D1281796

CHEMICAL ANALYSIS

Vol. 1. **The Analytical Chemistry of Industrial Poisons, Hazards, and Solvents.** *Second Edition.* By Morris B. Jacobs
Vol. 2. **Chromatographic Adsorption Analysis.** By Harold H. Strain (*out of print*)
Vol. 3. **Colorimetric Determination of Traces of Metals.** *Third Edition.* By E. B. Sandell
Vol. 4. **Organic Reagents Used in Gravimetric and Volumetric Analysis.** By John F. Flagg (*out of print*)
Vol. 5. **Aquametry: Application of the Karl Fischer Reagent to Quantitative Analyses Involving Water.** By John Mitchell, Jr. and Donald Milton Smith (*temporarily out of print*)
Vol. 6. **Analysis of Insecticides and Acaricides.** By Francis A. Gunther and Roger C. Blinn (*out of print*)
Vol. 7. **Chemical Analysis of Industrial Solvents.** By Morris B. Jacobs and Leopold Scheflan
Vol. 8. **Colorimetric Determination of Nonmetals.** Edited by David F. Boltz
Vol. 9. **Analytical Chemistry of Titanium Metals and Compounds.** By Maurice Codell
Vol. 10. **The Chemical Analysis of Air Pollutants.** By Morris B. Jacobs
Vol. 11. **X-Ray Spectrochemical Analysis** *Second Ed'tion.* By L. S. Birks
Vol. 12. **Systematic Anlaysis of Surface-Active Agents.** *Second Edition.* By Milton J. Rosen and Henry A. Goldsmith
Vol. 13. **Alternating Current Polarography and Tensammetry.** By B. Breyer and H. H. Bauer
Vol. 14. **Flame Photometry.** By R. Herrmann and J. Alkemade
Vol. 15. **The Titration of Organic Compounds** (*in two parts*). By M. R. F. Ashworth
Vol. 16. **Complexation in Analytical Chemistry: A Guide for the Critical Selection of Analytical Methods Based on Complexation Reactions.** By Anders Ringbom
Vol. 17. **Electron Probe Microanalysis.** *Second Edition.* By L. S. Birks
Vol. 18. **Organic Complexing Reagents: Structure. Behavior, and Application to Inorganic Analysis.** By D. D. Perrin
Vol. 19. **Thermal Methods of Analysis.** *Second Edition.* By Wesley Wm. Wendlandt
Vol. 20. **Amperometric Titrations.** By John T. Stock
Vol. 21. **Reflectance Spectroscopy.** By Wesley Wm. Wendlandt and Harry G. Hecht
Vol. 22. **The Analytical Toxicology of Industrial Inorganic Poisons.** By the late Morris B. Jacobs
Vol. 23. **The Formation and Properties of Precipitates.** By Alan G. Walton
Vol. 24. **Kinetics in Analytical Chemistry.** By Harry B. Mark, Jr. and Garry A. Rechnitz
Vol. 25. **Atomic Absorption Spectroscopy.** By Walter Slavin
Vol. 26. **Characterization of Organometallic Compounds** (*in two parts*). Edited by Minoru Tsutsui
Vol. 27. **Rock and Mineral Analysis.** By John A. Maxwell
Vol. 28. **The Analytical Chemistry of Nitrogen and Its Compounds** (*in two parts*). Edited by C. A. Streuli and Philip R. Averell
Vol. 29. **The Analytical Chemistry of Sulfur and Its Compounds** (*in three parts*). By J. H. Karchmer
Vol. 30. **Ultramicro Elemental Analysis.** By Gunther Tolg
Vol. 31. **Photometric Organic Analysis** (*in two parts*), By Eugene Sawicki
Vol. 32. **Determination of Organic Compounds: Methods and Procedures.** By Frederick T. Weiss
Vol. 33. **Masking and Demasking of Chemical Reactions.** By D. D. Perrin
Vol. 34. **Neutron Activation Analysis.** By D. De Soete. R. Gijbels, and J. Hoste
Vol. 35. **Laser Raman Spectroscopy.** By Marvin C. Tobin
Vol. 36. **Emission Spectrochemical Analysis.** By Morris Slavin
Vol. 37. **Analytical Chemistry of Phosphorus Compounds.** Edited by M. Halmann
Vol. 38. **Luminescence Spectrometry in Analytical Chemistry.** By J. D. Winefordner. S. G. Schulman, and T. C. O'Haver
Vol. 39. **Activation Analysis with Neutron Generators.** By E. P. Przybylowicz and S. S. Nargolwalla
Vol. 40. **Determination of Gaseous Elements in Metals.** Edited by Laben M. Melnick, Lynn L. Lewis, and Ben D. Holt

CHEMICAL ANALYSIS

A SERIES OF MONOGRAPHS ON ANALYTICAL CHEMISTRY AND ITS APPLICATIONS

Editors

P. J. ELVING · I. M. KOLTHOFF

VOLUME 40

A WILEY-INTERSCIENCE PUBLICATION

JOHN WILEY & SONS

New York/London/Sydney/Toronto

Determination of Gaseous Elements in Metals

Edited by

LABEN M. MELNICK

Research Laboratory
United States Steel Corporation
Monroeville, Pennsylvania

LYNN L. LEWIS

Research Laboratories
General Motors Corporation
Warren, Michigan

BEN D. HOLT

Chemistry Division
Argonne National Laboratory
Argonne, Illinois

A WILEY-INTERSCIENCE PUBLICATION

JOHN WILEY & SONS

New York/London/Sydney/Toronto

Library of Congress Cataloging in Publication Data:

Melnick, Laben Morton.
Determination of gaseous elements in metals.

(Chemical analysis, v. 40)
"A Wiley-Interscience publication."
Includes bibliographical references.
1. Metals—Analysis. 2. Gases—Analysis. I. Lewis, Lynn L., joint author. II. Holt, Ben D., joint author. III. Title. IV. Series.

QD132.M44 546′.3 74-768
ISBN 0-471-59328-1

Printed in the United States of America

10 9 8 7 6 5 4 3 2 1

PREFACE

This book represents, in a sense, an American sequel to an earlier Russian text, *Analysis of Gases in Metals*, by Z. M. Turovtseva and L. L. Kunin, published by the Publishing House of the Academy of Sciences of the USSR in 1959 and translated by James Thompson, Consultants Bureau Enterprises, Inc. of New York. Much significant research has been published in the intervening years, and the discussions and treatments of the subjects covered here have been expanded accordingly.

The first two chapters of the present book are basic in nature. Chapters 3–6 treat the fundamentals of various techniques, and certain ones for determining hydrogen, nitrogen, and oxygen are surveyed in Chapters 7, 8, and 9, respectively. All the common metals and many of the less common ones often analyzed for gases are treated in Chapters 10–19. Appropriately placed last is Chapter 20 on surface analysis by ion beam techniques, which are relatively new and provide information not obtainable by those discussed earlier. Extensive efforts have been made to give an integrated treatment of the subject of determining gaseous elements in metals as presented by the various authors.

Although the rapidity with which new developments are introduced and applied make it impossible for any book of this type to be completely current, emphasis has been placed on bringing subject coverage up to date as much as possible. This has been quite difficult because during the lengthy preparation of this book, three of the authors retired and three others left this field of research. In addition, Bernard D. La Mont, a coauthor of Chapter 9, passed away. His courageous efforts to continue his work during his protracted terminal illness were a source of inspiration to those of us who knew him.

Chapters written by a number of authors require careful editing to assure uniformity of text organization, use of units, and citation of literature. The editors have endeavored to accomplish this task, and hope that any deviations will cause no difficulties. With regard to articles in foreign literature, many of the references were obtained from the original journals and some were culled from *Chemical Abstracts*, *Analytical Abstracts*, or similar sources, and are so noted. However, some of the information obtained from abstracts and various articles read in translation has been identified only by the original journal reference.

The editors are indeed grateful to the many people who had so much to do with preparing several parts of the book:

At U. S. Steel: We wish to thank Mrs. Linda Bracco, who did part of the typing of both text and correspondence and who diligently coordinated much of the work; Ms. Marilyn Byron, Patricia Coleman, Dolores Davis, Rose Farkas, Joan Fennell, and Barbara Smith for typing; Messrs. A. J. Cornelissen and D. H. Voegeli for supervising the preparation of drawings and photographs; and Mr. R. O. Davis and Ms. Jean Frisinger for editing and review. In addition, the encouragement provided by Dr. Waldo Rall, Chief of the Physics and Analytical Chemistry Division of the U. S. Steel Corporation Research Laboratory, is sincerely appreciated.

At General Motors: Grateful acknowledgment is extended to Ms. Joann Gilbert, Janice Sarka, and Eileen Stempin for typing assistance. Of particular importance was the encouragement provided by Mr. M. D. Cooper, Head of the Analytical Chemistry Department, Research Laboratories, General Motors Corporation.

At Argonne: We are indebted to Dr. D. C. Stewart, Associate Director of the Chemistry Division, Argonne National Laboratory, and to Mr. E. E. Voiland, Manager of Analytical Services, for their encouragement to produce the book. The typists to whom we are especially grateful are Ms. JoAnn DiBuono, Brenda Grazis, Joanne Harmon, Dorothy Maes, Sandra Tasharski, and Edith Veal.

Permission to reproduce various illustrations and data from technical journals has kindly been given by authors, publishers, and manufacturers; reference to these are given in the text.

Finally, we acknowledge with much gratitude the exemplary patience of our respective wives, Bernice, Ruth, and Louise, and families with our intermittent but prolonged occupation with this book.

LABEN M. MELNICK
LYNN L. LEWIS
BEN D. HOLT

Monroeville, Pa.
Warren, Mich.
Argonne, Ill.

CONTENTS

CHAPTER

1

BEHAVIOR OF GASEOUS ELEMENTS IN METALS

LYNN L. LEWIS

Research Laboratories
General Motors Corporation
Warren, Michigan

CONTENTS

I. INTRODUCTION

Like Alice trying to enter a strange new territory in Wonderland, you may feel the need of a special talisman or a secret formula as you seek to understand the behavior of gaseous elements in metals. Unlike Alice, however, we can draw on extensive scientific studies to provide insight for our understanding. At least two groups have been engaged in these studies of the behavior of hydrogen, nitrogen, and oxygen in metals: the theorists primarily interested in the "harmonious order of the parts" and the applied scientists concerned with relationships on a macroscopic scale. As a result, the principles underlying the solubility, diffusion, and thermochemistry of gases in metals are well understood [1–7]. Furthermore, these principles are being applied widely because of the effect these gaseous elements have on the physical and mechanical properties of metals.

As requirements on the properties of metals continue to become more stringent, so will needs continue to arise for new and improved methods for determining the gas content of metals and identifying the phases that are present. An increased understanding and application of theory and basic data is required to meet these demands, with the realization that such applications will often be limited to the better known gas–metal systems. It is the purpose of this chapter to introduce some of the physical and chemical principles involved in the determination of gases in metals and to show how these principles can be applied to obtain reliable analytical information.

As a practical matter, the effect of gases on a metal can range from decidedly beneficial to extremely deleterious. The sources and losses of a gas may occur in many ways. Gas–metal reactions occur when metals are exposed to gases, for example, while being melted or heat-treated during processing operations. In some reactions the gaseous reactant is removed from the atmosphere, as in the direct solution of the gas in a metal. In other reactions one gas is exchanged for another; this occurs, for example, when water vapor reacts with a metal to form the metal oxide and hydrogen or when ammonia reacts with a metal to form the metal nitride and hydrogen. At other times a gas is formed, as in the thermal decomposition of metal oxides, nitrides, and hydrides.

II. SOLUBILITY

A. INTRODUCTION

Gas–metal interactions have been classified as being of four general types: physical adsorption, chemical adsorption (chemisorption), solution formation, and bulk compound formation [8]. Adsorption reactions are confined to the surface, while in the last two the gas is absorbed in the metal. Although solution formation is of principal interest here, all are discussed because most clean metal surfaces are highly reactive. The reaction products, whether those of adsorption, absorption, compound formation or a combination thereof, are of direct concern because of their effect on analytical results.

1. Physical Adsorption

The first step in a gas–metal reaction involves the gas molecules striking a metal surface and being adsorbed by physical or chemical forces. In physical adsorption, the gas molecules are adsorbed weakly by physical (van der Waals) forces when the conditions of temperature and pressure are favorable. The thickness of the adsorbed layer is limited to a few monolayers, with the quantity of gas adsorbed being proportional to the area of the metal surface. The adsorption process is reversible, and the exothermic heat of adsorption is low, usually less than 5 kcal/mole. Gases that are adsorbed physically on a metal surface may be removed by reducing the pressure; the process can be accelerated by raising the temperature. The interaction of a metal with a rare gas such as argon is an example of physical adsorption.

2. Chemical Adsorption

In chemical adsorption, usually called chemisorption, chemical interaction between the adsorbed gas and the metal results if bonding to the surface is stronger than that within the gaseous molecule. The bonds formed are almost as stable as those existing in stoichiometric compounds. (The chemisorbed layer may be covered by gas molecules bound by physical forces.) Chemisorption occurs at higher temperatures than physical adsorption, and the heats of adsorption are usually 20–100 kcal/mole. Chemisorption may be spontaneous at ambient temperature and pressure on an active surface, but activation energies of more than 20 kcal/mole are sometimes required. Vigorous treatments are often required to remove a chemisorbed layer from a metal surface, particularly for systems of high reactivity and strong bond formation.

The adsorption of gases on the clean surface of a number of metals is classified according to chemical reactivity in Table 1.1 [8]. The metals are

TABLE 1.1. Classification of the Metals and Semimetals According to the Chemical Reactivity of Their Surfaces[a]

Li	Be											B	C
Na	Mg							D				Al	Si
K	Ca	(Sc)[b]	Ti	(V)	Cr	Mn	Fe	Co	Ni	Cu	Zn	Ga	Ge
Rb	Sr	(Y)	Zr	Nb	Mo	(Tc)	(Ru)	(Rh)	Pd	Ag	Cd	In	Sn
Cs	Ba	La	(Hf)	Ta	W	Re	(Os)	Ir	Pt	Au	Hg	Tl	Pb
	C			A		B		E					

Metals	Reacting Gases						
	O_2	C_2H_2	C_2H_4	CO	H_2	CO_2	N_2
Group A	3	3	3	3	3	3	3
Group B	3	3	3	3	3[c]	3	2
Group C	3	3	3	3	2 or 3	3	2
Group D	3	3	3	3	3	3	1
Group E	3	3	3	3	3	1 or 0	1 or 0
Cu	3	3	3	1	0	?	0
Ag	2 or 3	1	1	0	0	?	0
Au	0	3	3	1	0	?	0
B	3 or 2	?	?	3 or 2	3 or 2	?	3 or 2
Al	3	2	2	3	0	?	0
Si, Ge	3	?	?	0	2	2 or 3	0
K	3	3	0	0	0	?	0
Other metals	3	0	0	0	0	?	0

[a] Reproduced by permission from Academic Press [8].
[b] Parentheses indicate that there are insufficient data to make a definite classification, but the metal is presumed to have properties similar to its neighbors.
[c] The adsorption of H_2 on Mn is activated at 300°K.
3—A rapid and probably nonactivated uptake of gas detectable at 300°K and 10^{-4} torr pressure.
2—A slower and sometimes activated uptake of gas detectable at 300°K and 10^{-4} torr pressure.
1—A detectable uptake of gas at 195°K, but not at 300°K.
0—No detectable uptake of gas at 195°K or 300°K and 10^{-4} torr pressure.
?—No data.

grouped in order of decreasing activity, with group A being the most reactive. The group-A metals chemisorbed all the gases studied, the group-B metals chemisorbed all gases rapidly except for nitrogen, and so on. Trapnell [9] stated that there is a general order of affinity of gases for metal surfaces, with the order being

$$O_2 > C_2H_2 > C_2H_4 > CO > H_2 > N_2.$$

If a metal chemisorbs a particular gas, it will also chemisorb all gases higher in the scale; if it does not chemisorb the gas, it will not chemisorb gases lower in the scale. With regard to the analysis of metals for gases, chemisorption of oxygen rather than that of nitrogen predominates when specimens are exposed to the atmosphere during surface preparation.

When the surfaces of most metals are carefully prepared for analysis, the relative quantity of gaseous element(s) on the surface is usually insignificant when compared with that in the bulk of the metal. Problems from chemisorption of gases are encountered, however, in the analysis of metals with a low bulk gas content and those with a high surface–volume ratio. Reference is commonly made to the "cleaning" of metal surfaces by acid etching, careful filing, and such, but, in reality, clean surfaces are prepared and maintained only in the absence of reactive gases. To illustrate, an atomically clean surface has been defined as "one free of all but a few percent of a single monolayer of foreign atoms, either adsorbed on or substitutionally replacing surface atoms of the parent lattice" [10]. In work of the tungsten–nitrogen system [11], a tungsten sample was degassed at 2500°C, cooled to 25°C, and exposed to nitrogen; when reheated to 2500°C, the quantity of nitrogen evolved (assumed to be complete; see Section III.C) was 4×10^{14} molecules/cm^2, a quantity corresponding to approximately one monolayer of adsorbed nitrogen. For samples exposed to oxygen, one monolayer of chemisorbed oxygen can be calculated to contribute at least 0.01 μg oxygen/cm^2 of surface. This estimate is, of course, too low for oxide films of greater thickness and surfaces that actively absorb moisture from the atmosphere. For example, carefully etched and dried copper specimens were determined to have about 0.24 μg oxygen/cm^2 of surface [12].

To reduce surface contamination by oxygen, Hill et al. [13] cleaned specimens by cathodic etching, using an argon ion source. Uranium samples cleaned by acid etching and filing were determined to "contain" a total oxygen content of 7.6 ppm, whereas samples that were acid-etched and then cathodically etched "contained" 4.3 ppm. In similar work with iron, the cathodic etch treatment resulted in an oxygen decrease of from about 4.9 to 3.8 ppm. Greater differences would probably have been observed if exposure of the cathodically etched specimens to the atmosphere could have been avoided before vacuum fusion analysis. A promising alternate approach to surface cleaning would be flash evaporation of the surface by heating to high temperatures (about 5000°C) by means of a xenon flash discharge lamp [14].

Following a different approach to the problem of surface contamination, Guardipee [12] developed and applied two techniques for differentiating between oxygen on the surface and in the bulk of the metal. In one, specimens were heated in two stages to provide for separate evolution (and determination) of the two sources of oxygen. In the other, samples of varying surface–volume ratio were carefully analyzed to obtain total oxygen by plotting a regression line of parts per million total oxygen versus surface area per gram of sample; the quantity of bulk gas was determined from the intercept. His thorough investigation revealed that surface oxygen accounted for an appreciable fraction of the total oxygen content of the copper, chromium, and tungsten that was analyzed.

3. Absorption

All gas–metal reactions have in common a series of consecutive steps that occur during the absorption of the adsorbed gas into the bulk of the metal by diffusion into the metal lattice:

1. The dissociation of the gas molecules that are adsorbed on the surface of the metal
2. The transfer of the adsorbed gas from the surface of the metal to true solution within the metal lattice
3. The diffusion of the gas in the atomic form through the metal lattice.

Any one of these steps may control the rate at which saturation equilibrium is approached, depending on the gas–metal system and the environmental conditions [15, 16]. The three steps as listed above present an oversimplified concept of the absorption process, because of the formation of surface films. For example, in oxidation, the chemical and physical properties of the gas–oxide interface, the oxide, the oxide–metal interface, and the metal all contribute to the nature and extent of the gas–metal reactions [17]. The classification of oxidation reactions based on film thicknesses, Fig. 1.1, indicates that the characteristics of oxide films are continually changing as oxidation proceeds [18]. As a further note, gases that have been dissociated are absorbed by metals much more rapidly than those that are in molecular form initially, and the saturation limit is higher. The degree of dissociation of a gas increases with the temperature, and the rate of its absorption by the metal increases correspondingly.

4. Interstitial Phases

When an atom of one element enters the metal lattice of another, the added atom (e.g., zinc added to copper) may take the place of a parent metal

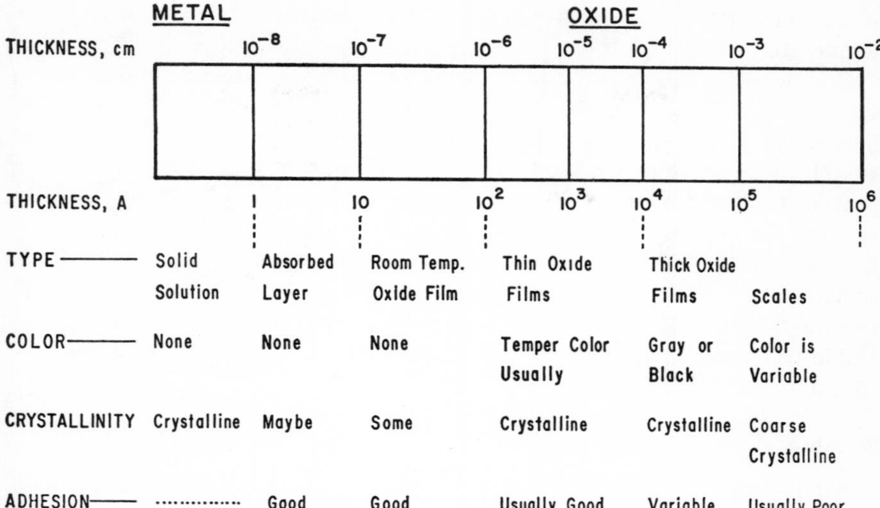

Fig. 1.1 Classification of extent of oxidation reactions based on film thickness. (Reproduced by permission from *Corrosion* [18].)

atom in the lattice to form a substitutional solid solution. However, some nonmetallic elements (hydrogen, nitrogen, oxygen, boron, and carbon) form interstitial solid solutions in certain metallic crystals by entering into the vacant spaces, or interstices, between metal atoms of the lattice; the metal atoms are not displaced from their positions in the lattice, although the original arrangement of the metal atoms may be distorted or expanded.

The solution formed by an interstitial element in a solid metal can exist over a range of concentration. This solution is called a *primary solid solution* when this range extends to the composition of the pure metal. The concentration limits of the phases usually are small at room temperature and broaden with increasing temperature. When the solubility is exceeded at a given temperature, a *secondary solid solution* (hydride, nitride, or oxide phase) is formed. The crystal structure of the secondary solid solution usually differs from that of the primary solid solution.

To illustrate the phase changes occurring during the solution of a gas in a metal, the phase diagram for the iron–nitrogen system is given [19] in Fig. 1.2. In this example, at 400°C nitrogen forms an interstitial solid solution in iron in concentrations up to 0.4 atom of nitrogen/100 atoms of iron. When the nitrogen content is in the primary-solid-solution range, the terminal solubility for nitrogen has not been exceeded and the one phase present is body-centered cubic, which is the structure of the solvent metal, iron, at that temperature. On the addition of more nitrogen, the terminal solubility is

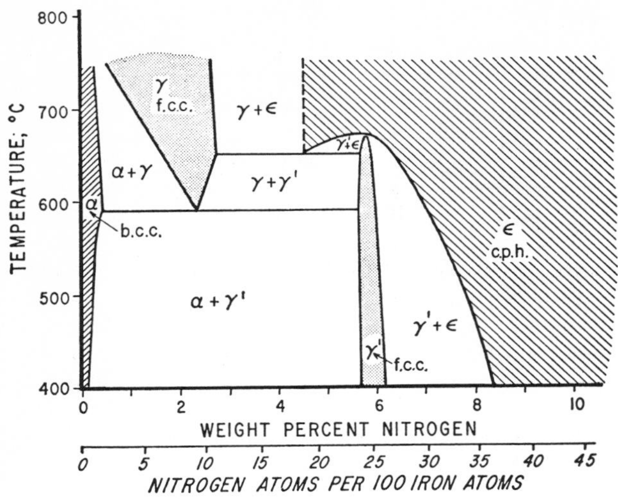

Fig. 1.2 Iron–nitrogen system. (Reproduced by permission from the *Discussions of the Faraday Society* [19].)

exceeded and a second phase, Fe_4N (γ'), is formed that contains 24.1–25.8 atoms of nitrogen/100 atoms of iron; the second phase, which is nonstoichiometric, is face-centered cubic in structure. The close-packed hexagonal ε phase is formed when the nitrogen content reaches 35.5 nitrogen atoms/100 atoms of iron. Thus, it can be seen that the addition of the gas modifies the packing of the metallic atoms.

As might be expected, the size relationship and the number of potential bond-forming contacts are important in the formation of interstitial phases [20–22]. The size relationship is given as the ratio of the atomic diameter of the interstitial atom to that of the solvent atoms. Barrer [23] states, "In a close-packed array of metal atoms two kinds of interstitial position occur: 'octahedral' sites (coordination number 6); and 'tetrahedral' sites (coordination number 4). Which kind of site X (nonmetallic element) will occupy depends on the rule that it attains the highest coordination number compatible with its size. Larger atoms X tend then to occupy octahedral sites and smaller ones the tetrahedral sites, the maximum radius ratio being about 0.59 for coordination number 6 and 0.41 for coordination number 4. There is no need for all the sites of a given type to be filled because the metal lattice is stable in the absence of any interstitial atoms. However, since there is only a limited number of sites the composition cannot be past a specific upper limit." Hydrogen tends to occupy the interstitial tetrahedral (coordination number 4) sites, and the larger nitrogen and oxygen atoms tend to occupy the octahedral sites (coordination number 6) in a close-packed metallic lattice (face-centered cubic or close-packed hexagonal).

Data on some interstitial phases, as classified by Barrer [23], are given in Table 1.2. These interstitial phases may be nonstoichiometric in composition, as is the case with many metal hydrides, nitrides, and oxides; equilibrium conditions for compounds of this type have been discussed [24].

5. Effect of Temperature and Pressure

Abrupt changes occur in the solubility of a gas in a metal at the temperatures of phase transformation and melting, as illustrated in Fig. 1.3 for the solubility of hydrogen and nitrogen in iron [25]. The alpha and delta phases of iron are body-centered cubic in structure, whereas γ-iron is face-centered cubic. Solubility for interstitial elements is greater in the face-centered cubic lattice, which allows for a larger interstitial body of spherical shape.

Pressure as well as temperature govern the solubility relationships for a gas–metal system, so phase diagrams are usually drawn from data collected at constant pressure, which is often 760 torr of the gas. At any given temperature the equilibrium partial pressure of the gaseous constituent of a gas–metal system will vary with the composition of the gas–metal system. When the partial pressure of the gas equals or exceeds the decomposition pressure of a gas–metal compound, the compound is formed.

The amount of hydrogen, nitrogen, or oxygen that can dissolve in a solid or a liquid metal is proportional to the square root of the partial pressure, p,

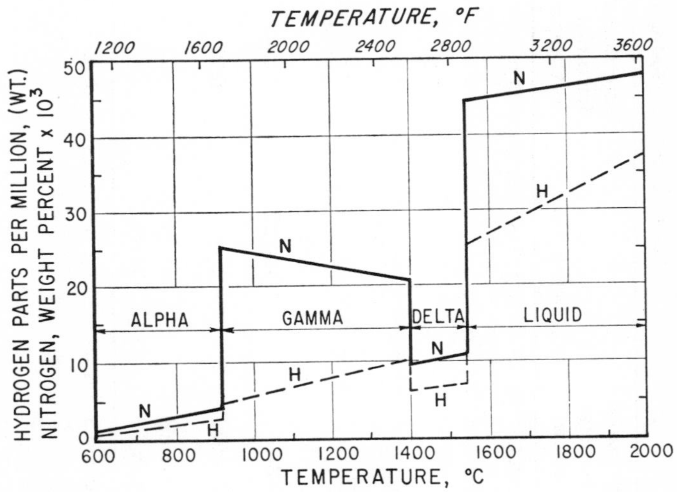

Fig. 1.3 Effect of temperature on the solubility of hydrogen and nitrogen at 760 torr in iron. (Reproduced by permission from John Wiley & Sons, Inc. [25].)

TABLE 1.2. Crystal Chemical Data on Some Interstitial Solutions[a]

Coordination of Interstitial Element	Radius Ratio	Proportion of Interstitial Sites Occupied	Arrangement of Metal Atoms In Phase	Structure of Phase	Examples
Sixfold	0.41–0.59	1.0	Cubic close-packed	Rock salt	TiN, ZrN, ScN, VN, CrN
		0.5	Cubic close-packed	Rock salt	W_2N, Mo_2N
		0.25	Cubic close-packed	Rock salt	Mn_4N, Fe_4N
Fourfold	0.23–0.41	1.0	Cubic close-packed	Fluorite	TiH_2
		0.5	Cubic close-packed	Zinc-blende	ZrH, TiH
		0.25	Cubic close-packed	Zinc-blende	Pd_2H
		0.125	Cubic close-packed	Zinc-blende	Zr_4H
Sixfold	0.41–0.59	0.25	Hexagonal close-packed	—	Mn_2N, Cr_2N, Fe_2N
Fourfold	0.23–0.41	0.25	Hexagonal close-packed	Wurtzite	Zr_2H, Ta_2H, Ti_2H
Fourfold	>0.29	—	Body-centered cubic	—	TaH
Sixfold	>0.58	—	Simple hexagonal	—	MoN

[a] Reproduced by permission from the *Discussions of the Faraday Society* [23].

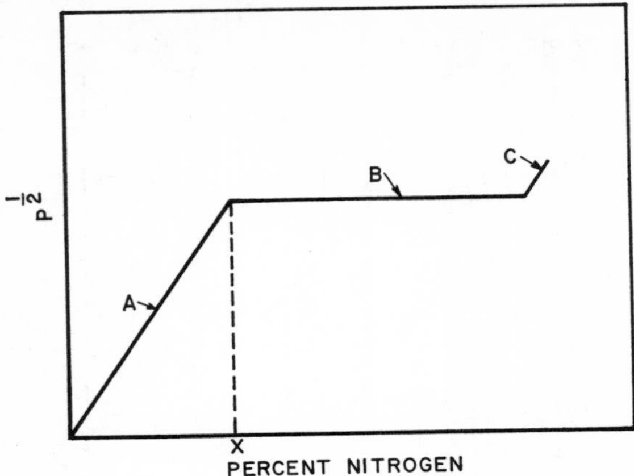

Fig. 1.4 General representation of $P^{1/2}$ versus percent nitrogen plot at constant temperature. A, Solid solution region where Sievert's law holds. B, Solid solution plus nitride. C, Nitride homogeneity range. X, Saturation solubility.

of the gas in the surrounding atmosphere (Sievert's law) [26]. In the case of nitrogen, for example,

$$N = k(p_{N_2})^{1/2}, \tag{1}$$

where N is the concentration (or more correctly, activity) of the dissolved nitrogen. This square-root relationship is given as evidence for the atomic solution of a diatomic gas such as hydrogen, nitrogen, or oxygen, as indicated by

$$\tfrac{1}{2}N_2(\text{gas}) = N \text{ (in the metal).} \tag{2}$$

Fig. 1.4 represents the nitrogen content of a metal at constant temperature increasing with the square root of the nitrogen equilibrium pressure. The gas is dissolved atomically at concentrations shown by line A until the terminal solubility is reached at X. As soon as the terminal solubility is reached, the pressure remains constant with increasing concentration during nitride formation (line B) until the nitride homogeneity range is reached (line C). This absorption process is reversible. In theory, the gas (e.g., nitrogen) content of a metal sample may be determined by extracting it from the heated metal sample in an evacuated system. The residual gas content after extraction will be along line A at a point determined by the lowest gas pressure that can be obtained in the analytical system.

The amount of gas in true solution in a metal under equilibrium conditions is also temperature dependent. The general equation expressing the concentration as a function of temperature and pressure for the solution of a diatomic

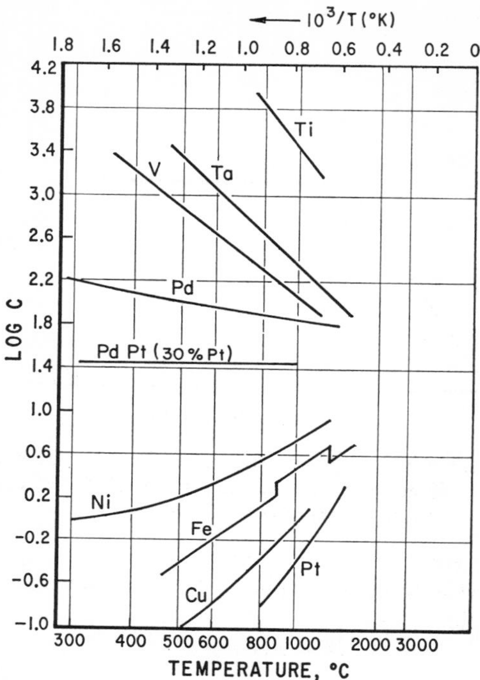

Fig. 1.5. Effect of temperature on the solubility of hydrogen in various metals at 760 torr of hydrogen. Hydrogen concentration is given in atoms per 10^4 metal atoms (Reproduced by permission from the Academic Press, Inc. [4].)

gas is

$$s = k'(p)^{1/2} \exp\left(\frac{-\Delta H}{RT}\right), \tag{3}$$

where s is the solubility of the gas in the metal, ΔH is the heat absorbed by the system during the solution of one mole of the gas, R is the gas constant, and T is the temperature in degrees Kelvin. The equation applies to many gas–metal systems at lower gas concentrations. An isobaric plot of hydrogen solubility versus reciprocal temperature (Fig. 1.5) shows solubility increasing with increasing temperature for the endothermic solution of hydrogen in nickel, iron, copper, and platinum and the solubility decreasing with increasing temperature for the exothermic solution of hydrogen in titanium, tantalum, vanadium, and palladium [4, 27].

6. Solubility Measurements

Solubility data are obtained usually by measuring the amount of gas absorbed by a known weight of a metal at a given temperature under a known pressure of the gas or by quenching the sample and determining the gas content by a chemical method. An example of the gravimetric method is given in the determination of the solubility of nitrogen in niobium–molybdenum solid solutions [28]. Indirect methods based on changes in the physical properties of solid metals are also used. The latter include measurement based on relaxation phenomena (internal friction) [1, 29–31], X-ray diffraction (lattice parameter) [1, 32, 33], and electrical resistivity [1, 32]; these methods generally are limited to metals of known composition and thermal history. The discrepancies in solubility data that are observed are often attributed to the following: measurement errors; the state of strain within the metal; the effect of impurity elements present; and nonequilibrium conditions at the time of measurement, which may be caused by the presence of an impervious oxide or nitride film on the metal surface.

7. Units of Measurement

The gas content of metals has been reported in a variety of units, the most common of which is in terms of cubic centimeters of gas per 100 grams of metal. Table 1.3 gives the units of measurement in use and conversion factors in terms of other units. The volumes given are those at standard temperature and pressure (STP), 0°C, and 760 torr.

TABLE 1.3. Units of Measurement

S = cubic centimeters of gas per 100 grams of metal = $22.42\ X/M$
V_g = cubic centimeters of gas per cubic centimeter of metal = $0.01\ SD$
X = milligrams of gas per 100 grams of metal = $0.04461\ SM$
C = milligrams of gas per 1000 grams of metal = $10\ X = 0.4461\ SM$ (parts per million, by weight)
F = atoms of gas per atom of metal[a] = $(8.922 \times 10^{-7})\ SA$
p_w = weight percent of gas = $10^{-3}\ X = (4.461 \times 10^{-5})\ SM$

$$p_a = \text{atomic percent of gas}^a = \frac{\dfrac{100p_w}{M/2}}{\dfrac{p_w}{M/2} + \dfrac{100 - p_w}{A}}$$

[a] For a diatomic gas.
D = density of the metal, M = molecular weight of the gas, and A = atomic weight of the metal.

B. HYDROGEN

The interaction between hydrogen and metals results in the formation of solid solutions and metal hydrides of three types that have been classified as ionic hydrides, covalent hydrides, and transition metal hydrides [3, 34–37].

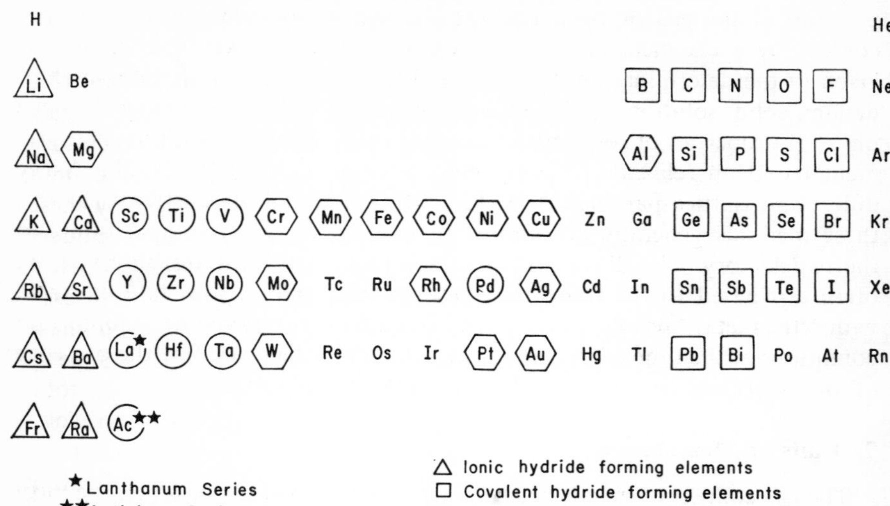

Fig. 1.6. Schematic representation of the type of reaction of the elements with hydrogen. (Reproduced by permission from Pergamon Press, Inc. [36].)

A schematic representation of the reaction types for the elements with hydrogen, as presented by Cotterill [36], is given in Fig. 1.6. Hydrogen evidently has no appreciable solubility in zinc, gallium, cadmium, indium, mercury, or thallium. No data are available presently for hydrogen in beryllium, technetium, ruthenium, rhenium, osmium, and iridium.

The ionic (saline) hydrides are formed with the alkali and the alkaline-earth metals. They are generally crystalline, nonvolatile compounds with stabilities that increase as the atomic weight of the metallic element decreases. These compounds with their negative hydrogen or hydride ions are analogous to the corresponding halides. Covalent hydrides analogous to the hydrocarbons are formed by the elements boron, silicon, germanium, tin, lead, arsenic, antimony, bismuth, selenium, and tellurium. They are gaseous at atmospheric temperatures. Familiar compounds of this type are arsine (AsH_3) and stannane (SnH_4).

The transition metal hydrides are generally classified as either endothermic or exothermic occluders. Endothermic occluders absorb hydrogen and form only solid solutions and/or metastable hydrides. Because simple solution is an endothermic process, the solubility of hydrogen increases as the temperature increases. As Figs. 1.5 and 1.6 indicate, nickel, iron, copper, and platinum are among the endothermic occluders. Hydrogen is dissolved atomically, resulting in the solubility increasing with the square root of the pressure (Sievert's law). The solubility of hydrogen is usually less than 0.1 at. % at hydrogen pressures of less than 760 torr.

The exothermic occluders, which are known as *pseudo-hydride formers* [3], include titanium, tantalum, and vanadium (Figs. 1.5 and 1.6). These metals absorb a total quantity of hydrogen that decreases as the temperature increases and varies as the square root of the pressure only over certain pressure ranges. At saturation, the amount of hydrogen absorbed by these metals is generally 2–10 at. %. Hydrogen dissolves in exothermic occluders to form a single phase up to a certain hydrogen content, beyond which a second phase (a hydride phase) forms. The solubility of hydrogen in exothermic occluders is given properly as that solubility at which a second phase (the hydride phase) appears in the hydrogen–metal system at a given temperature. This solubility (the terminal solubility) will always be less than the total amount of hydrogen that can be absorbed. In exothermic occluders, the solid solution of hydrogen proceeds endothermically and the solubility as defined above generally increases with temperature. However, hydride formation occurs exothermically. As hydrides are formed with titanium or tantalum, for example, the exothermic heat of hydride formation exceeds the endothermic heat of dissociation into the atomic state, and the overall exothermic reaction results in hydrogen solubility decreasing with increasing temperature.

Trends in the solubility of hydrogen in the elements of the periodic table have been given by Cotterill [36]: "The extreme difficulties involved in obtaining solubility data from one investigation to another (e.g. impurity variations, etc.) make a similar comparison of the solubility limits of the occluding metals qualitative rather than quantitative. However, a brief examination of the relevant data suggests that the solubility of hydrogen tends to decrease as the atomic number of the solvent increases, in any one group: for example, chromium, molybdenum, and tungsten of group VIA; or titanium, zirconium, and hafnium of group IVA. The solubility of hydrogen in the exothermic occluders is greater than that in the endothermic occluders. Hence, examination (Fig. 1.6) shows that (with the notable exception of palladium) the solubility of hydrogen decreases as the position of the solvent moves across the periodic table from left to right." Additional solubility data are presented by Dushman [3], Fast [4], Smialowski [38], and Eastwood [39].

The effect of alloying elements on the solubility of hydrogen in a metal may result in increased or decreased hydrogen solubility. This effect in liquid binary copper alloys [26] is shown in Fig. 1.7. The solubility of hydrogen in liquid iron alloys [40] has been studied (see Fig. 1.8); the solubility of hydrogen has been related to the electronic structure of the solute metal ion [40] and to the atomic number of the alloying metal with respect to three periods in the periodic table [41]. The solubility and diffusion of hydrogen in

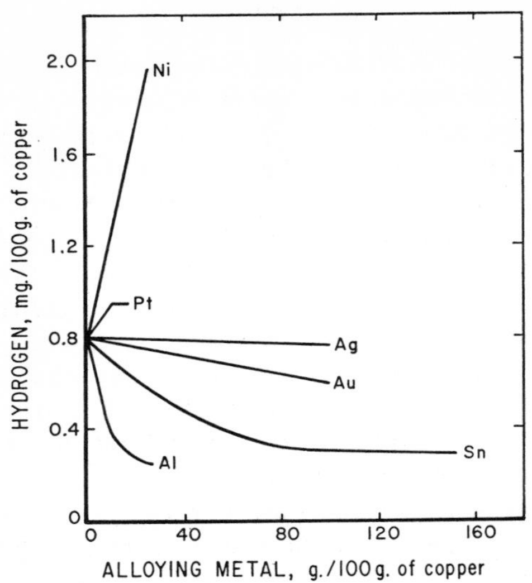

Fig. 1.7. Solubility of hydrogen in liquid copper alloys at 760 torr hydrogen pressure and 1220°C. (Reproduced by permission from *Zeitschrift fuer Erzbergbau und Metallhuetten wesen* [26].)

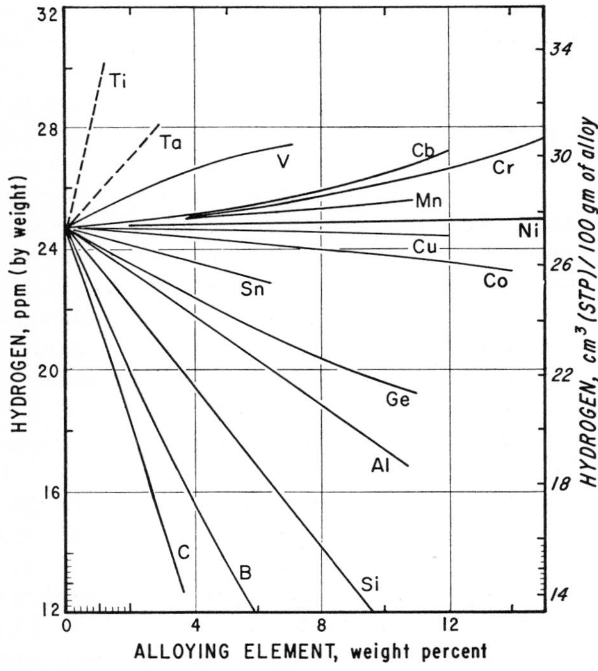

Fig. 1.8. Solubility of hydrogen in liquid iron alloys at 760 torr hydrogen pressure and 1600°C. (Reproduced by permission from Addison-Wesley [40].)

solid iron alloys increases with increasing temperature and the quantity of the alloying elements carbon, silicon, chromium, manganese, and nickel [42]. For iron–nickel alloys, Beck et al. [43] related the variation in hydrogen solubility and diffusivity (70°C, $P_{H_2} = 1$ atm) as a function of composition to the crystallographic and electronic properties of the alloy.

Metals that have been obtained by electrodeposition processes may contain much greater quantities of hydrogen than can be added by thermal means. For example, tin and zinc can occlude appreciable quantities of hydrogen when electroplated, although they absorb but little, if any, hydrogen by thermal processes. The hydrogen contents of some electrodeposited metals [34] are given in Table 1.4. The quantity of hydrogen occluded with an electrodeposited metal will vary with the composition, current density, and temperature of the bath [44]. Metals exposed to acid solutions in pickling operations may absorb large quantities of hydrogen because they are exposed, in effect, to high external pressures of atomic hydrogen. Very large hydrogen pressures would be required by thermal charging techniques to add hydrogen to metals in the quantities that can be added by electrolytic or chemical action; for example, about 10^4 atm of hydrogen would be required to be in equilibrium with the hydrogen that has been charged cathodically in palladium [45].

Although the importance of the effect hydrogen in metals is well recognized, there is relatively little information available on the solubility of hydrogen in the alloys that are manufactured and used commercially. However, interest has been shown in the distribution of hydrogen among the phases that constitute these alloys. For example, studies [46] of hydrogen in α-iron and the Laves phases of composition $TiFe_2$ and $NbFe_2$ (Fig. 1.9) show that the hydrogen concentration decreases with a decrease in temperature for α-iron and $NbFe_2$, whereas in $TiFe_2$ it increases from a minimum at 400°C to 14 ppm at ambient temperature. Thus, it appears that $TiFe_2$ when present could act as a hydrogen sink during the cooling of ferrous materials. In other work, based on an autoradiographic technique, differences in tritium concentration in zirconium were observed from grain to grain and within areas of a grain [47].

Chemical methods for distinguishing between atomic, molecular, and oxidized hydrogen in steel have been described [48] in which the oxidized hydrogen was presumably present as water or hydroxyl groups associated with inclusions. Direct evidence for the presence of hydrogen as methane in microvoids in solid steel has been obtained by dissolving samples in a bromine–methanol solution and analyzing the gases present [49]. Internal friction measurements have shown that only a small fraction of the hydrogen present in steel at temperatures less than approximately 200°C is in interstitial solution [50].

TABLE 1.4. Hydrogen Contents of Electrodeposited Metals[a]

Metal	Hydrogen Content (cc/100 g)
Sn	1
Cu	3
Ni	7×10^1
Zn	1×10^2
Co	3×10^2
Fe	8×10^2
Mn	1×10^3

[a] Reproduced by permission of the University of Chicago Press [34].

Sievert's law, Eq. 1, has an important application in the consideration of the conditions necessary for the quantitative determination of hydrogen in metals by vacuum extraction techniques. If we assume that Eq. 1 can be extended to low hydrogen concentrations and pressures without serious error, the value of k as determined at higher pressures can be used to calculate equilibrium hydrogen pressures for dissolved hydrogen at low concentrations. Some hydrogen pressures that would be in equilibrium with dissolved hydrogen at low levels have been calculated (Table 1.5; the constant k has been calculated from data given in the third and fourth columns). The data show that in the case of thorium, for example, a vacuum of 5.7×10^{-4} torr would

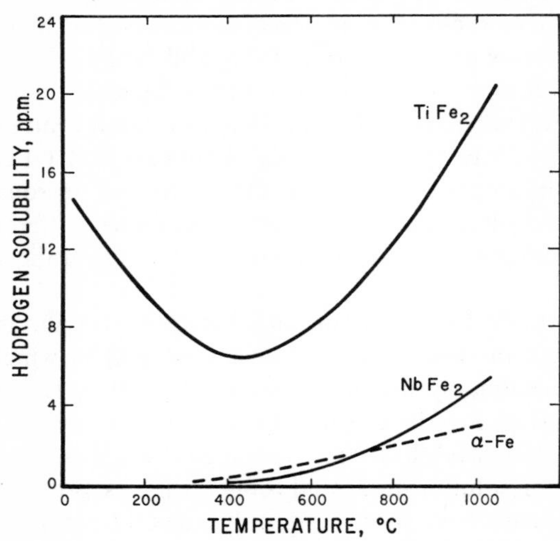

Fig. 1.9. Solubility of hydrogen in $TiFe_2$, $NbFe_2$, and α-Fe at 760 torr hydrogen pressure. (Reproduced by permission from the Iron and Steel Institute, London [46].)

TABLE 1.5. Equilibrium Hydrogen Pressures for Hydrogen in Various Metals

				Calculated	
Element	Temperature, °C	Dissolved Hydrogen, (wt %)	Gaseous Hydrogen Pressure, torr	Dissolved Hydrogen, (wt %)	Gaseous Hydrogen Pressure, torr
Mo[3]	1000	4.5×10^{-5}	760	1×10^{-5}	37
Fe[3]	1000	5.0×10^{-4}	760	1×10^{-4}	30
				1×10^{-6}	3
Ni[3]	1000	8.7×10^{-4}	760	1×10^{-4}	10
Th[51]	850	2.2×10^{-2}	32	1×10^{-4}	5.7×10^{-4}
Pr[52]	800	3.1×10^{-1}	3	1×10^{-4}	3.2×10^{-7}
Zr[53]	600	—	—	1×10^{-4}	4.2×10^{-7}
Zr[53]	700	—	—	1×10^{-4}	1.2×10^{-6}

be necessary at 850°C to reduce hydrogen to a concentration of 1×10^{-4} wt % (1 ppm). Pressures of 5.7×10^{-4} torr and lower are readily obtainable, and the calculation is verified by the fact that hydrogen in thorium has been determined by extraction at 840°C [54].

The calculated equilibrium hydrogen pressures for zirconium and praseodymium at the temperatures given are so low that hydrogen could not be extracted to the 1 ppm level with the type of hot extraction equipment now in general use. Higher extraction temperatures are therefore required to provide favorable equilibrium conditions and to reduce analysis time by means of higher diffusion rates. Zirconium has been successfully analyzed for hydrogen at the 53 ppm level in the 800–1500°C temperature range; the extraction times required were reduced from 30 min at 800°C to 5 min at 1500°C [55]. A 10-fold reduction in the hydrogen pressure in the analytical system will theoretically result in a 100-fold decrease in the hydrogen content of the metal at equilibrium. Thus, calculations show that the hydrogen content of iron at 1000°C could be reduced from 1×10^{-6} wt % at 3 torr to 1×10^{-8} wt % at a hydrogen pressure of 0.3 torr.

C. NITROGEN

Nitrogen is reported to be "insoluble" in ruthenium, osmium, cobalt, rhodium, iridium, nickel, palladium, platinum, copper, silver, gold, zinc, cadmium, mercury, aluminum, thallium, tin, lead, antimony, and bismuth [1, 3, 5, 6, 56]. Solid solutions are formed by nitrogen in titanium, zirconium, hafnium, thorium, uranium, vanadium, niobium, tantalum, chromium, molybdenum, tungsten, manganese, and iron [1, 3, 5, 6, 57]. A periodic table of the nitrides is reproduced in Table 1.6, which is taken from a survey on the preparation, crystallography, and thermochemistry of binary nitrogen

TABLE 1.6. Periodic Table of Nitrides[a]

Li	Be											B	C	N	O
Li_3N LiN (g)	Be_3N_2 BeN (g)											BN BN (g)	C_2N_2 (g) CN (g) C_4N_2	N_2 (g)	N_2O (g) NO (g) N_2O_3 (g) NO_2 (g) N_2O_4 (g) N_2O_5
Na	Mg											Al	Si	P	S
Na_3N NaN (g)	Mg_3N MgN (g)											AlN AlN (g)	SiN SiN (g) Si_2N_3 Si_3N_4 Si_xN	PN P_3N_5 PN (g)	S_5N_2 $(SN)_x$ S_4N_4 S_4N_2 SN (g) S_2N_2
K	Ca	Sc	Ti	V	Cr	Mn	Fe	Co	Ni	Cu	Zn	Ga	Ge	As	Se
K_3N KN (g)	Ca_3N_2 CaN (g)	ScN	TiN	V_2N VN V_3N	Cr_3N_2 Cr_2N CrN	MnN (g) Mn_3N_2 Mn_4N Mn_5N_2 Mn_2N	Fe_8N Fe_6N $Fe_{16}N_2$ Fe_3N Fe_4N Fe_2N	Co_3N Co_2N	Ni_3N	Cu_3N	Zn_3N_2	GaN (g) GaN (GaN) (g)	Ge_3N_4 Ge_3N_2	AsN AsN (g)	SeN Se_4N_4
Rb	Sr	Y	Zr	Nb	Mo	Tc	Ru	Rh	Pd	Ag	Cd	In	Sn	Sb	Te
Rb_3N RbN (g)	SrN (g) Sr_2N Sr_3N_2 Sr_3N_4	YN	ZrN	Nb_2N NbN	Mo_5N_4 Mo_2N MoN				PdN	Ag_3N	Cd_3N_2	InN InN (g)	Sn_3N_4	SbN SbN (g)	Te_3N_4
Cs	Ba	La	Hf	Ta	W	Re	Os	Ir	Pt	Au	Hg	Tl	Pb	Bi	Po
Cs_3N CsN (g)	BaN (g) Ba_2N Ba_3N_2 BaN_2	LaN	HfN	Ta_2N TaN Ta_3N_5	WN W_2N	$ReN_{0.43}$	Os_2N				Hg_3N	TlN TlN (g)	Pb_3N_4	BiN	
Fr	Ra	Ac	Th	Pa	U	Np	Pu	Am	Cm	Bk	Cf	E	Fm	Mv	No
			Th_3N_4 Th_2N_3 ThN	PaN	UN U_2N_3 U_3N_4 UN_2	NpN	PuN								
		Ce	Pr	Nd	Pm	Sm	Eu	Gd	Tb	Dy	Ho	Er	Tm	Yb	Lu
		CeN	PrN	NdN		SmN	EuN	GdN	TbN	DyN	HoN	ErN	TmN	YbN	LuN

[a] Reproduced by permission of the authors [58].

compounds of the elements [58]. Not all of the metal nitrides listed in Table 1.6 are formed directly by reacting the metal with nitrogen. For example, the explosively unstable silver nitride Ag_3N is obtained by crystallization from an ammoniacal solution of silver oxide [59].

Much of the work on the solubility of nitrogen in metals needs to be checked with metals of higher purity and with improved analytical techniques; some of the results are questionable because an impervious layer may have been formed on the metal surface that prevented the diffusion of nitrogen into the metal.

For information about the interaction of nitrogen with a particular element, phase diagrams of binary nitrogen–metal systems have been collected and reviewed [60]. Regarding ternary systems that contain nitrogen, the solubility of nitrogen in a metal may be changed markedly as an alloying element is added. For example, nitrogen solubility is increased when the alloying elements chromium, copper, molybdenum, niobium, tantalum, tungsten, and vanadium are added to liquid cobalt (1600°C), while aluminum, iron, nitrogen, and silicon decrease the nitrogen solubility [61]. Results obtained with liquid iron alloys are shown in Fig. 1.10 [62]. A method has been described for predicting the solubility of nitrogen in complex iron–base alloys from solubility data on binary alloy systems [63].

The hot-extraction method for determining nitrogen in metals is based on the fact that at a given temperature the quantity of nitrogen dissolved in a metal is proportional to the square root of the equilibrium nitrogen pressure (Sievert's law). The constant k in Eq. 1 can be calculated when the nitrogen concentration is known at a given equilibrium pressure. If we assume that the expression is correct for low nitrogen concentrations and pressures, the calculated k-value can be used to determine the equilibrium nitrogen pressure required to reduce the nitrogen concentration to the desired concentration level. Goward [64] has done this for several metals (Table 1.7); the equilibrium nitrogen pressure is that in equilibrium with 1 ppm of nitrogen in solution in the metal. His data show that the hot extraction of nitrogen from iron, molybdenum, and tungsten should be feasible. This has been verified for molybdenum and tungsten [11], which were heated to 2400°K at 10^{-6} torr in measurements on the solubility of nitrogen in these metals. As shown in Table 1.7, extraction of nitrogen from niobium (2420°C) and tantalum (2960°C) to the 1-ppm level could not be achieved readily because of the higher temperatures required. [Additional data [65] on the niobium–nitrogen and the tantalum–nitrogen systems have been presented as equilibrium (pressure–temperature–concentration) diagrams.] The previous discussion has been limited to temperatures obtainable with conventional equipment. To obtain higher temperatures (5000–8000°C) for the analysis of thin tantalum films, Guldner [14] used a xenon flash discharge to vaporize

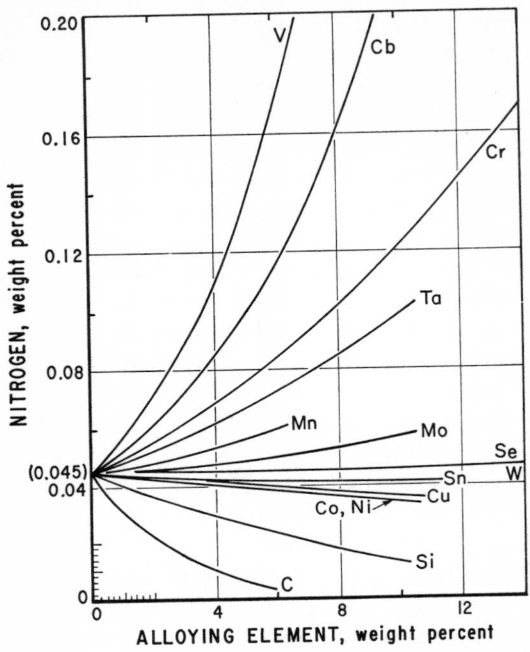

Fig. 1.10. Solubility of nitrogen in liquid iron alloys at 760 torr pressure of nitrogen and 1600°C. (Reproduced by permission from John Wiley & Sons, Inc. [62].)

TABLE 1.7. Calculated Equilibrium Pressures for Various Metals Containing 1×10^{-4} wt % Nitrogen[a]

Element	Temperature, °C	Equilibrium Nitrogen Pressure, torr	$t_{1/2}$[b]
Fe	1500	4.8×10^{-2}	1.6
Fe_l	1600	4.8×10^{-3}	1.4[c]
Cr	1300	9.9×10^{-6}	—
Cr_l	1600	4.6×10^{-7}	—
Mo	2000	6.9×10^{-2}	7
W	2000	6.3×10^{-2}	31
Nb	2000	1.2×10^{-9}	1.9
	2420	3.8×10^{-9}	—
Ta	2000	2.7×10^{-9}	2.8
	2960	1.4×10^{-6}	—

[a] Reproduced by permission from *Analytical Chemistry* [64].
[b] Extraction time, minutes, for removal of one-half of the nitrogen present.
[c] 1.6-cm crucible diameter.

the specimens for the extraction and determination of hydrogen, nitrogen, and oxygen.

The completeness of nitrogen recovery from a metal by a hot-extraction method may be affected by other elements present. Thus, although the data of Table 1.7 indicate that nitrogen could be extracted from pure iron, an element that forms a stable nitride may be present in a trace concentration great enough to prevent quantitative nitrogen extraction. Studies on the kinetics of the hot extraction of nitrogen from a steel at different temperatures show rate differences that are probably associated with the diffusion of dissolved nitrogen and the decomposition of a nitride [66].

Theoretical considerations of the determination of nitrogen in metals by the vacuum-fusion and hot-extraction methods usually have been based on the thermochemistry of the metal nitrides. Goward [64] has emphasized that the conclusions regarding the equilibrium pressures from such calculations are often in error because the substances being analyzed are not metal nitrides but metals containing dissolved nitrogen. In the niobium–nitrogen system, for example, the reactions for the removal of nitrogen are as follows: nitride (e.g., Nb_2N) $\rightarrow$ α solid solution + nitrogen; and α solid solution $\rightarrow$ metal + nitrogen when the partial pressure of nitrogen in the system was less than the equilibrium nitrogen pressure at the temperature and nitrogen concentration of the specimen. To remove nitrogen from the nitride, the nitrogen pressure in the analytical system need only be as low as the nitride dissociation pressure (line *B*, Fig. 1.4). If, for example, a metal nitride contains 30% nitrogen and the terminal solubility is 0.01%, the nitrogen content will be reduced from 30 to 0.01% at the nitride dissociation pressure and nitrogen recovery will be essentially complete. However, if the metal being analyzed contains only 0.009% nitrogen (in solution), no nitrogen will be extracted at the nitrogen equilibrium pressure represented by line *B*, and the sample will appear to contain no nitrogen. The dissolved nitrogen could be extracted, however, by reducing the equilibrium nitrogen pressure to an appropriate level as determined from Sievert's law. Thus, calculations on the equilibrium pressures required for extracting nitrogen from metals should not be based on the dissociation pressures of the nitrides.

D. OXYGEN

Many metals in the molten state dissolve large quantities of oxygen, most of which precipitates as oxides when the metal solidifies. The measurement of the terminal solubility of oxygen in many metals is difficult because an oxide phase is formed at low oxygen concentrations. The solubility of oxygen in osmium, iridium, platinum, gold, and mercury is so low that it has not been determined [1].

In contrast to the many instances in which the terminal solubility of oxygen is low, solid solutions containing approximately 30 at. % oxygen may be formed in titanium and zirconium before an oxide phase is formed [60]. The next largest quantities of oxygen that can be absorbed in solid solution are in vanadium, niobium, and tantalum. Unlike hydrogen and nitrogen, which dissolve interstitially in the transition elements, oxygen is known to dissolve interstitially in vanadium [67] and tantalum [68] only. The solubility of oxygen in transition metal alloys has been related to the electron–atom ratio of the alloy [69]. Information on the solubility of oxygen in metals has been collected and evaluated by Hansen and Anderko [60], who give phase diagrams as well as crystal-structure data on binary oxygen–metal systems.

Important information is often obtained by isolating, identifying, and determining quantitatively the oxide phases that are formed in metals. These analyses are difficult for several reasons:

1. The oxide in equilibrium with a metal may be of the type, $M_xO_{(y\pm a)}$, that is not represented by a simple chemical formula. The composition of lead oxide, for example, can vary between $PbO_{1.00}$ and $PbO_{1.10}$.
2. Studies on metal–oxygen systems are often complicated by the formation of a series of oxide phases (e.g., TiO, Ti_2O_3, Ti_3O_5, and TiO_2). The stoichiometry and structure of metal oxides have been discussed [59, 70].
3. Complex oxides of the type $A_xB_y \cdots O_n$ may be formed that contain two or more different kinds of atoms in addition to oxygen [71]. Examples of these are $MnSiO_3$, $MgCr_2O_4$, and $CaTiO_3$.

Thus, alloys that contain several metallic elements may combine with oxygen to form a number of simple and complex oxides. Because these oxides have varying chemical and thermal stability, their quantitative separation from metals and identification is difficult.

E. INERT GASES

The inert gases (helium, neon, argon, etc.) are formed in metals in situ as a result of nuclear disintegration processes; for example, helium is produced from both beryllium and boron by nuclear reactions, and krypton and xenon are produced from uranium by fusion. At temperatures too low for the inert gas to diffuse through the metal, supersaturation can occur. At higher temperatures, the inert-gas atoms tend to diffuse through the lattice and form gas bubbles, sometimes at pressures high enough to modify the properties of the material [72–75].

The inert gases evidently have no appreciable equilibrium solubility in metals in either the solid or liquid state [72, 76, 77]. When xenon was equilibrated at a pressure of 760 torr with three metals in the liquid state [78], the atom fraction of xenon present in each metal was 3.4×10^{-5} for sodium (147°C), 4.1×10^{-9} for bismuth (491°C), and 9.4×10^{-10} for mercury (22°C). The inert gases have been made to penetrate a metal lattice by means of an electrical discharge or ion bombardment; by these means, gas concentrations near the surface of the metal have approached 1 at. % [79].

III. DIFFUSION

A. INTRODUCTION

The diffusion of a single atom of hydrogen, nitrogen, or oxygen in a gas–metal system will occur as a series of random motions that will lead under fixed conditions to a uniform concentration of the gas throughout the metal. When the gas near the surface of the metal is removed by imposing non-equilibrium conditions upon the system, there will be a net transfer of the gas atoms from the region of higher to that of lower concentration. This transfer process is an integral part of many of the techniques used for determining gases in metals.

1. Fick's Laws

The mathematical expression describing diffusion in the steady state of flow shows that the rate of transfer of a diffusing substance through unit area of a section is proportional to the concentration gradient measured normal to the section,

$$J = -D\frac{\partial c}{\partial x}, \tag{4}$$

where J is the rate of transfer per unit area of section, c is the concentration of diffusing substance, x is the direction of diffusion normal to the section, and D is the diffusion coefficient. This is Fick's first law. The flux, J, of a material usually has the units of moles per second per square centimeter, with concentration c in moles per cubic centimeter and length x in centimeters. The diffusion coefficient, D, is independent of the units of concentration and is usually given in square centimeters per second. In the steady state the concentration gradient at point x, $\partial c/\partial x$, is constant throughout the object. (The negative sign is required because the diffusion occurs in the direction opposite to that of increasing concentration.) It is assumed that the diffusion coefficient is independent of concentration in the dilute solutions, the rate of flow of gas through the metal is independent of direction, and only one constituent, the gaseous atoms, moves in the system.

For the nonsteady state (when the concentration gradient changes with time) Fick's second law is applied

$$\frac{\partial c}{\partial x} = D \frac{\partial^2 c}{\partial x^2}. \tag{5}$$

This equation is applicable in integrated form to various boundary conditions. Solutions to the equation for diffusion into or out of objects of certain geometric shapes are given in Table 1.8. For each shape, the fraction of gas extracted is a function of Dt/L^2, where L is the radius of an infinite cylinder or a sphere, or the half-thickness of an infinite slab.

2. Fractional Extraction

Table 1.8 is useful for determining the time required for the fractional extraction (or fractional saturation) of a gas from samples of different diffusivity, D, and dimensions, L. The data for the slab and the cylinder apply where diffusion is negligible through the faces of the cylinder or the edges of the slab. The data given for the cylinder apply when the length of the specimen is (roughly) ten times more than that of the radius, in which case diffusion is mainly radial. It can be shown [80] that the function, Dt/L^2, for a cylinder of length L and radius 0.5 L approximates that given for a sphere, which is given in Table 1.8. From additional tables [2, 81], the spatial distribution of the diffusing species can be calculated. Gas–metal diffusion theory is treated in detail elsewhere [2, 57, 82, 83].

The data of Table 1.8 are used with the assumptions that initially the gas is distributed evenly throughout the metal and the diffusion step is rate-controlling. In practice, however, the desorption of the gas from the metal surface is sometimes slow in comparison with diffusion to the surface. In analytical work, extraction times longer than indicated by the above considerations will be required because of the time needed for the diffusion gradient between axis and surface to be formed. In addition, time must be allowed for the sample to reach the temperature of extraction.

The effect of temperature on diffusion follows an exponential law

$$D = D_0 \exp\left(\frac{-E_D}{RT}\right), \tag{6}$$

where D and D_0 are expressed in the same units (area per time), E_D is the activation energy of the diffusion process, R is the gas constant, and T is the absolute temperature. The activation energy, E_D, is determined from the slope of the line obtained by plotting the logarithms of the experimentally determined D-values versus the reciprocal of the absolute temperatures. The slope of the line and the activation energy will change at temperatures of allotropic transformation of the metal. Hobson [84] has discussed data on the

TABLE 1.8. Fractional Extraction (Average Composition of Slab, Cylinder, or Sphere During Diffusion[a])

Dt/L^2	Fraction Extracted = $(C_m - C_0)/(C_s - C_0)$		
	Slab	Cylinder	Sphere
0.005	0.078	0.157	0.226
0.01	0.110	0.216	0.310
0.02	0.161	0.302	0.421
0.03	0.195	0.360	0.500
0.04	0.227	0.412	0.560
0.05	0.251	0.452	0.604
0.06	0.275	0.488	0.648
0.08	0.320	0.550	0.720
0.10	0.357	0.606	0.774
0.15	0.438	0.708	0.861
0.20	0.503	0.781	0.916
0.25	0.560	0.832	0.948
0.30	0.612	0.878	0.969
0.40	0.702	0.932	0.988
0.50	0.764	0.962	0.996
0.60	0.816	0.979	0.998
0.70	0.856	0.988	0.999
0.80	0.887	0.993	
0.90	0.912	0.996	
1.00	0.931	0.998	
1.50	0.980		
2.00	0.994		

[a] Reproduced by permission of the McGraw-Hill Book Co. [2].
C_m = mean concentration at time t.
C_0 = uniform initial concentration.
C_s = constant surface concentration.
D = diffusion constant.
L = the radius of the cylinder or sphere, or the half-thickness of the slab.

diffusion of hydrogen from steel in the 100–200°C range when the exponential law was not followed.

3. Rate-Controlling Factors

The rate of diffusion of a solute atom in a metal is controlled by the allotropic form of the metal, the temperature, the presence and concentration of alloying elements, and the heat treatment and cold work the metal has received. Diffusion may proceed along the grain boundaries as well as through the grains; as a result, the grain structures may affect the rate of migration. In the removal of gases from metals by evolution, the following consecutive steps occur: (1) the lattice diffusion of the gas (in the atomic form) to the metal surface; (2) the transfer of the absorbed atom at the gas–metal interface to give adsorbed molecules of gas; and (3) desorption of the molecules into the gas phase. The last step is frequently rate-controlling.

Certain metals are used as solid semipermeable membranes for the separation or purification of a gas. The best known are the palladium and the 75% palladium–25% silver systems that have been used to provide hydrogen with an impurity content estimated to be less than 1 part in 10^{10} [85, 86]. Heated palladium tubes are used routinely to separate hydrogen from other gases for the manometric determination of hydrogen (see Chapter IV, Section I.B.3.a). Oxygen of high purity can be admitted into an evacuated system by heating an attached silver tube that is exposed to the atmosphere [87].

The permeation of gases through the fused silica furnace tubes used in the determination of gases in metals is usually not considered. Interesting measurements have shown that with a fused silica capsule 1 cm in diameter, 10 cm long, and walls 0.1 cm thick held in air at 1000°C, the helium equilibrium pressure of 4×10^{-3} torr will be reached within an hour [88]. Oxygen permeation is much less, with a pressure of 2×10^{-4} torr being reached after 10 h. If the silica capsule were not degassed, the oxygen pressure would be about 1×10^{-2} torr in 10 h. Although these pressures are relatively low, they represent potential sources of error when small quantities of gas are being determined.

B. HYDROGEN

Hydrogen diffuses readily through most metals, partly because of the small size of the hydrogen atoms as compared to the metal atoms. For those conditions under which hydrogen is determined, it can probably be assumed that rapid desorption occurs at the gas–metal interface and that the diffusion step is rate-determining [89]. The extraction time will vary with the square of the dimensions, L, for similarly shaped specimens of the same material at a given temperature. For example, if at a given temperature 10 min are required for extracting hydrogen from a sheet of 2-mm thickness (1-mm half-thickness), 40 min will be required for a sheet 4 mm thick, and 160 min will be required for a sheet 8 mm thick. Some hydrogen-extraction-versus-time curves are shown in Fig. 1.11. The calculations are based on diffusivity data for γ-Fe (18% Cr, 9% Ni) at 1000°C [90]. Curves 1, 2, and 3 are for samples of the same volume (0.79 cm^3) but of different geometries. Hydrogen determinations are generally made on samples of cylindrical shape where the height is approximately twice the radius; curves 3 and 4 illustrate the effect of doubling these dimensions on the extraction time. Swinburn has provided theoretical curves for the evolution of hydrogen from ferritic steel cylinders of $\frac{1}{4}$- and $\frac{1}{2}$-inch diameter at temperatures of 500, 600, and 700°C [91].

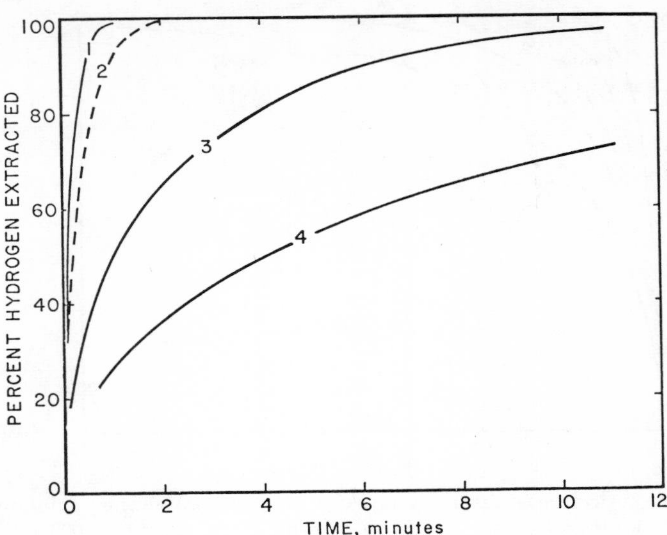

Fig. 1.11. Percent hydrogen extracted (average composition) during diffusion from samples of different geometry. 1, Cylinder: $h = 6.3$, $r = 0.20$, $v = 0.79$. 2, Sheet: $l = 3.0$, $w = 1.5$, $h = 0.17$, $v = 0.79$. 3, Cylinder: $h = 1.0$, $r = 0.5$, $v = 0.79$. 4, Cylinder: $h = 2.0$, $r = 1.0$, $v = 3.1$.
Height (h), radius (r), width (w), length (l) in cm, volume (v) in cm^3.

The time required to extract a given percentage of the hydrogen from a metal sample of a simple geometric shape can be calculated when the diffusion constant is known. For example, determine the time required for the vacuum extraction of 99% of the hydrogen from a sample of 3.1% Si-iron at 500°C where the sample is 1.0 cm long and 1.0 cm in diameter. The diffusion constant is 4.16×10^{-5} cm^2/sec:

$$\text{Fraction extracted} = \frac{C_m - C_0}{C_s - C_0} = 0.99. \tag{7}$$

From Table 1.8, $Dt/L^2 = 0.40$ (the cylindrical sample is of dimension $R = 0.5L$, and we use the date given for a sphere, as noted in Section III.A.2):

$$t_{\text{sec}} = \frac{(0.40)(0.50 \text{ cm})^2}{4.16 \times 10^{-5} \text{ cm}^2/\text{sec}} \tag{8}$$

$$= 2400 \text{ sec, or } 40 \text{ min.}$$

The diffusion rate of hydrogen through a metal will depend on the allotropic form of the metal. In the case of iron, for example, hydrogen diffuses much more rapidly through the alpha (ferritic) form than the gamma

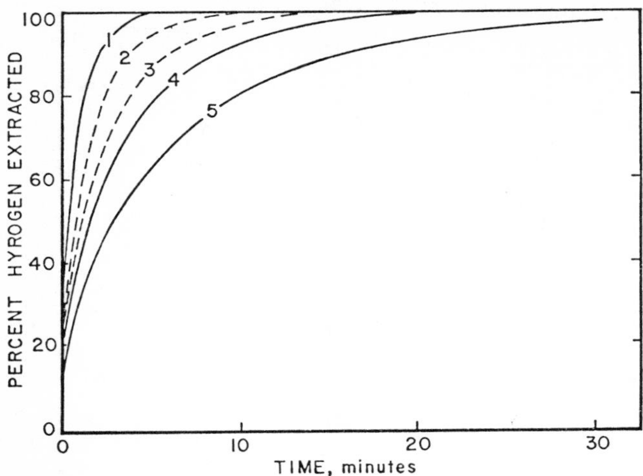

Fig. 1.12. Percent hydrogen extracted (average composition) during diffusion from cylindrical samples (h = 1.0 cm, r = 0.5 cm). 1, α-Fe, 600°C. 2, γ-Fe, 1000°C. 3, γ-Fe (18% Cr, 9% Ni), 1000°C. 4, α-Fe (27.5% Cr), 600°C. 5, α-Fe (4.3% Si), 600°C.

(austenitic) form. For this reason, hydrogen is often extracted from ferritic material at temperatures just below the temperature of the alpha–gamma transformation to take advantage of the maximum rate of diffusion of the hydrogen. Materials that are austenitic at room temperature (for example, 18% Cr, 8% Ni stainless steels) are usually analyzed for hydrogen at approximately 1000°C. Although the rate of diffusion increases exponentially with temperature (Eq. 6), curves 1 and 2 of Fig. 1.12 show that hydrogen can be extracted more readily from the alpha form at 600°C than from the gamma form at 1000°C.

Diffusion rates in alloys are complicated by the effects of alloying elements and lattice structure. Diffusivity data [90] were used to calculate the extraction curves of Fig. 1.12, which show that at 600°C the time required for the extraction of hydrogen from α-iron $<$ α-iron (27.5% Cr) $<$ α-iron (4.3% Si); and at 1000°C γ-iron $<$ γ-iron (18% Cr, 9% Ni). Diffusion rates have been correlated with changes in the hardness and density of steel [92]. The variability in the observed diffusion constants for hydrogen in iron and ferritic steels has been attributed to the interaction of hydrogen with trapping sites in the metal lattice and to composition [93–97]. The effects of isothermal transformation characteristics on the removal of hydrogen from various steels have also been studied [98].

Diffusion studies on some metals, such as steel [49, 99–102], are complicated by the fact that hydrogen is apparently present in different forms, namely, hydrogen in the atomic form in solution interstitially or in lattice defects and molecular hydrogen and/or methane in discontinuities. As a result, several temperature-dependent processes may contribute to the quantity of hydrogen evolved at any given extraction temperature, and the rate of hydrogen removal will not be that for hydrogen when present in a simple solid solution. It is therefore evident that attempts to separate the different forms of hydrogen present by a fractional extraction procedure at a given temperature will not be quantitative. Nevertheless, fractional extractions have been reported for steel, usually at temperatures of about 25°C, 650°C, and 1000°C.

C. NITROGEN

Nitrogen diffuses through some metals fast enough for the determination of nitrogen by heating the metal in an evacuated system and determining the nitrogen evolved. Data presented by Goward [64] (Table 1.7) include the calculated extraction times required for the removal of one-half of the nitrogen, $t_{1/2}$, from cylindrical metal samples 0.4 cm in diameter. The extraction times are short enough to be practical; analytical results that have been obtained on the determination of nitrogen in molybdenum and tungsten [103, 104] verify these calculations. In some instances the rate of evolution of nitrogen is controlled not by the rate of diffusion but by the rate of combination of the nitrogen atoms at the solid surface to form molecular nitrogen. Thus, the rate of denitriding ε-iron nitride has been determined to be 10^4 greater in hydrogen than under vacuum [105].

D. OXYGEN

As is the case with nitrogen, the diffusion rate of oxygen in solid metals is much less than that of hydrogen. Therefore, oxygen usually cannot be determined in a reasonable period of time by extraction from a solid sample at a high temperature. Figure 1.13 illustrates the difficulty that would be encountered in the determination of oxygen in iron by deoxidation in hydrogen; after the sample had been heated at 1000°C for 100 h in pure dry hydrogen, oxygen removal was still far from complete [106]. Deoxidation will proceed much faster at higher temperatures with the liquid metal. Studies with H_2:H_2O mixtures and oxygen in iron, cobalt, and nickel at 1550°C showed that equilibrium was reached within 1 h [107]. Even at high temperatures, a significant amount of the oxygen present may be combined as refractory oxides (SiO_2, Al_2O_3, etc.) that are not reduced by hydrogen (see Section IV.D).

When solid metals containing oxygen and carbon are heated under vacuum, they diffuse independently to the metal surface, where they combine and pass

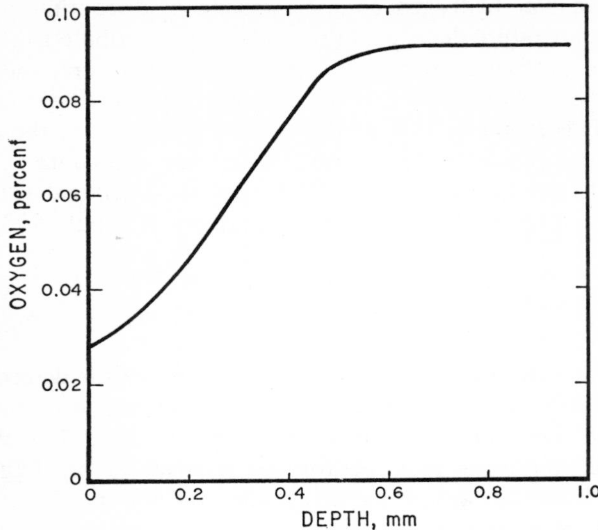

Fig. 1.13. Deoxidation of Armco iron in hydrogen for 100 h at 1000°C. (Reproduced by permission from *Transactions of the Faraday Society* [106].)

into the gas phase as carbon monoxide and carbon dioxide [3]. When the metal is heated to a still higher temperature, an additional burst of gas is often evolved because of the faster rates of diffusion. For the removal of most of the oxygen, carbon must be present in excess and sufficient time must be allowed for completion of the reaction. When nickel essentially free of carbon was heated under vacuum, no oxygen was evolved; the oxygen that diffused through the nickel remained adsorbed on the surface [108].

E. INERT GASES

The observed rate of evolution of an inert gas from a solid metal will depend on whether the gas is in solution or is present as gas bubbles and on whether the gas is concentrated near the surface (see Section II.E). According to Norton and Tucker [109], three types of release of rare gases from metals can be defined, depending on how the gases were placed in the metal. Placement by ion bombardment, fission processes, or gas discharge leads, respectively, to easy release at low temperatures, difficult release at high temperatures, or retention in the sputtered metal up to the point of evaporation of the metal.

The rate of diffusion of an inert gas in a solid metal is relatively low. Therefore, metals should be melted for a rapid gas extraction. For example, helium has been removed rapidly from beryllium (melting point 1283°C) by heating to 1300°C [110]. Alternatively, the metal sample may be dissolved in an acid solution and the liberated inert gas then determined.

IV. THERMOCHEMISTRY

A. INTRODUCTION

The accurate determination of the gas concentration in a metal depends essentially on the separation of the gaseous elements from the metals. Once separated, the quantity of each gas present can usually be determined reliably without difficulty. As can be noted from the numerous methods described in later chapters, various reactions are used to make these separations. All these reactions have a driving force, a free-energy change, that accompanies the reaction. The thermochemistry of these reactions should be considered when evaluating existing methods of analysis and considering new ones [2, 7, 111, 112].

1. Free-Energy Change and Equilibrium Constants

When a reaction occurs spontaneously at constant pressure and temperature, a decrease in free energy results. The greater the decrease in free energy, the greater is the tendency for the reaction to proceed. Because the standard free-energy change of a reaction is related to the equilibrium constant, free-energy data are helpful in determining how complete the reaction will be when equilibrium conditions are reached. Although the standard free-energy change for a reaction is negative, the reaction may not proceed at a favorable rate under a given set of conditions. For quantitative chemical analysis, reactions are sought that go to completion. Quite often reactions with an unfavorable equilibrium constant may go to completion, or essentially to completion, when the conditions of the reaction are changed or when one (or more) of the reaction products is removed as it is formed.

A type of reaction of interest here can be written as

$$m\mathrm{M(s)} + n\mathrm{X_2(g)} = \mathrm{M}_m\mathrm{X}_{2n}\mathrm{(s)}, \tag{9}$$

where M is a metal and X is hydrogen, nitrogen, or oxygen, and (s) and (g) signify solid and gas, respectively. For a given temperature, the equilibrium constant K for Eq. 9 can be written from the law of mass action as

$$K = \frac{[\mathrm{M}_m\mathrm{X}_{2n}]}{[\mathrm{M}]^m[\mathrm{X}_2]^n} \tag{10}$$

when the activity of each component is equal to its concentration. Equilibrium constants of interest are often not available.

Standard-free-energy data are available, however, from which the equilibrium constant for many reactions can be calculated by the well-known expression

$$\Delta G^\circ = -RT \ln K = -RT \ln \frac{[M_mX_{2n}],}{[M]^m[X_2]^n} \tag{11}$$

where ΔG° is the standard free energy of formation of one mole of M_mX_{2n} from its elements in their standard states at T° Kelvin and R is the gas constant. The substances are generally defined to be in their standard states (activity is unity) when present as a pure solid or liquid or as the gas at 1 atm pressure. When mM and M_mX_{2n} exist in their standard states at 1 atm pressure,

$$K = \frac{1}{[X_2]^n}, \tag{12}$$

and the equilibrium dissociation pressure at a particular temperature can be calculated by the relationship

$$\ln P_{X_2} = \frac{1}{n}\frac{\Delta G^\circ}{RT}. \tag{13}$$

Equilibrium data have been summarized as diagrams of $\Delta G^\circ = RT \ln P_{X_2}$ versus temperature. They are given for metal–nitrogen–metal-nitride systems in Fig. 1.14 [113] and for metal–oxygen–metal-oxide systems in Fig. 1.15 [2, 114]. Similar diagrams have been made for the metal–carbon–metal-carbide [115] and the metal–sulfur–metal-sulfide systems [116]. For a series of compounds at a given temperature, the relative stabilities will increase (decomposition pressures will decrease) as their respective ΔG°-values become more negative. The relative order of the stability of the compounds is given in the diagrams in terms of their standard free energy of formation at that temperature; at another temperature, the relative order of the stabilities of the compounds may not be the same. The use of these diagrams will be discussed later.

The simplifying assumptions leading to Eq. 13 cannot always be made, as when condensed phases of variable composition are formed. In that case, Darken and Gurry [2] have written,

> "The equilibrium constant may readily be evaluated even though the phases may vary in composition if sufficient data are available to establish the standard free energy change; however, the activity of the metal and of the oxide may not be equated to unity if the condensed phases are not of fixed composition. Thus P_{O_2} may not be evaluated even though ln K is known unless information is available not only as to the equilibrium compositions but also as to the activity of the metal and of the oxide as functions of composition. Hence one method of determining the relation between the partial pressure and the temperature is from a knowledge of the standard free energy change and the activities of the metal and of the oxide."

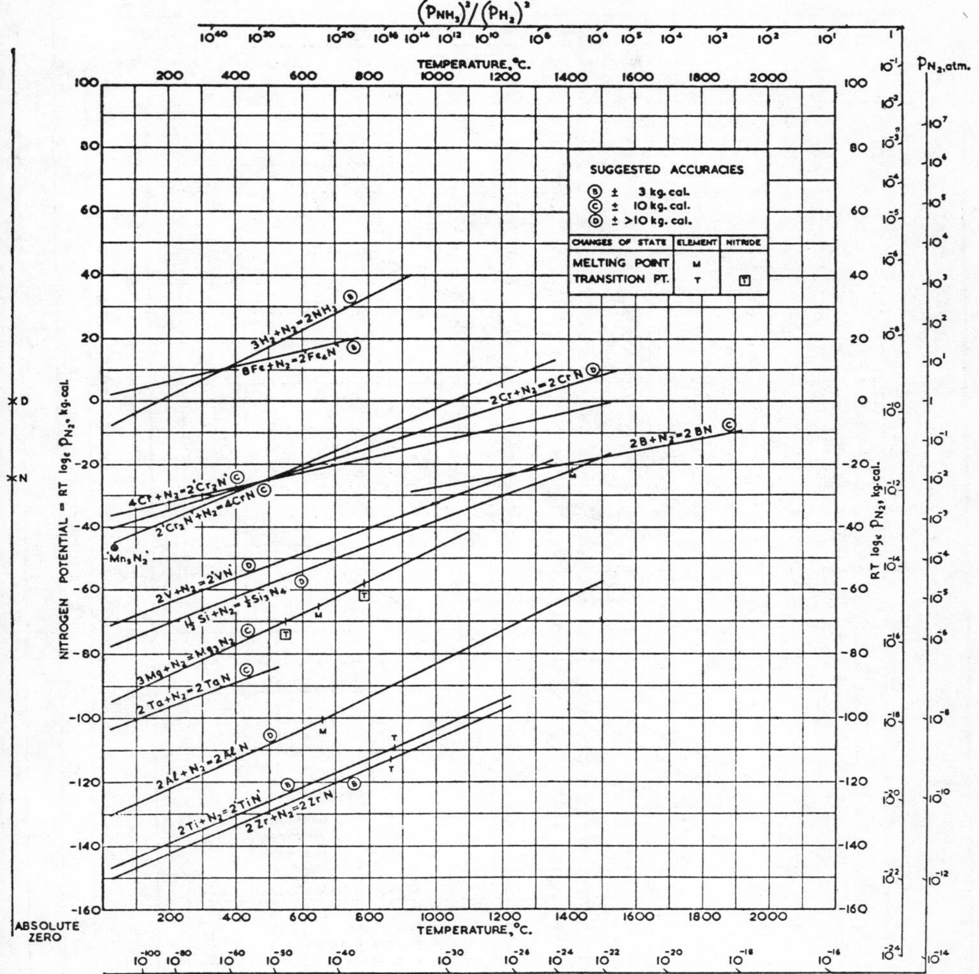

Fig. 1.14. Standard free energy of formation of nitrides as a function of temperature. (Reproduced by permission from the *Journal of the Iron and Steel Institute* [113].)

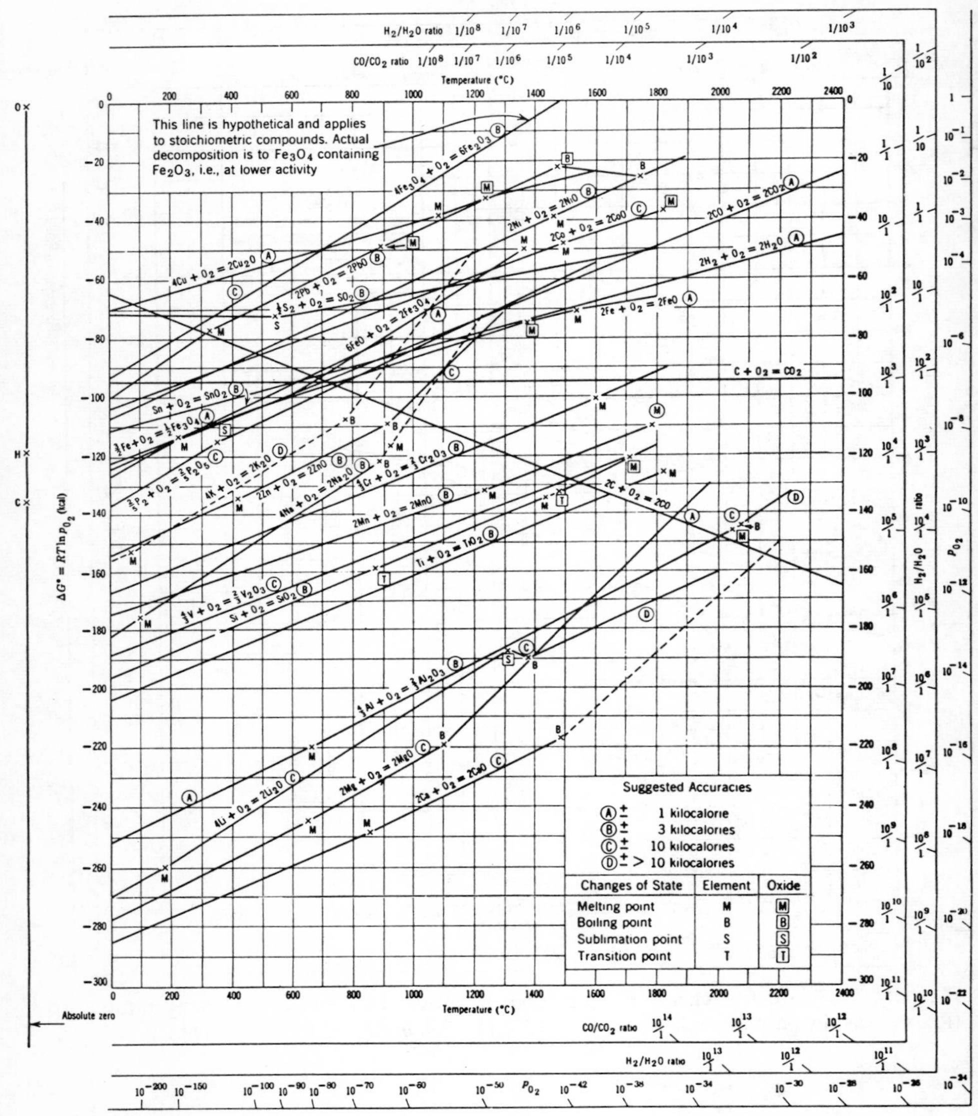

Fig. 1.15. Standard free energy of formation of oxides as a function of temperature. (Reproduced by permission from the McGraw-Hill Book Co. [2].)

The lack of free-energy and activity data for many gas–metal systems limits the use of thermochemical considerations in the appraisal of reactions for determining gases in metals.

The equilibrium pressure at a given temperature can be determined by Eq. 13 when vaporization occurs only by decomposition of the compound into its elements, as in

$$CuO(s) = Cu(s) + \tfrac{1}{2}O_2(g). \qquad (14)$$

However, vaporization also occurs by three other ways, namely:

1. Vaporization to gas molecules of the same composition,

$$ZrO_2(s) = ZrO_2(g) \qquad (15)$$

2. Vaporization by disproportionation to dissimilar molecules that are not both elements,

$$Al_2O_3(l) = 2AlO(g) + O(g) \qquad (16)$$

3. Vaporization with formation of a gaseous polymer,

$$3MoO_3(s) = (MoO_3)_3(g). \qquad (17)$$

Thus, a compound may vaporize by several ways simultaneously, in which case the pressure of the gas over the compound will be the sum of the partial pressures of all the gaseous molecules produced. There are no quantitative data on many of the vapor species of interest.

2. Thermochemical Calculations

Many of the methods used to determine gases in metals involve displacement-type reactions, some of which are represented by the following:

$$TiO_2 + C = TiC + 2CO \qquad (18)$$

$$2NbN + C = 2NbC + N_2 \qquad (19)$$

$$2CrN + 3H_2SO_4 = Cr_2(SO_4)_3 + 2NH_3 \qquad (20)$$

$$Si_3N_4 + 6NaOH + 3H_2O = 3Na_2SiO_3 + 4NH_3. \qquad (21)$$

Reactions typified by Eqs. 18 and 19 occur in the determination of oxygen and nitrogen by high-temperature fusion methods; they are discussed in Chapter 2. The reactions showing the formation of ammonia from the dissolution of chromium nitride in sulfuric acid (Eq. 20) and silicon nitride in sodium hydroxide (Eq. 21) are typical of established empirical methods for which no thermochemical data are available. Metal specimens are sometimes reacted with sulfur, hydrogen sulfide, or sulfur monochloride for the determination of oxygen, as discussed in Chapters 9 and 12.

The application of thermochemical data to an analytical problem can be illustrated as follows. Suppose that an independent method is needed to check upon the accuracy of the determination of oxygen in a calcium–calcium oxide mixture by the carbon-reduction method,

$$CaO + C = Ca + CO. \tag{22}$$

It can be shown that the hydrogen-reduction method,

$$CaO + H_2 = Ca + H_2O, \tag{23}$$

is not applicable, because the $H_2{:}H_2O$ ratio would have to be very high (larger than 10^{13} at 1000°C) for the reaction to proceed. The sulfur-reduction method, Eq. 27, is next considered. (In this method, the quantity of oxygen present is determined by an iodatimetric titration of the sulfur dioxide that is formed.) By algebraic addition of Eqs. 24–26 and their corresponding values of $\Delta G°$ [117],

$$\begin{array}{lll} 2CaO = 2Ca + O_2 & \Delta G° = +244{,}200 & (24) \\ 2Ca + S_2 = 2CaS & \Delta G° = -202{,}000 & (25) \\ \tfrac{1}{2}S_2 + O_2 = SO_2 & \Delta G° = -\ \ 65{,}450 & (26) \\ \hline 2CaO + \tfrac{3}{2}S_2 = 2CaS + SO_2 & G° = -\ \ 23{,}250 & (27) \end{array}$$

the $\Delta G°$ value for Eq. 27 is seen to be $-23{,}250$ calories at the chosen temperature of 1200°K (927°C). By substituting into Eq. 11,

$$-23{,}250 = -(4.574)(1200) \log K. \tag{28}$$

If we make the broad assumptions that all phases are in their standard states,

$$K = \frac{P_{SO_2}}{P_{(S_2)^{\frac{3}{2}}}} = 1.7 \times 10^4. \tag{29}$$

The equilibrium for the reaction is favorable, and the reaction appears to be possible. To ensure the complete removal of oxygen from the sample, a carrier gas should be used to sweep the sulfur dioxide away as it is formed and thus force the reaction to completion. Oxygen in magnesium oxide could not be determined by sulfur reduction [118], and thermochemical calculations indicate that the method would not work for magnesium oxide.

The importance of thermodynamics as a source of information on the equilibria involved in the various reactions used for the determination of gases in metals is apparent. The rates at which the reactions proceed are also important. The two most important rate-controlling processes in the reactions are probably those of transport within a solid or liquid phase, and reaction at a phase boundary. To increase reaction rates, reactions are usually carried out at temperatures as high as is practical. Because diffusion rates are much higher in a liquid than in a solid, a fluxing agent is sometimes added to the material being analyzed. Material transport by convection is considerably more rapid than by diffusion; for this reason and because induction heating of the liquid causes stirring, induction heating is better than resistance heating in some applications.

The thermochemical data of interest to those determining gases in metals will include data not only on the metal nitrides and oxides but also data on the compounds that are formed in the chemical reactions used, such as the metal carbides, chlorides, and sulfides. Data on the latter are not included here because several compilations are available [117, 119, 120]. One source [117] presents a particularly convenient series of tables and charts on the physical and thermochemical properties of inorganic substances that includes a tabulation of standard-free-energy data over a wide range of temperature.

Data on the vapor pressures of the elements and their compounds are important, as when separations are based on the differences between the vapor pressures of the metals and their compounds. For example, the oxygen content of magnesium has been determined after subliming magnesium from the oxide [120]. Data on the vapor pressures of the solid and liquid elements are given in Table 1.9, as presented by Honig [122].

Nonmetallic phases are often separated from the metal by halogenating the sample and subliming or distilling the metal halide formed from the nonmetallic phase. Chlorination techniques are often chosen because the chlorination proceeds readily and because many of the chlorides formed have relatively high vapor pressures. Vapor-pressure data on some of the metal chlorides [123] are given in Fig. 1.16.

B. HYDROGEN

Hydrogen can be separated readily from most metals by hot-extraction methods. When extraction temperatures are required at which the metal has an appreciable vapor pressure, results may be low because of sorption of the evolved hydrogen by the metal vapor condensing in a cooler region of the analytical apparatus. This problem is overcome by reducing the vapor pressure of the metal by (1) dissolving the sample in a metal having a low melting point and a low vapor pressure, such as tin; (2) heating the sample in a closed capsule through which the hydrogen diffuses; or (3) heating the metal in dry oxygen to form the less volatile metal oxide and water, which is measured. The rate of hydrogen removal is of concern in all of these methods. The conditions required for extraction are a function of sample size and geometry and are determined empirically. Some metal hydrides have been analyzed for hydrogen by combustion in oxygen, the conditions being 30 min at 900°C for titanium, 30 min at 800°C for zirconium, 15 min at 1000°C for vanadium, and 30 min at 1100°C for niobium [124].

Hydrogen has been determined to be present in the fractional parts-per-million level in some materials by the hot-extraction method. The oxide

TABLE 1.9. Vapor Pressure Data for the Solid and Liquid Elements[a]

Symbol[b]	Element	Melting Point, °K	Boiling Point, °K	Data Temp. Range, °K	Temperatures (°K) for Vapor Pressures (torr)											
					10^{-8}	10^{-7}	10^{-6}	10^{-5}	10^{-4}	10^{-3}	10^{-2}	10^{-1}	1	10^{1}	10^{2}	10^{3}
Ac	Actinium	1320	3470	1873, Est.	1230	1305	1390	1490	1605	1740	1905	2100	2350	2660	3030	3510
Ag	Silver	1234	2435	958–2200	847	899	958	1025	1105	1195	1300	1435	1605	1815	2100	2490
Al	Aluminum	932	2736	1220–1468	958	1015	1085	1160	1245	1355	1490	1640	1830	2050	2370	2800
Am	Americium	<1103	2880	1103–1453	848	905	971	1050	1140	1245	1375	1540	1745	2020	2400	2970
As_4	Arsenic (s)	1090	886		377	400	423	447	477	510	550	590	645	712	795	900
At_2	Astatine	(575)	(610)	Est.	252	265	280	296	316	338	364	398	434	480	540	620
Au	Gold	1336	3081	1073–1847	1080	1150	1220	1305	1405	1525	1670	1840	2040	2320	2680	3130
B	Boron	2300	3950	1781–2413	1555	1640	1740	1855	1980	2140	2300	2520	2780	3100	3500	4000
Ba	Barium	983	1895	1333–1419	545	583	627	675	735	800	833	984	1125	1310	1570	1930
Be	Beryllium	1556	2757	1103–1552	980	1035	1105	1180	1270	1370	1500	1650	1830	2080	2390	2810
ΣBi	Bismuth	545	1852		602	640	682	732	790	860	945	1050	1170	1350	1570	1900
ΣC	Carbon		4130	1820–2700	1930	2030	2140	2260	2410	2560	2730	2930	3170	3450	3780	4190
Ca	Calcium	1123	1756	730–1546	555	590	630	678	732	795	870	962	1075	1250	1475	1800
Cd	Cadmium	594	1040	411–1040	347	368	392	419	450	490	538	593	665	762	885	1060
Ce	Cerium	1077	(3740)	1611–2038	1245	1325	1420	1525	1650	1795	1970	2180	2440	2780	3220	3830
Co	Cobalt	1768	3174	1363–1522	1195	1265	1340	1430	1530	1655	1790	1960	2180	2440	2790	3220
Cr	Chromium	2176	2938	1273–1557	1110	1175	1250	1335	1430	1540	1670	1825	2010	2240	2550	3000
ΣCs	Cesium	302	955	300–955	257	274	297	322	351	387	428	482	553	643	775	980
Cu	Copper	1357	2846	1143–1897	905	1060	1125	1210	1300	1405	1530	1690	1890	2140	2460	2920
Dy	Dysprosium	1680	2710	1258–1773	898	955	1020	1090	1170	1270	1390	1535	1710	1965	2300	2780
Er	Erbium	1770	(2850)	1773, Est.	922	981	1050	1125	1220	1325	1450	1605	1800	2060	2420	2920
Eu	Europium	1099	1764	696–900	556	592	634	682	739	895	884	981	1100	1260	1500	1800
Fr	Francium	(300)	(950)	Est.	242	260	280	306	334	368	410	462	528	620	760	980
Fe	Iron	1809	3148	1356–1889	1165	1230	1305	1400	1500	1615	1750	1920	2130	2390	2740	3200
Ga	Gallium	303	2676	1179–1383	892	950	1015	1090	1180	1280	1405	1555	1745	1980	2300	2730
Gd	Gadolinium	1585	(3000)	Est.	1035	1100	1170	1250	1350	1465	1600	1760	1955	2220	2580	3100
ΣGe	Germanium	1210	3100	1510–1885	1085	1150	1220	1310	1410	1530	1670	1830	2020	2320	2680	3180
Hf	Hafnium	>2400	4690	2035–2277	1760	1865	1980	2120	2270	2450	2670	2930	3240	3630	4130	4780
Hg	Mercury	234	630	193–575	201	214	229	246	266	289	319	353	398	458	535	642

Ho	Holmium	1734	2842	923–2023	922	981	1050	1125	1220	1325	1450	1605	1800	2060	2410	2910
In	Indium	429	2364	646–1348	761	812	870	937	1015	1110	1220	1355	1520	1740	2030	2430
Ir	Iridium	2727	4810	1986–2600	1850	1960	2080	2220	2380	2560	2770	3040	3360	3750	4250	4900
K	Potassium	336	1031	373–1031	294	315	338	364	396	434	481	540	618	720	858	1070
La	Lanthanum	1193	3610	1655–2167	1295	1375	1465	1570	1695	1835	2000	2200	2450	2760	3150	3680
Li	Lithium	454	1597	735–1353	508	541	579	623	677	740	810	900	1020	1170	1370	1620
Lu	Lutetium	1925	(3300)	Est.	1185	1260	1345	1440	1550	1685	1845	2030	2270	2550	2910	3370
Mg	Magnesium	923	1376	626–1376	458	487	519	555	600	650	712	782	878	1000	1170	1400
Mn	Manganese	1517	2309	1523–1823	778	827	884	948	1020	1110	1210	1335	1490	1695	1970	2370
Mo	Molybdenum	2890	4924	2070–2504	1865	1975	2095	2230	2390	2580	2800	3060	3390	3790	4300	5020
Na	Sodium	371	1156	496–1156	347	370	396	428	466	508	562	630	714	825	978	1175
Nb	Niobium	2770	4640	2304–2596	2035	2140	2260	2400	2550	2720	2930	3170	3450	3790	4200	4710
Nd	Neodymium	1297	3335	1240–1600	1000	1070	1135	1220	1320	1440	1575	1770	2000	2300	2740	3430
Ni	Nickel	1725	3159	1307–1895	1200	1270	1345	1430	1535	1635	1800	1970	2180	2430	2770	3230
Os	Osmium	3318	5260	2300–2800	2170	2290	2430	2580	2760	2960	3190	3460	3800	4200	4710	5340
P_4	Phosphorus	870	704		327	342	361	381	402	430	458	493	534	582	642	745
Pb	Lead	601	2016	1200–2028	615	656	702	758	820	898	988	1105	1250	1435	1700	2070
Pd	Palladium	1823	3310	1294–1640	1115	1185	1265	1355	1465	1590	1735	1920	2150	2450	2840	3380
ΣPo	Polonium	527	1220	711–1286	384	408	432	470	494	537	588	655	743	862	1040	1250
Pr	Praseo-dymium	1208	3295	1423–1693	1070	1140	1220	1315	1420	1550	1700	1890	2120	2420	2820	3370
Pt	Platinum	2043	4097	1697–2042	1565	1655	1765	1885	2020	2180	2370	2590	2860	3190	3610	4170
Pu	Plutonium	913	3508	1392–1793	1105	1180	1265	1365	1480	1615	1780	1975	2230	2550	2980	3590
Ra	Radium	973	(1800)	Est.	520	552	590	638	690	755	830	920	1060	1225	1490	1840
Rb	Rubidium	312	974		271	289	312	336	367	402	446	500	568	665	802	1000
Re	Rhenium	3453	5960	2494–2999	2220	2350	2490	2660	2860	3080	3340	3680	4080	4600	5220	6050
Rh	Rhodium	2239	4000	1709–2205	1550	1640	1745	1855	1980	2130	2310	2520	2780	3110	3520	4070
Ru	Ruthenium	(2700)	4392	2000–2500	1780	1880	1990	2120	2260	2420	2620	2860	3130	3480	3900	4450
ΣS	Sulfur	388	718		263	276	290	310	328	353	382	420	462	519	606	739
ΣSb	Antimony	903	1908	693–1110	552	582	618	656	698	748	806	885	1030	1250	1560	1960
Sc	Scandium	1811	3280	1301–1780	1045	1110	1190	1280	1380	1505	1650	1835	2070	2370	2780	3360
ΣSe	Selenium	490	952	550–950	336	356	380	406	437	516	570	636	719	719	826	972
ΣSi	Silicon	1685	3418	1640–2054	1265	1340	1420	1510	1610	1745	1905	2090	2330	2620	2990	3490

TABLE 1.9. *(continued)*

Symbol[b]	Element	Melting Point, °K	Boiling Point, °K	Data Temp. Range, °K	Temperatures (°K) for Vapor Pressures (torr)											
					10^{-8}	10^{-7}	10^{-6}	10^{-5}	10^{-4}	10^{-3}	10^{-2}	10^{-1}	1	10^{1}	10^{2}	10^{3}
Sm	Samarium	1345	2076	789–833	644	688	738	790	853	926	1015	1120	1260	1450	1715	2120
Sn	Tin	505	2891	1424–1753	955	1020	1080	1170	1270	1380	1520	1685	1885	2140	2500	2960
Sr	Strontium	1043	1640		514	546	582	626	677	738	810	900	1005	1160	1370	1680
Ta	Tantalum	3270	5510	2624–2948	2230	2370	2510	2680	2860	3080	3330	3630	3980	4400	4930	5580
Tb	Terbium	1638	(3295)	Est.	1070	1140	1220	1315	1420	1550	1700	1890	2120	2420	2820	3370
Tc	Technetium	(2400)	(4900)	Est.	1840	1950	2060	2200	2350	2530	2760	3030	3370	3790	4300	5000
Te_2	Tellurium	723	1267	481–1128	428	454	482	515	553	596	647	706	791	905	9065	1300
Th	Thorium	1968	5020	1757–1956	1705	1815	1935	2080	2250	2440	2680	2960	3310	2750	4340	5130
Ti	Titanium	1940	3575	1510–1822	1335	1410	1500	1600	1715	1850	2010	2210	2450	2760	3130	3640
Tl	Thallium	577	1710	519–924	556	592	632	680	736	803	882	979	1100	1255	1460	1750
Tm	Thulium	1873	2005	809–1219	731	776	825	882	953	1030	1120	1235	1370	1540	1760	2060
U	Uranium	1406	4090	1630–2071	1405	1495	1600	1720	1855	2010	2200	2430	2720	3080	3540	4180
V	Vanadium	2190	3652	1666–1882	1435	1510	1605	1705	1820	1960	2120	2320	2560	2850	3220	3720
W	Tungsten	3650	5800	2518–3300	2390	2520	2680	2840	3030	3250	3500	3810	4180	4630	5200	5900
Y	Yttrium	(1773)	3570	1774–2103	1230	1305	1390	1490	1605	1740	1905	2105	2355	2670	3085	3650
Yb	Ytterbium	1097	(1800)	Est.	520	552	590	638	690	755	830	920	1060	1225	1490	1840
Zn	Zinc	693	1184	422–1089	396	421	450	482	520	565	617	681	760	870	1010	1210
Zr	Zirconium	2128	4747	1949–2054	1755	1855	1975	2110	2260	2450	2670	2930	3250	3650	4170	4830

[a] Reproduced by permission from the RCA Review [122].
[b] Those elements whose gas phase is made up of two or more species are identified by a Σ preceding the symbol. For those elements that consist mostly of one molecular species, the appropriate subscript is added to the chemical symbol, as in As_4.

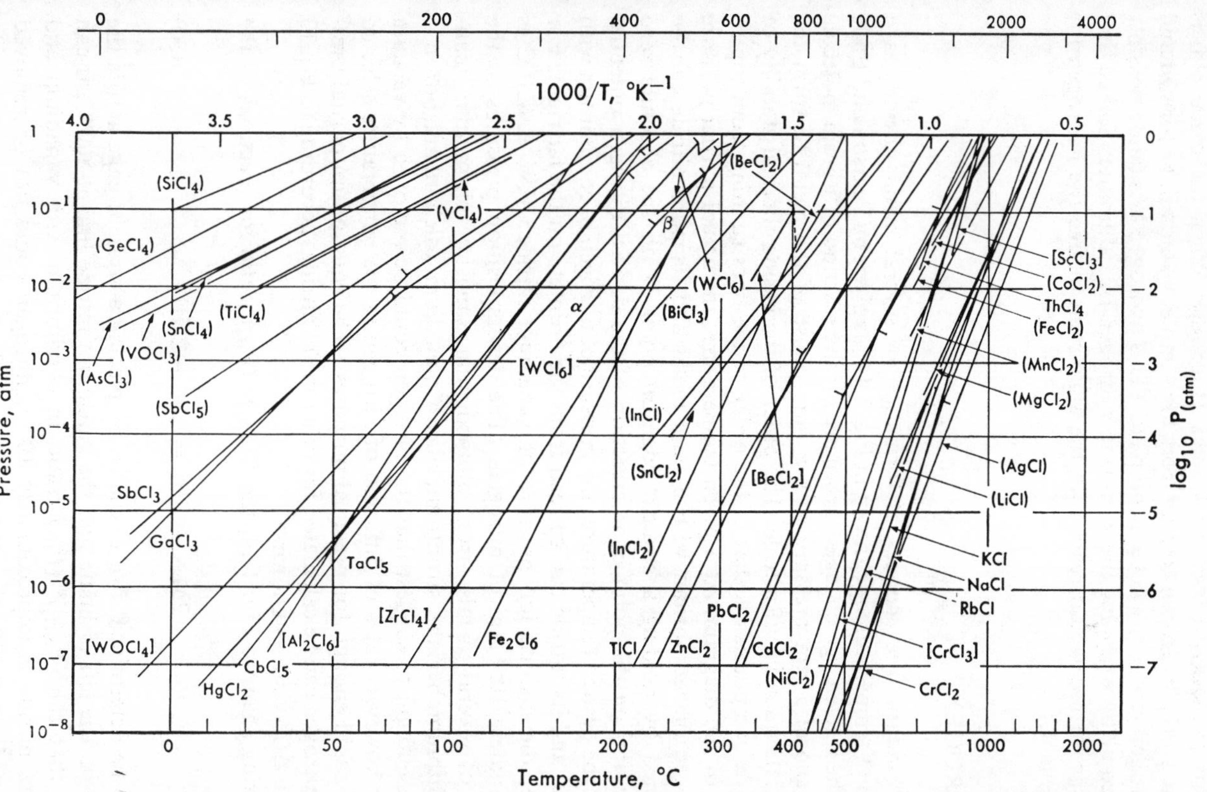

Fig. 1.16. Vapor pressures of the metal chlorides. (Reproduced by permission from *Zeitschrift fuer Erzbergbau und Metallhuettenwesen* [123].)
The formula for the chloride is enclosed in square brackets where the vapor pressure is for the solid only and in parentheses where the vapor pressure is for the liquid only. Where the curve is for both the solid and the liquid, the formula is not enclosed. A short, straight line cutting a curve indicates a phase change.

coating that forms spontaneously on samples prepared in the most careful manner is probably at least 10 A thick (Fig. 1.1). When the oxide film is reducible by the hydrogen coming from the sample, water will be formed and the hydrogen recovery will be incomplete. (This hydrogen loss could possibly be overcome by wrapping the sample in hydrogen-free aluminum foil so that any water formed would react with the aluminum to free the hydrogen for measurement.)

C. NITROGEN

The standard free energies of formation of several metal nitrides as a function of temperature are shown in Fig. 1.14, as given by Pearson and Ende [113]. The free-energy changes shown are for the reaction of 1 gram molecule of nitrogen at 1 atm pressure with a pure phase of the metal to form the appropriate quantity of nitride. The order of increasing thermal stability of the nitride compounds is that of the order of the lines running from top to bottom of the diagram. At a given temperature, the actual partial pressure of nitrogen in equilibrium with a pure metal and its pure nitride can be determined by holding a straight edge on point D on the absolute zero ordinate and on the point of intersection of the desired nitrogen potential line and the desired temperature; the pressure of nitrogen in equilibrium with the metal and its nitride can be read from the ordinate on the far right. For example, the dissociation pressure of vanadium nitride, VN, at 800°C is approximately 10^{-8} atm of nitrogen. (The decomposition pressures of the nitrides of several metals—iron, molybdenum, tungsten, chromium, manganese, lithium, barium, strontium, silicon, vanadium, calcium, and boron—are high enough to suggest that the nitrogen could be removed by vacuum heating; however, rates of diffusion of nitrogen at the temperature selected must be rapid enough for practical purposes.) The equilibrium partial pressure of nitrogen can be calculated for other metal nitrides according to Eq. 13 when the $\Delta G°$-value is known.

Referring again to Fig. 1.14, the $(P_{NH_3})^2:(P_{H_2})^3$ ratio for reactions of the type

$$2MN + 3H_2 = 2M + 2NH_3 \tag{30}$$

can also be determined. For example, by placing a straight edge across the point N on the left ordinate and the vanadium nitride line at 800°C, it can be seen that the $(P_{NH_3})^2:(P_{H_2})^3$ ratio in equilibrium with vanadium and vanadium nitride is approximately 10^{-15}; nitrogen could not be removed from vanadium nitride by treatment with hydrogen to form ammonia because of the unfavorable equilibrium.

The foregoing discussion of metal nitrides is relevant when metal nitrides are being considered, but nitrogen in solution must be considered when nitrogen is being determined. Regarding the latter, thermochemical calculations can seldom be made because of the paucity of data on the partial molar free energy of solution of nitrogen in metals. As a result, general differences in the behavior of metal–nitrogen systems are used to advantage in chemical analysis. For example, Oelsen and Sauer [125] based their determination on the "free" nitrogen content of steel on the nitrogen removed (as NH_3) in hydrogen at 750°C. In other examples, differences in the reactivity of metal nitrides to water, acid and alkaline solutions, halides, and organic reagents are the basis of procedures used to isolate and determine nitrides in metals.

D. OXYGEN

The variation of the standard-free-energy change with temperature in the conversion of 1 gram molecule of oxygen at 1 atm pressure into the metal oxide is shown in Fig. 1.15, which is reproduced from Richardson and Jeffes [114] as modified by Darken and Gurry [2]. Changes in the slopes of the lines appear where phase transformations take place.

When a metal oxide dissociates with the formation of separate pure phases of the metal and metal oxide, the partial pressure of oxygen in equilibrium with the metal and its oxide can be determined from Fig. 1.15. To do this, place a straight edge on point O of the ordinate on the left and the point where the line for the reaction crosses the temperature of interest; read the pressure (in atmospheres) from the nomograph on the far right. For example, at 1000°K (727°C) the pressure of oxygen in equilibrium with the metal oxide–metal phase according to the reactions given in Fig. 1.15 would be approximately 10^{-13} for PbO:Pb; 10^{-21} for FeO:Fe; 10^{-36} for SiO_2:Si; and 10^{-56} for CaO:Ca. If one of these metals is in an atmosphere (or vacuum) where the partial pressure of oxygen is greater than the value given, oxidation will occur. Because of the low partial pressure of oxygen over hot calcium metal, pure gases that will not react with calcium are sometimes passed over hot calcium to remove any traces of oxygen and water vapor present.

The more negative the standard free energy of formation of an oxide at any temperature, the lower is the equilibrium oxygen dissociation pressure and the more stable is the oxide. As a result, an oxide is theoretically capable of being reduced by a metal that forms an oxide with a more negative standard free energy of formation, and the reducing power of an element increases as the distance between the lines on the diagram increases. (All reactions are referred to 1 mole of oxygen so that oxygen vanishes when two reactions are "subtracted" from one another.) Thus, if pure zinc oxide is reduced to pure zinc metal, any oxides of iron, tin, cobalt, nickel, and copper present will be reduced under the same conditions, while more stable oxides, such as those of aluminum or magnesium, may not be reduced, depending on the reducing agent. Similarly, Fig. 1.15 indicates that FeO can be reduced by silicon and aluminum; of the two, aluminum is known to reduce iron–oxygen systems to a greater degree than silicon. Some of the methods used for determining oxygen in certain metals depend on reacting the sample with aluminum and calculating the oxygen content from the quantity of aluminum oxide formed [126].

Metal oxides are reducible by hydrogen and carbon monoxide by the reactions

$$M_mO_{2n}(s) + 2_nH_2(g) = mM(s) + 2_nH_2O(g) \tag{31}$$

$$M_mO_{2n}(s) + 2_nCO(g) = mM(s) + 2_nCO_2(g). \tag{32}$$

The ratios of $H_2:H_2O$ and $CO:CO_2$ in equilibrium with the pure metal and metal oxide phases at various temperatures can be determined from Fig. 1.15. The $H_2:H_2O$ ratio for the reaction of interest can be determined by placing a straight edge over H on the ordinate at the left and the reaction line at the desired temperature, and by reading the ratio on the $H_2:H_2O$ nomograph. Similar information on reactions involving $CO:CO_2$ equilibria are obtainable by using point C on the ordinate and reading the $CO:CO_2$ ratios from the appropriate nomograph.

For the hydrogen reduction of cobalt oxide, for example,

$$CoO + H_2 = Co + H_2O, \tag{33}$$

the $H_2:H_2O$ ratio at equilibrium with the pure phases is approximately 1:30 at 1100°C. The oxide would be reduced if the ratio were greater than 1:30, and the cobalt metal would be oxidized if the ratio were less. Thus, the reduction of cobalt oxide at 1100°C by a stream of dry hydrogen is practical. On the other hand, the hydrogen reduction of aluminum oxide at 1100°C is not practical, because the $H_2:H_2O$ ratio must exceed 5×10^9. Extremely dry hydrogen would be needed, and in large volumes that would have to be circulated to keep the water formed below the equilibrium concentration. Equilibrium data have been listed for the hydrogen reduction of the oxides of copper, iron, nickel, cobalt, manganese, chromium, molybdenum, and tungsten [3].

In the above discussion it is assumed that the reactions proceed as written; however, an oxide may not dissociate into its elements. It is also assumed that the reactants and products are in their standard states as pure phases, whereas mutual solubility of the phases may occur. In the latter case, the activities of the reactants and products must be taken into consideration, as recently discussed in the reduction of titanium and zirconium oxides by calcium in systems containing impure phases [111, 112].

The above considerations pertain to metal–metal oxide equilibria, and no consideration was given to the oxygen remaining in solution after the oxide phase has been removed. The amount of oxygen in solution at temperatures used for analysis may represent a significant portion of the total oxygen present. (For example, the solubility limit of oxygen in molybdenum is 0.0065 wt % at 1700°C and 0.015 wt % in iron at 1450°C.) The reduction of a metal oxide (FeO, for example) by a H_2:H_2O (or CO:CO_2) mixture leads to a solid solution of oxygen in iron with the oxide phase absent. When the solid solution and gaseous phases are in equilibrium, the H_2:H_2O (or CO:CO_2) ratio at a given temperature is related to the dissolved oxygen content of the solid phase. A variable concentration of oxygen may be present, with the oxygen concentration being determined by the gas mixture in equilibrium with the metal at a given temperature; this is illustrated by Fig. 1.17 [127]. Calculations show that a H_2:H_2O mixture containing less than 0.03% water could be used to reduce the oxygen content of iron at 1600°C to 1 ppm.

Among the examples appearing in later chapters for determining oxygen by hydrogen reduction are molybdenum, tungsten, and rhenium, 1200°C [128]; lead and lead–tin alloys, 600°C, and copper and copper–base alloys, 1150°C, [129]; and bismuth, 900°C [130]. Dry hydrogen (dew point, −80°F) has been used to reduce the oxygen content of chromium to about 0.001 wt %, but a reaction time of about 20 h at 1500°C was required [131]. Because of the slow rate of oxygen diffusion, reaction times are long, and reactions of this type for solid metals are seldom used in analysis.

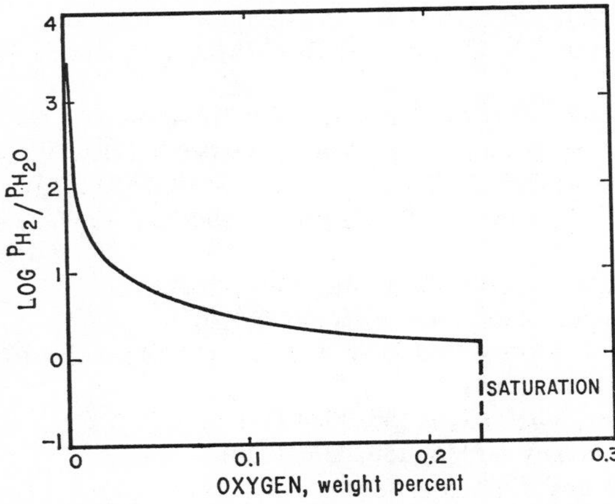

Fig. 1.17. P_{H_2}/P_{H_2O} in equilibrium with dissolved oxygen in liquid iron at 1600°C. (Reproduced by permission from the *Transactions of the American Institute of Mining, Metallurgical, and Petroleum Engineers* [127].)

The value of Fig. 1.15 lies in determining if the metal oxide can be reduced, in which case it may be possible to reduce the dissolved oxygen content to a satisfactory level if the equilibria appear to be quite favorable. Although the conclusions reached may be questionable, no better answer may be available because of a lack of thermochemical data pertaining to dissolved oxygen in metals [7]. As may be apparent, thermochemical data are lacking in part because of the difficulty of obtaining reliable oxygen analyses at low concentrations. In the absence of such data, analysis based on independent methods will be required to provide the best estimate. Methods of satisfactory accuracy are available for most materials, but further improvements are required, particularly for new materials.

REFERENCES

1. C. R. Cupp, "Gases in Metals," in B. Chalmers, Ed., *Progress in Metal Physics*, Vol. IV, Pergamon, New York, 1953, pp. 105–173.
2. L. S. Darken and R. W. Gurry, *Physical Chemistry of Metals*, McGraw-Hill, New York, 1953.
3. S. Dushman, *Scientific Foundations of Vacuum Technology*, 2nd ed., J. M. Lafferty, Ed., Wiley, New York, 1962.
4. J. D. Fast, *Interaction of Metals and Gases*, Vol. I, Academic, New York, 1965.
5. D. P. Smith, L. W. Eastwood, D. J. Carney, and C. E. Sims, *Gases in Metals*, American Society for Metals, Cleveland, 1953.
6. L. I. Sokol'skaya, *Gases in Light Metals*, Trans. by G. F. Modlen, Pergamon, New York, 1961.
7. O. Kubaschewski, A. Cibula, and D. C. Moore, *Gases in Metals*, Elsevier, New York, 1970.
8. D. O. Haywood, "Gas Adsorption," in J. R. Anderson, Ed., *Chemisorption and Reactions on Metallic Films*, Vol. I, Academic, New York, 1971, p. 231.
9. B. M. W. Trapnell, *Proc. Roy. Soc., Ser. A*, **218,** 566 (1953).
10. F. G. Allen, J. Eisinger, H. D. Hagstrum, and J. T. Law, *J. Appl. Phys.*, **30,** 1563 (1959).
11. R. Frauenfelder, *J. Chem. Phys.*, **48,** 3966 (1968).
12. K. W. Guardipee, *Anal. Chem.*, **42,** 469 (1970).
13. J. H. Hill, C. J. Morris, and J. W. Frazer, *AEC Accession No. 40817*, Rept. No. UCRL-14959 (1966).
14. W. G. Guldner, *Anal. Chem.*, **35,** 1744 (1963).
15. F. N. Rhines, *Trans. AIME*, **156,** 336 (1944).
16. J. D. Fast, *Philips' Tech. Rundschau.*, **7,** 73 (1942); through *Chem. Abstr.*, **38,** 2911 (1944).
17. K. Hauffe, *Oxidation of Metals*, Plenum, New York, 1965.
18. E. A. Gulbransen, *Corrosion*, **21,** 76 (1965).
19. C. Goodeve and K. H. Jack, *Disc. Faraday Soc.*, **No. 4,** 83 (1948).

20. L. Pauling, *The Nature of the Chemical Bond*, Cornell University Press, Ithaca, N.Y., 1960.
21. J. Stringer and A. R. Rosenfield, *Nature*, **199,** 337 (1963).
22. D. A. Robins, *J. Less-Common Metals*, **1,** 396 (1959).
23. R. M. Barrer, *Disc. Faraday Soc.*, **No. 4,** 68 (1948).
24. *Nonstoichiometric Compounds*, Advances in Chemistry Series No. 39, American Chemical Society, Washington, D.C., 1963.
25. J. Chipman and J. F. Elliott, "Physical Chemistry of Liquid Steel," in C. E. Sims, Ed., *Electric Furnace Steelmaking, Theory and Fundamentals*, Vol. II, Interscience, New York, 1963, p. 101.
26. A. Sieverts, *Z. Metallk.*, **21,** 37 (1929).
27. R. H. Fowler and C. J. Smithells, *Proc. Roy. Soc., Ser. A*, **160,** 38 (1937).
28. G. Hörz and E. Steinheil, *J. Less-Common Metals*, **21,** 84 (1970).
29. A. S. Nowick, "Internal Friction in Metals," in B. Chalmers, Ed., *Progress in Metal Physics*, Vol. IV, Pergamon, New York, 1961, pp. 1–70.
30. K. M. Entwistle, *Metall. Rev.*, **7,** 175 (1962).
31. G. M. Leak, "Application of Internal-Friction Measurements to the Study of Gases in Metals," in *Spec. Rept. No. 68*, Iron and Steel Institute, London, 1960, pp. 270–95.
32. J. C. Barton and F. A. Lewis, *Talanta*, **10,** 237 (1963).
33. P. Schwarzkopf and R. Kieffer, *Refractory Hard Metals*, Macmillan, New York, 1953, pp. 228–254.
34. D. P. Smith, *Hydrogen in Metals*, University of Chicago Press, Chicago, 1948.
35. D. T. Hurd, *Chemistry of the Hydrides*, Wiley, New York, 1952.
36. P. Cotterill, "The Hydrogen Embrittlement of Metals," in B. Chalmers, Ed., *Progress in Materials Science*, Vol. IX, No. 4, Pergamon, New York, 1961, pp. 201–301.
37. W. M. Mueller, Ed., *Metal Hydrides*, Academic, New York, 1968.
38. M. Smialowski, *Hydrogen in Steel*, Pergamon, New York, 1962.
39. L. W. Eastwood, *Gas in Light Alloys*, Wiley, New York, 1946.
40. J. F. Elliott and M. Gleiser, *Thermochemistry for Steelmaking*, Vol. II, Addison-Wesley, Reading, Mass., 1963.
41. P. J. Depuydt and N. A. D. Parlee, *Met. Trans*, **2,** 612 (1971).
42. W. Schwarz and H. Zitter, *Arch. Eisenhuttenw.*, **36,** 343 (1965).
43. W. Beck, J. O'M. Bockris, M. A. Genshaw, and P. K. Subramanyan, *Met. Trans.*, **2,** 883 (1971).
44. C. A. Zapffe and M. E. Haslem, *Plating*, **37,** 610 (1950).
45. R. M. Barrer, *Trans. Faraday Soc.*, **36,** 1235 (1940).
46. P. D. Blake, M. F. Jordan, and W. I. Pumphrey, "The Solubility of Hydrogen in α-Iron of High Purity and in the Laves Phases $TiFe_2$ and $NbFe_2$," in *Spec. Rept. No. 73*, Iron and Steel Institute, London, 1962, pp. 76–82.
47. C. R. Cupp and P. Flubacher, *J. Nucl. Mater.*, **6,** 213 (1962).
48. Yu. A. Klyachko and O. D. Larina, *Stal'*, **21,** 604 (1961); through *Chem. Abstr.*, **55,** 25583e (1961).
49. H. H. Podgurski, *Trans. AIME*, **221,** 389 (1961).
50. J. Hewitt, "The Study of Hydrogen in Low-Alloy Steels by Internal Friction

Techniques," in *Spec. Rept. No. 73*, Iron and Steel Institute, London, 1962, pp. 83–89.

51. M. W. Mallett and I. E. Campbell, *J. Am. Chem. Soc.*, **73,** 4850 (1951).
52. R. N. R. Mulford and C. E. Holley, *J. Phys. Chem.* **59,** 1222 (1955).
53. E. A. Gulbransen and K. F. Andrew, *Trans. AIME*, **203,** 136 (1955).
54. D. T. Peterson and D. G. Westlake, *Trans. AIME*, **215,** 444 (1959).
55. R. K. McGeary, *Zirconium and Zirconium Alloys*, American Society for Metals, Cleveland, 1953, p. 172.
56. C. J. Smithells, *Metals Reference Book*, Vol. II, 2nd ed., Interscience, New York, 1955.
57. R. M. Barrer, *Diffusion in and Through Solids*, Cambridge University Press, New York, 1951.
58. R. G. Ehl, R. J. Sime, and J. L. Margrave, "Binary Nitrogen Compounds of the Elements: A Literature Survey," *U.S. Dept. Com., Office Tech. Service, PB Rept. 161,315*, 1959.
59. A. F. Wells, *Structural Inorganic Chemistry*, 2nd ed., Clarendon, Oxford, 1950.
60. M. Hansen and K. Anderko, *Constitution of Binary Alloys*, McGraw-Hill, New York, 1958.
61. R. G. Blossey and R. D. Pehlke, *Trans. AIME*, **236,** 28 (1966).
62. J. Chipman and J. F. Elliot, "Physical Chemistry of Liquid Steel," in C. E. Sims, Ed., *Electric Furnace Steelmaking, Theory and Fundamentals*, Vol. II, Interscience, New York, 1963, p. 102.
63. F. C. Langenberg and M. J. Day, "Application of Nitrogen Solubility Data to Alloy Steelmaking," in *Proc. 15th Ann. Elec. Furnace Steel Conf. AIME*, 1957, pp. 7–15.
64. G. W. Goward, *Anal. Chem.*, **37,** 117R (1965).
65. E. Geghardt, E. Fromm, and D. Jakob, Plansee Proc., 5th Seminar, 1964, Reutte/Tyrol, 1965, pp. 421–437; through *Chem. Abstr.*, **64,** 17150g (1964).
66. G. Subat, *Schweiz. Arch. Angew. Wiss. Tech.*, **29,** 338 (1963).
67. C. W. Tucker, A. U. Seybolt, and H. T. Sumsion, *Acta Met.*, **1,** 390 (1953).
68. T. Kê, *Phys. Rev.*, **74,** 9 (1948).
69. R. T. Bryant, *J. Less-Common Metals*, **4,** 62 (1962).
70. L. Brewer, *Chem. Rev.*, **52,** 1 (1953).
71. T. R. Allmand, *Microscopic Identification of Inclusions in Steel*, British Iron and Steel Research Association, 1962.
72. D. E. Rimmer and A. H. Cottrell, *Phil. Mag.*, **2,** 1345 (1957).
73. R. S. Barnes and D. J. Mazey, *Proc. Roy. Soc., Ser. A*, **275,** 45 (1963).
74. D. Kramer, K. R. Garr, C. G. Rhodes, and A. G. Pard, *J. Iron Steel Inst.* (London), **207,** 1141 (1969).
75. K. C. Russell, *Acta Met.*, **20,** 899 (1972).
76. R. S. Barnes and D. J. Mazey, *Proc. Roy. Soc., Ser. A*, **275,** 47 (1963).
77. A. W. Castleman, F. E. Hoffman, and A. M. Eshaya, "Diffusion of Xenon Through Aluminum and Stainless Steel," *U.S. At. Energy Comm. BNL-624*, 1960.
78. C. Mitra, "Solubility of Xenon in Liquid Metals," Columbia University, Univ. Microfilms, 1961; *Diss. Abstr.*, **22,** 100 (1961).

79. G. Brebec, V. Levy, and Y. Adda, *Compt. Rend.*, **252,** 722 (1961).
80. A. Demarez, A. G. Hock, and F. A. Meunier, *Acta Met.*, **2,** 214 (1954).
81. J. D. Hobson, *J. Iron Steel Inst.* (London), **191,** 342 (1959).
82. C. J. Smithells, *Gases and Metals*, Wiley, New York, 1937.
83. A. D. Le Claire, "Diffusion in Metals," in B. Chalmers, Ed., *Progress in Metal Physics*, Vol. IV, Pergamon, New York, 1961, pp. 265–332.
84. J. D. Hobson, *J. Iron Steel Inst.* (London), **189,** 315 (1958).
85. J. R. Young, *Rev. Sci. Instrum.*, **34,** 891 (1963).
86. D. N. Jewett and A. C. Makrides, *Trans. Faraday Soc.*, **61,** 932 (1965).
87. J. R. Young and N. R. Whetten, *Trans. Natl. Vacuum Symp.*, **8,** 625 (1961).
88. F. J. Norton and A. U. Seybolt, *Trans. AIME*, **230,** 595 (1964).
89. P. L. Chang and W. D. G. Bennett, *J. Iron Steel Inst.* (London), **170,** 205 (1952).
90. W. Geller and T. Sun, *Arch. Eisenhuttenw.*, **21,** 423 (1950).
91. D. G. Swinburn, *J. Iron Steel Inst.* (London), **209,** 620 (1971).
92. U. V. Bhat and H. I. Lloyd, *J. Iron Steel Inst.* (London), **165,** 382 (1950).
93. A. McNabb and P. K. Foster, *Trans. AIME*, **227,** 618 (1963).
94. F. R. Coe and J. Moreton, *J. Iron Steel Inst.* (London), **204,** 366 (1966).
95. D. G. Swinburn, *J. Iron Steel Inst.* (London), **208,** 508 (1970).
96. R. A. Oriani, *Acta Met.*, **18,** 145 (1970).
97. G. M. Evans and E. C. Rollason, *J. Iron Steel Inst.* (London), **207,** 1484 (1969).
98. J. H. Andrew, A. K. Mallik, and A. G. Quarrell, *J. Iron Steel Inst.* (London), **153,** 67p (1946).
99. M. L. Hill and E. W. Johnson, *Trans. AIME*, **221,** 622 (1961).
100. Iron Steel Inst. (London), *Spec. Rept. No. 73*, 1962.
101. B. A. Shmelev, *Zavodskaya Lab.*, **23,** 263 (1957); through *Chem. Abstr.* **52,** 972g (1958).
102. P. K. Foster, A. McNabb, and C. M. Payne, *Trans. AIME*, **233,** 1022 (1965).
103. M. W. Mallett and C. B. Griffith, *Trans. ASM*, **46,** 375 (1954).
104. J. E. Fagel, R. F. Witbeck, and H. A. Smith, *Anal. Chem.*, **31,** 1115 (1959).
105. C. Goodeve and K. H. Jack, *Disc. Faraday Soc.*, **No. 4,** 82 (1948).
106. A. Bramley, F. W. Haywood, A. T. Cooper, and J. T. Watts, *Trans. Faraday Soc.*, **31,** 707 (1935).
107. E. S. Tankins, N. A. Gokcen, and G. R. Belton, *Trans. AIME*, **230,** 820 (1964).
108. C. J. Smithells and C. E. Ransley, *Proc. Roy. Soc., Ser. A*, **155,** 195 (1936).
109. F. J. Norton and C. W. Tucker, *J. Nucl. Mater.*, **2,** 350 (1960).
110. L. W. Hillen and M. T. Thackray, *J. Chromatogr.* **10,** 309 (1963).
111. A. K. Biswas and G. R. Bashforth, *The Physical Chemistry of Metallurgical Processes*, Chapman and Hall, London, 1962.
112. C. Bodsworth, *Physical Chemistry of Iron and Steel Manufacture*, Longmans, London, 1963.
113. J. Pearson and U. J. C. Ende, *J. Iron Steel Inst.* (London), **175,** 53 (1953).
114. F. D. Richardson and J. H. E. Jeffes, *J. Iron Steel Inst.* (London), **160,** 261 (1948).
115. F. D. Richardson, *J. Iron Steel Inst.* (London), **175,** 33 (1953).
116. F. D. Richardson and J. H. E. Jeffes, *J. Iron Steel Inst.* (London), **171,** 167 (1952).

117. J. F. Elliott and M. Gleiser, *Thermochemistry for Steelmaking*, Vol. I, Addison-Wesley, Reading, Mass., 1960.
118. A. K. Babko, K. E. Kleiner, and L. V. Markova, *Zavodskaya Lab.*, **22,** 640 (1956); through *Chem. Abstr.*, **50,** 15339g (1956).
119. O. Kubaschewski and E. Ll. Evans, *Metallurgical Thermochemistry*, Pergamon, New York, 1958.
120. C. E. Wicks and F. E. Block, "Thermodynamic Properties of 65 Elements—Their Oxides, Halides, Carbides, and Nitrides," *Bureau of Mines Bulletin 605*, U.S. Govt. Printing Office, Washington, 1963.
121. B. D. Holt and H. T. Goodspeed, *Anal. Chem.*, **34,** 374 (1962).
122. R. E. Honig, *RCA Review*, **23,** 567 (1962).
123. R. Hörbe and O. Knacke, *Z. Erzbergbau Metallhuttenw.*, **8,** 556 (1955).
124. M. M. Antonova, *Zh. Analit. Khim.*, **19,** 1408 (1964); through *Chem. Abstr.*, **62,** 5881 (1965).
125. W. Oelsen and K. H. Sauer, *Arch. Eisenhuttenw.*, **38,** 141 (1967); through *Anal. Abstr.*, **15,** 3332 (1968), and *Chem. Abstr.*, **67,** 29039p (1967).
126. N. Gray and M. C. Sanders, *J. Iron Steel Inst.* (*London*), **143,** 321P (1941).
127. T. P. Floridis and J. Chipman, *Trans. AIME*, **212,** 549 (1958).
128. R. Geyer and K. Friedrich, *Z. Anal. Chem.*, **213,** 259 (1965); through *Chem. Abstr.*, **63,** 15549 (1965).
129. W. A. Baker, *Metallurgia*, **40,** 188 (1949).
130. E. S. Funston and S. A. Reed, *Anal. Chem.*, **23,** 190 (1951).
131. A. U. Seybolt and R. A. Oriani, *J. Metals*, **8,** 556 (1956).

CHAPTER

2

VACUUM FUSION FURNACE REACTIONS

WILLIAM F. HARRIS

Youngstown Sheet and Tube Company
Youngstown, Ohio

CONTENTS

I. INTRODUCTION

The vacuum-fusion and inert-gas-fusion methods have been used extensively for the determination of the gas content of metals. Of the two, the reactions under vacuum conditions have been studied the more extensively and will be discussed here. The same general theory applies because they are both based on heating the sample in a graphite crucible and removing the evolved gases for measurement. The oxygen present in solution in the metal or as oxide reacts with the carbon from the crucible and is liberated as carbon monoxide. Hydrogen and nitrogen, which may be present in solution or in separate phases, are liberated as elemental hydrogen and nitrogen, respectively. Despite the fact that the fusion techniques can be described so simply, a complete theoretical treatment of the subject is very complex.

Walker and Patrick [1] suggested the vacuum fusion method in 1912, and for 30 or more years it was used primarily for the analysis of gases in iron and steel. As nonferrous metallurgy became more important, attempts were made to apply the vacuum-fusion technique to a variety of metals. As a result of these attempts, which were of varying success, it became apparent that in order to successfully apply vacuum fusion to a large number of metals and alloys, it would be necessary to develop a theory that could be used to calculate the approximate conditions necessary for the complete removal of gas from the metal being analyzed.

In 1952 Sloman et al. [2] used thermodynamic calculations to determine the correct temperature conditions for the recovery of oxygen and nitrogen from eight high-melting metals. In this work it was assumed that an iron bath would have to be used for most nonferrous metals. The four mechanisms postulated for the reduction of the oxides were:

1. Reduction of oxide with carbon to form carbon monoxide and very slow solution of the liberated metal in the iron bath
2. Reduction of oxide with carbon to form carbon monoxide and solution of the liberated metal in the iron bath
3. Reduction of the oxide with carbon to form carbon monoxide and a reaction of the liberated metal with carbon to form the metal carbide
4. A combination of mechanisms 2 and 3 depending on the relative affinity of the liberated metal for iron and carbon.

The authors had established that Al_2O_3 could be quantitatively reduced at 1600°C. It was then calculated that for the equation

$$3C + Al_2O_3 \rightarrow 2Al + 3CO \tag{1}$$

the pressure of CO in equilibrium with the Al_2O_3 at 1600°C is 2.4 torr. With the above data, the decomposition temperatures of MoO_2, SiO_2, UO, ZrO_2, UO_2, TiO, and ThO_2 were calculated for the four postulated mechanisms. A comparison of the calculated temperatures with those experimentally determined showed that postulates 1 and 2 agreed closely with the experimental data. It was assumed that in most cases postulate 1 was too simple to explain the mechanism of vacuum fusion and that postulate 2 was probably the correct mechanism.

A similar procedure was followed for nitrogen. The postulated mechanisms for the release of nitrogen during vacuum fusion were as follows:

5. Thermal decomposition of the nitride without solution of the released metal
6. Thermal decomposition of the nitride with solution of the released metal in the iron bath
7. Reaction of the nitride with carbon to form the metal carbide and nitrogen
8. A combination of mechanisms 6 and 7 depending on the relative affinity of the metal for iron or carbon.

The authors determined that the nitrogen was evolved rapidly and completely from TiN at 1800°C. On the basis of this information the decomposition temperatures were calculated for the four postulated mechanisms for Mo_2N, Si_3N_4, VN, AlN, UN, ZrN, and Th_3N_4. When the calculated temperatures are compared to the experimentally determined temperatures, the best agreement is with postulated mechanisms 5 and 6. The authors state that the most probable mechanism is number 6, which is analogous to the mechanism for the reduction of oxides because both involve solution of the liberated metal in the iron bath and neither involves carbide formation.

This first attempt to determine the temperature conditions necessary for vacuum-fusion analysis was an excellent start towards an understanding of the vacuum-fusion process.

II. REACTION MECHANISMS

To understand fully the vacuum-fusion process, the factors necessary for the quantitative removal of the gases from the metal in the vacuum-fusion system must be considered. It is necessary that the oxides be brought into intimate contact with carbon at a temperature high enough to cause reduction to carbon monoxide. While the carbon is probably not necessary for the elimination of nitrogen, the temperature must be high enough to thermally decompose the nitrides. It is also necessary that the metal be in a form through which the gases can rapidly pass. This may be either a mobile metal melt or, in some cases, solid metal. The conditions that meet the above requirements are as follows:

1. Extraction from molten sample contained in a graphite crucible
2. High-temperature extraction in a graphite crucible from solid metal
3. Extraction from the sample after alloying with a suitable molten metal.

All of these methods have been used successfully for the determination of gases in metals, although number 3, which is the well-known bath technique, is the most widely used.

The extraction of gases from a molten metal, which essentially is using the molten metal as its own bath, is limited to metals that have melting points below 1800°C, oxides that are reducible at the desired temperature, and either low vapor pressures or insignificant reaction rates between the metal vapor and the gases released. This type of extraction has been used in the analysis of gases in iron and mild steels, tin, copper, and a few other metals that meet the requirements for the technique.

It is apparent that one of the conditions for the successful removal of gases is not necessarily a large quantity of carbon dissolved in the sample. For example, at 1200°C copper neither dissolves appreciable quantities of carbon nor does it wet the crucible, yet carbon monoxide is quickly and quantitatively evolved.

The high-temperature extraction from solid metal specimens contained in a graphite crucible holds promise for a few special melts. It has been applied to massive samples of molybdenum and tungsten [3] and niobium [4], and to powdered tantalum [3]. In this method the specimen is heated in a graphite crucible at 2000°C for an appropriate length of time, 15–20 min for molybdenum and tungsten and 30–45 min for niobium. The results obtained for oxygen, nitrogen, and hydrogen by high temperature extraction agree well with those obtained by bath techniques. This method is applicable mainly to the pure metals; small alloying additions may cause incomplete extraction of the gases.

The exact mechanism of the liberation of the gases from the metal by high-temperature extraction is not known. It is known, however, that the carbon content of the tungsten and molybdenum specimen is no greater at the end of the extraction than at the beginning; thus, diffusion of the carbon into the specimen is evidently not necessary. The relatively high vapor pressures of the oxides and nitrides of molybdenum and tungsten at 2000°C would seem to indicate that thermal decomposition may be a controlling factor.

III. APPLICATION OF BATH TECHNIQUES

For the successful determination of the oxygen and nitrogen content of most metals, the bath technique is used. The metal sample is fused in a bath of molten metal, usually iron or platinum. The main purposes of the bath are as follows:

1. To bring the oxides into intimate contact with carbon
2. To lower the temperature range of liquids by the formation of low-melting alloy
3. To lower the partial vapor pressure of reactive metal vapors
4. To increase equilibrium pressure of carbon monoxide and nitrogen over the melt.
5. To prevent viscous or solid carbide formation.

While intimate contact with carbon is evidently not a necessary condition for the quantitative extraction of oxygen from all metals as carbon monoxide, this condition is necessary in many cases. For this purpose, iron is widely used as a bath material because the solubility of carbon in iron at normal vacuum-fusion temperatures (about 1650°C) is 5–6%. The solubility of carbon in platinum is somewhat less, about 1.5% at 1900°C, but is quite effective in bringing oxides of alloyed metals into intimate contact with sufficient quantities of carbon to promote rapid reduction.

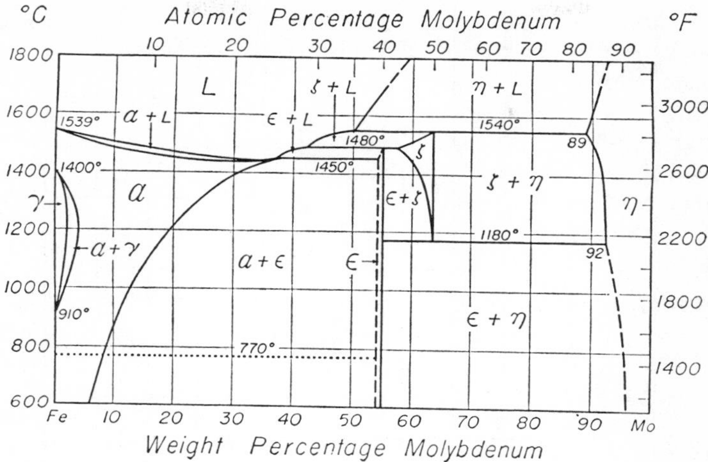

Fig. 2.1. Iron-molybdenum phase diagram. (Reproduced by permission from American Society for Metals [5].)

Iron is a very useful bath material because of the low-melting alloys that it forms with many high-melting metals. For example, when molybdenum (melting point 2625°C) is alloyed with iron, the melting point of the alloy containing up to 30% molybdenum is about 1450°C [5]. The phase diagram is shown in Fig. 2.1. Since an iron bath contains several percent carbon, further lowering of the melting point of the molybdenum iron alloy is obtained. The phase diagram of the iron–molybdenum–carbon system shown in Fig. 2.2 [5] is for the alloy containing 20% molybdenum. It shows the melting point depression of the alloy to about 1200°C at carbon contents of above 3.3%.

Similar effects can be seen by examining the iron-tungsten (Fig. 2.3) and the carbon–iron–tungsten (Fig. 2.4) phase diagrams. Tungsten, which has a melting point of 3410°C, can be alloyed with iron in quantities of up to 30% and yield an alloy that melts at 1450°C [5]. Again the addition of carbon to this alloy further lowers the melting point. The phase diagram of the carbon–iron–tungsten system containing 20% tungsten, Fig. 2.4 [5], shows the melting point of the alloy to be below 1100°C when the melt contains 3.7% carbon. The phase diagrams graphically illustrate the advantages of using an iron bath to determine the gas content of high-melting-point metals.

More recently platinum baths have been used for the determination of gases in refractory metals. A platinum bath has the advantages of a very low vapor pressure at high temperatures and a resonably high solubility for most

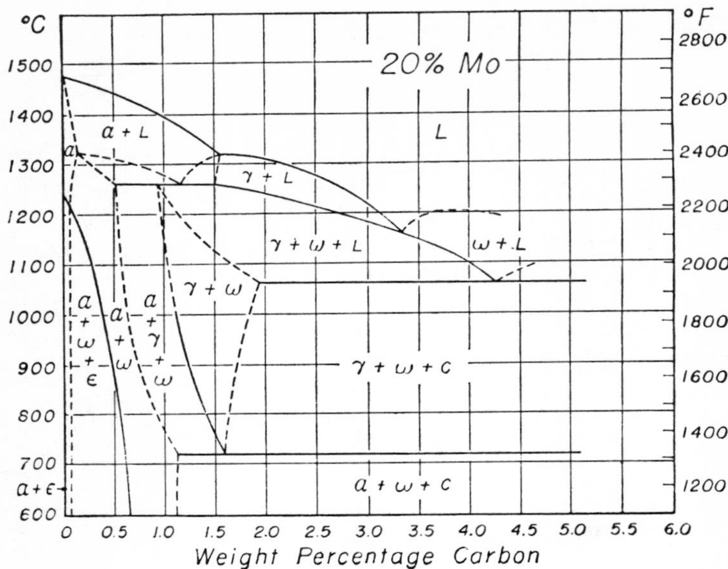

Fig. 2.2. Iron–carbon–20% molybdenum phase diagram. (Reproduced by permission from American Society for Metals [5].)

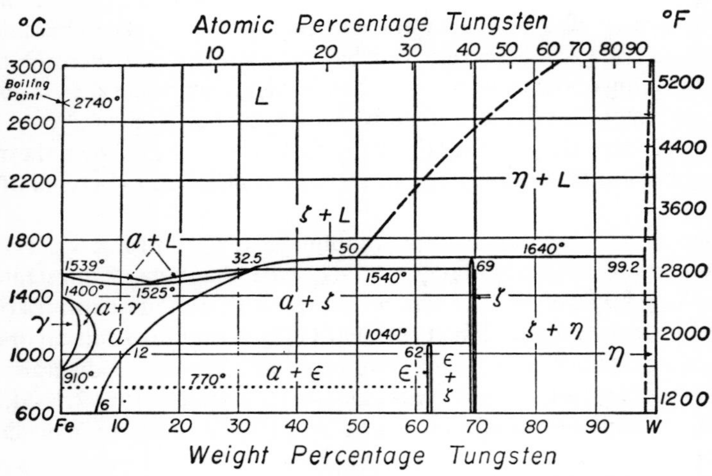

Fig. 2.3. Iron–tungsten phase diagram. (Reproduced by permission from American Society for Metals [5].)

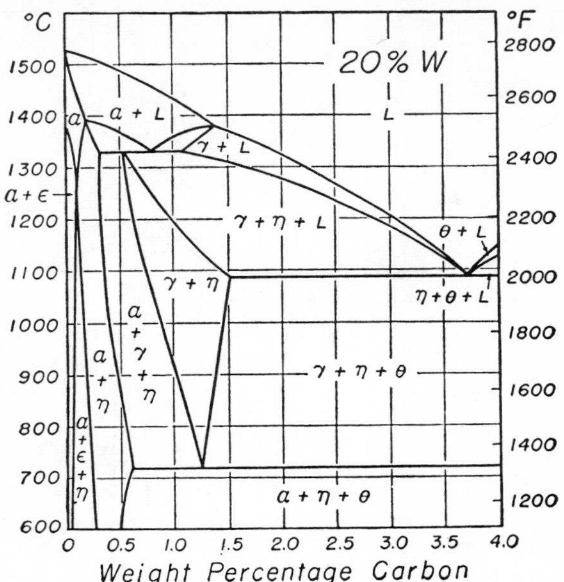

Fig. 2.4. Iron–carbon–20% tungsten phase diagram. (Reproduced by permission from American Society for Metals [5].)

metals. However, there are certain instances where the use of a platinum bath can lead to complications.

The phase diagram of the molybdenum–platinum system is shown in Fig. 2.5 [6]. At the normal temperature of a platinum bath, 1950°C, the solubility of molybdenum is about 14%. Therefore, samples of a very small weight must be used to keep the bath fluid. To dissolve more than 15% molybdenum in platinum, it is necessary to go to temperatures above 2100°C. This usually leads to an increased blank rate and, in the case of the molybdenum–platinum system, a bath with increased viscosity. In the iron-bath systems it was noted that the addition of several percent carbon to the bath lowered the melting point by over 200°C. This does not seem to be true of the platinum–carbon system. The melting point of platinum is 1774°C, and the eutectic at 1.2% carbon is reported at 1734°C. It seems reasonable to assume that the molybdenum solubility will not be changed greatly by the addition of carbon.

The platinum–tungsten system, Fig. 2.6 [6], also shows a low solubility of tungsten in platinum at the usual vacuum-fusion temperatures. At 1950°C the solubility of tungsten in platinum is about 15%. In an actual vacuum-fusion crucible, however, the effective concentration of tungsten or molybdenum may be lower than is shown on the phase diagram because of an

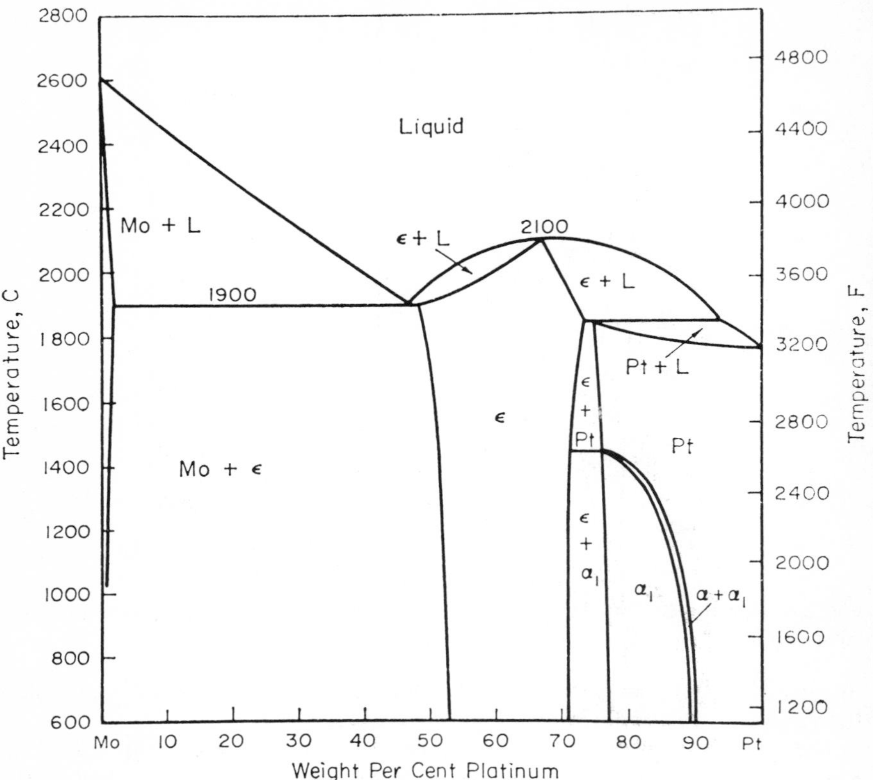

Fig. 2.5. Molybdenum–platinum phase diagram. (Reproduced by permission from Battelle Memorial Institute [6].)

uneven distribution of the solute metal in the melt. An investigation of platinum baths has shown that the solute metal tends to be more concentrated at the top of the bath; saturation takes place when the surface of the bath freezes. The inhomogeneity of the platinum bath is caused in part by the difference in density between the platinum and the solute metal. Although very little mixing of the melt is obtained with 450-kHz radio-frequency heating, the concentration gradient of the solute may not be noticed when generators of 10 kHz are used, because of the stirring action imparted to the melt.

Certain metals have high vapor pressures at the temperatures necessary to reduce their oxides with carbon to carbon monoxide. The vapor of the metal is detrimental to the analysis if the metal vapor or the deposited metal film sorbs ("getters") the liberated gas. Gettering in the vacuum-fusion process will be discussed more fully later, but in general, it can be said that the partial

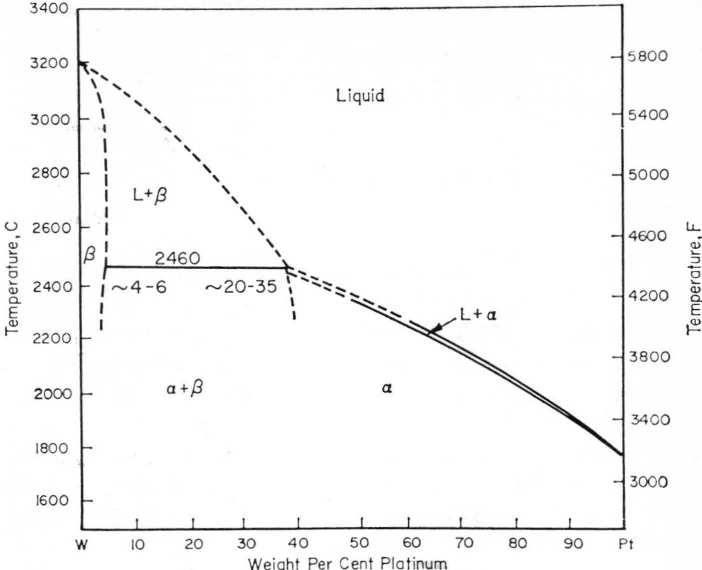

Fig. 2.6. Tungsten–platinum phase diagram. (Reproduced by permission from Battelle Memorial Institute [6].)

pressure of the metal must be lowered to reduce gettering. The partial pressure of the offending metal can generally be reduced somewhat by the use of a bath.

Raoult's law states that in an ideal solution, the partial pressure of the solute is proportional to the relative concentrations of the solute and solvent. The partial pressure of solute metal in an iron bath at a given temperature can be represented by

$$p_{A[Fe]} = n_A p_A^\circ, \tag{2}$$

where $p_{A[Fe]}$ is the partial pressure of solute metal in the iron bath, n_A is the mole fraction of solute metal, and p_A° is the vapor pressure of the pure solute metal at the given temperature. The term p_A° can be calculated from tables prepared by Dushman [7]. Although Eq. 2 applies only to dilute solutions or perfect solutions where the activity is equal to the mole fraction of solute, it is useful for obtaining the approximate vapor pressure of the solute. Table 2.1 shows the reduction in the vapor pressure of aluminum by the use of an iron bath at 1650°C.

While a bath serves to lower the partial pressure of volatile metals that may getter the liberated gases, it also raises the partial pressure of the carbon monoxide and nitrogen over the melt above that which would exist if the

TABLE 2.1. Partial Pressure of Aluminum in an Iron Bath at 1650°C

Weight % Al	Partial Pressure, torr
1	0.063
5	0.284
10	0.591
20	1.078
30	1.485
100	3.160

sample were melted without the use of a bath. This means that the same pressure of carbon monoxide and nitrogen in equilibrium with a pure metal can be obtained at a lower temperature if the metal is alloyed with an iron bath. This is illustrated in Tables 2.2 and 2.3. The increase in the equilibrium pressure of carbon monoxide when an iron bath is used is caused by the free-energy contribution to the total reaction of the solution of the liberated metal in the iron. The increase in the nitrogen pressure above the bath is caused by a similar effect.

It can be seen from Table 2.3 that the extraction of nitrogen from titanium containing TiN would be impossible at 1600°K. With the use of an iron bath, however, the pressure is high enough for nitrogen to be extracted. In addition, the removal of the nitrogen from the molten bath would be much quicker than by diffusion through the solid titanium specimen. It is assumed that these remarks apply to platinum baths also; however, because of the lack of good data for heats of solution of metals in platinum, quantitative calculations are difficult to make.

The success of the platinum bath in vacuum-fusion analysis may be caused by more than merely the higher melting point and lower vapor pressure of the metal. It has been observed that when certain metals, in particular Nb, Ta, Ti, and Zr, are alloyed with a platinum bath, a flash of light is observed that indicates the release of large quantities of energy and an instantaneous

TABLE 2.2. Pressure of Carbon Monoxide in Equilibrium with Al_2O_3 and Carbon With and Without an Iron Bath

	CO Pressure, torr	
Temp., °K	Without Fe Bath	With Fe Bath
1400	2.6×10^{-4}	3.4×10^{-2}
1600	3.0×10^{-2}	2.3
1800	1.3	60.4
2000	24.0	821

TABLE 2.3. Pressure of Nitrogen in Equilibrium with TiN With and Without an Iron Bath

	N_2 Pressure, torr	
Temp., °K	Without Fe Bath	With Fe Bath
1400	6.5×10^{-13}	3.8×10^{-7}
1600	9.0×10^{-10}	1.2×10^{-4}
1800	2.3×10^{-7}	1.2×10^{-2}
2000	2.3×10^{-5}	0.5

increase in bath temperature. This has been observed in both vacuum fusion and the arc-chamber–spectrographic technique. This momentary increase in temperature has been attributed to the formation of a platinum–metal intermetallic compound. It is quite possible that the additional energy contributed by this compound formation materially contributes to the decomposition of oxides and nitrides in the metal. It is interesting to note that the use of a platinum bath for the extraction of gases has not been of any particular advantage for metals, such as Mo and W, that do not give a light flash when alloyed with platinum.

IV. THERMODYNAMIC CONSIDERATIONS

In pure metals oxygen exists as a precipitated oxide or in solid solution in the metal. When alloys or impure metals are considered, the oxygen may exist in forms making a quantitative theoretical treatment difficult. In addition, the removal of the oxygen can be accomplished by several mechanisms, including simple reduction of the oxide and solution of the metal or carbide formation by the metal before, after, or along with the reduction of the oxide.

The simplest case, and that for which the best and most abundant thermodynamic data exist, is that of the reduction of the metal oxide and the subsequent solution of the sample metal in the bath. The reduction of oxides as a function of temperature and pressure has been treated by Darling [8]. Unfortunately, in many of the refractory metals we are not dealing with the reduction of oxides because of the high solubility of oxygen in the metals; however, qualitatively these calculations may be of value.

If it is assumed that the reaction involved in extracting the oxygen from the metal is a simple reduction, the equation for the reaction can be written as

$$MO_2 + 2C \rightarrow M + 2CO. \quad (3)$$

The equilibrium constant of the reaction is related to the free energy of the reaction by the equation

$$\Delta F^\circ = -RT \ln K = -4.574\text{T} \log K. \tag{4}$$

Because the carbon monoxide is the only noncondensed phase, Eq. 4 becomes

$$\Delta F^\circ = -4.574\text{T} \times 2 \log P_{\text{CO}}. \tag{5}$$

Since free-energy functions are additive, the free energy for a specific system can be calculated from the sum of the free energy of the individual reactions comprising the system

$$\text{M} + \text{O}_2 \rightarrow \text{MO}_2 \qquad \Delta F_6^\circ \tag{6}$$

$$2\text{C} + \text{O}_2 \rightarrow 2\text{CO} \qquad \Delta F_7^\circ. \tag{7}$$

Therefore, the equation for the free energy of the reaction expressed in Eq. 3 becomes

$$\Delta F^\circ = \Delta F_7^\circ - \Delta F_6^\circ. \tag{8}$$

The difference in the free energy of reaction for the carbon reduction of metal oxides, Eq. 8, is plotted against temperature in Fig. 2.7 [8]. Shown also in Fig. 2.7 are the carbon monoxide equilibrium pressures for the metal–metal oxide systems. Thus, for the reduction of Al_2O_3 at 1600°C, the partial pressure of carbon monoxide must be less than 3×10^{-3} atm (2.5 torr). In vacuum-fusion analysis this is done by pumping the carbon monoxide from the furnace into the analytical section of the system. In inert-gas fusion the carbon monoxide is swept out with helium or argon, but in both, the carbon monoxide is removed from the system so that the reaction may go to completion. As indicated in Fig. 2.7, the reduction of all of the metal oxides with carbon will occur at the temperatures and pressures commonly used in vacuum fusion. These calculations, however, apply only to the pure oxide. They are based on the assumption that the oxide, metal, and carbon exist in pure phases and that the equilibrium conditions can be defined in terms of the carbon monoxide pressure above the melt.

Based strictly on thermodynamic considerations, Klyachko et al. [9] have shown that the oxides of molybdenum, silicon, vanadium, and aluminum should be reduced during melting under conditions found in vacuum-fusion analysis without the use of an iron bath. Their calculations indicate that an iron bath is necessary to reduce the oxides of zirconium, uranium, and titanium.

Based on thermodynamic considerations, the fractional vacuum-fusion technique was developed in an attempt to differentiate between the various oxide species that are present in iron. The analysis is based on the theory that each oxide species has a definite decomposition temperature; therefore, if the reduction is carried out at a series of successively increasing temperatures, the various oxide species can be decomposed stepwise.

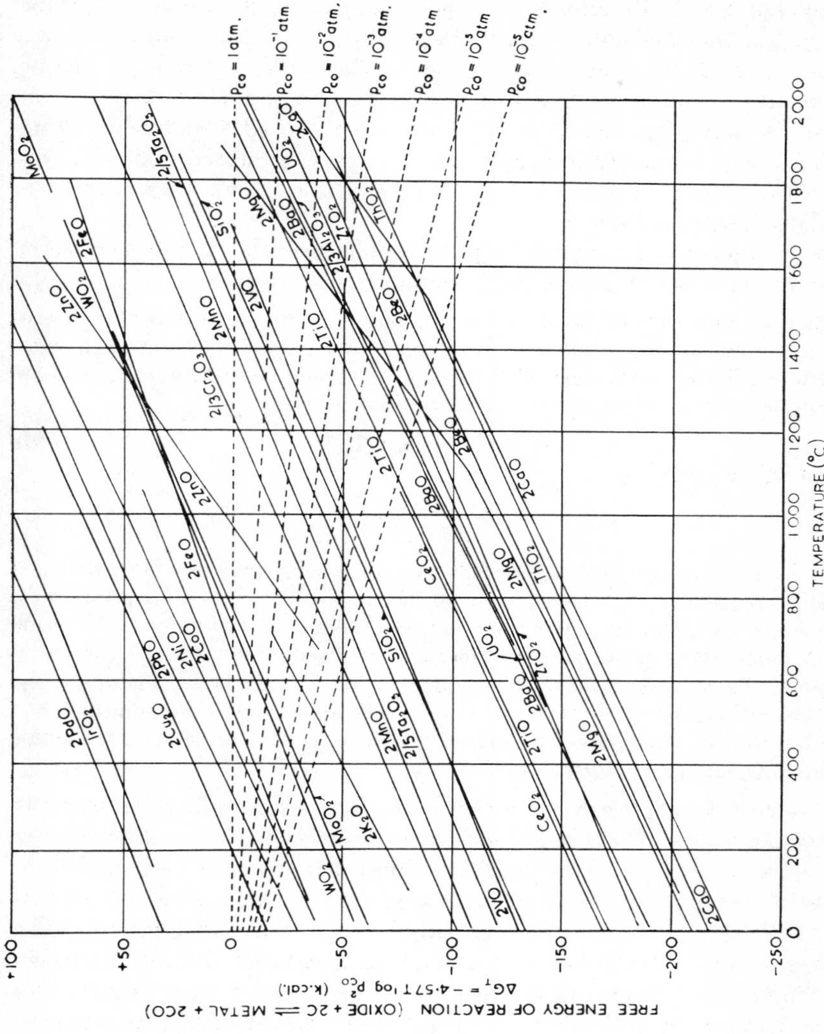

Fig. 2.7. Free energy of reaction of carbon reduction of metal oxides, with equilibrium carbon monoxide pressure (Reproduced by permission from *Metallurgia* [8].)

Hoyt and Scheil [10] reported very good agreement between the sum of the oxygen fractions and total oxygen determined at one temperature on three types of steel. In this work tin was added to the sample to lower the melting points and to reduce, as much as possible, the gettering effect due to manganese. A very slow reduction of the oxides was used by allowing about $\frac{1}{2}$–1 h between temperature increments. This work indicated that FeO was reduced at 1050°C, MnO at 1170°C, SiO_2 at 1320°C, and Al_2O_3 at 1570°C under the bath conditions used.

The comments of Phragmen [11] concerning this paper point out some of the problems involved in fractional vacuum-fusion analysis:

"The fractional vacuum fusion method is founded on the assumption that there is a definite reduction temperature for every oxide. The oxides, however, form compounds or solutions with each other. Even a pure oxide has no definite reduction temperature. For example, the equilibrium of the reaction

$$SiO_2 + 2C \rightarrow Si + 2CO \tag{9}$$

is given by the equation

$$f(T) = \frac{[Si][CO]^2}{[SiO_2][C]^2}. \tag{10}$$

Thus, the reduction temperature depends on the pressure in the reaction vessel and on the silicon content of the molten metal. The apparent reduction temperature is, of course, higher than the equilibrium temperature and the difference is not fixed.

"The reduction temperature observed may be influenced by the composition of the steel in the same degree as by the composition of the inclusions. Of course when the fractional vacuum fusion method is used, only one sample is to be melted in a graphite crucible. If, for instance, some pure iron is melted first in the crucible, the reduction temperature will be lower."

In later work [12] the reduction behavior under vacuum-fusion conditions of several oxides and combined oxides was investigated. This work indicates that silicates and aluminates require higher temperatures than had been previously reported. Reduction curves were obtained that showed that there is no definite decomposition temperature for an oxide or series of oxides but a temperature range over which the oxide is reduced. Because the reduction temperature ranges of the various oxides overlap to a considerable degree, separations of the oxides are not possible with any degree of certainty. A two-step fractional vacuum-fusion approach was suggested in which reductions at 1150°C and 2000°C were employed. At 1150°C the ferrous and manganese oxides are decomposed, and the more stable oxides of silicon, aluminum, titanium, and calcium are decomposed at 2000°C. It is interesting to note that tin was not used in this work because its effect was considered to be possibly deleterious. It appears that fractional vacuum fusion can be used to make only general separations of oxides; at times the conclusions reached may be subject to doubt.

When the metal forms a carbide before the reduction of the oxide, a different mechanism than that presented in Eq. 3 probably takes place. Titanium and zirconium are two metals that readily form carbides under the conditions normally used for vacuum-fusion analysis. Equation 11 is the expected reaction for the reduction of TiO based on Eq. 3:

$$TiO + C \rightarrow Ti + CO. \tag{11}$$

At 1727°C the free energy of reaction of TiO is −21,900 cal/mole, which corresponds to an equilibrium pressure for CO of 250 torr. At the same temperature, the free energy of formation of TiC is −37,500 cal/mole, and this reaction will occur in preference to the oxide reduction. When carbide formation of the metal takes place before complete reduction of the oxide, the reduction of the remaining oxide would have to take place by the reaction

$$TiO + TiC \rightarrow 2Ti + CO. \tag{12}$$

The free energy of this reaction at 1727°C is +11,000 cal/mole; therefore, this reaction will not proceed.

The platinum flux technique is generally more successfully used than the platinum-bath technique for the determination of oxygen in zirconium and titanium by vacuum and inert-gas fusion. In the platinum flux technique, the sample is either wrapped or encapsulated in platinum or a quantity of platinum is added to the crucible with the sample. The platinum to sample ratio is usually above 6:1. This technique apparently gives better results than the bath techniques because, in the light of the previous remarks concerning carbide formation, titanium and zirconium are very strong carbide formers.

In many vacuum-fusion baths the distribution of the solute (sample) is not homogeneous throughout the bath. This is particularly true in the case of platinum baths with 450-kHz heating that provides little or no mixing of the bath. The concentration of the solute is much higher at the surface than in the lower levels of the bath. Consequently, the advantages of the bath are diminished with the addition of each succeeding sample. In addition, the bath contains an equilibrium quantity of carbon in solution.

When the flux technique is used, the equivalent of a new bath is prepared with each sample, and nonequilibrium conditions for carbon exist. The optimum ratio of platinum to sample metal can be controlled, so that any extra energy provided by the formation of intermetallic compounds with the platinum can be utilized. The carbon content of the new portion of the bath is also kept much lower than the equilibrium concentration to minimize carbide formation. Because of the gas content of the flux metal, the blanks obtained with this technique are not as low as those of the bath technique.

V. CONDITIONS FOR THE REMOVAL OF OXYGEN

Although proper thermodynamic conditions are necessary for the reduction of the oxides, physical conditions in the bath may influence the completeness of reduction and the extraction of the carbon monoxide from the bath. Some factors that can be considered as affecting the removal of oxygen from molten metals are the mechanism and kinetics of reduction of the oxides, solubility of carbon monoxide in the sample or bath metal, and the viscosity of the bath. Because complete information is not available the effect of these factors cannot be completely evaluated; however, some general conclusions can be drawn.

In vacuum-fusion analysis the reduction of the oxygen-containing species by carbon is usually accomplished in an iron bath in which the carbon and oxygen are present in dilute solution. The following reaction takes place:

$$[\mathrm{C}]_{\mathrm{Fe}} + [\mathrm{O}]_{\mathrm{Fe}} \rightleftharpoons \mathrm{CO}. \tag{13}$$

From this equation it might be assumed that the [C] × [O] product would be constant for a given equilibrium pressure of carbon monoxide. It has been determined, however, that the presence of carbon in molten iron reduces the activity of oxygen [13]. While an increase in carbon content lowers the oxygen activity in molten iron it also substantially lowers the solubility of carbon monoxide in molten iron. Under normal vacuum-fusion conditions the oxygen content of an iron bath should be below measurable amounts [8].

With regard to the mechanism of oxide reduction during vacuum-fusion analysis, Birks and Booth [13a], studied the kinetics of carbon monoxide evolution from aluminum-killed steel samples in a carbon-saturated iron bath. The rate-controlling step is evidently the dissolution of alumina particles in the bath. Thus, diffusion of dissolved carbon and oxygen, their reaction to give carbon monoxide, and the evolution of carbon monoxide were not rate-controlling.

The carbon content of an iron bath can cause conditions that prevent proper alloying of the sample with the iron bath and complete removal of the carbon monoxide from the melt. Smith [14] found that if iron baths were heated in a graphite crucible at 1800° or 2000°C for 2 h, graphite flakes were observed on the surface. This graphite "kish" forms a viscous zone that prevents proper alloying of the samples with the bath and probably causes incomplete carbon monoxide recovery from the melt. The kish cannot be dissolved by raising the temperature of the vacuum-fusion bath. This problem is apparent particularly when an iron bath is degassed at a higher temperature than that at which it is used. Consequently, in actual practice the bath is usually degassed at the temperature at which the oxygen determinations will be made, not a higher temperature.

Quantitative measurements of this effect have been made by Shvidkovskiy [15] in an investigation of the effect of alloy composition on the viscosity of molten metals. Some of the determinations of viscosity were made on ferrochrome in graphite crucibles at temperatures up to 1700°C and therefore were similar to the conditions found in a vacuum-fusion crucible. It was found that when the melt was cooled 20° or 30° below the maximum temperature used during the test, the viscosity of the melt increased greatly. In this case the increase in viscosity was attributed to the formation of graphite flakes in the melt, forming a heterogeneous system. Further work showed that alloying constituents did not appreciably affect the viscosity of the melt.

VI. CONDITIONS FOR THE REMOVAL OF NITROGEN

The models proposed by Sloman et al. [2] for the removal of nitrogen from metals by vacuum fusion were based on nitrogen present as a nitride. The reaction of the matrix metal was not treated. However, if the matrix were iron or steel, some of their proposed models are adequate. The principal conclusion of this work was that the primary reaction involved in the vacuum-fusion extraction of nitrogen from metals was thermal decomposition of the nitride and solution of the liberated metal in the iron bath. It appeared that carbide formation played little or no part in the reaction.

In a complete review of the problems associated with the removal of nitrogen from metals by vacuum fusion, Goward [16] emphasized that nitrogen in many of the refractory metals exists in the metals in solid solution rather than as a precipitated nitride. Based on the theories of Chipman [17] and Wagner [18], Goward developed equations that allowed the equilibrium pressure of nitrogen over an iron–metal bath to be calculated. For these calculations, the activity coefficients for nitrogen in iron–metal systems as presented by Pehlke and Elliott [19] were used. A graphical summary of the log of the activity coefficient of nitrogen in an iron bath at 1600°C versus percent of the other metal present is given in Fig. 2.8 [16]. This figure shows the marked effect that a few percent of a metal can have on the activity coefficient of nitrogen in an iron bath. This is particularly true when the metal being added is one of the refractory metals such as Nb or V.

Using the information in Fig. 2.8 and the equations developed from Chipman's and Wagner's work, Goward calculated the equilibrium nitrogen content in molten iron–metal systems at various nitrogen pressures. These are tabulated in Table 2.4. All calculations were based on a 10-g iron bath and a 1-g sample heated at 1600°C.

In most vacuum-fusion systems the partial pressure of residual nitrogen in contact with the melt is of the order of 10^{-6} torr. From the information

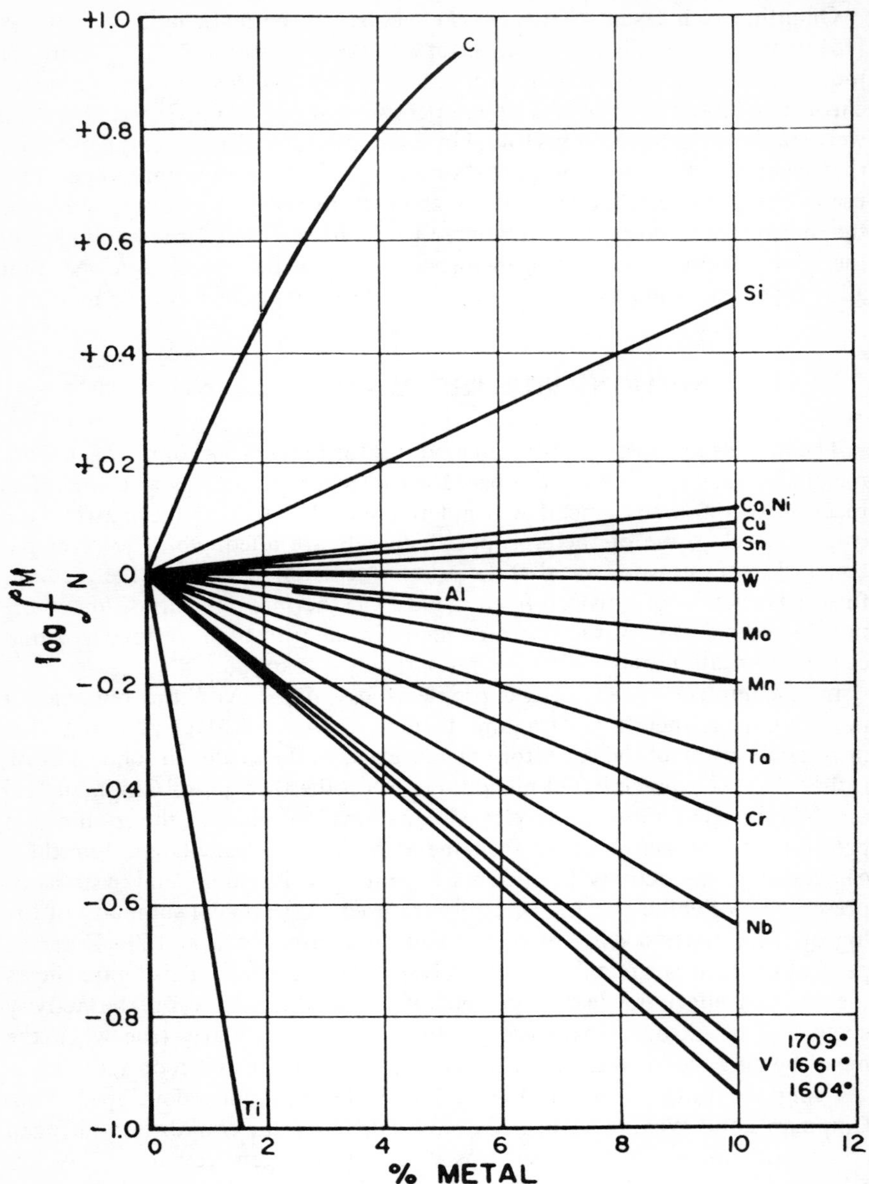

Fig. 2.8. Log activity coefficient of nitrogen versus percent metal in iron at 1600°C. (Reproduced by permission from *Analytical Chemistry* [16].)

TABLE 2.4. Equilibrium Nitrogen Contents in Molten Fe–M Solutions at Various Nitrogen Pressures[a]

System	% Metal in Bath	% C in Bath	Equil. N_2 Content in Bath (ppm) at Indicated Pressure		
			10^{-4} torr	10^{-5} torr	10^{-6} torr
Fe–V	5	0	0.48	0.15	0.048
	5	5.3	0.050	0.016	0.0050
	10	0	1.32	0.42	0.132
	10	5.3	0.15	0.048	0.015
	1700°C {10	0	1.07	0.34	0.107
	1700°C {10	5.3	0.12	0.038	0.012
Fe–Nb	5	0	0.33	0.10	0.033
	5	5.3	0.035	0.011	0.0035
	10	0	0.70	0.22	0.070
	10	5.3	0.073	0.023	0.0073
Fe–Ta	5	0	0.23	0.074	0.023
	5	5.3	0.023	0.0073	0.0023
	10	0	0.34	0.11	0.034
	10	5.3	0.037	0.012	0.0037
Fe–Ti	2	0	2.8%	0.87%	0.28%
	2	5.3	0.32	0.10	0.032
	5	5.3	23	7.4	2.3

[a] All at 1600°C unless otherwise indicated.

contained in Table 2.4, the following conclusions can be drawn:

1. Reducing the nitrogen pressure over the melt by a factor of 100 will lower the equilibrium nitrogen content of the melt by a factor of 10.
2. For most metals, the equilibrium nitrogen concentration in the bath increases as the metal concentration increases. In a carbon-saturated bath, as long as the solute metal concentration is under 10%, this increase should not affect nitrogen results (except in the case of Ti). As the solute metal concentration increases above 10%, the equilibrium nitrogen concentration can be expected to approach significant concentrations.
3. The effect of carbon in reducing the equilibrium nitrogen content of a bath is significant. In most cases an iron–metal bath containing no carbon will have an equilibrium solubility 10 times that of a carbon-saturated bath of the same iron–metal composition. For metals that form carbides, the carbon helps to reduce the effective reactive metal concentration and thus tends to lower the equilibrium nitrogen concentration.

These conclusions are valid only for thermodynamic considerations and undoubtedly must be modified by kinetic considerations. Drawing on the work of Pehlke and Elliott [20], Goward [16] calculated that it would require 12.5 min to reduce the nitrogen content of a bath from an initial concentration of 50 ppm to a final value of 0.09 ppm. If the bath contained 10% Nb,

TABLE 2.5. Dissociation Pressures of Various Nitrides[a]

Nitride	Temperature (°C) for Several Equilibrium Nitrogen Pressures		
	1 torr	10^{-2} torr	10^{-3} torr
TiN	2912	2489	2331
ZrN	2603	2205	2046
VN	1242	1024	933
(NbN)	(1946)	(1620)	(1491)
Nb_2N	2050	1747	1614
TaN	1921	1599	1460
"Cr_2N"($Cr_2N_{0.76}$)	1050	782	675
Mo_2N	329	243	209
W_2N	344	256	221
AlN	1487	1275	1187
Si_3N_4	1421	1164	1060
BN	1340	1075	927
Th_3N_4	2401	2037	1890

[a] Reproduced by permission from *Analytical Chemistry* [16].

it would take 25 min to effect the same removal of nitrogen. These times are in approximate agreement with the few successful reports in the literature of nitrogen determination by vacuum fusion. One other point that could affect the extraction of nitrogen from the bath would be the formation of carbonitrides at the same time the metal carbides are being formed. The removal of nitrogen from the solid carbide would be a very slow process and could cause an unfavorable extraction rate. The literature contains very little information on this possibility.

Goward [16] assembled data relating to the quantitative hot extraction of nitrogen [3, 21] from metals below their melting point. The data presented in Table 2.5 indicate that the nitrides of Cr, Mo, W, V, Nb, Ta, Al, Si, and B should be thermally unstable at temperatures that are attainable in the usual vacuum-fusion system. The data also indicate that for thermal decomposition of the nitrides of Ti, Zr, and Th, a much higher temperature than usually used in vacuum-fusion analysis would be necessary, which is basically in agreement with findings of Sloman et al. [2].

Despite the fact that the nitrides may be unstable at hot extraction temperatures, a successful extraction can be obtained only if the diffusion through the sample is rapid enough. Using published diffusion data, Goward [16] calculated $t_{1/2}$-values for a series of metals in the form of a cylinder of 0.4-cm diameter. The $t_{1/2}$-values in Table 2.6 indicate that, if it is assumed that diffusion of the nitrogen through the metal is the limiting factor, the hot-extraction process is theoretically feasible for many metals.

TABLE 2.6. Calculated Equilibrium Pressures for 1 ppm Nitrogen for Various Metals[a]

Element	Temp., °C	Equilibrium N_2 Pressure (torr)	$t_{1/2}$ (min)[b]
Fe	1500	4.8×10^{-2}	1.6
Fe(l)	1600[c]	4.8×10^{-3}	1.4
Cr	1300	9.9×10^{-6}	—
Cr(l)	1600	4.6×10^{-7}	—
Mo	2000	6.9×10^{-2}	7
W	2000	6.3×10^{-2}	31
Nb	2000	1.2×10^{-9}	1.9
	2420	3.8×10^{-9}	—
Ta	2000	2.7×10^{-9}	2.8
	2960	1.4×10^{-6}	—

[a] Reproduced by permission from *Analytical Chemistry* [16].
[b] 0.4-cm-diameter cylinder.
[c] Molten Fe, 1.6-cm crucible diameter.

VII. INCOMPLETE RECOVERY DUE TO UPTAKE OF GASES

For a vacuum-fusion analysis of metals to be successful, the gases have to be not only liberated from the sample and extracted from the bath but also transferred from the furnace to the analytical system without loss. During outgassing of the bath and during analysis of the samples, considerable quantities of metal are vaporized and deposited on the walls of the furnace. According to Ehrke and Slack [22], two main types of gettering exist: (1) dispersal gettering, in which clean-up takes place while the sorbing substance is being dispersed or volatilized in the presence of the gas; and (2) contact gettering, in which the previously deposited metal surface removes gas by contact (this also applied to sorption that may continue after volatilization ceases in cases of dispersal gettering).

Their study was confined to the contact gettering of bright and diffuse metal surfaces. The bright surfaces were produced at pressures of about 0.01 torr, while the diffuse surfaces were produced at pressures of 1–3 torr. In general, the diffuse layer picked up more hydrogen and nitrogen than did the bright layer, probably because of a greater effective surface area. They reported that hydrogen and nitrogen were not adsorbed by films of aluminum and magnesium, while films of thorium, uranium, barium, and misch metal did adsorb these gases. It was also noted that in the presence of mercury the gettering power of some metals is diminished and that metals that amalgamate with mercury getter very little, if at all. This finding may apply to many vacuum-fusion systems where mercury diffusion pumps are used and where mercury lifts are used to insert the samples. In these systems, the pressure of mercury can reasonably be expected to be 3–5 torr.

A study of the activity of evaporated metal films in gas chemisorption was made by Trapnell [23]. This was a general study not related directly to vacuum fusion. The adsorption characteristics of the metal films with nitrogen, hydrogen, and carbon monoxide were investigated between 0 and −183°C at pressures of 10–100 torr. This subject is also discussed in Chapter 1, Section II.A.2. The recovery of nitrogen and oxygen (as carbon monoxide), during analysis by fusion techniques, is treated in Chapter 4, Section I.A.4.

REFERENCES

1. W. H. Walker and W. A. Patrick, *Orig. Com. 8th Intern. Congr. Appl. Chem.*, **21,** 139–148 (1912); through *Chem. Abstr.* **6,** 3380 (1912).
2. H. A. Sloman, C. A. Harvey, and O. Kubaschewski, *J. Inst. Metals Bull. Met. Rev.*, **80,** 391 (1952).
3. J. E. Fagel, R. F. Witbeck, and H. A. Smith, *Anal. Chem.*, **31,** 1115 (1959).
4. W. R. Hansen and M. W. Mallett, *Anal. Chem.*, **29,** 1868 (1957).
5. T. Lyman, Ed., *Metals Handbook*, American Society for Metals, Cleveland, 1948.
6. DMIC Report 152, Battelle Memorial Inst., 1961.
7. S. Dushman, *Scientific Foundations of Vacuum Techniques*, 2nd ed., J. M. Lafferty, Ed., Wiley, New York, 1962, Chapter 10.
8. A. S. Darling, *Metallurgia*, **64,** 7 (1961).
9. Yu A. Klyachko, L. L. Kunin, and E. M. Chistyakova, *Tr. Komissii Anal. Khim., Akad. Nauk SSSR, Inst. Geokhim. i Anal. Khim.*, **10,** 10 (1960).
10. S. L. Hoyt and M. A. Scheil, *Trans. AIME*, **125,** 313 (1937).
11. G. Phragmen, *Trans. AIME*, **125,** 298 (1937).
12. H. Schenk and K. Zinger, *Arch. Eisenhuttenw.*, **30,** 549 (1959).
13. S. Marshall and J. Chipman, *Trans. Amer. Soc. Metals*, **30,** 695 (1942).
13a. N. Birks and D. Booth, *J. Iron Steel Inst.* (London), **204,** 340 (1966).
14. W. H. Smith, *Anal. Chem.*, **27,** 1636 (1955).
15. Ye. G. Shvidkovskiy, Nekotaryye Nyazkosti rasplovlennykh metallov, State Publishing House for Technical and Theoretical Literature, Moscow, 1955 (NASA Technical Translation F-88).
16. G. W. Goward, *Anal. Chem.*, **37,** 117R (1965).
17. J. Chipman, *J. Iron Steel Inst.* (London), **180,** 97 (1955).
18. C. Wagner, *Thermodynamics of Alloys*, Addison-Wesley Press, Cambridge, 1962.
19. R. D. Pehlke and J. F. Elliott, *Trans. AIME*, **218,** 1088 (1960).
20. R. D. Pehlke and J. F. Elliott, *Trans. AIME*, **227,** 844 (1963).
21. M. W. Mallett and C. B. Griffith, *Trans. Amer. Soc. Metals*, **46,** 375 (1954).
22. L. F. Ehrke and C. M. Slack, *J. Appl. Phys.*, **11,** 129 (1940).
23. B. M. W. Trapnell, *Proc. Roy. Soc., Ser. A*, **218,** 566 (1953).

CHAPTER

3

ACTIVATION ANALYSIS

EVERETT E. WICKER

United States Steel Corporation
Research Laboratory
Monroeville, Pennsylvania

CONTENTS

I. HISTORICAL DEVELOPMENT

Since the discovery of the neutron by Chadwick in 1932, a number of techniques that make use of the properties of this particle for chemical analysis have been devised and tested [1]. Many of these are based on measurements of the transmission, scattering, and absorption of neutrons either at selected single energies or as a function of energy over a broad spectrum. Another useful property of the neutron is that it can be diffracted by a crystal lattice, and this property has also been used in chemical analysis. However, the greatest contribution of neutrons to chemical analysis has been through the development of activation analysis. In this technique the neutron is used to produce a radioactive nuclear species in the sample, and the radioactivity of the sample is measured for qualitative and quantitative analysis. Several excellent books [2–5] describe various features and details of this powerful analytical technique.

The first use of this technique was reported by de Hevesy and Levi in 1936 [6]. These investigators used an isotopic source (Ra–Be) of neutrons to detect and measure quantitatively the dysprosium content of yttrium. They found that concentrations of dysprosium as small as 0.1% could be detected. They also used this isotopic source to determine europium in gadolinium samples. These two cases were exceptionally favorable in that the elements of interest were very readily activated, so that even the weak sources available at that time proved to be satisfactory. With such weak sources, however, there were very few analytical possibilities available, and consequently, the technique was not developed further for some time. With the development of nuclear reactors during World War II, large intense sources of neutrons became available, and the activation of most elements became practical. Thus, the technique was applied to many different analytical problems, and its great sensitivity became the chief motivation for its further development.

Nuclear reactors, of course, are very large and very expensive pieces of experimental apparatus—certainly too expensive to be considered for general use by most analytical laboratories. Therefore, it seemed that neutron-activation analysis would be beyond the general reach of most industrial and academic laboratories. This situation has been changed, however, largely by the advent of relatively inexpensive electrical accelerators capable of producing rather large fluxes of fast neutrons. The analytical techniques to be discussed in this chapter are based primarily on the use of such accelerators.

II. THEORY AND PRINCIPLES OF THE METHOD

A. GENERAL

The method of activation analysis consists generally of the bombardment of a target material by nuclear particles and the transmutation of that target material to a radioactive species. The type and energy of radiation emitted by the radioactive nucleus and the half-life for its decay are properties that can be used to identify the radioactive species and thus provide a qualitative chemical analysis. The intensity of the emitted radiation in terms of events detected per unit time (e.g., counts per second) provides the necessary information to make the analysis quantitative.

The bombarding particles used to produce a transmutation of the target nucleus may be neutrons, protons, deuterons, tritons, α-particles, or others. Although activation by all of these particles or atomic nuclei has proven to be of considerable value in many instances (see Chapter 20) for the present discussion neutrons will be the activating particles of greatest interest.

Neutrons may have widely varying energies depending on their source. The energy is usually expressed in units of *electron volts* (eV) or in multiples such as *thousand* (*kilo-*) *electron volts* (keV) or *million* (*mega-*) *electron volts* (MeV). Neutrons are frequently referred to as *slow* or *fast*. By *slow* it is usually meant that the neutron is in thermal equilibrium with its surroundings; it is therefore also called *thermal*. This is an energy of about 0.025 eV. Fast neutrons have energies in excess of 1 MeV. (In familiar cgs units, 1 MeV $= 1.602 \times 10^{-6}$ erg.)

There is a wide variety of radiation detection equipment that may be used, depending on the type and energy of radiation emitted by the activated material. The present discussions will be concerned primarily with scintillation detectors for energetic gamma rays.

B. NUCLEAR REACTIONS

Various nuclear reactions are possible when a target nucleus is struck by a neutron. The extent to which one type of nuclear reaction may predominate over the others is primarily a function of the energy of the incoming neutron and the mass of the target nucleus. Nuclear reactions are discussed in detail in most nuclear physics textbooks [7], and only the most basic features will be mentioned here.

The most commonly used nuclear reaction is one known as radiative capture and is usually expressed symbolically as (n, γ). In this reaction the

target nucleus absorbs the neutron and emits a γ-ray almost instantaneously. The new nucleus that is formed therefore has the same atomic number as the target nucleus but has an increase of 1 in mass number. The new, heavier nucleus of the same chemical species of the same atomic number is normally, although not always, radioactive. It may be identified by the type and energy of particles emitted during radioactive decay. Nuclear reactions of this type usually take place with slow, or thermal, neutrons and may occur for a wide range of atomic mass and atomic number in the target nucleus. A few examples of this type of reaction are written below in the normally accepted terminology; the resultant products are radioactive and have the half-lives shown:

$$^{23}\text{Na}(\text{n}, \gamma)\,^{24}\text{Na}; \qquad t_{1/2} = 15.0 \text{ h}$$

$$^{51}\text{V}(\text{n}, \gamma)\,^{52}\text{V}; \qquad t_{1/2} = 3.75 \text{ min}$$

$$^{197}\text{Au}(\text{n}, \gamma)\,^{198}\text{Au}; \qquad t_{1/2} = 2.698 \text{ days.}$$

Another type of reaction to be considered is one in which a neutron is absorbed by the target nucleus and a proton emitted. This is usually referred to, and written as, an (n, p) reaction. Because the neutron that is absorbed and the proton that is emitted have the same mass, the product nucleus has the same atomic mass as the target nucleus; however, because the proton that is emitted carries 1 esu of positive charge, the atomic number will be decreased by 1. Reactions of this type frequently require that the incoming neutron have a fairly high threshold energy of the order of a few megaelectron volts. Furthermore, these reactions tend to occur more frequently among light nuclei than among heavy ones. Examples of the (n, p) reaction with the threshold energy and half-life of the product are listed below:

$$^{16}\text{O}(\text{n}, \text{p})\,^{16}\text{N}; \qquad 10.2 \text{ MeV}; \qquad t_{1/2} = 7.14 \text{ sec}$$

$$^{19}\text{F}(\text{n}, \text{p})\,^{19}\text{O}; \qquad 4.2 \text{ MeV}; \qquad t_{1/2} = 29.1 \text{ sec}$$

$$^{27}\text{Al}(\text{n}, \text{p})\,^{27}\text{Mg}; \qquad 1.9 \text{ MeV}; \qquad t_{1/2} = 9.46 \text{ min}$$

$$^{28}\text{Si}(\text{n}, \text{p})\,^{28}\text{Al}; \qquad 4.0 \text{ MeV}; \qquad t_{1/2} = 2.31 \text{ min.}$$

The next type of reaction to be considered is one in which a neutron is absorbed by the target nucleus and an α-particle is emitted. This is usually written as an *(n, α)* reaction. The (n, α) reaction is also a threshold reaction, taking place primarily when the incoming neutron has an energy of the order of a few megaelectron volts and the target nucleus is of light or intermediate mass such that the mass number is generally less than 80. The α-particle emitted has a mass of 4, whereas the incoming neutron has a mass of only 1; therefore, the product nucleus has a mass 3 less than the target nucleus.

Furthermore, the α-particle has a double positive charge, so that the atomic number is decreased by 2. Some examples of the (n, α) reaction and their thresholds are

$$^{27}Al(n, \alpha)\,^{24}Na; \qquad 3.25 \text{ MeV}$$
$$^{55}Mn(n, \alpha)\,^{52}V; \qquad 0.65 \text{ MeV}.$$

The half-lives of the ^{24}Na and ^{52}V are, of course, the same as when these isotopes are produced by (n, γ) reactions.

A fourth type of nuclear reaction that is frequently of use for activation analysis is one in which the neutron is absorbed by the target nucleus and two neutrons are emitted; that is called an (n, 2n) reaction. In this reaction there is a decrease of 1 in the mass number of the target nucleus, and there is no gain or loss of charged particles, so that the atomic number or chemical species does not change. The (n, 2n) reaction is, like the two preceding nuclear reactions, a threshold reaction requiring that the incoming neutron possess energy greater than a certain minimum value. The threshold for these reactions is usually somewhat larger, frequently in the range 10–20 MeV. These reactions are also somewhat more predominant among intermediate and heavy nuclei than are the two preceding types of nuclear reaction. Examples of potentially useful (n, 2n) reactions with light or intermediate nuclei and the thresholds for the reactions are

$$^{63}Cu(n, 2n)\,^{62}Cu; \qquad 11.0 \text{ MeV}; \qquad t_{1/2} = 9.76 \text{ min}$$
$$^{19}F(n, 2n)\,^{18}F; \qquad 11.0 \text{ MeV}; \qquad t_{1/2} = 109.7 \text{ min}$$
$$^{54}Fe(n, 2n)\,^{53}Fe; \qquad 13.9 \text{ MeV}; \qquad t_{1/2} = 8.51 \text{ min}.$$

There are, of course, other nuclear reactions that may be quite valuable in neutron-activation analysis but are not widely used at present. The most common nuclear reaction is simply scattering of the incoming neutrons either elastically or inelastically. Some very interesting applications for the use of this reaction have been proposed but are not pertinent to the present subject. Perhaps the best-known nuclear reaction is that of fission, or what might be written as an (n, f) reaction. This reaction, while not of practical use as a means of chemical analysis, is of concern as a possible source of interference in the determination of oxygen.

Although several nuclear reactions have been suggested and used for the activation analysis for oxygen, many of these require activation by charged particles such as deuterons or α-particles, which are difficult to obtain inexpensively in large quantities and yield only surface information. Others are used to activate the oxygen isotopes with mass numbers 17 and 18, which are present in such small abundance that the sensitivity of the method is greatly limited. Therefore, the most desirable nuclear reaction for the analysis for bulk oxygen in metals is the one listed above under (n, p) reactions. Specifically, the reaction is $^{16}O(n, p)^{16}N$.

The ^{16}N product nucleus has a half-life of only 7.14 sec, and therefore, its γ-ray activity must be measured very promptly after activation. This also means, however, that a relatively short activation time of perhaps 20 to 30 sec is sufficient to approach the saturation activity within the sample. The radioactive ^{16}N product nucleus decays by the emission of high-energy β-particles.

A distinct advantage of this activation process is that the γ-ray photons emitted are of very high energy (6.14 and 7.11 MeV). This energy range is greater than any commonly found in radioactive decay, so that interferences are minimized. The most important interference is from fluorine, because the reaction $^{19}F(n, \alpha)^{16}N$ produces the same radioactive isotope as does the oxygen reaction. Because of the relative cross section for these reactions, a correction becomes necessary only if the fluorine concentration exceeds about 50% of the oxygen concentration—a relatively uncommon situation in metals. Boron may also produce an interference but only if the oxygen content does not exceed 30% of the boron present—again an unusual circumstance in metals. The only other possible interference would come from the presence of fissionable material, such as thorium or uranium, that might emit high-energy γ-rays and delayed neutrons. Therefore, interferences by other chemical elements can usually be ignored.

The $^{16}O(n, p)\,^{16}N$ reaction is very widely used and will be the center of interest in this chapter. The determination of oxygen based on this reaction has been the subject of numerous technical papers (Ref. 8 contains a good bibliography). A brief discussion of the physical properties and relationships that provide a quantitative basis for all neutron-activation analysis may be useful and will be presented here.

C. QUANTITATIVE ASPECTS

1. Cross Section and Flux

A nuclear property of fundamental concern in activation analysis is the reaction cross section. This quantity may be considered variously as a reaction rate or, perhaps more commonly, as the probability that a certain nuclear reaction will take place. To get an understanding of the meaning of the cross section, consider a uniform beam of neutrons with a density in the beam of n particles per cubic centimeter traveling with a velocity v centimeters per sec. If this beam impinges on a very thin sheet of target material, the number of reactions taking place per square centimeter per unit time will be proportional to the number of nuclei per unit area of the target and to the number of particles incident on the target according to the following equation:

$$\frac{\text{No. of reactions}}{\text{cm}^2 \times \text{sec}} = \sigma \times \frac{\text{No. of target nuclei}}{\text{cm}^2} \times nv \frac{\text{particles.}}{\text{cm}^2 \times \text{sec}} \quad (1)$$

The proportionality constant, σ, in this equation is called the *microscopic cross section* for the reaction of interest. Equation 1 indicates that σ must have units of square centimeters. Because of this dimension of square centimeters, or area, the name *cross section* is used for this constant. For most nuclear reactions the magnitude of σ is between 10^{-27} and 10^{-24} cm^2, and for convenience the unit of 10^{-24} cm^2 has been given the name *barn*, so that cross sections are usually expressed in terms of barns or millibarns. A very important characteristic of the cross section, but one that is beyond the present discussion, is that the σ for a given reaction in any isotope varies greatly and irregularly with the energy of the incident particle.

In the above equation, the product nv, the density of neutrons in the beam multiplied by the velocity of the neutrons, is usually referred to as *flux* and has units of neutrons per square centimeter per second. This quantity occurs very frequently in activation-analysis equations and is generally given the symbol ϕ.

2. Formation and Decay of Radioactive Species

If a target material is exposed to a nuclear reaction that results in a radioactive product, this chain of events may be represented as follows:

$$\mathrm{X} + \mathrm{n} \rightarrow \mathrm{Y}^* \underset{\lambda}{\rightarrow} \mathrm{Z}$$

This indicates that the stable isotope X is bombarded by neutrons, and when it absorbs a neutron it becomes the radioactive material Y, which then decays with a disintegration constant λ to the isotope Z (which may or may not be radioactive).

The rate of formation of radioactive isotope Y will depend on the number of atoms of X present (N_{X}); the activation cross section of isotope X (σ_{act}); and on the neutron flux, ϕ. While isotope Y is being formed, it is also decaying to isotope Z, so that the number of atoms of Y(N_{Y}) is changing at a rate, λN_{Y}, where λ is the disintegration constant and will be equal to 0.693 divided by the half-life and will have units of reciprocal time. Therefore, the overall rate of change of N_{Y} (the number of Y atoms present at any time) may be expressed as

$$\frac{dN_{\mathrm{Y}}}{dt} = \phi \sigma_{\mathrm{act}} N_{\mathrm{X}} - \lambda N_{\mathrm{Y}}. \quad (2)$$

This equation may be integrated to give the number of radioactive atoms present at the end of an irradiation time t_i. If it is assumed that the number of

Y atoms is always small compared with the number of X atoms so that N_X may be considered a constant and if it is further assumed that the number of Y atoms is equal to zero at the beginning of the irradiation time, the result of this integration will be

$$N_Y = \frac{\phi\sigma_{act}N_X}{\lambda}[1 - \exp(-\lambda t_i)]. \tag{3}$$

The rate of disintegration, or the activity (A), of the radioactive material produced by this reaction will be given as

$$A_Y = \lambda N_Y = \phi\sigma_{act}N_X[1 - \exp(-\lambda t_i)]. \tag{4}$$

This equation represents the amount of radioactivity present at the end of a given irradiation period, t_i. If there is a delay between the end of the irradiation and measurement of this radioactivity, the activity (A_Y) will be decreased by the exponential factor $e^{-\lambda t_d}$, where t_d is the delay time.

In Eq. 4, the factor in brackets is usually referred to as the *saturation factor*, which may be used to examine quantitatively the approach to the maximum attainable activation as a result of irradiation. The form of the activation curve in terms of the number of half-life periods is shown in Fig. 3.1, as is the form of the radioisotope decay curve which shows the rate at which the radioactive material would decay after removal from the neutron flux.

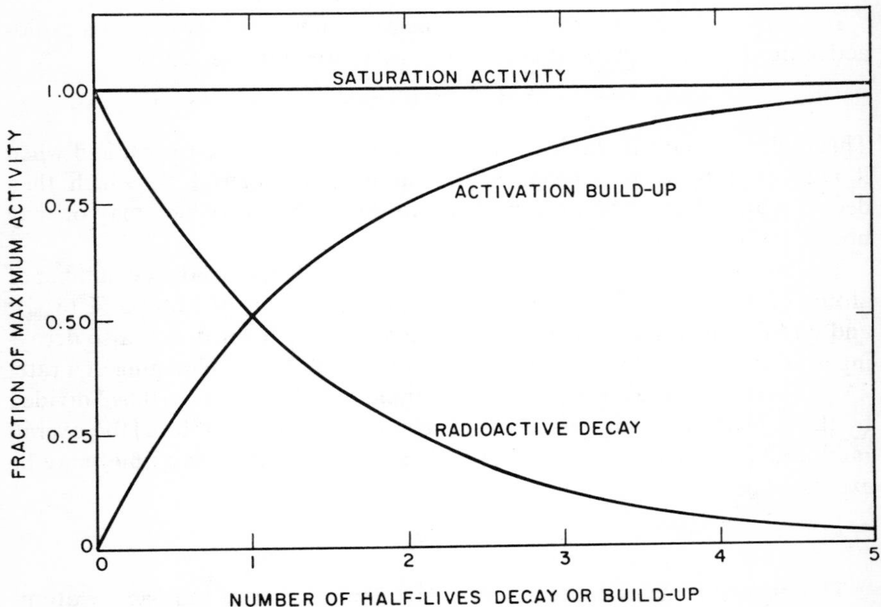

Fig. 3.1. Radioactive build-up and decay.

The quantity of interest in activation analysis is usually the weight of an element that was originally present in the target material; in the example we have been considering here, it could be the weight of X or W_X. This quantity can be related to the number of atoms of X by the following equation:

$$N_X = \frac{W_X f}{M} 6.02 \times 10^{23}, \tag{5}$$

where f is equal to the isotopic abundance of the nucleus producing the desired reaction, M is equal to the atomic weight of that isotope, and 6.02×10^{23} is Avogadro's number. If this expression is substituted for N_X, Eq. 4 can be rewritten and solved for W_X

$$W_X = \frac{M_X A_Y}{\phi \sigma_{act} f (6.02 \times 10^{23})(1 - e^{-\lambda t_i})}. \tag{6}$$

In principle it should be possible to measure the activity of the radioactive species and thus determine the value for A_Y. With all other factors on the right-hand side of the equation known or constant, it should be possible to calculate the weight W_X of the target nucleus X present in the initial sample.

While it is true that the constants in Eq. 6 would seem to permit a direct calculation of the weight of element X in the sample, closer examination reveals that this is not necessarily true in practice. For example, the flux, ϕ, at the sample is difficult to measure with precision and may vary during irradiation. Therefore, it can rarely be substituted directly into a calculation. Furthermore, the activation cross section, σ_{act}, is sometimes known only with limited accuracy, so that substitution of published values for this constant might lead to errors. The disintegration rate of the radioactive isotope Y is also subject to errors and uncertainties in its experimental measurement. Therefore, the general practice in activation analysis is to compare the activity induced in a standard sample with that induced in the material under investigation.

The standard sample should be chosen to have nearly the same density and composition as the unknown sample under investigation. If the standard and unknown are then irradiated for the same time in the same flux and their activities are measured with the same efficiency, the weight of element X in the unknown will be given by the equation

$$W_{Xu} = \frac{A_{Xu}}{A_{Xs}} W_{Xs}, \tag{7}$$

where W_{Xu} is the weight of element X in the unknown; W_{Xs} is the weight of element X in the standard; A_{Xu} is the activity of element X in the unknown at the end of irradiation; and A_{X_s} is the activity of element X in the standard at the end of irradiation.

If this procedure is followed, the uncertainties in flux, cross section, and decay constant should not affect the accuracy of the results. It is important to note, however, that identical irradiation and counting of the sample and standard are essential and may be difficult to attain.

The standard material should be thoroughly documented, resistant to radiation damage, stable over long time periods, and of suitable physical and chemical form.

III. EQUIPMENT

To measure the oxygen content of a material by the nuclear reaction $^{16}O(n, p)\,^{16}N$, it is necessary to have equipment for production of fast neutrons and for measurement of the radioactive decay of the resultant ^{16}N. Furthermore, it is necessary to be able to transfer the samples between the irradiation position and the counting station reliably and rapidly and to determine the neutron flux.

Detailed discussions of equipment, which are beyond the scope of the present book, appear in the published literature [9, 10, 11]. However, the various components and some of the desired capabilities are listed here:

1. A neutron generator that produces 14-MeV neutrons by the deuterium–tritium reaction, $^{3}H(d, n)\,^{4}He$. The output should be at least 10^{10} n/sec. Continuously evacuated Cockroft-Walton accelerators, such as is shown in Fig. 3.2, have been used. Recently, however, small, sealed-tube sources, as shown in Fig. 3.3, have demonstrated great reliability, long life, excellent stability, and high neutron output (more than 10^{11} n/sec). Both of these generators were manufactured by Kaman Sciences Corp., Colorado Springs, Colorado. Similar devices are produced by other companies in the United States and abroad.
2. One or more sodium iodide scintillation counters, large enough to absorb a significant fraction of the 6.13–7.11-MeV γ-rays from the decay of ^{16}N. The counting geometry is optimum with two crystals 5-in. long by 5-in. in diameter.
3. Electronic components such as amplifiers, pulse-height analyzers, and scalers to indicate the number of γ-ray photons at the desired energy.
4. A system for transferring the sample rapidly between the neutron source and the radiation detector, positioning the sample reproducibly, and controlling the times during which the sample is irradiated, counted, and in transit.
5. Some device for monitoring the neutron flux at the sample position during irradiation.

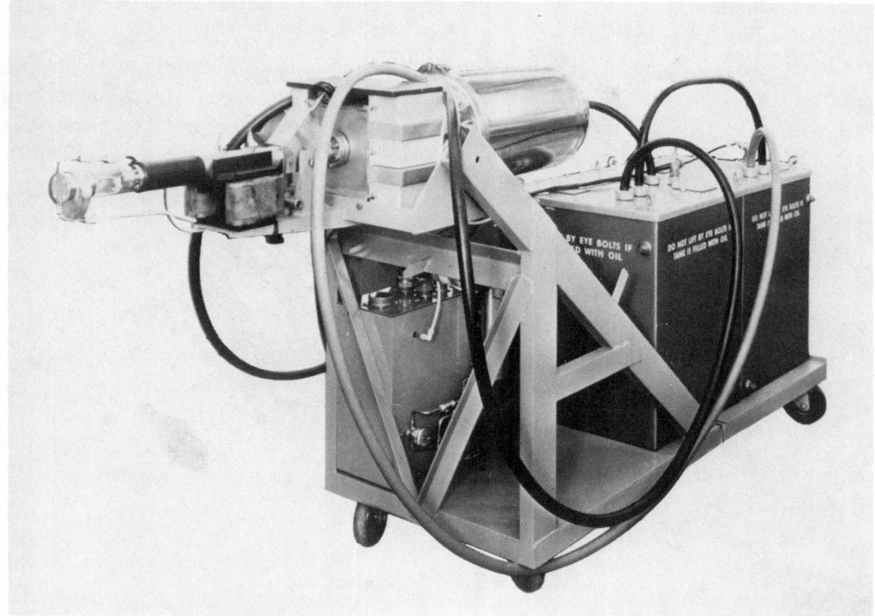

Fig. 3.2. Cockroft-Walton generator for 14-MeV neutrons.

Most laboratories, both in the United States and abroad, have assembled their analytical systems either from commercial or "home-made" components. Two industrial embodiments have been made and thoroughly tested, one in England and one in Belgium. The first of these was described by Gray [12] in 1964 and given the commercial name Analox. This equipment has been in use for several years at a British steelworks and has recently determined the oxygen content of 100 samples per eight-hour shift [13]. A second system for use in steel production facilities was developed by Hoste and co-workers [14] under the sponsorship of Euratom and was described in a 1967 report. This equipment too has been tested and used extensively in actual steel-mill operations with excellent results [15].

Both of these systems use neutron generators of the deuterium–tritium type described above, and both are continuously pumped rather than being of the sealed-tube types. The generator used in the Euratom system operates at somewhat higher voltage (150 kV versus 100 kV) and has a correspondingly higher neutron output. In other aspects of the system, the differences are greater and more significant. The Euratom system, for example, irradiates an unknown and a standard sample simultaneously, with the standard sample being positioned behind the unknown. This provides not only a standard for

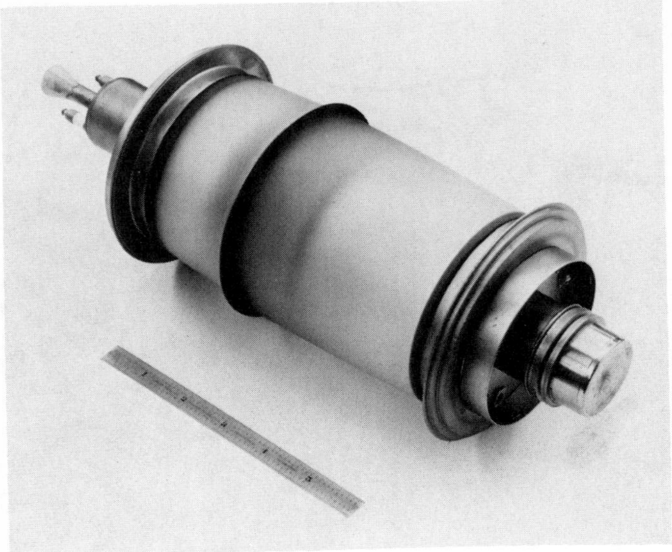

Fig. 3.3. Sealed tube for neutron production.

ratio comparison with each measurement but also a flux monitor. The two pieces, each about 26 mm in diameter and 9 mm thick, are counted simultaneously in separate, but equivalent, radiation-measuring systems. The British apparatus, on the other hand, irradiates only the unknown sample and uses a BF_3 neutron counter embedded in the shielding wall as a flux monitor. The gamma activity of the sample is then determined by a pair of sodium iodide scintillation crystals 3 in. in diameter and 3 in. long, and the pulses are processed through a suitable amplifier and a single channel analyzer to the digital read-out tubes of a counter.

Computation of the oxygen content is based on a function of the ratio of gamma count to neutron (BF_3) count in the British apparatus and on a function of the ratio of gamma counts from the sample to gamma counts from the standard in the Euratom method.

Although the differences between these two systems are significant, they may be compensating because excellent results have been demonstrated for both systems.

IV. NEUTRON-ACTIVATION PROCEDURES AND RESULTS

A. INTRODUCTION

As is indicated by the dates of the commercial assemblages described in the previous section, the determination of oxygen, particularly in steel, is by now a well-established technique and, in fact, the subject of a test method established in 1969 by the American Society for Testing and Materials [16]. This method, designated as E385-69T and entitled, "Method of Test for Oxygen Content Using a 14-MeV Neutron Activation and Direct-Counting Technique," describes in somewhat general terms the required apparatus and desirable procedure for oxygen determination in almost any matrix.

Even though this procedure has become routine in some laboratories, there are details of it that should be considered in a work of this kind and will be discussed here.

B. SAMPLE SIZE AND PREPARATION

Neutron activation, in common with most analytical techniques that depend on atomic or nuclear properties, is independent of the state of combination of the element of interest. Specifically, this means that an analysis of this type will yield a measurement of all the oxygen present whether it is chemically part of the sample, adsorbed on the surface, adhering as a corrosion product, or simply trapped in a sample container. Consequently, allowance must be made for this fact, and all extraneous forms of oxygen must be either eliminated or accounted for by correction factors. Furthermore, any other isotope that can be converted to ^{16}N by transmutation under bombardment with fast neutrons will also be measured as oxygen unless corrections are applied. Such interferences fortunately are usually inconsequential but will be discussed shortly.

Examination of Eq. 4 reveals that the amount of radioactivity of a given type induced in a specimen is directly proportional to the actual number of atoms of the pertinent isotope present. Thus, in the case of oxygen analysis the activity determined by γ-ray counting depends on the total weight of ^{16}O in the sample. This means that the lower detectable limit will be perhaps a few tenths of a milligram of ^{16}O, and whether this is of the order of parts per million or percent depends on the sample size. Therefore, the sample should be as large as possible if low concentrations are to be measured.

The principal limit on sample size is imposed by the sample transfer system. This system must move the sample from the irradiation site to the counting position and back to the loading port very rapidly and position it reproducibly. Obviously, large samples with their greater inertia make this more difficult. In fact, a steel sample weighing 100 g and moving at a speed of a few hundred feet per second becomes a hazardous missile.

Although one German report [17] refers to steel samples as heavy as 200 g being transported pneumatically, samples are generally much less massive than that. The two commercial systems referred to earlier use samples of 40–50 g of steel. One of the most commonly used sample containers in this country is the 2-dram polyethylene vial, whose inside dimensions are 1.5 cm (diameter) and about 4.5 cm (length). A steel sample this size would have a mass of about 60 g. The maximum sample accommodated by the sample carriers described by Pasztor and Wood [18] is about 8 mm by 38 mm and represents a steel mass of about 15 g.

In regard to sample size, it should be noted that frequently the sampling method provides a limitation as well as the transfer system. For example, one of the most widely accepted methods for sampling molten steel and other metals is the casting of a small pin by sucking the melt into an evacuated tube. The maximum homogeneous sample size for such pins is about 6 mm in diameter and 25–30 mm in length. Thus, in larger sample carriers there may be air-filled space that must be purged or otherwise compensated for.

A sample for the determination of oxygen by neutron activation may be either a liquid, solid, or powder; however, the use of a solid sample is preferable, particularly for metals. When a solid sample is not available, there is no inherent reason why a powdered sample should not be used if proper cognizance is taken of the oxygen contained in the air space surrounding the particles. If a solid sample is used, the first requirement is that the surface be as clean as possible so that no oxide film will interfere with the analysis. The cleaning of the surface may be accomplished by either mechanical, chemical, or electrochemical means.

Remarkably few authors have commented on the actual method of preparation of the surface of the sample to be analyzed, even though this may be of critical importance from the standpoint of both analytical accuracy and time consumed per analysis. Samples are customarily submitted to the author's laboratory in a freshly machined condition and stored under methanol. The samples are visually inspected for surface discoloration or blemishes, for cracks, gouges, gas pockets, or large inclusions at the surface during the weighing process. These are removed by careful filing, not grinding, before activation analysis. Even though the importance of large inhomogeneities in the sample is minimized by rotation during irradiation and "good" geometry counting, the effect cannot be ignored.

Powdered samples must be encapsulated during transfer, and solid samples usually are encapsulated to protect both the sample and the components of the transfer system. Many encapsulating materials have been tried and reported in the literature; however, the most commonly used material is polyethylene.

Although the use of a sample container ("bunny") is desirable, it is also a potential source of difficulty and requires that certain precautions be exercised. The most serious problem is that the encapsulating material may contain large amounts of oxygen, thus providing a high blank value and restricting the sensitivity of the analysis. Anders and Briden [19] have reported a method for selecting suitable capsule materials for the analysis of the reactive metal cesium, for which a capsule was obviously required and for which the maximum sensitivity was desired.

A further problem encountered in the use of a capsule is that usually the sample does not completely fill it, so that some oxygen will be retained in the capsule and will also contribute to the blank value. For solid samples, the ends of the container may be perforated so that the activated air is removed during return of the samples to the counter. The use of capsules perforated on all sides has been recommended by Nargolwalla and co-workers [20–22] in a classic series of papers on neutron and γ-ray attenuation in both the sample and the container.

It may be possible to use no capsule or perhaps to use only a portion of a capsule such as polyethylene rings surrounding a solid sample to aid in moving the sample through the transfer system when a very low blank value is required. Another means of reducing the capsule blank is to keep the sample encapsulated during transfer and irradiation but to strip rapidly and mechanically the capsule from the sample prior to counting at the scintillation detector.

An additional factor that may influence the blank value and thus limit the sensitivity of the analysis has been pointed out by Anders and Briden [19]. They observed that recoiling ^{16}N nuclei produced in air in the vicinity of the specimen may be driven onto the specimen or capsule surface and adhere there during counting. Since this ^{16}N will decay back to ^{16}O, it also represents an oxygen contamination that may build up with successive use of the sample container as long as air is used as the transfer medium. While this effect may be small, it is certainly significant where the maximum sensitivity is desired, and the use of nitrogen (rather than air) for the transfer gas is desirable for very low oxygen concentrations.

C. STANDARD SAMPLES

Because the neutron-activation analysis method does not permit the absolute determination of the oxygen content of a sample without precise knowledge of the constants in Eq. 6, the determination is usually based on comparison with standard samples. In principle, these standard samples can be made from any material of known oxygen content, such as pure metal oxides or pure organic chemicals of known stoichiometric oxygen content. A wide variety of materials has been used for this purpose. The principal requirement is that the standard sample must have the same physical size and shape as the sample to be analyzed and must have as nearly as possible the same overall density if computational corrections are to be avoided.

One suggested method of accomplishing this in the preparation of standard samples for the determination of oxygen in steel is to fill a polyethylene capsule with successive layers of low-oxygen steel and Mylar. Because Mylar is readily available in a wide range of thicknesses and because its oxygen content is known to be quite constant from one lot to another (33.2%), this procedure would seem to be quite suitable if the steel used had an oxygen content as near zero as possible. This procedure permits the establishment of a variety of standards covering the oxygen-concentration range of interest, and they are essentially absolute standards in that they do not depend on any other analytical method for calibration. An alternative method of standard sample preparation, using compressed powders of known composition, has been described in considerable detail for use with the commercial system of Hoste [23]. When standards are to be used for interlaboratory analyses, it should first be ascertained that they are compatible with each sample transfer system.

D. FLUX-MONITORING

Probably the most important variable, and the one most difficult to control during the determination of oxygen in metals by neutron-activation analysis, is the flux from the neutron generator. The neutron flux incident on the sample to be analyzed will vary with such obvious factors as the position of the specimen, the accelerating voltage applied to the generator, and the beam current that is striking the target. In addition, the neutron flux will vary with such less obvious factors as the age of the tritium target and its surface cleanliness, the atomic and molecular components of the deuteron beam in the generator and the sharpness of focus of the beam, and the position of the beam on the target. Because it is essentially impossible to control all these variables during any given irradiation period, it becomes necessary to devise a system for measuring the flux to which the sample is exposed during the irradiation period. The importance of this factor is amply attested to by the variety and ingenuity of flux-monitoring systems described in the literature.

Probably the simplest and most commonly used method for monitoring the flux of a 14-MeV-neutron generator is to arrange a neutron counter tube somewhere in the proximity of the generator and read the output from this tube either continuously with a rate meter or integrated over the irradiation period by a scaler. The neutron detector frequently used for this application is a proportional counter with boron trifluoride (BF_3) as the filling gas. The counter tube is usually located at some distance from the actual target position, perhaps in the shielding wall surrounding the generator or in a large volume of moderator material, such as paraffin, located within the shielded chamber [24]. This is necessary because the BF_3 counter is sensitive only to neutrons that have been reduced in energy to less than about 30 keV. This is a serious disadvantage because the neutrons of interest have an energy of 10 MeV or greater, which is required to activate oxygen. Thus, the quantity measured by this type of flux monitor is only indirectly related to the neutron flux of interest.

To avoid this energy problem with BF_3 counters, several investigators have used fast-neutron counters. For example, Anders and Briden [25] have described the use of plastic scintillator beads viewed by a photomultiplier tube located behind the specimen being irradiated. This counter, which detects recoil protons generated in the plastic scintillator by fast neutrons, also indicates the extent to which neutrons are removed from the beam by the sample. A similar fast-neutron counter is based on the principle of the proton-recoil telescope described by Bame et al. [26] and has been used for neutron-activation flux-monitoring by Benjamin et al. [27]. Counters of this type, if used with circuitry that will permit integration of the pulses during the irradiation time, should provide a much more reliable indication of the actual neutron flux.

Anders and Briden [25] have included an ingenious modification in the use of the neutron counter tube for monitoring the flux from the 14-MeV-neutron generators. They have used an integrating count–rate meter with an RC time constant of 10.6 sec, the same as the mean life ($t_{1/2}/0.693$) of the ^{16}N activation product. They have demonstrated that if the count–rate meter has this time constant, then the instantaneous rate meter reading at any time will be proportional to the buildup of activity in the sample due to ^{16}N. Thus, if the rate-meter potential is applied to a strip chart recorder, the reading on the recorder will be proportional to the flux seen by the specimen. Furthermore, they show that the accuracy of the analytical result is enhanced by placing the flux-monitor tube behind the sample, so that the flux measured is dependent upon the sample absorption.

Another scheme for monitoring the neutron flux, described by Steele and Meinke [28], consists of monitoring the radioactivity in the cooling water circulating past the target of the neutron generator. All Cockroft-Walton generators using the (d–T) reaction require that the target be cooled, and obviously some of the oxygen in the cooling water will be activated to ^{16}N because of exposure to the intense field of 14-MeV neutrons. If the water-cooling lines are wrapped around either a Geiger tube or a scintillation counter, it is possible to monitor continuously the radioactivity of the water and correlate this with the neutron output. A reliable correlation requires that the counting system be calibrated against activation foils of metal such as copper, the $^{63}Cu(n, 2n)^{62}Cu$ reaction being used for absolute measurement of 14-MeV-neutron fluxes.

A flux-monitoring technique that has more recently come into almost universal use is based on the reference sample. In this technique, a known sample is sent to the irradiation position at the same time as the sample to be analyzed. The reference sample is irradiated simultaneously and for the same irradiation period. The activity induced in the reference sample is therefore proportional to the flux and can be used to normalize the induced activity in the sample to an activity per unit of flux.

It is of course important that the reference sample be exposed to the same beam as the sample to be analyzed, and therefore they should ideally occupy the same geometrical position. This can best be approximated by moving both samples within the beam area. One common way is to rotate both the sample and the reference material about an axis parallel to the deuteron beam and in a plane parallel to the target [8, 29]. It is also necessary that both the sample and the reference have the same physical size and geometry.

After irradiation, the sample to be analyzed and the reference material may be counted either simultaneously in separate counting facilities or sequentially in the same counting facility provided that the ^{16}N activity induced in the reference material is sufficiently great that a high counting rate persists even after three or four half-lives. This will be true if the reference contains about 1 g of oxygen. The reference-sample technique obviously makes the system more complex in that essentially two transfer systems are required; however, this may well be justified by the increased precision of the analysis that results from careful flux-monitoring.

One additional flux-monitoring technique that has been reported is the use of an internal standard in much the same manner that internal standards have been used in emission spectrographic analysis for many years. Twitty and Fritz [30] have used ^{25}Na, ^{56}Mn, and ^{48}Sc as internal standards in the determination of oxygen in magnesium, steel, and titanium, respectively. This technique is obviously limited to cases for which a suitable internal standard can be found in the matrix in sufficiently large quantities to provide good counting statistics; however, the examples they have used cover three important cases and it seems quite likely that suitable standards could be found for other oxygen analyses of interest.

Several of the flux-monitoring techniques described above have been considered in detail and have been compared and evaluated in a study by Iddings [31].

E. OPERATION

The actual operating procedure used in determining oxygen in a metal sample should be as rapid and as automatic as possible. The need for speed is obvious from what has been said earlier in this chapter regarding the 7.14-sec half-life of ^{16}N, and the requirement of automation results from the desire to maintain the maximum possible reproducibility of the various steps of the analytical procedure. Consequently, it would be desirable simply to insert a prepared and encapsulated sample into the analytical apparatus, initiate the analytical procedure from a single switch, and have all sample motion, irradiation, and counting performed and timed automatically. The result would be two numbers, one corresponding to the radioactivity in the specimen due to high-energy γ-rays and the other a similar number corresponding to the same activity in a reference material—or, preferably, one result in terms of percent or parts per million of oxygen in the sample.

Many different times for the various phases of the operation have been reported. It would seem that irradiation times longer than 15–20 sec would be of little value, since after two half-lives of the ^{16}N isotope (about 15 sec) the ^{16}N activity in the sample would be about 75% of the saturation value as indicated by Eq. 4. Similarly, counting times longer than three or four half-lives would contribute very little to the reliability of the analysis of samples with low oxygen content. If the oxygen content is too small to achieve adequate counting statistics with a 15–20-sec irradiation time followed by a 20–25-sec counting time, it would be advisable to use multiple determinations on the same sample to enhance the statistical accuracy.

F. CALIBRATION

As explained earlier in this chapter, the number of atoms of a given type can be calculated, in principle at least, from Eq. 4. However, this activation equation contains several factors that are either not known or known with inadequate accuracy. Therefore, it is advisable to use standard samples to prepare a calibration curve relating oxygen weight to a simple function of radioactivity.

Standards are prepared with a known amount of oxygen distributed fairly homogeneously throughout a sample of the same physical dimensions and density as the sample to be analyzed, as indicated earlier. If the standard samples are irradiated simultaneously with a flux monitor or reference sample, a plot can be made of the ratio between the activity of the standard and the activity of the reference versus the weight of oxygen in the standard sample (Fig. 3.4). This plot is a straight line with the equation

$$\frac{A_s}{A_r} - B = KW_s, \tag{8}$$

where B is the background activity ratio due to the sample container, A_s is the total count from the standard sample in the oxygen energy range, A_r is the similar total count from the reference or flux monitor sample, W_s is the weight of oxygen added to the standard sample (milligrams), and K is the

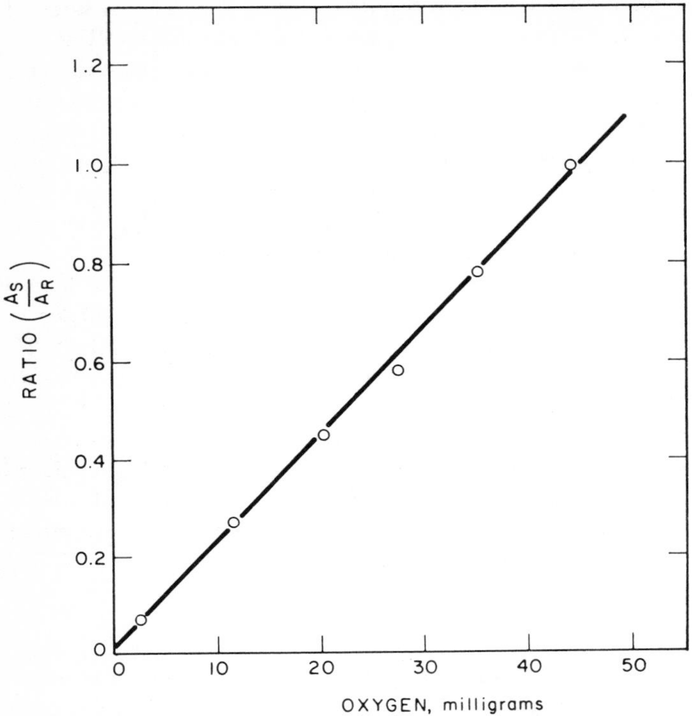

Fig. 3.4. Calibration curve for oxygen in steel.

calibration factor in counts per milligram of oxygen. Once the factor K has been determined for a given size, density, experimental cycle, and geometry, it should only be necessary to run standards at relatively infrequent intervals (perhaps once or twice a day) to determine that there has been no change of the calibration curve. However, changes in sample size or density will cause changes in K and may require correction of the activity ratio for absorption effects as described by Nargolwalla [21].

Using this factor, K, the weight of oxygen in an unknown sample (W_X) may be calculated if the activity ratio (A_X/A_r) is measured from the equation

$$\frac{(A_X/A_r) - B}{K} = W_X \text{ (mg oxygen)}, \tag{9}$$

and the concentration in parts per million is

$$\frac{W_X}{M} \times 10^6 = \text{ppm oxygen}, \tag{10}$$

where M is the mass, in milligrams, of the sample being analyzed. If the background count due to natural radioactivity, residual activated air, or other causes is a significant fraction of the count (A_X) from the sample being analyzed, it should be subtracted from A_X before the computation is completed. (The background is assumed negligible in relation to the reference count A_r.)

G. SENSITIVITY, ACCURACY, AND PRECISION

Virtually every author who has written about the neutron-activation technique has investigated the precision of his results and, where possible, compared them with oxygen analyses performed by more conventional methods, particularly the vacuum-fusion and inert-gas fusion methods.

Studies by Alpat'ev and co-workers [32] in the Soviet Union resulted in an approximate formula for determination of minimum sensitivity. They state that the limiting sensitivity may be estimated within 30% from the formula

$$Q \geq 5 \times 10^6 \frac{1}{\phi P_s} \% \qquad (11)$$

where Q is the concentration of oxygen, ϕ is the neutron flux at the sample (neutrons per square centimeter per second), and P_s is the sample mass in grams. Thus, for a 50-g sample in a flux of 10^9 n/cm²-sec the sensitivity would be limited to concentrations greater than 0.001% or 10 ppm. It should be noted, however, that the constant 5×10^6 in Eq. 11 depends on such experimental factors as geometry of irradiation and counting and the various cycle times and, so, will vary for different experimental conditions.

As with any radioactivity measurement, the sensitivity and precision depend ultimately on the number of radioactive decay events that are detected. Since they are random events following a Poisson distribution, the standard deviation (σ) of any measurement is equal to the square root of the total number of events detected ($\sqrt{N}$). In addition to the oxygen content itself, a number of instrumental factors can affect the total count, N, the most significant of which are the neutron flux, the sample size, the efficiency of radiation detection, and the various times involved in the analytical cycle. These may, of course, be optimized for the equipment available and provide an irreducible precision error. It is always the goal of the analyst to reduce all other errors such as "blank" content, reproducibility of positioning and timing, sample heterogeneity, uncontrolled flux variation, and radiation absorption to an insignificant level, so that the precision depends solely on the number of counts detected.

Studies of the accuracy and precision of instrumental neutron-activation analysis for oxygen were carried out by Anders and Briden [25], Mott and

Orange [24], Wood [8], Fujii and co-workers [33, 34], and Nargolwalla et al. [20–22], among others. All these authors examined the causes of imprecision or irreproducibility and indicated how they could be controlled or compensated for by sample rotation [29], flux-monitoring, and absorption corrections. In general, these authors showed that proper experimental conditions, equipment, and procedures permit the relative standard deviation to approach the value predicted by counting statistics above; that is, 1–2% for oxygen in steel, beryllium, and most other metals.

Of the published studies, probably the most extensive and detailed experimental investigation of the accuracy of activation analysis relative to gas fusion for the determination of oxygen in carbon steel has been conducted by Hoste [14]. In his report, results for 60 samples that had been repeatedly analyzed over a period of eight days were compared with results from two, or sometimes three, laboratories using fusion methods. The comparative results are quite good even though the samples show considerable inhomogeneity on the basis of measurements on opposite sides of disc samples. The inhomogeneity of the oxygen content in a steel bar has been studied by van Wyk [35], and the results from various neutron-activation analysis systems were compared with each other as well as those obtained by vacuum fusion.

Van Grieken [36] recently studied the precision or reproducibility of oxygen analysis in steel and cast iron with Hoste's apparatus. As a result of some 500 analyses of 100 samples spread over several months, he and his co-workers concluded that both statistical and instrumental "errors" affected the precision. The statistical errors should be less than 2% for oxygen contents greater than 100 ppm. The instrumental error varied widely between apparent limits of 5 ppm oxygen for low concentration and 1.7% of the amount present for high concentrations.

Results of oxygen determination by neutron-activation analysis in metals other than steel have also been reported. Hoste [14] included precision data on analysis of bismuth, cadmium, cobalt, niobium, titanium, tantalum, zinc, aluminum, copper, lead, and zirconium. Fujii et al. [37] and Brune and Jirlow [38] have reported on the analysis of aluminum and aluminum alloys for oxygen content.

Because the accuracy can only be compared with that of destructive methods using a much smaller sample and because precision depends so strongly on instrumental factors, generally applicable values for these parameters are not available. However, the ASTM procedure [16] states that "for metal samples in the range from 100 to 200 ppm oxygen, the reproducibility has been found to be about 3 percent." In general, this statement would seem to be conservative for modern experimental facilities that can analyze a sample weighing 15 g or more.

H. ADDITIONAL FAST-NEUTRON ANALYSES

Although oxygen is by far the element most widely determined by fast-neutron-activation analysis, it is not the only one. Other reactions may be used for the determination of gases.

Nitrogen can be activated by fast neutrons through the reaction

$$^{14}N(n, 2n)^{13}N.$$

Nitrogen-13 has a half-life of about 10 min and decays by emission of a positron. This radioactivity may be detected by counting the β-rays directly, by counting the annihilation γ-rays at 0.511 MeV, or by coincidence counting, since the γ-rays are emitted in pairs. This analysis has been discussed by Andersen and Algots [39] and by Tsuji [40]. Guinn [10] has stated that the interference-free sensitivity for nitrogen determination is in the range 0.1–0.3 mg; however, because many fast-neutron reactions lead to positron emitters, the interference-free condition is improbable. Most successful experimental work has been done either at very high concentrations, as in metal nitrates or nitrides, or in hydrocarbons where interferences are minimal.

As discussed earlier, fluorine can be activated by the $^{19}F(n, \alpha)^{16}N$ reaction that interferes with oxygen analysis. However, fluorine also undergoes two other reactions that may make the separation of the oxygen and fluorine possible. One of these reactions, $^{19}F(n, p)^{19}O$, yields γ-rays with energies of 0.197 and 1.37 MeV with a decay half-life of 29.4 sec. This reaction was used by Steele and Meinke [28] to show that 4 ppm fluorine could be detected in a 100-g sample. The other fluorine reaction, $^{19}F(n, 2n)^{18}F$, leads to a product with a long enough half-life (110 min) to suggest the possibility of chemical separation prior to measurement of the positron or annihilation radiation.

These and all other fast-neutron activation-analysis studies are referenced in a very valuable bibliography prepared by Van Grieken and Hoste and published by the Eurisotop Bureau of the Commission of the European Communities [41].

V. ADDITIONAL ACTIVATION PROCEDURES

A. INTRODUCTION

The preceding portions of this chapter have dealt almost exclusively with the use of 14-MeV neutrons for the determination of oxygen in metals. The emphasis on that analysis is justified by the fact that it is more widely used, particularly in industry, than any other activation procedure and by the fact that the general principles and many of the specific details are common to all activation analysis. Nevertheless, there are many other possible ways to activate the nuclei of the elements and to measure the radioactivity. Some of these that are pertinent to the determination of gases in metals will be considered in this final section; some others which use radiation techniques, but not strictly activation, are covered in Chapter 20.

Neutrons, and particularly fast neutrons, were considered as the bombarding or activating particles in the earlier parts of this chapter. They have certain advantages as noted there, but they certainly are not the only particles usable in activation analysis. Nuclear reactions leading to radioactive isotopes can also be brought about by bombardment with protons, deuterons, tritons (^{3}H nuclei), α-particles, ^{3}He nuclei, γ-photons, and others. In a survey paper published in 1963, Bate [42] listed 25 nuclear reactions for oxygen. These are shown in the conventional concise notation in Table 3.1 with the

TABLE 3.1. Nuclear Reactions of Oxygen[a]

Reaction	Half-Life of Induced Nuclide
$^{16}O(n, p)^{16}N$	7.35 sec
$^{16}O(t, n)^{18}F$	112 min
$^{16}O(\gamma, n)^{15}O$	124 sec
$^{16}O(^{3}He, p)^{18}F$	112 min
$^{16}O(t, p)^{18}O$	Stable[b]
$^{16}O(n, \alpha)^{13}C$	Stable[c]
$^{16}O(^{6}Li, n)^{21}Na$	23 sec
$^{16}O(^{6}Li, ^{4}He)^{18}F$	112 min
$^{16}O(n, 2n)^{15}O$	124 sec
$^{16}O(d, n)^{17}F$	66 sec
$^{16}O(^{3}He, n)^{18}Ne$	1.46 sec
$^{16}O(d, p)^{17}O$	Stable[d]
$^{16}O(p, n)^{16}F$	Short
$^{16}O(\alpha, n)^{19}Ne$	18 sec
$^{16}O(d, \gamma)^{18}F$	112 min
$^{16}O(\alpha, pn)^{18}F$	112 min
$^{16}O(p, \alpha)^{13}N$	10 min
$^{16}O(p, 3p)^{14}C$	5770 yr
$^{16}O(d, \alpha)^{14}N$	Stable[c]
$^{16}O(p, pn)^{15}O$	124 sec
$^{16}O(p, 2p2n)^{13}N$	10 min
$^{16}O(p, 3p3n)^{11}C$	20.5 min
$^{16}O(^{3}He, \alpha)^{15}O$	124 sec
$^{16}O(d, t)^{15}O$	124 sec
$^{16}O(\alpha, p)^{19}F$	Stable[b]

[a] Reproduced by permission from McGraw-Hill, Inc. [42].
[b] Proton is measured.
[c] Alpha is measured.
[d] 0.875-MeV prompt gamma is measured.

half-life of the resulting nuclide. Obviously, these are not all of equal utility since some lead to products with either very short or very long half-lives or even nonradioactive species. Furthermore, the cross sections for these reactions vary widely, so that some of them might be quite improbable. The few that have been found most useful will be discussed shortly.

A more extensive list showing the nuclear reactions for all gaseous elements of low atomic mass number is given in Table 3.2. This table was adapted from an article by Moiseev et al. [43] and includes the reaction cross section at the indicated incident particle energy. The reactions are arranged according to the product nucleus, and its half-life ($t_{1/2}$) is shown in parentheses. Most of the radioactive isotopes produced—^{11}C, ^{13}N, ^{18}F, ^{15}O—decay by emission of a positron (β^+-particle), so that the resulting annihilation γ-ray at 0.511 MeV will be subject to a great deal of interference, and either the half-life or some other distinguishing characteristic must be used to identify the radiation source properly. A more practical way of avoiding interferences, when it is possible, is to select the energy of the bombarding particles in such a way that the threshold energy for the undesired reactions exceeds the particle energy.

It is clear that there are nuclear reactions initiated by particles other than neutrons that may be useful for determining the concentration of gases in metals. These may be classified generally as either photon reactions or charged-particle reactions. The photon reactions are usually initiated by the high-energy X-rays, known as *bremsstrahlung*, generated when fast electrons are decelerated in the coulomb field of the atomic nuclei of a solid target. The useful electron energies may vary from 2 to 3 MeV in a small Van de Graaf accelerator to an order of magnitude greater than that in a betatron or linear accelerator. The bremsstrahlung photons are emitted in a continuous spectrum of energy ranging from zero to a maximum of the kinetic energy of the electron. The radiation is quite comparable to the "white," or continuous, radiation produced in an X-ray tube. The electrons that have been accelerated to very high, relativistic energies are stopped in solid metal "conversion" targets (used to convert one type of radiation to another) of Al, Cu, Ag, Pt, or other metals. These conversion targets could be expected to have an optimum thickness when used for activation analysis, and Engelmann [44] shows this to be true. However, the optimum thickness is independent both of the material being activated and the target material but dependent on the electron energy. Thus, for 35-MeV electrons, the optimum thickness was 4 g/cm^2 for aluminum, copper, or platinum when activating carbon, but the optimum thickness of a platinum conversion target increased linearly from about 2.5 to 4.5 g/cm^2 as the electron energy increased from 25 to 45 MeV.

Charged particles, such as protons, deuterons, or ^{3}He nuclei used for activation analysis are also generated in Van de Graaf accelerators for the lower energies, but cyclotrons are usually used to impart higher energies to

TABLE 3.2. Nuclear Reactions Involving Light-Element Isotopes[a]

Original Isotope	Isotope Content in Natural Mixtures, at. %	Type of Reaction	Reaction Cross Section, mb
	Formation of the Isotope ^{11}C($t_{1/2}$ = 20.34 *min*)		
^{14}N	99.6	p, α	100(6.7 MeV)
^{14}N	99.6	α, αt	16(30 MeV)
^{14}N	99.6	d, αn	20(15 MeV)
^{14}N	99.6	t, α 2n	—
^{14}N	99.6	n, tn	7(14 MeV)
^{16}O	99.8	n, α 2n	—
^{14}N	99.6	γ, t	5(23 MeV)
^{16}O	99.8	γ, α n	
	Formation of the Isotope ^{13}N($t_{1/2}$ = 9.96 *min*)		
^{14}N	99.6	p, pn	65(20 MeV)
^{16}O	99.8	p, α	40(14.6 MeV)
^{19}F	100	p, Li	—
^{14}N	99.6	α, αn	40(30 MeV)
^{16}O	99.8	α, αt	7(30 MeV)
^{14}N	99.6	^{3}He, α	15(8 MeV)
^{14}N	99.6	d, t	60(15 MeV)
^{16}O	99.8	d, αn	—
^{16}O	99.8	t, α 2n	—
^{14}N	99.6	t, tn	—
^{14}N	99.6	n, 2n	19(14.1 MeV)
^{16}O	99.8	n, p 3n	—
^{19}F	100	n, α 3n	—
^{16}O	99.8	γ, t	12(20 MeV)
^{14}N	99.6	γ, n	5.8(20 MeV)
	Formation of the Isotope ^{18}F($t_{1/2}$ = 1.83 *h*)		
^{16}O	99.8	α, pn	200(35 MeV)
^{16}O	99.8	α, pn	200(35 MeV)
^{19}F	100	α, αn	45(35 MeV)
^{15}N	0.37	α, n	108(10 MeV)
^{14}N	99.6	α, γ	2.1(21 MeV)
^{19}F	100	n, 2n	60(14 MeV)
^{18}O	0.2	n, n	500(6 MeV)
^{19}F	100	p, pn	175(21 MeV)
^{20}Ne	90.9	p, 2pn	—
^{16}O	99.8	^{3}He, p	480(8 MeV)
^{19}F	100	^{3}He, α	60(8 MeV)
^{15}N	0.37	^{3}He, γ	35(8 MeV)
^{16}O	99.8	t, n	129(2.4 MeV)
^{19}F	100	t, tn	30(2.7 MeV)
^{20}Ne	90.5	t, αn	—

TABLE 3.2 (*contd.*)

Original Isotope	Isotope Content in Natural Mixtures, at. %	Type of Reaction	Reaction Cross Section, mb
	Formation of the Isotope $^{15}O(t_{1/2} = 2.07$ *min*)		
^{16}O	99.8	n, 2n	20(14 MeV)
^{16}O	99.8	γ, n	16(30 MeV)
^{15}N	0.37	p, n	40(20 MeV)
^{16}O	99.8	α, αn	—
^{16}O	99.8	d, t	—
^{14}N	99.6	d, n	30(15 MeV)
^{16}O	99.8	^{3}He, α	—
	Formation of the Isotope $^{16}N(t_{1/2} = 7.14$ *sec*)		
^{16}O	99.8	n, p	90(14 MeV)
^{19}F	100	n, α	—
^{15}N	0.37	d, p	18(15 MeV)
^{16}O	99.8	d, 2p	—
^{14}N	99.6	α, 2p	15(35 MeV)
^{14}N	99.6	t, p	—

[a] Adapted from Moiseev et al. [43].

the heavier particles. The very versatile Van de Graaf generator with its intense, stable beam at low energy costs only about twice as much as the Cockroft-Walton fast-neutron generator discussed earlier; however, larger accelerators such as the linear accelerator or cyclotron may cost several hundred thousand dollars.

Although these costs limit the general applicability of both photon-activation and charged-particle-activation analysis, extensive research is under way in this field, and much progress has been made in recent years in understanding and solving the problems involved.

B. PHOTON-ACTIVATION ANALYSIS

Photon-activation-analysis techniques have been described frequently in the scientific literature over the past 35 years, approximately the same period covered by neutron-activation analysis. Recent excellent survey articles by Engelmann [44] and Lutz [45] describe both the theory and applications of this technique. Since the interest in this book is on the determination of gases, much of the photon-activation work will not be discussed, as it applies to heavier elements.

The principal advantage of photon activation over neutron methods is that the sensitivity may be much higher, particularly for the light elements carbon, nitrogen, and oxygen. Additional advantages may accrue in certain specific cases in which the radioisotopic species produced by neutron bombardment may have undesirable properties, such as a very short half-life, a mode of decay not easily measured with suitable accuracy, interference from other neutron products, or excessive radiation buildup in the matrix material.

Most photon activation procedures depend on the production of a new nuclear species by the absorption of γ-ray energy followed by the emission of a neutron or, in some cases, a proton. Thus, the reactions may be indicated as (γ, n) or (γ, p). Most of these reactions have energy thresholds of 10 MeV or more and consequently can only be produced by photons from fairly large expensive accelerators. Two notable exceptions to this rule are reactions of beryllium and deuterium, which have thresholds of 1.67 MeV and 2.23 MeV, respectively. The reactions are $^{9}Be(\gamma, n)^{8}Be$ and $^{2}H(\gamma, n)^{1}H$. Beryllium-8 decays very rapidly ($t_{1/2} \cong 2 \times 10^{-16}$ sec) to two α-particles. Hydrogen (^{1}H) is stable, and consequently the radiation best suited to measurement is the neutron emitted by the compound nucleus. These neutrons are moderated and detected by BF_3 counters, by scintillation detectors, or by activating an intermediate material whose activity is then determined.

Because these thresholds are so low, the reactions can be brought about by gamma rays from a certain few radioisotopes, and historically, this was the original photon-activation technique. Antimony-124 with a maximum γ-ray energy of 2.09 MeV can initiate the beryllium reaction but not the deuterium one. Only ^{24}Na and ^{208}Tl have sufficient energy and long enough half-lives to be useful for the deuterium reaction. Consequently, even for these low-threshold reactions, accelerators are the most desirable.

A significant experiment on the photodisintegration of deuterium was reported by Guinn and Lukens [46]. They used a Van de Graaf accelerator as a photon source and irradiated their samples for 1 h with a beam intensity of 3.0 MeV and 1 mA for deuterium. They detected the neutrons by the activation of manganese in the reaction $^{55}Mn(n, \gamma)^{56}Mn$. The γ-rays emitted by the ^{56}Mn were then determined by scintillation counting. The sample and the MnO_2 detector were held in concentric cylindrical cells during photon irradiation. Using this technique, Guinn and Lukens demonstrated a sensitivity for deuterium of about 0.12 ppm in a 70-ml sample or about $\frac{1}{500}$ of the concentration in natural water. Lutz [45] reported that the Russian workers, Mazukevich and Shkoda-Ul'yanov estimated that hydrogen could be determined in metals by this technique at a concentration of 1 ppm with an accelerator beam of 50 mA at 4.5 MeV.

The photon reaction most commonly used for determination of oxygen is $^{16}O(\gamma, n)^{15}O$. The feasibility of this reaction for determining oxygen was first demonstrated by Basile et al. [47] with γ-rays from a betatron. The product nucleus, ^{15}O, has a half-life of 2.07 min and emits a positron with maximum energy of 1.74 MeV. Because the reaction has a threshold of 15.7 MeV, bremsstrahlung from electrons accelerated to energies of at least 25 MeV are necessary, and maximum sensitivity requires even greater energy. Englemann [44] has shown that the ^{15}O activity increases from about 600 disintegrations per minute per microgram (dpm/μg) of oxygen at 25 MeV to about 20,000 dpm/μg at 40 MeV.

Because ^{15}O is a positron emitter, the 0.511-MeV annihilation γ-ray is available for measurement. As noted above, however, many elements, particularly low-atomic-weight elements, also form positron emitters, so that γ-ray spectroscopy is not as effective as it is with the fast-neutron method. The usual technique is to determine the radioactive decay curve at the energy of 0.51 MeV and resolve it into its components either graphically or by more sophisticated computer analysis. The elementary principles involved in this procedure are discussed in many books on radiochemistry, such as that by Friedlander, Kennedy, and Miller [48].

From the intensity of annihilation radiation with a half-life of 2.07 min, it is possible to estimate the oxygen content. However, much depends on the interfering elements present, and frequently chemical separations are necessary. Baker and Williams [49] showed that by fast chemical separation of the ^{15}O before counting, a sensitivity of 0.1 mg may be obtained. Using completely instrumental methods, Albert [50] reported that theoretically 1 ppm of oxygen can be determined in beryllium without chemical separation but that interferences prevent the instrumental determination of normal concentrations of oxygen in iron and zirconium. Generally speaking, the determination of oxygen in metals by purely instrumental photon-activation techniques is successful only on rare occasions, so that resort must usually be made to the chemical separation techniques mentioned above.

Nitrogen may also be determined by photon-activation methods through the reaction $^{14}N(\gamma, n)^{13}N$. The ^{13}N isotope is also a positron emitter and decays with a half-life of 9.96 min. Unfortunately, the use of this reaction for completely instrumental analysis is hampered by interferences from iron and copper. The photon reactions $^{63}Cu(\gamma, n)^{62}Cu$ and $^{54}Fe(\gamma, n)^{53}Fe$ also lead to positron emitters, and their half-lives are, respectively, 9.9 and 8.5 min so that decay-curve resolution is also impossible. Thus, even a small impurity of iron or copper would interfere with the nitrogen determination by photon activation.

Successful determinations have been made by use of chemical separations after irradiation. Since the half-life of the ^{13}N is about 10 min, it is sometimes possible to extract it from metals before excessive decay has taken place. Albert [50] used the Kjeldahl method for distillation of NH_3 in determining the nitrogen content of aluminum, zirconium, and beryllium. With this technique, he found, for example, a nitrogen content of less than 0.1–0.2 ppm in double electrolytically or zone-refined aluminum. Iron, unfortunately, reacts so slowly with hydrochloric acid that the 10-min half-life is too short to permit the determination.

Mention should be made of the special interferences by neighboring nuclei that may lead to production of the radioactive isotope of interest. For example, when oxygen is being determined by the $^{16}O(\gamma, n)^{15}O$ reaction, interference may arise from the reactions $^{19}F(\gamma, tn)^{15}O$ and $^{20}Ne(\gamma, \alpha n)^{15}O$, both of which obviously yield ^{15}O as the reaction product. Similarly, the determination of nitrogen by the reaction $^{14}N(\gamma, n)^{13}N$ may encounter interference from the reactions $^{16}O(\gamma, t)^{13}N$ and $^{19}F(\gamma, \alpha 2n)^{13}N$. Fortunately, the reactions in which larger or multiple particles are emitted have higher threshold energies (20–30 MeV versus 10–15 MeV) than do the (γ, n) reactions and advantage may be taken of this fact. Table 3.3 [44] shows how electron energy affects the amount of oxygen that must be irradiated to produce the same amount of ^{13}N activity as produced by 1 μg of nitrogen. Evidently this interference can be avoided at lower electron energy.

Fluorine may also be determined by the photonuclear method through the reaction $^{19}F(\gamma, n)^{18}F$. This reaction has a threshold energy of 10.5 MeV, and the ^{18}F reaction product is, like so many other products, a positron emitter. However, ^{18}F has a relatively long half-life of about 110 min, so that the ^{13}N, ^{15}O, and many other positron emitters will decay in time for measurement of the desired radiation. Interferences of the type mentioned in the preceding paragraph are $^{20}Ne(\gamma, d)^{18}F$ and $^{23}Na(\gamma, \alpha n)^{18}F$, but these have thresholds about 20 MeV, so that electron beams of about 25 MeV will excite the ^{19}F reaction much more efficiently than either of the interferences.

Engelmann [44] quotes sensitivities for the (γ, n) reactions based on irradiation for one half-life with a 100-μA beam current in the accelerator. He finds, for example, that with a beam energy of 35 MeV, N, O, and F

TABLE 3.3.[a] Relative Interference by Oxygen in Determination of Nitrogen

Energy of Electron Beam, MeV	Oxygen Producing the Same ^{13}N Activity as 1 μg N, μg
27	>3700
30	800
32	300
34	200
36	150
38	110
40	85
42	70

[a] Reproduced by permission from Academic Press [44].

may be detected at levels of 0.14, 0.0067, and 0.036 μg, respectively. These are indeed very good sensitivities. Of the other gases, only chlorine has been shown to have a reasonably sensitive (γ, n) reaction. The positron emitter, ^{34}Cl, can be produced by the reaction $^{35}Cl(\gamma, n)^{34}Cl$ and has a half-life of 32.0 min. With bremsstrahlung from a 100-μA beam at 30–40 MeV, the detection sensitivity should be a few tenths of a microgram.

Photon-activation methods are obviously quite sensitive for some of the gases (more sensitive, incidentally, for many metallic elements). These methods do require rather large, complex accelerators and more sophisticated treatment of data than the fast-neutron method for oxygen. Consequently, it is probable that the photon methods will continue to benefit research studies rather than to influence production methods.

C. CHARGED-PARTICLE-ACTIVATION ANALYSIS

Activation analysis methods using charged particles for the bombardment were apparently used first in 1938 by Seaborg and Livingood [51]. However, these methods were not developed to any great extent for nearly 20 years. Cyclotrons and other large accelerators were quite scarce prior to World War II, and immediately following the war, the research emphasis centered on making use of the very high neutron fluxes made available by nuclear reactors. In the early 1950s larger accelerators were developed, and the number of such machines and their technology advanced rapidly. As might be expected, this heightened activity brought about remarkable improvements in the stability and simplicity of cyclotrons and Van de Graaf accelerators, while simultaneously reducing their size and cost to bring them within the budget of many academic institutions and larger industrial and contract research establishments. Consequently, research papers in the field of charged-particle-activation analysis began to appear in the late 1950s, and the subject developed very rapidly during the 1960s. Some indication of the extent of the activity in this field is apparent from the two international symposia in the field in the successive years 1967 and 1968 [52, 53]. Some of the outstanding developments, particularly as they pertain to the determination of trace quantities of gases in metals, will be discussed here, and others will be discussed in Chapter 20. Excellent reviews of recent developments in this field may be found in the papers by Tilbury [54], Ricci [55], and De Soete et al. [56].

As was indicated earlier, the charged particles usually considered for activation analysis are protons, deuterons, tritons, ^{3}He nuclei, and ^{4}He nuclei (α-particles). Larger nuclei, such as ^{7}Si, ^{12}C, and ^{14}N, have been used, but they require much higher energies and lead to more complex reactions; these will not be discussed here.

The principal difference between neutron-, photon-, and charged-particle-activation analysis is in the strong electric-field interaction between the ions and the target nuclei, a feature totally absent in neutron and photon bombardment. Consequently, neutrons and photons penetrate deeply into most solids and can interact with a large volume of sample. This feature is particularly useful in the fast-neutron activation of oxygen, in which both the neutrons and the high-energy γ-rays are very penetrating. Charged particles, on the other hand, can penetrate only very limited thicknesses of metals.

Table 3.4 shows the range, or maximum penetration into iron, of various charged particles with an initial energy of 20 MeV [57]. As the table shows, the penetration varies with both the mass and the charge of the bombarding particle, and for α-particles it is less than 100 μm. For this reason, charged-particle-activation analysis is particularly useful for studying either very thin samples or the surface layers of thicker samples.

TABLE 3.4. Penetration of 20 MeV Particles in Iron ($\rho = 7.86$ g/cm^2)[a]

	Range	
Charged Particle	g/cm^2	mm
p	0.688	0.875
d	0.417	0.531
t	0.317	0.403
^{3}He	0.0793	0.101
α	0.0662	0.0842

[a] From Williamson et al. [57].

One of the most important aspects of charged-particle-activation analysis is that the bombarding particle must have sufficient energy to overcome the coulomb barrier surrounding the target nucleus. This is just the electrostatic repulsion between the positively charged particle and the positively charged target nucleus. It is expressed as

$$V = \frac{Z_1 Z_2 e^2}{R_1 + R_2}, \tag{12}$$

where Z_1e is the charge of the nucleus, Z_2e is the charge of the bombarding particle, and R_1 and R_2 are the respective radii.

If $R = 1.5 \times 10^{-13}(A^{1/3})$ cm, where A is the atomic mass number, Eq. 12 may be simplified to

$$V = \frac{0.96 Z_1 Z_2}{A_1^{1/3} + A_2^{1/3}} \text{ (MeV)}. \tag{13}$$

Thus, for the reaction $^{14}N(\alpha, p)^{17}O$, the coulomb barrier would be

$$V = \frac{0.96 \times 7 \times 2}{14^{1/3} + 4^{1/3}} = 3.4 \text{ MeV}. \tag{14}$$

In addition to this energy, classical mechanics requires that, for conservation of momentum, the minimum energy must be $(A_1 + A_2)/A_1$ times the potential barrier, so that in the present case, the actual minimum energy required for the α-particle is

$$3.4 \times \frac{18}{14} = 4.4 \text{ MeV}.$$

These relations establish threshold energies for charged-particle reactions that are exoergic—that is, where a positive energy is released for each interaction. For endoergic reactions, however, the required energy may be greater [$(A_1 + A_2)/A_1$ times the negative Q-value] if this is larger than the coulomb barrier.

Another factor that greatly complicates charged-particle reactions and has been the subject of extensive research is the variation of the reaction cross section (σ) with the energy of the particle (E). The energy itself varies with the distance traveled in the target. The reaction cross section also is a function of energy for neutron and photon reactions, of course, but the variation in particle energy within the sample is small enough that the cross section may be assumed to be constant for activation-analysis purposes. Such simplification is not possible with charged particles, and a variety of methods have been proposed to account for this complex variation. Two of these methods are in widespread use and have contributed greatly to the establishment of charged-particle analysis as a practical technique.

The first of these methods is a totally experimental procedure, devised by Englemann [58], in which samples of various thickness from zero (sample absent) to the total range of the particle are irradiated between two very thin layers of a standard material (for example, nylon) if oxygen is to be determined by the $^{18}O(p, n)^{18}F$ reaction. From these data, a curve of specific count–rate ratio as a function of thickness can be developed. Integration of this curve permits the determination of a single constant known as the *equivalent thickness*, a fictitious thickness throughout which the saturation activity would be uniformly distributed at the energy of the undegraded beam. The activation equation in terms of disintegrations (D) from the target is then

$$D = In\sigma_0 e, \tag{15}$$

where I is the charged-particle flux, n is the number of target nuclei per milligram of target, e is the equivalent thickness, and σ_0 is the cross section at the energy of the undegraded beam.

An alternative means for assigning a constant cross section is the *average-cross-section method* developed by Ricci and Hahn [59, 60], who claim that it "reduces charged-particle activation analysis of thick samples to almost the simplicity of neutron activation." These authors define an average cross section

$$\bar{\sigma} = \frac{\int_0^R \sigma_t \, dt}{\int_0^R dt}, \tag{16}$$

where σ_t is the variation of the cross section with thickness t and R is the range of the particle.

These two integrals can be related to the *excitation function* and *stopping power*; the first can be experimentally determined, and the second can be calculated. Thus, the constant $\bar{\sigma}$ can be evaluated. Furthermore, Ricci and Hahn [60] have shown that it is approximately independent of the target material, and constant for a given reaction and bombarding-particle energy.

Both the average-cross-section method and equivalent-thickness method are described clearly and with good detail in the recent survey paper by Ricci [55]. Both are entirely suitable, and the choice between them would probably be based on a preference for experimental or computational procedure.

Table 3.1 shows that many of the charged-particle reactions with oxygen result in the formation of ^{18}F. These are probably the most useful reactions, because the resulting radionuclide has a suitably long half-life (112 min) to permit removal from the vacuum system and any necessary physical or chemical surface treatment before the radioactivity is measured. Furthermore, the fact that ^{18}F decays with the emission of a positron means that sensitive, coincidence-counting techniques may be applied to the 0.511-MeV annihilation γ-rays. Interference by other positron emitters is a potential problem but may be minimized or eliminated by chemical treatment or by half-life analysis of the decay radiation.

The most useful reactions with ^{16}O are $^{16}O(t, n)^{18}F$; $^{16}O(^{3}He, p)^{18}F$; $^{16}O(p, n)^{18}F$; $^{16}O(d, \gamma)^{18}F$; and $^{16}O(\alpha, d)^{18}F$. Reactions starting with ^{18}O, such as $^{18}O(p, n)^{18}F$ and $^{18}O(d, 2n)^{18}F$, are also useful, but the fact that the natural abundance of ^{18}O is only 0.2% limits the sensitivity unless isotopically enriched oxygen is available.

Osmond and Smales [61] reported on the use of the first reaction above to determine the oxygen content of powdered beryllium metal. The beryllium powder was mixed with lithium fluoride, and the mixture was irradiated in the neutron flux of a nuclear reactor. The reaction $^{6}Li(n, \alpha)^{3}H$ produced tritons that could activate oxygen to ^{18}F by the reaction above. Born and Riehl [62] reported a sensitivity of 10 parts per billion for the use of this technique in the determination of the oxygen content of iron. This is, however, a surface method, since the range of the 2.7-MeV tritons would be no more than 30 μm in solids.

Tritons can be produced directly by ionization and acceleration of tritium in an accelerator. With modern experimental techniques, this has several advantages over the $^{6}Li(n, \alpha)^{3}H$ mode of production. Barrandon and Albert [63] used tritons from a 3-MeV Van de Graaf generator to study the surface of zirconium and aluminum. They were readily able to measure the effect of surface treatment on surface oxygen concentration. They stated that a purely instrumental method is possible for iron, chromium, and nickel as well as for aluminum and zirconium. The applicable range for surface oxygen is from 100–10^{-3} μg/cm^2. As noted earlier, all charged-particle methods lead essentially to a surface analysis. This is particularly true of the ^{3}He methods developed by Markowitz and Mahony [64] and by Ricci and Hahn [59, 60]. Consequently, further discussion of applications of these methods is presented in Chapter 20.

REFERENCES

1. T. I. Taylor and W. W. Havens, in W. G. Berl, Ed., *Physical Methods in Chemical Analysis*, Vol. III, Academic, New York, 1956, p. 449.
2. W. S. Lyon, Jr., Ed., *Guide to Activation Analysis*, Van Nostrand, Princeton, N.J., 1964.
3. Denis Taylor, *Neutron Irradiation and Activation Analysis*, Van Nostrand, Princeton, N.J., 1964.
4. H. J. M. Bowen and D. Gibbons, *Radioactivation Analysis*, Oxford University Press, London, 1963.
5. J. Hoste, J. Op de Beek, R. Gijbels, F. Adams, P. Van den Winkel, and D. De Soete, *Activation Analysis*, CRC, Cleveland, 1971.
6. G. de Hevesy in *Proceedings of the Symposium on Radiochemical Methods of Analysis*, Vol. I, Int. Atom. Energy Agency, Vienna, 1965, p. 8.
7. I. Kaplan, *Nuclear Physics*, Addison-Wesley, Cambridge, Mass., 1955, p. 360 ff.
8. D. E. Wood, in J. L. Duggan, Ed., *Proceedings of the Conference on the Use of Small Accelerators for Teaching and Research*, Oak Ridge, Tenn., USAEC Conf-680411, 1968, p. 56.
9. J. E. Strain, in H. A. Elion and D. C. Stewart, Eds., *Progress in Nuclear Energy, Analytical Chemistry*, Vol. IV, Pt. 3, Pergamon, 1965.
10. V. P. Guinn, in J. L. Duggan, Ed., *Proceedings of the Conference on the Use of Small Accelerators for Teaching and Research*, Oak Ridge, Tenn., USAEC Conf-680411, 1968, p. 1.
11. D. E. Wood, in J. M. A. Lenihan, S. J. Thomson, and V. P. Guinn, Eds. *Advances in Activation Analysis*, Academic, London, Vol. II, 1972, p. 265.
12. A. L. Gray, *Nuclear Engineering*, **9** (97), 205 (1964).

13. F. J. Armson and H. L. Bennett, *J. Iron Steel Inst.* (London), **208,** 748 (1970).
14. J. Hoste, D. DeSoete, and A. Speecke, *The Determination of Oxygen in Metals by 14-MeV Neutron Activation Analysis*, Euratom Report EUR-3565e, 1967.
15. P. C. Van Erkelens, *Euratom*, **7,** 59 (1968).
16. *Annual Book of ASTM Standards*, Pt. 30, American Society for Testing and Materials, Philadelphia, 1972, p. 1218.
17. H. J. Kopineck, G. Sommerkorn, R. Bass, and G. Presser, *Archiv. Eisenhuttenw.*, **35** (10), 987 (1964).
18. L. C. Pasztor and D. E. Wood, *Talanta*, **13,** 389 (1966).
19. O. U. Anders and D. W. Briden, *Anal. Chem.*, **37,** 530 (1965).
20. S. S. Nargolwalla, M. R. Crambes, and J. R. DeVoe, *Anal. Chem.*, **40,** 666 (1968).
21. S. S. Nargolwalla, M. R. Crambes, and J. E. Suddueth, *Anal. Chim. Acta*, **49,** 425 (1970).
22. S. S. Nargolwalla, E. P. Przybylowicz, J. E. Suddueth, and S. L. Birkhead, in *Proceedings of the Conference on Modern Trends in Activation Analysis*, Vol. II, Gaithersburg, Md., NBS Spec. Publ. 312, 1968, p. 879.
23. R. Gijbels, A. Speecke, and J. Hoste, *Anal. Chim. Acta*, **43,** 183 (1968).
24. W. E. Mott and J. M. Orange, *Anal. Chem.*, **37,** 1338 (1965).
25. O. U. Anders and D. W. Briden, *Anal. Chem.*, **36,** 287 (1964).
26. S. J. Bame, Jr., E. Haddad, J. E. Perry, and R. K. Smith, *Rev. Sci. Instrum.*, **29,** 652 (1958).
27. R. W. Benjamin, K. R. Blake, and I. L. Morgan, *Anal. Chem.*, **38,** 947 (1966).
28. E. L. Steele and W. W. Meinke, in *Proceedings of the Conference on Modern Trends in Activation Analysis*, College Station, Texas, 1961, p. 161.
29. F. A. Lundgren and S. S. Nargolwalla, *Anal. Chem.*, **40,** 672 (1968).
30. B. L. Twitty and K. M. Fritz, *Anal. Chem.*, **39,** 527 (1967).
31. F. A. Iddings, *Anal. Chim. Acta*, **31,** 206 (1964).
32. Y. S. Alpat'ev, E. G. Boreisha, A. G. Gordienko, V. M. Zelenin, and M. I. Korobko, *Zavodsk. Lab.*, **32,** 1492 (1966).
33. I. Fujii, H. Muto, and K. Miyoshi, *Japan Analyst*, **13,** 249 (1964).
34. I. Fujii, K. Miyoshi, H. Muto, and K. Shimura, *Anal. Chim. Acta*, **34,** 146 (1966).
35. J. M. van Wyk, M. Y. Cuypers, L. E. Fite, and R. E. Wainerdi, *Analyst*, **91,** 316 (1966).
36. R. Van Grieken, A. Speecke, and J. Hoste, *Anal. Chim. Acta*, **52,** 275 (1970).
37. I. Fujii, K. Takada, and H. Muto, *Japan Analyst*, **15,** 1239 (1966).
38. D. Brune and K. Jirlow, *J. Radioanal. Chem.*, **2,** 49 (1969).
39. G. H. Andersen and J. M. Algots, *J. Radioanal. Chem.*, **3,** 261 (1969).
40. H. Tsuji, *Japan Analyst*, **15,** 263 (1966).
41. R. Van Grieken and J. Hoste, *Annotated Bibliography on 14-MeV Neutron Activation Analysis*, Eurisotop Office Information Booklet 65, Bureau Eurisotop, 1972.
42. L. C. Bate, *Nucleonics*, **21** (7), 72 (1963).
43. L. I. Moiseev, V. I. Blokhin, and V. K. Bogatyrev, *J. Anal. Chem., USSR*, **32,** 1492 (1968).

44. C. Engelmann, in J. M. A. Lenihan, S. J. Thomson, and V. P. Guinn, Eds., *Advances in Activation Analysis*, Vol. II, Academic, New York, 1972, p. 1.
45. G. J. Lutz, *Anal. Chem.*, **43,** 93 (1971).
46. V. P. Guinn and H. R. Lukens, *Trans. Amer. Nuc. Soc.*, **9,** 106 (1966).
47. R. Basile, J. Hure, P. Leveque, and C. Schuhl, *Compt. Rend.*, **239,** 422 (1954).
48. G. Friedlander, J. W. Kennedy, and J. M. Miller, *Nuclear and Radiochemistry* 2nd ed. Wiley, New York, 1964.
49. C. A. Baker and D. R. Williams, *Talanta*, **15,** 1143 (1968).
50. P. Albert, in *Proceedings of the Conference on Modern Trends in Activation Analysis*, College Station, Texas, 1961, p. 78.
51. G. T. Seaborg and J. J. Livingood, *J. Amer. Chem. Soc.*, **60,** 1784 (1938).
52. Proceedings Euratom Symposium, *Practical Aspects of Activation Analysis With Charged Particles*, Liege, Belgium, 1967.
53. Proceedings Euratom Symposium, *Second Conference on Activation Analysis With Charged Particles*, Liege, Belgium, 1968, Rept. Eur. 3896.
54. R. S. Tilbury, *Activation Analysis With Charged Particles*, Nat. Acad. Sci.–Nat. Res. Council Report, NAS-NS, 3110.
55. E. Ricci in J. M. A. Lenihan, S. J. Thomson, and V. P. Guinn, Eds., *Advances in Activation Analysis*, Vol. 2, Academic, New York, 1972, p. 221.
56. D. DeSoete, R. Gijbels, and J. Hoste, in *Proceedings of the Conference on Modern Trends in Activation Analysis*, Vol. II, Gaithersburg, Md., NBS Spec. Publ. 312, 1968, p. 699.
57. C. F. Williamson, J. P. Boujot, and J. Picard, *Tables of Range and Stopping Power of Chemical Elements for Charged Particles of Energy* 0.05 *to* 500 *MeV*, CEA Report 3042, Commissariat A L'energie Atomique (France), 1966.
58. C. Englemann, in *Proceedings of Symposium on Radiochemical Methods of Analysis*, Vol. I, IAEA, Vienna, 1965, p. 405.
59. E. Ricci and R. L. Hahn, *Anal. Chem.*, **37,** 742 (1965).
60. E. Ricci and R. L. Hahn, *Anal. Chem.*, **39,** 794 (1967).
61. R. G. Osmond and A. A. Smales, *Anal. Chim. Acta*, **10,** 117 (1954).
62. H. J. Born and N. Riehl, *Angew. Chem.*, **16,** 559 (1960).
63. J. N. Barrandon and P. H. Albert, in *Proceedings of the Conference on Modern Trends in Activation Analysis*, Vol. II, Gaithersburg, Md., NBS Spec. Publ. 312, 1968, p. 794.
64. S. S. Markowitz and J. D. Mahony, *Anal. Chem.*, **34,** 329 (1962).

CHAPTER

4

VACUUM AND INERT-GAS FUSION

JOHN F. MARTIN AND LABEN M. MELNICK

United States Steel Corporation
Research Laboratory
Monroeville, Pennsylvania

CONTENTS

2. Specific Systems
   a. *Dallman-Fassel*
   b. *Hanin-Villeneuve*
   c. *LECO Oxygen Determinator*
   d. *LECO TN-14 Automatic Nitrogen Analyzer*
   e. *LECO TC-30 Simultaneous, Automatic Nitrogen–Oxygen Determinator*
   f. *LECO RO-16 Automatic Oxygen Determinator*
   g. *Strohlein Dinometer*

During the past 10 years there has been a gradual decrease in the use of vacuum fusion coupled with an increase in the use of inert-gas fusion. It appears that there are now many more users of the latter technique, at least in the United States, than there were in the past of the former. This is probably caused in part by the lower cost of some inert-gas fusion systems, since vacuum pumps and gauges are not needed with this technique. Also, the detection limit of the inert-gas fusion method should inherently be better than that of vacuum fusion, because with the former there is less gettering (loss of evolved gas by reaction with metal vapor or with metal or carbon films). In addition to the accelerated use of inert-gas fusion, separate fusion techniques for determining oxygen and nitrogen have been introduced. These have come about mainly because of the need for a rapid, sensitive, universally applicable technique for determining nitrogen in steel.

The choice of a gas analysis system will depend upon the following factors:

1. Capital investment
2. Space available
3. Other uses for apparatus
4. Metal to be analyzed
5. Sample size
6. Gases to be determined
7. Speed of analysis
8. Detection limit required

For example, for the determination of oxygen in steel, activation analysis, because it is nondestructive, would be advantageous if the sample must also be analyzed for the various oxides. (This may be accomplished by a chemical-extraction–spectrographic-analysis technique. The cationic elements are determined spectrographically, and the sum of the calculated oxygen equivalents is then compared with the activation-analysis result. Because of segregation of oxide phases in steel, it is desirable to perform both the oxygen and cationic-element determinations on the same sample.) In addition, at the present time, activation analysis requires about 70 sec as compared to several minutes or more, depending on the metal, for fusion analysis. Neutron-activation analysis, however, is presently limited to oxygen concentrations of 10 ppm or higher. Also, the cost of the neutron activation facility, while it may be used for other applications, is considerably higher than that for a fusion apparatus. Further, maintenance of an activation-analysis system requires a high degree of precaution and expertise.

To assist the reader in locating the discussions of certain parts of fusion systems, a word of explanation is in order. Where components, such as sample entry devices and oxidation catalysts, can be used in both vacuum and inert-gas fusion systems, the components are generally treated in the section dealing with the type of fusion in which they first appeared.

During the past 10–15 years, there have been a number of reviews written on determining gases in metals but only a few are cited here [1–8]. In addition, the authoritative work of Turovsteva and Kunin [9] and a collection of symposium papers [10] have appeared in book form.

I. VACUUM FUSION

The vacuum-fusion method has been widely used for the determination of oxygen and, to a lesser extent, of nitrogen and hydrogen in a number of metals. The initial development of this method has been variously attributed to Walker and Patrick [11], Oberhoffer and co-workers [12, 13], and Jordan and Eckman [14]. Considerable research in vacuum-fusion analysis has been performed by Sloman [15–24] and his associates in England under the auspices of the Oxygen Subcommittee of the Committee on the Heterogeneity of Steel Ingots.

The fundamentals of operation of vacuum pumps, vacuum gauges, and similar apparatus will not be presented here. Excellent discussions of vacuum techniques and the apparatus used in such work have been published [25–29]. A bibliography of publications on vacuum pumps, vacuum gauges, leak-detection equipment, and other applications of vacuum techniques in metallurgy has been compiled [30]. A tabulation of vacuum-equipment manufacturers, services, and repair facilities is published yearly [31].

There may probably be a number of investigators in the field of gas analysis of metals who will either disagree with some of the statements contained here or who will be unable to reproduce satisfactorily the work of others. At this point a quotation from a paper by Booth, Bryant, and Parker [32] is therefore appropriate: "It is a noticeable feature of vacuum fusion work that the optimum conditions for the analysis of a particular metal are by no means agreed upon by the various workers in the field. There have even been cases in which experiments repeated by other workers under what appear to be identical conditions have given unsatisfactory results."

A. EXTRACTION

1. Heating

Heating of the sample in vacuum-fusion analysis may be accomplished by electrical-resistance or radio-frequency-induction methods. (For a discussion of pulse heating, see Section II.C.2.d. Melting by d–c arc techniques is covered in Chapter 5.) Resistance heating requires simpler and less expensive apparatus. Also, for the same power, the cost of resistance heating is roughly one-half that of induction heating [33]. Resistance furnaces may be operated up to 1800–2000°C by using molybdenum radiation heaters and shields. With graphite heater rods and radiation shields, this temperature limit may be extended to 2200°C. The maximum temperature necessary in vacuum-fusion analysis will, of course, depend on the metal being analyzed.

Induction heating is preferred by many operators, having been used for gas analysis as early as 1935 [15]. With this method heating and cooling are rapid, the apparatus blank is lower because the quantity of graphite that must be degassed is less than that for resistance heating, and a stirring action of the molten metal is achieved, which accelerates outgassing. Temperatures as high as 2800°C have been obtained by induction heating.

The design of the induction coil is important. To obtain the maximum heating efficiency the material to be heated must be matched with the coil. Although coil design has been thoroughly treated [34], it is recognized as being somewhat empirical. Sloman [15] has described in detail the construction of a coil for a particular apparatus, and certain general recommendations have been made [35]. The length of the induction coil should be such that the height of the assembled coil will be at least twice the height of the material to be heated. To obtain more turns per unit height for the same area of bore, tubing of rectangular (not circular) cross section should be used. The oscillating current ideally should be carried completely by the outer skin of the conductor. Therefore, to prevent large electrical losses, the thickness of the conductor must be greater than some minimum value, which is a function of the frequency of the current. To obviate the problems encountered with thick-walled copper tubing (difficulty of production and lower number of turns per unit height), it may sometimes be necessary to braze a solid copper conductor to the hollow copper tubing, the solid conductor being on the inside of the coil. After winding of the coil, it should be acid-dipped and cleaned. The turns of the coil should then be insulated; this may be accomplished with braided Fiberglas sleeving, Glyptal paint, or certain clear organic spray paints manufactured for this purpose.

The frequency of the induction field may be varied over a wide range, depending upon the generating device used [36]. A frequency range of 3–10 kHz, 50–150 kHz, and 100–500 kHz or higher, may be obtained with motor–generator sets, spark-gap converters, and electron-tube oscillators, respectively. Exceptionally high humidity in a laboratory may cause the spark gap in the converter to short. The lower the frequency, the greater is the stirring action in the vacuum-fusion bath, but at frequencies below 10 kHz, increasing difficulties may be encountered in obtaining good coupling between the induction field and the sample.

In a study of the analysis for nitrogen in metals by vacuum fusion, Goward [37] determined that the promotion of stirring in the melt by use of low-frequency heating (30 kHz) would have no direct effect on the rate of evolution of nitrogen. However, it was noted that such low-frequency heating can accelerate the dissolution of the sample, the rate of solution of carbon from the crucible walls, and diffusion of this carbon to the reaction area to replace that consumed by carbide formation and oxide reduction. The formation of carbides enhances the rate of decomposition of nitrides.

The use of higher frequencies (for example, 500 kHz) results in the tendency of the graphite to pick up induced currents, which give rise to hot spots and therefore higher blank rates [38], although with a frequency of 900 kHz, small crucibles have been rapidly heated to high temperatures without the complication of insulating powders or heat shields [39]. Where radio-frequency power cannot be loaded into the sample, it will of course be necessary to use resistance heating exclusively or possibly to preheat the sample by this method to lower its resistivity sufficiently so that it can be heated inductively.

The ionization of vapor or glow discharge in the furnace by action of the high-frequency, electromagnetic field of the induction coil should be prevented or at least minimized. Such ionization may result in adsorption and desorption of metal vapors and gases on the furnace walls, thus invalidating the results. This ionization effect is most pronounced in the pressure range 0.01–0.05 torr [40]. Burden [41] has found that by plating silver strips on the inner wall of the furnace water-cooling jacket, the glow discharge could not be observed, even at maximum output of the generator. This quenching is apparently caused by a shielding of the electric field, so that the field strength is spread over a greater area of the furnace tube. The field strength is thus reduced below that necessary to produce electrons.

Another interesting means of minimizing the glow discharge has been reported by Everett and Thompson [42]. In their apparatus the induction coil was immersed in the water jacket surrounding the furnace tube to ensure a low radio-frequency voltage across the coil. Also, the power setting of the radio-frequency generator was not changed during a run. Gregory et al. [43] have found that with an 8-kW, 450-kHz Radyne heater, specially designed for heating graphite, no difficulty was encountered by water flowing freely over the induction coil.

2. Pumps

a. Fore Pumps

There are many mechanical vacuum pumps that may be used to produce the required vacuum before a diffusion pump can operate. Operational details of such pumps have been adequately described [44, 45]. The fore pump does not have to produce a high vacuum; however, the higher the vacuum, the better is the operation of the diffusion pump. Fore pumps that produce a vacuum of 10^{-3} or 10^{-4} torr are commonly used. Of great importance is the speed of the fore pump in removing the exhaust of the diffusion pump. This speed must be sufficiently high so that the critical backing pressure of the diffusion pump is not exceeded. If this were to happen, the diffusion pump would not function properly. Speeds of 70–100 liters/min or greater in the 10^{-3}–10^{-1} torr range are preferred. The pressure at which a specific pumping speed is obtained should be known in order to make proper use of the pumping speed.

Another desirable feature of fore pumps is that they be of the gas-ballasting type developed by Gaede so that condensable vapors are purged from the oil by admission of air from the intake side of the pump after the chamber has been shut off. Condensable vapors reduce the vacuum attainable by the pump, because the vapor pressure of the oil–water system is higher than that for the oil alone. These vapors may also corrode the metal parts of the pump or cause the oil to break down, thus impeding the operation of the pump.

Oil vapors should be prevented from entering the vacuum-fusion system. This can be accomplished by inserting in the line from the fore pump a trap containing molecular sieve 5A [46a], followed by a liquid-nitrogen trap. A protective screen should be placed in the line to prevent the molecular sieve particles from entering the fore pump. During analysis, both traps are used; overnight, when only the fore pumps are in operation, the molecular sieve trap alone provides sufficient protection for the system and liquid nitrogen is not needed. Both activated alumina and molecular sieve have been evaluated in their ability to prevent backstreaming from rotary pumps [46b], and an easily removable trap has been designed [46c].

A rotary vane pump without oil circulation and without ejector valve is described in Section I.L.5. This pump is used to collect the gas delivered from the vacuum-fusion furnace by a diffusion pump.

b. Diffusion Pumps

The choice of diffusion pumps for a vacuum-fusion unit is quite important. A vacuum of at least 5×10^{-5} torr should be attainable with the main diffusion pump. Although oil-diffusion pumps are less expensive and pump faster than mercury-diffusion pumps for the same operating pressure range, the latter are widely used in the United States. The former, however, are more often made of metal and these will, therefore, require glass-to-metal seals (as will metal mercury-diffusion pumps), which increase the intricacies of construction. Oil-diffusion pumps may also be more difficult to maintain than mercury-diffusion pumps. Mercury is relatively stable, but certain oils will decompose if air is inadvertently admitted to the pump or if the cooling water stops flowing. Cleaning of an oil-diffusion pump after such an occurrence is tedious. Further, the oil may decompose slowly over a period of time with back-diffusion of the decomposition products into the vacuum-fusion system. Pump oils that consist of polyphenyl ethers or liquid silicones having vapor pressures of less than 10^{-9} torr at room temperature are available [47, 48]. With polyphenyl ether no refrigeration other than cooling water is needed for the diffusion pump. (The advantages and disadvantages of both types of diffusion pump have been tabulated by Pirani and Yarwood [49].) Vaporization of mercury into the vacuum-fusion system can be greatly reduced by installing water condensers both before and after the mercury-diffusion pumps. Liquid-nitrogen cold traps on both the furnace and the evacuation pumps are recommended. A gold-foil trap [50] for collecting mercury vapor is therefore unnecessary. Moreover, copper-foil traps have been found ineffective for removing mercury vapor [51].

To assure a high pumping speed, especially in the vicinity of the furnace, the cross-sectional area of the throat should be as large as possible, the nozzle should be directed straight into the pumping direction, and the walls enclosing the pumping space should be cooled [52]. The furnace pump should have a high pumping speed and a high critical backing pressure. Because the efficiency of a pump will begin to fall at some high backing pressure before the critical backing pressure is reached, it is also recommended that the gas-collecting system should have a large volume [53]. Such properties are necessary to minimize the glow discharge and to effect a high throughput of gas, thereby reducing the possibility of loss of hydrogen or carbon monoxide (see Section I.A.4) in the furnace [42]. The previously mentioned nitrogen cold trap on the furnace pump will permit the obtaining of a pressure of less than 10^{-5} torr. The connection between the furnace and the pump should be as short as possible and should have a diameter of at least 1.5 in. with no sharp bends and a minimum number of gentle changes of direction [52] so that extraction of the gas may be performed as quickly as possible, preferably in less than 15 min.

A recommended arrangement for evacuation of the furnace is to have a single-stage pump, capable of pumping 40–50 liters/sec in the range 10^{-4}–10^{-5} torr and having a critical backing pressure of 0.02–0.03 torr, backed by another single-stage pump containing a positive cutoff (see below). This second pump should have a lower pumping speed, about 10–15 liters/sec at about 10^{-2} torr, but a higher critical backing pressure, about 10 torr. The backing pump with the positive cutoff can also be used as a circulating pump for the analytical train. For evacuating the vacuum-fusion system, a single-stage pump followed by a two-stage pump, with the stages in parallel may be used. With this combination a pumping speed of 25–30 liters/sec in the range 10^{-2}–10^{-4} torr and a critical backing pressure of 0.5–1 torr can be obtained. Covington and Bennett [54] have used a three-stage metal oil-diffusion pump for evacuation. A three-stage oil-diffusion pump is used satisfactorily on the Balzers Exhalograph EA-1 (see Section I.L.5).

The stated speed of a diffusion pump can be misleading. As mentioned in Section I.A.2.a, the pressure should be noted at which the speed was determined. Ideally, all conditions affecting the pumping speed should be specified, because the speed is a function of the molecular weight of the gas being pumped and the radius and length of the inlet port. Also, most pumping speeds are probably measured before the cold trap on the high-vacuum side of the pump. Pumping speeds measured after the cold trap would be considerably lower. Other factors affecting the speed are power input to the heater, temperature of the cooling water in the condenser, and pressure of the fore-pump manifold. Values of heating power for diffusion pumps, usually supplied by the pump manufacturer, should be adhered to. A detailed treatment of diffusion or vapor pumps is beyond the scope of this book. Discussions on this topic are available [55, 56].

Although mercury-diffusion pumps are usually constructed of glass, some have been made of metal. Speight and Gill [57] used a four-stage metal pump capable of evacuating at a rate of 75 liters/sec with a critical backing pressure of 35 torr. With such a high pumping speed, the blanking operation was fast, and reaction of metal vapors with carbon monoxide, hydrogen, or nitrogen was reduced. Sterling [58] has used a Gaede, Type E, all-metal, mercury-diffusion pump. This three-stage pump has a pumping rate of 15–20 liters/sec at 10^{-3} torr with a critical backing pressure of 20 torr. A stainless steel, mercury-diffusion pump is described in Section I.L.2.

The diffusion pump used for collecting the gases in the analytical system should contain a positive cutoff so that the measuring volume of the system is constant. Otherwise, the volume of the pump will change with gas pressure because the mercury-vapor atoms will condense at different points, depending on their velocities. Such a pump has been designed by Naughton and Uhlig [59] (see Fig. 4.1). A water-cooled cutoff (cold finger) is sealed into the pump so that the mercury atoms will condense at a specific point. The calibrated volume of the system is thus independent of the mercury-distillation rate or heater output. A Toepler pump may be used instead of the Naughton-Uhlig

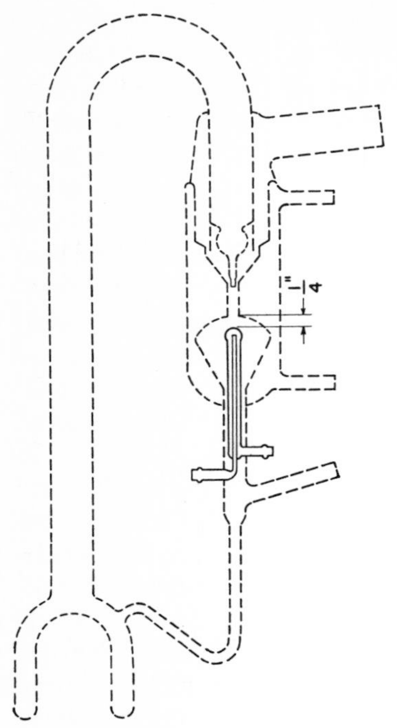

Fig. 4.1. Naughton-Uhlig diffusion pump. (Reproduced by permission from *Analytical Chemistry* [59].)

pump to obtain a positive cutoff. The use of the former, however, results in increased analysis time. In general, mercury-diffusion pumps are used for circulating gases through the analytical train. To transfer a gas onto a chromatographic column where a slug of uniform composition is required, a Toepler pump in conjunction with a gas buret has been used [60] (see also Section I.L.6). Automatic Toepler pumps have been commercially available for some time, and several have been described [61–68]. Under specific conditions, a pump cycle as low as 10 sec has been achieved [65]. A transistorized relay control circuit with a photocell sensor has been incorporated in an automatic Toepler pump [66] so that no sparking occurs as in the pumps where mercury makes electrical contact with tungsten wires. An automatic Toepler pump was developed by McLaren and Williams [69] that reportedly was an improvement over previous designs [65, 68, 70]. With this pump solenoid valves and repeat cycle timers are not required. Good reliability, safety and ease of operation are obtained over a wide range of gas sample sizes. The surging of mercury on the upstroke is minimized, and gas transfer is quantitative.

3. Furnaces

Furnace design differs widely among users of vacuum fusion. A distinction may be made, however, between macrofurnaces used for samples weighing 0.5 g or more and microfurnaces used for sample weights in the range of 2–100 mg. Regardless of the furnace size, high-purity graphite should be used for crucibles, funnels, and insulation material. In the United States such graphite may be purchased from U ion Carbide Corporation, Carbon Products Division, New York, New York; Ultra Carbon Company, Bay City, Michigan; and Ringsdorff Carbon Corporation, East McKeesport, Pennsylvania. Other grades of graphite are mentioned in the ensuing discussion. If the graphite powder is prepared in the laboratory by crushing and grinding, the material should be passed over a magnet to remove any particles of iron or other magnetic material [71].

To decrease the amount of residual gas in the system a d also the outgassing time, Gokcen and Tankins [72] recommended the use of pyrolytic graphite. Although this material is more expensive than crucible grade graphite, it is nonporous and contains about one-fifth as much gas. The preparation and properties of pyrolytic graphite have been discussed [73–79]. The use of ultra-pure graphite [38] (no. 208 spectrographic grade) from Le Carbone Ltd., London, England, has also been recommended for faster degassing and lower ultimate blank rates. Pyrolytic boron nitride has been recommended [80a] as a replacement for the quartz crucible envelope in inert-gas-fusion analysis to obtain lower oxygen blanks at temperatures above 2000°C (see Sections II.C.1 and II.C.2.a). With quartz the higher oxygen blank is caused by the high temperature reaction of carbon with silica to yield carbon monoxide and silicon [80b]. Below 2000°C the oxygen and nitrogen blanks of quartz are lower. From 2000–2200 C the nitrogen blanks of the two materials are similar, while the oxygen blank for quartz increases rather sharply. The properties of pyrolytic boron nitride have been summarized [80c]. Thimbles of this material are available from Union Carbide Corporation, Carbon Products Division [80d].

As a substitute for graphite insulation around the crucible, graphite cloth has been recommended [81a] (see Section I.A.3.a). This material has been described by Rohl and Robinson [81b]. It has been manufactured by the Carborundum Company and by the Carbon Products Division of Union Carbide Corporation. Originally developed for ablative uses in aerospace, graphite cloth is made by a combination of heating and chemical treatment, so that the oxygen and hydrogen disappear leaving only carbon [82]. It is a good heating element and has double the tensile strength at 1650°C that it has at room temperature. Care must be taken when using it, as it has been found to couple with induction fields (but only with certain heaters) at about

450 kHz. Graphite felt, which is more dense, is also available from Union Carbide.

For oxygen determinations fresh graphite crucibles frequently require "conditioning" with one sample or with bath or flux metal before valid results can be obtained. This should be ascertained with the metal in question before proceeding with the analyses.

a. Macrofurnaces

One of the most popular vacuum-fusion furnaces that has been used in the United States was designed by Guldner and Beach [83]. This furnace is shown in Fig. 4.2 (all dimensions are in inches.) The graphite crucible is contained in a clear quartz tube suspended in a glass envelope made of Corning 774 Pyrex. The quartz tube is inserted into the envelope at the bottom through a ground-glass joint that is made vacuum-tight by applying a thin coat of sealing wax to the ground surface. (The joint is cool during furnace operation, and therefore a number of waxes are suitable.) The outer glass surface is then heated with a Bunsen burner, and the plug is rotated until it is seated properly. Attached to this plug is a glass disk that acts as a heat shield and protects the joint if any molten metal is spattered. The temperature of the metal bath is measured with an optical pyrometer through the glass flat at the top of the furnace. The lower surface of the glass flat is protected from metal vaporized from the crucible by magnetically moving the glass-enclosed iron slug in the top side-arm. Samples are admitted through another side-arm.

The quartz crucible holder is 2 in. o.d. and 6.5 in. long. It is suspended from the two glass hooks by 0.050-in platinum wire. By thus minimizing the mass of material heated at high temperature, very low furnace blanks are obtained. The graphite crucible is 0.75 in. o.d., 0.5 in. i.d., and 3 in. long. A graphite funnel, split lengthwise to avoid coupling with the induction field, is inserted into the crucible as shown.

The quartz tube is filled about one-third full with graphite powder, −200 mesh or less. Then the crucible and funnel are inserted, the funnel is covered with a close-fitting metal cap [71], and powdered graphite is added up to the top of the funnel. The graphite powder should be loosely packed to provide maximum heat insulation and optimum conditions for outgassing without danger of lifting the powder from the assembly. A $\frac{1}{4}$-in. layer of −20 mesh graphite can be used on top of the −200 mesh graphite to reduce the possibilities of blowing out the fine powder during initial evacuation. The metal lid is then removed, and any powdered graphite in the crucible is ejected by blowing out with a low-pressure air stream. It is most important to remove this powder because it does not reach the temperature of the crucible during outgassing and therefore causes high blanks. Also, the gas content of the powder could cause the molten metal to be spattered out of the crucible. A

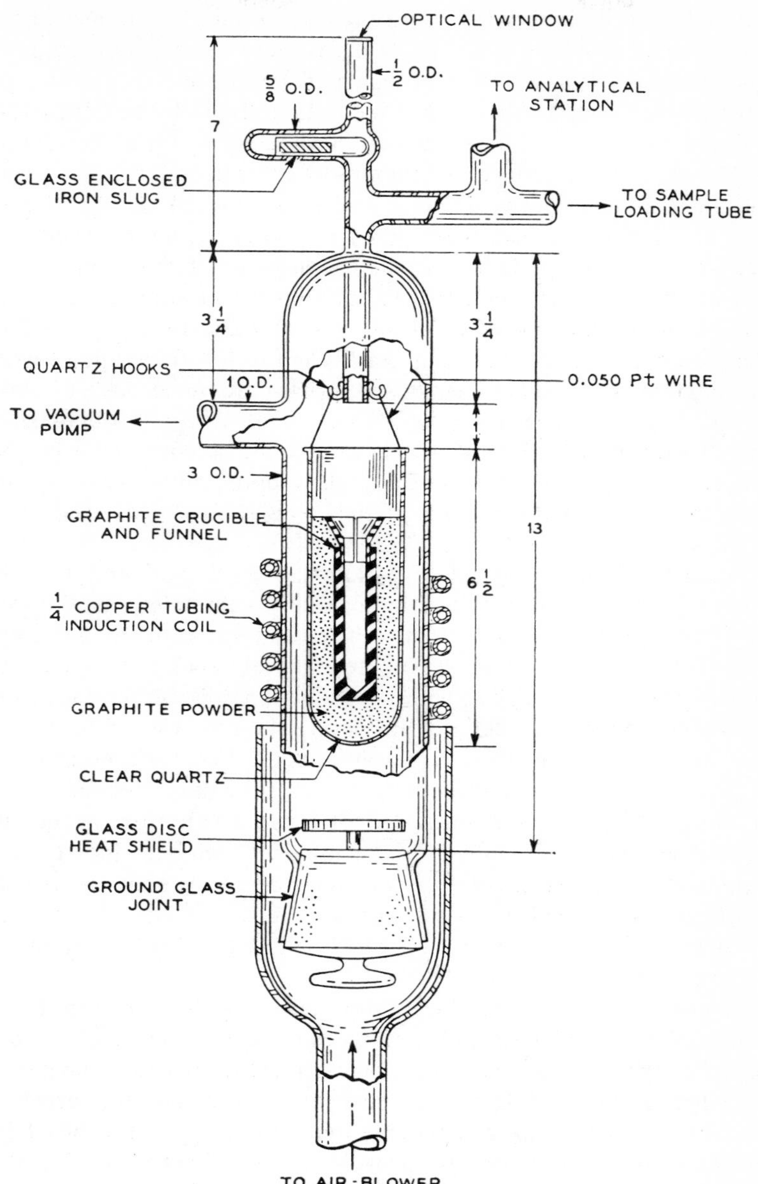

Fig. 4.2. Guldner-Beach furnace. (Reproduced by permission from *Analytical Chemistry* [83].)

high-velocity air stream, directed through a special glass funnel and uniformly about the furnace walls, cools the furnace. This air stream may be obtained from an electric blower of regulated speed.

Outgassing of such a crucible may be accomplished by heating at 2400–2600°C for two hours. Some workers have used a lower temperature, about 2000°C [32], to reduce the amount of graphite volatilized. However, this results in a longer outgassing time and a higher blank. Longer degassing times and/or higher temperatures will be necessary for larger crucibles and greater amounts of graphite insulation. The degassing temperature and time will depend in part on the temperature and time used during actual analysis. With a 2-h outgassing period at 2400°C, Guldner and Beach [83] obtained, in 30 min, a carbon monoxide blank equivalent to less than 2 μg of oxygen.

The advantages of the Guldner-Beach furnace are numerous. It is easy to construct, assemble, dismantle, and clean. Full vision is obtained, and because the quartz tube is suspended in vacuum, massive metal heads are eliminated, as are also quartz to metal or glass joints, rubber or lead gaskets, cements for attaching dissimilar materials, radiation shields, and ceramic supports.

The quartz furnace shown in Fig. 4.3*a* has been used at the authors' laboratory [81a]. To facilitate rapid removal of the extracted gas, the volume of the furnace was minimized. It has little dead space below the graphite crucible. The compact water jacket on the furnace provides adequate cooling and allows very close coupling with the induction coil. Replacement of the conventional graphite powder, for insulating the crucible, with three continuous wraps of Union Carbide graphite cloth, Fig. 4.3*b*, ensures a low blank by reducing the mass of graphite that must be outgassed and provides better insulation so that hot spots do not form. The bottom of the quartz envelope is covered to a depth of about one-half inch with loosely packed graphite cloth. The quartz envelope is connected by a standard-taper 60/50 joint, which is air-cooled, to a glass cap. Connection is made to the sample inlet system (see Section I.A.5) via a 40/25 ball and socket joint to facilitate interchange of furnaces.

The upper $\frac{1}{2}$ in. of the crucible (Ultra Carbon A 3878) is $\frac{1}{8}$ in. i.d.; the remainder of the crucible is $\frac{1}{4}$ in. i.d. A split graphite shield $6\frac{3}{4}$ in. long is connected to the crucible. This shield is machined from 1-in. diameter, high purity, Ultra Carbon rod. It provides less surface area than the surrounding glass for evaporated metals to condense upon and react with evolved gases. The shield or funnel extension also provides a more direct "line of sight" for pumping and results in faster removal of the extracted gases from the furnace. The lower $\frac{1}{2}$-in. of the shield is machined to fit the crucible. The upper $\frac{1}{2}$-in. of the crucible is machined to $\frac{1}{16}$-in. wall thickness to accommodate the funnel (Ultra Carbon A 3874), which is outside the periphery of the shield.

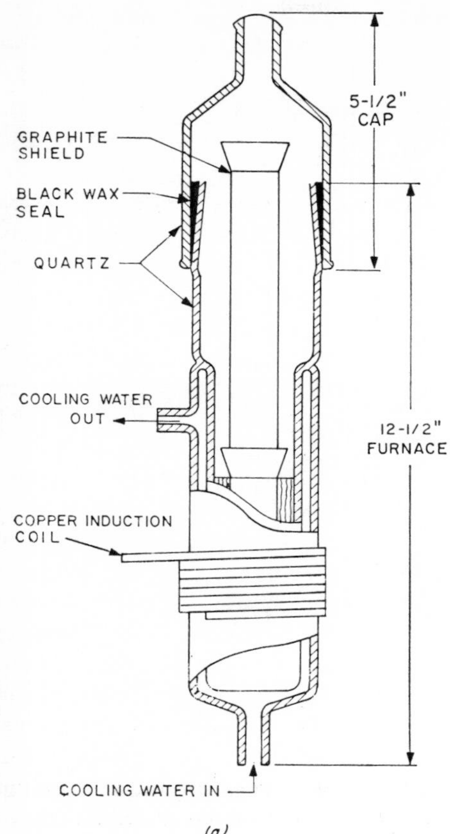

Fig. 4.3*a* and *b*. Martin Furnace [81a].

To avoid overheating the black wax seal on the standard-taper 60/50 joint, outgassing of the crucible is accomplished by cycling the crucible temperature between 1500° and 2000°C. After 90 min of outgassing, the nitrogen and oxygen blanks, respectively, are 5 and 10 μg for the 5-min extraction period. A 2-μg oxygen blank may be achieved by outgassing for 3 h.

The need for a mobile, gas-analysis system with less fragile components led to the use of a nickel-plated steel head for the furnace section [57]. This head is connected to the silica furnace tube by a thick, neoprene-rubber ring compressed against the outer wall of the tube by means of a screwed collar. The same type of seal is used for the glass water jacket. The two tubes are therefore individually water-tight and vacuum-tight.

With a resistance-heated quartz vacuum furnace having graphite-electrode heating elements, difficulties may be encountered when using a cooling coil.

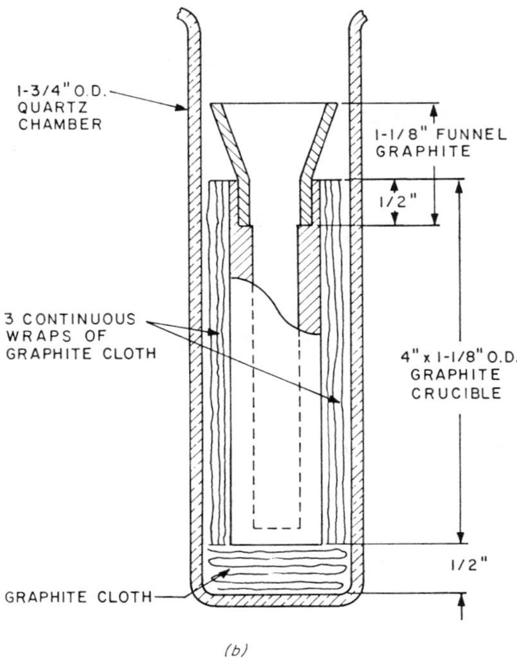

Fig. 4.3*b*.

Fluctuations in water pressure may result in local overheating and weakening of the furnace wall, followed by its collapse because of the external atmospheric pressure. Also, at temperatures above 800°C, quartz glass gradually becomes permeable to gases. Klyachko et al. [84] obviated these difficulties by substituting a low-carbon steel tube of the same dimensions as the quartz tube. The stronger and more durable metal tube is cooled by means of a metal water jacket. The cooling was sufficiently effective to result in a reduced degassing period. Accuracy of results was thereby improved, and the cost of the apparatus was reduced.

Cook and Speight [85] have described a carbon resistance furnace contained in a stainless-steel tube (Fig. 4.4*a*). The graphite crucible assembly (Fig. 4.4*b*) consists of a crucible mounted concentrically, with a $\frac{1}{32}$-in. annular gap, inside a cylindrical graphite heating element. This element is split longitudinally to within $\frac{1}{2}$ in. of the top, thus providing a resistive path for the current between two copper conductors to which the heating element is bolted. There is no difficulty due to expansion and contraction of the graphite on heating and cooling. The water-cooled conductors are secured to the

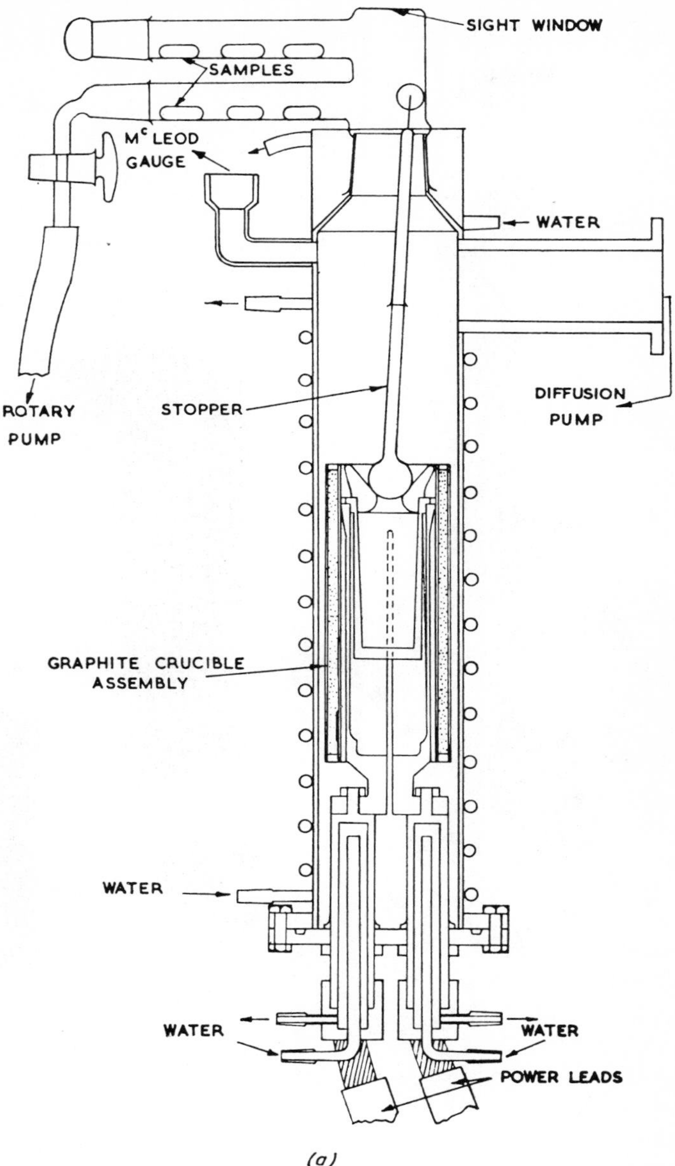

Fig. 4.4*a* and *b*. Cook-Speight Furnace. (Reproduced by permission from the *Journal of the Iron and Steel Institute* [85].)

Fig. 4.4*b*.

stainless-steel base flange by insulated vacuum-tight seals. To minimize gas adsorption, the conductors are chromium-plated.

Two concentric thin-walled graphite tubes, containing −300 mesh graphite powder in the annular space, comprise the radiation shield. This method of shielding is reportedly more effective with a resistance furnace than with a high-frequency induction furnace.

The melting crucible will accommodate up to eight 15-g steel samples. A graphite cone and stopper at the mouth of the crucible prevent excessive splashing of high-oxygen steels. The stainless-steel furnace chamber is 3 in. in diameter and 15 in. long and is cooled by water circulating through a copper coil soldered to the outside. A glass head, containing a sight window and two sample arms, is connected to the furnace tube by a cone and socket joint that is water-cooled. The bottom of the furnace tube ends in a stainless-steel flange to which the base flange of the crucible assembly is bolted. A greased rubber ring fitted into the groove in the base flange ensures a vacuum-tight joint.

A step-down transformer is used to provide the power to the furnace. The primary winding is connected to a 240-V, single-phase supply; a variable resistance is used to control the voltage. The secondary winding is connected through an ammeter to the copper conductors. For degassing at 2100°C the power requirements are 16 V and 250 A; for melting at 1650°C, 13 V and 200 A are required.

Klyachko et al. [86] have discussed the design of resistance furnaces for vacuum-fusion analysis. A resistance furnace designed by Covington and Bennett [54] is an integral part of the LECO vacuum-fusion analyzer described in Section I.L.4. In addition, a resistance furnace is used in the Balzers vacuum-fusion apparatus described in Sections I.L.5, 7, and 8.

b. Microfurnaces

The micro method has been applied to the analysis of the rarer metals to minimize consumption of material. The use of small samples in turn expedites analysis. Further, with microfurnaces a less-expensive heating unit is required in comparison to that used with larger furnaces. In addition, blank values are lower because of the reduced amount of graphite used. A vacuum-fusion apparatus suitable for microanalysis has been described by Everett and Thompson [42]. This apparatus, which is a modification of that designed by Gregory and co-workers [43, 87] has been used for the analysis of beryllium, niobium, thorium, uranium, zirconium, and other metals. The furnace and crucible assembly are pictured in Fig. 4.5*a*–*b*.

The crucible and other graphite components were made from either Acheson AGT or Morgan EY9166 grades of graphite and were vacuum-degassed at 2200°C for several hours before use. This reportedly resulted in

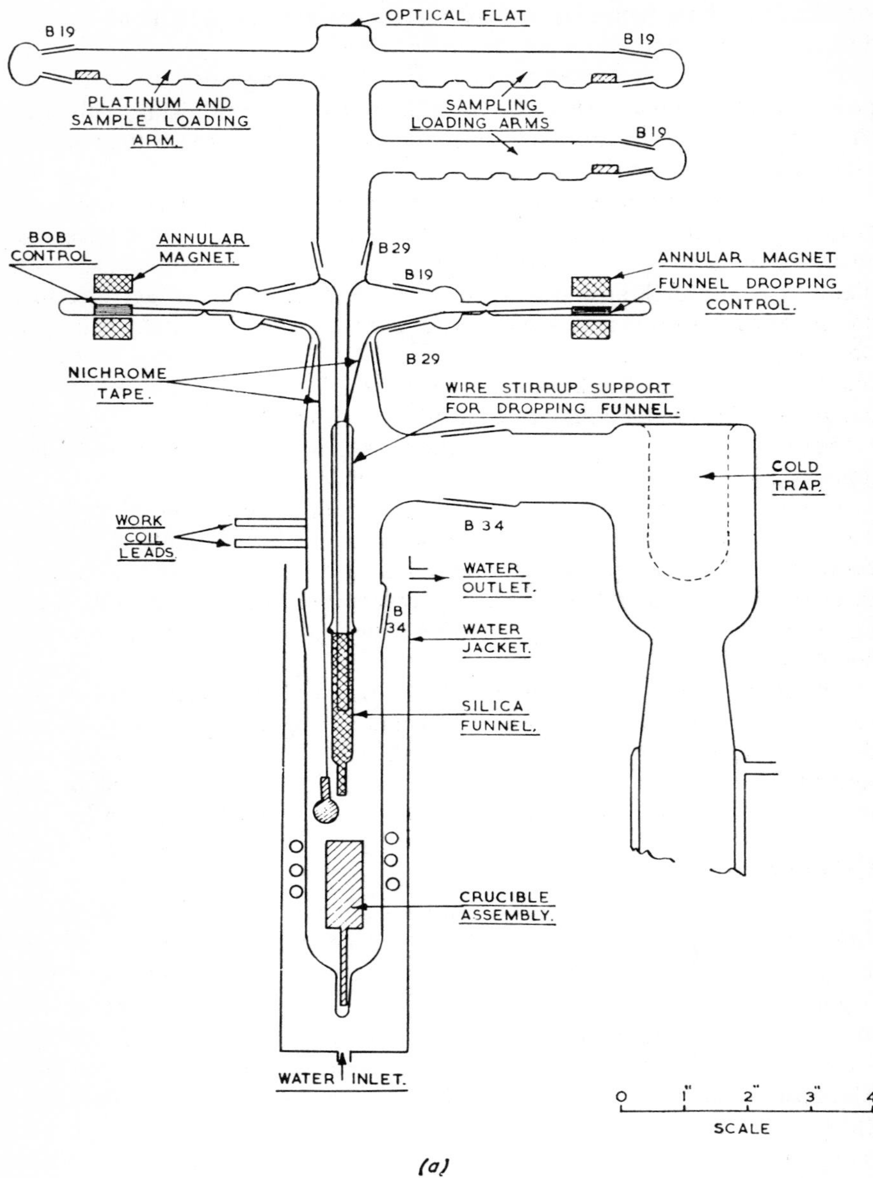

Fig. 4.5*a* and *b*. Everett-Thompson microfurnace. (Reproduced by permission from The Society of Analytical Chemistry [42].)

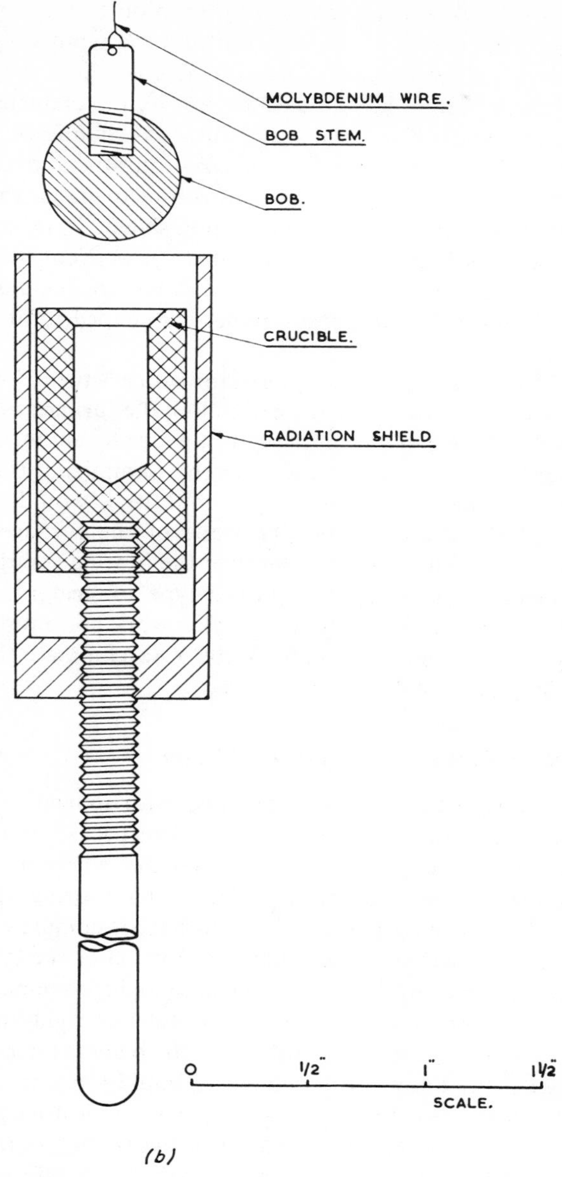
MOLYBDENUM WIRE.
BOB STEM.
BOB.
CRUCIBLE.
RADIATION SHIELD
0
1/2"
1"
1 1/2"
SCALE.
(b)

Fig. 4.5*b*.

a lower apparatus blank and reduced film formation in the microvacuum-fusion apparatus. Also, to obtain a low blank rate, attempts were made to minimize the weight of graphite used. The crucible is $1\frac{3}{16}$ in. long externally and $\frac{5}{8}$ in. in diameter. Internal dimensions are $\frac{5}{16}$-in. diameter and $\frac{3}{4}$-in. depth; a $\frac{7}{16}$-in. diameter spherical lid sitting on a 45° chamfered edge is used to close the crucible. To minimize the condensation of evaporated graphite on the silica furnace tube, which may cause this tube to crack, the crucible is surrounded by a 0.04-in.-thick graphite radiation shield, $\frac{13}{16}$ in. o.d. and 2 in. long. Four vertical slots $1\frac{7}{8}$ in. long and spaced at 90° were cut to reduce high-frequency power absorption. The crucible and radiation shield are mounted on a $\frac{1}{4}$-in.-diameter threaded graphite support peg shaped to fit a $\frac{1}{4}$-in. bore extension of the furnace tube.

A 15-kW, 450-kHz radio-frequency generator with a 3-turn $1\frac{3}{4}$-in.-diameter work coil immersed in a water jacket surrounding the furnace tube was used to heat the crucible. The coil was immersed in water to obtain a low radio-frequency voltage across the work coil, thereby minimizing the possibility of vapor ionizing in the furnace tube. Such ionization would result in gas adsorption on, or desorption from, the walls of the furnace tube, thus invalidating the results. The flow of water over the heating coil does not give rise to any problems, providing the proper type of heater is used [43]. Bach et al. [88] found that the voltage across a work coil immersed in the cooling water had to be reduced to about 100 V to eliminate the glow discharge when a frequency of 400 kHz was used.

c. *Measurement of Furnace and System Vacuum*

Many vacuum-fusion furnaces are equipped with ionization gauges, of either the indicating or the recording type, to detect leaks and to monitor the pressure during extraction. For all-metal systems, where it is impossible to see if the sample has entered the crucible, the monitoring application is useful. To determine when outgassing of the furnace is complete, the ionization gauge should be inserted in the calibrated portion of the system. When mercury-diffusion pumps are used, the ionization-gauge connection to the vacuum-fusion system should be made via a U-tube or right-angle bend to aid in condensing mercury vapor that will cause the readings to be erroneous. If condensable gases are known to be absent, a liquid-nitrogen trap may be used for this purpose [89]. If a spark-gap converter is used as the high-frequency source, it should be turned off when taking the ionization-gauge readings, because the converter emits radiation that may affect the reading on the gauge.

Waldron [90a] has described the use of a strain gauge to compare the pressure in the exhaust line from a vacuum-fusion furnace with the pressure in a reference chamber as a function of the position of a thin diaphragm that separates the two chambers. This gauge, similar to a micromanometer, is a temperature-compensated pressure transducer, Model PM5TCd ± 0.15–350, manufactured by Statham Instruments, Inc., Los Angeles, California. The thin diaphragm is attached to one end of each of four matched wire resistors that have their opposite ends attached to rigid supports. The resistors are connected in a Wheatstone bridge arrangement, in which the resistances are a function of the stresses applied to the wires.

The transducer of Waldron's gauge is sealed into the vacuum system near the exhaust port of the circulation pump to monitor the gas pressure during evolution from the sample and during all separation steps. This protects the transducer from an excessive pressure differential because the maximum differential is limited by the mercury cutoffs to about 50 torr. The manufacturer's specified maximum differential across the diaphragm is about 180 torr, above which rupture may occur. A low, stable reference pressure is obtained by connecting the reference chamber to the fore pump. The reference pressure must be low because the maximum allowable pressure differential below which accurate measurements may be made is about 7.8 torr.

A circuit diagram of the gauge is shown in Fig. 4.6. Low-voltage input is obtained from a 1.34-V mercury cell. The signal output is amplified by a Kintel Model 111 BF amplifier and recorded by a Leeds and Northrup Speedomax H recorder with a 0–10-mV range. A 50-fold amplification produces a full-scale recorder deflection equivalent to 750 μ. Adjustable zero

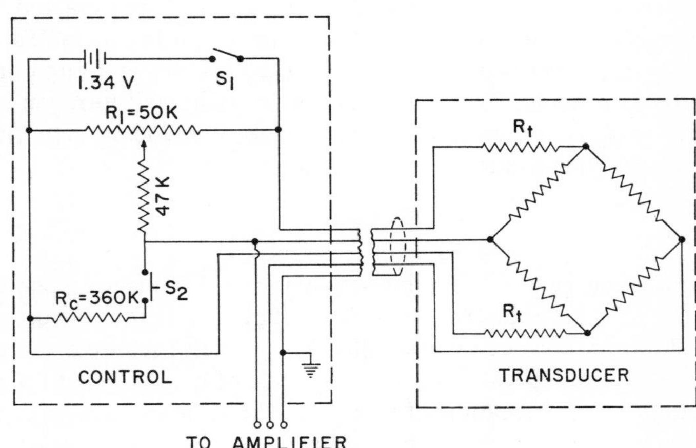

Fig. 4.6. Waldron temperature-compensated pressure transducer. (Reproduced by permission from the author [90a].)

suppression is obtained with the 10-turn Helipot, $R_{\dagger}$. Input voltage and other electrical parameters may be checked by means of a calibrating resistor, R_c, without actual establishment of a known pressure differential. The response of this gauge is continuous, linear, and independent of the type of gas present. A broad adjustable pressure range of about 7.5–750 μ can be covered. The instrument has been proven reliable for several years. An elaborate sensitive capacitance manometer is manufactured by MKS Instruments, Burlington, Massachusetts [90b]. This instrument (Baratron) is applicable over the range 10^{-5}–5000 torr. Pressure measurement is independent of gas composition.

4. Recovery of Nitrogen, Oxygen, and Hydrogen

The recovery of oxygen (as carbon monoxide) and nitrogen from vacuum-fusion baths has been the subject of numerous investigations. It is unfair to evaluate critically some of these investigations, because certain facts and techniques are available today that were not available earlier. For example, as noted above, it is now possible to analyze the same specimen by both neutron-activation and vacuum-fusion techniques, to determine the amount of oxygen lost by gettering. Further, it has been shown that with the LECO Automatic Nitrogen Analyzer (see Section II.C.2.d.) it is possible, on certain silicon steel samples, to get values higher than obtainable by usual chemical techniques, indicating decomposition of silicon nitrides with the LECO analyzer.

The degree of recovery of hydrogen and carbon monoxide from a vacuum-fusion bath depends to a great extent on the volatility and reactivity of the metallic elements in the sample and the reactivity of the sublimed carbon. Bath temperature, area of melt exposed to the vacuum, and pumping speed must also be considered. The mechanism of gettering and nitrogen recovery are covered in Chapter 2, and specific means of reducing gettering when analyzing certain metals are described in later chapters.

a. Nitrogen

Little nitrogen is gettered by sublimed carbon or by various metal films, as noted in the next section. It is generally believed that low nitrogen values are often obtained in vacuum-fusion analysis of steels because erosion of the graphite in the crucible by iron produces flakes of graphite that float on the melt and impede the evolution of nitrogen. The formation of viscous baths and the nature of the carbon precipitate have been discussed, respectively, by McDonald et al. [91] and Smith [92]. Both investigations, however, were in connection with the determination of oxygen. Smith [92] noted that the largest concentration of graphite flakes was at the surface of the melt. Low nitrogen recovery may also occur because the solubility of nitrogen in molten iron increases with temperature [93]; further, the solubility increases as the Mn, Cr, Mo, and V contents increase from 0 to 5% [94]. Ihida [95] has found that nitrogen is expelled from the melt on cooling and that a residual amount remains in the melt. The extensive British Iron and Steel Research Association study on the determination of nitrogen in steel [53] showed that the last traces of nitrogen were indeed difficult to extract especially in the presence of strong nitride-forming elements.

Although the techniques of ejecting each sample from the crucible after analysis of nitrogen (Balzers apparatus, Section I.L.5) or of analyzing only one sample for nitrogen per crucible (single crucible technique used in the LECO and Strohlein systems, Section II.C.2.d and g, respectively) have been extensively used, it is of interest to consider other alternatives suggested earlier. Niebuhr [96] controlled the viscosity of an iron bath to allow escape of gas bubbles by adding tantalum, a strong carbide former, to the bath. The tantalum content must be less than 30%. Severus-Laubenfeld et al. [97], in a detailed study of the determination of nitrogen in steel by vacuum fusion, concluded that in conventional vacuum-fusion analysis the extraction of nitrogen from the melt was incomplete. By increasing the carbon concentration in the melt, they obtained decreased nitrogen recoveries. These investigators also added tantalum to the bath (tantalum–iron weight ratio = 0.8/1: 1.5/1). Quantitative nitrogen recovery was achieved for aluminum-, titanium-, and zirconium-bearing steels. The analysis time was 7 min, but there was no indication of the number of samples that could be analyzed in a single crucible. The tantalum carbide that is formed is dense. It sinks to the bottom of the bath, and therefore, each sample melts in the neighborhood of the surface of the bath, with the result that the diffusion path for nitrogen is short and is not obstructed by a layer of graphite. The use of tantalum was also considered advantageous because it does not form a carbonitride; however, the use of this element precludes the simultaneous determination of oxygen, because with tantalum the supply of carbon to form carbon monoxide is deficient.

Gerhardt [98] and Gerhardt et al. [99–101] critically evaluated published analytical data for nitrogen in steel covering a 35-year period and showed that most vacuum-fusion results were low in comparison to chemical values. Gerhardt [98] showed that low nitrogen results are caused by increased viscosity of the melt resulting from precipitation of flake graphite and are not caused by incomplete decomposition of nitrides or by increased solubility of nitrogen in the melt. The solution to this problem (see Section I.L.5, but note also Section I.L.8) was the addition of a carbon-saturated alloy of nickel and 1% cerium to a crucible that can be emptied after each analysis by rapid rotation about its axis. The cerium causes the graphite to be precipitated in spheroidal form, as noted by Morrogh and Williams [102], instead of flake form. With this method, using the Balzers Exhalograph EA-1, complete extraction of nitrogen from 1.5-g steel samples at 1600°C was achieved. The use of cerium, a strong oxide-former, precludes the simultaneous determination of oxygen. Jaudon [103], in using the single crucible technique for determining nitrogen in steel by inert-gas fusion, stated that vacuum fusion could also be used but showed no results for the latter.

Fassel et al. [104] obtained complete nitrogen and oxygen recovery simultaneously from a variety of steel samples, including many NBS certified standards using a platinum bath (35 g minimum added initially), platinum flux (0.5 g added with each sample) technique. Up to twelve consecutive 1-g samples or twenty to thirty-five 0.5-g samples could be analyzed in a single crucible.

b. Oxygen and Hydrogen

Beach and Guldner [105] investigated the effect of thin films of carbon and carbon plus iron, nickel, aluminum, manganese, germanium, platinum, or platinum–titanium on the recovery of oxygen (as carbon monoxide), hydrogen, and nitrogen. Only pure gases were used in this work; from 2 to about 150 cc-torr of each gas were admitted to the system. Recovery was measured at 5-min intervals for a total of 30 min. Usually maximum sorption occurred in 15 min. Bath temperatures were 1550°C for C plus Al; 1650°C for C and C plus either Fe, Ni, Ge, or Mn; and 2000°C for C plus Pt and C plus Pt–Ti. Above 20 cc-torr there was less than 7% nitrogen loss to any of the films. Results for loss of hydrogen and carbon monoxide are shown in Table 4.1. Also shown are results obtained by Izmanova et al. [106a] on a variety of sublimate films. In this study, a Balzers Exhalograph EA-1 was used and the crucible temperature was 1650°C. Karpov et al. [106b] found little or no sorption of carbon monoxide on graphite with the crucible temperature at 1800°C. Also, little loss of the compound was experienced for platinum at 1900°C, iron plus tin at 1700°C, and iron plus tin plus 5% titanium at 1850°C. A slight loss was noted for iron and for iron plus 10% niobium, both at 1700°C. Considerable carbon monoxide was lost with titanium in a dry crucible at 1850°C and also with an iron bath at 1700°C. No operating details were given for the Giredmet-911M apparatus, presumably manufactured in Russia. However, Beach and Guldner [105] pointed out that the quantitative conclusions are a function of surface area of furnace, pumping speed, and bath temperature. Other factors, such as the amount of gas admitted, are also involved. Gettering of hydrogen by calcium has been observed [107–109].

Yoneda [110, 111] has stated that carbon monoxide is adsorbed by certain

TABLE 4.1. Sorption Loss of Hydrogen and Carbon Monoxide to Various Sublimate Films

Film	Loss %			
	Beach-Guldner [105]		Izmanova et al. [106]	
	H_2	CO	H_2	CO
C	—	—	—	5–6[b]
C	—[a]	<±3	20	10[c]
C + Fe	<10	<3	10–18	10–15
C + Ni	<±13	<±17[d]	11–16	4–10
C + Al	<10[e]	60–80	—	—
C + Mn	<14[f]	100	—	—
C + Ge	<9[e]	<2[g]	—	—
C + Pt	<14[h]	<5[i]	—	—
C + Co	—	—	8–15	4–10
C + Sn	—	—	—	3–8
C + Pt + Ti	<2[e]	<13[j]	—	—
C + Co + 17% Sn	—	—	~20	—
C + Ni + 30% Cr	—	—	—	10–13[b]
C + Ni + 30% Cr	—	—	—	30[c]
C + Co + Ni + Cr	—	—	—[k]	—[k]

[a] Sorption increases with amount added.
[b] Larger gas volume.
[c] Smaller gas volume.
[d] For >30 cc-torr of CO added. Apparently, appreciable CO blank.
[e] For >40 cc-torr of H_2 added.
[f] For >60 cc-torr of H_2 added.
[g] For >80 cc-torr of CO added.
[h] For >30 cc-torr of H_2 added.
[i] For >70 cc-torr of CO added.
[j] For >20 cc-torr of CO added.
[k] Sorption increases with increasing temperature.

metals when they are at 300–500°C. Lounamaa et al. [112] essentially confirmed this in the analysis of steels for oxygen. They provided a specially designed stopper funnel above the crucible and maintained the funnel temperature during analysis at 900–950°C. This temperature was low enough to condense evaporated metals on the funnel but high enough to minimize the reaction of carbon monoxide with the metal film. Koizumi et al. [94], in the analysis of steel for nitrogen and oxygen, obtained almost 100% recovery of oxygen by use of a molybdenum hood above the crucible. The hood temperature was about 900°C. Holt and Goodspeed [113] used a movable, split tantalum cylinder on which to condense magnesium vapors prior to the reduction of magnesium oxide. The collection of magnesium on the cylinder was facilitated by the use of argon at 100 torr during analysis. After removal of the cylinder from the crucible area, the temperature was raised and the oxide was reduced. Similarly, Wood and Oliver [114], in the analysis of manganese-bearing titanium alloys, obtained improved recovery of oxygen by volatilizing manganese at 1500°C, followed by reduction of oxide at 1880°C in the presence of platinum. Gettering of carbon monoxide by the volatilized manganese was prevented by adding tin to the bath, the tin presumably depositing over the manganese.

Conflicting opinions appear in the literature regarding the use of crucible lids as a means of condensing metal vapors to prevent gettering. Many investigators [32, 42, 50, 88, 115] cited the need for lids, especially in the analysis of beryllium [32, 42, 50]. In the analysis of titanium, Walter [116] used a cover to minimize the ejection of graphite powder from the crucible. Guernsey and Franklin [117], however, reported that there was no clear-cut evidence that a cover improved performance of their vacuum-fusion apparatus in the analysis of steel, titanium, molybdenum, chromium, copper, silver, and high-temperature alloys. The lids may even stick to the crucible during analysis of steel, because of splashing of the metal. Lids may be magnetically moved [42, 50], and this has been accomplished automatically [88].

Other gettering problems have been described. In a micro-vacuum-fusion furnace, a graphite heat shield may be used in place of powdered graphite insulation to reduce outgassing time. This requires a shield of proper design to minimize gettering problems. Without the heat shield (or the powdered graphite insulation), evaporated graphite might condense on the furnace walls as a film, absorb radiant heat, and cause excessive thermal gradients that could crack the walls [43]. Extensive study showed that the top of the circular heat shield should be flush with the top of the crucible. During the analysis of uranium it was determined that if the heat shield extended above the level of the crucible, uranium vapor condensed on this upper portion, and the hot uranium gettered carbon monoxide and nitrogen. Any uranium or graphite that condensed on a cold portion of the system did not appear to getter the gas. To minimize high-frequency heating of the shield, two fine vertical saw cuts 7–8 cm long were made diametrically opposite each other. In contrast to the height of this shield, Everett and Thompson [42] extended the height of their radiation shield about $\frac{1}{4}$ in. above the level of the crucible. The use of a heat shield is dependent on the metal being analyzed, the temperature, and the frequency of the induction heater (see Section I.A.1).

To enhance the evolution of hydrogen and carbon monoxide from the molten metal and to enable more samples of certain metals to be analyzed between crucible changes, various metal bath materials have been used. The main function of such materials is to obtain a more fluid melt into which the samples are dropped. Bath and flux materials best suited for specific metals are noted in later chapters (see, for example, Chapter 19, Section VI.B.1, for an excellent discussion on the use of tin). The desirable characteristics of a metal bath are as follows [32]:

1. It must remain liquid and not form a carbide at the operating temperature
2. It must be more noble toward carbon and oxygen than the metal being analyzed
3. It must have a low vapor pressure and be a poor getter
4. It must readily dissolve the sample metal
5. It must dissolve carbon only to a slight degree.

Bath materials that have been used include platinum, nickel, cobalt, iron, and platinum plus iron. Platinum and tin have been added as fluxes with each sample. Platinum has the advantage of low vapor pressure; further, carbon is only slightly soluble in it. Its use has been described [118–120]. The cost of platinum is somewhat overcome by the fact that it is about 80% recoverable from the bath. In the United States recovery of platinum is usually performed by the suppliers of the metal. However, a laboratory chemical scheme can be worked out for this purpose.

To avoid high and erratic blanks, flux and bath materials must be homogeneous and of low gas content, especially where small amounts of hydrogen and oxygen are to be determined. Booth and Parker [121] used high-purity tin rod, heating the rod at 600°C for 3–4 h at less than 10^{-5} torr to achieve homogeneity of the residual oxide. The surface was abraded immediately before insertion into the vacuum-fusion apparatus, where the tin was then preheated at 1400°C for 1 min before use. The oxygen content was thereby lowered from 80–100 ppm to 30 ± 6 ppm at the 95% confidence level.

In cooperation with American Society for Testing and Materials [122], the Baker Platinum Division of Englehard Industries, Inc., has improved the homogeneity of both platinum rod and foil and has reduced the oxygen content of these materials. Previously, in one lot of platinum rod the oxygen concentration varied from 9.6 to 16.4 ppm. In one lot of commercial platinum used as a fluxing agent, the oxygen content varied from 5 to 50 ppm. The average oxygen content of eight specimens of $\frac{1}{8}$-in. diameter, low-oxygen platinum rod, commercially produced in 1961, was 2.1 ± 1.0 ppm at the 95% confidence level. For the same number of specimens of low-oxygen 10-mil (0.254-mm) foil, the average oxygen content was 5.2 ± 1.9 ppm at the 95% confidence level.

It has been reported that the formation of graphite flakes in a platinum bath after several analyses can prevent proper alloying of the sample and quantitative recovery of the gas [123]. This can be obviated by adding platinum as a flux with each sample, either alone or in conjunction with tin;

iron together with tin has also been used [124]. The tin acts not only to increase the fluidity of the bath but also to effect stirring [125] during its boiling, thus minimizing entrapment of the gas in the melt. Further, the volatilized tin condenses over films of active metals and reduces gettering. In addition, tin apparently causes desorption of carbon monoxide from active films of beryllium [42]. Sawa [126], however, found that while the use of tin during steel analysis diminished gettering when a resistance furnace is used, the opposite can occur with an induction furnace because of the discharge occurring during evaporation of the tin, the induction field supposedly causing the tin to be more reactive. Donovan et al. [38] analyzed chromium, silicon, copper, and thorium in an iron bath; iron, ferrosilicon, ferromolybdenum, and ferroaluminum wrapped in nickel foil in an iron bath; and silicon and molybdenum in an iron bath with a tin flux. Karpov et al. [106b] cautioned that while tin can almost completely reduce the amount of gas sorbed in a sublimate, the use of this element can cause an increased blank. Therefore, tin should only be used when a sublimate shows a strong gettering effect.

In the work of Fassel et al. [104] quantitative recovery of oxygen was obtained for steel samples in the form of both cubes and compressed chips. Both low-alloy and high-alloy steels were analyzed using the platinum-bath–platinum-flux technique. Kraus et al. [127] recommended the addition of nickel to the bath to recover oxygen from aluminum-killed steels. The use of nickel increases the oxygen activity and hinders the formation of a graphite layer on the melt. Chistyakova and Stepanov [128a] in determining oxygen in stainless steels with high manganese contents found that cobalt provided a more fluid bath than nickel. Also, tin was added to prevent gettering by covering sublimed metal films. In a study with carbon monoxide labeled with ^{18}O, Kamada and Furuya [128b] noted that with tin or platinum as the bath metal and with gas circulation times up to 5 min, 90% or greater and 82% or greater, respectively, of the carbon monoxide was recovered. Addition of a steel sample caused little change in recovery with the tin bath but resulted in a greatly reduced recovery with the platinum bath.

Several precautions must be taken when using bath and flux techniques. Samples must dissolve in the bath and not float on top. To accomplish this in the analysis of titanium and zirconium [129], the samples were placed in steel capsules. This also served to dilute the volatile components. Platinum capsules containing degassed tin have been wrapped with platinum foil to achieve this purpose [123]. Capsules have also been made of graphite. Although this material is a reactant and not a flux, it is of interest to consider it here. Use of graphite in capsule form provides for prolongation of the time period that the sample is intimately exposed to the graphite during decomposition or reaction. Beck and Clark [130] used this technique when determining oxygen in aluminum, aluminum nitride, and various metal oxides by the inert-gas fusion method. Capsules were prepared from preformed spectroscopic electrode cups and were sealed with a graphite plug. Metallic films are deposited on the walls of this miniature reaction chamber and not on the walls of the crucible, the carbon monoxide diffusing through the walls of the chamber. Presumably, gettering was minimized in this manner. Oxygen was successfully determined over the range 0.05–40%. However, gettering of small amounts of oxygen at these concentration levels would not be noticed. On the other hand, Zakharov [131] used graphite capsules in the analysis of titanium for hydrogen and oxygen, and Columbo and Rodari [132] used copper filings in graphite capsules in the analysis of aluminum for oxygen. In an elegant effort to determine parts-per-billion gases in copper and to eliminate sorption effects, Bohm et al. [133a] placed the crucible inside a quadrupole mass spectrometer, where the sample was heated by electron bombardment.

5. Sample-Storage and -Entry Systems

The type of sample-storage or -entry system that is most practical will depend partly on (1) the material being analyzed, (2) whether or not hydrogen is to be determined, and (3) the frequency at which samples are submitted. Metals that will amalgamate with mercury cannot be passed through a mercury lift. If ferritic steels are to be analyzed for hydrogen, which diffuses at an appreciable rate from this material even at room temperature, a multiple-sample-storage system will not be suitable. This latter type of system also will not be practicable if samples are submitted periodically throughout the day and if rapid analyses are required.

Conventional barometric-length mercury lifts [133b] have been used for many years for introducing steel samples into vacuum-fusion systems. Yanagisawa and Seki [134], however, attributed the poor accuracy and reproducibility of their oxygen results to contamination of the steel samples with a mercury oxide present in the mercury lift. (There was good agreement between nitrogen values obtained with and without the mercury lift, so any air admitted with the sample was presumably evacuated prior to analysis.) Another disadvantage of the mercury lift is that if it should break, a large amount of mercury can be pulled into the furnace, resulting in explosion of the furnace and volatilization of mercury into the laboratory atmosphere at toxic concentrations. To prevent the influx of mercury into the furnace, a self-locking check valve (see Fig. 4.7) has been constructed [135]. This valve is a modified standard-taper ball-and-socket joint placed in the inlet system between the mercury and the furnace. A sudden entry of mercury or air causes this valve to close.

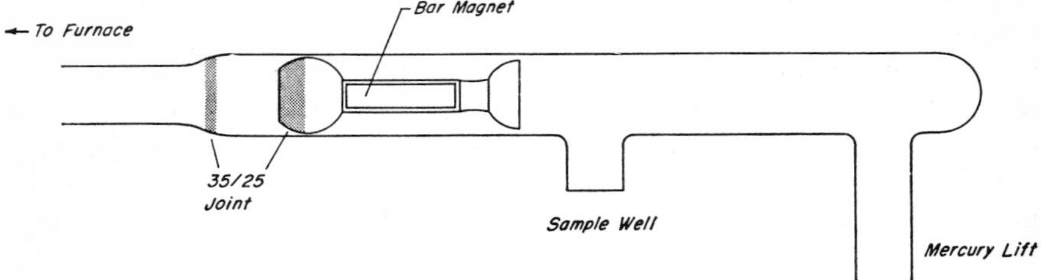

Fig. 4.7. Self-locking check valve. (Reproduced by permission from *Transactions of the Metallurgical Society of AIME* and The Metallurgical Society of AIME [135].

Considerable water vapor and some air have been detected mass spectrometrically in the authors' laboratory with samples brought in through the mercury lift. This necessitated a pump-down time of 5 min to regain a pressure of 5×10^{-5} torr. To minimize the amount of mercury used and the amount of air entrained with the sample, the inlet system shown in Fig. 4.8*a* was developed in 1963. In this system two Jamesbury Corporation (Worcester, Massachusetts) ball valves, type 1213, $\frac{3}{4}$ in. in diameter, are connected to a

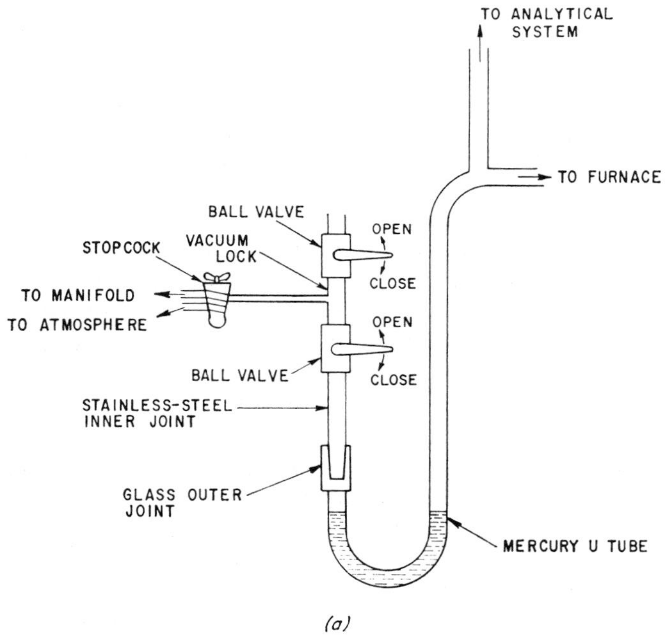

Fig. 4.8*a* and *b*. Sample-inlet systems.

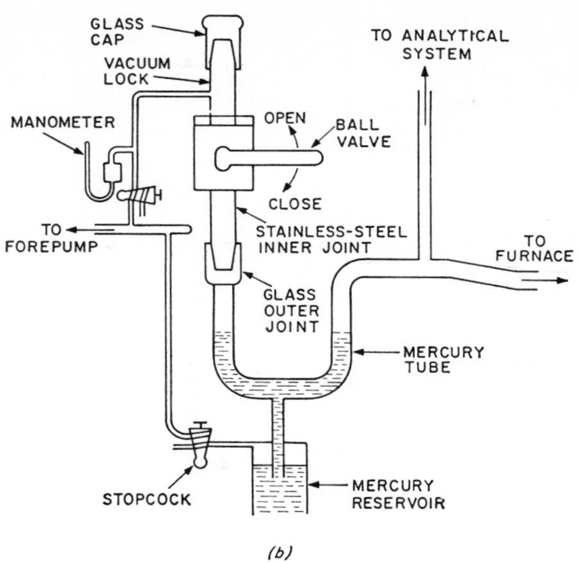

Fig. 4.8*b*.

stainless-steel tee to form a vacuum lock that can be pumped down separately. (The vacuum lock can also be constructed using a metal standard-taper joint with a side-arm and a glass cap at the upper end, as shown in Fig. 4.8*b*). A small amount of mercury is used to isolate the ball valves from the high vacuum and also to serve as a cushion to protect the glass when the metal sample drops through the lower valve. By isolating the high-vacuum side, pump-down time after sample admission can be held to a minimum. When the ball valves were used with mercury and each sample was held in the vacuum lock for a pump-down period of 1 min, no gas was detected and pump down to 5×10^{-5} torr required less than 1 min. By omitting the mercury, samples that amalgamate with it can be analyzed.

The heart of the ball valve is a finely machined sphere through one axis of which has been drilled a large-diameter hole. The sphere is held at each end by two specially shaped, annular seats made of neoprene or Teflon. When the valve is in the fully open position, the hole in the ball lines up with the path through the valve body. Turning the sphere 90° completely closes the valve and enables it to hold vacuum in either direction. Gate valves also were tried, but samples tended to lodge in the valve housing. This resulted not only in loss of the sample but often prevented closing of the valve, thus requiring dismantling of the valve with considerable loss of time.

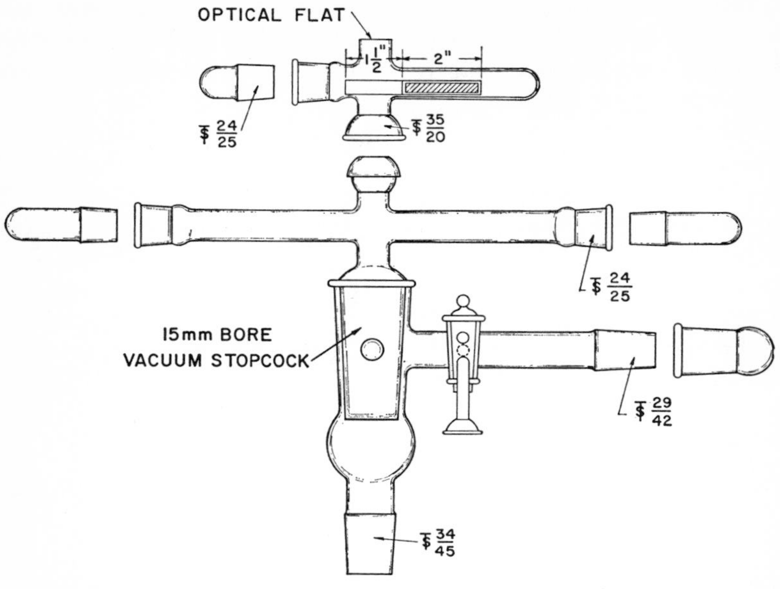

Fig. 4.9. McKinley sample-inlet system. (Reproduced by permission from the author [136].)

McKinley [136] has used the system shown in Fig. 4.9 for charging nonmagnetic samples. The top section contains a magnetically operated shield for keeping the optical flat clean. The middle section serves as both a handle for the 15-mm-bore stopcock and as a storage chamber for samples, flux and bath materials, and a magnetic pusher (the system may be modified to omit the middle section). These sections are evacuated simultaneously with the vacuum-fusion system. The bottom section affords access for samples inserted singly. The sample is placed in the open cup of the sample introductory assembly, which is then sealed. After evacuation of this section via the three-way stopcock, the 15-mm-bore stopcock is rotated, and the sample is pushed through the bore so that it falls into the crucible.

A similar entry port for the introduction of samples singly into a vacuum-fusion system has been described by Bennett and Covington [137]. The sample is introduced through a standard-taper joint, which is then capped (see Fig. 4.10). The chamber containing the sample is then evacuated through one arm of a three-way stopcock. After evacuation the sample is admitted to the furnace through a vacuum valve (Veeco T-62-S, manufactured by Vacuum-Electronics Corporation, Plainview, New York). This is a high-vacuum bellows valve machined from square brass stock, thus allowing mounting on any of the flat faces. Synthetic rubber or Teflon gaskets are obtainable with this device. The bellows valve and other similar metal valves may be sealed to the system by means of standard-taper or ball-and-socket metal–glass connections. Such connections are preferable to mercury-seal joints and glass-to-metal graded seals.

A sample-storage system designed for micro-vacuum-fusion analysis of beryllium is shown in Fig. 4.5*a* [42]. The $1\frac{1}{2}$-in.-diameter silica furnace tube is connected by an Apiezon W40 (W100 would be better) waxed B34 standard-taper joint to a glass furnace head that accommodates up to 12 samples and 3–10 g of platinum beads in individual depressions in the side-arms. Samples of platinum are moved in the side-arms by magnetically operated rods so that when the crucible lid is removed, the samples fall through the central sample-dropping tube and the lowered silica extension funnel into the crucible.

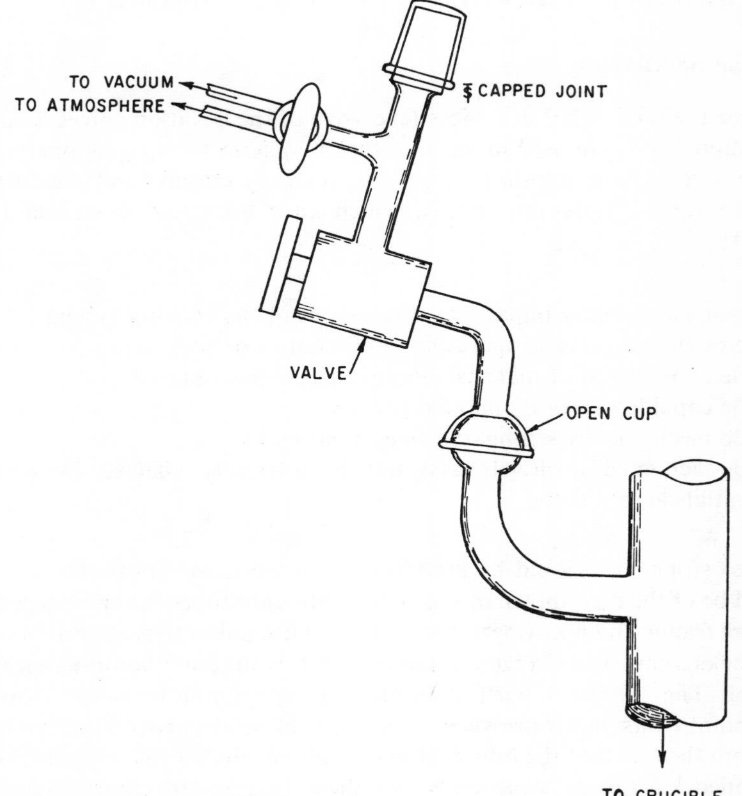

Fig. 4.10. Bennett-Covington sample-inlet system. (Reproduced by permission from *Analytical Chemistry* [137].)

Other sample-storage systems for micro-vacuum-fusion analysis have been described that incorporate hinged trip buckets [43, 87] and a 24-compartment tree arrangement [138]. The former are unduly complicated. Advantages of the latter and the similar systems described earlier must be equated with the disadvantages when analyzing specific metals. In general, the disadvantages of such systems are that (a) they have several cone-and-socket joints that are sealed with wax each time a group of samples is analyzed, a time-consuming operation, and (b) they add considerably to the furnace volume, making it more difficult to pump down. Also, although the magnetically operated plunger that is used for sample-release is made to fit loosely, difficulty may be encountered in evacuating each sample-compartment.

B. ANALYSIS BY PARTIAL PRESSURE MEASUREMENT

1. Isolation Devices

Three types of isolation devices (exclusive of the isolation valves described in Section I.A.5) are used in vacuum-fusion systems to separate one portion of the system from another: stopcocks, mercury cutoffs, and mercury-less vacuum valves. Isolation devices, which have been well described [139], should:

1. Not significantly impair the throughput of the vacuum system
2. Not evolve gases at operating temperatures or pressures
3. Be constructed of materials inert to gases encountered
4. Be capable of being operated rapidly
5. Be mechanically strong and free of gas leaks
6. Be activated positively and not be adversely affected by pressure differentials across them.

Glass stopcocks should be used as little as possible. Those that are used should be of the type shown in Fig. 4.11. Although stopcocks are inexpensive, they are fragile and may freeze after extended use unless they are relubricated. Stopcocks may also freeze in laboratories with poor room-temperature control. The lubricant itself presents a problem not so much from the standpoint of its vapor pressure, which should be about 10^{-6} torr or lower, but from the fact that the lubricant may evolve occluded gas, especially when the stopcock is turned. Apiezon N and silicone greases are commonly used as lubricants. When using stopcocks, the bore should be large enough to provide good conductance.

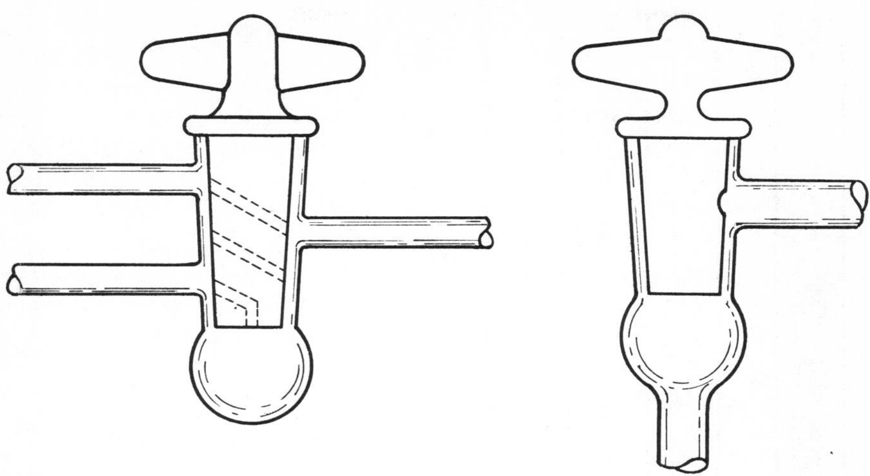

Fig. 4.11. Vacuum stopcocks.

The U-type mercury cutoff, such as that shown in Fig. 4.12, is quite common. This is a float valve containing spherically ground seats and valve plungers with glass-enclosed metal cores; it requires a mercury reservoir. Actuation of the cutoff may be accomplished manually or semiautomatically by means of a motor-operated solenoid. Little difficulty is encountered in the operation of U-type cutoffs. However, in rare instances where the pressure surge is too great, the cutoff may leak. Mercury cutoffs should be able to withstand a pressure differential of 10 torr.

Solenoid-operated glass–mercury valves [140–142] have been used in place of the U-type mercury cutoff (see Fig. 4.13). In the design of McKinley [142] the seal is a nickel cup attached to a solid iron stem. Horton and Brady [141] controlled their valve by means of a coil that had a dc resistance of 285 Ω and an inductance of 1.25 H with the iron slug out and 1.55 H with the iron slug in. With $\frac{1}{4}$ in. of mercury pressure, differentials up to 2 torr could be tolerated. Greater differentials tended to hold the inverted cup down against the pull of the pickup coil or blow it open. The main advantage of this type of valve is that it does not require a mercury reservoir.

A somewhat different type of solenoid-operated glass–mercury valve has been described by Lench and Martin [143]. This valve also does not require a mercury reservoir. It is constructed so as to be normally held in the open position by the tension of a light spring. When closed, a maximum pressure differential of 10 torr can be maintained. The depth to which the cap is drawn into the mercury in the annulus and the rate at which the valve is closed are regulated by the variable autotransformer that controls the power to the solenoid.

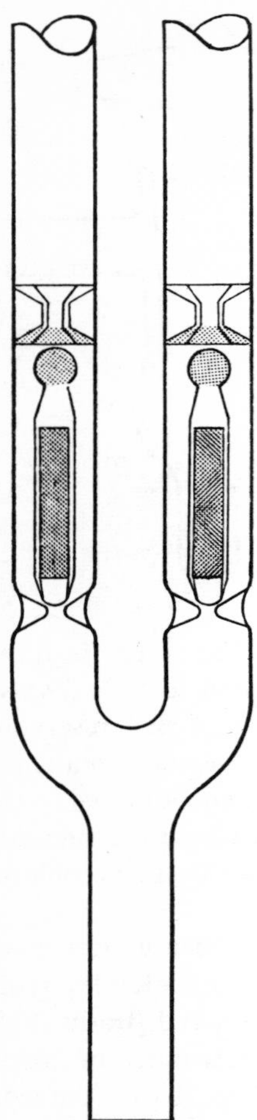

Fig. 4.12. Mercury cutoff.

Lench and Martin [143, 144] eliminated difficulties encountered with a large-bore stopcock by substituting a stainless-steel O-ring vacuum valve. (The stopcock had been connected to a vacuum-fusion furnace, and the difficulties included sticking in cold weather, gas adsorption on the large area of grease exposed, and atmospheric leakage.) With the new valve the possibility of ejection of the graphite insulation powder from around the crucible on sudden application of high vacuum was avoided by adjusting the aperture of the valve and therefore the degree of vacuum applied. A tee piece 2 in. in diameter was modified to take the vacuum valve by adding a 1-in.-diameter, 3-in.-long section directly opposite the vertical part of the tee

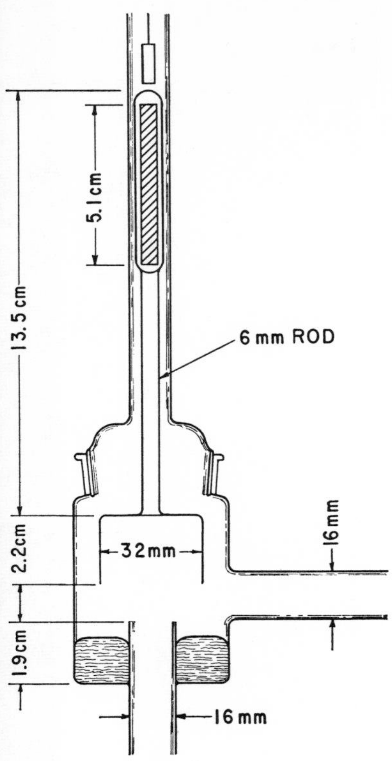

Fig. 4.13. Solenoid-operated mercury cutoff. (Reproduced by permission from the author [142].)

(see Fig. 4.14) [144]. The body (G) of the valve was attached to the 1-in.-diameter pipe flange by a modified form of Quickfit Visible Flow (QVF) fitting (from Quickfit and Quartz Company, Ltd.) in which the projecting metal flange on the body replaces the usual metal backing flange. A vacuum-tight seal is ensured by the O-ring (L) between the flat ground end of the flange and the body of the valve. The valve seat (B) is held between the flange of the stem (F) and the flange of the piping (A) by QVF pipe union fittings; the valve seat is separated from each of the flanges by a neoprene gasket (P), forming a vacuum-tight seal. Another O-ring (C) set in the valve seat ensures a vacuum-tight seal between valve (D) and the valve seat. Valve (D) is loosely attached to the spindle by means of a split ring (E) and a steel ball (O). The loose fitting allows the valve to tilt a few degrees from the horizontal plane in all directions to accommodate any small misalignment of valve and valve seat that may exist. The ball is attached to the spindle by a threaded length of rod that permits, in conjunction with the use of a locknut, a lengthwise adjustment of the spindle up to 1 in. The piston slides in the tube

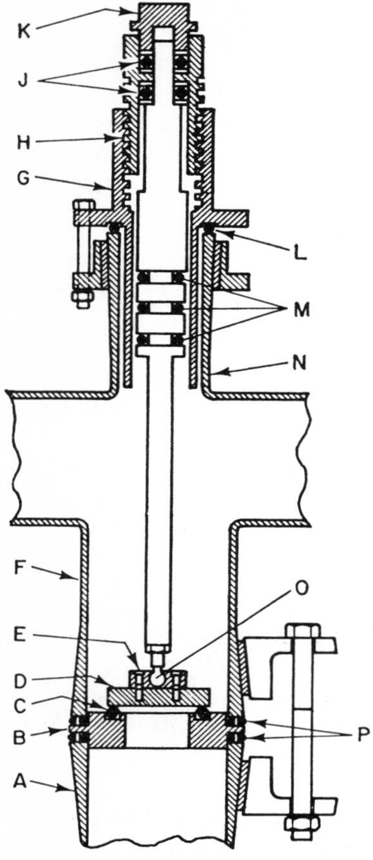

Fig. 4.14. Lench-Martin vacuum valve. (Reproduced by permission from The Institute of Physics and The Physical Society [144.]

on three O-rings (M) that provide a vacuum seal and hold the piston centrally in the tube. The internal flange on the bonnet (H) and the flange on the locking nut (K) attached to the end of the piston ensure that movement of the bonnet in or out of the body will cause a corresponding movement of the valve. Threading of the body and bonnet enables the size of the valve aperture to be finely adjusted by rotating the bonnet. Ball bearings and washers (J) enable the bonnet to be rotated without rotating the piston.

2. Pressure Gauges

The pressure gauge in the analytical part of a vacuum-fusion system must operate independently of the specific gaseous elements or compounds in the system and must be able to measure pressures generally over the range of

10^{-3}–10 torr. Mercury, with its high specific gravity, cannot be used in a manometer in this application because the lowest pressure recordable with this liquid is 1 torr. (See, however, Section I.L.5.) U-tube manometers [145a] containing a low vapor pressure oil such as *n*-amyl sebacate or dibutyl phthalate have been used, but they are not recommended for accurate gas analysis. Although they are simple, are inexpensive, and can provide pressure readings accurate down to 7×10^{-3} torr with a cathetometer, oil manometers suffer from several disadvantages. Oil exposed to air at atmospheric pressure will dissolve some air that will be evolved as bubbles, thus hindering pressure measurements. Also, the oil may wet the glass and creep into other parts of the system. Replenishment of oil in the gauge requires dismantling of a section of the apparatus. Sellenger [145b] has presented an excellent discussion of the operation, application, and calibration of mechanical manometers and thermal conductivity and ionization gauges.

The McLeod gauge best serves the purpose for measuring pressures of gases extracted by vacuum fusion. It is the only gauge other than the complex and fragile Knudsen gauge that operates independently of noncondensable gas composition. The McLeod gauge is normally applicable over the range 10^{-5}–10 torr. Special modifications permit its use at lower pressures. Because time is required for the system and gauge pressures to equalize, instantaneous measurements are impossible with any type of McLeod gauge. In addition, the McLeod gauge cannot be used with condensable gases—that is, those gases having a critical temperature above room temperature. Water vapor is the only such offender encountered in vacuum fusion. Ideally, the room in which the gas analysis is performed should be thermostatically controlled. If not, correction should be applied to the measured pressure of the extracted gas.

In the construction of a McLeod gauge, the capillary should be closed with a tapered glass plug so that the end of the capillary is flat on the inside and nearly so on the outside. Such construction permits complete observation of a line on the scale behind the closed end of the capillary. To diminish breaking and sticking of mercury threads in the capillaries, the bore should be ground with an abrasive such as aluminum oxide by drawing a lubricated brass wire back and forth through it [146]. Further, the bore should be of uniform diameter.

Calibrated McLeod gauges are available commercially. Such gauges are at best accurate to within 1% of the pressure for the range 10^{-2}–1.5 torr, which is typical for vacuum-fusion analysis. A double-scale gauge [147] is available that has a capillary of fine bore in the top half for accurate reading of low pressures and a capillary of coarse bore in the lower half to increase the range. Fine- and coarse-bore comparison tubes are to the left and right of the capillary.

A tilting or rotating McLeod gauge, which is much smaller and less expensive than the gauges described above, has been used on several commercially available gas-analysis systems (both vacuum fusion and hot extraction). This type of gauge contains much less mercury than the stationary McLeod gauge and does not necessitate a pump. However, the tilting gauge also utilizes a smaller, compressed scale and is therefore less accurate; it is not recommended where high accuracy is desirable. Tilting McLeod gauges are available in several pressure ranges.

The mercury in the capillaries of a stationary McLeod gauge will not ordinarily become contaminated with dirt, because the mercury enters the gauge through a tube extending nearly to the bottom of the large reservoir. However, in the tilting type, dirt can present a problem. The cleaning of the precision-bore capillary in this gauge has been facilitated by closing the end of the capillary with a Teflon plug containing a standard O-ring seal [148]. This seal may be easily opened or closed and will remain vacuum-tight for pressures as low as 10^{-6} torr. In all types of McLeod gauges, only triply-distilled mercury should be used.

A rotary McLeod gauge [149] has been modified to allow recording of pressures [150] but has not been found to be completely satisfactory. A precision McLeod gauge that provides greater accuracy than conventional McLeod gauges has been developed by Podgurski and Davis [151a]. Work and Hawk [151b] have developed a McLeod gauge with automatic zeroing. This gauge is manufactured by Hastings-Raydist, Inc., Hampton, Virginia.

3. Analysis of Gases

The method chosen for analysis of the gas mixture will depend on the gases to be determined and the sensitivity required. Complete recovery of nitrogen from the nitride of interest should be confirmed prior to analysis, especially in view of conflicting data in the literature. For example, a British Iron and Steel Research Association report [152] states that the vacuum-fusion method is universally applicable for the determination of nitrogen in steel, except where certain types of carbon-resistance apparatus are used. However, Karp et al. [153] have shown that with certain samples of $3\frac{1}{4}\%$ silicon steel, higher nitrogen results were obtained by a chemical method than by the vacuum-fusion procedure using an iron bath.

The sequence in which the gases are determined and in which the oxidation of hydrogen and carbon monoxide are performed is important. For example, when Hopcalite is used to oxidize carbon monoxide, hydrogen should be removed first, as it can be slowly oxidized by this substance [115, 154, 155]. Also, the efficiency of oxidizers and gas absorbers should be checked daily.

a. Diffusion of Hydrogen Through Palladium

The high diffusion rate of hydrogen through palladium has been known for many years. Fleiger [156] and Ransley [157] were among the first to report the use of a palladium tube in an analytical system; many workers have since utilized palladium for the admission of hydrogen to vacuum systems. Within the past few years, however, the palladium tube has been increasingly used in the determination of hydrogen in metals after removal of this gas by diffusion. In such systems the inside of the tube is connected to a diffusion pump to maintain a lower partial pressure of hydrogen on the exit side. (In a system containing hydrogen at about 100 torr, this gas has been caused to diffuse to the atmosphere through a palladium tube simply by heating the tube.)

Prior to analysis, the palladium tube must be outgassed. For this operation, temperatures of 600–750°C have been used [158–160] and times of 10–15 h have been recommended [159, 160]. Operating temperatures vary from 350–600°C [32, 158, 161–163]. It has been reported [155] that at 600°C none of the gases extracted from metals interferes if oxygen is absent; the diffusion rate of hydrogen may be low when oxygen is present [158].

Although minor objections have been raised concerning the slowness [72] and the incompleteness (94–96%) [164] of hydrogen diffusion, this technique has been shown [32] to be rapid, comparatively inexpensive, and quantitative. Further, on analysis of the undiffused gas by gas chromatography, no hydrogen was detected [161].

Although the diffusion of hydrogen through palladium has been the subject of many investigations, and comprehensive studies of the metallurgical properties of palladium have been performed [165, 166], comparatively few publications have appeared describing the use of the palladium tube in a vacuum-fusion system. In 1956 Turovtseva et al. [167] and, later, Still [115] and Mallett et al. [154] discussed the application of this technique. One deterrent to its use has been that the palladium tube may crack if cooled below 150°C in an atmosphere containing hydrogen [150]. (Heating and cooling involves transformations between the α- and the β-phase of palladium.) In another study [165], the critical temperature range was stated to be between 100° and 250°C. Martin et al. [161] eliminated the cracking difficulty by maintaining the palladium tube at 400°C after degassing. If the proper precautions are taken, the palladium tube can be operated in a reproducible manner with a minimum of maintenance for many years [168, 169]. Hunter [170] has shown that by alloying palladium with 10–50% silver the embrittlement is prevented. This discovery has been utilized by Serfass in vacuum-fusion analysis [171]. Hydrogen containing only about 10^{-8}% impurities has been obtained by diffusion through palladium and also through palladium–25% silver [172a]. High hydrogen-diffusion rates may be obtained by using a tube made of palladium containing boron, silver, or boron plus silver [172b].

Regardless of the precautions observed with the palladium tube, it is still advisable to check its condition periodically by injecting a known amount of hydrogen into the system followed by measurement of the pressure of the diffused gas. For this purpose, gas mixtures containing either 50% nitrogen, 30% carbon monoxide, and 20% hydrogen or 50.0% nitrogen, 49.5% carbon monoxide, and 0.5% hydrogen have been used in the authors' laboratory.

b. Oxidation Catalysts

At least four catalysts have been used in vacuum-fusion systems for oxidizing both hydrogen to water and carbon monoxide to carbon dioxide: (1) a platinum filament heated in the presence of a known amount of oxygen, (2) heated copper oxides, (3) heated copper oxide containing 1% iron oxide, and (4) a heated mixture of copper oxide and oxides of rare-earth elements consisting mainly of cerium. Hopcalite, a mixture of cobalt, copper, manganese, and silver oxides has been used for the oxidation of carbon monoxide. To reduce the outgassing time of such catalysts, the system should be so constructed that the catalyst chamber can be isolated during regeneration of the catalyst by atmospheric air.

The method of adding a known amount of oxygen at a heated platinum filament [50] is not entirely satisfactory, because some oxygen is lost by adsorption on the filament [117], thus necessitating the determination of the unadsorbed oxygen by the addition of hydrogen and further combustion [32]. Such extra operations are time-consuming and provide additional sources of error. Also, dilution of the extracted gas by excess oxygen may result in erroneous nitrogen values when the nitrogen content is low.

Copper oxide at 320–360°C [72, 124, 173] has been used as an oxidation agent by a number of investigators. (To oxidize methane and higher-molecular-weight organic gases, temperatures above about 360°C are needed. This is discussed in Sections I.B.3.f and I.I.) Copper oxide is subject to poisoning by chlorinated solvents, such as carbon tetrachloride, that are used in cleaning stopcocks. Copper oxide may be made [72] by repeated slow air-oxidation of $\frac{1}{2}$-in. balls of 30-gauge pure copper wire (presumably at 320°C) and reduction in a 5% hydrogen–95% nitrogen mixture to form finally a layer of copper oxide. The copper oxide blank may consist of an appreciable quantity of oxygen. To obtain a much lower blank than that normally encountered with copper oxide, a catalyst composed of copper oxide and oxides of rare-earth elements has been used [116, 135, 174]. This catalyst, developed at the National Bureau of Standards and described by Walter [116], is prepared in the following manner.

Dissolve separately 23.5 g of rare-earth oxides, containing at least 35% cerium, and 150 g copper in nitric acid, and mix the two solutions. Evaporate the solution to dryness, heat to 800°C, and maintain this temperature for 24 h to remove the acid fumes. Cool the material, crush, and put through an 80-mesh screen. Mix the powder with 36 g of pure kaolin, and moisten with water to form a plastic mass. Dry the mixture, and bake at 800°C for 4 h. Then break the mixture into small lumps, reduce in a stream of hydrogen at 300–400°C, and reoxidize at the same temperature. Repeat the reduction–oxidation cycle. Crush the material, and put through a 10-mesh screen. Retain the material that is caught on a 20-mesh screen, and place in a catalyst U-tube. Use the catalyst at 320°C.

Copper oxide containing 1% iron oxide has been used at 400°C [141] and 600°C [85]. In the latter study, pure copper oxide, ferric oxide, and kaolin were mixed and ground in the proportions 99:1:20. After addition of water, the paste was squeezed through a 2-mm-diameter orifice. The resultant threads were dried, broken into suitable lengths, fired at 600°C, and then packed loosely in a fused silica U-tube. Reduction in hydrogen and re-oxidation in oxygen at 600°C were then carried out. It was noted that this catalyst was more reactive than pure copper oxide and that during analysis, recirculation of gas over the catalyst was unnecessary.

The use of Hopcalite as an oxidant for carbon monoxide in a vacuum-fusion system was originated by Davis and Gray [175]. (In the United States, Hopcalite may be purchased from Mine Safety Appliance Company, Pittsburgh, Pennsylvania, and in England from Siebe Gorman, Ltd., Davis Road, Chessington, Surrey.) Oxidation of carbon monoxide is performed at room temperature and requires from 2–3 min. Hopcalite has no measurable vapor pressure or reaction products under these conditions [137]. It has been found necessary, however, to heat Hopcalite before using it for the first time; otherwise, a gas, consisting mainly of water vapor, is given off even after 8 h of evacuation [115]. Heating at 80–100°C for about 1 h under vacuum is recommended; heating above 120°C impairs the oxidation efficiency of the Hopcalite. The catalyst may be regenerated with air at 15–20°C [176]. As mentioned in Section I.B.3, hydrogen can be slowly oxidized by Hopcalite and should therefore be removed prior to oxidation of carbon monoxide.

c. Water Traps

Several means of isolating water include absorption in magnesium perchlorate or phosphorus pentoxide or trapping in tubes cooled by mixtures of dry ice with ethylene glycol monoethyl ether acetate [177], ethylene glycol-acetone, or trichloroethylene; or liquid nitrogen with methanol or ethanol. A dry-ice–acetone mixture supercooled to −95°C by subjecting it to reduced pressures has also been used [116].

Various opinions and recommendations concerning water traps appear in the literature. For example, several workers [85, 117, 178] have used phosphorus pentoxide, which has two disadvantages [179]. One is that during use, a syrupy coating forms on the surface, thus reducing the drying power of the oxide. The second disadvantage is that the presence of phosphorus trioxide affects the efficiency of the desiccant by reacting with water vapor to form hydrogen phosphide, which reacts with mercury to form mercury phosphide and hydrogen [180]. The trioxide can be converted to the pentoxide by heating the material in a bath to 300°C in a stream of ozone; alternatively, the more volatile trioxide can be removed by subliming in a current of dried air [181]. Gokcen and Tankins [72] preferred using a refrigerated U-tube instead of phosphorus pentoxide. The U-tube was immersed in ethanol cooled to −100°C by adding liquid nitrogen. This mixture was found to be superior to dry-ice–acetone, dry–ice–trichloroethylene, and methanol–liquid-nitrogen mixtures because of the lower water vapor pressure that it maintains. However, the mixture of methanol and liquid nitrogen should be adequate because at the temperature of this mixture, about −95°C, the vapor pressure of water is approximately 0.03 μ [141]. Care must be taken not to freeze the alcohol entirely, because temperatures cold enough to trap carbon dioxide may be attained and the expansion of alcohol on freezing may result in breakage of the Dewar flask.

All of the various cold traps cited will absorb water from the atmosphere with eventually an increase in temperature, and therefore, periodic temperature measurements should be made. The methods of the American Society for Testing and Materials [174] for oxygen in molybdenum and titanium, and the recommended methods of the Panel on Methods of Analysis, Metallurgical Advisory Committee on Titanium [182], for oxygen in titanium specify magnesium perchlorate as the water absorber with dry ice–acetone as an acceptable substitute (vapor pressure of water is about 0.5 μ with this trap). The anhydrous magnesium perchlorate must be reagent grade and screened to remove fines. Material passing a no. 20 (840-μ) sieve and retained on a no. 200 (74-μ) sieve is recommended. A magnesium perchlorate trap may be conveniently attached to a system using ball-and-socket joints sealed with Apiezon W high-vacuum wax. This trap is evacuated at about 240°C prior to analysis. Reactivation is performed similarly. With the system under vacuum the magnesium perchlorate is kept in condition by maintaining this temperature. During analysis this absorbent is kept at room temperature. The ease with which magnesium perchlorate may be regenerated and the high efficiency and capacity of this reagent make it attractive for removing water vapor. Smith and Diehl [183] have developed a magnesium perchlorate desiccant containing potassium permanganate as indicator, which is light purple in the dry condition and dark brown when it is spent.

d. Carbon Dioxide Traps

Carbon dioxide has been removed during vacuum-fusion analysis by freezing out in a trap immersed in liquid nitrogen or by absorption in an inert material impregnated with sodium hydroxide (Ascarite, Caroxite, or Mikohbite). Condensation by liquid nitrogen is the recommended practice of two cooperative groups [174, 182]. Solid absorbents for carbon dioxide cannot be reactivated. If they are used, the ball-and-socket arrangement mentioned in Section I.B.3.c is convenient.

e. Sequence of Gas Removal

Trapping of the gases has been performed in various sequences. The order of removal and pressure measurement depends upon the gases being determined. After a pressure measurement of the total gas mixture, water and carbon dioxide can be frozen out in liquid nitrogen and the nitrogen determined. By substituting a dry-ice–acetone trap for the liquid nitrogen trap, after removal of nitrogen, a pressure measurement of the carbon dioxide can be made. Evacuation of this gas followed by removal of the second cold trap provides for a measurement of water. If the determination of hydrogen is of secondary or no importance, the water can initially be absorbed in magnesium perchlorate and determined by difference. Freezing out of carbon dioxide then allows a pressure measurement of nitrogen.

f. Treatment of Miscellaneous Gases

The measurement of oxygen evolved during vacuum fusion as sulfur dioxide or carbonyl sulfide will be dealt with only briefly here because the presence of such compounds depends on the material being analyzed; a detailed discussion of this topic is in later chapters. Suffice it to say, analysis for such compounds should be performed when new materials are being investigated. Analysis may be performed by gas chromatography, providing the retention times are known, or with a mass spectrometer. Where the metal sample contains little or no hydrogen or nitrogen, a simplified vacuum-fusion unit can be used for determining oxygen evolved primarily as carbon monoxide and secondarily as sulfur dioxide. Harris and Hickman [184] used such a unit in the analysis of copper (see Chapter 16). In their system, which is essentially a furnace attached to a mercury manometer, the gases are not removed from the furnace and any sulfur dioxide evolved is converted to carbon monoxide according to the following equation:

$$SO_2 + 2C \rightarrow 2CO + S. \tag{1}$$

In a conventional vacuum-fusion unit containing a copper-oxide–rare-earths-oxide catalyst, erroneous oxygen results are obtained if sulfur dioxide is present; this compound appears to be absorbed by the catalyst and then to be slowly released over a long period of time.

Carbon disulfide and carbonyl sulfide will cause oxygen results to be high. Hamner and Fowler [185] hydrogenated these compounds, obtained during the analysis of high-sulfur irons, by mixing the gas with an equal volume of hydrogen and passing it over a platinum spiral heated to 1050°C until the reaction appeared to be complete. Then the gas was passed through a pipet containing lead acetate or iodine to remove hydrogen sulfide. Presumably, carbon monoxide was released from the carbonyl sulfide during hydrogenation.

Any methane in steel is normally decomposed during vacuum-fusion analysis [186]. However, Turovsteva and Kunin [187] state that any methane that passes into the analytical system can be determined by combustion over copper oxide heated to 850°C (see also Section I.I). The methane content is calculated from the amount of additional carbon dioxide formed after the initial oxidation of hydrogen and carbon monoxide at 300°C. The determination of methane in steel by other means is discussed in Chapter 7. Methane has also been determined by gas chromatography [188]. Hydrocarbons (apparently including methane) have been found by mass spectrometry in the vacuum-fusion analysis of silver and tin [189a] (see Section I.I).

C. CALIBRATION OF APPARATUS

The volume of the analytical portion of the system must be accurately known. It can be determined by admitting a known volume of dry air at atmospheric pressure and measuring the pressure in the system at a known temperature. From Eq. 2 the volume (V_2) of the system can be calculated.

$$V_2 = \frac{P_1 V_1}{P_2} \tag{2}$$

where P_1 is the atmospheric pressure, V_1 is the volume of dry air at atmospheric pressure, and P_2 is the pressure in the system expressed in the same units as P_1.

The dry air (or other gas where desirable) may be admitted to the vacuum system via a standard-taper joint, by means of a dosing stopcock or a gas pipet having a known volume between two stopcocks. The volume of this admitting device is determined by filling it with mercury and then weighing the mercury to the nearest tenth of a milligram after removal. Several calculations of both the volume of air taken and the analytical volume of the fusion system should be made and the average values calculated. Although calculated to the nearest milliliter, the analytical volume is accurate only to about 2%. During these measurements care should be taken to ensure the cleanliness of the mercury and the admitting device. All air bubbles must be liberated from the mercury contained in the device, and all traces of mercury must be removed prior to weighing the empty device. Also, stopcock grease must not be allowed to alter the volume of the device when the stopcocks are turned.

Gas-dosing valves that permit the introduction of accurately known amounts of gases to vacuum systems are available from Ströhlein and Company, 4 Dusseldorf 1, West Germany [189b]. These valves are available in nominal volumes of 0.015, 0.05, 0.1, 0.3, 0.5, 0.8, 1.2, 2.0, 3.0, and 5.0 ml. (All dosing valves are marked with the exact volume, which in fabrication may differ by up to 10% from the nominal value.) The inaccuracy in the specified volume is less than 1%, except for the smallest size, where an error of up to 3% may be expected. The leakage of these valves is less than 10^{-7} liter-torr/sec.

The Strohlein gas-dosing valve is shown in Fig. 4.15. A plunger rod, D, with seals E, F, and G, slides in the precision-bore glass tube, A. The rod, D, is moved by a manually operated rotary drive. The seal, E, closes off the dosing valve toward the outside. When the rod, D, is moved from its lower end position into the position shown, a quantity of gas corresponding to the dose volume is separated from chamber B, which is connected to a gas cylinder, and trapped between seals F and G. As the rod, D, moves into the upper end position, this quantity of gas is introduced into chamber C, which is connected to the fusion apparatus.

The Laboratory Equipment Corporation, St. Joseph, Michigan, manufactures a gas-dosing device, X707–100 [189c]. Extremely accurate aliquots of gas reportedly may be introduced to an analyzer with this device. Also, dosing valves are supplied with the Balzers apparatus (see Sections I.L.5, 7, and 8); these can be removed and used on other instruments (see Section I.L.2.c).

D. DETERMINATION OF THE OPERATING BLANK

Results obtained by vacuum-fusion analysis must be corrected for appreciable contributions by the blank. Classical furnace systems are outgassed for at least 2 h at a temperature considerably higher than the operating temperature but low enough to minimize volatilization of graphite (as

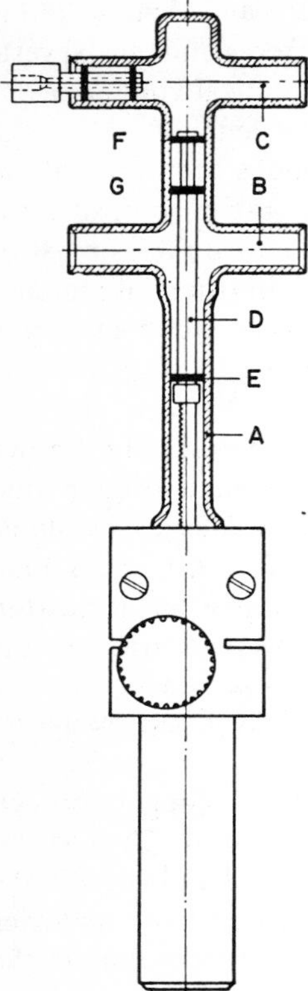

Fig. 4.15. Strohlein gas-dosing valve. (Reproduced by permission from Strohlein and Co. [190a].)

described in Section I.A.3), the temperature is reduced to that used during analysis and a simulated run is carried out. Analysis of the gas provides blank values for hydrogen as the elemental gas or as H_2O, oxygen as CO or CO_2, and nitrogen. By mass-spectrometric analysis of the blank gas over a period of several years, it has been determined in the authors' laboratory that with the low blank value (usually less than 8 micron-liters in 10 min) and the constancy of its composition (about 80% H_2, 8% CO, 8% N_2, and lesser concentrations of other gases), only the total pressure need be measured, and the contribution of the constituents can be calculated.

If the furnace is conditioned with a sample of the metal being analyzed or if a bath material is used, determination of the blank should be performed after addition and outgassing of this material. On the other hand, if platinum or some other foreign material is used as a flux (that is, added with each sample), it is recommended that several replicate determinations of the gas content of the flux material be made [182] and that an average value be added to that of the blank. The total then constitutes the blank for the apparatus and the flux material. Comparison of blank rates of various systems should be made only when the blanks have been obtained in the same manner. For commercially obtained systems, blanking instructions are usually provided.

E. CALCULATIONS

The total pressure measured is caused by hydrogen, carbon monoxide, and nitrogen. With catalytic oxidation there is no pressure change in converting hydrogen to water or carbon monoxide to carbon dioxide. To calculate the percent of hydrogen, nitrogen, and oxygen, the pressure of each gas must first be corrected for its contribution to the blank (see Section I.D). The percent of the various gases may then be determined by using the appropriate pressure values and the following equations:

$$PV = \frac{g}{M} RT \qquad \text{or} \qquad g = \frac{MPV}{760\,RT}, \tag{3}$$

where P is the net pressure in torr, V is the volume in milliliters, g is the weight of gas in grams, M is the number of grams per mole of the gas, R is the gas constant, 82.06 ml-atm/(deg)(mole), and T is the ambient temperature in degrees Kelvin. For hydrogen (converted to and removed as water or diffused through palladium),

$$\%\,\mathrm{H} = \frac{g}{W} \times 100 = \frac{2.016\,PV}{760\,RTW} \times 100, \tag{4}$$

where W is the sample weight in grams. For oxygen (converted to CO and measured as CO_2),

$$\%\,\mathrm{O} = \frac{16\,PV}{760\,RTW} \times 100. \tag{5}$$

For nitrogen,

$$\%\,\mathrm{N} = \frac{28\,PV}{760\,RTW} \times 100. \tag{6}$$

To convert the percent composition to parts per million, multiply by 10,000.

F. DETECTION LIMITS

The usual limit of detection of the vacuum-fusion method with pressure measurement of the gases is about 10 μg for nitrogen and oxygen and about 1 μg for hydrogen. Lower limits for all three gases may be obtained by using mass-spectrometric or gas-chromatographic techniques (see Sections I.I and I.J; for inert-gas fusion, see the various systems described in Section II.C.2).

G. SAFETY FEATURES

In addition to those features cited in previous sections, several other points are noted to provide for safety and efficiency, mainly in operation of a glass vacuum-fusion unit.

The operator should wear safety glasses. Also, it is recommended that a shield of Plexiglas or equivalent material be placed in front of the furnace and expansion volume sections of the apparatus to provide protection in case of implosion. Alternatively, the expansion bulbs may be enclosed with wire mesh [89]. Shields must be easily movable to provide ready access to the equipment. Belt guards should be installed on all fore pumps.

To protect glass mercury cutoffs from sudden mercury surges the manifold system should be connected through a stopcock to the main exhaust system so that the same degree of vacuum is maintained on both sides of the mercury pots. Also, a room-temperature mercury trap should be installed on this manifold line to protect the fore pump. This trap should be easily removable; connection with Apiezon W grease is recommended. Where an air-cooled furnace is used, the power supplies to the blower and furnace should be interlocked so that the furnace power cannot be turned on unless the blower has previously been put into operation.

A flow meter containing a make-or-break relay should be installed on the exhaust side of the cooling water of a diffusion pump so that the electric power to the diffusion pumps (where the cooling water lines are in series) will be turned off if the water pressure drop is too great. Such devices have been described for both electrical [190, 191] and gas supplies [192]. An audio alarm can be wired into the circuit to provide a warning to the operator. Also, open or sight drains should be used for the cooling water.

Safety devices for a vacuum-fusion unit have been described [115]. These devices can automatically switch off the induction heater, rotary oil pump, Pirani gauge, and the gas supply to the gas burners for the mercury-diffusion pumps. Thus, the system is guarded in case of failure of cooling water and air supplied and leakage in the vacuum system. Among other things the Pirani gauge is protected so that the filaments are not heated in air. This could also be accomplished by electrically connecting the Pirani gauge to the induction heater so that any leakage in the vacuum system would initiate a signal from the gauge that would switch off the heater.

Where a barometric-length mercury lift is used for introducing samples, a self-locking check valve should be installed so that in case of a break in the lift, the entrance of mercury into the furnace is prevented (see Section I.A.5).

Similar check valves consisting of a steel ball cemented to a piece of glass or metal have been used [193].

With a Guldner-Beach type of furnace [83] (see Section I.3.a), the pump-down operation should be carried out very slowly to prevent displacement of the powdered graphite. This may be accomplished by initially evacuating the furnace to about 10^{-2} torr through the manifold. After closing the mercury cutoff between the furnace and manifold, the furnace may be evacuated further by means of the diffusion pump. Pump-down by this technique requires about 30 min.

H. LEAK TESTING

There are a number of instruments for detecting leaks in vacuum systems, and these have been well described [194, 195a, 195b]. The first step in locating leaks in a glass system is usually to determine in which portion the leak occurs. This may be accomplished with a multistation hot-wire vacuum gauge by successively measuring the pressure in each part of the system after it has been isolated from the other parts of the system. The leak may then be pinpointed with a Tesla coil. However, the coil should be of sufficiently high frequency, about 10^6 Hz, or the spark discharge may puncture the glass.

The Tesla coil can be used for testing at pressures down to about 10^{-2} torr. For smaller leaks a Pirani gauge or, better yet, an ionization gauge may be used in conjunction with butane or carbon dioxide as the search gas [194]. The Tesla coil cannot be used where the leak occurs in a mercury-diffusion pump because a glow is normally obtained with this device when the pump is leak-tight because of the presence of mercury. Further, the Tesla coil cannot be used on metal systems or glass-to-metal seals. For these applications, Pirani and ionization gauges and helium and halogen (Freon-type gases) leak detectors may be utilized. With the Pirani and ionization gauges and butane as the search gas, the minimum detectable air leaks are 5×10^{-6} and 10^{-7} liter-torr/sec, respectively [194]. (The use of flammable gases, such as butane, is not recommended). With the helium and halogen leak detectors, the sensitivity is even greater, especially with the former.

I. VACUUM FUSION–MASS SPECTROMETRY

The gaseous compounds evolved during fusion of a metal with graphite in a vacuum may be qualitatively examined and quantitatively determined by mass spectrometry. This has been performed, for example, in the analysis of copper for oxygen [196]; lanthanum [197], silver, and tin [189a] for nitrogen and oxygen; cast iron for nitrogen [198]; and steel [135], zirconium, uranium, and plutonium [199] for hydrogen, nitrogen, and oxygen. Aspinal [200] and

Aspinal and Hazelby [201] used vacuum fusion–mass spectrometry to determine less than 2 ppm oxygen in copper and to obtain a display of gas evolution rates; oxygen, nitrogen, and/or hydrogen were also determined in molybdenum, zirconium, tantalum, and steels (see Section I.L.1). Iron and steel were analyzed by this technique by Yavoiskii et al. [202] and Kammori et al. [203].

In all cases the evolved gases were analyzed directly without catalytic oxidation. If catalytic oxidation is utilized, difficulty is encountered in pumping water vapor [135]. In the event that a large amount of gas is evolved, it may be necessary to analyze an aliquot portion. Where the mass spectrometer shows that only one constituent is evolved during fusion, a simpler means may be substituted for final determination of this gas [196].

Pure samples of each component of the gas mixture evolved during fusion must be used to calibrate the mass spectrometer. From a total pressure measurement of the gas evolved by fusion and from the compositional analysis of the gas by mass spectrometry, the partial pressure of each gas may be calculated. (The fundamentals of mass spectrometry are beyond the scope of this book; they may be found elsewhere [204, 205]. Also, the techniques for interpreting mass spectrometric patterns are usually described in detail in the instruction manuals provided with the various commercially available mass spectrometers.) The percent of each constituent in the original sample can then be calculated using Eqs. 3–6 in Section I.E.

Kammori et al. [202] obtained the following limits of detection: 0.07 μg for oxygen, 0.1 μg for nitrogen, and 0.005 μg for hydrogen. Analysis time was 30 min. Conzemius and Svec [189a], using a Johnson-Nier double-focusing high-resolution mass spectrometer, obtained sensitivities of 0.45 μg and 0.08 μg, respectively, for oxygen (as CO) and nitrogen in silver. The results for these gases by conventional vacuum fusion were considerably higher than those obtained by mass spectrometry. With the latter technique, in addition to scanning other peaks, the 28 $\pm$ 0.10 mass region was cyclically scanned for CO, N_2, and $C_2H_4^+$ with each scan cycle requiring about 10 sec. A significant $C_2H_4^+$ ion current was noted. The authors concluded that the hydrocarbons present were not oxidized over CuO at 325°C in the conventional vacuum-fusion system. Methane (which presumably was also detected by mass spectrometry) passed through the cold trap and was measured as nitrogen by pressure measurement. Higher hydrocarbons condensed out in the liquid-nitrogen cold trap and were determined as oxygen (CO_2). Conzemius and Svec [189a] noted that in designing vacuum-fusion–mass-spectrometry systems the flow rate due to outgassing of the furnace must be within the capability of the mass-spectrometer–vacuum system to maintain normal operating pressures. Their vacuum-fusion system contained a mercury-diffusion pump in series with a dry-ice cold trap before the mass spectrometer, which itself was differentially pumped with an ion pump in the source chamber and a liquid-nitrogen cold trap followed by a two-stage mercury-diffusion pump and mechanical pump in the analyzer chamber.

Aspinal and Hazelby [201] compared the mass-spectrometric technique with other methods for analyzing gases evolved during vacuum fusion (see Table 4.2). Their analytical system is discussed in detail in Section I.L.1.

TABLE 4.2. Methods of Determining Gases Obtained by Vacuum Fusion (N_2, H_2, CO, CH_4)[a]

Method	Analysis	Remarks
Chemical	Measures H_2 by diffusion through Pd thimble	N_2 by difference
	CO by oxidation (Hopcalite) and absorption, or by oxidation and conductimetric finish	No allowance made for methane
Physical	H_2 by thermal conductivity	N_2 by difference
	CO by infrared	No allowance made for methane
Gas chromatographic	Can measure positively all gases likely to be liberated	Not able to follow evolution of constituent gases during evolution
		Unexpected peaks could be difficult to identify
Mass spectrographic	Can measure positively all gases likely to be liberated	Can follow the evolution of any constituent gas
		Can readily identify unexpected peak

[a] Reproduced by permission from Associated Electrical Industries, Ltd. [201].

The limits of detection for oxygen, nitrogen, and hydrogen with their system were 0.1, 0.1, 0.01 μg, respectively.

J. VACUUM FUSION–GAS CHROMATOGRAPHY

A method of equal or greater sensitivity than that described in Section I.I and requiring much simpler and less expensive instrumentation is that in which the gases evolved during vacuum fusion are analyzed by gas chromatography. The fundamentals of this method, beyond the scope of this book, have been adequately covered [206–208]. The gas-chromatographic method does not provide the positive identification of mass spectrometry. However, once the retention times for the various gases have been determined, no further identification is necessary as the possibility of some foreign gas occurring in the specific metal and being eluted simultaneously with one of the normally occurring gases is very remote.

The gas-chromatographic separation and analysis of mixtures containing hydrogen, nitrogen, carbon monoxide, and methane have been studied [209–213], and various workers [214–230b] have used these techniques to analyze the gases evolved during vacuum fusion. Analysis is usually carried out directly without catalytic oxidation. (This principle has been used in the Heraeus-Feichtinger instrument and one laboratory analyzer described in Section I.L.) The gas sample must be introduced to the analytical column as a slug of uniform composition. Some workers have accomplished this by compressing the gases from the vacuum apparatus with a piston directly into the stream of the carrier gas [215], by using a sampling bulb [216], or by injecting the gases into the carrier gas stream with a Toepler pump [220, 223]. By using a gas buret and Toepler pump in combination, the evolved gas may be trapped in a known volume and then aliquoted [223]. Hydrogen, carbon monoxide, and nitrogen were adsorbed at $-196°C$ in one loop of a gas-sampling valve [221, 227, 229]. After the valve was rotated, the loop was heated to desorb the gases into a carrier gas stream, and a chromatogram was obtained (see also Section I.L.2). Temperatures of 200°C [224, 229] and 300°C [221, 227] were used to remove carbon monoxide, hydrogen, and nitrogen from the chromatographic columns specified. Coe [230c] quantitatively transferred gas from a vacuum-fusion system to a gas chromatograph by means of a specially made glass stopcock used in conjunction with a Toepler pump. Different proportions of the total gas mixture could be selected and transferred to the chromatograph without the need for intermediate storage bulbs. The volume of the gas sample for analysis could thus be adjusted to the requirements of the detector.

In most chromatographic analyses of the permanent gases, Linde molecular sieve 5A [231] (manufactured by Linde Air Products Company, a division of Union Carbide and Carbon Corporation, New York) the calcium zeolite that will admit molecules up to about 5 Å in diameter, has been used as the stationary phase. (Porapak Q, manufactured by Dow Chemical Company, Inc., Midland, Michigan may also be used, but the separation time for nitrogen and carbon monoxide is somewhat longer, about 8 min [732].) Before a molecular sieve column can be used, it must be activated. This has been accomplished by heating the column to 350°C overnight while maintaining a flow of helium through the column. Shorter times may be used. Activation results in the removal of adsorbed carbon dioxide, occluded water, and water of constitution [233] and increases the adsorptive capacity of the column. According to Janak et al. [212], activation of molecular sieves should be effected between 300° and about 600°C. Heating above 600°C causes the molecular sieve lattice to break down. For molecular sieve columns that will no longer separate oxygen from nitrogen, Jeffery and Kipping [234] state that it is often more convenient to repack the column with new material than to regenerate the old material. However, this necessitates recalibration for the constituents being determined. Regeneration can be achieved, however, by heating the column at 350–400°C for 1 h with argon or helium flowing through. After activation the molecular sieve column should not be exposed to air more than is necessary [221]. Activation should not be performed in vacuo, as this will make the molecular sieve too active and will result in an increased retention time for carbon monoxide [235]. Water preadsorbed in a molecular sieve column will cause a decreased adsorptive capacity for nitrogen [236]; when the amount of water exceeds 7% by weight, the capacity is almost completely suppressed.

Other factors to be considered when preparing the chromatographic column are its length and inner diameter and the particle size of the stationary phase. The column dimensions will depend mainly on the activity and particle size of the stationary phase. Smaller particle sizes will often provide sharper separations and allow shorter columns to be used. Several workers have used approximately 20–60-mesh molecular sieve on columns 4.8 m long by 5 mm i.d. [209], 1.52 m long by 6.4 mm i.d. (213), 1.83 m long by 3.2 mm i.d. [221], and 1.00 m long by 10.5 mm i.d. [223]. Recently, however, 100–120-mesh molecular sieve has been made available. Activated charcoal, 30–60-mesh, has been used in the determination of hydrogen, nitrogen, methane, and oxygen (as carbon monoxide) in steel [237]. The column was 3 mm in diameter and 1.5 m long. Argon was used as the carrier gas.

High-purity helium or argon can be used as the carrier gas without further purification. Where detection is accomplished with a thermal conductivity cell, it is desirable to use a carrier gas having a thermal conductivity greatly different from that of the gas being determined. From Table 4.3 it can be seen

TABLE 4.3. Thermal Conductivity[a] ($\times 10^5$) of Various Gases at 100°C[b]

Argon	5.2
Carbon monoxide	7.2
Carbon dioxide	5.3
Helium	41.6
Hydrogen	53.4
Methane	10.9
Nitrogen	7.5
Oxygen	7.6

[a] All thermal-conductivity values are expressed in cal cm^{-3} sec^{-2} (°C $cm^{-1})^{-1}$.

[b] Reproduced by permission from John Wiley and Sons, Inc. [208].

that helium is the preferable carrier where oxygen (as carbon monoxide) and nitrogen are to be determined and argon where hydrogen is to be determined. However, argon is more readily available in some areas and less expensive than helium. Either carrier gas may be used with radioactive or argon ionization-type detectors. However, thermal-conductivity detectors are considerably less expensive, but also less sensitive, than ionization detectors; the former, though, may be obtained with sensitivities that are presently more than sufficient for gas analysis of metals. For example, $\frac{1}{3}$ full-scale deflection can be obtained for 0.1 ppm carbon monoxide, using a 1-g sample and a 1-mV recorder. Such sensitivity therefore permits the use of small samples and the aliquoting of the evolved gas as mentioned earlier. Actually the practical lower limit of the amount of oxygen and nitrogen that can be determined depends more so on the magnitude of the blank. In one study with a thermal-conductivity detector [221], no carbon monoxide blank was noted, and the nitrogen blank was always 0.5 μg or less. Over the range of 0.002–0.017% oxygen and 0.004–0.014% nitrogen, these gases can be determined with a standard deviation of 5% of the amount present. Measurement of peak area rather than peak height is preferred to obtain increased precision and sensitivity. It is recommended that the calibration curves be checked each time a new cylinder of carrier gas is used [213].

Lilburne [228, 229] has used a molecular sieve column and a micro-ionization detector of the Lovelock-type with helium as the carrier gas in the determination of hydrogen, nitrogen, and oxygen in various metals. This detector has approximately equal sensitivity for all permanent gases. The linear range of the ionization detector is three orders of magnitude, and the maximum concentration of permanent gases in the helium carrier gas that can be handled is 1 in 10^{-3}. Therefore, to use this detector over its full range requires that the carrier gas have a total impurity content of no more than 1 ppm by volume. This was accomplished with a helium-diffusion cell similar to that designed by McAfee [238]. It was obtained from Electron Technology, Inc., Kearny, New York, and consists of a large bundle of capillary silica tubes, housed in a steel tube, and capable of withstanding high pressures and temperatures. When the cell was operated at a pressure of 700 psig and a temperature of 350°C, a helium flow rate of 100 cm^3/min at 10 psig was obtained on the low-pressure side. Only small amounts of hydrogen and neon diffused with helium, but with the helium supplies commonly available, the concentration of these impurities never reached 1 ppm by volume. The blank contribution was less than 10% of the total volume of the gas usually collected, and the blank gas, before addition of bath material, consisted of 83% H_2, 4% CH_4, and 13% CO. Samples containing as little as 0.2 ppm H_2, 8 ppm N_2, and 10 ppm O_2 were analyzed.

Gas-chromatographic analysis of permanent gases can be performed in 2

[239] to 3 [221] min. Total analysis time for vacuum fusion–gas chromatography has been accomplished in 5 [227], 10 [229], and 12 [224] min, and this time is partly dependent on the material being analyzed.

In an effort to shorten the analysis time required in a previous vacuum-fusion–gas-chromatographic investigation [215], Feichtinger et al. [240] used the argon-surge technique, in which a small amount of argon, about 1 cc, is introduced into the furnace chamber. The tremendous expansion of the argon to 10–100 liters reportedly results in a powerful flushing effect that causes rapid and complete transfer of the gases to the chromatograph. Minimum analysis times for hydrogen, nitrogen, and oxygen (as carbon monoxide) were 12, 19, and 28 min, respectively. However, it was stated that these times could be reduced.

K. VACUUM FUSION–INFRARED SPECTROMETRY

The determination of oxygen in metals by infrared-absorption measurement of the carbon monoxide produced during vacuum fusion has been applied in the Balzers Exhalograph EA-1 [241, 242] and EAO-202 apparatus. This is described in detail in Sections I.L.5 and 7.

Briefly, the pressure and then the thermal conductivity of the gas mixture evolved during vacuum fusion are measured. The gas is then transferred to a nondispersive infrared analyzer, where air is admitted until the gas mixture is at atmospheric pressure. The amount of carbon monoxide present is indicated on a strip-chart recorder. A number of papers have been published describing the operation of the Balzers Exhalograph EA-1 and the results obtained; a few of these papers are cited [243–247a].

L. VACUUM-FUSION SYSTEMS

The following descriptions represent a sampling of systems that were commercially available or had been constructed in laboratories during the past 10–20 years. All of these systems contain unique contributions to the art, although they may not be used today for the purposes originally specified. This also applies to the inert gas fusion systems described in Sections II.C.2.a–g. Vacuum fusion for hydrogen only is described in Chapter 7.

1. GEC/AEI Metallurgy MS10 Vacuum-Fusion—Mass-Spectrometric Analyzer

Aspinal [200] and Aspinal and Hazelby [201] have designed and constructed a vacuum-fusion–mass-spectrometer system at the Central Research Laboratory of the Associated Electrical Industries Division of GEC/AEI

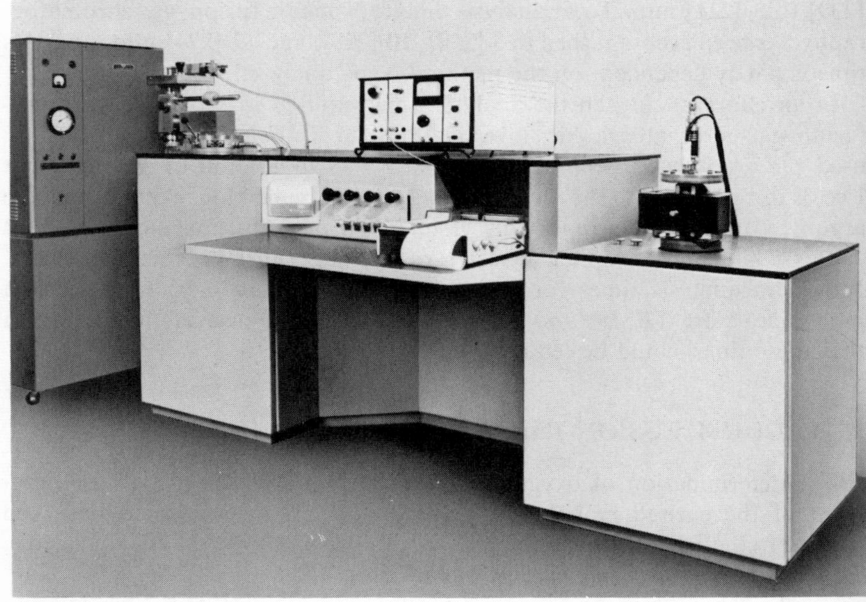

Fig. 4.16. GEC/AEI MS10 analyzer. (Reproduced by permission from Associated Electrical Industries, Ltd. [201].)

in Rugby, England. This system, the GEC/AEI Metallurgy MS10, was particularly designed for determining oxygen contents below 10 ppm in metals and utilizes an AEI MS10, 180°-deflection mass spectrometer.

A general view of the apparatus is shown in Fig. 4.16. The apparatus consis basically of three parts:

1. The furnace section containing a vacuum lock for the insertion of samples into the vacuum, a sample storage facility, and the crucible itself. Figure 4.17 shows this part of the equipment. Gases released from the crucible are pumped away by an 80-liters/sec oil-diffusion pump to reduce the possibility of gettering by evaporated films. The pumping speed at the mouth of the crucible is approximately 50 liters/sec.
2. A reservoir volume in which the gases from the furnace are collected. This volume can be varied between 1 and $2\frac{1}{2}$ liters. It is previously evacuated.
3. The continuously pumped mass spectrometer, which is joined to the reservoir section by one of two capillary leaks. The conductance of the leaks is chosen so that only a negligible quantity of gas is removed from the reservoir during the time required to take readings of the ion current of the mass spectrometer.

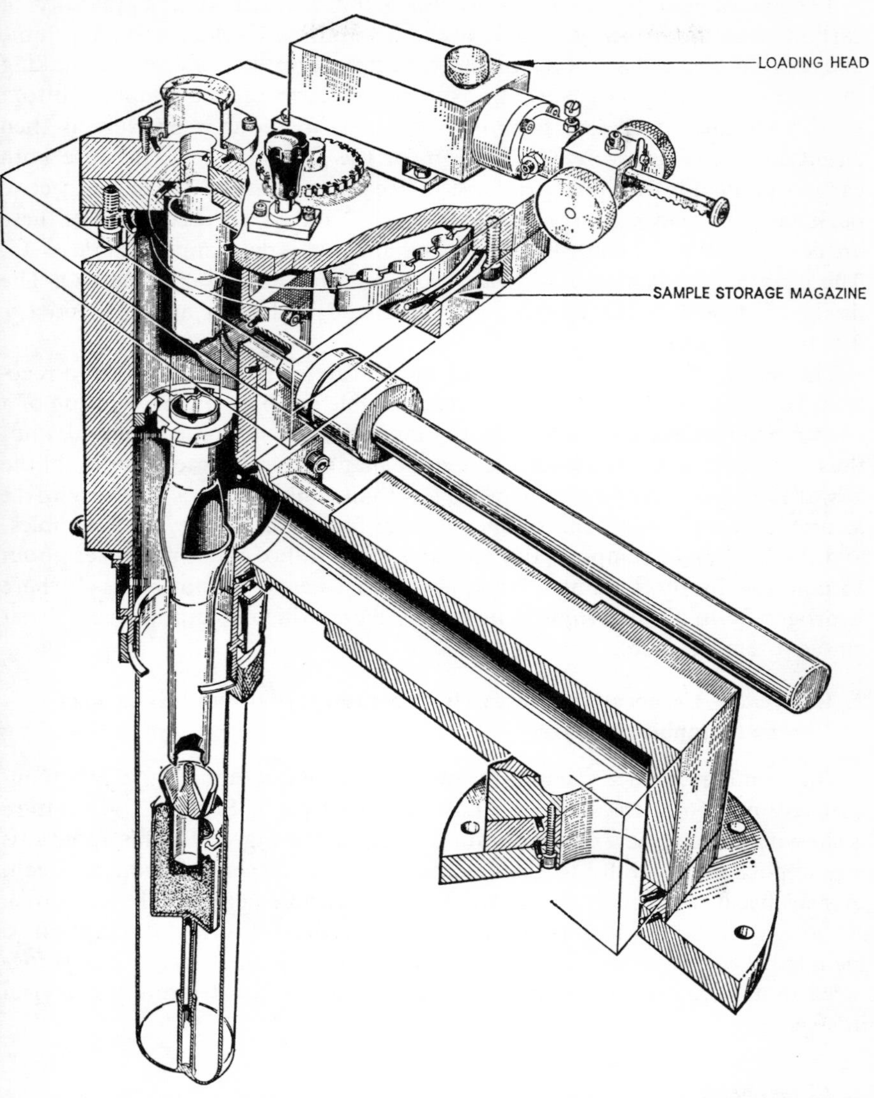

Fig. 4.17. GEC/AEI MS10 furnace assembly. (Reproduced by permission from Associated Electrical Industries, Ltd. [201].)

The instrument is calibrated by admitting a small known quantity of carbon monoxide, nitrogen, hydrogen, or methane to the reservoir volume and noting the mass spectrometer readings. This is achieved by taking the known quantity of gas at atmospheric pressure and expanding it into a known volume. A second portion of this gas at reduced pressure is then admitted to the reservoir. Calibration for the gases is linear. Because both carbon monoxide and nitrogen have molecular weights of 28, the respective peak heights at mass–charge (m/e) ratios of 12 and 14 are used. Both of these are corrected for any contribution of methane that is determined at m/e = 15. The peak height at m/e = 14 is also corrected for doubly charged CO. The limits of detection for oxygen, nitrogen, and hydrogen are, respectively 0.1, 0.1, and 0.01 μg.

The speed of an analysis is limited by the rate of the vacuum-fusion reaction. However, with the AEI MS10 it is possible to observe the evolution of a chosen product gas and, therefore, to limit the collection time to a minimum, thus speeding the analysis and keeping the blanks low. After extraction of the gas, the analysis may be completed in less than 3 min. It is possible to analyze a sample every 5 min. The sample storage facility holds up to 18 samples, and the loading of samples (and also any bath or flux materials) takes about 15 min. Obviously, loading of samples under vacuum cannot be used where hydrogen is to be determined in metals from which it diffuses readily at ambient temperatures.

2. U.S. Steel Corporation, Research Laboratory, Vacuum-Fusion–Gas-Chromatographic Analyzer

The semiautomatic vacuum-fusion–gas-chromatographic analyzer constructed in 1966 at the Research Laboratory of the U.S. Steel Corporation is shown in Fig. 4.18; a flow diagram is presented in Fig. 4.19. This apparatus was applicable generally to the determination of oxygen, but not nitrogen, in steel because it incorporated the multiple-sample-per-crucible technique. Although the apparatus is no longer used, the following description is included, because at the time it contained some unique designs. It is recognized that there are less-expensive systems available for determining oxygen in steel.

a. Components

(1) FURNACE (see Fig. 4.4).

(2) INDUCTION HEATER. Induction, rather than resistance, heating was selected because of the more-rapid heating rates and higher temperatures obtainable with the former method. Further, with an electrical frequency of only 3 kHz, it was believed that the increased stirring of the bath would effect

Fig. 4.18. Vacuum-fusion–gas-chromatographic analyzer.

evolution of nitrogen from the bath. A motor-generator set (Tocco Division, Ohio Crankshaft Company, Cleveland, Ohio) was used as a power supply. With a 15-turn coil fabricated from $\frac{1}{4}$-in. square copper tubing, temperatures in excess of 2500°C were easily attained in the graphite crucible.

(3) FURNACE PUMP. An Edwards High Vacuum, Inc. (Grand Island, New York), 2M4A mercury-diffusion pump was used to effect rapid removal of the gas from the furnace. This pump, constructed of stainless steel, has four stages and is especially designed to operate with a high critical backing pressure (about 35 torr). The unbaffled speed of this pump is 80 liters/sec with an ultimate vacuum of 10^{-3} torr. However, to keep the mercury in the pump and to achieve a lower ultimate vacuum of 10^{-7} torr, a refrigerated baffle was used. This baffle, which has a conductance of 35 liters/sec, uses thermoelectric cooling to maintain a cooled surface between the diffusion pump and the vacuum system. Four pairs (positive and negative in each pair) of semiconducting bismuth telluride elements were arranged in electrical series in a $\frac{1}{2}$-inch cube, and a current of 10–15 A at about 1 V was passed through the elements. In this way, a temperature difference of 25–30°C was

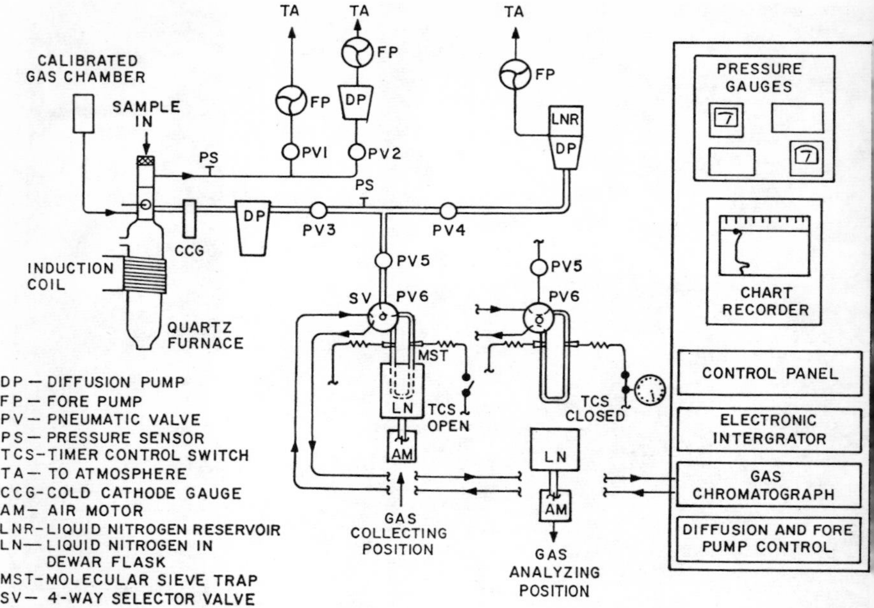

Fig. 4.19. Vacuum-fusion–gas-chromatographic Analyzer Flow Diagram.

maintained between two opposite faces of the cube. The lower temperature face was placed in good thermal contact with a dish-type baffle and thermally insulated from the main body; the opposite face was placed in contact with a finned copper block. Facilities were provided on the block for water cooling or air cooling, the final temperature of the baffle depending on the temperature of the block. For example, if the cooling water temperature was 15°C, the baffle temperature would be approximately −10°C. If the block were air cooled, the baffle temperature would probably be several degrees higher than this, depending on the ambient temperature. Water cooling was used here.

To isolate the furnace from the pumping system, a quarter-swing butterfly valve was used. This compact, high-conductance valve is capable of resisting a pressure differential of 760 torr in either direction. When graphite crucibles were changed in the furnace, this valve was used to maintain ultimate vacuum in most of the system, while atmospheric pressure prevailed in the furnace section.

The pump, baffle, and butterfly valve formed one compact unit and were fabricated from stainless steel. With this pumping system a vacuum of 5×10^{-5} torr could be maintained in the furnace at the operating temperature of 1850°C.

(4) SAMPLE-INLET SYSTEM. A system for rapidly introducing both magnetic and nonmagnetic samples to the furnace was constructed with an inlet system that included a 29/42 standard-taper glass cap connected to a stainless steel, Jamesbury Corporation type 1213 ball valve. The glass cap was removed to insert the sample. The vacuum lock between the glass cap and ball valve was evacuated through a Veeco Instruments, Inc., Plainview, N.Y. type R38S bellows valve. The vacuum lock was evacuable to 10^{-5} torr with a fore-pump–silicone-oil-diffusion-pump combination. The latter pump has a pumping speed of 2 liters/sec at 10^{-3} torr and less, and an ultimate vacuum of 5×10^{-7} torr. To achieve rapid evacuation, 1-in. lines were used throughout this system even though the volume to be pumped was less than 100 cc. An advantage of this inlet system was that the operating vacuum of 10^{-5} torr was attained in the inlet vacuum lock. Thus, no extraneous gas was carried into the analytical part of the system, and little time was spent in transferring the sample to the graphite crucible.

(5) GAS-COLLECTION SYSTEM. The gas-collection system consisted of two parts, a volume between valves (between PV5 and PV6 in Fig. 4.19) and a molecular sieve trap column made of stainless steel that was directly heated electrically. All the gas extracted from a sample was initially adsorbed in the molecular sieve trap at liquid-nitrogen temperature. The Dewar flask containing the liquid nitrogen was raised and lowered by a jack operated by an air-motor. (The use of the air-motor-operated jack and several other ideas for automating the apparatus were obtained from T. D. McKinley, Pigments Division, E. I. duPont de Nemours, Inc., Wilmington, Delaware, who had previously applied these in different form to a vacuum-fusion unit where the component gases were determined by pressure measurement.) The trap column was electrically separated from the copper tube inlet by Hoke, Inc., Cresshill, New Jersey, nylon connectors. Since the gas could not be adsorbed on the sieve as rapidly as it was extracted from a sample, a volume in which the gas could accumulate was necessary. This volume could be held to a minimum to simplify construction, because the furnace diffusion pump had a high (35 torr) critical backing pressure.

To ensure reliability, speed of operation, and ease of maintenance, in-line, high-vacuum pneumatic valves of stainless-steel construction (Thermionics Laboratory, Inc., Hayward, California) were used. To provide flexibility in the system, all valves were interchangeable and were incorporated into the systems with QF (Quick-Flange) connectors. These connectors, manufactured by Leybold in West Germany and obtained from Leybold-Heraeus, Inc., Monroeville, Pennsylvania, were modified as noted below. The QF connectors are mating flanges sealed with a neoprene O-ring and centering ring and joined to provide a vacuum-tight seal with a clamp. The tubing of the entire vacuum system was of stainless steel. Successive portions were joined with QF connectors to provide easy access for cleaning, replacing, or modifying any part of the system.

(6) GAS-ANALYSIS SYSTEM. The gas-analysis system was made up of those components through which the helium carrier gas flowed. The main component was the gas chromatograph. The link between the gas-collection and gas-handling systems was a four-way selector valve (SV in Fig. 4.19), no. Pl-418, manufactured by Circle Seal Products Company, Inc., Pasadena, California. During a fusion, the extracted gas went directly to the molecular sieve trap column, while the other two parts of the valve were connected to the analytical system. After the gas was adsorbed, the valve was switched from vacuum to a helium stream at 30 psi, and the trap column was heated electrically to about 300°C to effect desorption of the gas and its transferral to the chromatograph.

A certain amount of helium remained in the sample collection loop after transferral of the gas to the chromatographic column. This helium had to be evacuated prior to adsorption of the gas from the next sample. Therefore, after each analysis, the sample loop was rotated back to the vacuum system and the remaining helium was evacuated to the atmosphere.

A Beckman Instruments, Inc. (Fullerton, California), Model GC-1 Gas Chromatograph was used for the gas analysis. This instrument contained a Linde 5A molecular sieve (−30 mesh) chromatographic column $\frac{5}{16}$-in. i.d. and $3\frac{1}{2}$-ft long for separating the different gases. A thermal-conductivity cell and recorder were used to monitor the amount of each gas. The chromatographic column was maintained at a temperature of 50–60°C. Detector current was 300 mA. To maintain a constant sample weight (about 1 g) while simultaneously accommodating the wide range of oxygen concentrations encountered in steels and titanium, a Leeds and Northrup Company (Philadelphia) Speedomax G recorder with a wide signal acceptance range (variable from 1 mV full scale to 50 mV full scale, with a 1-sec, full-scale response time) was selected. Since it was not always possible to preselect the magnitude of the signal, an automatic attenuator was added. This attenuator or voltage divider kept the recorder pen on the chart scale at all times. Thus, no information was lost, and the signal was recorded at the optimum sensitivity.

To further automate the system, a printing integrator was attached to the recorder. This integrator automatically printed cumulative integrals of the recorder pen position with respect to time, to provide peak areas for quantitative interpretation. Incorporated in the integrator was a valley sensor, so that an area count was provided for every signal received from the gas chromatograph. Thus, although it did not provide a direct concentration readout system, the integrator provided a number that could be readily converted to concentrations. With this system of interpretation of the gas chromatograms, linear calibration of peak area versus oxygen concentration was obtained.

(7) GAUGES. The vacuum gauges in the system served two purposes: first, to follow the various steps in the analysis as they occurred, and, second, to locate leaks in the system. Two different types of Heraeus gauges were used. One was a three-position thermocouple gauge; pressures in the range of $1–10^{-3}$ torr were indicated on this gauge. Gauge sensing tubes for this type were mounted by means of QF connectors in the pumping system, in the gas-collection system, and in the sample-inlet system (vacuum lock). The other type of vacuum gauge was a cold cathode ionization gauge, which operated in the range of $10^{-2}–10^{-6}$ torr. This gauge was mounted in the furnace section.

(8) SYSTEM VACUUM PUMP. Analytical requirements dictated an operating pressure near 10^{-5} torr. Therefore, the system pump had to be able to attain this vacuum after each analysis. The diffusion pump selected as the system pump was an Edwards High Vacuum, type 2M3B, three-stage, water-cooled, mercury-diffusion pump constructed of stainless steel. The prime factors in selecting the pump were the pumping speed of 150 liters/sec at 10^{-3} torr and less and the ultimate vacuum of about 10^{-7} torr attainable by using a matching liquid-nitrogen cold trap. With the cold trap, the critical backing pressure of this pump was 0.5 torr. The fore pump was a standard two-stage pump with a displacement of 58 liters/min. With this pumping system less than 1 min was required to reduce the pressure in the vacuum system to 10^{-5} torr between samples. Also, this system provided adequate pumping capacity for rapid initial pump-down during bake-out of a fresh crucible.

b. Assembling the Unit

This vacuum-fusion apparatus was designed to be flexible and to provide for easy replacement of components. These features also provided for rapid initial construction. It was originally planned to use only commercially available QF connectors. However, because many of the necessary components were available only in cast aluminum having a rough interior and because the size of these cast-aluminum pieces was not compatible with the balance of the vacuum system, the connections in the system were made by butt-welding QF connectors to 1-in. o.d. stainless-steel tubing. The centering rings, the O-rings, and the clamps were standard items. In the places where it was necessary to use glass (for the sight port on the sample-entry system and for the furnace body and cap), standard-taper joints (metal and glass) were used. The metal standard-taper joints were welded to the metal portion of the system. The glass furnace was sealed to the system with the mating joints by using Apiezon W (melting point of 100°C) vacuum wax. The standard taper joints of the sight port used for the sample-entry system were sealed with Apiezon L (melting point of 40°C) vacuum grease.

Each vacuum component was leak-checked with helium before installation. In order to provide some degree of versatility in adding new components or modifying the system, several extra QF connections were incorporated both in the furnace section and in the gas-collection section. These connections were sealed by using blind flanges. The system was quite compact. For safety purposes, a single, clear-plastic guard was placed in front of part of the apparatus to provide protection against implosion of the glass furnace when under vacuum and to shield the induction coil and also the gas-collection loop, which, as noted earlier, was heated directly by electricity.

Except for dropping the metal sample into the crucible, the complete analytical cycle was semiautomatic, the operation being controlled by the push-button system shown in the middle of the cabinet in Fig. 4.18. This cabinet also contained all the vacuum gauges, the strip-chart recorder, the automatic attenuator, and the peak-area integrator.

c. *Calibration*

In many systems for gas analysis of metals a calibrated volume and a pressure-measuring device are used. However, with the chromatograph, it was possible to eliminate both a McLeod gauge and a calibrated volume in this instrument. To permit calibration of the instrument with pure gases, the Balzers device, described in Section I.L.5, for admitting known quantities of pure gases was incorporated.

Although the chromatographic column always quantitatively separated carbon monoxide, nitrogen, and hydrogen, the rate at which the gases were desorbed changed with age of the column. The calibration curves obtained by using pure gases were confirmed by analysis of National Bureau of Standards metal standards. The chromatographic column was considered usable as long as resolution between nitrogen and carbon monoxide was obtainable. Reactivation of this column was accomplished by heating it to 350°C overnight, while maintaining a flow of helium through it.

Total analysis time was about 5 min for plain carbon and low-alloy steels and about 8 min for high-alloy steels.

3. NRC Model 912-D Vacuum-Fusion Gas-Analyzer

The NRC Model 912-D Vacuum-Fusion Gas-Analyzer has been widely used. It was formerly manufactured by National Research Corporation (Newton Highlands, Massachusetts), which is now part of Varian Associates. The apparatus is shown in Fig. 4.20. A Model 912-S was also manufactured, this instrument having the features of the 912-D but also a 3500-ml

Fig. 4.20. NRC 912D vacuum-fusion–gas-analyzer. (Reproduced by permission from NRC Equipment Corp.)

gas-storage volume and an extra circulating pump so that gas from one sample could be stored while the previous gas was being analyzed. In these analyzers a graphite crucible insulated with graphite powder is contained in a Guldner-Beach, Pyrex glass furnace (see Section I.A.3.a). Heating is effected by induction and requires a 220-V, single-phase, 60-Hz, ac power source with a current of 25 A at full load and a 2-gal/min water flow. By this means, temperatures of at least 2400°C may be obtained.

The system is evacuated by means of a three-stage oil-diffusion pump of 70 liters/sec pumping speed at pressures of 0.5 μ and less. Samples are introduced through a modified Jamesbury Corporation ball valve, and gases are removed from the furnace by a 30-liters/sec (at 0.5 μ and less), two-stage mercury-diffusion pump. A three-stage mercury-diffusion pump having a speed of about 15 liters/sec at a few microns pressure and lower circulates the gases to a manually operated Toepler pump, which packs the total gas into a McLeod gauge having three linear ranges. After a pressure measurement the gases are circulated through a heated copper oxide–ceric oxide catalyst. Water and carbon dioxide are removed in a magnesium perchlorate trap and a liquid-nitrogen trap, respectively. After measurement of the nitrogen pressure, the carbon dioxide is warmed to room temperature for a pressure measurement.

Incorporated in the vacuum-fusion system are water-flow and power-failure safety devices. The former is comprised of a water-flow switch and a relay to shut off power to the diffusion pumps if water flow ceases. The latter consists of a pair of solenoid-operated vacuum valves to isolate both mechanical pumps in case of power failure, thus maintaining the vacuum and preventing mechanical-pump oil backsurge. Also, provision is made in the pumping system for inclusion of thermocouple and/or ionization gauges.

Analysis time with the Model 912-D is 40–50 min/sample depending on the type of material. This time can be shortened by 40% by use of the Model 912-S analyzer. An advantage of both of these analyzers is that the NRC hot-extraction furnace used for determination of hydrogen (see Chapter 7) can be interchanged with the vacuum-fusion furnace.

4. Laboratory Equipment Corporation (LECO) Vacuum-Fusion Analyzer

The LECO Model 578-000 Vacuum-Fusion Analyzer, manufactured by the Laboratory Equipment Corporation, St. Joseph, Michigan (see Fig. 4.21), was designed originally by Covington and Bennett [54] for use mainly in the analysis of titanium (see Chapter 17). With minor modifications it can be used for steel analysis. This apparatus consists of a resistance-heated furnace connected to a double analytical section. Gases are determined by pressure measurement before and after oxidation coupled with selective absorption.

Power requirements for this apparatus include a 230-V, 60-Hz ac, three-phase source and a 115-V, 60-Hz ac, single-phase, 15-A line. A 5-gal/min water supply is also necessary. The 230-V power source serves as the input to a water-cooled, three-phase, 6-kVA ac transformer. The output from this transformer can be varied from 2 to 14 V at a maximum of 360 A by means of a saturable reactor that regulates the three-phase input. The furnace heating element consists of a graphite cylinder composed of three equal segments connected by a $\frac{1}{2}$-in. ring at the top. Each segment is attached to the leads from the output of the transformer. The graphite crucible is suspended inside the heating element and is further surrounded by two graphite heat shields. The whole is contained in a gas-tight, water-cooled metal housing. Temperatures up to 2300°C may be obtained with the furnace. The outgassing time is only about 1 h because there is no graphite packing around the crucible. The furnace is evacuated by means of a two-stage mercury-diffusion pump (of 8-liters/sec pumping speed) with a cold finger that serves as one boundary of the collection volume. Evacuation of the system is accomplished with a metal, three-stage oil-diffusion pump of 60-liters/sec pumping speed. The sample is introduced into a port fabricated from a metal vacuum

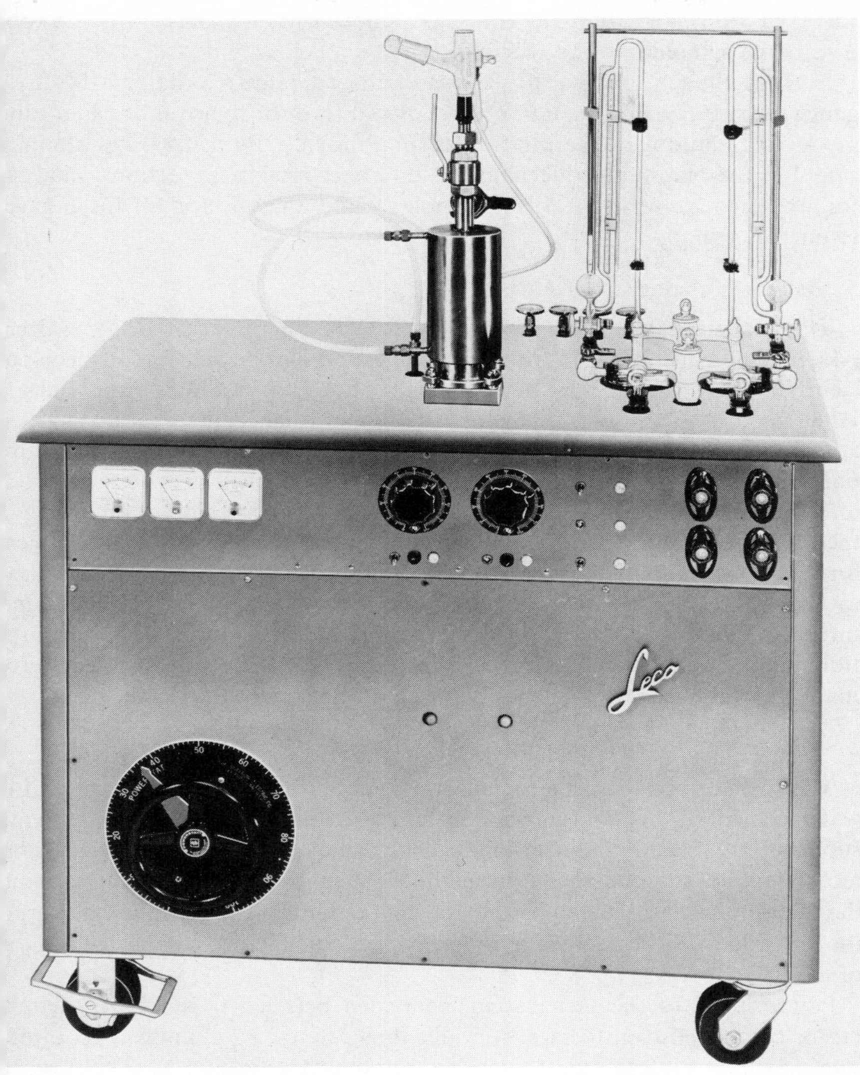

Fig. 4.21. LECO 578-000 vacuum-fusion analyzer. (Reproduced by permission from Laboratory Equipment Corp.)

valve, and any air admitted with it is removed immediately. A guide tube directs the sample into the crucible.

Each of the two analytical sections of the apparatus contains a McLeod gauge, a copper oxide oxidation tube, a cold trap for removal of water, and an Ascarite bulb for absorption of carbon dioxide. By using 0.1-g titanium samples, the platinum-flux technique, and both analytical sections, analysis for oxygen requires only 5 min/sample, while analysis for all three gases requires 8 min.

5. Balzers Exhalograph EA-1

The Exhalograph EA-1 [241–244, 247a] is manufactured by Balzers Aktiengesellschaft für Hochvakuumtechnik und Dünne Schichten, Liechtenstein. This instrument may be obtained in the United States from Balzers High Vacuum Corporation, Santa Ana, California. This apparatus, which is essentially constructed of metal, is shown in Fig. 4.22. The principal components are a furnace with a sample lock, a calibration system, an oil diffusion pump, a gas-collecting pump, a measuring block containing a thermal conductivity cell, a nondispersive infrared analyzer, a mechanical fore pump, a central unit, and a compensating recorder. In an analysis the total gas pressure is measured, and the hydrogen, oxygen (as CO), and nitrogen contents are determined by thermal conductivity, infrared absorption, and mathematical difference, respectively. Values are plotted automatically on the strip chart within 10 sec of the actual determinations.

a. Furnace

The furnace body is constructed of a double wall of stainless steel and is water-cooled. Two concentric resistance-heated graphite tubes provide support and heat for the graphite crucible. A graphite heat shield and molybdenum radiation shield are used. Sight ports are included for viewing the melt in the crucible and for measuring the temperature on the bottom of the crucible with the built-in pyrometer. The crucible has a capacity of 60 g of iron.

Power input to the furnace can be varied between 0 and 9 kW with a motor-controlled transformer. A water flow rate of 3.2 gal/min is required. The maximum (intermittent) temperature obtainable is 2200°C. The furnace may be operated for prolonged periods at 2000°C. Resistance heating, in contrast to induction heating, permits an all-metal furnace to be used. Thus a shorter outgassing time is achieved than with a glass furnace [26].

b. Sample Lock

An all-metal sample-lock system consisting of two valves keeps the furnace under high vacuum when a sample is admitted.

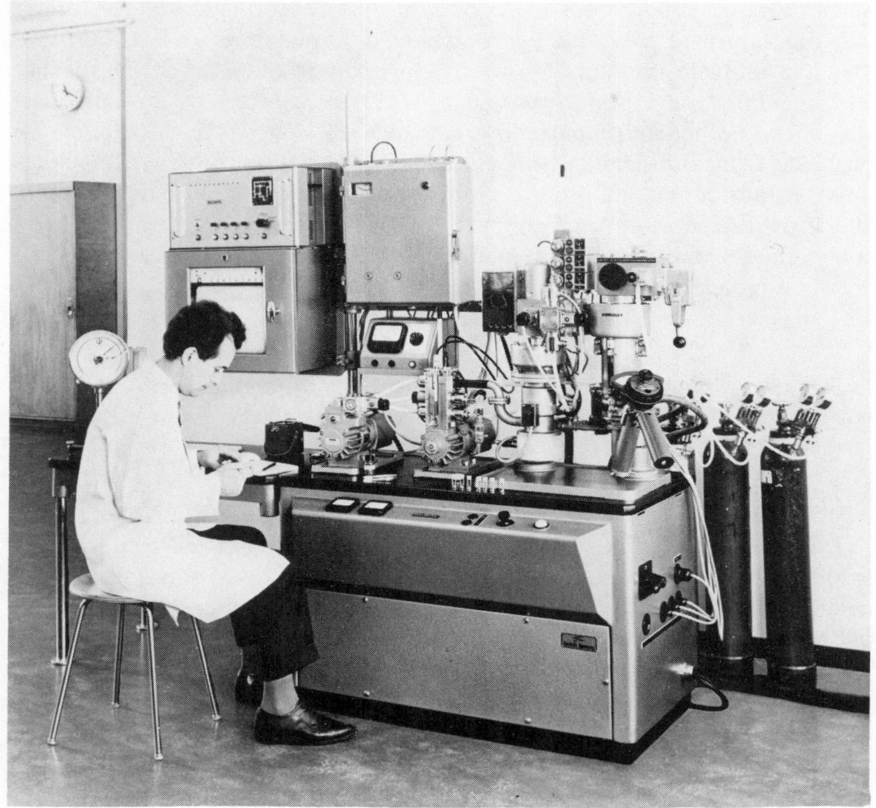

Fig. 4.22. Balzers Exhalograph EA-1. (Reproduced by permission from Balzers Aktiengesellschaft für Hochvakuumtechnik und Dünne Schichten.)

c. Calibration System

This system consists of three similar units, each comprised of a pair of solenoid valves with a cavity of 10 mm^3 between them. These valves are controlled by a variable-impulse transmitter programmed to open alternately the valves on opposite sides of each of the three cavities; each pair of valves is controlled individually. A counter is provided to record the number of times a valve is open to the vacuum. Calibration gases from cylinders are passed into plastic surge bags so that the gases can be stored at atmospheric pressure prior to their entrance into the system.

d. Pumping System

In addition to the conventional fore pump, there are a high-speed (170 liters/sec from 10^{-3}–10^{-5} torr) oil-diffusion pump to evacuate the furnace and a collector pump to pack these gases into the measuring block. The former contains saturated paraffinic hydrocarbons of vapor pressure less than 10^{-8} torr. An ultimate vacuum of 1×10^{-6} torr can be obtained in the furnace within 2 min. The collector pump is a two-stage, oil-free, rotary-valve type with a speed of 0.3 liter/sec from 760 torr to 200 μ. Therefore, contamination problems from pumping fluid are eliminated. The collector pump can quantitatively transfer and compress gas in the measuring block up to a pressure of 100 torr. Pressures in various parts of the system are determined by a cold-cathode ionization gauge and three thermocouple gauges using one common indicator.

e. Measuring Block

This unit is composed of a U-type pressure gauge and a thermal conductivity cell. The former has a sensing coil to determine the height of the mercury in the closed end; it is accurate to better than 10^{-1} torr.

f. Infrared Analyzer

This instrument (LIRAS I, manufactured by Hartmann and Braun A. G., Frankfurt, West Germany) is applicable over the 2–15-μm range. It is connected to the measuring block. To avoid the effect of intensity fluctuations of the radiation source, both a measuring beam and a reference beam are used. The beams are emitted by two electrically heated coils and are collimated by two parabolic mirrors. The sealed reference gas is 10% carbon monoxide in air. The detector is selective and is sealed on one end. During analysis, it is evacuated, and then all of the gas extracted from the metal is transferred to it. To prevent errors due to pressure differences, air is bled into the detector, so that all measurements are made at atmospheric pressure.

g. Control Unit

All of the steps of the automatic analysis are controlled by this unit. In addition, the sensitivity of each analyzer and the zero point of the total pressure can be adjusted here.

h. Operation

Oxygen in steel has been determined in this apparatus at the authors' laboratory for several years with little down-time. To determine nitrogen in steel with this apparatus, a spinning crucible has been devised [98–100]. The crucible, mounted on a water-cooled vertical shaft, is rotated at high speed after each analysis by an electric motor actuated by a signal from the recorder. The centrifugal force ejects the melt from the crucible onto a water-cooled graphite ring, where it solidifies. Because this technique requires the use of a nickel–cerium alloy and because cerium is a strong oxide former, nitrogen and oxygen cannot be determined simultaneously. Further, because high hydrogen blanks are usually obtained in systems containing substantial amounts of graphite, it is recommended that the Exhalograph EA-1 not be used for determining hydrogen in steel. Samples can be weighed on the torsion balance with electrical scanning, and the sample weight is then stored as an analogue value, which is later recorded automatically on the strip chart. Extraction of gas from the sample is carried out until the blank rate is again obtained. Total pressure and results for carbon monoxide, hydrogen, and nitrogen appear as peaks on the strip-chart recorder. The mean relative error of an individual measurement for each of the three respective gases is 0.5, 0.9, and 1.0%. For a 1-g steel sample, analysis time for oxygen is about 5 min, which includes time for two calibration checks. The analysis time for nitrogen, including calibration checks, is only slightly higher.

6. Heraeus-Feichtinger Vacuum Analyzer VH-8

This instrument (Fig. 4.23) is manufactured by W. C. Heraeus GMBH, Hanau, West Germany, and is based in part on the design of Feichtinger

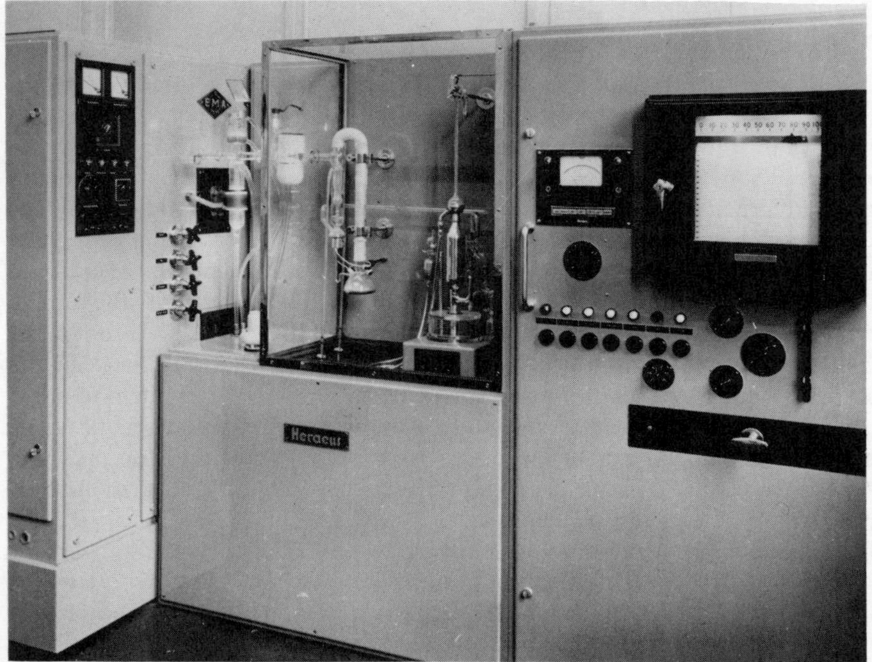

Fig. 4.23. Heraeus-Feichtinger vacuum analyzer VH-8. (Reproduced by permission from W. C. Heraeus GMBH.)

et al. [215, 240]. It may be obtained in the United States from Leybold-Heraeus, Inc., Monroeville, Pennsylvania. Analysis for hydrogen, oxygen, and nitrogen is automatic.

The vacuum system contains both resistance (for warm extraction) and induction (for vacuum fusion) furnaces, two fore pumps, three diffusion pumps, and a Feichtinger pump. The furnaces are so attached as to make possible the use of one or the other by operating certain valves. The temperature of the furnaces may be automatically programmed, continuously or stepwise. The change in pressure during degassing of a sample can be monitored and recorded.

The Feichtinger pump is similar to a Toepler pump in that on each stroke of the pump, a portion of the gas sample is collected; it is then measured in a gas buret. A liquid-nitrogen cold trap is used to remove any water and carbon dioxide from the gas before analysis. The gas sample is analyzed with a chromatograph that has thermistor detectors and control of both column and detector temperature.

The limit of detectability for both oxygen and nitrogen, with helium as the carrier gas, is approximately 0.5 μg; for hydrogen with argon as the carrier gas, the limit of detectability is 0.1 μg.

7. Balzers Exhalograph EAO 202

The Balzers Exhalograph EAO 202 (Fig. 4.24), manufactured by Balzers Aktiengesellschaft für Hockvakuumtechnik und Dünne Schichten, Leichtenstein, is an automatic vacuum-fusion analyzer capable of producing an oxygen result in about 90 sec [247b]. In analysis, the sample (about 1 g) is placed in a graphite crucible, which is inserted in the furnace chamber and sealed. The charging lid is closed, and this starts the automatic analysis cycle. For 39 sec the system is pumped down by a two-stage rotary-valve pump. Then, with the crucible acting as a resistance heater, an electrical pulse melts the sample during a 25-sec period. Oxygen in the sample diffuses as CO through the crucible wall, and the latter gas is pumped into a chamber between the second stage of the collector pump and a valve. The valve then opens to allow the CO to enter an evacuated mixing chamber, whereupon a second valve opens to allow air to enter and mix with the CO at atmospheric pressure. After a third valve opens, the gas mixture expands into the evacuated measuring cuvette of an infrared analyzer. The quantity of CO is indicated on a linear scale calibrated from 0 to 1 mg oxygen.

The EAO 202 contains a calibration system by which measured quantities of CO can be admitted to the analyzer by alternately opening and closing two valves. The required opening and closing frequency can be set digitally on the instrument. The standard deviations for admitting equal quantities of calibration gas are between 5×10^{-7} g and 1×10^{-5} g oxygen at the low and high points of the linear scale, respectively. Maximum sample dimensions are a 7-mm diameter and a 9-mm length. Steel, beryllium, cobalt, lead, nickel, niobium, tantalum, zirconium, noble metals, and other materials may be analyzed with the EAO 202.

Heater power can be set at five different positions with a maximum temperature of about 2200°C; the maximum power is 4.5 kVA. The instrument may be wired for a variety of three-phase voltages at 50 or 60 Hz. Approximately 5 liters/min cooling water at 1.5×10^3 to 7.6×10^3 torr is required.

8. Balzers EAN 202

The Balzers EAN 202 is also manufactured by Balzers AG, Liechtenstein [247c]. In appearance and basic construction, it is almost identical with the EAO 202 (Fig. 4.24). The EAN 202 is an automatic analyzer capable of

Fig. 4.24. Balzers EAO 202. (Reproduced by permission from Balzers Aktiengesellschaft für Hochvakuumtechnik und Dünne Schichten.)

yielding a result for nitrogen in less than 3 min. Both melting of the sample and calibration are effected as in the oxygen-analyzer. During analysis, the gas mixture, containing nitrogen, carbon monoxide, and hydrogen, is transported by a pump to the water-cooled reagent, Reaktit 300, where most of the carbon monoxide is adsorbed. The hydrogen and any remaining carbon monoxide are then oxidized to water and carbon dioxide, respectively, over a heated oxidizer, and then also adsorbed. The amount of nitrogen is then determined by a thermal-conductivity measurement, and the result is displayed on a linear scale that covers the range 0–100 μg. Each scale division corresponds to 1 μg of nitrogen. The average error of a single measurement at the lower detection limit is 0.1 μg. The average relative error of a single measurement at the 100-μg level is 0.5%. The steel sample weight is about 0.5 g. Maximum sample dimensions and electrical-power and cooling-water requirements are the same as those for the Balzers EAO 202 analyzer.

The oxidation and adsorption reagents are regenerated by heating and outgassing during pump-down of the system. The adsorption reagent is then automatically moved to a water-cooled zone, where, during extraction, the gases are adsorbed.

II. INERT-GAS FUSION

In recent years there has been a plethora of articles on the inert-gas-fusion method for determining oxygen in metals. Some papers have also discussed the determination of nitrogen alone or both oxygen and nitrogen simultaneously. Hanin and Jaudon [248] and Dallman and Fassel [249] have reviewed inert-gas-fusion analysis, and the latter authors also conducted an extensive investigation. In addition, many articles on inert-gas-fusion analysis have been cited in some of the annual reviews sections of *Analytical Chemistry* and in review articles mentioned in Section I of this chapter. Although early workers in the inert-gas-fusion determination of oxygen in metals oxidized carbon monoxide to carbon dioxide and determined the latter gravimetrically, conductometrically, or coulometrically, such investigations will not be described here except for certain specific details. The methods that will be described are those utilizing mainly thermal-conductivity measurement of constituents. Inert gas fusion for hydrogen is described in Chapter 7.

Certain facets of inert-gas-fusion methods are common to vacuum fusion and have already been described in Section I of this chapter. Arc fusion is described in Chapter 5; it is often coupled with measurement techniques applicable to the inert-gas-fusion method.

Inert-gas fusion possesses several advantages over vacuum fusion. Analysis by inert-gas fusion is faster, requiring no pump-down and consequently no elaborate vacuum pumps. (A good fore pump is required, however, where carbon dioxide is determined by the capillary-trap [250, 251] or vapor-pressure-measurement technique.) Inert-gas fusion utilizes simpler apparatus in general and is, therefore, less expensive than vacuum fusion. Also, because the inert carrier gas suppresses the rate of vaporization of metals that may act as getters for oxygen, this method can be applied to samples that cannot be analyzed by vacuum fusion [252]. Although the evolution of gas from the melt by bubble formation is effected faster by application of vacuum than by use of an inert carrier gas [241], the viscosity of the bath also plays an important part in recovery of the gas. This has already been discussed with regard to the problems encountered in the determination of nitrogen in steel. These problems have been solved by the use of the single-crucible technique discussed in Section I.A.4. This technique is further described in Sections II.C.2.d–g. Inert gas fusion for hydrogen is described in Chapter 7.

A. INERT GAS

The term *inert gas* is used here to include (1) the flushing gas utilized to remove air from the sample-inlet system, (2) the fusion-carrier gas for the fusion process, and (3) the chromatographic-carrier gas for the reference and sample streams of the thermal-conductivity cell. Often, only one source of gas is utilized for all three purposes. Ideally, three separate gas sources should be used to maintain the proper flow rate through the furnace and through the thermal-conductivity measuring system, and for the most precise work, all three streams should be purified before entering the system. Helium and argon are readily available in the United States in grades of 99.995% minimum purity, and higher purities are available at increased costs. With these gases fairly simple purification systems may be used that require only infrequent regeneration or replacement. Common reagents and methods for oxidizing or removing impurities from helium and argon are given in Table 4.4. Where the inert gas is not purified a variety of problems may be encountered. For example, the fusion-carrier gas might contain enough oxygen and nitrogen to contribute significantly to the blank. Also, when a short chromatographic column is used to expedite analysis, the nitrogen peak may appear as a shoulder on the oxygen peak if the latter has not been removed. The removal of air from pressure regulators is discussed in Chapter 5, Section II.B.1.b.

Various fusion-carrier gases have been used without purification or with varying degrees of purification. Nitrogen was used in the original work by Singer [253], who determined oxygen in steel. Bobalek and Shrader [254] used helium for the gravimetric determination of both hydrogen and oxygen in magnesium. Yoneda [255] and Shanahan and Cook [256] also used nitrogen. Holt and Goodspeed [257] tried nitrogen to convert magnesium

TABLE 4.4. Common Reagents and Methods for Oxidizing or Removing Impurities from Helium and Argon

Reagent or Method	Usual Impurity Oxidized (*a*) or Removed (*b*)
Heated CuO	Organic gases[b] (*a*), H_2 (*a*), CO (*a*)
Heated Schutze reagent	CO (*a*)
Hopcalite	CO (*a*)
Soda asbestos	CO_2 (*b*)
Silica gel	H_2O (*b*)
$Mg(ClO_4)_2$	H_2O (*b*)
Heated Ti, Zr	N_2 (*b*), O_2 (*b*)
Heated Ca, U	O_2 (*b*)
Liquid-N_2 trap[a]	H_2O (*b*), CO_2 (*b*)
Molecular sieve at −196°C[a]	Organic gases (*b*), CO (*b*), CO_2 (*b*), H_2O (*b*), O_2 (*b*)

[a] Not to be used when Ar is the carrier gas, as Ar liquifies at −185.7°C and solidifies at −189.2°C.
[b] Higher temperatures are needed to oxidize organic gases than to oxidize H_2 or CO.

to the nitride in order to prevent gettering of oxygen by this metal. However, cyanogen formed that interfered in the manometric measurement of carbon dioxide.

Hancart and Marot [258] were among the first to couple gas-chromatographic analysis to inert-gas fusion. They determined both oxygen and nitrogen in steel using argon as both the fusion- and chromatographic-carrier gas. Their work has been followed by a number of studies. For the determination of oxygen in metals, many investigators have used argon [250, 251, 257, 259–271a].

Argon has also been used in the determination of nitrogen [271b–c]. Helium has been used in the determination of hydrogen, nitrogen, and oxygen [271d], nitrogen and oxygen [249, 272–275a], nitrogen only [103, 275b–278], and oxygen only [275b, 279]. Hydrogen has been mentioned as a carrier gas [280], but this presents a problem of safety. It is of interest to note that the use of argon was specified with the obsolete conductometric oxygen-analyzer manufactured by LECO [281] but that helium is now specified with this manufacturer's latest oxygen-analyzer [279], where carbon dioxide is measured by thermal conductivity.

Several examples of methods of purifying the fusion-carrier gas are cited. In the determination of nitrogen, Vasserman and Turovsteva [268] purified "technical" argon by passing it successively through Ascarite, anhydrone, and finally calcium turnings at 650°C. For determining oxygen in zirconium, argon was passed through an elaborate system consisting of heated CuO–CO_2 absorbent–$Mg(ClO_4)_2$–H_2SO_4–$Mg(ClO_4)_2$–CO_2 absorbent-titanium sponge at 600°C [265]. In place of titanium, zirconium chips at 800°C [260], uranium turnings at 625°C [250] or 300°C [182] and turnings of zirconium–titanium alloy at 850°C have been used [182]. Schutze reagent (a special preparation of iodic acid) [263] and Hopcalite at 125°C [182] have also been utilized in the purification train to oxidize any carbon monoxide. Schutze reagent has been heated from 120–160°C for this purpose [182, 260, 261, 265].

Goldbeck et al. [272], in determining oxygen and nitrogen, dried helium over silica gel and then passed the gas through zirconium sponge at 700°C and then 5A molecular sieve at −196°C. Dallman and Fassel [249], in determining the same constituents, purified helium by passing it through 20–50 mesh, 13X molecular sieve at −196°C. Escoffier [271d] used a similar method for purifying helium in determining hydrogen, nitrogen, and oxygen in iron, steel, niobium, tantalum, tungsten, molybdenum, and chromium and noted that trace amounts of neon were not removed but were chromatographically determined with hydrogen.

B. SEPARATION AND DETERMINATION OF DESIRED CONSTITUENTS

Although a few specific analytical systems will be described in detail in Section II.C, certain general aspects are described here. For the determination of oxygen (as carbon monoxide) alone or oxygen and nitrogen, some investigators adsorb the gas on molecular sieve at −196°C [249, 253, 272], desorb the gas by heating the column in a stream of helium, and finally chromatographically determine the concentrations of carbon monoxide and nitrogen. The use of helium as the chromatographic carrier gas yields greater sensitivity for carbon monoxide and nitrogen than does argon and also allows the use of the liquid-nitrogen cold trap. Molecular sieve is also used for the chromatographic separation. Mosen et al. [282] used silica gel in the cooled trap column in their work.

In the LECO apparatus for oxygen [279], helium sweeps the carbon monoxide over heated copper oxide and the resulting carbon dioxide is adsorbed on molecular sieve at ambient temperature. After the sieve is heated, helium sweeps the desorbed carbon dioxide into a thermal-conductivity cell. In the LECO nitrogen-analyzer [277], interfering gases are removed by oxidation and absorption, and nitrogen is then directly determined in helium. In the Strohlein apparatus for hydrogen, oxygen, and nitrogen [283], the helium carries the gas from the furnace through an infrared analyzer where carbon monoxide is determined. The gas then passes over Schutze reagent, and thus, carbon monoxide is oxidized to carbon dioxide, which is then absorbed on molecular sieve at ambient temperature. The sieve also serves to separate hydrogen from nitrogen for the thermal-conductivity measurement of the latter.

C. APPARATUS

1. General Discussion

The direction of flow of inert gas through the furnace portion of the system will depend in part on the geometry of the system and the number of separate sources of inert gas used. For example, Smiley [250] passed argon upward through the furnace and sample-inlet to prevent any air that enters with the sample from reaching the crucible. In the Strohlein apparatus [283] helium enters in the side of the furnace and exits at the top. In the LECO nitrogen-analyzer [277] the flow is from top to bottom, while the reverse direction has been used by Dallman and Fassel [249].

When choosing the type of induction heating unit for the fusion, consideration should be given to the frequency used. High frequencies in the vicinity of 13 MHz require the use of carbon black as insulating material [264]. Carbon black may be difficult to blank out and has a tendency to be blown into the crucible when the flow of inert gas is increased, especially if it is excessively compacted [265]. To minimize this problem, Banks et al. [261] and Kallman and Collier [264] used graphite crucibles longer than usual. The use of carbon black in the form of granules rather then flocculent particles also helped [261]. The granular material was heated before being placed in the furnace; it could be reused three or four times before it agglomerated and no longer served as a good insulator. Powdered carbon black can be reused also. Where entrainment of particulate matter in the gas stream may occur, the latter should be passed through glass wool immediately upon emerging from the furnace area.

The substitution of a 2.5-kW, 450-kHz radio-frequency generator [260] for the higher-frequency generator mentioned earlier permitted the use of graphite insulation, which is preferable to carbon black [264]. With this power generator in conjunction with graphite, the blank amounted to 10 μg oxygen for 10 min. Blanks of 2–7 μg oxygen for 10 min were obtained by using an uninsulated crucible with a 10-kHz radio-frequency generator [250]. With pile-grade graphite insulation and a 10-kW generator with a frequency of 400 kHz [251], blanks were as low as 7–10 μg.

The blank in inert-gas fusion (excluding special techniques such as pulse heating) is caused mainly by the high-temperature reaction of graphite or carbon black with the quartz thimble for the crucible [80b]:

$$SiO_2 + 2C \underset{\Delta}{\rightarrow} Si + 2CO. \qquad (7)$$

Smith and co-workers [284, 285] reduced the blank by eliminating the insulation and the quartz thimble, and they used an elaborate current-concentrator furnace to obtain temperatures of 2400–2800°C. Dallman and

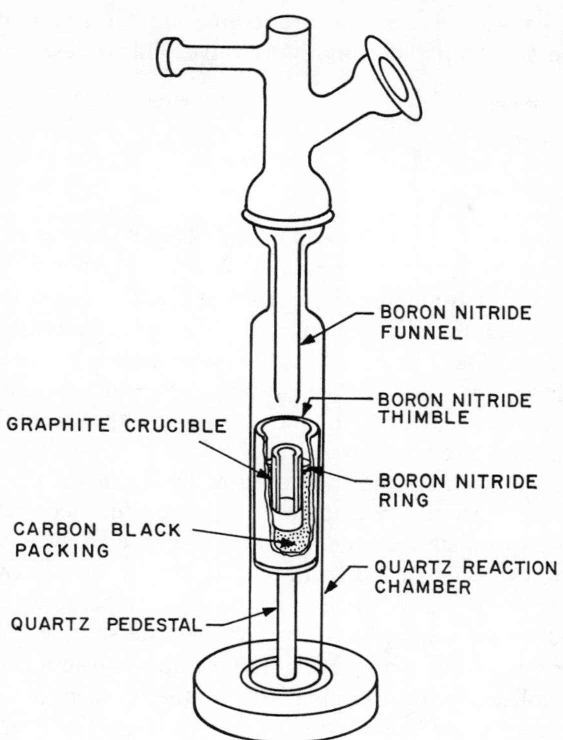

Fig. 4.25. Pyrolitic boron nitride crucible. (Reproduced by permission from *Analytica Chimica Acta* [267].)

Fassel [80a, 249] substituted pyrolytic boron nitride (PBN), as described in Section II.C.2.a, for quartz in the thimble. Beck and Chambers [267] also used PBN in constructing a thimble (see Fig. 4.25) for the LECO No. 534–300 Oxygen-Analyzer. In addition, a PBN ring was placed on the carbon-black insulation to prevent it from being swept out of the crucible. It was noted further that the blank could be increased by diffusion of alkali metal (from samples) to the quartz and reaction with it. Also, any tin in the crucible may deposit on the quartz and couple with the induction field, causing local hot spots.

The blank and gas-recovery problems involved in analyzing several samples per crucible by inert-gas fusion have been overcome by use of a fresh crucible for each sample. This work was pioneered by Vasserman and Turovtseva [268] and Lemm [276]. As noted earlier, similar work was performed by Gerhardt et al. [98–101] in vacuum.

Many inert-gas-fusion systems are connected to gas chromatographs through a collection trap and a multiport valve. This is described in detail in the following.

2. Specific Systems

a. Dallman-Fassel

The system devised by Dallman and Fassel [249] at the Institute for Atomic Research and Department of Chemistry at Iowa State University, Ames, Iowa, has been used for determining nitrogen and oxygen in 19 different base metals. Values for those elements were then compared with those obtained by the Kjeldahl chemical and vacuum-fusion procedures, respectively. A diagram of the inert-gas-fusion–gas-chromatographic analysis apparatus is shown in Fig. 4.26, and pertinent experimental details and operating conditions are summarized in Table 4.5. According to the procedure, purified helium enters the furnace zone by means of the gas-manifold assembly connected to the bottom of the Vycor furnace jacket. Samples contained in the storage arms are introduced into the crucible by means of a large ball valve. Gases extracted from the sample are carried by the purified helium into a collection trap. At liquid-nitrogen temperature the molecular sieve contained in the trap effectively removes the sample gases from the carrier-gas stream. After completing the sample-extraction and the gas collection, the collection trap is switched from the furnace carrier-gas stream into the chromatographic helium stream, and the liquid nitrogen is removed from the trap. The collection trap is then heated for 12–15 sec, whereupon the collected gases are immediately released and are flushed onto the chromatographic separation column. Just prior to elution of the nitrogen from the separation column, the collection trap is switched back into the furnace carrier-gas stream. This action returns the recorder base line to zero from the deflection induced by pressure disturbances during the trap switching and heating procedures. The gases are separated and detected by conventional gas-chromatographic procedures. The areas under the recorded peaks are automatically integrated and printed out.

Integrator response to nitrogen and carbon monoxide was calibrated directly in terms of counts per microgram of the specific gas. The calibrations were achieved by introducing known amounts of nitrogen or carbon monoxide into the carrier-gas stream ahead of the furnace assembly. The calibrated volumes were measured in the gas-manifold assembly. Calibration data were obtained at different crucible temperatures with the furnace assembly in various states of cleanliness. These data demonstrated that the various crucible temperatures and furnace conditions did not influence the calibrations in any systematic manner; the collection of the nitrogen in the trap

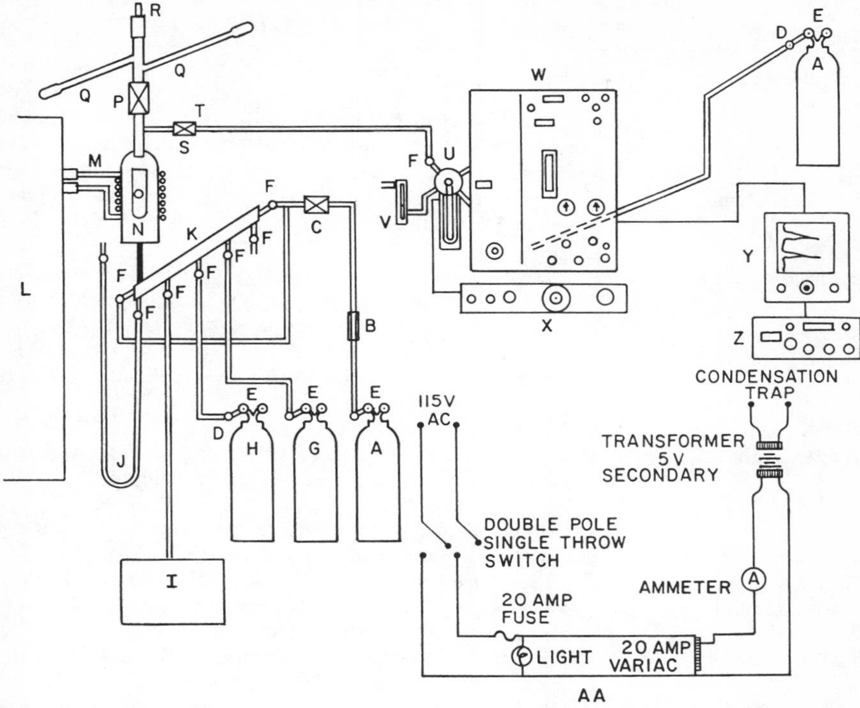

Fig. 4.26. Dallman-Fassel inert-gas-fusion–gas-chromatographic system. (Reproduced by permission from Analytical Chemistry [249].)

A, Helium supplies; Bureau of Mines, Grade A
B, Helium purification column
C, Hoke flow-control needle valve.
D, Matheson shut-off valves
E, Matheson two-way pressure regulators
F, Hoke toggle valves
G, Calibration dry-nitrogen supply; Linde Air Products
H, Calibration carbon monoxide supply; Matheson, CP
I, Welch mechanical vacuum pump
J, Mercury manometer
K, Gas manifold
L, Induction generator
M, Work coil
N, Vycor furnace jacket
O, Crucible assembly
P, 1-in. Jamesbury ball valve; NRC Equipment Corp.
Q, Sample-storage arms
R, Sighting prism and optical flat
S, $\frac{1}{4}$-in. Jamesbury ball valve; NRC Equipment Corp.
T, Glass-wool filter plug
U, Precision gas-sampling-valve–collection-trap assembly
V, Flow meter
W, Gas chromatograph
X, Collection-trap heater unit
Y, Recorder
Z, Printing integrator
AA, Electrical circuit for collection-trap heater

TABLE 4.5. Experimental Apparatus and Procedures for Simultaneous Determination of Oxygen and Nitrogen in Metals by Inert-Gas Fusion[a]

Heating unit	Lepel High Frequency Laboratories, Inc., induction generator, Model T-5N-1, with nominal power output of 5 kW and frequency of 400 ± 100 kHz. Maximum power output coupling to work load achieved with maximum tank-coil setting and an 8-turn, 4-in.-diameter work coil.
Crucible assembly	Ultra Carbon Corp., graphite crucible C-625 and graphite funnel F-703, or equivalent, cut down to dimensions specified [249].
Furnace assembly	Design based on Guldner-Beach furnace [83]. Crucible assembly components floated on Ultra Carbon Corp., UCP-2, −200-mesh graphite powder insulation, or equivalent, within specially fabricated "pyrolytic" boron nitride thimble [80a] that is suspended by platinum wire hooks within air-cooled Vycor jacket.
Furnace carrier gas	Purified helium, 275–300 cc/min.
Helium purification	Furnace carrier-gas purified by passing it through a 12-ft coiled length of $\frac{1}{4}$-in. o.d. copper tubing packed with approximately 50 g of 20–50 mesh 13× molecular sieve maintained at liquid-nitrogen temperature. Sieve originally activated by heating in vacuo at 180–190°C for 24 h.
Crucible degassing and bath addition	Crucible temperature raised to 2450–2500°C over 1 h period and degassed for 45 min. Temperature reduced to 1750°C and platinum–tin bath materials added slowly and intermittently to crucible. Crucible and bath then gradually brought to the desired operating temperature.
Bath conditioning and degassing	Conditioning metal added to bath when operating temperature is reached. When operating temperature exceeds 1900°C, conditioner is added at 1900°C, after which desired temperature is gradually attained. Degassing of entire composition at operating temperature is then conducted for 30–45 min.
Sample preparation	Carefully filed and cleaned samples fluxed with 12-gauge platinum wire, platinum capsules formed from 1-mil foil, or both.
Platinum-flux blanks	Platinum wire: less than 2 ppm nitrogen; 5–8 ppm oxygen. Platinum foil: less than 2 ppm nitrogen; 10–15 ppm oxygen.
Furnace blanks	Furnace blanks somewhat temperature-dependent: 0.1–0.8 μg nitrogen/min.; 0.2–0.6 μg oxygen/min.
Collection of evolved gases	Achieved by passing furnace carrier-gas stream through a stainless-steel collection trap packed with approximately 2 g of 20–50 mesh 13× molecular sieve, which is maintained at liquid-nitrogen temperature for the gas collection. Collected gases released from the trap by 12–15-sec heating period. Sieve originally activated by heating in an oven at 180–190°C for 24 h.
Gas chromatograph	Perkin-Elmer Corp., Vapor Fractometer, Model 154D, including a Precision Gas-Sampling Valve, of which the collection trap is an integral part.

TABLE 4.5 (*contd.*)

Separation column	Perkin-Elmer Corp., synthetic zeolite column I. Column periodically reactivated in situ for 8–12 h at 190°C, under a 100-cc/min flow of helium.
Chromatograph-carrier gas	Helium, 92 cc/min
Column and detector temperature	80°C
Detector voltage	8.0 V
Recorder	Leeds & Northrup Co., Speedomax "H," AZAR
Recorder sensitivity	0–2 mV, full scale
Recorder response	1 sec, full scale
Recorder-chart speed	$\frac{1}{2}$ in./min
Integrator	Perkin-Elmer Corp., Printing Integrator, Model 194B
Integrator calibration	Specified as integrator counts/μg of nitrogen or carbon monoxide, obtained by analysis of known quantities of nitrogen and carbon monoxide.

[a] Reproduced by permission from *Analytical Chemistry* [249].

from the helium-carrier stream was quantitative and the blank corrections employed were accurate. The calibrations were absolute and had a high degree of sensitivity under the experimental conditions used, amounting to approximately 85 counts/μg of nitrogen. The carbon monoxide calibrations were nearly coincident with those for nitrogen, because the calibrations were based on thermal-conductivity changes, and the thermal conductivities of the two gases are nearly identical. Syringe injections of the calibration gases directly onto the separation column yielded calibration values concordant with those obtained by the procedure described and therefore could be used to evaluate the integrator response daily.

b. Hanin-Villeneuve (IRSID)

Hanin and Villeneuve [271b, 271c] have developed an inert-gas-fusion–gas-chromatographic system for use at the Institute de Recherches de la Siderurgie (IRSID), St. Germain-en-Laye, France. The system is shown in Fig. 4.27.

The furnace is composed of a clear quartz tube inside of which the graphite crucible rests on a hollow graphite column, which in turn is supported by a Pyrex tube. The quartz tube is held between two pieces of duralumin by water-cooled joints. The fusion temperature of 1750°C is attained by induction heating using a 500-kHz, 2-kVA generator. Samples are introduced through a movable head cover under argon with the gas flowing from bottom to top through the furnace. This permits nitrogen determinations to be made directly on filings or powders without the use of a capsule for sample containment.

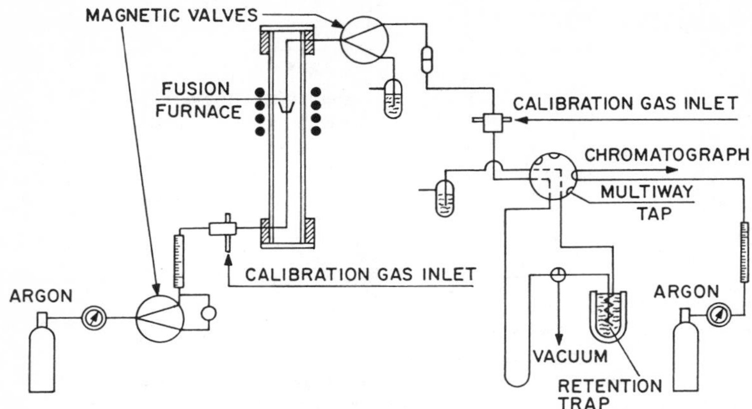

Fig. 4.27. Hanin-Villeneuve inert-gas-fusion–gas-chromatographic system. (Reproduced by permission from Institut de Recherches de la Siderurgie Francaise [271b–c].)

The graphite crucibles used are of two sizes. The smaller crucible weighs about 0.8 g and is for a 1-g sample in which only nitrogen or both oxygen and nitrogen are to be determined. The larger crucible weighs about 4.5 g and is capable of holding a dozen 1-g samples for the determination of oxygen only.

Argon is used as both the fusion- and chromatographic-carrier gas. Both streams of argon are purified by passage of the gas through 5A molecular sieve cooled by liquid oxygen. During fusion of the steel sample at 1750°C for 3 min, the argon carries the carbon monoxide, nitrogen, and hydrogen to a 4A molecular-sieve trap column immersed in liquid oxygen. This trap column consists of a Pyrex coil about 4–6 mm i.d. and 55 cm long, containing about 5 g of molecular sieve. It is connected by means of a multiposition valve to either the system or the chromatograph. Carbon monoxide and nitrogen are retained on this column while hydrogen passes through with argon. The trap column is also connected to a vacuum line, because when the column is cooled, a certain amount of argon condenses on it. Heating of the trap to 100°C would lead to an excessively high pressure if part of this argon were not previously removed by a vacuum pump. After evacuation of some argon and after heating the molecular-sieve trap with boiling water, the carbon monoxide and nitrogen are swept by the chromatographic-carrier gas at a reduced flow rate to the chromatographic column and the katharometer.

The chromatographic column consists of 5A molecular sieve contained in a copper column 4 mm i.d. and 300 cm long kept at 83°C. This column is regenerated after about 200 separations by heating at 200°C under a pressure of argon for 6 h, followed by cooling in the same gas.

The system contains two inlets for injecting carbon monoxide and nitrogen calibration gases; one is before the furnace and the other is before the chromatograph. The latter is used for calibration purposes. The former is used to study the effect of extraction temperature, condensed metal films, and other factors on the recovery of the two gases.

The determination of both carbon monoxide and nitrogen is based on peak height measurement. This method of quantitation is made possible by the use of a mercury-level pressure regulator upstream from the thermal-conductivity cell and a diaphragm pump downstream from the cell to maintain a sufficiently constant gas flow rate (6 liters/hr). In addition, to ensure that the gases do not undergo a preliminary separation on the trap column but are injected simultaneously to the chromatographic column, a certain amount of moisture is allowed to remain on the former column when it is installed. Presumably, this moisture blocks enough sites in the 4A molecular sieve to prevent chromatographic separation of the two gases here.

c. LECO Oxygen Determinator

The LECO Oxygen Determinator [279] (Model 734-300) is manufactured by Laboratory Equipment Corporation, St. Joseph, Michigan, and is shown in Fig. 4.28. This is an automatic instrument with which oxygen has been determined in steel, copper, titanium, zirconium, and other metals in about 3 min, including sample-preparation time, when a series of samples is analyzed. In analysis, the sample is introduced to the graphite crucible by means of a special loading device. Helium, purified by passing it through anhydrone, Ascarite, and hot zirconium sponge, carries the carbon monoxide and other gases given off during fusion through hot copper oxide where carbon monoxide is oxidized to carbon dioxide as noted in Section II.B. The latter is then sorbed in 5A molecular sieve at ambient temperature. The carbon dioxide is then desorbed in a stream of helium from a second tank while the molecular sieve is heated to about 350°C. Any other impurities are separated from the carbon dioxide by briefly retaining this compound on a silica-gel column at 40°C; this column also aids in equilibrating the gas flow to the thermal-conductivity detector. The amount of oxygen in percent is provided on a digital voltmeter.

During the sorption of carbon dioxide on the molecular-sieve column, the furnace carrier gas passes through this column and through the sample-loading chamber to the atmosphere, while the dried chromatographic-carrier gas passes through the reference side of the detector, through the silica-gel column, through the measuring side of the detector, and then to the atmosphere. During the desorption of the carbon dioxide from the molecular sieve, the furnace carrier gas goes from the furnace to the sample-loading chamber.

Fig. 4.28. LECO 734-300 oxygen determinator. (Reproduced by permission from Laboratory Equipment Corp.)

The crucible holds approximately 25 g of sample (depending on the density of the material). As little as 0.0005 mg oxygen may be detected. Calibration may be accomplished by using certified steel samples or by introducing aliquots of carbon dioxide with the LECO gas-dosing device. Degassing of the sample is carried out at approximately 2700°C. The furnace operates on 155/230-V, 60-Hz, single-phase or 220-V, 50-Hz grounded neutral current, 4.5 kVA. It uses 0.4-gal/min water flow for cooling. An automatic safety device will prevent operation of the furnace until water is flowing.

d. LECO TN-14 Automatic Nitrogen-Analyzer

The LECO TN-14 Automatic Nitrogen-Analyzer [277] is manufactured by the Laboratory Equipment Corporation, St. Joseph, Michigan. With this instrument, steel samples of widely varying composition may be analyzed for nitrogen in about 2.5 min. The limit of detection is about 0.0005 mg nitrogen and the standard deviation, as a measure of repeatability, is 0.0001% nitrogen at the 0.004% level. Titanium alloys cannot be analyzed with this instrument. The instrument is shown in Fig. 4.29. Cooling water and high-purity helium are required.

Fig. 4.29. LECO TN-14 automatic nitrogen analyzer. (Reproduced by permission from Laboratory Equipment Corp.)

The instrument may be powered by 208-, 220-, or 115/238-V, single-phase, 60-Hz current or 220-V, single-phase, 50-Hz current, 6 kVA. The electronic components are solid-state, plug-in type. The basic idea for this instrument stems from the work of Vasserman and Turovsteva [268], who established the use of pulse heating for extracting oxygen from metals and certain compounds. The advantage of this method, described below, is that the heating period is of short duration and therefore evaporation of graphite and metals is insignificant. Goldbeck et al. [272] developed a modified pulse heating furnace. An evaluation of an early model LECO nitrogen-analyzer was performed by Jecko and Touvenin [275c]. A later model was evaluated by Cline et al. [275d].

(1) HELIUM PURIFICATION. Any traces of hydrogen and carbon monoxide can be oxidized by passing the helium through a hot mixture of rare-earth oxides and copper oxide. The resulting carbon dioxide and water are removed by Ascarite and anhydrone, respectively. The helium cylinder is connected to a manifold system of at least two such cylinders. This permits changing of a cylinder while maintaining a helium atmosphere for the thermistor without shutting down the instrument. If the instrument is completely shut down, it requires about 4 h to stabilize the thermistor before operation can be resumed.

(2) FURNACE ASSEMBLY. The furnace (left side of chassis, Fig. 4.29) consists of two copper electrodes between which the graphite crucible is located during analysis. The crucible, used without a lid, is seated in the lower electrode, which is attached to a movable rod for raising the crucible into the furnace. After the crucible is raised into position, the power supply is activated. The graphite crucible acts as the primary resistance to current flow in the secondary circuit of a transformer that is briefly overloaded. The very high current pulses heat the crucible and sample to temperatures in the 2500°C range. A series of pulses are delivered during the 10-sec fusion cycle. The duration of the pulses is sufficiently large that the transformer does not reach excessive temperatures. The water-cooled furnace is protected by a switch that prevents current flow whenever there is not enough cooling water flowing.

The sample is loaded into the crucible from the sample loading plunger, which rides between six O-rings. The furnace is sealed at the bottom by two O-rings located just below the lower electrode. Only one sample is analyzed per crucible so that there is no buildup of carbon in the molten metal that can hinder the evolution of the nitrogen.

Four solenoids direct the helium flow during the idling (between analyses), outgassing, and analyzing cycles.

(3) ANALYTICAL PURIFICATION TRAIN. The analytical gas stream passes through the following prior to the thermal-conductivity cell: (1) glass-wool trap to remove carbon and other solids, (2) heated rare-earth oxide–copper oxide, (3) activated charcoal to adsorb any organic gases, (4) Ascarite, and (5) magnesium perchlorate.

(4) ANALYZER. A thermistor thermal-conductivity cell is maintained at constant temperature. The response for nitrogen is integrated and read directly as percent nitrogen on a digital voltmeter. A binary coded decimal output is available.

(5) SAMPLE-WEIGHT COMPENSATOR. The instrument is equipped with sample-weight-compensation controls. For sample weights between 0.300 and

0.699 g the weight-compensation dial is adjusted accordingly, and the nitrogen result is read from the digital voltmeter. A sample of weight outside this range is analyzed with the controls set at 0.50 g, and the percent nitrogen is calculated by multiplying the digital voltmeter reading by the ratio of the sample-weight to 0.5. It is recommended by the manufacturer that weights greater than 0.699 g not be taken, to avoid the possibility of molten metal being ejected from the crucible and welding to the upper electrode.

(6) ANALYSIS. After the crucible has been baked out, the sample is loaded and the analysis is automatically performed. If the sample consists of millings or similar small particles, the crucible must be removed from the furnace after bake-out to permit loading of the sample. On replacing the crucible, it is necessary to purge air from the furnace prior to analysis.

(7) TROUBLE-SHOOTING. Occasionally the digital voltmeter will advance during analysis and then roll back to a lower number. This may be caused by a dirty solenoid, depleted Ascarite or magnesium perchlorate, a faulty O-ring seal around the bottom electrode, or some electronic failure. Rollback may also be experienced in the analysis of fine iron powders. With such samples, the problem may be caused by the large amount of surface oxide, which is first converted to carbon monoxide. The concentration of this compound may be too high for complete oxidation to be accomplished, pure carbon monoxide therefore passing through the system and affecting detector response. A temporary solution to this particular problem is simply to note the highest nitrogen value obtained in the digital voltmeter.

A decrease in current through the crucible during analysis can be encountered. This has been found to be caused by a coating formed on the high-voltage cable connected to the electrode. After disconnecting the power and removing the cable the affected area should be abraded.

e. LECO TC-30 Simultaneous, Automatic Nitrogen–Oxygen Determinator

The LECO TC-30 Determinator, manufactured by the Laboratory Equipment Corporation, St. Joseph, Michigan, is shown in Fig. 4.30. The sample is fused in purified helium in a single-use crucible in a manner similar to that described in the section above. (Ferrous and certain nonferrous metals may be analyzed.) The helium carries the evolved gases over hot rare-earth oxide–copper oxide where carbon monoxide and hydrogen are oxidized, respectively, to carbon dioxide and water. After absorption of the water in magnesium perchlorate, nitrogen and carbon dioxide are separated chromatographically on silica gel and are determined in a thermistor cell. Successive bridge outputs for nitrogen and oxygen are integrated and displayed on separate digital voltmeters. A binary coded decimal output is available for both gases. Analysis time is about $3\frac{1}{2}$ min/sample.

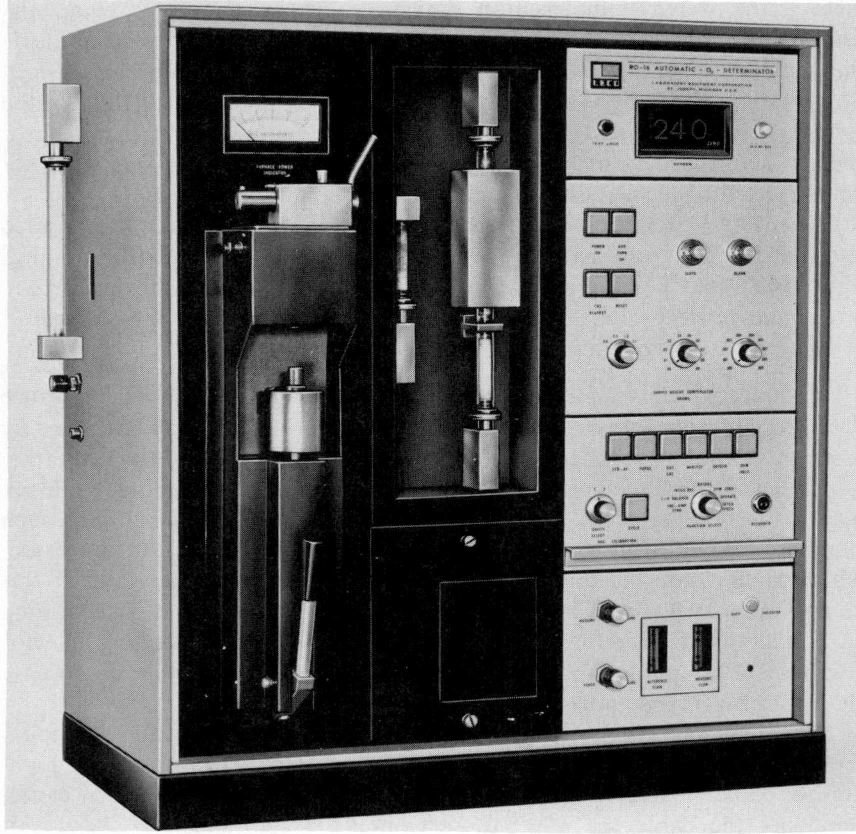

Fig. 4.30. LECO TC-30 simultaneous automatic nitrogen–oxygen determinator. (Reproduced by permission from Laboratory Equipment Corp.)

Over the range of 1–1000 ppm with 0.5-g samples, nitrogen can reportedly be determined to ±2 ppm or ±2% of the amount present, whichever is greater. For oxygen, the corresponding numbers are ±2 ppm or ±3%.

The analytical cycle consists of (1) inserting the crucible, (2) loading the sample in the entry position, (3) outgassing the crucible, (4) introducing the sample to the crucible, and (5) analysis. Steps 3 and 5 are initiated by depressing a button. Helium back-flushes the sample cavity during sample-introduction to prevent contamination by air. Calibration may be achieved with standard samples, recommended for the initial work, followed by standard gases for periodic checks. With the latter, the LECO gas-dosing system is used.

The furnace can be built for 60-Hz, 6-kVA, 208-, 220-, or 230/115-V, single-phase or 50-Hz, 6-kVA, 220-V, single-phase operation. All electronic components are solid state. A series of check points are used in conjunction with one of the digital voltmeters to check these components for proper output. Also provided are a safety switch to turn off the furnace if cooling water flow is insufficient, and a micro-switch to turn off the furnace power if the furnace is opened at an improper time. All filters and chemical reagents that require periodic replacement are contained in glass tubes held in place by spring-loaded quick-disconnect seals. The electronic components are solid state and are mounted on removable boards.

f. LECO RO-16 Automatic Oxygen Determinator

The LECO RO-16 Determinator (Fig. 4.31), applicable to many ferrous and nonferrous metals, is manufactured by the Laboratory Equipment Corporation, St. Joseph, Michigan. The heating principle and crucible are the same as those described in the two preceding sections. The electronic components are solid state and are mounted on removable boards. Purified nitrogen is used as the carrier gas. Hydrogen is oxidized to water and absorbed in anhydrone, while carbon monoxide is oxidized to carbon dioxide and measured by a thermistor cell. Detector output is integrated and displayed directly for 1–1000 ppm oxygen for samples weighing between 0.8 and 1.2 g, because of the digital weight compensator. A binary coded decimal output is available.

The system blank is about 3 ppm. For homogeneous samples, results in the above range are accurate to within 2 ppm or 3%, whichever is greater. The system may be wired for 208-, 220-, or 115/230-V, 60-Hz, single-phase, 6-kVA or 220-V, 50-Hz, single-phase, 6-kVA operation. The instrument is protected from inadvertent overloading if cooling water flow is insufficient or if the furnace is opened at an improper time.

g. Strohlein Dinometer

The Strohlein Dinometer (Fig. 4.32) was developed by Koch and Lemm [274, 275a–b] and by Lemm [278]. It is manufactured by Strohlein and Company, Dusseldorf, West Germany [283]. The basic operation of the Dinometer is briefly described in Section II.B.

This instrument requires the use of a new crucible for each determination. The infrared and thermal-conductivity outputs for carbon monoxide and for hydrogen and nitrogen, respectively, are integrated during continuous flow of the gas mixture. Results for carbon monoxide, on the one hand, and for hydrogen and nitrogen, on the other, are displayed on two strip-chart recorders and presented digitally on two counters. Analysis time for oxygen and nitrogen determined simultaneously is about 3 min.

Fig. 4.31. LECO RO-16 automatic oxygen determinator. (Reproduced by permission from Laboratory Equipment Corp.)

This apparatus has several unique features. The Strohlein gas-dosing valve [189b], described in Section I.C, has been incorporated into the system so that pure carbon monoxide and nitrogen can be admitted to the system both for the purpose of calibration and for checking the response of the gas-analyzers. Further, the furnace section includes the following:

1. An electric elevator for raising and lowering the crucible to facilitate changing of crucibles and to exactly locate the crucible during heating

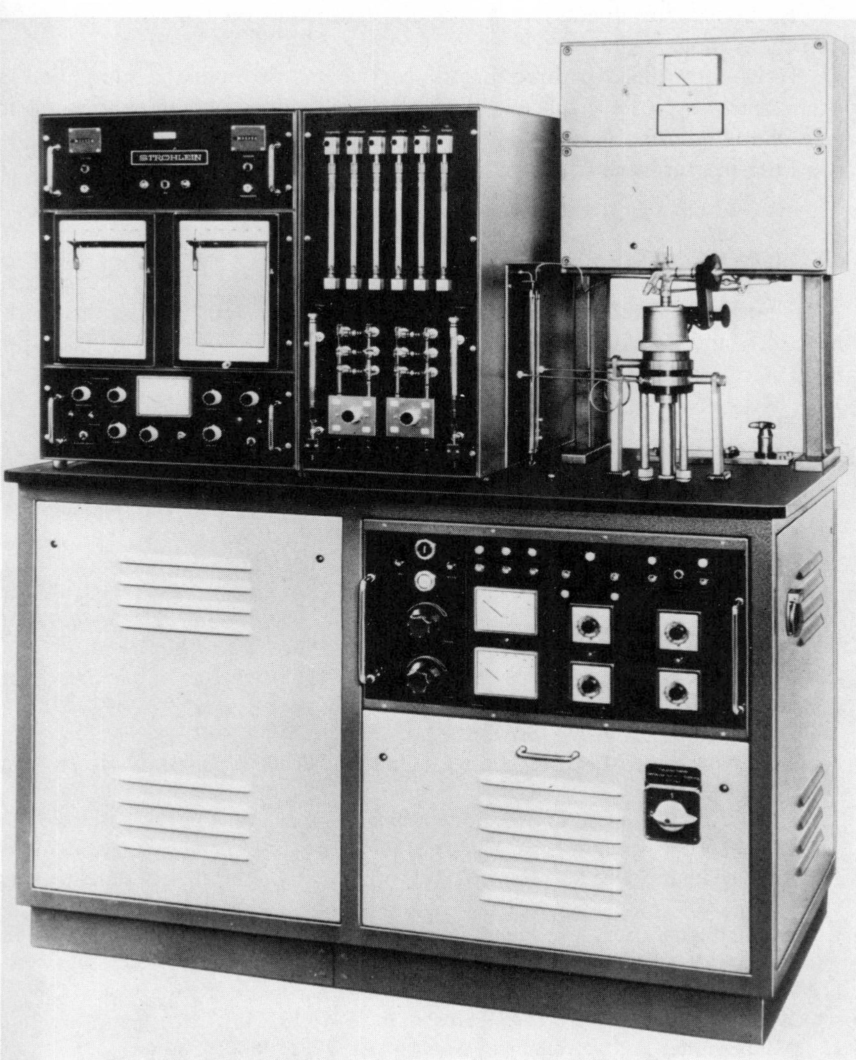

Fig. 4.32. Strohlein dinometer. (Reproduced by permission from Strohlein and Co.)

2. Electroprogrammer to vary the temperature while extracting the gases from the sample
3. Flow-controller for directing the carrier gas through the sample-entry port and through the furnace section so that air does not enter the system when changing crucibles
4. Plug-in graphite resistance heating elements.

ACKNOWLEDGMENTS

The authors gratefully acknowledge the many helpful discussions with, and contributions from H. A. Barnett, H. S. Karp, F. J. Malloy, R. Rapp, and R. C. Takacs of the U.S. Steel Corporation Research Laboratory.

REFERENCES

1. M. LePape and R. Platzer, *Comm. Energie At. Serv. Doc. Ser. Bibliograph*, No. 23 (1962); through *Chem. Abstr.*, **61,** 4931e (1964).
2. R. Kern, *Phys. Kondensierten Materie*, **1** (2), 105 (1963).
3. J. A. James, *Met. Rev.*, **9,** 93 (1964); through *Chem. Abstr.*, **61,** 12599g (1964).
4. Yu. A. Klyachko and O. M. Gorlova, *Zhur. Vses. Khim. Obsch. im. D. I. Mendeleeva*, **9** (2) 205 (1964); through *Anal. Abstr.*, **12,** 3161 (1965).
5. J. Montuelle, *ATB Met.*, **5** (8), 343 (1964–1965); through *Iron Steel Inst. Abstr. Book Title Svce.*, **543.27**, 61022.
6. K. Zimmermann, *Stahl Eisen*, **86** (25), 688 (1966).
7. S. Yanagisawa, M. Ihida, E. Kato, and Y. Abe, *Japan Analyst*, **15** (11), 1277 (1966).
8. S. Ikeda, J. Kashima, and E. Sudo, *Japan Analyst*, **16,** 135 (1967).
9. Z. M. Turovtseva and L. L. Kunin, *Analysis of Gases in Metals*, Academy of Sciences of the USSR, Moscow, 1959, Trans. by Consultants Bureau, New York, 1959.
10. Iron and Steel Inst., *The Determination of Gases in Metals*, Spec. Rep. No. 68, Iron and Steel Institute, London, 1960.
11. W. H. Walker and W. A. Patrick, *Orig. Com. 8th Intern. Cong. Applied Chem.*, **21,** 139 (1912); through *Chem. Abstr.*, **6,** 3381 (1912).
12. P. Oberhoffer and H. Schenck, *Stahl Eisen*, **47,** 1526 (1927).
13. W. Hessenbruch and P. Oberhoffer, *Arch. Eisenhuttenw.*, **1,** 583 (1938).
14. L. Jordan and J. R. Eckman, *Sci. Pap. Natl. Bur. Std.* (U.S.), **20,** 445 (1925).
15. H. A. Sloman, *Spec. Rep. No. 9*, Iron and Steel Institute, London, 1935, p. 71.
16. H. A. Sloman, *Spec. Rep. No. 16*, Iron and Steel Institute, London, 1937, p. 82.
17. H. A. Sloman, *Spec. Rep. No. 25*, Iron and Steel Institute, London, 1939, p. 43.
18. H. A. Sloman, *J. Iron Steel Inst.*, **43,** 298 (1941).

19. H. A. Sloman, *J. Inst. Metals Bull. Met. Rev.*, **71,** 391 (1945).
20. H. A. Sloman, *Engineering*, **160,** 385 (Nov. 9, 1945).
21. H. A. Sloman, *Engineering*, **160,** 404 (Nov. 16, 1945).
22. H. A. Sloman, *Engineering*, **160,** 419 (Nov. 23, 1945).
23. H. A. Sloman, *Report No. 5066*, National Physical Laboratory, Metallurgy Division, (Oct. 9, 1947); through Ref. 173.
24. H. A. Sloman, C. A. Harvey, and O. Kubaschewski, *J. Inst. Metals, Bull. Met. Rev.*, **80,** 391 (1951–1952).
25. J. Yarwood, *High Vacuum Technique*, 3rd ed., Wiley, New York, 1955.
26. M. Pirani and J. Yarwood, *Principles of Vacuum Engineering*, Chapman and Hall, London, 1961.
27. S. Dushman, *Scientific Foundations of Vacuum Technique*, 2nd ed., J. M. Lafferty, Ed., Wiley, New York, 1962.
28. A. Guthrie, *Vacuum Technology*, Wiley, New York, 1963.
29. C. M. Van Atta, *Vacuum Science and Engineering; Properties of Gases at Low Pressure, Vacuum Measurements, Design and Operating Features of Vacuum Pumps and Systems*, McGraw-Hill, New York, 1965.
30. R. F. Bunshah and B. S. Bunshah, Bibliography in R. F. Bunshah, Ed., *Vacuum Metallurgy*, Reinhold, New York, 1958, p. 443.
31. *Vacuum Technology, Directory and Specifications Catalog*, Thompson, Chicago.
32. E. Booth, F. J. Bryant, and A. Parker, *Analyst*, **82,** 50 (1957).
33. See Ref. 26, p. 467.
34. P. G. Simpson, *Induction Heating*, McGraw-Hill, New York, 1960.
35. D. H. Templeton and J. I. Watters, "Low Pressure Methods," in C. J. Rodden, Ed., *Analytical Chemistry of the Manhattan Project*, McGraw-Hill, New York, 1950, p. 644.
36. F. W. Curtis, *High-Frequency Induction Heating*, 2nd ed., McGraw-Hill, New York, 1950.
37. G. W. Goward, *Anal. Chem.*, **37,** 117R (1965).
38. P. D. Donovan, J. L. Evans, and G. H. Bush, *Analyst*, **88,** 771 (1963).
39. AGARD Materials Group of the Structure and Materials Panel, *Working Paper M33, Analysis of Refractory Metals*, NATO Advisory Group for Aeronautical Research and Development, Paris, 1963, pp. 21, 24.
40. J. O'M. Bockris, J. L. White, and J. D. Mackenzie, *Physicochemical Measurements at High Temperatures*, Academic, New York, 1959.
41. J. P. Burden, *Nature*, **188,** 221 (1960).
42. M. R. Everett and G. E. Thompson, *Analyst*, **87,** 515 (1962).
43. J. N. Gregory, D. Mapper, and J. A. Woodward, *Analyst*, **78,** 414 (1953).
44. See Ref. 27, p. 118.
45. See Ref. 26, p. 29.
46. (a) M. A. Biondi, in C. R. Meissner, Ed., *Proceedings of the Seventh National Symposium of Vacuum Technology Transactions*, Pergamon, New York, 1960, p. 24; (b) M. J. Fulker, *Vacuum*, **18,** 445 (1968); and (c) W. W. Roepke and K. G. Pung, *ibid.*, 457.
47. H. A. Steinherz and P. A. Redhead, *Scientific American*, March 1962, p. 78.

48. M. L. Olashkevich and V. I. Mirimanova, *Pribory i Tekhnika Eksperimenta*, **6** (Nov.–Dec. 1966).
49. See Ref. 26, p. 71.
50. J. N. Gregory and D. Mapper, *Analyst*, **80,** 225 (1955).
51. R. H. McFarland and D. G. McDonald, *Rev. Sci. Instrum.* **29,** 530 (1958)
52. See Ref. 27, p. 149.
53. British Iron and Steel Research Association, *Spec. Rep. No. 62*, Iron and Steel Institute, London, 1958, p. 44.
54. L. C. Covington and S. J. Bennett, *Anal. Chem.*, **32,** 1334 (1960).
55. See Ref. 26, p. 50.
56. See Ref. 27, p. 141.
57. K. Speight and G. M. Gill, *Metallurgia*, **55,** 155 (1957).
58. J. T. Sterling, in *Ductile Chromium and Its Alloys*, American Society for Metals, Cleveland, 1957, p. 188.
59. J. J. Naughton and H. H. Uhlig, *Ind. Eng. Chem., Anal. Ed.*, **15,** 750 (1943)
60. R. Lesser, in L. E. Preuss, Ed., *Transactions of the Eighth National Vacuum Symposium and Second International Congress on Vacuum Science and Technology*, (1961), Vol. II, Pergamon, New York, 1962, p. 782.
61. G. Urry and W. H. Urry, *Rev. Sci. Instrum.*, **27,** 819 (1956).
62. R. M. Roberts and J. J. Madison, *Anal. Chem.*, **29,** 1555 (1957).
63. E. Altmann, *Dechema Monograph*, **31,** 91 (1959); through *Chem. Abstr.*, **53**, 20938b (1959).
64. R. Gilissen, P. J. Demeester, and A. J. Filipot, *Rev. Sci. Instrum.*, **35,** 855 (1964).
65. J. A. Rodder, *Rev. Sci. Instrum.*, **36,** 867 (1965).
66. D. A. Rice and J. Roach, *J. Sci. Instrum.*, **44,** 473 (1967).
67. M. Bufalini and J. E. Todd, *J. Chem. Ed.*, **44,** 425 (1967).
68. See Ref. 9, p. 129.
69. K. G. MacLaren and W. T. Williams, *J. Sci. Instrum.*, **44,** 1033 (1967).
70. F. Vratny and B. Graves, *Rev. Sci. Instrum.*, **30,** 597 (1959).
71. *Methods for Chemical Analysis of Metals*, American Society for Testing and Materials, Philadelphia, 1960, pp. 28, 298.
72. N. A. Gokcen and E. S. Tankins, *J. Metals*, **14,** 584 (1962).
73. General Electric Co., *Pyrolytic Graphite*, *Preliminary Engineering Data*, Metallurgical Products Dept., Specialty Alloys Section, Detroit, 1962; through F. J. Miller, *Anal. Chem.*, **35,** 929 (1963).
74. High Temperature Materials, Inc., *Pyrolytic Graphite*, *Property Data* Brighton, Mass. Oct. 6, 1961; through F. J. Miller, *Anal. Chem.*, **35,** 929 (1963)
75. P. L. Walker, Jr., *Amer. Scientist*, **50,** 259 (1962).
76. C. A. Klein, *International Science and Technology*, August 1962, p. 60.
77. J. Pappis and S. L. Blum, *J. Amer. Cer. Soc.*, **44,** 592 (1961).
78. W. A. Robba, *High Frequency Heating Review*, **2** (2), 4.
79. W. F. Knippenberg, B. Lersmacher, H. Lydtin, and A. W. Moore, *Philips Tech. Rev.*, **28** (8), 231 (1967).
80. (a) W. E. Dallman and V. A. Fassel, *Anal. Chem.*, **38,** 662 (1966); (b) J. L. Potter, J. E. Murphy, and H. H. Heady, *Anal. Chem.*, **34,** 1635 (1962); (c)

Technical Information Bulletin No. 713-204EF, Union Carbide Corp., Carbon Products Division, New York; and (d) *Technical Information Bulletin No. 712-201GG*, Union Carbide Corp., Carbon Products Division, New York.
81. (a) J. F. Martin and L. M. Melnick, ASTM Committee E3, Symposium on Recent Advances in the Determination of Gases in Metals, Atlantic City, N.J., June 28, 1966; and (b) C. W. Rohl and J. H. Robinson, *Research/Development*, July 1966, pp. 21–24.
82. T. H. Malim, *Iron Age*, **200** (12), 61 (1967).
83. W. G. Guldner and A. L. Beach, *Anal. Chem.*, **22,** 366 (1950).
84. Yu. A. Klyachko, A. G. Atlasov, and E. M. Chistyakova, *Zavodskaya Lab.*, **16,** 17 (1950); through *Chem. Abstr.*, **44,** 6741c (1950).
85. R. M. Cook and G. E. Speight, *J. Iron Steel Inst.*, **176,** 252 (1954).
86. Yu. A. Klyachko, A. G. Atlasov, and M. M. Shapiro, *The Analysis of Gases and Inclusions in Steel*, Metallurgy, Moscow, 1953; through Ref. 9, p. 117.
87. J. N. Gregory and D. Mapper, *Analyst*, **80,** 225 (1955).
88. B. B. Bach, J. V. Dawson, and L. W. L. Smith, *J. Iron Steel Inst.*, **176,** 257 (1954).
89. NRC Equipment Corporation (now part of Varian Associates), *Operating Instructions for Vacuum Fusion Apparatus Type 912C*, Newton Highlands, Mass. November 27, 1957, p. 14.
90. (a) H. F. Waldron, Pittsburgh Conference on Analytical Chemistry and Applied Spectroscopy, 1964; and (b) Bulletins 77 and 90, Feb. 9, 1966 and June 25, 1971, MKS Instruments, Inc., Burlington, Mass.
91. R. S. McDonald, J. E. Fagel, and E. W. Balis, *Anal. Chem.*, **27,** 1632 (1955).
92. W. H. Smith, *Anal. Chem.*, **27,** 1636 (1955).
93. C. J. Smithells, *Metals Reference Book*, Vol. II, 3rd ed., Butterworths, Washington, 1962, pp. 576–577.
94. T. Koizumi, F. Tsugane, and M. Kamakura, *Tetsu To Hagane*, *Overseas*, **4,** 376 (1964).
95. M. Ihida, *Japan Analyst*, **8,** 786 (1959).
96. J. Niebuhr, *Plansee Proc. 3rd Seminar*, *Reutte/Tyrol 1958*, *1959*, 313–323; through *Chem. Abstr.*, **54,** 16271d (1960).
97. H. Severus-Laubenfeld, J. Steiger, and R. Stahel, *Z. Anal. Chem.*, **218,** 241 (1966).
98. A. Gerhardt, doctoral thesis, Technishce Universitat, Berlin, 1965.
99. A. Gerhardt, T. Kraus, and M. G. Frohberg, *Giess. Techn. Wiss.*, **17,** 203 (1965).
100. A. Gerhardt, T. Kraus, and M. G. Frohberg, *Neue Hutte*, **12,** 115 (1967).
101. A. Gerhardt, T. Kraus, and M. G. Frohberg, Pittsburgh Conference on Analytical Chemistry and Applied Spectroscopy, Pittsburgh, Pa., 1966.
102. H. Morrogh and W. J. Williams, *J. Iron Steel Inst.*, **155,** 321 (1947).
103. E. Jaudon, *Rev. Met.*, **63,** 1025 (1966).
104. V. A. Fassel, F. M. Evens, and C. C. Hill, *Anal. Chem.*, **36,** 2115 (1964).
105. A. L. Beach and W. G. Guldner, *Anal. Chem.*, **31,** 1722 (1959).
106. (a) T. A. Izmanova, E. M. Chistyakova, and A. Ya. Bessmertnaya, *Zavodsk.*

Lab., **32,** 1336 (1966); and (b) Yu. A. Karpov, G. G. Glanin, and L. L. Kunin, *Zh. Anal. Khim.*, **24,** 276 (1969).

107. N. Christensen and K. Gjermundsen, *J. Iron Steel Inst.*, **190,** 248 (1959).
108. W. F. Murphy, *Trans. Amer. Soc. Metals*, **41,** 888 (1949).
109. D. T. Peterson and V. G. Fattore, *Anal. Chem.*, **34,** 579 (1962).
110. N. Yoneda, *Tetsu To Hagane*, **43,** 395, 949 (1957).
111. N. Yoneda, *Hitachi Hyoron*, **40,** 8, 1019 (1958).
112. N. Lounamau, J. U. Aass, J. Kuhne, and T. Perrson, *Anal. Chim. Acta*, **29,** 267 (1963).
113. B. D. Holt and H. T. Goodspeed, *Anal. Chem.*, **34,** 374 (1962).
114. D. F. Wood and J. A. Oliver, *Analyst*, **84,** 436 (1959).
115. J. E. Still, *Spec. Rep. No. 68*, Iron and Steel Institute, London, 1960, p. 43.
116. D. I. Walter, *Anal. Chem.*, **22,** 297 (1950).
117. D. L. Guernsey and R. H. Franklin, *Spec. Tech. Publ. No. 222*, American Society for Testing and Materials, Philadelphia, 1957, p. 3.
118. J. N. Gregory and D. Mapper, *Analyst*, **80,** 230 (1955).
119. D. H. Wilkins and J. F. Fleischer, *Anal. Chim. Acta*, **15,** 334 (1956).
120. W. G. Smiley, *Spec. Tech. Publ. No. 222*, American Society for Testing and Materials, Philadelphia, 1957, p. 25.
121. E. Booth and A. Parker, *Analyst*, **84,** 546 (1959).
122. T. D. McKinley, *Englehard Industries, Inc., Tech. Bull.* **2** (4), 140 (1962).
123. W. R. Hansen, M. W. Mallett, and M. J. Trzeciak, *Anal. Chem.*, **31,** 1237 (1959).
124. A. L. Beach and W. G. Guldner, *Spec. Tech. Publ. No. 222*, American Society for Testing and Materials, Philadelphia, 1957, p. 15.
125. W. M. Albrecht and M. W. Mallett, *Anal. Chem.*, **26,** 401 (1954).
126. S. Sawa, *Tetsu To Hagane*, **38,** 567 (1952); through *Henry Brutcher Tech. Trans. No. 3647.*
127. T. Kraus, M. Frohberg, and A. Gerhardt, *Arch. Eisenhuttenw.*, **35,** 39 (1964); through *Brit. Iron Steel Inst. Translation 3677*, April 1964.
128. (a) E. M. Chistyakova and V. I. Stepanov, *Novye Metody Ispit. Metal., Tsentr. Nauch.–Issled. Inst. Chernoi Met.*, **No. 49,** 98 (1966); and (b) H. Kamada and K. Furuya, *Bull. Chem. Soc. Jap.*, **48,** 1256 (1968).
129. Yu. A. Klyachko and E. M. Chistyakova, *Tr. Komissii po Anal. Khim., Akad. Nauk SSSR, Inst. Geokhim. Anal. Khim.*, **12,** 126 (1960); through *Chem. Abstr.*, **54,** 20644e (1960).
130. E. J. Beck and F. E. Clark, *Anal. Chem.*, **33,** 1767 (1961).
131. E. L. Zakharov, *Byull' Ts IIN*, **1957**; through Ref. 9.
132. A. Columbo and E. Rodari, *Anal. Chim. Acta*, **42,** 133 (1968).
133. (a) H. Bohm, K. G. Gunther, and W. Kuhl, *Z. Anal. Chem.*, **209,** 198 (1965); and (b) G. Derge, W. A. Peifer, and B. Alexander, *Trans. AIME*, **162,** 361 (1945).
134. S. Yanagisawa and M. Seki, *Bunseki Kagaku*, **9,** 176 (1960); through *Chem. Abstr.*, **57,** 2834h (1962).
135. J. F. Martin, J. E. Friedline, L. M. Melnick, and G. E. Pellissier, *Trans. AIME*, **212,** 514 (1958).

36. T. D. McKinley, private communication, January 18, 1963.
37. S. J. Bennett and L. C. Covington, *Anal. Chem.*, **30,** 363 (1958).
38. D. A. Swann and D. A. Williams, *Analyst*, **83,** 113 (1958).
39. Reference 26, p. 214.
40. S. Jnanananda, *High Vacua*, Van Nostrand, New York, 1947, p. 256.
41. W. S. Horton and J. Brady, *Anal. Chem.*, **25,** 1891 (1953).
42. T. D. McKinley, private communication, 1963.
43. A. Lench and G. S. Martin, *Anal. Chem.*, **31,** 1726 (1959).
44. G. S. Martin and A. Lench, *J. Sci. Instrum.*, **36,** 141 (1959).
45. (a) See Ref. 26, p. 81; and (b) F. R. Sellenger, *Vacuum*, **18,** 645 (1968).
46. P. Rosenberg, *Rev. Sci. Instrum.*, **9,** 258 (1938).
47. H. S. Martin & Co., *Catalog*, Evanstown, Ill., 1964, p. 71, Section D.
48. R. Gilmont and M. C. Parkinson, *Research/Development*, **13,** (11), 50 (1962).
49. M. Axelbank, *Rev. Sci. Instrum.*, **21,** 511 (1950).
50. S. Yanagisawa, M. Seki, Y. Watanabe, and S. Nakamura, *Mikrochim. Acta*, **No. 1,** 1 (1959).
51. (a) H. H. Podgurski and F. N. Davis, *Vacuum*, **10,** 377 (1960); and (b) R. H. Work and C. E. Hawk, *Research/Development*, June 1966, pp. 79–81.
52. British Iron and Steel Research Association, *Spec. Rep. No. 62*, Iron and Steel Institute, London, 1958, p. 51.
53. H.S.Karp, L. L. Lewis, and L. M. Melnick, *J. Iron Steel Inst.*, **200,** 1032 (1962).
54. M. W. Mallett, D. F. Kohler, R. B. Iden, and B. G. Koehl, *Tech. Rep. WAL TR 823/5*, Office of Technical Services, Washington, 1962.
55. R. Eborall, *Spec. Rep. No. 68*, Iron and Steel Institute, London, 1960, p. 192.
56. A. G. Fleiger, *Ind. Eng. Chem., Anal. Ed.*, **10,** 544 (1938).
57. C. E. Ransley, *G.E.C. Journal*, **11,** 135 (1940).
58. J. R. Young and N. R. Whetten, *Rev. Sci. Instrum.*, **31,** 1112 (1960).
59. R. W. Crompton and M. T. Elford, *J. Sci. Instrum.*, **39,** 480 (1962).
60. See Ref. 9, p. 138.
61. J. F. Martin, R. C. Takacs, R. Rapp, and L. M. Melnick, *Trans. AIME*, **230,** 107 (1964).
62. See Ref. 9, p. 133.
63. C. E. Ransley, *Analyst*, **72,** 504 (1947).
64. T. A. Izmanova, Yu. A. Klyachko, and N. S. Larichev, *Tr. Komissii Anal. Khim., Akad. Nauk SSSR, Inst. Geokhim. Anal. Khim., im V. I. Vernadskogo*, **10,** 267 (1960); through *Chem. Abst.*, **55,** 6946c (1961).
65. A. S. Darling, *Platinum Met. Rev.*, **2,** 16 (1958).
66. M. van Swaay and C. E. Birchenall, *Trans. AIME*, **218,** 285 (1960).
67. Z. M. Turovtseva, N. F. Litvinova, G. V. Mikhailova, A. S. Noskov, and R. Sh. Khalitov, *Zh. Anal. Khim.*, **12,** 208 (1957).
68. E. A. Gulbransen, *Trans. Electrochem. Soc.*, **81,** 327 (1942).
69. E. A. Gulbransen and K. F. Andrew, *Trans. AIME*, **203,** 136 (1955).
70. J. B. Hunter, U.S. Patent No. 2,773,561 assigned to Atlantic Refining Co., December 11, 1956.
71. E. J. Serfass, *Operating Instructions, Serfass Gas Analyzer, Bulletin FS-270*, Fisher Scientific Co., Pittsburgh, 1957.

172. (a) J. R. Young, *Rev. Sci. Instrum.*, **34,** 891 (1963); and (b) A. S. Darling, British Patent 956,176, April 22, 1964.
173. R. A. Yeaton, *Vacuum*, **2** (2), 115 (1952).
174. *Methods for Chemical Analysis of Metals*, American Society of Testing and Materials, Philadelphia, 1960, pp. 298, 524, 558.
175. H. C. Davis and J. A. Gray, *Royal Aircraft Establishment Report Met.*, **1955,** 86; through Ref. 32.
176. V. V. Nedin, D. Z. Gel'man, A. P. Yanov, and A. K. Matvienko, USSR Pat. 128,450, May 15, 1960; through *Chem. Abstr.*, **54,** P23125h (1960).
177. W. G. Guldner, private communication, 1963.
178. V. S. Baikov, *Primenenie Vakuuma v Met., Akad. Nauk SSSR, Inst. Met. im A. A. Baikova*, **1960,** 320; through *Chem. Abstr.*, **55,** 12145c (1961).
179. A. A. Morton, *Laboratory Technique in Organic Chemistry*, McGraw-Hill, New York, 1938, p. 7.
180. J. J. Manley, *J. Chem. Soc.* (*London*), **121,** 331 (1922).
181. W. F. Hillebrand, G. E. F. Lundell, J. I. Hoffman, and H. A. Bright, *Applied Inorganic Analysis*, 2nd ed., Wiley, New York, 1953, p. 48.
182. S. Vigo, Chairman, *Panel on Methods of Analysis*, *Metallurgical Advisory Committee on Titanium*, *Information Bulletin No. T8*, *Part III*, AD277228, Office of Technical Services, Washington, D.C., 1962.
183. G. F. Smith and H. Diehl, *Talanta*, **3,** 107 (1959).
184. W. F. Harris and W. M. Hickam, *Anal. Chem.*, **31,** 281 (1959).
185. H. L. Hamner and R. M. Fowler, *Trans. AIME*, **194,** 1313 (1952).
186. D. I. Walter and G. A. Picklo, Jr., Pittsburgh Conference on Analytical Chemistry and Applied Spectroscopy, Pittsburgh, 1959.
187. See. Ref. 9, p. 126.
188. A. Fuchs, H. Reinhard, J. Niebuhr, and R. Beck, *Arch. Eisenhuttenw.*, **34,** 361 (1963).
189. (a) R. J. Conzemius and H. J. Svec, *Anal. Chim. Acta*, **33,** 145 (1965); (b) Leaflet No. 884-E, Strohlein & Co., Dusseldorf, West Germany, 1968; and (c) Form 184b, Laboratory Equipment Corp., St. Joseph, Michigan.
190. G. Houghton, *J. Sci. Instrum.*, **33,** 199 (1956).
191. G. W. Green, *J. Sci. Instrum.*, **35,** 147 (1958).
192. C. Steel, R. F. Smith, and B. Sunners, *J. Sci. Instrum.*, **34,** 125 (1957).
193. B. B. Bach, J. V. Dawson, and L. W. L. Smith, *J. Sci. Instrum.*, **31,** 343 (1954).
194. See Ref. 26, p. 331.
195. (a) See Ref. 27, p. 353; and (b) Ref. 29, pp. 133–168.
196. W. M. Hickam, *Anal. Chem.*, **24,** 362 (1952).
197. D. T. Peterson and D. J. Beernsten, *Anal. Chem.*, **29,** 254 (1957).
198. E. Kato, *Rept. Castings Research Lab.*, *Waseda Univ.* (Tokyo), **No. 8,** 81 (1957); through *Chem. Abstr.*, **52,** 7064f (1958).
199. R. E. Taylor, *Anal. Chim. Acta*, **21,** 549 (1959).
200. M. L. Aspinal, *Analyst*, **91,** 33 (1966).
201. M. L. Aspinal and D. Hazelby, Int. Vac. Met. Conf., New York, June 15, 1967, and AEI Research Publication CRL 131, 1967.

202. V. I. Yavoiskii, L. B. Rosterev, V. L. Safonov, and M. I. Afanas'ev, *Izv. Vyssh. Ucheb. Zavedo, Chern. Met.*, **10,** 39 (1967).
203. O. Kammori, N. Yamaguchi, and H. Kanno, *Nippon Kinzoku Gakkaishi*, **31,** 679 (1967).
204. A. J. B. Robertson, *Mass Spectrometry*, Wiley, New York, 1954.
205. G. P. Barnard, *Modern Mass Spectrometry*, Institute of Physics, London, 1953.
206. A. I. M. Keulemans and C. G. Verver, *Gas Chromatography*, Reinhold, New York, 1957.
207. R. L. Pecsok, Ed., *Principles and Practice of Gas Chromatography*, Wiley, New York, 1959.
208. S. Dal Nogare and R S. Juvet, Jr., *Gas-Liquid Chromatography*, Wiley, New York, 1962.
209. G. Kyryacos and C. E. Boord, *Anal. Chem.*, **29,** 787 (1957).
210. K. Wencke, *Chem. Tech.* (Berlin), **9,** 404 (1957).
211. J. Janak, M. Krejci, and H. E. Dubsky, *Chem. Listy*, **52,** 1099 (1957); through *Chem. Abstr.*, **52,** 15331h (1958).
212. J. Janak, M. Krejci, and H. E. Dubsky, *Collect. Czech. Chem. Commun.*, **24,** 1080 (1959).
213. J. F. Ellis and C. W. Forrest, *Anal. Chim. Acta*, **24,** 329 (1961).
214. J. Vaclavinek, *Hutnicke Listy*, **14,** 905 (1959); through *Anal. Abstr.*, **7,** 3556 (1960).
215. H. Feichtinger, H. Bachtold, and W. Schuhknecht, *Schweiz. Archiv. Angew. Wiss. Tech.*, **25,** 426 (1959).
216. C. Baque and L. Champeix, *Rev. Met.*, **57,** 919 (1960).
217. R. Lesser and H. Gruber, *Z. Metallk.*, **51,** 495 (1960).
218. R. Lesser, *Angew. Chem.*, **72,** 775 (1960).
219. W. Hein and K. Lohberg, *Giesserei Tech.-Wiss. Beih.*, **13,** 221 (1961); through *J. Iron Steel Inst.*, **200,** 685c (1962).
220. R. Lesser in L. E. Preuss, Ed., *Transactions of the Eighth National Vacuum Symposium and Second International Congress on Vacuum Science and Technology*, (1961), Vol. II, Pergamon, New York, 1962, p. 782.
221. L. L. Lewis and L. M. Melnick, *Anal. Chem.*, **34,** 868 (1962).
222. B. Marincek, *Neue Zurcher Zeitung* (Switzerland), No. 675, Feb. 2, 1962; through *Technological Digests*, **7,** (12) 77 (1962).
223. J. A. Roff, *Chem. Ind.* (London), March 30, 1963, p. 537.
224. Yu. A. Klyachko, E. M. Chistyakova, and Yu. D. Labut'ev, *Sb. Tr. Tsentr. Nauchm.-Issled. Inst. Chernoi Met.*, **No. 31,** Pt. 1, 87 (1963); through *Chem. Abstr.*, **59,** 5761h (1963).
225. F. Sperner and K. H. Koch, *Metallurgia*, **18,** 701 (1964).
226. F. Sperner and K. H. Koch, *Metallurgia*, **19,** 742 (1965); through *Chem. Abstr.*, **63,** 11103b (1965).
227. J. F. Martin and L. M. Melnick, American Society for Testing and Materials, Comm. E-3, Symposium on Recent Advances in the Determination of Gases in Metals, Atlantic City, N.J., 1966.
228. M. T. Lilburne, *Analyst*, **91,** 571 (1966).
229. M. T. Lilburne, *Talanta*, **14,** 1029 (1967).

230. (a) J. L. Botts, H. G. Davis, and W. R. Laing, Eleventh Conference on Analytical Chemistry in Nuclear Technology, Gatlinburg, Tenn., 1967; (b) W. T. Barnes, J. E. Maurits, and D. E. Wilson, *Lab. Management*, July, 1966, pp. 16–18, 51; and (c) F. R. Coe, *Analyst*, **92,** 199 (1967).
231. Linde Company, Linde Molecular Sieves, *Form 9947-A*, New York, 1959.
232. E. L. Obermiller and G. O. Charlier, *J. Gas Chromatogr.*, **6,** 446 (1968).
233. J. Janak, M. Krejci, and H. E. Dubsky, *Ann. N.Y. Acad. Sci.*, **72,** 731 (1959).
234. P. G. Jeffery and P. J. Kipping, *Gas Analysis by Gas Chromatography*, Pergamon, Oxford, 1964.
235. A. Weinstein, *Chem. Ind.* (London), Oct. 24, 1959, p. 1347.
236. S. A. Stern and F. S. DiPaolo, *J. Vac. Sci. Tech.*, **4** (6), 347 (1967).
237. S. Sawa, T. Mori, and H. Tsumita, *Tetsu To Hagane*, **49,** 643 (1963).
238. K. B. McAfee, private communication to M. T. Lilburne, noted in Ref 228.
239. R. Berry, *Nature*, **188,** 578 (1960).
240. H. Feichtinger, H. Bachtold, and K. Brauner, *Schweiz. Arch. Agnew. Wiss. Tech.*, **28,** 125 (1962).
241. T. Kraus, *Arch. Eisenhuttenw.*, **33,** 527 (1962).
242. T. Kraus and O. Winkler, *Rev. Met.*, **61,** 87 (1964).
243. G. Ramsey, O. Winkler, and T. Kraus, "A New Instrument for Determination of Gas Content in Metals," in M. A. Cocca, Ed., *Transactions Vac. Met. Conf.*, Amer. Vac. Soc., Boston, 1964, pp. 421–31.
244. T. Kraus, *Z. Anal. Chem.*, **209,** 206 (1965).
245. N. N. Timoshenko, T. A. Izmanova, and E. M. Chistyakova, *Zavod. Lab.*, **31,** 1068 (1965).
246. K. Styblo, F. Ermis, P. Pivoda, and M. Kovarik, *Hutnicke Listy*, **20,** 288 (1965); through *Anal. Abstr.*, **13,** 4149 (1966).
247a. A. Gerhardt, T. Kraus, and M. G. Frohberg, "New Methods of Oxygen and Nitrogen Control During Steel Production," in M. A. Orehoski and R. F. Bunshah, Ed., *Transactions Vac. Met. Conf.*, Amer. Vac. Soc., Boston, 1966, pp. 82–97.
247b. Circular P51-76 7108e, Exhalograph EAO 202, Balzers Aktiengesellschaft fur Hockvakuumtechnik und Dunne Schichten, Liechtenstein, 1971.
247c. Circular P51-78 7108d, Exhalograph EAN 202, Balzers Aktiengesellschaft fur Hockvakuumtechnik und Dunne Schichten, Liechtenstein, 1971.
248. M. Hanin and E. Jaudon, *Rev. Met.*, **62,** 37 (1965).
249. W. E. Dallman and V. A. Fassel, *Anal. Chem.*, **39,** 133R (1967).
250. W. G. Smiley, *Anal. Chem.*, **27,** 1098 (1955).
251. J. I. Peterson, F. A. Melnick, and J. E. Steers, Jr., *Anal. Chem.*, **30,** 1086 (1958).
252. W. G. Guldner, *Talanta*, **8,** 191 (1961).
253. L. Singer, *Ind. Eng. Chem., Anal. Ed.*, **12,** 127 (1940).
254. E. G. Bobalek and S. A. Shrader, *Ind. Eng. Chem., Anal. Ed.*, **17,** 544 (1945).
255. N. Yoneda, *Nippon Kinzoku Gakkaishi*, **21,** 392 (1957); through *Chem. Abstr.*, **56,** 7993d (1962).
256. C. E. A. Shanahan and F. Cooke, *J. Iron Steel Instr.*, **188,** 138 (1958).
257. B. D. Holt and H. T. Goodspeed, *Anal. Chem.*, **34,** 374 (1962).

258. J. Hancart and J. Marot, *Rev. Met.*, **57,** 911 (1960).
259. K. Abresch and H. Lemm, *Arch. Eisenhuttenw.*, **30,** 1 (1959).
260. P. Elbling and G. W. Goward, *Anal. Chem.*, **32,** 1610 (1960).
261. C.V.Banks, J.W.O'Laughlin, and G. J. Kamin, *Anal. Chem.*, **32,** 1613 (1960).
262. T. A. Sullivan, B. J. Boyle, A. J. Mackie, and R. A. Plott, *Rept. Invest. No. 5834*, U.S. Bur. Mines, 1961; through *Chem. Abstr.*, **55,** 26837c (1961).
263. V. A. Fassel, W. C. Dallmann, R. Skogerboe, and V. M. Horrigan, *Anal. Chem.*, **34,** 1364 (1962).
264. S. Kallmann and F. Collier, *Anal. Chem.*, **32,** 1616 (1960).
265. American Society for Testing and Materials, *1968 Book of ASTM Standards*, Pt. 32, Philadelphia, 1968, pp. 685–92.
266. H. Goto, S. Ikeda, A. Onuma, *Sci. Rep. Res. Inst. Tohoku Univ.*, **17,** 318 (1965).
267. E. J. Beck and W. E. Chambers, *Anal. Chim. Acta*, **43,** 348 (1968).
268. A. M. Vasserman and Z. M. Turovtseva, *Zh. Anal. Khim.*, **20,** 1359 (1965).
269. G. J. Kamin, J. W. O'Laughlin, and C. V. Banks, *Anal. Chem.*, **35,** 1053 (1963).
270. J. Kashima and T. Yamazaki, *Bull. Chem. Soc. Jap.*, **39,** 1448 (1966).
271. (a) J. Kashima and T. Yamazaki, *Jap. Analyst*, **15,** 9 (1966); (b) M. Hanin and D. Villeneuve, *Chim. Anal.* (Paris), **48,** 442 (1966); (c) M. Hanin and D. Villeneuve, *ibid.*, **47,** 634 (1965); and (d) P. Escoffier, *ibid.*, **49,** 208 (1967).
272. C. G. Goldbeck, S. P. Turel, and C. J. Rodden, *Anal. Chem.*, **40,** 1393 (1968).
273. C. G. Goldbeck, S. P. Turel, and C. J. Rodden, Eleventh Conference on Analytical Chemistry in Nuclear Technology, Oak Ridge, Tenn., 1967.
274. W. Koch and H. Lemm, Pittsburgh Conference on Analytical Chemistry and Applied Spectroscopy, Cleveland, Ohio, 1968.
275. (a) W. Kock and H. Lemm, *Arch. Eisenhuttenw*, **38,** 881 (1967). (b) W. Koch and H. Lemm, *Z. Anal. Chem.*, **215,** 377 (1966). (c) G. Jecko and R. Touvenin, *Rev. Met.*, **66,** 823 (1969). (d) R. W. Cline, H. S. Karp, and L. M. Melnick, Pittsburgh Conference on Analytical Chemistry and Applied Spectroscopy, Cleveland, Ohio, 1970.
276. H. Lemm, *Z. Anal. Chem.*, **209,** 114 (1965).
277. Form No. C258, Laboratory Equipment Corp., St. Joseph, Michigan, 1969.
278. H. Lemm, Pittsburgh Conference on Analytical Chemistry and Applied Spectroscopy, Pittsburgh, Pa., 1967.
279. Form No. 184D, Laboratory Equipment Corp., St. Joseph, Michigan, 1967,
280. P. Tyou, *Inst. Hierro y. Acero*, **13,** 383 (1960); through *Chem. Abstr.*, **54,** 19290h (1960).
281. Form No. 152A, Laboratory Equipment Corp., St. Joseph, Michigan.
282. A. W. Mosen, R. E. Kelley, and H. P. Mitchell, *Talanta*, **13,** 371 (1966).
283. Form No. 510-E, Strohlein and Co., Dusseldorf, West Germany, 1967.
284. M. E. Smith, J. M. Hansel, R. B. Johnson and G. R. Waterbury, *Anal. Chem.*, **35,** 1502 (1963).
285. M. E. Smith, J. M. Hansel, and G. R. Waterbury, *Anal. Chem.*, **37,** 782 (1965).

CHAPTER

5
ELECTRICAL DISCHARGE EXTRACTION METHODS

ROYCE K. WINGE AND VELMER A. FASSEL

Ames Laboratory-USAEC
and
Department of Chemistry
Iowa State University
Ames, Iowa

CONTENTS

I. INTRODUCTION

A. SCOPE

Most methods for determining the gaseous elements in metals require that they first be extracted in one form or another from the metal. In general, therefore, the apparatus consists of two main sections: the extraction system, which converts the oxygen, nitrogen, and hydrogen contents to CO, N_2, and H_2, and the analysis section, which separates, if necessary, and quantitatively measures each component.

This chapter will be subdivided in like manner, with the extraction and analytical systems discussed in separate sections. Only those techniques in which the electrical discharge is used to melt a relatively large sample (approximately 0.1–2 g) and to extract completely its gaseous element content will be discussed. Discharges of this type include the dc-arc, the hollow cathode, and, if the electrical discharge is broadly defined, even the electron beam.

B. HISTORICAL SURVEY

As is evident from the other chapters in this book, various modifications of the vacuum-fusion, carrier-gas-fusion, hot-extraction, and Kjeldahl chemical techniques have been used extensively for determining oxygen, nitrogen, and hydrogen in metals. Often these techniques have required considerable time per analysis and have involved numerous procedural subtleties. Spectral methods, involving arc or spark excitation sources, have been widely used for the rapid determination of alloying elements and traces of metallic impurities in metals. It was therefore natural to apply these methods to determining the nonmetallic impurities as well, and thus was started the utilization of electrical discharges for the determination of gaseous elements in metals.

1. High-Voltage-Spark and High-Energy-Impulse Discharges

High-voltage-spark and high-energy-impulse discharges may be used for the detection and determination of oxygen, nitrogen, and hydrogen in the sample vaporized by these discharges. These techniques are discussed in detail in Chapter 6.

2. Other Discharge Types

Other discharge types capable of melting a more massive sample may allow, under appropriate experimental conditions, extraction of the total oxygen, nitrogen, or hydrogen content of metals. With the greater quantities of gases obtained through complete melting of a sample, other analytical techniques are applicable. Yudowitch [1] explored the possibility of determining oxygen in titanium by arcing self-electrodes in an argon atmosphere and observing the TiO band spectra but was unable to detect less than 2–3% oxygen. Oxygen in steel was successfully determined by Rosen [2–4], who used a hollow-cathode discharge for melting the metal and exciting a carbon monoxide band-head. Concentrations as low as 0.0010% could be determined if samples weighing 10 g were used. Because of the large samples, long extraction periods were required and the procedure as a whole was slow. Fassel and Tabeling [5] were the first to apply successfully a dc-arc discharge to extraction and spectral excitation of trace concentrations of gases in metals. Since that time, the dc-arc extraction technique, in combination with several other analytical schemes, has been successfully applied to determining the gaseous element content of an extended list of metals and alloys. To reduce the time required per analysis, Evens and Fassel [6] replaced the photographic spectral method with a gas-chromatographic technique for the simultaneous determination of oxygen and nitrogen in steels. This procedure was extended to the determination of oxygen and nitrogen in a number of refractory metals [7]. Matsumoto et al. [8] later developed a direct-reading spectrometric method of determining oxygen in steels.

Grieser et al. [9] have investigated the determination of oxygen in sodium metal by a dc-arc spectrometric technique. To reduce quenching of the excitation process by sodium vapor, the metal was vaporized from oxide residuals by resistance heating of the electrode prior to the excitation step. Oxygen emission could not be obtained, however, from the residual of sodium metal samples (if any remained), a phenomenon the authors attributed to difficulties in the heating system for volatilizing the sodium. Grieser et al. did obtain oxygen–argon intensity ratios proportional to the oxygen content of calcium, magnesium, and sodium compounds.

Ivanova et al. [10] have used a dc arc for the epuilibration and excitation steps of a spectral-isotopic method for determining oxygen and nitrogen in titanium and niobium.

Fromm [11] extracted gases from large samples (10–20 g) of niobium, tantalum, and vanadium by means of an electron beam. The gases were analyzed gas chromatographically. The purpose of this work was not to determine quantitatively the gaseous element content but to determine the composition of gases evolved from these metals during the electron-beam purification procedure. Guenther et al. [12–13] used an electron beam to extract gases from copper. A quadrupole mass spectrometer was used as the analysis system. With gradual heating of the sample the surface gases could be distinguished from the bulk gases. The method was reported to have a detection limit of a few picograms per gram of sample.

Table 5.1 summarizes the essential information of published work involving electrical discharges for the extraction and determination of the gaseous element content of metals.

II. EXTRACTION SYSTEMS

A. EXPERIMENTAL FACILITIES AND PROCEDURES

1. Extraction Chamber

The extraction chamber serves two principal purposes: it prevents atmospheric gases from contaminating the extracted gases, and it confines the extracted gases until a gas sample can be taken or until an analysis can be completed. The typical chamber, shown in Fig. 5.1, consists of a vacuum-tight enclosure containing a horizontal, rotating electrode stage on which a number of sample electrodes can be rotated into arcing position. Regardless of whether a gas chromatographic or spectroscopic analytical method is used, it is advantageous to keep the chamber volume small to reduce dilution of the extracted gases. Chamber volumes as low as 25 cm^3 were evaluated by the Applied Research Laboratories [39], but this small size proved to be

TABLE 5.1. Summary of Electrical-Discharge Extraction Methods

Method	Extraction: Atmosphere	Extraction: Flux	Analytical Method	Elements Determined	Sample	Reference Literature
DC arc	Argon	Pt	Spectroscopy	O	Zr, Zircalloy	14
	Argon	None, Pt	Spectroscopy	O	Nb	15
	Helium	None	Gas chromatography	O	Steels	6
	Argon	None	Spectroscopy	O	V	16
	Argon	None, Pt	Spectroscopy	O	Steel, V, Nd, Ta, Y, Th, Ti, Zr, Hf	17
	Argon	Pt	Spectroscopy	O	Ti, Ti alloys	18
	Argon	None	Spectroscopy	O, N	Steel	19
		Pt		O	Ti, Zr, Nb, Y	
		None		H	Ti	
	Argon	None	Spectroscopy	O	Steel	20
		Pt		O	Ti	
		Ni		O	La	
	Argon	None	Spectroscopy	O	Steel	5
	Argon	None	Spectroscopy	O	Steel	21–22
	Argon	Pt	Spectroscopy	O	Ti	23
	Argon	None	Infrared absorption	O	Steel	24
	Argon	None	Gas chromatography	O, H, N	Steel	25
	Argon	Pt	Condensation-volumetric	O	Y	26
	Argon	None	Condensation-volumetric	O	Steel	27
	Argon	None	Spectroscopy (direct reading)	O	Steel	8
	Argon	None	Spectroscopy	O, H	Fe–Al	28
		Pt		O, H	Zr, U	
	Helium	Pt	Gas chromatography	O	Cr, Hf, Mo, Nb, Y, Zr, Steel, Ta, Tb, Th, Ti, V	7
				N	Cr, Nb, Steel, Tb, Th, Ti, V, Y	

TABLE 5.1. *(continued)*

Method	Extraction: Atmosphere	Extraction: Flux	Analytical Method	Elements Determined	Sample	Reference Literature
	Helium	None, Pt Ag, Sn	Gas chromatography	O, N	Steel	29
	Helium	None	Spectral-isotope dilution	O, N	Ti, Nb	10
	Argon	None	Spectroscopy	O	Na, Ca, Mg	9
	Argon	None	Spectroscopy	O	Zr, UC	31
	Argon		Spectroscopy	H	Zircalloy	32
	Argon	None	Emission spectroscopy	O	SiC	90
	Argon	None	Emission spectroscopy	O	TiC	91
	Argon	None	Emission spectroscopy	O	HfC	92
	Helium	Pt	Gas chromatography	O, N	Zr, Th, Cu, Steel, Ti	30
	Argon	Pt	Gas chromatography	H	Ti, Zr	40
	Helium	None	Emission spectroscopy	None	Steel	41
	Argon	None	Infrared absorption; emission spectroscopy	O N	Steel	83
	Argon	None	Emission spectroscopy	O	Nb	84
	Argon	None	Emission spectroscopy	H	Ti powder	85
	Argon	Ni	Emission spectroscopy	O	Ti	86–89
Electron beam	None (high vacuum)	None	Gas chromatography	O, H, N	Nb, Ta, V	11
	None (high vacuum)	None	Mass spectroscopy	H, CO, CO_2	Cu	12, 13

Hollow cathode	Argon	None	Spectroscopy	O	Steel	2–4
	Helium	None	Spectroscopy	O, N	Steel	33–36
		Ni		O, N	UC, W, Cu	
		None		H	Zr, Zircalloy	
	Helium	None	Spectral-isotope	O	Sn, Pb, Te, Al, Ge, Cu, Nb, Mo, W	93
	Helium	None	Spectral-isotope	O	Si	94
Impulse capsule	Argon	None	Infrared absorption	O	TaC, NbC, ZrN, Ta, Nb, V	37
	Helium	Sn with Be	Condensation-gas chromatography	O, N	B, BN, Be, BeO, U, UN_x, Th, ThO_2, Ilmenite	38

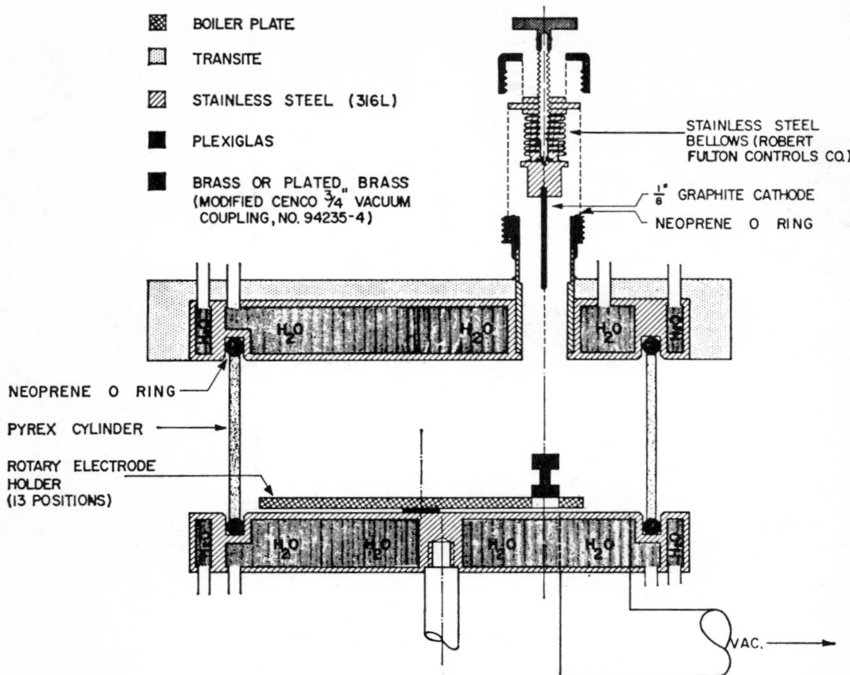

Fig. 5.1. Extraction chamber of Evens and Fassel. (Reproduced by permission of *Analytical Chemistry* [15].)

impractical because of a tendency of the arc to strike to the chamber wall. This problem was eliminated by increasing the chamber volume to 50 cm³, but another problem became apparent with this chamber. More than 50% of the carbon monoxide and nitrogen added to the chamber was lost by adsorption on the walls or graphite deposits or by gettering by vaporized metal [39] (see Section II.B.3 and Chapter 4 for discussions of gettering). Internal volumes of other chambers have ranged from 50 cm³ [25, 27] to approximately 2 liters [14].

Kashima [27] and Kamada et al. [25] used small quartz arc chambers having a single sample electrode. The chamber of Kamada is shown in Fig. 5.2. Vasserman and Turovtseva [37] have used an interesting variation of the dc-arc technique for achieving extraction temperatures up to 3000°C. In their procedure a small carbon capsule, shown in Fig. 5.3, is resistance-heated by an impulse current of about 400 A for 3 sec. The liberated carbon monoxide diffuses through the heated walls, while vaporized metal is retained as the metal carbide within the capsule. The chamber used in this work is shown

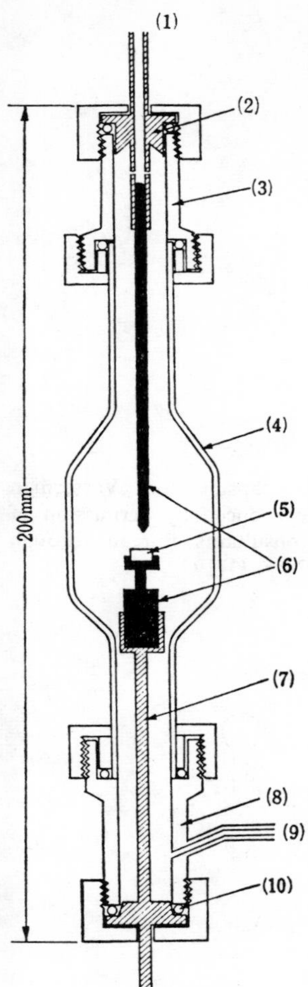

Fig. 5.2. Extraction chamber of Kamada: (1) gas outlet; (2) stainless steel; (3) brass; (4) quartz; (5) metal sample; (6) graphite electrode; (7) stainless steel; (8) brass; (9) gas inlet; (10) O-ring. (Reproduced by permission of the authors and *Bunseki Kagaku* [25].)

in Fig. 5.4. Webb and Webb [33–36] used a hollow cathode as an extraction furnace. A schematic diagram of this chamber is shown in Fig. 5.5. The chamber contains a turret assembly, so that up to 9 samples can be added to the extraction crucible cathode. A shutter system separates the sample-turret from the extraction chamber. Thus, samples may be added without exposing the chamber to the atmosphere. The hollow-cathode discharge serves as the excitation source for spectrometric analysis of the extracted gases. A compact arc chamber with a capacity for 28 electrodes and designed for the spectrographic determination of the gaseous elements in metals was developed by Spitz and Van Danh [31]. It is shown in Fig. 5.6.

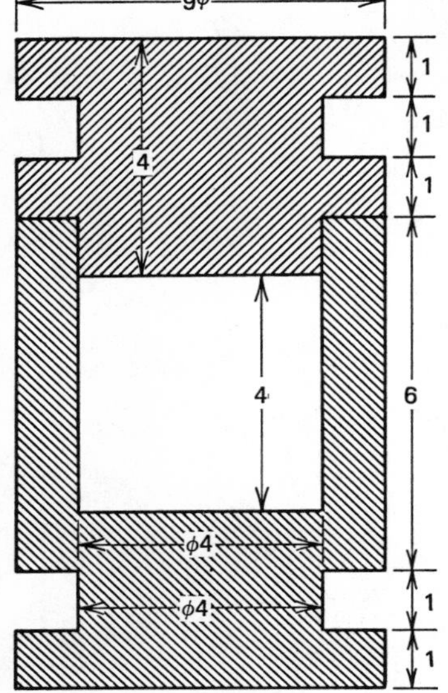

Fig. 5.3. Extraction capsule of Vasserman and Turovtseva. (Reproduced by permission of copyright holder, Consultants Bureau through Plenum Publishing Crop. [37].)

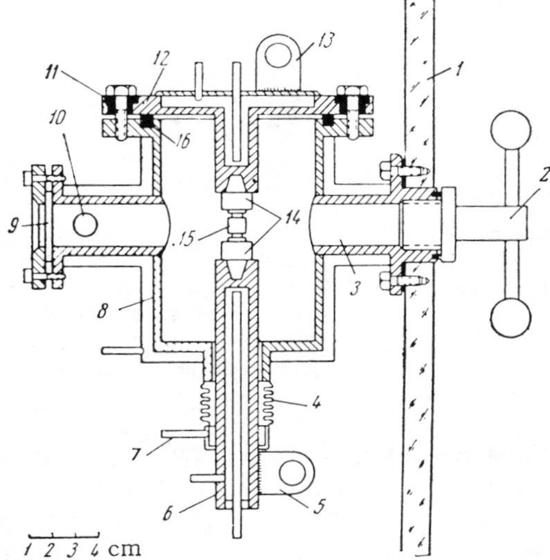

Fig. 5.4. Extraction chamber of Vasserman and Turovtseva: (1) plexiglas glovebox front; (2) stopper; (3) charging hatch; (4) bellows; (5, 13) current-carrying busbars; (6) water-cooled movable contact; (7) argon supply fitting; (8) water-cooled chamber; (9) viewport; (10) vacuum port; (11) texolite plugs insulate lid from chamber; (12) water-cooled lid with stationary contact; (14) molybdenum terminals; (15) heated capsule; (16) rubber seal. (Reproduced by permission of copyright holder, Consultants Bureau through Plenum Publishing Corp. [37].)

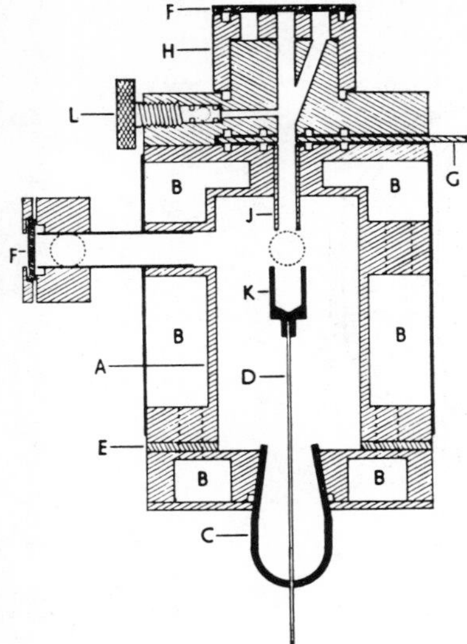

Fig. 5.5. Extraction chamber of Webb and Webb: (a) stainless-steel cathode chamber; (b) water jacket; (c) Pyrex glass joint; (d) 2 mm tungsten rod; (e) sheet rubber gasket; (f) quartz windows; (g) shutter; (h) turret sample-head; (j) anode (at earth potential); (k) cathode; (l) valve. (Reproduced by permission of the authors and Elsevier Scientific Publishing Co. [34].)

Another chamber, designed by the authors [29] and shown in Fig. 5.7, has a short cylindrical form (1) with a horizontal axis. In contrast to previous chambers, "dead volume appendages" such as vacuum gauges, extension tubes for viewports, valve body volumes, and the associated manifold were either eliminated or separated from the chamber by valves. The resulting consolidated volume of the chamber, with its maximum dimension in the vertical direction, allows the extracted gases to homogenize very rapidly with the extraction atmosphere through convection currents set up by the arc discharge. The internal volume of the chamber is approximately 650 cm^3. The internal diameter is 11.4 cm, and the depth is 6.35 cm. The electrode stage is rotated in a vertical plane through a gear set (2) (recessed into cover plate) and knob (3). Experimental use of first the 5-electrode stage (4) and then the 10-electrode stage (5) indicated that a 15-electrode stage could very probably be accommodated in the same compact chamber without danger of arc-over to adjacent electrodes.

A telescoping sample loading arm (6), which penetrates the rear of the chamber (7) at a 45° angle, provides storage for 10 samples. Each sample is loaded into the appropriate electrode just prior to its extraction step. A vacuum sample-introduction lock can be added to the loading arm to permit an analysis of a sample a very short time after its preparation. The vertical electrode stage allows the use of a novel method for minimizing carbon deposits on the viewport, which is especially important if spectroscopic measurements are to be made with the chamber. A quartz window is located in each

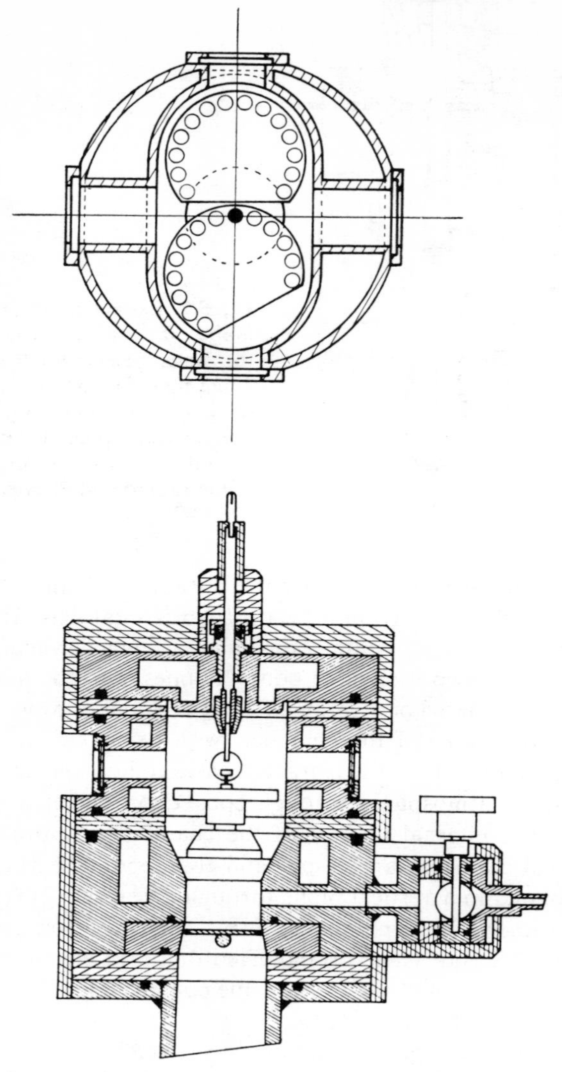

Fig. 5.6. Extraction chamber of Spitz and Van Danh. (Reproduced by permission of the authors and GAMS [31].)

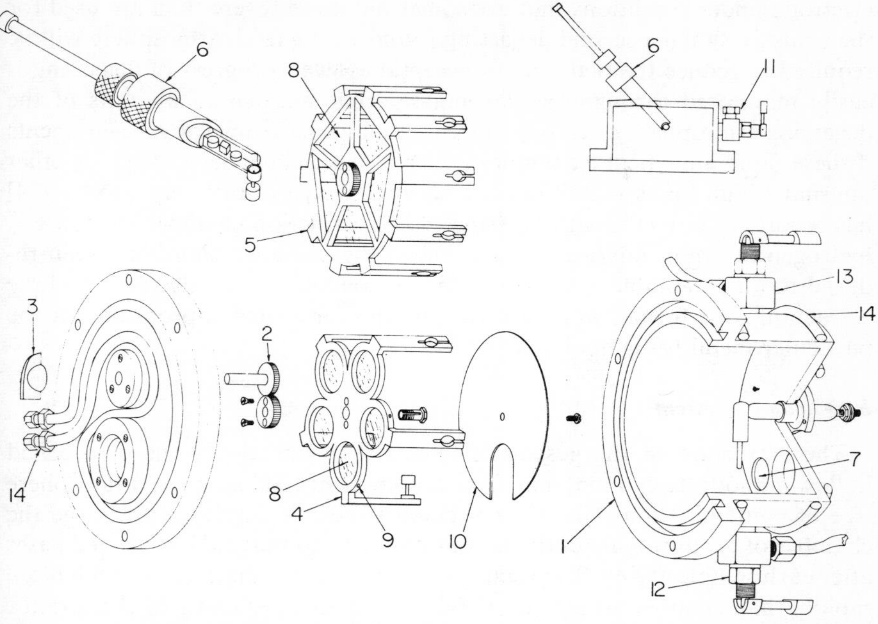

Fig. 5.7. Extraction chamber of Winge and Fassel. Legend numbers are described in text. (Reproduced by permission of *Analytical Chemistry* [29].)

of the five circular or trapezoidal ports (8) in the electrode stage and is retained by a Teflon-tipped setscrew (9). All windows, with the exception of the portion in line with the electrode in arcing position, are protected from deposits by a stainless steel shield (10). The inlet for the extraction atmosphere, the outlet to the gas-chromatography sampling valve, and the port to the high-vacuum system are controlled by valves (12, 13, and 11, respectively). The chamber is water cooled through copper cooling coils (14).

a. Operating Procedure

Operating procedures for the various extraction chambers differ according to the particular extraction technique and to the requirements of the sample metal being analyzed. Consultation with the original literature is recommended for specific operating details. In general, the operating procedure for an arc extraction chamber similar to that shown in Fig. 5.7 is as follows: After cleaning the interior surfaces of the chamber, the electrodes are placed, the samples are loaded into the storage arm, and the chamber is sealed, evacuated, and filled with the extraction atmosphere to approximately atmospheric pressure. The chamber and electrodes are then degassed by arcing each electrode under conditions that are somewhat more severe than are used for the analysis. Often a second degassing round with a fresh atmosphere will be required to reduce the blank to an acceptable level. Progress of degassing is easily monitored through gas chromatographic analysis of aliquots of the degassing atmosphere or through direct-reading spectrometric measurements if these rapid analytical techniques are employed. Chamber leakage or other unusually high blank conditions are easily detected in this way. Höller [24] has reported more efficient degassing of the extraction chamber when a 6% hydrogen in argon mixture is used. When the chamber blank has been reduced to an acceptable level, the sample is loaded into an electrode and the extraction is performed at an arc current and for a time appropriate for the particular metal being analyzed.

b. Vacuum System

The extraction of the gaseous elements from metals by the dc-arc and hollow-cathode techniques must be accomplished in an inert atmosphere free of contamination. Usually a vacuum system is employed to purge the chamber of air during its initial preparation and to purge the extracted gases after each sample arcing. The vacuum system removes these gases much more rapidly than simply flushing; also, evacuation aids degassing of the interior surfaces of the chamber. A conventional high-vacuum system employing an oil-diffusion pump and a mechanical vacuum pump was used in the authors' laboratory.

B. FACTORS INFLUENCING EXTRACTION AND MEASUREMENT OF GASES

1. Extraction Atmosphere

a. Choice of Atmosphere

In the dc-arc extraction technique the primary purpose of the extraction atmosphere is to support the electrical discharge. Argon and helium serve equally well in this capacity with the exception that somewhat higher currents are required in argon to achieve the same sample-electrode temperatures [42]. The most important consideration in choosing an extraction atmosphere is that it be compatible with the analytical method.

For the extraction of gases by the hollow-cathode discharge, Rosen [2–4] used an argon atmosphere at a few torrs pressure. Webb and Webb [34] reported that more reliable results and higher sensitivity were obtained and less metal sputtering occurred with an atmosphere of helium than with argon.

hese differences may have greatest importance in relation to the spectro-
netric analytical technique; for example, to reduce gettering by some active netals, the rate of sputtering must be reduced (to be discussed in a later ection). Argon is often used in conjunction with dc-arc spectrometric nethods, because this element provides useful reference lines [5].

With gas chromatographic analytical methods the choice of extraction tmosphere depends on the type of detector and on the gases for which ighest sensitivity is desired. This is explained in greater detail in Section II.C. Where infrared absorption is used to measure carbon monoxide, elium and argon should perform equally well.

. *Purity*

The purity of the extraction atmosphere is very important, because many f the common impurities in helium or argon contribute to the analytical lank. In many cases it is possible to obtain commercial argon or helium of ufficient purity. To maintain this purity, it is important to use only clean netal tubing and fittings from the gas cylinder to the extraction chamber and nalysis system. High-vacuum fittings containing swageable metal ferrules vork very well. Rubber tubing tends to contaminate a pure gas stream.

Pressure regulators containing rubber diaphragms are often a source of ontamination. Regulators with diffusion-resistant metal diaphragms should e used. Another source of contamination arises from failure to remove air rom the regulator before opening the tank valve. If the regulator outlet alve is closed when the tank valve is opened, the air trapped in the regulator s free to diffuse into the tank. This could easily account for a 5–10-ppm ontamination of the tank if the total regulator volume of air diffused into he tank. Even if the regulator is flushed first by opening its outlet valve efore opening the tank valve, air trapped in the gauges and other occluded paces within the regulator may bleed out for an extended period of time. This problem is best eliminated by evacuating the regulator and connecting ubing through the vacuum system of the extraction chamber before the ank valve is opened.

There are several relatively simple methods of purifying helium. Purer t al. [43] employed a cryogenically cooled (35°K) charcoal adsorption trap operated at high pressure to prepare helium containing only 2 parts neon per 10^9 and no detectable traces of other impurities. Limoncelli [44] obtained more than 99.9999% pure helium by passing the gas through a trap containing a synthetic zeolite and then through a column of titanium sponge heated to 830°C. Another method for purifying helium involves its selective diffusion through heated quartz [45, 46]. Escoffier [47] used only a synthetic zeolite trap cooled in liquid nitrogen for the purification of helium. The minimum impurity levels of hydrogen, oxygen, nitrogen, methane, and carbon monoxide from the zeolite trap are defined by the equilibrium vapor pressures of these gases above the zeolite at a given loading level and temperature. The liquid-nitrogen-cooled zeolite trap should produce very pure helium if the adsorption bed is long enough to contain the elution bands of the impurities for the duration of the experiment. Greene [48] suggests that the amount of zeolite correspond to three times that necessary to contain the mass-transfer zone of the impurities. If the capacity of the zeolite trap is exceeded, the order of appearance of the impurities in the helium will be hydrogen, oxygen, nitrogen, methane, and carbon monoxide.

Argon is not as easily purified as helium. Limoncelli [44] purified argon to less than 1 ppm impurity by the method, mentioned above, for purifying helium.

2. Extraction Mechanisms

a. Stirring Action, Temperature Gradient

The dc-arc discharge extracts the gaseous elements from metals much more rapidly than the usual furnace fusion techniques. Fassel et al. [19] have attributed the rapid extraction to the high temperature in the vicinity of the anode spot (the highly luminous site where the arc discharge strikes the anode) and to the stirring action thought to be caused by the high temperature gradient in the molten metal globule. The temperature of the anode spot reaches approximately the boiling point of the metal [49], whereas the temperature at the sample-electrode interface is thought to be at approximately the melting point of the metal.

That the dissolution of carbon and its subsequent reaction with oxygen occur much more rapidly at high temperature in the molten metal is well known. Not known, however, is the mechanism that causes the apparent stirring effect that can easily be observed in the metal globule during arcing. It is speculated that this stirring action is caused by the variation of surface tension of the melt with temperature. In general, the surface tension is lowest at the anode spot and increases as the temperature decreases toward the supporting electrode. The molten globule is therefore in a nonequilibrium state, and in attempting to equilibrate, the high-surface-tension areas draw molten metal away from the anode spot. The globule remains in a nonequilibrium state, with the arc supplying heat at the top, while the electrode removes heat from below by conduction. This process could effectively cause mass transfer throughout the globule and continually expose new surface where degassing occurs more rapidly than through diffusion from the interior. In addition, the small size of the dc-arc sample (in contrast to the typical furnace-fusion melt) results in a much larger surface–volume ratio and, therefore, a more rapid evolution of the gases.

b. Electrode, Flux

The electrode plays a particularly important part in the extraction of gases from a metal sample. It constitutes a major part of the total extraction environment, which must provide favorable conditions for the complete degassing of the sample.

One of the main functions of the electrode is to support the metal sample in such a way that it can be heated to a temperature high enough to cause the chemically combined gases to be evolved. The constricted neck of the electrode provides the thermal isolation necessary to reach the required temperature with moderate arcing currents. As the sample melts, carbon dissolves into the molten metal at the electrode-sample interface. The role of carbon in the degassing of metal samples in the dc arc is probably quite similar to its role in the typical vacuum-fusion furnace. It is important that sufficient carbon dissolve in the metal to promote rapid conversion of the sample oxygen to carbon monoxide and yet not so much that the melt becomes too viscous to permit the evolution of the gases.

The amount of carbon that dissolves in the sample can be controlled by several techniques. Fassel and Altpeter [16] have described the influence of electrode geometry on extraction efficiency for the determination of oxygen in vanadium. Fig. 5.8A shows the geometry of an electrode that allowed the extraction of the oxygen content of 0.5-g vanadium samples in about 30 sec. The electrode platform proved to be too small, however, because about one-fifth of the molten globules would fall off. Figure 5.8B shows an electrode geometry on which the molten vanadium would remain. Approximately 60 sec were required for the oxygen extraction with this geometry. Other geometries have been developed for the rapid and complete degassing of niobium [15] and steel [16].

For certain metals the formation of refractory metal carbides may prevent the complete evolution of the gaseous element content [18]. The platinum-flux technique, achieved with the electrode assembly shown in Fig. 5.9, apparently retards or limits the transfer of carbon to the sample metal. Platinum has a carbon-solubility of about 1% at 1750°C [50]. Platinum possesses several other distinctive advantages as a fluxing agent. The boiling point of platinum is high and, consequently, its partial pressure at extraction temperatures is low. Platinum also reduces by dilution the vapor pressure of the sample metal. Platinum readily and exothermically forms alloys with even the most refractory metals. The effect of this alloying reaction is

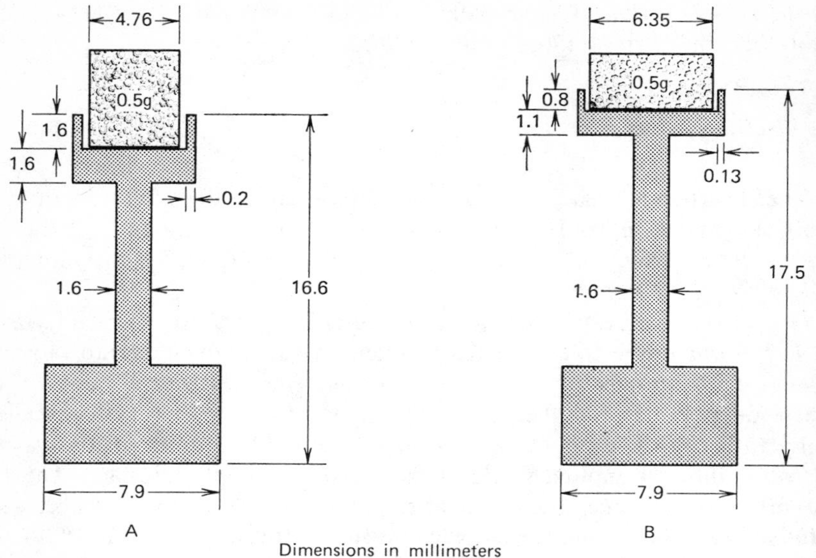

Fig. 5.8. Electrode assemblies used for vanadium. (Reproduced by permission of copyrigh holder, Microforms International [16].)

illustrated in Fig. 5.10 by the rapid evolution of the oxygen (evolved as carbon monoxide) and nitrogen contents of thorium. The alloying reaction occurs 3–5 sec after the arc is initiated. Within only a few seconds after this reaction the measureable carbon monoxide and nitrogen reach maximum values. These gases are evolved much more slowly when the thorium is arced without the platinum flux.

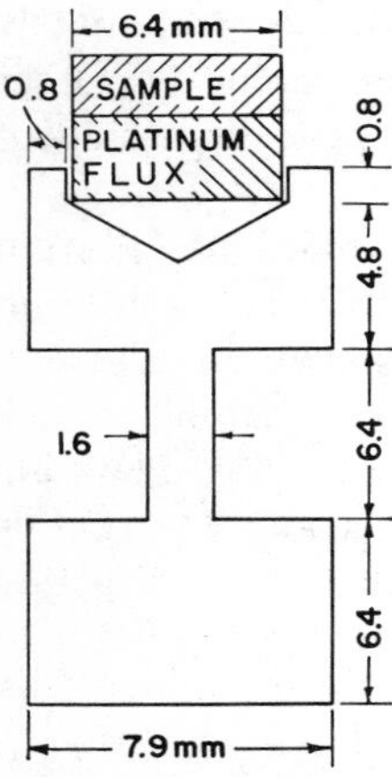

Fig. 5.9. Electrode assembly used for platinum-flux technique. (Reproduced by permission of *Analytical Chemistry* [7].)

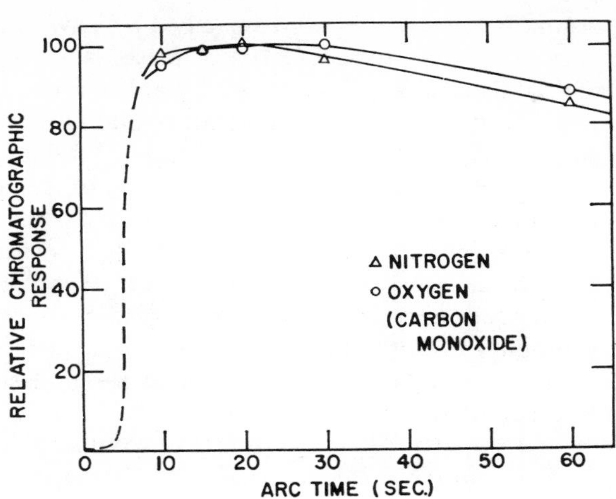

ig. 5.10. Gas-evolution curve for thorium with the platinum flux technique. (Reproduced y permission of *Analytical Chemistry* [7].)

. Limiting Conditions

Limitations to the measurement of low concentrations of gases in metals sually do not reside in lack of measurement sensitivity. Rather, the primary miting factors are (a) operating blank, (b) loss of gases to competing reacons, (c) incomplete extraction processes at low concentration levels, (d) ıck of precise standard samples for evaluating measurement accuracy, and e) sampling problems [51].

. *Blank, Positive, and Negative*

The term *blank*, in its usual sense, denotes that part of the analytical ignal that originates from sources other than the sample. In electrical ischarges this blank may result from outgassing of internal surfaces of the hamber, leakage of atmospheric gases by faulty joints or seals, contaminaion of the gas used as the extraction atmosphere, and residual gas content f the flux metal. The blank obtained from these sources is always of positive ign. It must be determined separately and subtracted from the total signal o obtain the net signal from the sample. In the authors' laboratory the ositive blank of the chamber has been reduced by employment of accepted igh-vacuum construction techniques, reduction of the number of O-ring eals, and the use of all-metal plumbing between the extraction atmosphere ylinder and the extraction chamber. With these precautions the chamber blank can be regularly reduced to approximately 1 μg oxygen for a 20-se 15-A arc between graphite electrodes. Under these same conditions th nitrogen blank is not detectable. Of the positive blank sources, the platinu blank is the most difficult to reduce. The 12-gauge platinum wire normall contains 5–10 ppm oxygen and no detectable nitrogen. With a 0.3-g sampl weight and a 1.0-g platinum flux, the 5–10 ppm residual oxygen is equivale to a 17–33-ppm blank referred to the sample. Even if the oxygen content of given lot of platinum wire is constant to within $\pm 10\%$, this variation is sti quite detrimental to the accurate determination of 10 ppm oxygen in sample. Surface adsorption or contamination is suspected to be a majc source of the platinum oxygen blank. Removal of surface oxygen from th platinum prior to its use as a flux may be possible by xenon-flash-desorptio [52, 53] or cathodic-sputtering [54] techniques.

The possibility of a "negative" blank must also be considered in extraction that are performed by an electrical discharge [29]. A negative blank ma arise when all of the analyte in the sample, although extracted, is nc quantitatively measurable. This situation occurs when a portion of the analyt is excluded from measurement through gettering, adsorption on the chambe surfaces, or chemical reaction to form a new species that the analytical metho is incapable of measuring. Incomplete extraction of the gas content of a meta could also appear as a negative blank if the equilibrium concentration of gas in a metal, under the extraction conditions, is a significant fraction of th original gas content of the sample. The actual blank, therefore, is the alge braic sum of a number of positive and negative components.

Winge and Fassel [29] have examined in detail the positive- and negative blank phenomena of the dc-arc method. When the arc is conducted to a empty electrode, the measurable carbon monoxide level changes little wit respect to arc time, as shown in Fig. 5.11. In contrast, the measurabl nitrogen level decreases with arc time, and the decrease is more severe, th smaller the starting quantity of nitrogen. This disproportionate loss of ni trogen then appears as a negative blank when the analyte response is plotte as a function of the quantity of analyte added to the extraction chamber. Th presence of a metal sample during arcing also affects the losses of carbo monoxide and nitrogen. Evens and Fassel [6] noted that the loss of nitroge was greatest when the arc was conducted to an empty electrode, but the los of carbon monoxide was greatest when a steel sample was arced. Thes observations suggest the reaction with carbon to form cyanogen [19] account for the loss of nitrogen, whereas the reaction with vaporized metal account for the loss of carbon monoxide.

The negative-blank concept implies that the loss of analyte is constant for given arc period regardless of analyte concentration. This is only a roug approximation of the truth. At a low starting concentration the loss of analyt

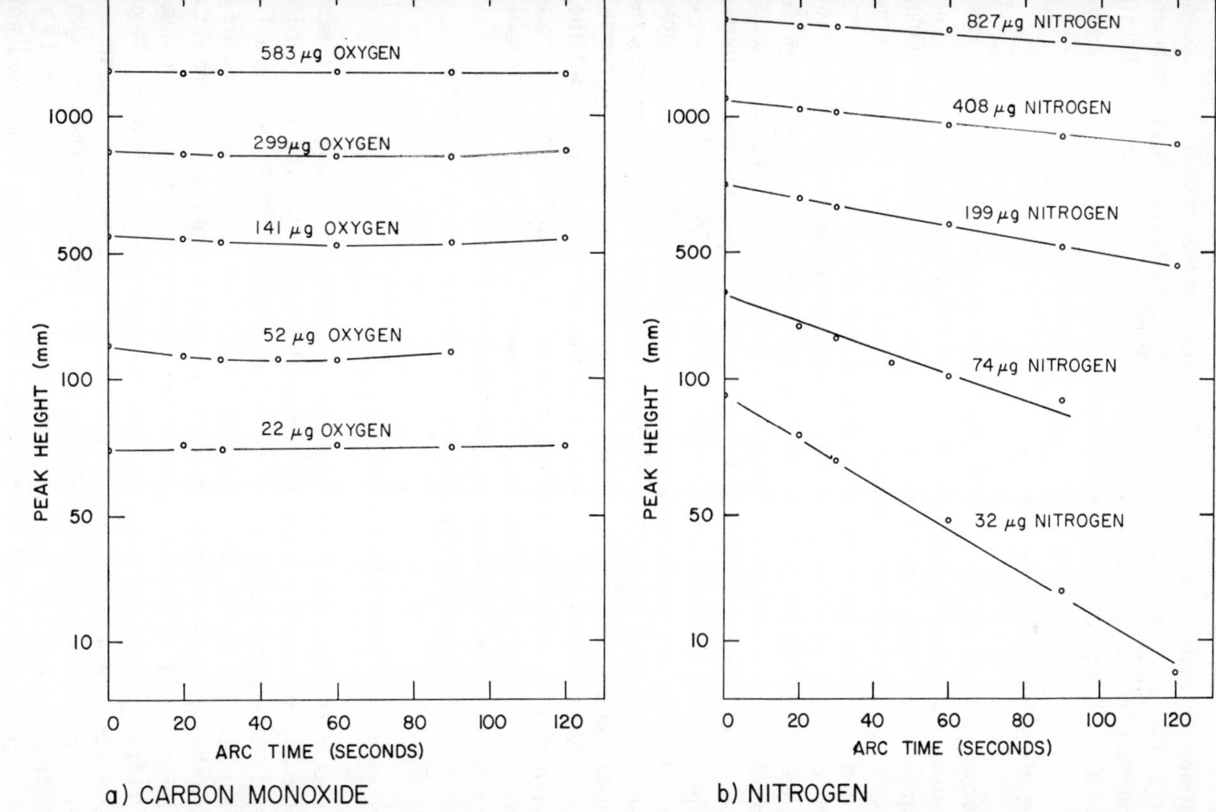

Fig. 5.11. Dependence of analytical response for carbon monoxide and nitrogen on arc time. (Reproduced by permission of *Analytical Chemistry* [29].)

appears to be limited by its concentration, whereas at higher starting concentrations the analyte loss appears to be limited by the amounts of reactive carbon and metal vapor made available by the arc discharge. This hypothesis is supported by the concordance of the data and the least-squares curves above 0.01–0.02 wt % oxygen and nitrogen.

b. Negative-Blank Compensation

It is evident from the above discussion that the negative blank phenomena could cause difficulties in determining oxygen and nitrogen in metals in the low-ppm range. What can be done to reduce this negative blank? One approach is to reduce the amounts of reactive carbon and metal vapor generated by the electrical discharge. A platinum flux does exactly that for reasons stated in Section II.B.2.b. Also, the platinum, because of its limited solubility for carbon, may reduce carbon vaporization, although this may be of secondary importance to the reduction caused by the presence of a metal sample in the electrode. Further and probably most important, the platinum flux allows most extractions to be completed in less than 20 sec [7] as opposed to the 60–90 sec arc time required without platinum. Another approach consists of extracting the gaseous elements from the metal sample into an atmosphere containing known amounts of carbon monoxide and nitrogen. The objective of this *negative-blank-compensation technique* is to put the measurement of small quantities of the extracted gases into the linear or most sensitive region of the analytical curve. Preliminary experiments [29] showed the reproducibility of this artificially created positive blank was adequate for the extension of the detection limit to about 5 ppm for a sample weight of 1 g.

c. Minimum Degassing for Nitrogen

Studies in the authors' laboratory have revealed that nitrogen degasses from the dc-arc extraction chamber very readily [29]. When the chamber is degassed immediately after preparation and sample loading, nitrogen is observed usually only in the first arcing cycle of the degassing procedure. Often, if the extraction chamber was loaded and left to pump down overnight, no nitrogen could be observed even in the products of the first degassing cycle. This suggested that the rigorous degassing normally required to reduce the oxygen blank would be unnecessary if only the nitrogen content of samples were to be determined. This proved to be true. With a single degassing cycle, consisting of a 60-sec, 15-A arc to a single electrode, a subsequent arc to a second electrode, under the normal sample arcing conditions, released no measurable nitrogen. The sample extractions that followed gave a linear analytical curve over the 0.0028–0.0490 wt % nitrogen range, and the negative blank (determined from concentration axis intercept, which was obtained by the least-squares technique) was the same as was obtained from the more rigorous degassing procedure.

d. *Effect of Reactive Metal Content of Samples*

Because of comparatively high vapor pressure at fusion temperatures, manganese and aluminum alloying constituents actively getter or chemisorb both oxygen and carbon monoxide [55] and will affect any fusion-extraction method in which appreciable quantities of these metals are vaporized [24, 56–58]. Greater recovery of both oxygen and nitrogen from a variety of metals (in particular, of oxygen from metals containing manganese) by the addition of tin to the reaction medium has been reported by a number of investigators using vacuum- and carrier-gas-fusion techniques [59, 60]. Jones found that in the dc-arc extraction technique, the addition of small quantities of tin or silver to certain steel samples caused a greater recovery of the oxygen content [39]. Figures 5.12 and 5.13 show the results of a study conducted in the authors' laboratory of the effect of manganese on the determination of oxygen and nitrogen in steel [29]. The oxygen data indicate that significantly more scatter of points occurs when no flux is employed (see Fig. 12b and d) and that those samples that fall farthest from the curve have the highest manganese content. Table 5.2 lists elemental compositions and reference oxygen and nitrogen values for the samples used in these studies. The curve shown in Fig. 5.12a is derived from the best linear relation obtained by the least-squares method for data of the platinum-flux technique. This same curve, or its linear coordinate transform, is reproduced in Figs. 5.12a, b, d as a reference for comparison of the alternate methods. Figure 5.12c indicates that the addition of 30–40 mg of either silver or tin to the manganese steel samples causes these data to very nearly equal that obtained by the platinum-flux technique. Figure 5.12d shows that the platinum-flux technique does not give acceptable results when the manganese content reaches 10%, as evidenced by the zero peak heights observed for samples 1, 2, and 23. This figure also shows that the silver or tin addition methods may give acceptable results up to approximately 10% manganese. More exact information on the capabilities of the various method modifications is restricted by the limited availability of high-manganese steel samples. Figures 5.13a, b, c compare the nitrogen data obtained under the same experimental conditions. Again, the curve of Fig. 5.13a was derived from the best linear relationship by the least-squares method, and this same curve is reproduced in Figs. 5.13b, c for comparison. These figures indicate that consistent nitrogen results can be obtained by any of the dc-arc extraction methods, regardless of the manganese content.

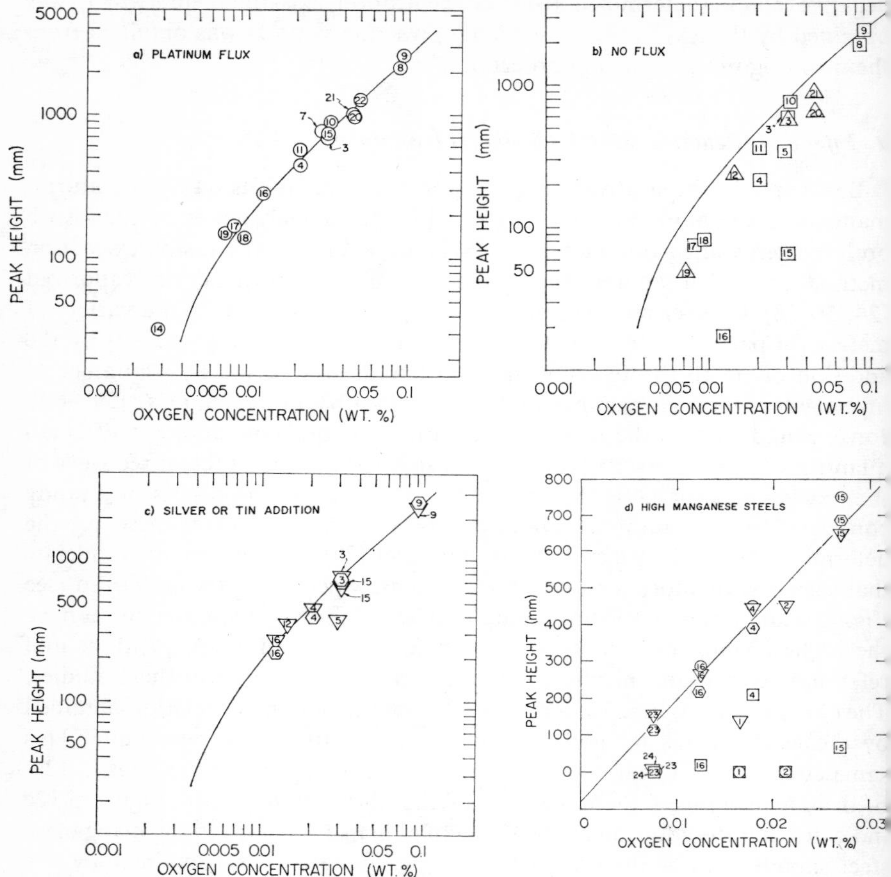

Fig. 5.12. Comparison of arc-extraction oxygen data from steels. ○: Platinum flux 20-sec extraction, 15 A arc, 0.3-g sample; □: No flux, 60-sec extraction, 15-A arc, 1.0-g sample; △: No flux, 90-sec extraction, 15-A arc, 1.0-g sample; ▽: Silver addition, 30–40 mg, 60-sec extraction, 15-A arc, 1.0-g sample; ⬡: Tin addition, 30–40 mg, 60 sec. extraction 15 amp. arc, 1.0-g sample. Samples are identified in Table 5.2. (Reproduced by permission of *Analytical Chemistry* [29].)

TABLE 5.2. Identification of Steel Samples

	Sample	Composition (%)	Concentration (wt %)	
			Nitrogen[a]	Oxygen[a]
1	High Alloy A2A	6.8 Cr, 27.4 Mn	0.75[b]	0.017[c]
2	High Alloy A9	12.2 Cr, 10.5 Mn	0.24[b]	0.021[c]
3	NBS 8i	0.5 Mn	0.018	0.031
4	NBS 101e	0.4 Si, 18 Cr, 1.8 Mn, 9.5 Ni, 0.4 Mo	0.040	0.021 0.018[c]
5	NBS 111b	0.3 Si, 0.7 Mn, 1.8 Ni, 0.3 Mo	0.0048	0.030
6	NBS 122d	3.3 C, 0.6 Si, 0.5 Mn	0.0044	0.052
7	NBS 125a	3.3 Si, 0.05 Mn	0.0037	0.028
8	NBS 129b	0.3 S, 0.8 Mn	0.014	0.086
9	NBS 133a	0.4 Si, 0.3 S, 12.9 Cr, 1.0 Mn, 0.2 Ni, 0.3 Mo	0.030	0.091
10	NBS 152	0.2 Si, 0.8 Mn	0.0035	0.032
11	NBS 343	15.8 Cr, 2.1 Ni	0.067	0.021
12	NBS 344	0.4 Si, 15 Cr, 0.6 Mn, 1.2 Al, 7.3 Ni, 2.4 Mo	0.027	0.014
13	NBS 425	Mn, Ni, Cr	0.0042	0.0024
14	NBS 427	0.4 Si, 1.0 Cr, 1.0 Mn, 0.3 Mo	0.0044	0.0028
15	NBS 443	0.2 Si, 18.5 Cr, 3.4 Mn, 9.4 Ni	0.023	0.031 0.027[c]
16	NBS 444	0.6 Si, 20.5 Cr, 4.6 Mn, 10.1 Ni, 0.2 Mo	0.049	0.012 0.012[c]
17	NBS 462	0.3 Si, 0.7 Cr, 0.9 Mn, 0.7 Ni	0.0092	0.0082
18	NBS 463	0.4 Si, 0.7 Cr, 1.2 Mn, 0.4 Ni, 0.1 Mo	0.0072	0.0095
19	Low Alloy 8	0.5 Si, 0.8 Cr, 0.4 Mn, 0.4 Ni	0.0068	0.0072
20	Low Alloy 11	0.6 Mn	0.0046	0.045
21	Low Alloy 28	0.3 Mn	0.0028	0.044
22	Low Alloy 40	0.6 Mn	0.027	0.049
23	High Alloy 65	0.5 Si, 15 Cr, 10 Mn, 22.5 Ni, 2.0 Mo	0.23	0.0076[c]
24	High Alloy	0.7 Si, 18.4 Cr, 14.3 Mn, 0.3 Ni	0.48	0.0074[c]

[a] All oxygen and nitrogen values, unless otherwise identified, were determined by the vacuum-fusion platinum-bath technique.
[b] These nitrogen values were determined by the Kjeldahl chemical technique.
[c] These oxygen values were determined by the neutron-activation technique.

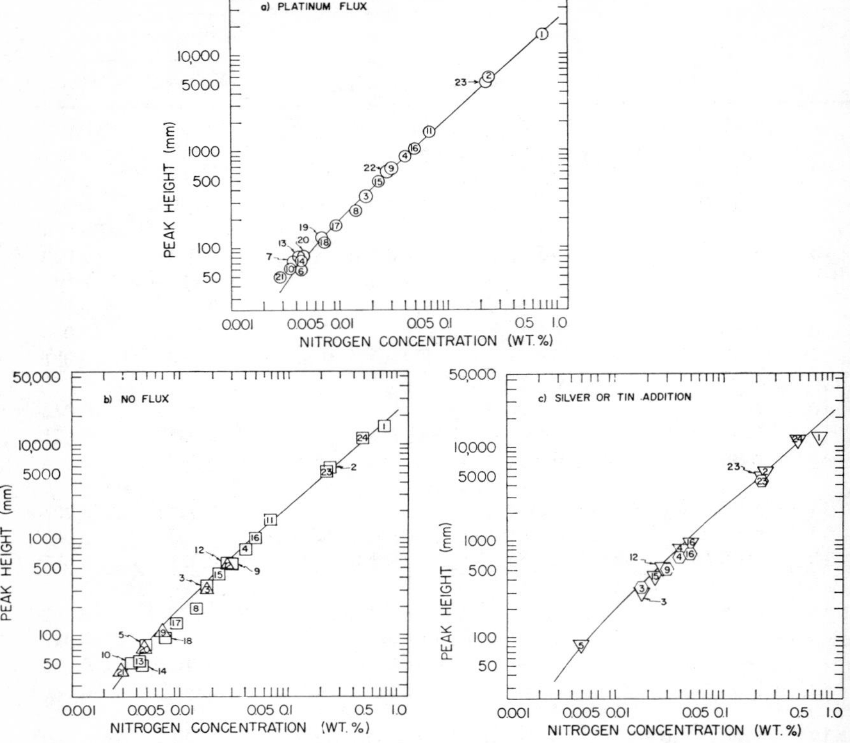

Fig. 5.13. Comparison of arc-extraction nitrogen data from steels. Symbol legend is the same as for Fig. 5.12. Samples are identified in Table 5.2. (Reproduced by permission of *Analytical Chemistry* [29].)

III. ANALYTICAL SYSTEMS

A. MANOMETRIC ANALYSIS

The manometric method, when used with the dc-arc extraction chamber, is one of the least expensive techniques for measuring the oxygen content of metals. In this technique the oxygen content is evolved from the metal as carbon monoxide, which is later oxidized to carbon dioxide by Schutze's reagent [61, 62]. The carbon dioxide is passed through a capillary trap cooled with liquid oxygen or liquid nitrogen. Liquid oxygen is preferred, especially if the trapping is to be done at atmospheric pressure, because its boiling point is slightly above that of argon. The carbon dioxide is retained in the trap while the argon extraction atmosphere flows on through. After all of the extracted gas has been transferred to the trap, the trap is closed off

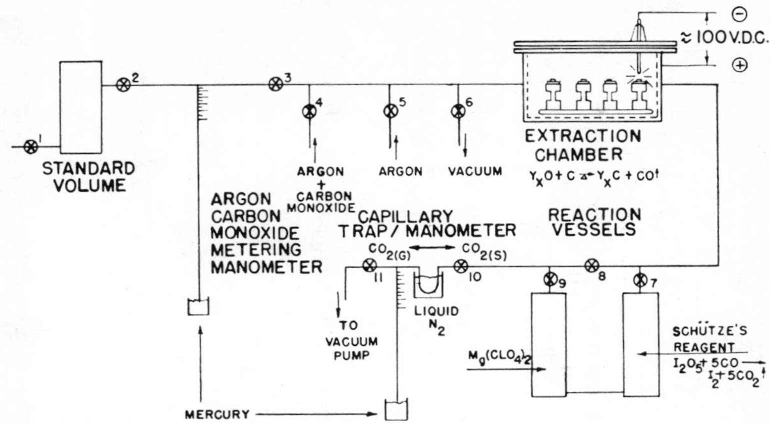

ig. 5.14. Schematic drawing of equipment used for the manometric determination of xygen in yttrium metal [26].

rom the system and is warmed to room temperature. The quantity of oxygen s then determined directly from the pressure–volume readings of the prealibrated trap-manometer.

A dc-arc manometric apparatus, shown in Fig. 5.14, was used by Karohl 26] for the determination of oxygen in yttrium. By evacuating the extraction hamber from 0.5 to 0.04 atm pressure two times, more than 99% of the xtracted gases theoretically would be transferred from the chamber to the rap. Smiley [62] used the carrier-gas fusion technique with an argon flow ate of 100 cm³/min through a simple U-tube trap. Hanin [63] was not able to rap all of the carbon dioxide with this U-tube trap and developed a doubleelix capillary-tube arrangement. A flow rate of 100 cm³/min was also used vith this trap. Transfer of the carbon monoxide from the chamber by flushing s a rather slow process, as can be seen from the data in Table 5.3. Kashima

TABLE 5.3. Times[a] Required to Flush 99% of Extracted Gases from Chamber

Flow rate of flushing gas (cc/min)	Chamber Volume (cm³)			
	1000	500	250	100
50	92 min	46 min	23 min	9.2 min
100	46	23	11.5	4.6
200	23	11.5	5.8	2.3
500	9.2	4.6	2.3	0.9

Flushing times were calculated from a linear differential equation describing the change n quantity of extracted gas in the chamber with time.

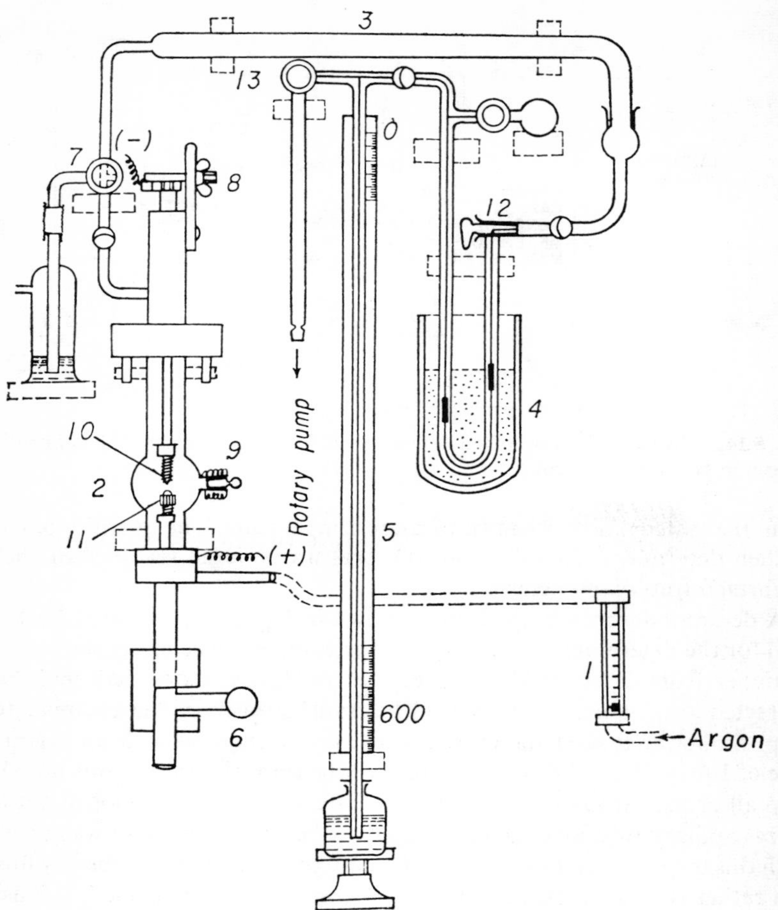

Fig. 5.15. Apparatus for the manometric method of Kashima: (1) flowmeter; (2) ar chamber; (3) Schutze's reagent; (4) liquid-oxygen-cooled trap for CO_2; (5) manometer (6) chamber closure; (7) three-way stopcock; (8) electrode gap adjustment; (9) sample inlet; (10) cathode; (11) anode; (12, 13) two-way stopcocks. (Reproduced by permissio of the author and *Nippon Kinzoku Gakkaishi* [27].)

[27] has developed a rapid procedure using a 500 cm³/min flow rate through chamber of small volume for the determination of oxygen in iron and stee The high flow rate was tolerated by a special trap containing glass woc plugs. The apparatus is shown in Fig. 5.15. Flushing of the extracted gase from the discharge region may be responsible for avoidance of the analyte consuming reactions experienced under static conditions in the Applie Research Laboratories test chamber of similar size (see Section II.A.1).

B. SPECTROSCOPY

1. Spectrographic Techniques

The basic principles behind spectrographic techniques are essentially the same for gases extracted from large and small samples. Generally, the same electrical discharge is used both for the extraction of the gases from the metal sample and for the spectral excitation of the extracted gases. The discussion of spectrographic techniques used in this manner will therefore be deferred to Chapter 6.

2. Infrared Absorption

The infrared-absorption technique for measuring oxygen (as carbon monoxide) is independent of the actual extraction process and is therefore appropriately discussed in this chapter along with the other independent analytical methods. Höller [24], Vasserman and Turovtseva [37], and Hilger and Watts, Ltd. [64] have used infrared absorption as a rapid analytical technique for their dc-arc and impulse-current extraction techniques. The oxygen content of the sample is extracted into an argon atmosphere. The resulting argon–carbon monoxide mixture is expanded into an infrared measuring cell from which the carbon monoxide absorption reading is converted to weight percent oxygen in the sample. The homonuclear diatomic molecules, hydrogen and nitrogen, cannot be measured by infrared-absorption techniques.

C. GAS CHROMATOGRAPHY

Gas chromatography is particularly appropriate for the determination of gases in metals because of its ability to measure quantitatively very minute quantities of the individual components in the gas mixture. The basic gas chromatographic apparatus consists of a carrier-gas supply, sample-introduction valve, separation column, and detector. This assemblage is shown in relation to the arc extraction chamber in Fig. 5.16.

1. Column

The synthetic zeolites (molecular sieves) are used almost exclusively as the column packing material for the separation of the gases normally extracted from metals—namely, hydrogen, nitrogen, and carbon monoxide. These zeolites are also capable of separating methane, which occurs occasionally in the extracted gases.

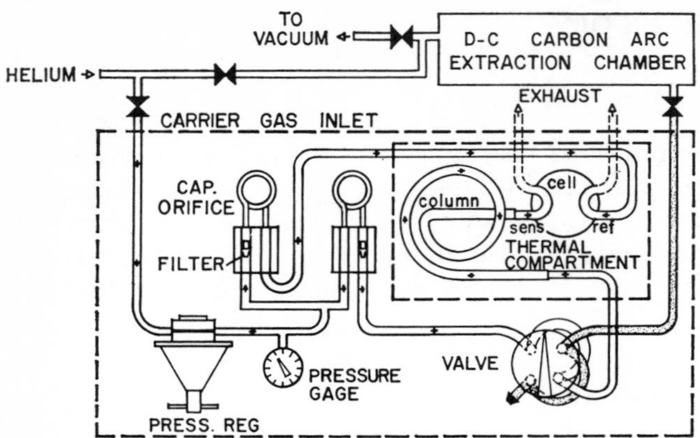

Fig. 5.16. Schematic drawing of arc-extraction chamber and gas-chromatograph assembly. (Reproduced by permission of *Analytical Chemistry* [6].)

2. Detector

Several types of detectors can be used to measure the permanent gases as they elute from the column. Only the most widely used types, thermal-conductivity and ionization detectors, will be discussed in this chapter.

a. Thermal Conductivity

To achieve high sensitivity with a thermal-conductivity detector, the thermal conductivity of the sample gas should be significantly different from that of the carrier gas. Thermal conductivities of gases that may be encountered in a gases-in-metals analysis system are listed in Table 4.3. These data clearly show that helium is the better carrier gas for the determination of nitrogen and carbon monoxide, whereas argon is the better choice for the determination of hydrogen. In addition, low concentrations of hydrogen in helium give an anomalous response from what would be expected from their thermal conductivities. The thermal conductivity of a hydrogen–helium mixture reaches a minimum at about 8 mole % hydrogen, reverses direction and equals the thermal conductivity of pure helium at about 17 mole % [65]. Purcell and Ettre [66] obtained normal peaks, regardless of the hydrogen concentration of the sample, when a commercially available 8.5 mole % hydrogen in helium mixture was used as the carrier gas.

Because the minimum thermal conductivity occurs at about 8 mole %, hydrogen should be determinable in helium up to about 5 mole % without a special carrier-gas mixture, polarity switching, or other special provision. With a chamber volume of 650 cm^3, 5 mole % would correspond to approximately 0.25 weight % hydrogen in a 1-g sample. With the high-sensitivity

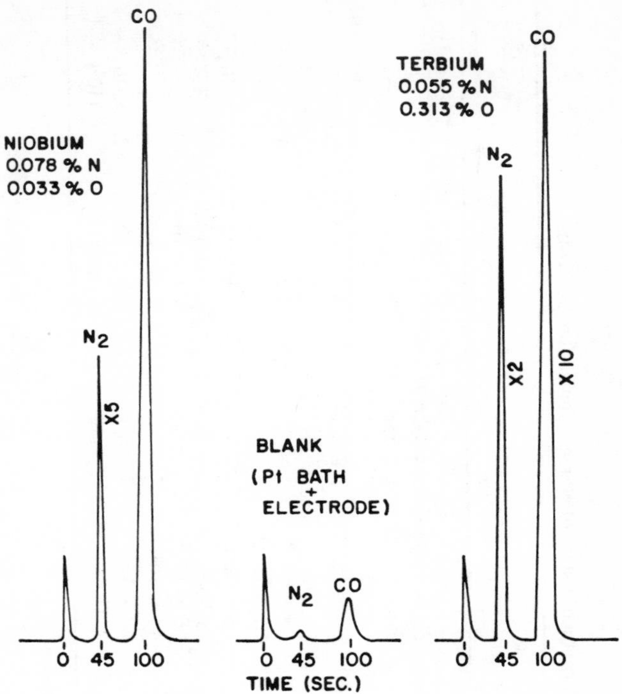

Fig. 5.17. Chromatograms of evolved gases. Numbers beside peaks indicate attenuation factors where used (e.g., ×10 indicates the signal was reduced by a factor of 10 before the peak was recorded). (Reproduced by permission of *Analytical Chemistry* [7].)

thermal-conductivity detectors presently available, the determination of hydrogen in metals by this analytical technique appears technically feasible.

Figure 5.17 shows typical chromatograms obtained with a thermal-conductivity detector and with helium as the carrier gas [7]. More recent results have shown sensitivities for both of about 0.4 μg/mm peak height with no measurable noise (1-mV detector signal provides 250-mm recorder response). This sensitivity is adequate for determining the oxygen and nitrogen content of samples down to approximately the 1-ppm level. The blank problems discussed in Section II.B.3.a, however, limit the lowest determinable concentration to somewhat higher values.

b. Ionization

Ionization detectors give very high sensitivity for the permanent gases when either argon or helium is used as the carrier gas. Table 5.4 summarizes reported detection limits obtained from ionization detectors for the permanent gases. For further information, the reader is referred to the numerous

TABLE 5.4. Detection Limits[a] for Permanent Gases of Various Ionization Detectors

Reference	Carrier gas flow rate (ml/min)	H_2	N_2	O_2	CO	CH_4
67	Helium purified, 75	5×10^{-7} ml	—	5×10^{-8} ml	—	5×10^{-8} ml
68	Helium purified, 85	6.3×10^{-11}	3.5×10^{-10}	1.4×10^{-10}	1.1×10^{-10}	5.1×10^{-11}
69	Argon	0.5 ppm	1 ppm	0.5 ppm	—	0.5 ppm
70	Helium, 50	1.5×10^{-12}	1.6×10^{-11}	3.6×10^{-12}	3.1×10^{-12}	2.3×10^{-12}
71	Argon, 25	4.8×10^{-1}	1.5×10^{-9}	3.8×10^{-11}	5×10^{-10}	8.8×10^{-11} $(\sim 10^4)$[b]
72	Krypton	2.6×10^{-11}	7.8×10^{-10}	1.6×10^{-10}	2.2×10^{-10}	1.3×10^{-10}
72	Xenon	1.8×10^{-12}	4.2×10^{-10}	3.1×10^{-10}	1.4×10^{-10}	2.9×10^{-11}
73	Commercial helium	3.6×10^{-9} (100)	4×10^{-8} (30)	2×10^{-8} (27)	1×10^{-7} (16)	7×10^{-8} (18)
74	Helium, 85	7.8×10^{-9} (10^2)	1.0×10^{-8} (5×10^2)	—	9.4×10^{-9} (10^3)	—
74	Helium, 80	6.4×10^{-9} (5×10^2)	8.8×10^{-9} (10^3)	3.5×10^{-9} (10^4)	8.8×10^{-9} (10^3)	2.0×10^{-9} (10^3)

[a] Detection limits in grams per second unless otherwise identified.
[b] Range of method shown in parenthesis.

books in the field and to reviews that have been written on ionization detectors [75, 76], trace analysis [77], and the gas-chromatographic determination of the permanent gases [78].

3. Sampling Systems

a. Aliquot

The aliquot sampling system is set up by simply connecting the extraction chamber to a chromatograph sample-valve, as is shown in Fig. 5.16. The most direct method of sampling of the chamber gases is by expansion into an evacuated sample-loop. The sample-valve must therefore have suitable seals, which prevent atmospheric leakage under high-vacuum conditions. A convenient sample-valve arrangement consists of two sample-loop volumes that can be interchanged between the chamber vacuum system and the chromatograph analysis system. This arrangement permits successive sampling of the same extraction atmosphere.

(1) HOMOGENIZATION. At the beginning of the extraction the arc supporting atmosphere of pure argon or helium is in the chamber. During the arc period, gases are extracted from the metal and are mixed with the supporting atmosphere. Because only an aliquot of the resulting gas mixture is analyzed, it is necessary that the extracted gases be homogeneously distributed throughout the supporting atmosphere to obtain a representative gas-chromatographic sample. The chamber design illustrated in Fig. 5.7 allows the extracted gases to homogenize very rapidly with the supporting atmosphere. Data in Table 5.5 [29] show little if any statistically significant difference in taking the sample immediately or in allowing a 30-sec interval after the arc is terminated. The immediate sampling technique is listed as having a homogenization

TABLE 5.5. Effect of Homogenization Period on the Measurement of Extracted Gases

Homogenization Period (sec)	Number of Samples	Average Peak Heights (mm) ± Coefficient of Variation	
		Nitrogen	Carbon Monoxide
1	7	100 ± 4.5%	206 ± 3.5%
30	6	103 ± 3.6%	207 ± 4.3%

Metal sample: NBS steel 8i
Weight: 0.3 g
Flux: Platinum
Weight: 1.0 g
Arc extraction: 20 sec, 15 A

period of 1 sec, because this is the approximate time required for the mechanical motions between termination of the arc and opening of the valve to the chromatograph sample-loop.

Ordinarily a 10-sec period is allowed between termination of the arc and sampling, because this is a convenient period for the necessary manipulations and allows a safety factor in case a significant fraction of the gas content of a sample evolves toward the end of the arc period. The vertical chamber configuration and the absence of dead volume appendages from the working volume of the chamber are credited for the rapid gas homogenization.

(2) TEMPERATURE CONTROL. Aliquot sampling of gases by expansion into an evacuated sample-loop is a common method in the field of gas chromatography. However, this method of sampling is sensitive to differences in temperature between the sample-valve and the gas source, and it is very difficult, if not impossible, to make the sample-valve follow exactly the temperature cycle of the gas in the extraction chamber. In the authors' apparatus (Fig. 5.7) the sample-valve is located a short distance from the extraction chamber and remains at room temperature. Also, the sample valve acts as a massive heat sink and is relatively unaffected by the heat content of the gas sample that expands into it.

Temperature effects on sampling result from the change in volume of the chamber and the change in pressure of the gas in the extraction chamber with temperature, the latter being of much the greater consequence. These two effects cannot be separated physically but can be separated mathematically.

With regard to the expansion of the chamber volume with temperature, a temperature rise of 100°C would decrease the amount of gas transferred to the sample loop by 0.3%, a negligible amount (assuming coefficient of expansion of stainless steel to be 1×10^{-5}/°C).

With regard to the effect of temperature on the pressure of the gas in the chamber, the fraction, F, of the total chamber gas expanded into the sample loop can be obtained from

$$F = \frac{V_s}{V_c(T_s/T_c) + V_s} \tag{1}$$

where V_s is the sample loop volume at temperature T_s and V_c is the extraction-chamber volume at temperature T_c. When both volumes are at the same temperature, the fraction of chamber gas that expands into the sample-loop is simply the ratio of the loop volume to the total volume of the system.

The temperature of the gas within the chamber is estimated to rise about 100°C during a normal arc cycle. From the above equation, if $V_c = 650\ \text{cm}^3$, $V_s = 5\ \text{cm}^3$, $T_c = 398$°K, and $T_s = 298$°K, the fraction of the total chamber gas expanded into the sample loop would be 33% greater than if the two volumes were at the same temperature. The chamber and sample valve could be thermostatted to keep both at the same temperature. However, the electrodes, electrode stage, and associated parts inside the chamber are not readily cooled, and a relatively long time period would be required to equilibrate these parts (and the chamber gas) with the temperature of the sample valve. As can be seen from Eq. 1, it is not necessary that the chamber gas and the sample valve be at the same temperature but only that the temperatures be reproduced in a constant ratio, T_s/T_c, to obtain reproducible aliquot sampling. Cooling of the chamber is still required, of course, to minimize thermal outgassing and to dissipate the heat created during each discharge cycle.

The following procedure is used to reproduce the T_s/T_c ratio. After the initial degassing, which is more rigorous than that in the normal analytical procedure, the chamber is allowed to cool a few minutes. Several electrode blanks are then determined under the same conditions as for a sample analysis. These determinations confirm that the blank has been reduced to an acceptable level and set up a reproducible temperature-cycle rhythm for the succeeding sample analyses.

The importance of reproducing temperature conditions is not only to obtain consistent analytical results from samples. These temperature effects must also be considered if quantitative calibrations are established by adding measured quantities of gases to the chamber. It is apparent that the same temperature cycle must be used with the calibration gases as with metal samples, to obtain the same fraction of chamber gas in the chromatographic sample.

If consistent results are expected from methods involving several different arc times, it would seem advisable to install a thermocouple in such a manner that it would measure the gas temperature inside the chamber. The thermocouple should therefore be protected from direct radiation from the arc and also from carbon deposits that would slow its response. A small heater may be installed in the chamber to bring calibration gas samples up to sampling temperature so that the chromatographic aliquot sample can be taken at the same chamber gas temperature regardless of the procedure.

b. Sampling of Total Extraction Products

Kamada [25] has used a small quartz chamber containing a single electrode for the arc extraction of the gaseous element content of steel. The chamber was located in the flow system of the chromatograph just prior to the column. The extracted gases were swept directly to the chromatograph column with no sample valve interface. An argon flow rate of 120 ml/min was used through the arc chamber and column during the extraction. The extraction was conducted for 60 sec at 20 A. The carbon monoxide was eluted from the column in 4 min. Because the peaks were rather broad and the peak shape depended on the gas-evolution rate, the analytical data were obtained as integrated peak areas. Calibration for oxygen was obtained by injecting pure carbon monoxide into the argon flow. The oxygen data by this arc-extraction technique were shown to correspond well with vacuum-fusion oxygen values. Neither analytical data nor corroborating evidence were presented for hydrogen or nitrogen. Extraction curves for oxygen, hydrogen, and nitrogen were obtained spectrally, however, from a radio-frequency discharge.

D. CALIBRATION

All analytical techniques used with electrical-discharge extraction methods, except the trap-manometric determination of oxygen, are nonabsolute and therefore must be calibrated with some type of standard sample.

1. Metal Standards

The most reliable method of calibrating a gases-in-metals analysis system is undoubtedly the use of metal samples that contain accurately known gas contents and that have a chemical composition similar to the samples that are to be analyzed. The National Bureau of Standards is a source for a number of cast iron and steel standards of known oxygen and nitrogen content [79, 80]. For metals other than iron and steel, commercially available standards are virtually nonexistent. For these reasons the most widely used calibration system involves intralaboratory comparison with other analytical techniques, as for example, vacuum fusion [7, 29], inert-gas fusion [27], and neutron activation [29]. Homogeneous samples having a suitable range of gaseous element contents are not always available for intralaboratory comparison.

Oxygen standards have been synthesized through blending, by arc melting known amounts of metal oxide with a relatively pure base metal and by high-temperature diffusion of measured amounts of molecular oxygen into a metal sample [81]. Evens and Fassel [15] found that the addition of molecular oxygen to niobium metal produced more concordant results than the arc-melting procedure. Fassel, Gordon, and Tabeling [20] also used this diffusion technique for the preparation of lanthanum–oxygen standards.

2. Chemical Standards

To avoid the difficulties of obtaining homogeneous metal samples of known gas content, several investigators have used chemical compounds containing the desired element as calibration standards. These standards have been prepared from measured amounts of copper oxide [14, 17, 28], stannic oxide [17, 28], zinc oxide [14, 17, 28], ferric oxide [34], uranium oxide [38], and zirconium hydride [28, 32] pelleted with powdered graphite. The results of Fassel and Goetzinger [17] showed good concordance of spectral data obtained from copper, stannic, and zinc oxides and nine different metals. Laboratory Equipment Corporation (LECO), St. Joseph, Michigan, offers tin capsules containing measured amounts of potassium acid phthalate for use as oxygen standards. No data are available on the application of these tin-capsule standards to the electrical-discharge extraction methods. Dukat and Marley [82] have reported the use of cadmium hydrogenhydroxyethylenediaminetriacetate as a calibration standard for the determination of oxygen, nitrogen, and carbon in metals by a carrier-gas-fusion method.

TABLE 5.6. References to Applications of Electrical-Discharge Extraction Techniques for the Determination of Oxygen, Nitrogen, and Hydrogen

Sample Material	Hydrogen	Nitrogen	Oxygen
Al			93
B, BN		38	38
Be, BeO			38
Cr		7	7
Cu	12, 13		12, 13, 35, 93
Fe, steel		6, 7, 19, 29, 30, 35, 41, 83	3–8, 17, 19–22, 24, 25, 27–29, 30, 33–36, 83
Ge			93
Hf, HfC			7, 17, 92
La			20
Mo			7, 93
Na			9
Nb, NbC	11	7, 10, 11	7, 10, 11, 15, 17, 19, 37, 84, 93
Pb			93
Si			90, 94
Sn			93
Ta, TaC	11	11	7, 11, 17, 37
Tb		7	7
Te			94
Th		7	7, 17, 30, 38
Ti, TiC	19, 40, 85	7, 10	1, 7, 10, 17–20, 23, 86–89
U, UC, UN		35, 38	28, 31, 35, 38
V	11	7, 11	7, 11, 16, 17, 38
W			35, 93
Y		7	7, 17, 19, 26
Zr, ZrN zircalloy	40	30	7, 14, 17, 19, 28, 30–32, 35, 38
Sulfides of Ti and Hf		29	29

3. Gas Standards

Probably the most simple of all calibration schemes is the addition of measured quantities of the appropriate gases to the extraction chamber. This technique has been used with varying degrees of success by a number of investigators [3, 14, 17, 25, 28, 38]. For the technique to be successful several complicating factors must be considered. The gas samples should be treated in such a manner that the same losses occur as in the normal extraction of gases from metals. Also, if an aliquot sampling technique is used, the ratio of temperatures of the extraction chamber and aliquot volume must be reproduced as explained in Section III.C.3.a.2.

IV. APPLICATIONS

Table 5.6 lists with appropriate references an extensive list of materials that have been analyzed for their oxygen, nitrogen, and hydrogen contents by electrical-discharge extraction techniques. By reducing the effects of volatility and reactivity of the sample metal, the platinum-flux technique has

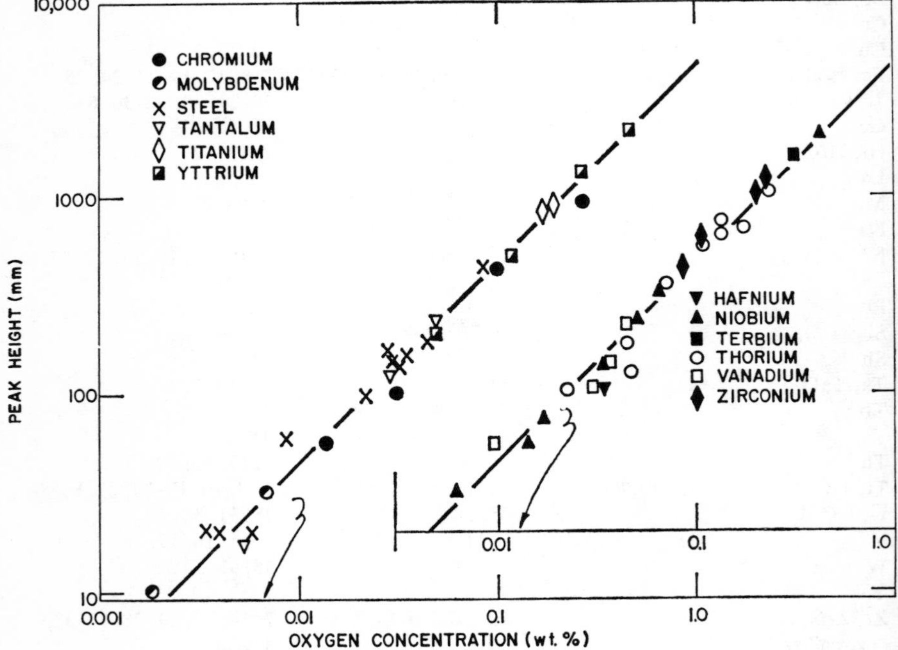

Fig. 5.18. Oxygen analytical curve of Winge and Fassel. (Reproduced by permission of *Analytical Chemistry* [7].)

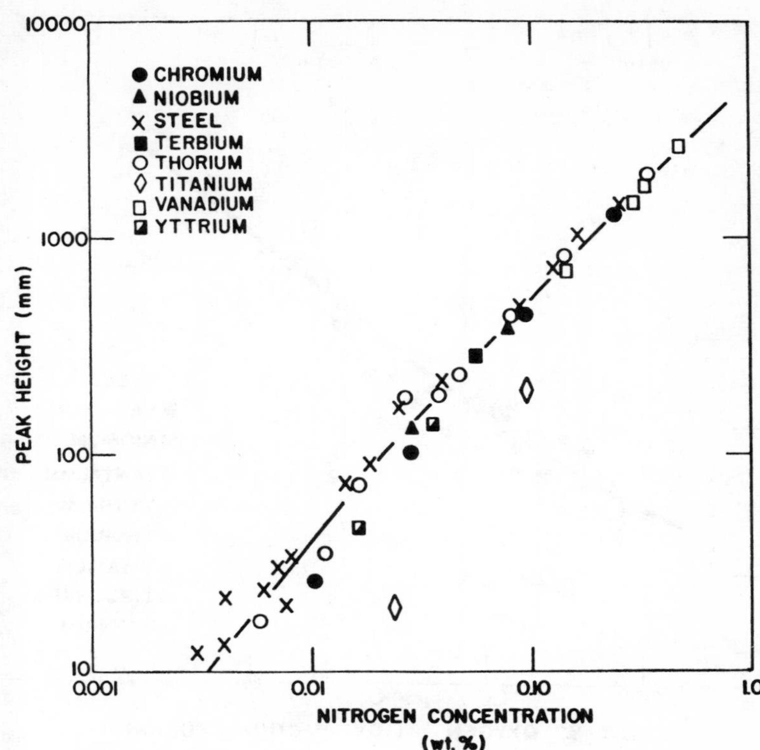

Fig. 5.19. Nitrogen analytical curve of Winge and Fassel. (Reproduced by permission of *Analytical Chemistry* [7].)

TABLE 5.7. Precision Data

Method	Sample	Concentration (wt %) Oxygen	Concentration (wt %) Nitrogen	Relative Standard Deviation (%)[a] Oxygen	Relative Standard Deviation (%)[a] Nitrogen
No flux (6)	Steel	0.037	0.017	2.6	2.7
	Steel	0.0085	0.0089	5.5	2.9
Platinum flux (7)	Thorium	0.110	0.016	7.0	5.5
	Vanadium	0.036	0.298	7.8	7.9
	Yttrium	0.266	0.034	6.4	7.2
	Zirconium	0.118	—	3.1	—

[a] Calculated from 10 individual determinations on as many different days.

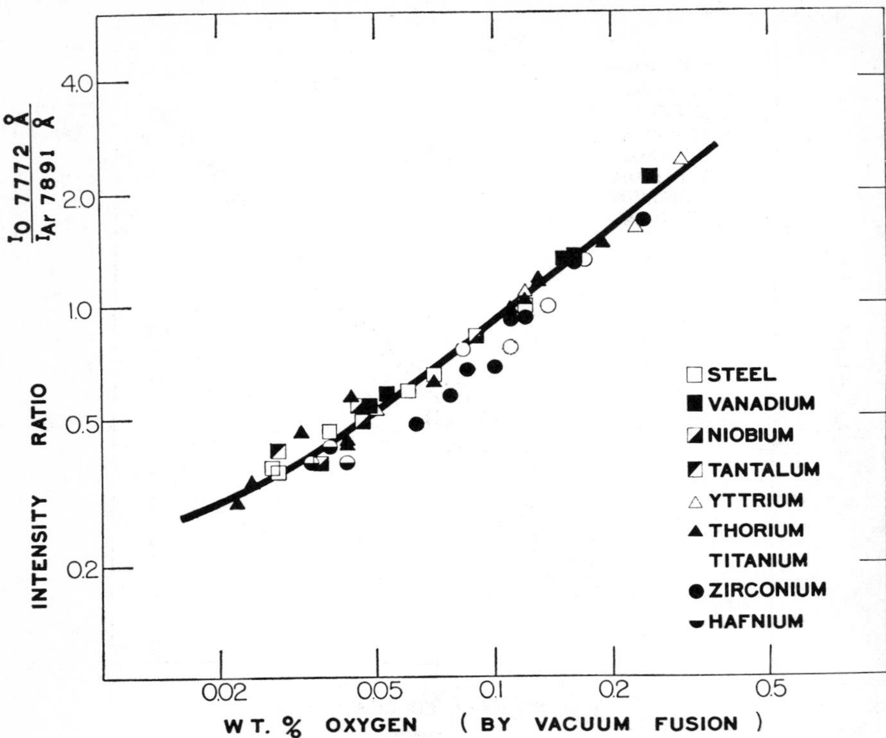

Fig. 5.20. Comparison of oxygen results from gas standards and metal samples by Fassel and Goetzinger. (Reproduced by permission of copyright holder, Microforms International [17].)

permitted the oxygen and nitrogen content of a variety of metals to be determined under the same operating conditions. Evidence of this is shown by the concordance of the data presented in Figs. 5.18, 5.19 [7], and 5.20 [17]. The concordance of the metal-sample data with those obtained from gas standards (heavy line) in Fig. 5.20 also suggests that quantitative evolution of the oxygen (carbon monoxide) is achieved by the dc-arc extraction technique.

In general, the precision of the analytical results obtained by electrical discharge extraction techniques is comparable with that obtained by other available methods. Table 5.7 shows precision data obtained without [6] and with [7] the platinum-flux technique. The variation in these precision data for the different samples suggest that at least a portion of the deviation is caused by true inhomogeneities of the gas content of the bulk samples or by differences caused by the sample preparation procedure.

REFERENCES

1. K. K. Yudowitch, *Rept. No. NP-4486*, Armour Res. Foundation, (1952).
2. B. Rosen and M. I. Ottelet, Colloq. Intern. Spectrog. Strasbourg, October, 1950, Group. pour l'Avan. des Methodes Spectrogr. des Prod. Met., Paris, 1950, pp. 155–167.
3. B. Rosen, *Rev. Universelle Mines*, **9,** 445 (1953).
4. B. Rosen, Congr. Group Avan. Methodes Anal. Spectrog. Prod. Met. XX, Paris, Group. pour l'Avan. des Methodes Spectrog., Paris, 1957, pp. 237–243.
5. V. A. Fassel and R. W. Tabeling, *Spectrochim. Acta*, **8,** 201 (1956).
6. F. M. Evens and V. A. Fassel, *Anal. Chem.*, **35,** 1444 (1963).
7. R. K. Winge and V. A. Fassel, *Anal. Chem.*, **37,** 67 (1965).
8. C. Matsumoto, V. A. Fassel, and R. N. Kniseley, *Spectrochim. Acta*, **21,** 889 (1965).
9. D. R. Grieser, G. C. Cocks, E. H. Hall, W. M. Henry, and J. McCallum, *Rept. No. 1538*, Battelle Memorial Institute, Columbus, Ohio, 1961.
10. T. F. Ivanova and V. V. Federova, *Zh. Anal. Khim.*, **23,** 1750 (1968).
11. E. Fromm, *Z. Metallk.*, **56,** 493 (1965).
12. H. Bohm, K. G. Guenther, and W. Kuhl, *Fresenius' Z. Anal. Chem.*, **209,** 198 (1965).
13. K. G. Guenther and H. Lamatsch, *Trans. Intern. Vacuum Congr. 3rd* (Stuttgart), **2** (1), 161 (1966).
14. J. Artaud and C. Berthelot, *Mem. Sci. Rev. Met.*, **57,** 338 (1960).
15. F. M. Evens and V. A. Fassel, *Anal. Chem.*, **33,** 1056 (1961).
16. V. A. Fassel and L. L. Altpeter, *Spectrochim. Acta*, **16,** 443 (1960).
17. V. A. Fassel and J. W. Goetzinger, *Spectrochim. Acta*, **21,** 289 (1965).
18. V. A. Fassel and W. A. Gordon, *Anal. Chem.*, **30,** 179 (1958).
19. V. A. Fassel, W. A. Gordon, and R. J. Jasinski, *Proc. 2nd Intern. Conf. Peaceful Uses of Atomic Energy* (Geneva) **28,** 583 (1958).
20. V. A. Fassel, W. A. Gordon, and R. W. Tabeling, *ASTM Spec. Tech. Publ. No. 221*, Symp. Spectrochem. Anal. Trace Elem., 1958, pp. 3–22.
21. H. Goto, S. Ikeda, K. Hirokawa, and M. Suzuki, *Sci. Rept. Res. Inst. Tohoku Univ.*, **S18,** 1 (1966).
22. H. Goto, S. Ikeda, K. Hirokawa, and M. Suzuki, *Fresenius' Z. Anal. Chem.*, **228,** 180 (1967).
23. R. E. Heffelfinger and W. M. Henry, *Titanium Met. Lab. Rept. No. 98*, Battelle Memorial Institute, Columbus, Ohio, 1958.
24. P. Höller, *Arch. Eisenhuettenw.*, **34,** 425 (1963).
25. H. Kamada, K. Iwata, and I. Ogahara, *Bunseki Kagaku*, **16** (11), 1203 (1967).
26. J. G. Karohl, thesis, Iowa State University, Ames, Iowa, 1959.
27. T. Kashima, *Nippon Kinzoku Gakkaishi*, **29,** 478 (1965).
28. G. Rossi and J. Melamed, *Comm. Energie At.* (*France*), *Centre d'Etudes Nucleaires*, Saclay, AEC Translation No. 6266, 1961.
29. R. K. Winge and V. A. Fassel, *Anal. Chem.*, **41,** 1606 (1969).

30. A. P. D'Silva, P. K. Wahi, and S. S. Biswas, *Bhabha Atomic Research Centre Report No. 336*, Bhabha Atomic Research Centre, Bombay, India, 1968.
31. J. Spitz and T. Van Danh, *Method. Phys. Anal.*, **4,** 371 (1968).
32. G. Rossi and Z. Hainski, *Met. Ital.*, **58,** 290 (1966).
33. M. S. W. Webb and R. J. Webb, *Nature*, **201,** 487 (1964).
34. M. S. W. Webb and R. J. Webb, *Anal. Chim. Acta*, **33,** 138 (1965).
35. M. S. W. Webb and R. J. Webb, *Anal. Chim. Acta*, **36,** 403 (1966).
36. M. S. W. Webb and R. J. Webb, Third National Meeting, Society for Appl. Spectry., Cleveland, 1964.
37. A. M. Vasserman and Z. M. Turovtseva, *Zh. Anal. Khim.*, **20,** 1359 (1965).
38. C. G. Goldbeck, S. P. Turel, and C. J. Rodden, *Anal. Chem.*, **40,** 1393 (1968).
39. J. L. Jones, Applied Research Laboratories, Hasler Research Center, Goleta, California, personal communication, 1964.
40. A. P. D'Silva, P. K. Wahi, and S. S. Biswas, *Bhabha Atomic Research Centre Report No. 337*, Bhabha Atomic Research Centre, Bombay, India, 1968.
41. K. Hirokawa and H. Goto, *Fresenius' Z. Anal. Chem.*, **234,** 340 (1968).
42. B. L. Vallee and M. R. Baker, *J. Opt. Soc. Amer.*, **46,** 77 (1956).
43. A. Purer, L. Stroud, and T. O. Meyers, *Advan. Cryog. Eng.*, **10** (A–L), 398 (1964).
44. E. A. Limoncelli, *Report TIM-824*, U.S. Atomic Energy Commission, 1965.
45. M. T. Lilburne, *Analyst*, **91,** 571 (1966).
46. J. R. Young and N. R. Whetten, *Rev. Sci. Instrum.*, **32,** 453 (1961).
47. P. Escoffier, *Rech. Aerospatiale*, **106,** 25 (1965).
48. K. D. Greene, Union Carbide Corporation, Linde Division, Buffalo, New York, personal communication, 1968.
49. L. B. Loeb, *Fundamental Processes of Electrical Discharge in Gases*, Wiley, New York, 1939, p. 616.
50. M. R. Nadler and C. P. Kempter, *J. Phys. Chem.*, **64,** 1468 (1960).
51. V. A. Fassel, *Report of the Panel on Analytical Problems in Refractory Metals, Rept. MAB-154-M(1)*, Vol. II, National Academy of Sciences, 1959, pp. 17–22.
52. W. G. Guldner, *Anal. Chem.*, **35,** 1744 (1963).
53. J. S. Hetherington and Y. E. Strausser, 14th National Symp. of the American Vacuum Society, Kansas City, October 1967.
54. J. H. Hill, C. J. Morris, and J. W. Frazer, *UCRL Report 70056*, Lawrence Radiation Laboratory, Livermore, California, 1966.
55. D. O. Hayward and B. M. W. Trapnell, *Chemisorption*, 2nd ed., Butterworths, Washington, 1964, pp. 73–78.
56. A. L. Beach and W. G. Guldner, *Anal. Chem.*, **31,** 1722 (1959).
57. S. Sawa, *Tetsu to Hagaen*, **38,** Pt. 1, p. 567; Pt. 2, p. 672 (1952). Translation available from Henry Brutcher, Technical Translations, P.O. Box 157, Altedena, California 91001.
58. T. Koizumi, F. Tsugane, and M. Kamakura, *J. Iron Steel Inst. Jap.*, **50,** 925 (1964).
59. V. A. Fassel, W. E. Dallmann, and C. C. Hill, *Anal. Chem.*, **38,** 421 (1966).
60. W. E. Dallmann and V. A. Fassel, *Anal. Chem.*, **39,** 133B (1967).
61. M. Schutze, *Ber.*, **77B,** 484 (1944).

62. W. G. Smiley, *Anal. Chem.*, **27,** 1098 (1955).
63. M. Hanin, *Rev. Met.*, **57,** 1133 (1960).
64. F. B. P. Gray and J. Skinner, Pittsburgh Conference on Analytical Chemistry and Applied Spectroscopy, Pittsburgh, March 1967.
65. L. J. Schmauch and R. A. Dinerstein, *Anal. Chem.*, **32,** 343 (1960).
66. J. E. Purcell and L. S. Ettre, *J. Gas Chromatogr.*, **3,** 69 (1965).
67. R. Berry, "Analysis of milli-microlitre quantities of permanent gas mixtures," in M. van Swaay, Ed., *Gas Chromatography*, Butterworths, London, 1962, pp. 321–330.
68. P. J. Bourke, R. W. Dawson, and W. H. Denton, *J. Chromatogr.*, **14,** 387 (1964).
69. V. Willis, *Nature*, **184,** 894 (1958).
70. C. H. Hartmann and K. P. Dimick, *J. Gas Chromatogr.*, **4,** 163 (1966).
71. S. R. Lipsky and M. M. Shahin, *Nature*, **197,** 625 (1963).
72. S. R. Lipsky and M. M. Shahin, *Nature*, **200,** 566 (1963).
73. R. A. Landowne and S. R. Lipsky, *Nature*, **189,** 571 (1961).
74. H. P. Williams and J. D. Winefordner, *J. Gas Chromatogr.*, **6,** 11 (1968).
75. A. Karmen, J. Calvin Giddings and R. A. Keller, Eds., *Advances in Chromatography*, Vol. II, Dekker, New York, 1966, pp. 293–336.
76. J. C. Sternberg, *4th International Symposium on Gas Chromatography*, Academic New York, 1963, pp. 161–191.
77. V. Svojanovsky, M. Krejci, K. Tesarik, and J. Janak, *Chromatographic Reviews*, Vol. VIII, M. Ledered, Ed., Elsevier, New York, 1966, pp. 91–171.
78. R. S. Juvet and F. Zado, *Advances in Chromatography*, Vol. I, J. Calvin Giddings and R. A. Keller, Eds., Dekker, New York, 1965, pp. 249–307.
79. *Catalog and Price List of Standard Materials Issued by the National Bureau of Standards, NBS Special Publ. 260*, U.S. Government Printing Office, Washington, D.C., 1970; *Supplement*, January 1972.
80. *Determination of Oxygen in Ferrous Materials SRM 1090, 1091, and 1092, NBS Misc. Publ. 260–14*, U.S. Government Printing Office, Washington, D.C. 1966.
81. D. I. Walter, *Anal. Chem.*, **22,** 297 (1950).
82. A. J. Dukat and J. L. Marley, Pittsburgh Conference on Analytical Chemistry and Applied Spectroscopy, Cleveland, Ohio, 1969.
83. P. Dickens, P. Loenig, K. H. Schmitz, and P. Jaensch, *Arch. Eisenhuettenw.*, **39,** 45 (1968).
84. V. N. Egorov and Y. N. Gryslov, *Zavod. Lab.*, **36,** 1342 (1970).
85. I. A. Grikit, E. G. Galusko, V. V. Polonik, R. K. Ognev, G. G. Kolomoets, and A. I. Polonik, *Sb. Tr., Uses. Nauchno.-Issled. Proekt. Inst. Titana*, **6,** 155 (1970).
86. I. A. Grikit and T. I. Rumyantseva, *Ab. Tr., Uses. Nauchno.-Issled. Proekt. Inst. Titana*, **6,** 146 (1970).
87. I. A. Grikit and T. I. Rumyantseva, *Zh. Prikl. Spektrosk.*, *12*, 602 (1970).
88. I. A. Grikit, T. I. Rumyantseva, and E. G. Galushko, *Sb. Tr., Uses. Nauchno.-Issled. Proekt. Inst. Titana*, **4,** 221 (1970).
89. I. A. Grikit, T. I. Rumyantseva, and E. G. Galushko, *Zh. Prikl. Spectrosk.*, **10,** 387 (1969).

90. L. F. Kravchenko and V. D. Kurochkin, *Zh. Anal. Khim.*, **27,** 398 (1972).
91. V. M. Plotnitskii, *Zavod. Lab.*, **36,** 1068 (1970).
92. V. M. Plotnitskii and A. N. Rudoi, *Zavod. Lab.*, **37,** 1324 (1971).
93. R. R. Shvangiradze, K. A. Oganezov, and B. Y. Chikhladze, *Zh. Prikl. Spektrosk.*, **7,** 265 (1967).
94. R. R. Shvangiradze, K. A. Oganezov, and B. Y. Chikhladze, "Metody Opred. Issled. Sostoyaniya Gazov Metal," *1968*, 83.

CHAPTER

6

OPTICAL AND MASS-SPECTROSCOPIC METHODS

ROYCE K. WINGE AND VELMER A. FASSEL

Ames Laboratory-USAEC
and
Department of Chemistry
Iowa State University
Ames, Iowa

CONTENTS

I. INTRODUCTION

Optical-emission spectroscopic methods of analysis have been used for over a hundred years for the determination of alloying elements and residual impurities in a variety of metals without requiring time-consuming separation or concentration procedures. It was not until 1950 that optical-emission spectroscopic techniques were first applied to determining the gaseous element content of metals [1, 2]. This tardy development was caused by several basic problems. First, it was necessary to volatilize the constituents to be determined and to excite their spectra in surroundings free from the

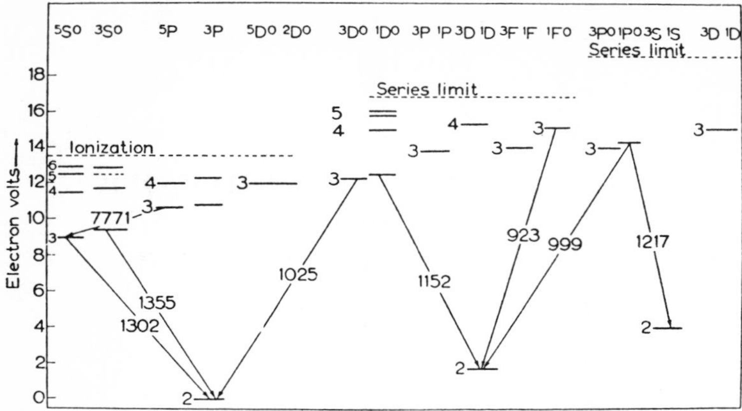

Fig. 6.1. Energy level diagram of oxygen. (Reproduced by permission from *Spectrochimica Acta* [57].)

atmospheric gases. Second, the sensitive resonance lines of oxygen, nitrogen, and hydrogen occur in the vacuum ultraviolet region of the spectrum, the analytical usefulness of which has received little study until recently. In addition, conventional excitation sources energetic enough to excite the emission lines of the gaseous elements normally volatilize little sample. As a consequence, adequate powers of detection may be difficult to achieve.

Mass spectroscopy may also be used for determining the gaseous elements in metals, either directly through spark-source volatilization and ionization or as an adjunct to the vacuum-fusion method (see Chapter 4).

II. ANALYTICAL TECHNIQUES

A. EMISSION SPECTROSCOPY

Various electrical-energy sources have been employed for the extraction and spectral excitation of the gaseous element content of metals. The relatively few spectroscopic techniques can be roughly categorized according to the country of origin by the type of excitation source used—that is, vacuum spark in France, spark and impulse discharge in Russia, and dc arc in the United States. Relatively high-energy excitation conditions are required to achieve the necessary sensitivity for determining these gases in the low parts-per-million range. Figure 6.1 shows excitation energies required for some of the most sensitive lines of oxygen [57]. It is apparent from this diagram that most of the sensitive lines lie in the vacuum ultraviolet region of the spectrum. This same generalization can be drawn for the atomic lines of nitrogen and hydrogen. For conventional spectrographic techniques the atomic line at 7771.9 Å allows oxygen to be determined to the low parts-per-million range.

1. Spark and Impulse Discharges

Following the work of Mandel'shtam and Fal'kova first reported in 1950 [1], many papers were published on the application of spark- and impulse-discharge methods for determining the gaseous element content of metals (3–56). To appreciate these excitation techniques, it is necessary to draw some distinctions between low- and high-voltage spark discharges. These operate at potentials below approximately 1000 and above 10,000 V, respectively. The main difference between spark and impulse discharges is the value of the circuit capacitance. In spark circuits, relatively low-valued capacitors can be charged to the breakdown potential of the combined analytical and auxiliary gaps many times per second. This type of discharge is sometimes referred to as an intermittent spark. In impulse-discharge circuits the capacitance is so high (up to 8000 μF [12]) that relatively long charging times are required with charging circuits of modest proportions. The impulse discharges are, therefore, initiated manually or by means of a timer, and often only one [56] or a few [37, 40, 42] discharges are required to obtain adequate intensity. The high energy attained in an impulse discharge can be estimated from the work of Sventitskii et al. [54], in which O II lines were used.

Normally, excitation by spark- or impulse-discharge techniques takes place in an evacuable chamber or bell jar in an atmosphere of helium, carbon dioxide, or nitrogen. Shubina et al. [49], however, have developed a method for determining the oxygen content of steels using only a stream of nitrogen to protect the impulse discharge from atmospheric oxygen. The authors claim that the protective nitrogen need not be of exceptionally high purity, because of the short period of the discharge and the directed action of the shock wave, which expands outward, impeding the ingress of gas from the outside atmosphere. Figure 6.2 shows the electrode arrangement, along with the optimum shape nozzle for directing the nitrogen flow. The blank contribution of the residual oxygen in the extraction atmosphere could be reduced further by time-resolving the emitted radiation in such a way as to reject the first 25% of each light pulse, which corresponds to the time period of maximum emission of the surrounding atmosphere [12].

Detection limits of spark- and impulse-discharge methods for oxygen and nitrogen occasionally have been reported as low as 10 or 20 ppm [39], more often in the range 50–100 ppm, and sometimes as high as 200–500 ppm [14, 29, 32]. The best detection limit reported for hydrogen has been 1 ppm [48]. The primary factor controlling this lower limit of detection appears to be the small amount of metal actually sampled by the spark or impulse

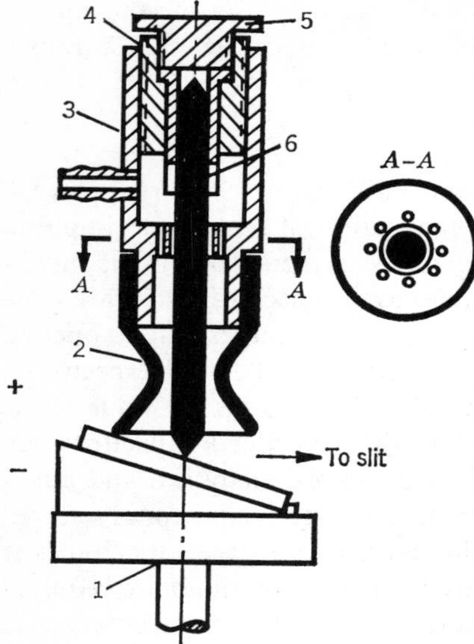

Fig. 6.2. Electrode arrangement [49] for isolating discharge gap from the atmosphere by a flow of neutral gas. (1) Lower electrode with wedge-shaped support to prevent screening of light beam; (2) glass nozzle; (3) body of upper electrode holder; (4) sleeve; (5) nut for aligning clamp; (6) clamp for holding tungsten electrode.

discharge. Vaporization rates from steel samples using high-frequency spark and unidirectional, overdamped condenser discharges were determined by Fassel and Tabeling [57], as shown in Tables 6.1 and 6.2. Berneron [58] estimated that only 1 or 2 mg of sample was affected by the approximately 100 sparks required for an analysis. The powerful impulse discharges are capable of higher vaporization rates, but considering the short exposure times, only a few milligrams of sample contribute to the analytical measurement.

In addition to limiting sensitivity, the small sample size is also responsible for several other problems. Because the bulk of the energy of the spark or impulse discharge is expended at or near the surface of the sample, surface conditions of the sample may affect the analytical results [57, 59]. Presparking has been used to clean the surface before making the analytical exposure [56]. This technique is not always applicable, however. Mandel'shtam and Fal'kova [1] reported a local impoverishment of the nitrogen content in the immediate region of previous spark sites on steel samples, which required moving of the sample between discharges for a multiple-impulse exposure.

The small amount of sample vaporized by the spark and impulse discharges also tends to magnify inhomogeneity problems. The gaseous element impurities are often segregated on a microscopic scale throughout the metal sample. The electrical discharge may selectively seek out or avoid the inclusions according to their electrical or geometric properties.

TABLE 6.1. Rate of Steel-Sample Weight-Loss by High-Frequency Spark Discharge[a]

Atmosphere	Weight Change (mg/min)
H_2	−0.3
He	−0.3
Ar	−0.4
O_2	−1.9
N_2	−1.8
Air	−1.8

[a] Reproduced by permission from *Spectrochimica Acta* [57].

Another problem is that different metal and alloy compositions require different excitation conditions [9, 28, 48, 51]. Structural changes brought about by heating and quenching a previously annealed sample have also been observed to cause a change in nitrogen "sensitivity." The same results were obtained after reannealing the quenched specimen, however, as were obtained from the original annealed sample [9]. Skotnikov [27, 30] has expressed the view that standard samples should be made of exactly the same grade of steel as the analyzed steel because the spectroscopic results depend on the nature of the alloying elements and the heat treatment. This is a rather optimistic viewpoint, because reliable standard samples are difficult to obtain even without the criterion of matching the composition exactly.

In general, the most sensitive spectral lines of an element are those originating from transitions of a sole electron between p- and s-states in the outer shell [63–65]. This situation corresponds to the electronic configuration of the alkali metals, which of course, have very sensitive spectral lines. For oxygen and nitrogen this alkali metal configuration would correspond to the O VI

TABLE 6.2. Rate of Steel-Sample Weight-loss as Anode and Cathode of Unidirectional Overdamped Condenser Discharge[a]

Atmosphere	Weight Change (mg/min)	
	Anode	Cathode
Ar	0.0	−12.7
H_2	0.0	−14.2
N_2	—[b]	−0.6
Air	+2.9	−4.0

[a] Reproduced by permission from *Spectrochimica Acta* [57].
[b] Not measured.

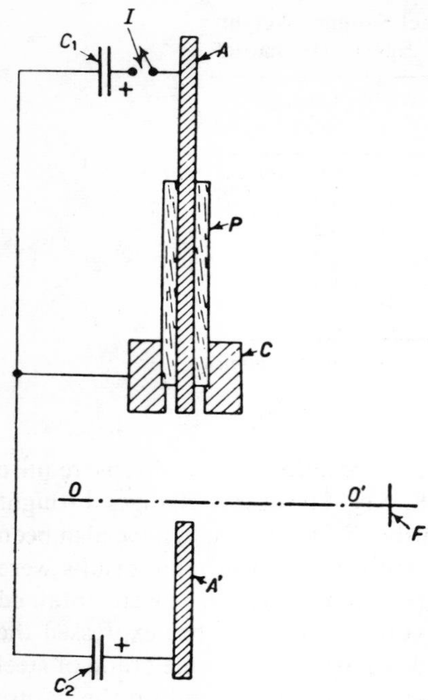

Fig. 6.3. Arrangement of the electrodes for the production of a vacuum spark with the aid of an auxiliary spark. (Reproduced by permission from *Spectrochimica Acta* [73].)

and N v lines. Excitation of these lines requires powerful electrical discharges (vacuum sparks), and the observation of these lines in the vacuum ultraviolet region requires special experimental techniques. These techniques have been developed primarily at the Centre National de la Recherche Scientifique, Bellvue, France and at the Institute de Recherche de la Siderurgie, Saint-Germain en Laye, France [58, 59, 63, 66–75]. The apparatus developed in these laboratories consisted of a spectrometer and a spark chamber, both evacuated to 10^{-4}–10^{-6} torr [72, 76]. A special electrode arrangement, shown in Fig. 6.3, was required to obtain a spark across the vacuum analytical gap [73]. As can be seen from this figure, C serves as the cathode for both anode A of the initiating circuit and anode A′ of the analytical-gap circuit. A′ is the sample electrode. A low-power initiating spark travels radially across the lower end of the alumina tube P between C and the center electrode A. This initiating spark produces ionization in the analytical gap and allows the powerful analytical discharge to occur between C and A′. The contamination caused by the alumina tube P is noticeable particularly in determining low concentrations of oxygen [66]. The spectral background has also been a problem in this far ultraviolet region but has been reduced by lowering the excitation source voltage from 17 kV to 8 kV [66]. Figures 6.4 and 6.5 show the results obtained for the determination of oxygen and nitrogen in niobium

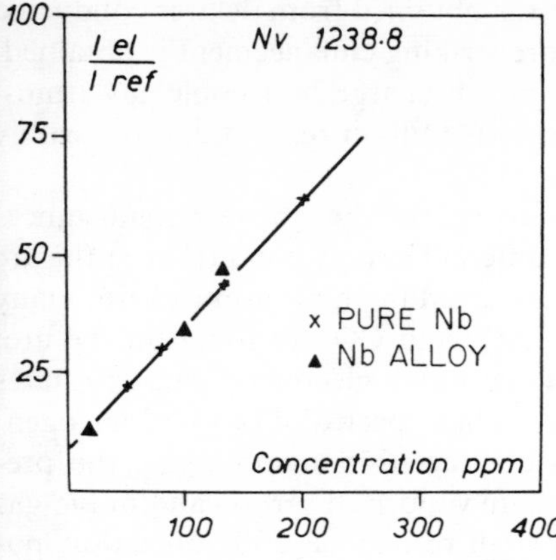

Fig. 6.4. Analytical curve for nitrogen in niobium and its alloys. (Reproduced by permission from Adam Hilger Ltd. [66].)

and its alloys [66]. These curves show the capabilities of the technique to give very linear relations of concentration and intensity even to quite low concentrations.

Buyanov et al. [59] and Lindstrem et al. [77] have also studied the determination of oxygen and nitrogen in metals by spectroscopic methods in the vacuum ultraviolet region, with the latter reporting studies of analytical curves made from eight oxygen lines in this region.

2. DC Arc

Useful analytical lines of oxygen, nitrogen, and hydrogen are barely visible, if at all, in dc-arc discharges between metal electrodes. Moderate

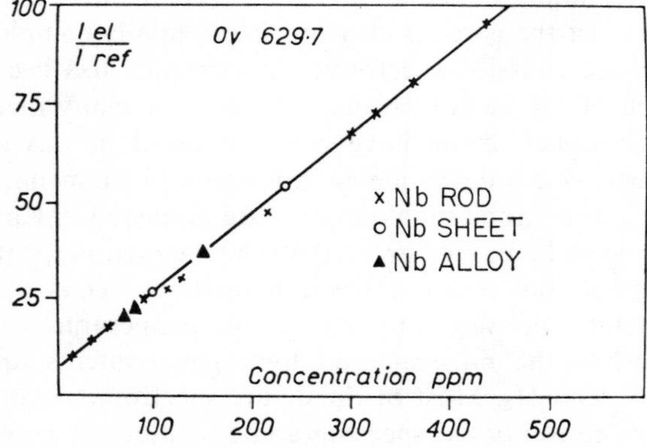

Fig. 6.5. Analytical curve for oxygen in niobium and its alloys. (Reproduced by permission from Adam Hilger Ltd. [66].)

intensity of certain lines of these elements is obtained from dc arcs conducted between graphite electrodes in air. A more striking enhancement is obtained, however, by conducting the dc-carbon-arc discharge in a noble-gas atmosphere. As an example, only a few parts per million oxygen can be readily detected in a dc-arc discharge in argon.

The difference in the intensity encountered in the above circumstances appears to be the result of considerably different energy conditions in the arc column. When the dc-arc discharge is conducted in an air atmosphere, many molecular species such as N_2, O_2, CN, NO, and CO are found in the arc. These species have rotational, vibrational, and electronic excited states below the energies required to excite the atomic spectra of oxygen, nitrogen, and hydrogen. When the carbon arc is conducted in a noble gas, the predominant species of the discharge column are vaporized carbon and noble-gas atoms, both of which possess relatively high resonance and ionization potentials. The mean energy and energy distribution of the carbon-arc argon discharge are therefore increased sufficiently over that of the other discharge conditions for efficient excitation of the emission spectra of oxygen, nitrogen, and hydrogen. With metal electrodes the arc column is flooded with metal atoms having ionization states and many excited states that lie at relatively low energy levels. Prevention of flooding of the arc column with sample vapor can be accomplished in several ways: (1) The excitation can be performed with the discharge passing to an empty electrode rotated into arcing position after the arc-extraction step [78]. (2) With many metals this change of electrode is not necessary, because the metal has a relatively low vapor pressure, which is further reduced by the carbon dissolved in the molten metal during the extraction process [79]. (3) A platinum flux may be used to reduce volatility of the sample metal (see discussion of platinum flux in Chapter 5). In all cases the beginning of the analytical measurement is usually delayed until the extraction of the gaseous elements is essentially complete.

The dc-carbon-arc emission-spectrometric technique has been applied to the determination of the gaseous element content of many metals [57, 60, 78–111]. Excitation phenomena have been discussed in this chapter; the extraction chamber, electrode geometry, extraction phenomena, calibration, bath and flux techniques, and blank control are discussed in Chapter 5. It is appropriate to recall here, however, that the extraction phenomena are similar in the dc-arc and fusion extraction methods. That is, the oxygen content of the metal is converted to CO by high-temperature carbon-reduction reactions, while the nitrogen and hydrogen contents are thermally decomposed to N_2 and H_2. Most of the dc-arc spectrometric methods have used more or less conventional spectrographic facilities. Two experimental arrangements merit special mention, however.

Goto et al. [92] used a dc-arc extraction method in conjunction with a 'acuum ultraviolet spectral determination of oxygen in iron and steel. The .rc extraction was performed in argon at 600 torr in a chamber separated rom the evacuated spectrometer by a calcium fluoride lens. The intensity atio O I 1302.17 Å/C II 1721.66 Å was determined photoelectrically and 'aried linearly with the oxygen content of the sample in the range 0.01–0.1%. Iirokawa and Goto [60] later reported the determination of nitrogen in steel .lso in the vacuum ultraviolet region. They used the line pair N I 1745 Å/C I 930 Å for the range of approximately 15–200 ppm nitrogen.

Matsumoto et al. [78] developed a direct-reading spectrometric technique or determining oxygen in steels using the oxygen line at 7772 Å. The experi- nental facility is shown in Fig. 6.6. The method takes advantage of the

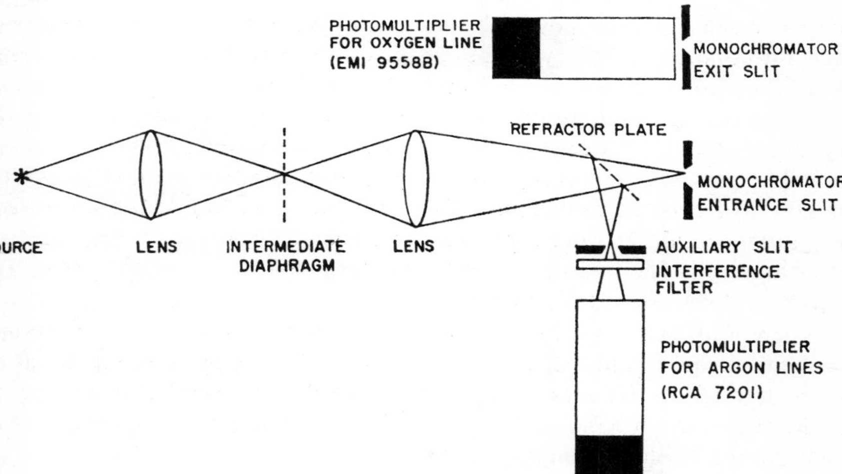

ig. 6.6. Schematic of direct-reading spectrometer. (Reproduced by permission from *'pectrochimica Acta* [78].

elatively large aperture of a 0.5-m Ebert-type spectrometer. The intermediate .iaphragm selects the central 3 mm of the analytical gap and thus blocks the ontinuum radiation of the electrodes from entering the spectrometer. The ortion of the discharge transmitted by the diaphragm is then focused on the lit of the spectrometer. A glass refractor plate, however, diverts a portion f this light at a right angle to an interference filter, which transmits a group f argon lines to a photomultiplier tube for use as an internal reference. An nterference filter with peak transmission at 7772 Å is also used at the entrance lit of the spectrometer to reduce stray light and thereby increase the signal o noise ratio of the oxygen line by approximately 2.5.

3. Discharge Tube Excitation

a. Electrodeless Discharges

Electrodeless-discharge excitation techniques have found limited use fo determining gases after they have been extracted by other means [112–119] This excitation technique is commonly used, however, in conjunction witl isotope-equilibration methods, as discussed in a later section.

Because of a large number of lines and an intense background, Berezi and Malyshev [115] chose electrodeless-discharge excitation rather tha spark excitation for the determination of hydrogen and oxygen in uranium The gases were extracted from the uranium in a resistance-heated graphit or molybdenum boat in an evacuated bell jar. The gases were then pumpe into a discharge tube powered with a 100-W, 5–10-MHz high-frequenc generator. Neon at 8 torr was used as the carrier gas. Hydrogen and oxyge were determined down to 10^{-4} and 10^{-2} wt %, respectively. The author reported that the same results could be obtained for oxygen from either th oxygen line at 7772 Å or the carbon monoxide bandhead at 4380 Å.

Babko and Getman [114] have developed a sensitive method for deter mining hydrogen in germanium. The germanium was burned in an oxyge jet, and the resulting water vapor was frozen and delivered to an electrodeless discharge cell. The hydrogen was then determined to 2×10^{-5} % from th hydrogen lines at 6563 Å and 4861 Å.

Koch et al. [116–119] have used discharge tube methods for determinin oxygen, nitrogen, and hydrogen in steels after these gases were separated b hot-extraction techniques. They used a 45-W, 40.68-MHz generator t excite the nitrogen bandhead at 3755 Å, the carbon monoxide bandhead a 4835 Å, and the hydrogen line at 4861 Å.

b. Hollow-Cathode Discharge

Some of the earliest studies on determining the gaseous elements in metal spectrometrically were performed by Rosen [2, 120, 121], who used a hollov cathode for the extraction and excitation of the oxygen content of steel Webb and Webb [122–125] have extended this approach to the determinatio of oxygen and nitrogen in steel and uranium carbide, oxygen in tungsten an copper, and hydrogen in zirconium and zircalloys. They modified a Czerny Turner-type spectrometer so that oxygen and nitrogen could be determine simultaneously. The necessary two channels were obtained by using th conventional and reverse paths through the spectrometer, as shown in Fig 6.7. One channel measured the oxygen triplet at 7772–7775 Å and th adjacent background, while the other channel measured the nitrogen band head 3914 Å in the second order. To determine hydrogen, the spectromete

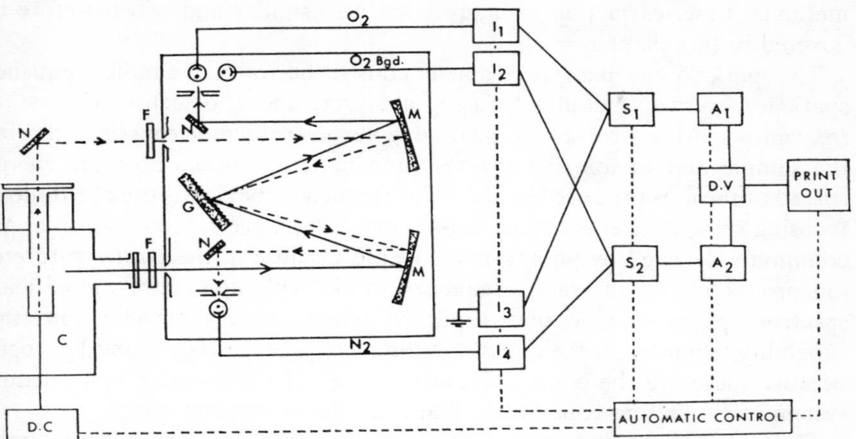

Fig. 6.7. Schematic of single monochromator for simultaneous determination of oxygen and nitrogen content of metals. (C) cathode chamber; (F) filters; (M) concave mirrors; (N) plane mirrors; (G) plane grating; (I) integrators; (S) selectors; (A) amplifiers; (D.V) digital voltmeter; (D.C) high-voltage dc source unit. (Reproduced by permission from *Analytical Chimica Acta* [124].)

was reset to measure the hydrogen line at 6563 Å. Sample weights as low as 4 mg were used, and it was suggested that sample weights as high as 100 mg could be used for the determination of hydrogen. Sample weights in the low milligram range may introduce problems arising from nonrepresentative samples and from surface adsorption and contamination. Also, of concern is the probability of metal vaporization and its attendant difficulties in the low-pressure discharge. The use of helium instead of argon as the discharge gas partially reduced the vaporization caused by sputtering. Limits of detection were listed in the fractional microgram range for oxygen, nitrogen, and hydrogen in most of the metals studied [124]. The hollow cathode has also been used for equilibration and spectral excitation in isotopic methods. This application is discussed in a later section of this chapter.

3. MASS SPECTROSCOPY

Mass spectroscopy, a highly sensitive analytical technique capable of analyzing very small quantities of sample, has only recently been applied to the determination of gases in metals [126–146]. Although vacuum-fusion-mass-spectroscopic techniques are discussed in Chapter 4, it is appropriate to discuss spark-source mass spectroscopy in this chapter because of its similarity in sampling characteristics to the spark-emission spectroscopic methods. Laser-extraction techniques are also similar and will therefore b covered in this chapter.

The spark-source mass spectrometer consists basically of a radio-frequenc spark ion source, a double-focusing analyzer, and a detector. The radio frequency spark source serves the double purpose of vaporizing or sputterin the sample and of ionizing the resulting material. Because of the energ spread of the ions produced by the radio-frequency spark, the use of a double focusing spectrometer is almost mandatory [126]. Spectral ion intensities ar commonly detected by photographic means because of the erratic nature o ion production by the radio-frequency spark. Although spark-source mas spectroscopy is capable of exceptionally high powers of detection, thi capability is limited in the determination of oxygen, nitrogen, and hydroge because these are the common residual gases of the spectrometer vacuun system and therefore constitute a blank for the determination.

The first studies using spark-source mass spectroscopy for determining th gaseous elements in metals were troubled with problems of accuracy, preci sion, and high limits of detection [127, 128]. Improvements in equipment an experimental technique have alleviated some of these problems. Stefan et al. [129, 130] studied the determination of both metallic and gaseou element impurities in highly purified isotopes and the determination o hydrogen in metals with a threshold of detection as low as 0.0001 ppm Harrington et al. [131] were able to determine low concentrations of oxyge and nitrogen in silver, iron, tungsten, and copper matrices. They estimate the experimental blank to be less than 0.5 ppm. The microsampling achieve by the radio-frequency spark permitted reliable determinations of localize concentrations, which were useful for determining distributions of impurities A sample-scanning technique was proposed for estimation of bulk concentra tions of impurities as well as for evaluating homogeneity.

Leipziger and Guidoboni [132] determined oxygen and nitrogen in steels gold, and platinum, and hydrogen in molybdenum, niobium, tantalum, an tungsten. When a prespark of 100 nC was used to remove the last traces o surface impurities, a blank level below one atom per million was achieved Vidal et al. [133] determined oxygen and nitrogen in niobium but reporte poor reproducibility, which they attributed to the "pin-point character of th analysis and microsegregation."

Arbuzova et al. [136, 137] traced the local hydrogen concentration i aluminum by scanning with a laser beam focussed to a 10^{-4}–10^{-3}-cm diam eter. These authors were able to determine that the hydrogen concentratio at the grain boundaries in an aluminum sample having an average hydroge concentration of 0.2 ppm was somewhat higher than within the grains, whil in a titanium alloy the hydrogen was concentrated predominantly in the β phase within the grains. Ivanovski et al. [142] reported the use of a laser t chieve a local rise in temperature in the range 6000–10,000°C with the nergy focused in a 20-μ–1.5-mm-diameter spot. Gaseous impurities of the rder 10^{-3}–10^{-4} wt % could be determined mass spectroscopically.

:. ISOTOPIC TECHNIQUES

Methods involving stable isotopes for determining gases in metals are ariously referred to by the designations *isotope dilution*, *isotope balancing*, r *isotope exchange*. These designations refer essentially to the same process, hat is, equilibration of a known quantity of an enriched isotope with the otope content of a metal sample. The term *isotope dilution*, however, usually restricted to procedures in which the enriched isotope is added to the eaction furnace in the same chemical form as exists in the analytical sample. 'he terms *isotope balancing* and *isotope exchange* refer to procedures in which ne enriched isotope is added in a chemical form (usually as a gas) different rom that in the analytical sample. After equilibration, the gaseous element ontent of the metal is calculated from the mass- or emission-spectroscopic- lly determined abundance ratio of the light and heavy isotopes.

The relatively complicated and time-consuming experimental procedure is ompensated for by several very significant advantages. These advantages are 1) that complete evolution of the gas content of the sample is not required, nd (2) that the method is absolute, requiring no standard samples.

The isotopic method was first applied to the determination of gases in ıetals by Kirshenbaum et al., who were concerned with the determination of xygen in iron [147], copper [148], titanium [149, 150], and zirconium [149]. ince that time, the method has been applied to the determination of hydro- en, nitrogen, and oxygen in an extended list of metals [101, 151–191].

The reaction vessel can be a simple resistance-heated quartz tube if only ıoderate reaction temperatures are required, as for example, in the deter- ıination of hydrogen in metals [179, 180]. The stronger metal–oxygen and ıetal–nitrogen bonding and the slower diffusion rates of oxygen and nitrogen ave necessitated higher equilibration temperatures for these gases than for ydrogen. The higher temperatures have been obtained by inductive [160], ollow-cathode [101], and dc-arc heating [157, 170–172, 188–191] and hrough laser vaporization [159, 168, 188]. Zakorina et al. [191] reported educed analysis times resulting from an apparently more intensive isotopic xchange in the hollow-cathode discharge. Ivanova and Federov [101] btained very rapid isotopic equilibration [5–10 sec for titanium and nio- ium] with a 150-A dc arc as a heat source. Zaidel [186] has summarized emperature and equilibration-time data for oxygen, nitrogen, and hydrogen n a number of metals, as shown in Table 6.3.

TABLE 6.3. Conditions for Isotopic Equilibration for the Determination of Hydrogen, Nitrogen, and Oxygen in Metals [186]

		Balancing Conditions		
Element to Be Determined	Metal Being Analyzed	Temperature (°C)	Time (min)	Specimen Shape
Hydrogen	Zn	400	20	~5-mm granules
	Cd	300	20	~5-mm granules
	Al	500	30	~5-mm cylinders
	Cu	900	20	~5-mm cylinders
	Zr	1100	15	~5-mm cylinders
	Ti	1100	20	~5-mm cylinders
	Fe	1100	30	~5-mm cylinders
	V	1100	10	~5-mm cylinders
	Ni	1100	10	~5-mm cylinders
	W	1100	30	~5-mm cylinders
	Mo	1100	30	~5-mm cylinders
	Co	1100	200	~5-mm cylinders
	Cr	1100	300	~5-mm cylinders
	Fe 60 V 40	1100	15	~5-mm cylinders
	Fe 50 Ni 50	1100	15	~5-mm cylinders
	Fe 50 Co 50	1100	60	~5-mm cylinders
	Fe 30 Cr 70	1100	300	~5-mm cylinders
	Cr 20 Ni 80	1100	200	~5-mm cylinders
	Pd	1100	20	~5-mm cylinders
	Y	1100	10	~5-mm cylinders
	Nd	1000	10	~5-mm cylinders
	Pr	1000	10	~5-mm cylinders
	Nb	800	20	~1-mm plates
	Ta	800	15	~1-mm plates
	U	1000	15	~1-mm plates
Nitrogen	Fe	1150	120	~0.2-mm shaving
	Mo	1150	250	~3-mm bits
	W	1150	200	~3-mm bits
	W	1800	30	~3-mm bits
	Cr	1150	120	~2-mm bits
	Ni	1150	200	~0.3-mm shavings
	Ni	1600	25	melt
Oxygen	Fe	1700	30	melt
	W	1900	35	~5-mm bits
	Mo	1900	20	~5-mm bits
	Ni	1600	15	melt
	Co	1600	20	melt

Although procedurally more simple, the addition of the enriched isotope gaseous form does not always give accurate results [156]. The enriched otope is therefore often added in combination with a metal referred to as the *aster alloy* or *dosage flux*. The discrepancy mentioned above is more rious for some metals than for others and is also dependent on the gas eing determined. Apparently sequestering of the gaseous enriched isotope y adsorption, gettering, or other chemical reactions) before complete uilibration has taken place [156, 163] is the cause of the problem. The ojective is that a complete statistical exchange occur for all chemical forms the analyte and for all phases in which the analyte element exists in the uilibration system.

The overall procedure for an isotopic dilution determination is as follows. fter equilibration has been achieved, the isotope mixture in the gas phase is panded into a discharge tube [159, 185] or a mass-spectrometer source 74, 176] for analysis. The quantity of gas in the sample can be calculated om the formula given by Zaidel [186]:

$$V_0 = (1 + a)\frac{\left[\dfrac{C_1}{C_2} - \left(\dfrac{C_1}{C_2}\right)_0\right] P T_0 V}{\left[1 - a\left(\dfrac{C_1}{C_2}\right)\right]\left[1 + \left(\dfrac{C_1}{C_2}\right)_0\right] P_0 T} - V_0', \qquad (1)$$

here V_0 is the volume (STP) of the analyte-gas content of the sample; $C_1/C_2)_0$ and C_1/C_2 are the relative concentrations of the light and heavy otopes, respectively, in the gas phase before and after balancing; P is the essure (at temperature T) of the initial mixture of light and heavy isotopes troduced into the exchanger; the term $1 + (C)/C_2)_0$ corrects for the degree initial dilution of the heavy isotope by the light in the mixture; P_0 and T_0 e the normal pressure and temperature; V is the exchanger volume; and a a constant that corrects for the natural content of the heavy isotope in the ght ($a_H = 0.00015$, $a_N = 0.0038$, $a_O = 0.002$). The blank is determined om a "dry run," that is, from a normal analytical sequence without the nalytical sample. The blank correction V_0' is then calculated from Eq. 1 by tting $V_0 = 0$. When the roles of the equilibration furnace and the spectral-citation source are combined, as in the hollow cathode, the blank can be etermined along with the gas content of the sample in a single analytical quence. The gas content of the sample is then obtained from measurements the isotopic ratio after the equilibrations before and after addition of the mple metal. By eliminating the error due to the normal scatter of blank lues as determined by the dry run, Zakorina et al. [191] have reported a vefold improvement in the detection limit with the hollow-cathode technique.

TABLE 6.4 Wavelengths and Isotope Shifts of Spectral Lines Used in Isotopic Analytical Techniques [186]

	λ	$\Delta\lambda$
$H^1_\alpha(H^2_\alpha)$	6562.8 Å	−1.78 Å
$H^1_\beta(H^2_\beta)$	4861.3	−1.33
$N^{14}N^{14}(N^{14}N^{15})$	2976.8	+5.9
	3159.3	+3.5
	3943.0	−10.2
$CO^{16}(CO^{18})$	4123.6	+8.5
	5198.2	−19

The diffusion rate of hydrogen is high enough in most metals for isotop equilibrium to be achieved in 10–30 min for solid samples of moderate siz Chromium and cobalt require much longer times [186]. Because of the muc slower diffusion of oxygen and nitrogen in metals, it is usual to emplo samples in powder, shaving, or chip form [186]. This is a serious disadva tage, especially at low analyte concentrations, because of the greater relativ effect of surface films or contamination. To avoid the problems associate with these granular sample forms, the dc arc seems particularly appropria as an equilibration heat source because of its localized high-temperatu capabilities and the apparent stirring action it creates in the molten samp [84].

Zaidel and Petrov [186] have suggested the spectral lines listed in Table 6. as suitable for determining the isotopic ratios. The isotopic wavelength shif are also shown in this table. The excitation of these spectral lines is usual accomplished with a high-frequency electrodeless discharge, as for exampl the 1–5-MHz, 0.5-1.0-kW generator used by Zaidel [186]. The lowe concentrations of hydrogen, nitrogen, and oxygen determinable (with le than 30% error) by isotopic methods have been estimated by Zaidel [186] shown in Table 6.5. These data show the determination limits to be margi ally adequate for low-weight samples but quite good for sample weights 10 g.

TABLE 6.5. Determination Limits for Hydrogen, Nitrogen, and Oxygen (wt % [186]

Mass of Weighed Specimen (g)	H_2	N_2	O_2
0.1	2×10^{-4}	5×10^{-3}	10^{-2}
1	2×10^{-5}	5×10^{-4}	10^{-3}
10	2×10^{-6}	5×10^{-5}	10^{-4}

REFERENCES

1. S. L. Mandel'shtam and O. B. Fal'kova, *Zavod. Lab.*, **16,** 430 (1950).
2. B. Rosen and M. I. Ottelet, Colloq. Intern. Spectrog., Strasbourg, October, 1950, Group. pour l'Avan. des Methodes Spectrogr. des Prod. Met., Paris, 1950, pp. 155–167.
3. M. S. Alpatov, K. A. Sukhenko, P. P. Galonov, and O. B. Fal'kova, *Tr. Komis po Analit. Khim., Akad. Nauk SSSR, Inst. Geokhim. i Analit. Khim.*, **12,** 288 (1960).
4. T. A. Bessonova, A. V. Galilova, N. I. Zhorova, N. M. Zolotareva, M. K. Korsunskaya, A. B. Shaevich, and S. B. Shubina, *Metody Opred. Issled. Sostoyaiya Gazov Metal*, **1968,** 146.
5. I. B. Borovskii and S. A. Skotnikov, *Fiz. Sb. L'vovsk. Gos. Univ.*, **No, 4,** 217 (1958).
6. I. B. Borovskii and S. A. Skotnikov, *Titan i Ego Splavy*, *Akad. Nauk SSSR, Inst. Met.* **No. 2,** 165 (1959).
7. I. B. Borovskii, S. A. Skotnikov, and I. F. Petrushin, *Tr. Inst. Met. im. A. A. Baikova, Akad. Nauk SSSR*, **No. 3,** 276 (1958).
8. N. V. Buyanov and L. M. Federova, *Sb. Tr. Tsentr. Nauchno.-Issled., Inst. Chern. Metal*, **60,** 122 (1968).
9. N. V. Buyanova, L. M. Federova, and V. F. Korotkov, *Izv. Akad. Nauk SSSR, Ser. Fiz.*, **23,** 1126 (1959).
10. N. V. Buyanov, O. I. Vashkov, V. K. Gavrilova, and V. F. Karotkov, *Titan i Ego Splavy, Akad. Nauk SSSR, Inst. Met.*, **No. 2,** 174 (1959).
11. D. Cartlidge and G. Sale, *J. Inst. Metals*, **98,** 169 (1970).
12. D. Cartlidge and G. Sale, *Spectrochim. Acta*, **27B,** 421 (1972).
13. O. B. Fal'kova, *Izv. Akad. Nauk SSSR, Ser. Fiz.*, **19,** 149 (1955).
14. O. B. Fal'kova, *Zavod. Lab.*, **21,** 1083 (1955).
15. O. B. Fal'kova. *Tr. Komis po Analit. Khim., Akad. Nauk SSSR, Inst. Geokhim. i. Analit. Khim.*, **10,** 278 (1960).
16. L. M. Federova, E. P. Zanina, and V. P. Korneenko, *Zavod. Lab.*, **31,** 1347 (1965).
17. L. N. Filimonov and N. M. Kagan, *Fiz. Sb. L'vovsk. Gos. Univ.*, **4,** 222 (1958).
18. I. A. Grikit, V. N. Vovk, V. S. Makarenko, L. I. Tsikora, S. I. Bubyr, and M. N. Petrun'ko, *Sb. Tr. Vses. Nauchno.-Issled. Proekt. Inst. Titana*, **1,** 277 (1967).
19. K. Hirokawa and H. Goto, *Fresenius' Z. Anal. Chem.*, **240,** 311 (1968).
20. N. M. Kagan and L. N. Filimonov, *Zavod. Lab.*, **23,** 185 (1957).
21. O. I. Nikitian, O. I. Gorebaya, V. P. Ryabeka, and I. S. Jharapov, *Zavod. Lab.*, **37,** 183 (1971).
22. V. M. Nizel and S. B. Shubina, *Metody Opred. Issled. Sostoyaniya Gazov Metal*, **1968,** 151.
23. A. Petrakev and L. Petkova, *Mashinostroene* (Sofia), **16,** 530 (1967).
24. S. A. Skotnikov, *Zh. Prikl. Spectrosk.*, **9,** 316 (1968).
25. S. A. Skotnikov, *Izv. Akad. Nauk SSSR, Ser. Fiz.*, **23,** 1128 (1959).

26. S. A. Skotnikov, *Tr. Komis po Analit. Khim., Akad. Nauk SSSR, Inst. Geokhim. i Analit. Khim.*, **10,** 281 (1960).
27. S. A. Skotnikov, *Fiz. Khim. Osnovy Proizv. Stali, Akad. Nauk SSSR, Inst. Met. Tr. 5-oi (Pyatoi) Konf.*, **1959,** 337.
28. S. A. Skotnikov, *Tr. Inst. Met. im. A. A. Baikova, Akad. Nauk SSSR*, **43** (1963).
29. S. A. Skotnikov, *Tr. Inst. Met. im. A. A. Baikova*, **8,** 117 (1960).
30. S. A. Skotnikov and L. M. Federova, *Zavod. Lab.*, **28,** 555 (1962).
31. S. A. Skotnikov, *Metody Opred. Issled. Sostoyaniya Gazov Metal*, **1968,** 155.
32. K. A. Sukhenko, P. P. Galonov, and T. V. Barasheva, *Izv. Akad. Nauk SSSR Ser. Fiz.*, **23,** 1123 (1959).
33. K. A. Sukhenko, V. S. Grigorova, I. S. Lindstrem, N. S. Sventitskii, and P. P. Galonov, *Izv. Akad. Nauk SSSR, Ser. Fiz.*, **23,** 1116 (1959).
34. K. A. Sukhenko, V. S. Grigorova, I. S. Lindstream, N. S. Sventitskii, and P. P. Galonov, *Materialy Tret'ego Ural'sk Soveshch. po Spektroskopii, Inst. Fiz. Metal, Akad. Nauk SSSR, Komis po Specktroscopii*, 3rd., Sverdlovsk 1960, **101** (1962).
35. K. A. Sukhenko, V. S. Grigorova, and I. S. Lindstrem, *Metody Opred. Issled. Sostoyaniya Gazov Metal*, **1968,** p. 142.
36. T. K. Boner, I. V. Matyugina, A. G. Nechaeva, N. K. Shadrina, A. B. Shaevich, and S. B. Shubina, *Mater. Ural. Soveshch. Spektrosk.*, 4th., 1963, Sverdlovsk, 1965, p. 91.
37. L. M. Federova, *Sb. Tr. Tsentr. Nauchno.-Issled., Inst. Chern. Metal*, **31,** 83 (1963).
38. P. P. Galonov, K. A. Sukhenko, N. S. Sventitskii, N. G. Isaev, I. G. Tishin, and T. V. Barasheva, *Tr. Komis po Analit. Khim., Akad. Nauk SSSR, Inst. Geokhim. i Analit. Khim.*, **10,** 190 (1960).
39. V. S. Grigorova, I. S. Lindstrem, N. S. Sventitskii, and K. A. Sukhenko, *Izv. Akad. Nauk SSSR, Ser. Fiz.*, **26,** 924 (1962).
40. V. F. Korotkov and P. A. Kondrat'ev, *Sb. Tr. Tsentr. Nauchno.-Issled., Inst. Chern. Metal*, **31,** 50 (1963).
41. E. S. Kudelya and O. P. Ryabushko, *Avtomat. Svarka*, **10,** 95 (1957).
42. E. S. Kudelya and O. P. Ryabushko, *Avtomat. Svarka*, **11,** 12 (1958).
43. E. S. Kudelya and O. P. Ryabushko, *Tr. Komis po Analit. Khim., Akad. Nauk SSSR, Inst. Geokhim. i Analit. Khim.*, **10,** 183 (1960).
44. E. F. Runge and F. R. Bryan, *Spectrochim. Acta*, **12,** 96 (1958).
45. A. B. Shaevich, V. V. Danilevskaya, N. I. Shorova, G. P. Kazarina, and A. G. Torovina, *Zavod. Lab.*, **30,** 1343 (1964).
46. S. B. Shubina, *Tr. Ural'sk. Nauchno.-Issled. Inst. Chern. Metal*, **1,** 244 (1961).
47. S. B. Shubina, S. I. Kilina, and L. A. Bazanova, *Zavod. Lab.*, **31,** 694 (1965).
48. S. B. Shubina, A. B. Shaevich, and V. G. Dement'eva, *Zavod. Lab.*, **29,** 552 (1963).
49. S. B. Shubina, A. B. Shaevich, S. I. Kilina, S. I. Mel'nikov, and L. A. Bazanova, *Zavod. Lab.*, **28,** 942 (1962).
50. S. B. Shubina and N. I. Zhorova, *Zavod. Lab.*, **30,** 1350 (1964).

51. N. S. Sventitskii, K. A. Sukhenko, O. B. Fal'kova, P. P. Galonov, K. I. Taganov, and M. S. Alpatov, *Fiz. Sb. L'vovsk. Gos. Univ.*, **4**, 225 (1958).
52. N. S. Sventitskii and K. I. Taganov, *Fiz. Sb. L'vovsk. Gos. Univ.*, **4**, 209 (1958).
53. N. S. Sventitskii, K. I. Taganov, and Z. I. Shlepkova, *Izv. Akad. Nauk SSSR, Ser. Fiz.*, **23**, 1118 (1959).
54. N. S. Sventitskii, K. A. Sukhenko, P. P. Galonov, O. B. Fal'kova, M. S. Alpatov, and K. I. Taganov, *Zavod. Lab.*, **22**, 668 (1956).
55. J. Vaclavinek, *Hutnik.*, **31**, 356 (1964).
56. N. M. Zolotareva and T. A. Bessonova, *Zavod. Lab.*, **33**, 175 (1967).
57. V. A. Fassel and R. W. Tabeling, *Spectrochim. Acta*, **8**, 201 (1956).
58. R. Berneron, *British Iron and Steel Institute Translation*, *No. 3291*, June 1963.
59. R. Berneron and J. Romand, *Rev. Met.*, **66**, 695 (1969).
60. K. Hirokawa and H. Goto, *Fresenius Z. Anal. Chem.*, **234**, 340 (1968).
61. F. Malamand, *Office Nat. Etud. Rech. Aerospatiales*, *Note Tech.*, **105**, 41 (1967).
62. F. Malamand, *Method. Phys. Anal.*, **5**, 227 (1969).
63. J. Romand, G. Balloffet, and B. Vogar, *Spectrochim. Acta*, **11**, 268 (1957).
64. W. F. Meggers and B. F. Scribner, *J. Res. Nat. Bur. Stand.*, **13**, 657 (1934).
65. W. F. Meggers, *J. Opt. Soc. Am.*, **31**, 605 (1941).
66. J. Romand and R. Berneron, XIII Colloq. Spectry. Intern., Ottawa 1967, Hilger, London, 1968, p. 434.
67. G. Balloffet, *Ann. Phys.* (Paris), **Ser. 13**, 1243 (1960).
68. G. Balloffet and J. Romand, *Compt. Rend.*, **246**, 733 (1958).
69. R. Berneron, Memoire presente pour l'examen d'ingenieur diplome, par l'etat, Inst. De Recherches de la Siderurgie France, Saint-Germain en Laye, France, p. 92.
70. R. Berneron and J. Romand, *Mem. Sci. Rev. Met.*, **61**, 209 (1964).
71. J. Romand, G. Balloffet, and B. Vodar, VI Colloq. Spectry. Intern., Amsterdam, 1956, Pergamon, London, 1957, p. 268.
72. J. Romand, G. Balloffet, and B. Vodar, VIII Colloq. Spectry. Intern., Lucerne, Switzerland 1959, H. R. Sauerlander and Co., Aarau and Frankfurt, 1960, p. 160.
73. J. Romand, G. Balloffet, and B. Vodar, *Spectrochim. Acta*, **15**, 454 (1959).
74. M. J. Romand and M. R. Berneron, *Publ. Group. Avan. Methodes Spectrog.*, **1963**, 327.
75. B. Vodar, *ARL Technical Report 60-318*, Aeronautical Research Lab., Air Force Res. Div., Wright-Patterson Air Force Base, Ohio, October 1960, p. 25.
76. F. Malamand, *U.S. Govt. Res. and Develop. Rept.*, *N67-33501*, 1967, p. 41.
77. I. S. Lindstrem, N. S. Sventitskii, and Z. I. Shelpkova, *Mater. Ural. Svoeschch. Spektrosk.*, 4th., 1963, Sverdlovsk, 1965, p. 85.
78. C. Matsumoto, V. A. Fassel, and R. N. Kniseley, *Spectrochim. Acta*, **21**, 889 (1965).
79. V. A. Fassel and W. A. Gordon, *Anal. Chem.*, **30**, 179 (1958).
80. J. Artaud and C. Berthelot, *Mem. Sci. Rev. Met.*, **57**, 338 (1960).
81. M. Chaput and J. N. Savarit, *Anal. Chim. Acta*, **31**, 563 (1964).

82. P. Dickens, P. Koenig, K. H. Schmitz, and P. Jaensch, *Arch. Eisenhuettenw.*, **39,** 45 (1968).
83. V. N. Egorov and Y. N. Gryslov, *Zavod. Lab.*, **36,** 1342 (1970).
84. F. M. Evens and V. A. Fassel, *Anal. Chem.*, **33,** 1056 (1961).
85. V. A. Fassel, *Iron and Steel Inst.* (London) *Spec. Rept.*, **68,** 103 (1960).
86. V. A. Fassel and L. L. Altpeter, *Spectrochim. Acta*, **16,** 443 (1960).
87. V. A. Fassel and J. W. Goetzinger, *Spectrochim. Acta*, **21,** 289 (1965).
88. V. A. Fassel, W. A. Gordon, and R. J. Jasinski, *Proc. 2nd Intern. Conf. Peaceful Uses of Atomic Energy*, Geneva 1958, United Nations, New York, 1958, pp. 28, 583.
89. V. A. Fassel, W. A. Gordon, R. J. Jasinski, and F. M. Evens, *Rev. Universelle Mines*, **15,** 278 (1959).
90. V. A. Fassel, W. A. Gordon, and R. W. Tabeling, Symp. Spectrochem. Anal. Trace Elem. *ASTM Spec. Tech. Publ., No. 221*, 1958, p. 3.
91. H. Goto, S. Ikeda, K. Hirokawa, and M. Suzuki, *Sci. Rept. Res. Inst., Tohoku Univ.*, **S18,** 1 (1966).
92. H. Goto, S. Ikeda, K. Hirokawa, and M. Suzuki, *Fresenius' Z. Anal. Chem.*, **228,** 180 (1967).
93. F. B. P. Gray and J. Skinner, Pittsburgh Conference on Analytical Chemistry and Applied Spectroscopy, Pittsburgh, March 1967.
94. D. R. Grieser, G. C. Cocks, E. H. Hall, W. M. Henry, and J. McCallum, *Rept. No. 1538*, Battelle Memorial Institute, Columbus, Ohio , 1961, p. 37.
95. I. A. Grikit, E. G. Galusko, V. V. Polonik, R. K. Ognev, G. G. Kolomoets, and A. I. Polonik, *Sb. Tr., Uses. Nauchno.-Issled. Proekt. Inst. Titana*, **6,** 155 (1970).
96. I. A. Grikit and T. I. Rumyantseva, *Sb. Tr., Uses. Nauchno.-Issled. Proekt Inst. Titana*, **6,** 146 (1970).
97. I. A. Grikit and T. I. Rumyantseva, *Zh. Prikl. Spectrosk.*, **12,** 602 (1970).
98. I. A. Grikit, T. I. Rumyantseva, and E. G. Galushko, *Sb. Tr., Uses. Nauchno.-Issled. Proekt. Inst. Titana*, **4,** 221 (1970).
99. I. A. Grikit, T. I. Rumyantseva, and E. G. Galushko, *Zh. Prikl. Spectrosk.*, **10,** 387 (1969).
100. R. E. Heffelfinger and W. M. Henry, *Titanium Met. Lab. Rept. No. 98*, Battelle Memorial Institute, Columbus, Ohio, 1958, p. 19.
101. T. F. Ivanova and V. V. Federov, *Zh. Anal. Khim.*, **23,** 1750 (1968).
102. H. Kamada and V. A. Fassel, *Spectrochim. Acta*, **17,** 121 (1961).
103. L. F. Kravchenko and V. D. Kurochkin, *Zh. Anal. Khim.*, **27,** 398 (1972).
104. V. M. Plotnitskii, *Zavod. Lab.*, **36,** 1068 (1970).
105. V. M. Plotnitskii and A. N. Rudoi, *Zavod. Lab.*, **37,** 1324 (1971).
106. G. Rossi and Z. Hainski, *Met. Ital.*, **58,** 290 (1966).
107. G. Rossi and J. Melamed, *Comm. Energie At.* (France), *Rappt. No. 2179*, 1963, p. 9.
108. G. Rossi and J. Melamed, *Comm. Energie At. (France), Centre d'Etudes Nucleaires, Saclay, AEC Translation No. 6266*, 1961, p. 9.
109. G. Rossi and Z. Hainski, *Met. Ital.*, **58,** 290 (1966).
110. J. Spitz and T. Van Danh, *Method. Phys. Anal.*, **4,** 371 (1968).

111. K. K. Yudowitch, *Rept. No. NP-4486*, Armour Res. Foundation, 1952.
112. G. Abravanel, *Bull. Soc. Chim. France*, **10,** 2306 (1963).
113. G. Abravanel, *Method. Phys. Anal.*, **2,** 289 (1966).
114. A. K. Babko and T. W. Get'man, *Tr. Komis po Analit. Khim., Akad. Nauk SSSR, Inst. Geokhim. i Analit. Khim.*, **12,** 36 (1960).
115. I. A. Berezin and V. I. Malyshev, *Zh. Anal. Khim.*, **17,** 1101 (1962).
116. W. Koch, S. Eckhard, and F. Stricker, *Arch. Eisenheuttenw.*, **30,** 137 (1959).
117. W. Koch, L. Eckhard, and F. Stricker, *Angew. Chem.*, **71,** 545 (1959).
118. Max Planck Institut fur Eisenforschung, German Patent No. 1,087,378, August 1960.
119. F. Wever and W. Koch, Max Planck Institut für Eisenforschung, U.S. Patent No. 2,991,684, July 1961.
120. B. Rosen, *Rev. Universelle Mines*, **9,** 445 (1953).
121. B. Rosen, Congr. Group Avan. Methodes Anal. Spectrog. Prod. Met. XX, Paris 1957, Group. pour l'Avan. des Methodes Spectrog., Paris, p. 237.
122. M. S. W. Webb and R. J. Webb, *Nature*, **201,** 487 (1964).
123. M. S. W. Webb and R. J. Webb, *Anal. Chim. Acta*, **33,** 138 (1965).
124. M. S. W. Webb and R. J. Webb, *Anal. Chim. Acta*, **36,** 403 (1966).
125. M. S. W. Webb and R. J. Webb, Third National Meeting, Society for Applied Spectroscopy, Cleveland, 1964.
126. *Mass Spectrometric Analysis of Solids*, A. J. Ahearn, Ed., Elsevier, New York, 1966, p. 5.
127. W. M. Henry, *Final Report No. AD-410234*, Battelle Memorial Institute, Columbus, Ohio, 1963, p. 42.
128. J. Roboz, in G. H. Morrison, Ed., *Trace Analysis: Physical Methods*, Interscience, New York, 1965, p. 497.
129. M. M. A. Cavard, A. M. Andreani, and R. Stefani, *Method. Phys. Anal.*, **3** 38 (1967).
130. R. Stefani, M. Desjardins, R. Bourguillot, and A. Cornu, *Method. Phys. Anal*, **1,** 21 (1965).
131. W. L. Harrington, R. K. Skogerboe, and G. H. Morrison, *Anal. Chem.*, **38,** 821 (1966).
132. F. D. Leipziger and R. J. Guidoboni, *Appl. Spectrosc.*, **21,** 165 (1967).
133. G. Vidal, P. Galmard, and P. Lanusse, *Chim. Anal.* (Paris), **50,** 369 (1968).
134. G. Vidal, P. Galmard, and P. Lanusse, *Rech. Aerosp.*, **122,** 29 (1968).
135. G. Vidal, P. Galmard, and P. Lanusse, *Anal. Chem.*, **42,** 98 (1970).
136. L. A. Arbuzova, V. A. Danilkin, Y. A. Imas, V. A. Molchanov, and A. G. Mileshkin, *Zavod. Lab.*, **34,** 1199 (1968).
137. L. A. Arbuzova, A. M. Bonch-Bruevich, V. A. Danilkin, Y. A. Imas, V. A. Molchanov, and A. G. Mileshkin, *Metody Opred. Issled. Sostoyaniya Gazov Metal*, **1968,** 269.
138. W. E. Anable and E. D. Calvert, *PB Rept. 169*, PB-186148, U.S. Clearinghouse Fed. Sci. Tech. Inf., p. 15; through *U.S. Govt. Res. Develop. Rep.*, **69,** 106 (1969).
139. J. B. Clegg and E. J. Millett, *Microelectronics and Reliability*, **10,** 397 (1971).
140. A. Cornu, R. Bourguillot, and R. Stefani, *Energ. Nucl.*, **9,** 386 (1967).

141. I. Hattori and S. Nakai, *Ishikawajima-Harima Giho*, **44,** 678 (1968).
142. G. F. Ivanovskii, L. M. Blyumkin, S. V. Varnakov, and L. P. Lisovskii, *Zavod. Lab.*, **34,** 1263 (1968).
143. F. Konishi, *Int. J. Mass Spectrom. Ion Phys.*, **9,** 33 (1972).
144. L. I. Levi, V. I. Yavoiskii, L. B. Kosterev, A. N. Aleksandrova, V. L. Safonov, R. S. Afanas'eva, and M. I. Afanas'ev, *Izv. Vyssh. Ucheb. Zaved. Chern. Metal*, **11,** 146 (1968).
145. S. Oda, Z. Ohashi, and K. Furuya, *Talanta*, **19,** 779 (1972).
146. A. J. Socha and R. K. Willardson, *Report ARL 68-0132*, Aerospace Research Laboratories, Pasadena, California, 1968, p. 99.
147. A. D. Kirshenbaum and A. V. Grosse, *Trans. Amer. Soc. Metals*, **45,** 758 (1953).
148. A. D. Kirshenbaum and A. V. Grosse, *Anal. Chem.*, **26,** 1955 (1954).
149. A. D. Kirshenbaum, R. A. Mossman, and A. V. Grosse, *Trans. Amer. Soc. Metals*, **46,** 525 (1954).
150. A. D. Kirshenbaum and A. V. Grosse, *Anal. Chim. Acta*, **16,** 225 (1957).
151. V. P. Bondarev, V. A. Zhabina, and N. I. Klochkov, *Metody Opred. Issled. Sostoyaniya Gazov Metal*, **1968,** 79.
152. V. A. Borgest, A. N. Zaidel, and A. A. Petrov, *Tr. Komis po Analit. Khim., Akad. Nauk SSSR, Inst. Geokhim. i Analit. Khim.*, **10,** 270 (1960).
153. V. P. Borisov, V. M. Nemets, and A. A. Petrov, *Zh. Prikl. Spektrosk.*, **7,** 305 (1967).
154. L. S. Dale and deJong, S., Australian Atomic Energy Commission Research Establishment, Lucas Heights, Australia, March 1971, p. 11.
155. G. K. Dudich, A. A. Petrov, and M. P. Favorskaya, *Zavod. Lab.*, **37,** 436 (1971).
156. K. Furuya, S. Okuyama, T. Tachikawa, and H. Kamada, *Talanta*, **15,** 327 (1968).
157. T. Hagiwara, M. Harada, and K. Tanaka, *Bunko Kenkyu*, **18,** 141 (1969).
158. T. F. Ivanova, M. E. Trentovius, and V. V. Federova, *Izv. Akad. Nauk. SSSR, Ser. Fiz.*, **23,** 1120 (1959).
159. G. S. Lazeeva, A. A. Petrov, and G. V. Scvortsova, *Vestn. Leningrad Univ., Ser. Fiz. Khim.*, **22,** 63 (1967).
160. G. S. Lazeeva, A. A. Petrov, and G. A. Yusupova, *Vestn. Leningrad Univ., Ser. Fiz. Khim.*, **20,** 141 (1965).
161. C. R. Masson, S. G. Whiteway, W. D. Jamieson, and C. A. Collings, *Can. Metal Quart.*, **5,** 329 (1966).
162. C. R. Masson, *Met. Rev.*, **12,** 147 (1967).
163. N. M. Orlova and A. A. Petrov, *Vestn. Leningrad Univ., Ser. Fiz. Khim.*, **19,** 69 (1964).
164. K. B. Orlova and E. N. Vitol, *Zh. Anal. Khim.*, **21,** 1263 (1966).
165. K. B. Orlova, *Metody Opred. Issled. Sostoyaniya Gazov Metal*, **1968,** 74.
166. M. L. Pearce and C. R. Masson, *Spec. Rept. No. 68*, British Iron Steel Inst., London, 1969, p. 121.
167. A. A. Petrov and A. N. Zaidel, U.S.S.R. Patent No. 142,811, December 1961.

68. A. A. Petrov and G. V. Skvortsova, *Zh. Prikl. Spektrosk., Akad. Nauk Belorussk. SSR* (Minsk), **14,** 793 (1971).
69. L. N. Saksonova, D. I. Maksimov, and V. T. Burtsev, *Zh. Anal. Khim.*, **26,** 1360 (1971).
70. R. R. Shvangiradze, K. A. Oganezov, and B. Y. Chikhladze, *Zh. Prikl. Spektrosk., Akad. Nauk BSSR*, **3,** 300 (1965).
71. R. R. Shvangiradze, K. A. Oganezov, and B. Y. Chikhladze, *Zh. Prikl. Spektrosk.*, **7,** 265 (1967).
72. R. R. Shvangiradze, K. A. Oganezov, and B. Y. Chikhladze, *Metody Opred. Issled. Sostoyaniya Gazov Metal*, **1968,** 83.
73. H. G. Staley and H. J. Svec, *Anal. Chim. Acta*, **21,** 289 (1959).
74. E. N. Vitol, *Zavod. Lab.*, **29,** 649 (1963).
75. E. N. Vitol, K. B. Orlova, and G. I. Nikolaev, *Zh. Anal. Khim.*, **26,** 777 (1971).
76. A. I. Vygodskii, V. G. Nesterenko, and D. G. Sherman, *Zavod. Lab.*, **29,** 1474 (1963).
77. A. N. Zaidel and A. A. Petrov, *Zh. Tekh. Fiz.*, **25,** 2571 (1955).
78. A. N. Zaidel, A. A. Petrov, and G. V. Veinberg, *Spectroscopic-Isotopic Determination of Hydrogen in Metals*, 1957; trans., Butterworths, Washington, 1961, p. 120.
79. A. N. Zaidel, *Spectrochim. Acta*, **10,** 369 (1958).
80. A. N. Zaidel, A. A. Petrov, and K. I. Petrov, *Fiz. Sb. L'vovsk. Gos. Univ.* **No. 4,** 206 (1958).
81. A. N. Zaidel, G. V. Ostrovskaya, and A. A. Petrov, *Opt. Spektrosk.*, **10,** 637 (1961).
82. A. N. Zaidel and A. A. Petrov, *Zavod. Lab.*, **28,** 552 (1962).
83. A. N. Zaidel, G. S. Lazeeva, and A. A. Petrov, *Vestn. Leningrad Univ., Ser. Fiz. Khim.*, **18,** 55 (1963).
84. A. N. Zaidel, T. F. Ivanova, A. A. Petrov, V. V. Federov, and N. M. Chumakova, *Zadov. Lab.*, **29,** 693 (1963).
85. A. N. Zaidel, G. S. Lazeeva, and A. A. Petrov, *Zh. Prikl. Spektrosk., Akad. Nauk BSSR*, **1,** 236 (1964).
86. A. N. Zaidel and A. A. Petrov, *Zh. Prikl. Spektrosk.*, **3,** 383 (1965).
87. A. N. Zaidel, T. F. Ivanova, G. S. Lazeeva, A. A. Petrov, and V. V. Federov, *Mater. Ural. Soveshch. Spektrosk.*, 4th., 1963 Sverdlovsk, 1965, p. 92.
88. N. A. Zakorina, G. S. Lazeeva, A. A. Petrov, G. V. Skvortzova, and M. P. Favorskaya, *Vestn. Leningrad Univ., Ser. Fiz. Khim.*, **20,** 152 (1965).
89. N. A. Zakorina, G. S. Lazeeva, and A. A. Petrov, *Vestn. Leningrad Univ., Ser. Fiz. Khim.*, **21,** 38 (1966).
90. N. A. Zakorina, G. S. Lazeeva, and A. A. Petrov, *Zh. Anal. Khim.*, **23,** 1688 (1968).
91. N. A. Zakorina, G. S. Lazeeva, and A. A. Petrov, *Spektrosk., Tr. Sib. Soveshch.*, 4th., 1965, Tomsk, 1969, p. 377.

CHAPTER

7

HYDROGEN

JOHN F. MARTIN AND LABEN M. MELNICK

United States Steel Corporation
Research Laboratory
Monroeville, Pennsylvania

CONTENTS

I. INTRODUCTION

The determination of hydrogen in metals is performed by a variety of methods. The choice of a method depends on a number of factors. Of importance are detection limit, time required for analysis, and cost of equipment. Where a number of different metals or alloys are analyzed, one method of broad applicability is usually desirable. Also, for thermal-extraction methods a practical combination of comparatively low extraction temperature, coupled with a reasonably short analysis time, is advantageous to minimize vaporization of metallic elements that may getter hydrogen and to maximize the hydrogen concentration in the gas removed from the metal.

Only the more common methods are discussed, which, except for the first-bubble method, yield quantitative results. In later chapters of this book, in certain cases where it is used mainly for one type of metal, the method is treated in greater detail. Eborall [1] and Kappes and Schiffers [2] have written comprehensive reviews on the determination of hydrogen in metals, and the analysis of a number of metals for hydrogen has been covered by Turovsteva and Kunin [3]. McKinley [4] has reviewed various methods specifically applicable to titanium, and Sokol'skaya [5] has discussed methods for determining hydrogen in aluminum, magnesium, and their alloys. Methods for determining hydrogen in steel have been evaluated by Suarez Acosta [6], Wojcik [7], and Martin et al. [8]. Grigorenko et al. [9], Smith et al. [10], and Dugain and Bril [11] have studied the determination of hydrogen in aluminum and aluminum alloys, cobalt, and aluminum, respectively. Biennial reviews in *Analytical Chemistry* on ferrous metallurgy and non-ferrous metallurgy (I) have covered the determination of hydrogen in steel and in aluminum, beryllium, titanium, and magnesium, respectively.

The diffusivity of hydrogen in metals is treated in Chapter 1, and the subjects of sampling, sample-storage, and sample-preparation for specific metals are discussed in the chapters dealing with these metals.

II. VACUUM-TIN FUSION

This method, basically described by Bennek and Klotzbach [12], has been applied mainly in steel analysis. In analysis the furnace is evacuated and the tin is degassed, then the sample is fused in the tin bath generally contained in a borosilicate or quartz crucible. The use of tin permits fusion to take place at a temperature considerably below the melting point of the sample. For example, with steel, analyses are usually performed at 1100–1500°C as compared to 1535°C, the melting temperature of iron. During fusion, all of the hydrogen together with some oxygen (as carbon monoxide) and nitrogen are extracted under vacuum and transferred to the measuring system. Hydrogen may then be determined by a variety of methods as noted later. The fusion temperature is determined by the material being analyzed; the time of analysis is, in part, a function of the amount of sample taken; and the method for determining hydrogen is dependent mainly on whether other gases are extracted simultaneously with hydrogen. This method has been used for the analysis of ferromanganese [13], steel [14–18], magnesium and magnesium alloys [19], aluminum [20, 21], and calcium [22].

For the analysis of steel, Carney et al. [15] placed 130 g of tin and 0.7 g of silicon in a quartz crucible contained in an alumina crucible, the latter acting as a heat insulator and also as a container for the molten mass in case the quartz cracked. Silicon was added to depress carbon monoxide evolution. The hydrogen extracted was oxidized to water by hot copper oxide, and the water was absorbed in a dry-ice–acetone trap. Hydrogen was determined by pressure measurement before oxidation and after removal as water. The evolved gas contained 50–90% hydrogen, and the balance was carbon monoxide and nitrogen. The analysis time was about 15 min.

The analysis time was reduced to about 10 min by Shields et al. [16], who determined hydrogen by thermal-conductivity measurement. By this technique a maximum error of 0.5 ppm is possible at the 3-σ level when using 2–3-g samples containing about 6 ppm hydrogen. The evolved gas must contain more than 50% hydrogen to achieve proportionality between the thermal conductivity and pressure readings.

For the analysis of 2–4-g samples of magnesium and magnesium–lithium alloys, Mallett et al. [19] used a borosilicate furnace tube containing 100 g of tin as the initial charge. A liquid-nitrogen cold trap was incorporated to remove water desorbed from the system. After fusion at 450°C for about 15 min, during which time about 95% of the hydrogen was evolved, an additional 15 g of tin was added to effect stirring. A degassed steel weight was also used for this purpose during fusion. An Orsat apparatus was used for the hydrogen determination. About 1 h was required for each analysis.

Griffith and Mallett used a system similar to that described for magnesium [19] for analyzing aluminum alloys [20]. Complete dissolution of the sample required 1–1.5 h, depending on the type of alloy. The pressure of the extracted gas was measured, hydrogen was diffused through palladium, and the pressure of the undiffused gas was measured. Brandt and Cochran [21] analyzed aluminum by placing the sample in a graphite crucible contained in a second graphite crucible, which, in turn, was insulated from the outer quartz envelope by boron nitride powder. Hydrogen was determined by pressure measurements before and after diffusion of the hydrogen through palladium. Analysis time was 10–12 min.

Peterson and Fattore [22] determined hydrogen in calcium by fusion of

2–200 mg of sample with tin at 670°C in a quartz crucible contained in a quartz tube. Because hydrogen was the only gas evolved under these conditions, a direct pressure measurement could be made to complete the analysis. About 30 min was required for each analysis. In the range of 0.02–4.8%, hydrogen was determined with a relative error of 0.4–2.5%. Vacuum extraction without tin was unsuccessful for this analysis, because hydrogen was evolved at a significant rate only at temperatures at which calcium volatilization was too rapid for accurate analysis.

The desirability of the vacuum-tin-fusion method is dependent on a number of factors. As noted earlier, the fusion temperature may be considerably lower than that required for conventional vacuum fusion. Also, in the absence of graphite, the blank is smaller. Moreover, extraction time should be less than that required for vacuum extraction of the hydrogen from the solid sample.

III. VACUUM EXTRACTION

A. DISCUSSION

The determination of hydrogen by extraction from a solid sample under vacuum at elevated temperature is a rather simple and accurate method for many metals (for sodium, the sample is liquid, as noted below). The basis of this method was described in 1872 [23]. Usually conditions can be found under which hydrogen is either the only gas extracted or is by far the major constituent. The time for analysis will depend mainly on the extraction temperature, sample-material and sample-dimensions, and technique for determining the hydrogen. Regarding sample material, for example, extraction of hydrogen from an aluminum sample $\frac{5}{8}$ in. in diameter by $1\frac{1}{2}$ in. long requires 3 h or more [21], while for a ferritic steel sample of similar dimensions, the time is about 30 min. Because the temperatures used in this method are low in comparison to those encountered in classical fusion methods, there is usually less difficulty in obtaining a low apparatus blank. Also, gettering problems are minimized. To minimize gettering even further, Pepkowitz and Proud [24] developed a clever encapsulation technique for determining hydrogen in sodium. The sample was placed in a previously dehydrogenated iron capsule either alone or with a small piece of hydrogen-free magnesium or sodium, and a lid was welded to the capsule. The capsule was then placed in a vacuum system and heated at 700°C. Where hydrogen-free sodium was used, the release of hydrogen (present as NaOH) was accelerated according to the equation

$$2Na + 2NaOH \rightarrow 2Na_2O + H_2. \quad (1)$$

In either case, the hydrogen, presumably in the atomic form, diffuses through the walls of the capsule and may be determined without any problems due to volatilization of sodium. Walker and Seed [25], who also analyzed sodium by this technique, used sucrose as a standard source of hydrogen.

Coe et al. [26] and Perriton and Coe [27] analyzed aluminum and aluminum–zinc–magnesium alloys in the same manner. Steel was also analyzed [27]. Further, they prepared and evaluated standard samples of hydrogen in steel [28] because it was felt that a determination of gaseous hydrogen added in known amounts to a vacuum system did not represent realistic analytical conditions and because hydrogen charging of steel to known levels yielded standards from which the hydrogen would diffuse with time unless the standards were stored at low temperature (e.g., in liquid nitrogen), in which case the distribution of the standards to other laboratories would be difficult or impossible.

The hydrogen standards [28] are based on the fact that molecular hydrogen can be retained in a steel capsule at ambient temperature. During analysis at 650°C or higher, this hydrogen dissociates and the product diffuses through the thin walls of the cylinder. In the making of hydrogen standards, low-alloy-steel capsules of precisely machined internal dimensions and lids are heated to remove residual hydrogen. Then each capsule and its lid are mounted on the electrodes of a conventional resistance welding machine in a closed system. The system is filled with pure hydrogen, the temperature and pressure of which are monitored, and the lid is welded in position in 40–100 msec, depending on the dimensions. Although a set-down of a few thousandths of an inch results from welding, the gas volume enclosed is that determined by the physical volume of the capsule at the instant load is applied. The volume error is less than 2%. Standards are 1 in. long and either $\frac{1}{4}$ or $\frac{1}{2}$ in. in diameter. Their utility is obviously dependent on the size of the sample that a particular analytical apparatus can accept, although standards of other dimensions could be made.

In vacuum extraction a simple resistance furnace will suffice for heating the sample. Induction heating, however, has been used in two commercial analyzers (see Sections III.B.1 and III.B.2). Heating is usually carried out in a quartz tube furnace. Vacuum-extraction analyses have also been performed in resistance-heated vacuum-fusion apparatus using the graphite crucible; however, this invariably results in increased degassing time or an increased blank with resulting poorer detection limit. Alternatively, vacuum-fusion systems have been modified to provide for vacuum-extraction analyses without the inherent problems of graphite. For example, the standard NRC Model 917 hot-extraction furnace (Section III.B.2) may be installed on the NRC 912D Vacuum-Fusion Gas-Analyzer described in Chapter 4, Section I.L.3.

Usually, one diffusion pump is necessary for evacuation of the furnace and packing of the gas into the measuring system; a second diffusion pump is required for exhausting the entire system (see, however, Section III.B.1). The pumping speeds are not critical for either pump. However, pumps with high critical-backing pressures (about 2–5 torr) should be used for circulating the gas through a catalytic oxidation and absorption system or for pumping the hydrogen to a McLeod gauge after diffusion of the hydrogen through palladium. Both of these analytical systems should be as large as is practicable to accommodate a wide range of hydrogen contents; further, the McLeod gauge is more accurate at lower pressures. An appropriate cold trap should be placed between the furnace and its diffusion pump and in the measuring system to prevent residual moisture from affecting the analyses.

Where only hydrogen is evolved, analysis may be accomplished by a direct pressure measurement using a McLeod gauge. This basic technique has been used by Andrew et al. [29], who collected hydrogen over paraffin after evolution of the gas from steel at room temperature. (Coe [30] has discussed the collection of hydrogen over paraffin, glycerine, and mercury; it appears that in most American Society for Testing and Materials (ASTM) and British Welding Research Association (BWRA) tests for weld metal, mercury is used in the collecting system. Glycerine, however, has been widely used in continental Europe.) If other gases are present, hydrogen may be separated by diffusion through palladium (see Chapter 4, Section I.B.3.a), by catalytic oxidation to water followed by removal of the water (see Chapter 4, Sections I.B.3.b and c), or determined by thermal-conductivity measurement. By use of this last technique, Carson [31] was able to perform an analysis of a 2-g steel sample in 15 min. The apparatus was simplified by using the furnace as part of the calibrated collecting volume. (In reality this is an equilibrium pressure measurement, but because of the relatively low pressure attained after extraction, only a negligible amount of hydrogen remains in the sample.) Standard deviation of 0.14 ppm in the 2–3 ppm range was obtained when compared to results obtained by conventional hot vacuum extraction. Gas-chromatographic and/or mass-spectrometric analysis should be utilized initially to determine the presence of other constituents in the extracted gas. Also, complete evolution of hydrogen at the operating temperature must be verified either by performing the hot-extraction analysis at higher temperatures or more preferably by using a different technique such as vacuum fusion.

Because of the advantages noted earlier, vacuum extraction has usually been the most precise method for determining hydrogen in many metals. After considerable investigation this method was adopted by ASTM as the standard procedure for the analysis of zirconium [32] and steel [33]. The apparatus in Fig. 7.1 shows two alternative methods developed for steel

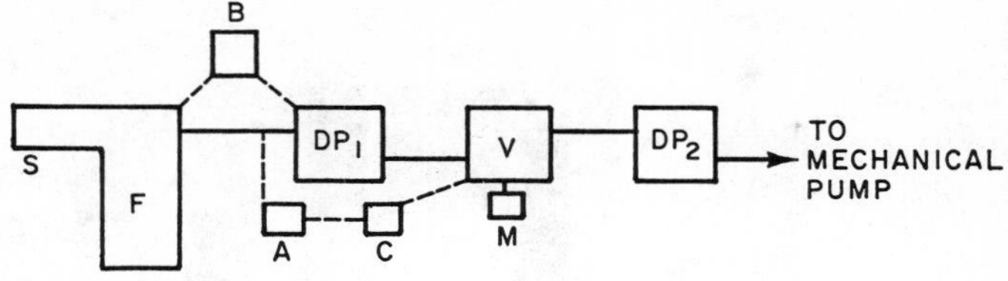

S	SAMPLE HOLDER OR VACUUM LOCK (ENTRY PORT)
F	FURNACE
DP_1	TRANSFER DIFFUSION PUMP
V	CALIBRATED ANALYTICAL VOLUME
M	McLEOD GAGE
DP_2	EVACUATING DIFFUSION PUMP
B	PALLADIUM TUBE
C	OXIDIZER
A	$Mg(ClO_4)_2$ TRAP

Fig. 7.1. Vacuum-extraction system for steel analysis. (Reproduced by permission from the American Society for Testing and Materials.)

analysis. If hydrogen is known to be the only gas evolved, a direct pressure measurement will suffice. However, if another gas is present, the hydrogen is determined by pressure measurement after diffusion through a palladium or palladium–silver tube into the calibrated analytical volume. Alternatively, the hydrogen can be determined by difference in pressure–volume measurements of the total gas evolved and the gas remaining after removal of hydrogen by oxidation to water and absorption in magnesium perchlorate.

When using the palladium tube, hydrogen passes through it while the direct line from the furnace is closed. After pressure measurement, the direct line is opened and the system is evacuated. When the hydrogen is to be removed as water, the evolved gases are collected in the analytical volume, the pressure is measured, and the gases are circulated through the oxidizer and water absorber and finally collected in the analytical volume. A second pressure measurement is then made.

In a comparative study of methods for determining hydrogen removed from steel by vacuum extraction [8], the completeness of diffusion of hydrogen through the palladium tube was verified by analysis of the undiffused gas by gas chromatography. Also, although the cost of the palladium thimble was about three times that required for catalytic oxidation and absorption, the analysis time, exclusive of extraction time, was considerably less with the former technique (6–8 min versus 20–30 min). The combination of the advantages of vacuum extraction and a simple measurement technique make such a method most suitable for the referee analysis of many metals. Lounamaa and Aass [34] also diffused hydrogen through palladium in analyzing

Fig. 7.2. LECO 534-600 hydrogen analyzer. (Courtesy Laboratory Equipment Corp.)

steels. Falecki [35] and Sannier and Leroy [36] used gas-chromatographic methods in determining hydrogen vacuum extracted from both ferrous and nonferrous materials.

In addition to the references cited above for applications of vacuum extraction, some others may be noted for the analysis of aluminum [37, 38] and steel [39–43a]. In all of these articles, descriptions of apparatus are included.

B. COMMERCIAL VACUUM-EXTRACTION ANALYZERS

1. LECO Hydrogen Analyzer

The LECO No. 534-600 Hydrogen-Analyzer, manufactured by Laboratory Equipment Corporation, St. Joseph, Michigan, is shown in Fig. 7.2. The sample is introduced through an air lock and greaseless stopcock arrangement directly into a quartz-enclosed graphite crucible, the bottom of which is supplied by the end of a movable rod. The sample is heated inductively, and after analysis this rod is lowered so that the sample can fall into a collection chamber from which it can later be removed. After extraction of hydrogen, a pressure reading is taken on the tilting McLeod gauge. Then hydrogen is catalytically oxidized to water, removed by absorption in magnesium perchlorate, and a second pressure reading taken. Hydrogen is determined from the difference in pressure readings. When only hydrogen is evolved a single pressure measurement will suffice. Analysis time is about 10 min.

The induction furnace requires 1.5 kW and is wired for 115-V, 50–60-Hz, ac power. Temperature control is effected with a variable transformer. An expansion volume is included in the vacuum system to accommodate large amounts of hydrogen. Sensitivity of the measuring system reportedly is 0.2 μg hydrogen. Only one mercury-diffusion pump, with a speed of 3–4 liters/sec is used. The water requirement for this pump is 0.7 liter/min.

2. NRC Vacuum Hydrogen Determinator–Titanium

The Vacuum Hydrogen Determinator–Titanium, Type 917B, has been widely used. It was formerly manufactured by the NRC Equipment Corporation, Newton Highlands, Massachusetts. This company is now part of Varian Associates. The analyzer was designed to be used for titanium and titanium-rich alloys from which only hydrogen is removed at 1400°C. This instrument, shown in Fig. 7.3, therefore, contains no oxidizing agents, cold traps, circulating pumps for handling gas mixtures, or palladium tube for removal of hydrogen. (A palladium leak is used, however, for admitting pure hydrogen to the system for calibration purposes.) With $\frac{1}{4}$-g samples, between 5 and 700 ppm of hydrogen may be determined.

Samples are charged into two horizontal loading arms, which are then sealed with glass caps. After a satisfactory blank has been obtained, a sample is moved along the loading arm by a magnetically operated pusher until it drops into a molybdenum crucible containing a thin thoria lining to prevent the titanium samples from welding to the crucible. The sample is heated inductively at 1400°C until the Pirani gauge, which is calibrated for hydrogen, indicates that no further gas is being evolved. The Pirani gauge is equipped with both an indicating meter and a strip-chart recorder. Hydrogen is collected in one or both of the volumes provided. The pressure is then read on a McLeod gauge. Finally, the crucible is lowered magnetically to dump the sample into a cold container below the furnace. (The manufacturer recommends that the crucible suspension magnets be remagnetized every third operating day.) Analysis time is about 15 min.

For the induction coil a 2-kW, 110-V, single-phase, 60-Hz, ac power supply is needed; the current is 20 A. A water supply of 7.5 liters/min is also required. Two diffusion pumps are contained in the system: (1) an oil-type

Fig. 7.3. NRC vacuum hydrogen determinator—titanium. (Courtesy NRC Equipmen Corp.)

pump of 70-liters/sec speed with a critical backing pressure of 0.2 torr and (2 a three-stage mercury-transfer pump of 10-liters/sec speed with a critica backing pressure of 5 torr. To operate any of the mercury cutoffs or th McLeod gauge, it is necessary to raise the mercury in its well column fron barometric level to the operating level by pressurizing the well with carbo dioxide. This gas is stored as a liquid in a 20-lb capacity cylinder equippe with a conventional regulator.

3. Serfass Hydrogen Analyzer

This instrument (Fig. 7.4), formerly sold by Fisher Scientific Company Pittsburgh, is no longer available but is still used in some laboratories (se Chapter 19). Single samples, magnetic or nonmagnetic, are introduced throug an air-lock, which is then closed and evacuated. (A multiple-sample inle tube was also available.) A large-bore stopcock is opened, and a magneti manipulator is used to drop the sample into the horizontal charging arm Another magnetic pusher moves it into the small quartz resistance furnace

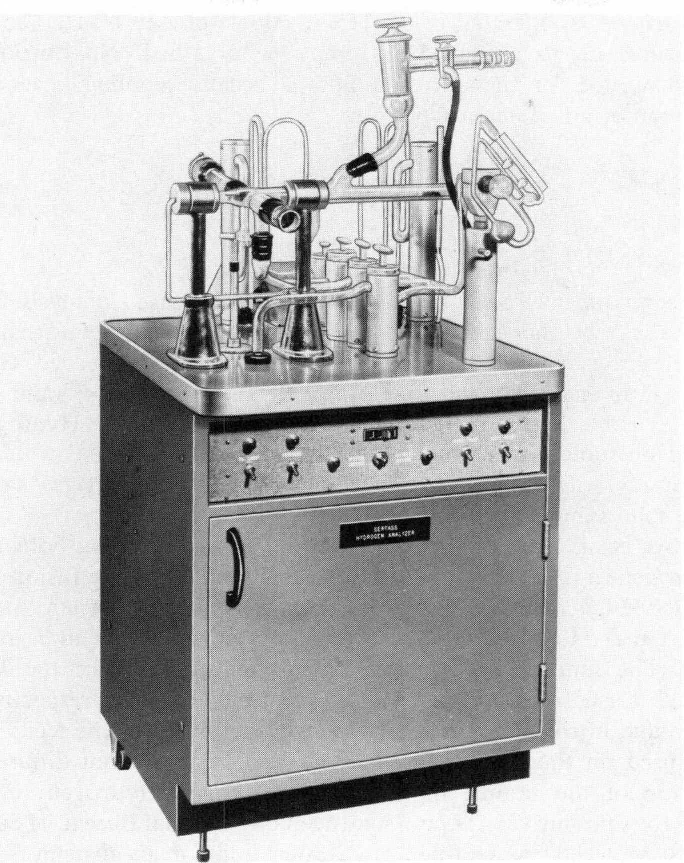

Fig. 7.4. Serfass hydrogen analyzer. (Courtesy Fisher Scientific Co.)

With the movable furnace tilted down, the sample slides into the heated end where the dissolved hydrogen is extracted. At the conclusion of the heating cycle, the furnace is tilted in the opposite direction, so that the analyzed sample slides into a waste trap at the other end of the tube.

The gases evolved are pumped into a glass bulb containing a palladium–silver-alloy tube (see Chapter 4, Section I.B.3.a) that is electrically heated. Only the hydrogen diffuses through the alloy into the tube and is pumped to a tilting McLeod gauge. A calibration chart is obtained by adding hydrogen from a calibrated volume in the system and plotting micrograms of hydrogen against the gauge readings. A 100-ml expansion volume permits the determination of large amounts of hydrogen. Precision reportedly is 0.2 ppm in the range 3–10 ppm hydrogen.

The furnace is operated with 115-V, single-phase, 60-Hz, ac power. Temperatures up to about 1100°C may be obtained. No outside water supply is needed for the diffusion pumps because cooling is effected by either Freon or air in heat exchangers.

IV. INERT-GAS FUSION

A. DISCUSSION

This technique has been utilized to obtain decreased analysis time for metals that can be analyzed in the solid state and also for metals that require fusion to release hydrogen. No high-vacuum equipment is needed. Goto and Hosoya [43b] analyzed steel and other metals in only 5–7 min, using a graphite crucible and an argon flow rate of 150 ml/min. Hydrogen was separated on molecular sieve 5A and determined by thermal conductivity. The accuracy was comparable to that obtained by inert-gas extraction from the solid sample.

Hargrove et al. [43c] developed a method for determining both nitrogen and hydrogen in milligram quantities of steel samples, using fusion in ultrapure (99.9999%) helium, followed by analysis with a Varian Aerograph (Walnut Creek, California) gas chromatograph with a helium ionization detector. The limiting sensitivities, corresponding to twice the base-line noise level, for a 1-mg sample were 0.02 and 0.03 wt ppm, respectively, for hydrogen and nitrogen. With regard to hydrogen in steel, the technique can only be used on those steels from which hydrogen will not diffuse during preparation of the sample in finely divided form. Hydrogen results are reported for titanium (28–39 ppm hydrogen on National Bureau of Standards Reference Material 352 certified at 32 ppm) and for an austenitic stainless steel (American Iron and Steel Institute Type 316).

B. LECO RH-1F (RH-1) HYDROGEN DETERMINATOR

This instrument, manufactured by the Laboratory Equipment Corporation, St. Joseph, Michigan, is shown in Fig. 7.5. Analysis is performed in a stream of purified argon with a single-use graphite crucible that is held between two water-cooled copper electrodes, the crucible and contents reaching temperatures near 2500°C when high-amperage current is applied (see Chapter 4). After introduction of the crucible, tin flux (for steel and titanium samples), and sample, the analysis is fully automatic.

In analysis, a tin pellet is placed in the crucible, which is then positioned in the furnace. The sample is introduced into the sample-loading device. Crucible outgassing is initiated by pushing a button. After a light indicates

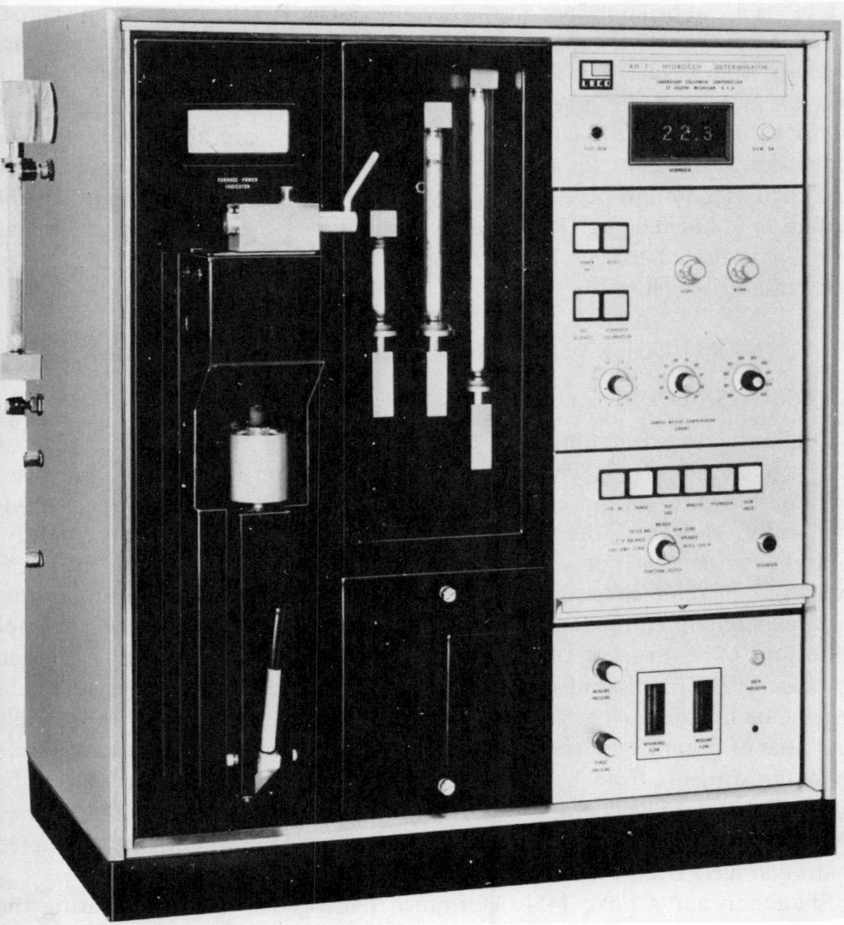

Fig. 7.5. LECO RH-1F hydrogen determinator. (Courtesy Laboratory Equipment Corp.)

conclusion of outgassing, the sample is transferred to the crucible, and pushing another button starts the analysis. Oxygen, as CO, is converted to CO_2 with Schutze reagent and absorbed in Ascarite. Nitrogen is separated chromatographically on a short molecular-sieve column. The output of the thermal-conductivity cell is then integrated and displayed digitally as ppm hydrogen.

Calibration of the RH-1F Hydrogen Determinator may be accomplished either with metal standards or with pure hydrogen introduced via the theoretical gas-dosing system supplied as a separate unit. Metals that can be analyzed include steel, tantalum, titanium, uranium, and zirconium. Up to

6 g of steel can be contained in the crucible. (The RH-1 system has a smalle crucible of about 1-g capacity.) Outgassing time is about 90 sec, and tota analysis time is about 150 sec. The latter can be reduced by outgassing a crucible and flux while gas evolved from the previous sample is being analyzed Blank values, which are corrected for electronically, are about 0.06 μg. The blank contribution of the argon, purified by passing over hot copper oxide and then magnesium perchlorate, is virtually nil. The detection limit for the system is about 0.06 μg. Reproducibility is 0.1 μg hydrogen for gas-dosing over the range 0.1–100 μg hydrogen. Accuracy is 0.3 μg or 3% of the hydro- gen present, whichever is greater.

V. INERT-GAS EXTRACTION FROM THE SOLID SAMPLE

A. DISCUSSION

An inert-gas extraction method for hydrogen in steel was attempted as long ago as 1903 by Heyn [44]. After the extraction in nitrogen, the gases were passed through hot copper oxide and the water was determined gravi- metrically. However, the sample and blank values differed only slightly, and the work was not continued. The main advantage of the inert-gas extraction method is the comparative simplicity and low cost of the apparatus no high-vacuum equipment is needed (except in the measuring system after diffusion of hydrogen through palladium). However, where hydrogen constitutes all or almost all of the extracted gas, a direct pressure measurement cannot be utilized as in vacuum hot extraction. Further, where appreciable amounts of other gases are coextracted, provision must be made to separate these constituents from hydrogen. If the proper precautions are taken, the time of analysis and accuracy of this method are comparable to those obtained by vacuum extraction. Various techniques for determining the extracted hydrogen have been used.

Shanahan and Cooke [45] determined hydrogen in steel by heating the sample in a horizontal silica combustion tube in dry nitrogen. The flow rate was 2 liters/h. Hydrogen was oxidized to water by copper oxide–ferric oxide at 550–600°C, and the water was determined either gravimetrically by absorption in anhydrone or titrimetrically with Karl Fischer reagent after absorption in absolute methanol. Any methane extracted from the sample was oxidized to carbon dioxide and water [46].

Coe [47, 48] and Coe and Jenkins [49] used argon as the carrier gas and detected the hydrogen by continuous thermal-conductivity measurement of the evolved gas, recording the measurements on a strip-chart recorder equipped with an automatic integrator. This provided greater sensitivity than the titrimetric method. As little as 0.005 ml hydrogen was determined; this corresponded to about 0.1 ppm hydrogen. The argon flow rate was 1

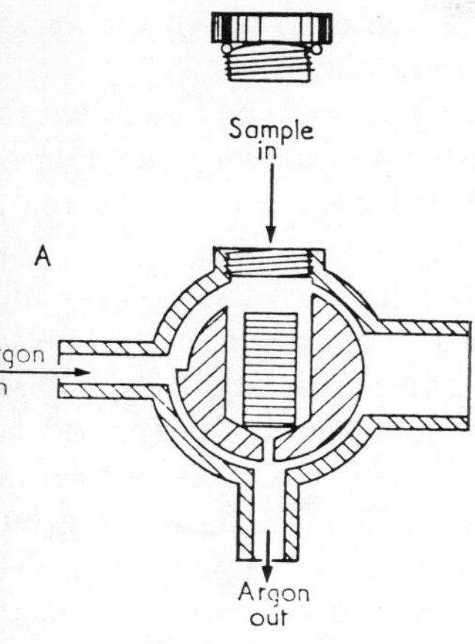

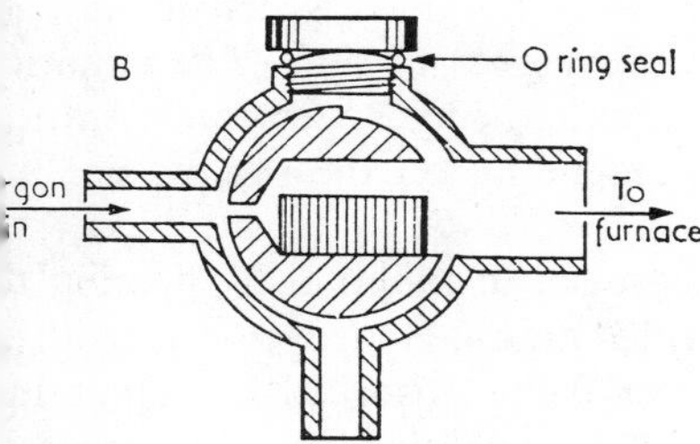

Fig. 7.6. Sample-loading lock. (Courtesy Journal Iron and Steel Institute.)

l/min. Approximately 30 min was required for each determination. Mois-re must be removed from the argon; otherwise, the reaction of the steel mples with moisture may yield hydrogen. Also, the argon flow rate and rrent supplied to the thermal-conductivity cell must be carefully controlled obtain precise results [50]. Further, rubber and plastic connections should reduced to a minimum, as they may be permeated by water vapor [51] d are subject to leaks.

Traces of air accidentally admitted with the sample produced thermal-nductivity peaks similar to those of hydrogen. This problem was solved use of a special loading lock shown in Fig. 7.6. It consists of a cylindrical ass block free to rotate on O-rings inside a close-fitting hollow brass linder. The center block is drilled to accommodate the sample. Gas inlet d outlet ports and a stoppered inlet-port connection to the furnace are ovided on the outer cylinder. The sample is loaded in position A, the stopper replaced, and argon flows and sweeps any entrained air to waste. The nter block containing the sample is rotated through 90°, allowing the mple to slide into the furnace (position B). Transfer of the sample into the furnace is accomplished by inclining the tilting board on which the sampl lock, furnace, and thermal-conductivity cell are mounted.

If other gases are extracted along with hydrogen, it was suggested that th could be separated with a molecular-sieve column prior to the therm conductivity cell, as described later. Any surface moisture evolved from t sample could be removed by a cold trap in a similar position.

Abresch et al. [52a] determined diffusible hydrogen in steel with hig purity argon as the inert carrier gas. (Nondiffusible hydrogen may be prese in steel as methane [52b].) Hydrogen was separated from argon by diffusi through a thin-walled palladium tube spiral of unique design and w collected over mercury in a measuring buret. The palladium spiral w connected to a standard ground socket by means of a standard grou stopper of silver that was water-cooled. Where rubber connections we needed, butyl rubber, instead of natural rubber, was used to minimize t blank. Analysis time for the diffusible hydrogen was about 30 min.

Fergason et al. [53] determined hydrogen in uranium by extraction 900°C in helium. Hydrogen was determined by mass spectrometry, integrating the rate of evolution curve with respect to time. The empiric number so obtained was directly proportional to the hydrogen concentratio Each determination required 3–10 min, depending on the sample-size a hydrogen concentration.

Walter and Offner [54] determined hydrogen in niobium by heating t sample at 1100–1150°C in a stream of purified argon. Hydrogen was oxidiz to water over hot copper oxide, the water being absorbed in magnesiu perchlorate and weighed. The argon (99.998% pure) was further purified passing it over hot copper to remove oxygen, over hot copper oxide decompose hydrocarbons, and over heated titanium to remove nitroge Oxide or nitride formation on the surface of niobium could impede t diffusion of hydrogen.

Giegerl [55] determined hydrogen in steel by heating the sample to 700 in a stream of argon. The evolved hydrogen was oxidized to water over h copper oxide, and the water in turn was reacted with hot graphite [1200° to form hydrogen and carbon monoxide. The carbon monoxide was th oxidized with Schutze reagent (I_2O_5), which does not oxidize the hydroge The resulting carbon dioxide was absorbed in barium hydroxide solutio with a resulting change in electrical potential between two electrodes in t solution. Hydroxyl ion was coulometrically generated until the origin potential was obtained, and the amount of carbon dioxide was determined b the time required at a predetermined current strength to bring the electroly back to the initial potential. Apparently, only low-carbon steels were analyz to prevent interference from any carbon monoxide evolved from the sampl during heating.

Kashima and Yamazaki [56a] determined hydrogen in ferrosilicon, mild steel, and cast iron by heating the sample in purified argon and analyzing for hydrogen by gas chromatography. After reducing the blank to less than 10^{-4} ml H_2/min, the sample was introduced to a quartz crucible and dried for several minutes in an argon stream at 150°C (no mention was made of investigation of hydrogen loss under these conditions). The sample was then heated to 1400°C for 4 min, during which time the evolved gas was collected on a silica-gel column at −70°C. The argon flow rate was increased from 10 ml/min to 60 ml/min, and without warming this column (presumably the cold trap was removed), hydrogen and other gases were transferred to a molecular-sieve 5A chromatographic column at 45°C. Here the hydrogen was separated in 2 min from nitrogen and carbon monoxide and finally determined by thermal conductivity. Amati et al. [56b] also have described a system for extraction of hydrogen from steel in argon followed by a chromatographic separation and thermal-conductivity measurement.

Goto and co-workers [57, 58] have used purified argon as the carrier gas in determining hydrogen both in iron and steel and in aluminum. The hydrogen extracted from the sample was diffused through a heated palladium thimble connected to a mercury-diffusion-pump–automatic-Toepler-pump combination and determined manometrically. Hydrogen was recovered from any methane present by maintaining the temperature of the palladium above 800°C. By mass-spectrometric analysis, the blank was shown to be caused by impurities in argon that decompose at the palladium thimble to yield hydrogen.

As can be seen, a variety of techniques have been used to determine hydrogen after inert-gas extraction. To minimize cost, manipulations, reagents, and instrument maintenance, the simpler techniques are desirable, such as the diffusion of hydrogen through palladium followed by a pressure measurement. Such techniques, coupled with the advantages of inert-gas extraction noted earlier, make this method very attractive for many metals.

B. BWRA HYTEST ANALYZER

The BWRA has developed an instrument for the rapid determination of hydrogen in metals utilizing the inert-carrier-gas method [27]. This instrument, called the Hytest Analyzer, is manufactured by De La Rue Frigistor, Ltd., Langley, Bucks, England.

In analysis the sample is introduced manually to a loading lock that prevents air from entering the furnace. The loading lock and the furnace are pivoted so that the sample falls into the latter. Purified argon sweeps over the heated sample and carries the evolved hydrogen to the thermal-conductivity detector, which has a sealed argon reference. The signal from the detector is amplified and presented on a four-figure digital counter. The only operations required for the analyst are sample-preparation, weighing, and loading, and balancing of the amplifier against the detector once every 24 h. The last requires 2 min.

To purify the argon carrier gas it is passed over titanium at 650°C, over Hopcalite at 650°C to oxidize hydrogen to water, and through molecular sieve to absorb the water. The argon flow rate is controlled to better than ±0.5%.

Iron, nickel, titanium, zirconium, and their alloys can be analyzed directly with the Hytest instrument. Aluminum and its alloys can also be rapidly analyzed, provided the samples are first placed inside capsules made of vacuum degassed mild steel. Hydrogen diffuses slowly from solid aluminum, and the use of these capsules permits a higher temperature, at which aluminum melts, to be utilized.

Analysis time varies in part with sample-size and sample-composition; in general, 4–15 min are required for samples of the above metals in diameters of $\frac{1}{8}$ in. or $\frac{1}{4}$ in. The Hytest Analyzer employs a range switch that changes the gain setting of the amplifier, so that it is a simple operation to analyze samples containing less than 1 ppm hydrogen and then samples containing more than 80 ppm. By using the range switch and changing the gas flow rate, up to 180 ppm hydrogen can be determined. In the range of 1–10 ppm hydrogen, contents are reportedly reproducible to within 1%. Above this concentration the relative reproducibility is 10%. By analysis of standard samples, it has been determined that the relative accuracies of the results are 5, 1, and 5%, respectively, for 0.1–1.0, 1.0–10, and 10–180 ppm. In addition to these features the Hytest instrument is constructed of metal and therefore avoids the limitations of glass systems.

VI. EQUILIBRIUM PRESSURE

A. SOLID SAMPLE

The equilibrium-pressure method is based on the amount of hydrogen evolved at constant temperature in a closed system of known volume. The hydrogen concentration in the original sample is then determined by summing up the amount of hydrogen in the gaseous phase and in the sample at equilibrium. The former value is obtained by calculation after measuring the pressure and room temperature at equilibrium (with the sample at the operating temperature); the concentration of hydrogen in the sample at equilibrium is taken from a calibration curve.

The apparatus used in this method consists of means for evacuating the system, for resistance-heating at a known temperature, and for measuring

TABLE 7.1. Probable Error of Single Determinations of Hydrogen in Titanium[a]

Range of Hydrogen Concentration (%)	Probable Error (%)
0.001–0.01	0.0003
0.01–0.02	0.0006
0.02–0.05	0.0012
0.05–0.1	0.0025
0.1–0.2	0.005
0.2–0.3	0.01

[a] Reproduced by permission of *Journal of the Electrochemical Society* [60].

the pressure. (In a commercial analyzer described later, an inert gas is used in place of the vacuum.) A distinct disadvantage of this method is that it can only be applied to metals for which an equilibrium hydrogen concentration–pressure relationship can be established. Also, the metal must have a high hydrogen solubility [59]. A very low hydrogen solubility introduces the problems of decreasing the internal volume of the apparatus and increasing the size of the specimen so that the amount of hydrogen required to set up the equilibrium pressure in the apparatus is small in comparison to the quantity dissolved in the specimen. Moreover, it is necessary to add known amounts of hydrogen to degassed samples in a known volume so that a curve correlating equilibrium pressure with hydrogen concentration may be obtained for the specific temperature. This method has been used in the analysis of titanium [4, 59, 60], and an apparatus for analyzing 200 g or more of this metal has been constructed.

A complete analytical cycle for 0.5-g samples takes about 15 min, and 20 samples may be analyzed in 8 h. The reproducibility is reportedly about twice that of the vacuum-fusion method. In Table 7.1 the probable errors of single hydrogen determinations in titanium are given.

The Cambridge Instrument Company, Ltd., London, developed an apparatus using the equilibrium-pressure method [61, 62]. The sample is heated at 650°C in argon in a closed system. When the mixture of argon and evolved hydrogen attains equilibrium, the thermal-conductivity cell or katharometer output, shown on an indicator, is proportional to the hydrogen content of the argon. The hydrogen retained in the sample after analysis is calculated from a provided formula. This instrument, which reportedly is rugged and easy to operate, is shown in Fig. 7.7. The galvanometer reading may be recorded on a strip chart. To prevent the reaction of moisture with the sample to produce hydrogen, the argon is dried immediately before use. Also, any water vapor admitted with the sample is absorbed in anhydrone placed between the bottom of the furnace and the katharometer. This anhydrone is adequate for at least 100 determinations; its exhaustion is indicated by the appearance of an oblique deflection on the recorder trace after evolution of the sample hydrogen. Analysis of a $\frac{1}{2}$-in.-diameter steel specimen requires about 30 min.

B. LIQUID SAMPLE

1. First-Bubble Method

A simple method for determining hydrogen for rapid quality control, but one less accurate than most of the other methods described in this chapter, has been developed by Dardel [63]. Considerable improvement has been made by Neil and Burr [64, 65], based on later work of Dardel [66] and on the work of Ransley and Neufeld [67], Lovtsov [68], and others [69a]. The applicability of the Dardel method is dependent upon the formation on the metal of a film that is impermeable to hydrogen and upon the release of hydrogen from the melt in the molecular form as noticeable bubbles. This method is applicable mainly to aluminum and its alloys; it has also been used for magnesium and magnesium alloys [5].

The determination is carried out by gradually reducing the pressure in a closed chamber in which the molten-metal sample is contained in a degassed crucible at constant temperature. The pressure and temperature when the first bubble appears are noted, and the hydrogen content is then calculated or determined from a plot relating pressure and hydrogen content. The solubility of hydrogen, S (from which the amount of hydrogen can be calculated), in milliliters per 100 g of metal, is calculated from

$$\log_{10} S = -\frac{A}{T} + B + \tfrac{1}{2}\log_{10} P, \tag{2}$$

where A and B are constants determined by the equilibrium solubility of the gas in the metal, T is the temperature in degrees Kelvin, and P is the corrected hydrogen pressure in torr Values for A and B, derived from log–log plots of solubility at 760 torr versus hydrogen pressure, for pure aluminum [67, 69b, 69c] and certain aluminum alloys [69c] have been published. If A and B are not known, chill-cast samples may be used to obtain corresponding pressure and temperature data using the initial bubble test. Analysis of the samples by an independent technique provides a hydrogen-content value. The observed pressure values are corrected according to

$$P = P_0 e \exp\left[2k\left(\frac{1}{T_r} - \frac{1}{T_0}\right)\right], \tag{3}$$

Fig. 7.7. Cambridge hydrogen-in-metals analyzer. (Courtesy Journal Iron and Steel Institute and Cambridge Instrument Co., Ltd.)

where P_0 is the observed pressure in torr, $K = 6355$, T_r is the reference temperature, and T_0 is the observed temperature. The plot of hydrogen contents determined independently against $\sqrt{P}$ is linear. A value of 943°K was chosen for T_r; all ordinary aluminum alloys are liquid at this temperature. For K, the value of 6355 was selected as reasonably applicable to commercial aluminum alloys; this was calculated according to

$$K = \frac{A}{\log_{10} e}. \tag{4}$$

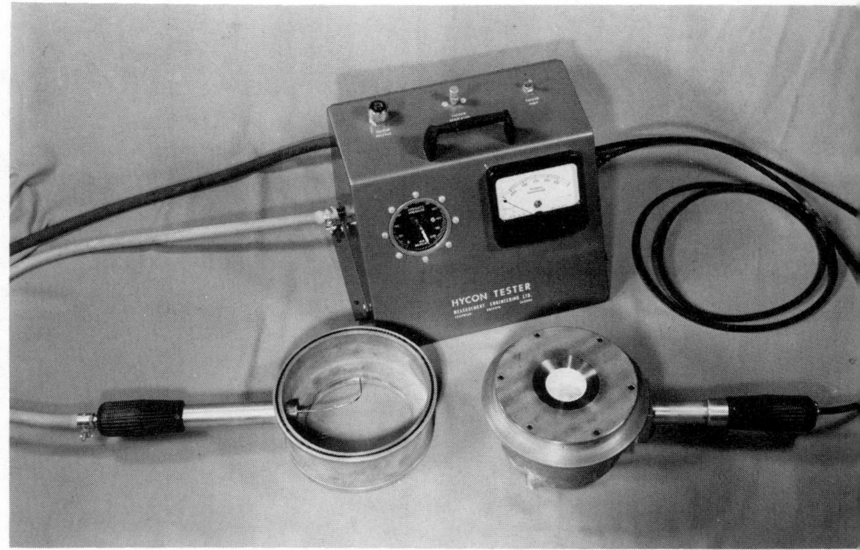

Fig. 7.8. Neil-Burr hydrogen apparatus (Courtesy *Revue de Metallurgie* (Paris) an *Transactions American Foundrymen's Society*.

The method as originally proposed by Dardel [63] aroused much criticisr [69]. The quantitativeness of the method was questioned. It was noted als that (1) the accuracy of the method decreases as the purity of aluminur increases, (2) at lower hydrogen concentrations there is difficulty in hydroge gathering and forming a bubble, (3) diffusion of hydrogen must be take into account, and (4) bubbles of gas may be trapped by advancing crystal: Also, if a volatile constituent is present in the aluminum, the vapor tensio of this constituent limits the range of the determinations [63].

Most of these and other objections appear to have been overcome [64, 65 with the development of the apparatus shown in Fig. 7.8. The sample i contained in a previously degassed boron nitride crucible of about 50–m capacity, on the outside of which is wound a chromel A heater of 325 W The crucible heater assembly is placed in a cast aluminum vessel with abou $1\frac{1}{2}$ in. of insulation between the two. The cover unit containing a glas viewing port is sealed in position with O-rings. The temperature is controlle by means of a chromel-alumel thermocouple, the leads of which pass throug the handle of the cover unit and inside the vacuum line and are connecte to a millivoltmeter calibrated for the 650–800°C temperature range. Th pressure is reduced with a rotary pump at a known and reproducible rat (about 5–10 torr/sec) controlled by a cutoff valve and is measured on manometer calibrated for the range of 0–200 torr. The apparatus is simple and rugged. Approximately 2 min are required for an analysis.

Samples must be as thick as is practicable, and the heating rate must be rapid. Because the volume–surface ratio must be large, this method is not applicable to thin sheets. The dispersion of results is very marked for aluminum alloys containing elements that produce a dull gray oxide film; apparently this color affects the perceiving of the first bubble. Other factors affecting results, such as inadequate cleaning of the crucible and carbon in refining agents, have been described by Plate [70]. In general, it is believed that the mean of two or more values will be precise to ± 0.05 ppm for the range of about 0.08–0.50 ppm. The convenience and rapidity of this method and the degree of precision reportedly make it preferable to the pouring, weight-ratio, Telegas, and Straube-Pfeiffer tests [70–73].

2. Telegas Method

An instrument for the direct measurement of the hydrogen content of molten aluminum alloys has been developed by Ransley and co-workers [74]. The measurement is based on the evolution of hydrogen into a recirculating stream of inert gas until an equilibrium is obtained between the hydrogen in the gaseous phase and in the sample. The instrument, called Telegas and manufactured by Edwards High Vacuum, Ltd., Crawley, Sussex, England, is composed mainly of a gas-admission valve, a circulating pump, a probe, a thermal-conductivity cell, and a millivoltmeter. In operation the probe is inserted into molten aluminum, a small volume of nitrogen is circulated through the system, and the amount of hydrogen extracted is determined by a thermal-conductivity measurement of the gas mixture. From this amount the hydrogen concentration in the original sample is calculated. With this instrument the response time varies from 2 min at the 0.1–0.2-ml hydrogen/100-g level to about 6 min at the 0.45-ml hydrogen/100-g level for commercial-purity aluminum [75]. With a duralumin-type alloy the response was more sluggish, the meter reading still climbing after 10 min. Another major problem has been the ceramic probe, which is fragile and sensitive to thermal shock. Also, the design of the probe is such that the efficiency of pick-up of the nitrogen–hydrogen mixture needs improvement.

VII. ISOTOPE DILUTION

In this method the sample is heated in an atmosphere containing a known amount of deuterium or tritium and equilibrated for a sufficient length of time so that isotopic exchange occurs between the hydrogen in the sample and the deuterium or tritium. Analysis of the atmosphere for deuterium or tritium then provides a means for calculating the hydrogen content of the sample. Alternatively, tritium has been determined in the gas removed from the doped sample by vacuum extraction [76]. Vygodskii et al. [77] have described a mass-spectrometric apparatus for analysis of HD and D_2 mixtures. With a 10-g sample and a hydrogen content of 5–10 ml/100 g the accuracy was 1–2%. Considerable work has been done by Zaidel in Russia utilizing optical-emission spectroscopic techniques for determining hydrogen–deuterium ratios, and this is discussed in Chapter 5.

It is assumed in this method that no isotopic fractionation occurs; that is, the proportions of hydrogen and its isotope are the same in both the gas phase and in the sample. However, this is often not the case. Sieverts and co-workers found that the solubility of deuterium in iron [78] and in nickel [79] was about 10% less than that of hydrogen. Temperature ranges of 500–1550°C and 200–1120°C, respectively, were covered in these studies. Even greater differences (up to 44%) were reported by Hawkins [80], who measured the solubilities of the three hydrogen isotopes in nickel and stainless steel from 300–750°C and noted that the solubilities decrease with increasing mass of hydrogen isotope. Frank et al. [81] determined hydrogen–deuterium solubility ratios for steel from 26–86°C; these varied from 3.2–7.8. Because the isotope-dilution method is subject to error and requires rather elegant equipment (optical-emission spectrograph, mass spectrometer, or counting equipment), it is used comparatively little. Extreme sensitivity may be obtained when using the radiochemical method with tritium [76], but as noted by Holt in Chapter 11, the use of tritium presents a potential health hazard. Also this isotope is more expensive than deuterium; therefore, its use in place of deuterium is justified only where counting analysis is more easily obtained than mass-spectrometric analysis.

Holt [82] utilized the isotope-dilution method for the analysis of sodium and sodium–potassium alloy, because the liberation of hydrogen by hot extraction was incomplete. Equilibration with deuterium was carried out for 5 min, and the hydrogen–deuterium ratio was determined mass-spectrometrically by

$$\text{H/D} = \frac{(2)/S_{\text{H}_2} + \frac{1}{2}(3)/S_{\text{HD}}}{\frac{1}{2}(3)/S_{\text{HD}} + (4)/S_{\text{D}_2}}, \tag{5}$$

where (2), (3), and (4) are the net outputs at these masses; S_{H_2}, S_{HD}, and S_{D_2} are the sensitivities, respectively, for the three gases involved; and (2) is the total output of 2 peak minus the outputs attributed to D_2 and HD patterns.

The hydrogen concentration is then calculated according to

$$\text{H, ppm} = \frac{[(\text{H/D})_s(d)_s - (\text{H/D})_b(d)_b](1.008)}{\text{grams of sample}}, \tag{6}$$

where s and b refer to sample and blank data and d is the number of micromoles of deuterium introduced. In practice d_s and d_b are identical.

Analysis of hydrogen-spiked, 2-g samples showed the standard deviation to be 2 ppm for the range 5–250 ppm hydrogen. It is necessary to apply a correction factor in the analysis of sodium–potassium to compensate for the effect of different rates of formation of the hydrides and deuterides. For sodium this effect is negligible. Total analysis time after obtaining the sample was about 45 min.

Hydrogen isotopes have been determined by Cercy et al. [83] by chromatographic separation at −196°C on a capillary glass column filled with alumina. Helium was used as the carrier gas and the determinations were made by thermal conductivity measurements. At STP the limits of detection for H_2, HD, and D_2, respectively, were 0.17, 0.1, and 0.08 micron-liter.

Gillespie [76] quenched liquid-aluminum samples in atmospheres of tritium–hydrogen and assayed them by vacuum extraction, followed by measurement of the tritium in an ionization chamber with a vibrating-reed electrometer. The time-consuming step was the extraction, which required 2–5 h. No isotope effects were noticed in this study. However, these might possibly be encountered in diffusion studies with solid samples. As mentioned earlier the tritium-tracer method provides greater sensitivity. Further, although surface gas corrections are necessary in the analysis of aluminum where gas-volume measurements are made, these corrections are reportedly not necessary in the tritium-tracer method when dealing with quenched liquid samples. This is especially advantageous at low hydrogen concentrations. Results obtained by vacuum extraction and the tritium-tracer method were comparable.

Evans and Herrington used the tritium-tracer method for determining hydrogen in solid sodium [84] and in uranium, beryllium, and aluminum [85]. Equilibration times were 10, 15, 20, and 30 min, respectively, for the four elements. Corrections had to be made for water adsorbed on the surface of the aluminum and uranium in order to obtain agreement between this method and vacuum extraction. This water is evolved on heating, partly unchanged and partly as hydrogen. The fraction evolved unchanged is not measured in the vacuum-extraction method but is measured by the isotopic-dilution method.

In the analysis, the metal was equilibrated with a known amount of tritium. A sample of diluted tritiated hydrogen was isolated and then purified by diffusion through palladium. This was followed by determination of the specific activity of the gas by Geiger counting. The volume, V, of the hydrogen in the metal was calculated from the formula

$$V = V_0\left(\frac{a_1}{a_2} - 1\right) - b, \tag{7}$$

where V_0 and b are the volumes of tritiated hydrogen introduced and apparatus blank, respectively, and a_1 and a_2 are the initial and final specific activities of the tritiated hydrogen. Little, if any, isotopic fractionation occurred during equilibration and gas-phase sampling. The mean and the relative standard deviation from the mean of 12 recovery experiments using hydrogen-spiked samples were 101 ± 3%.

VIII. COMBUSTION

A number of workers [86–89] have determined hydrogen in metals by combustion in a stream of oxygen, followed by determination of the water in a variety of ways. This method is susceptible to error from water adsorbed on the sample, the sample-container, and the various internal parts of the system. Because of the ubiquitous nature of water the blank may represent an appreciable fraction of the hydrogen content of the metal, thereby limiting the sensitivity of the method. Further, hydrogen results obtained by combustion of titanium are often higher than those obtained by vacuum extraction or equilibrium-pressure measurement. McKinley [4] has suggested that this is caused by an unmeasured blank, the basic blank being measured at the nominal operating temperature, while considerably higher temperatures are attained during the exothermic combustion of the sample. Another problem is that in most cases the sample must be in the form of chips or drillings; this requires additional preparation time and results in a larger surface area on which water vapor may collect. Also, the method is cumbersome in view of the many operations required in assembling and disassembling the apparatus and reagents. An advantage of the combustion method is that in some instances [90, 92, 94] carbon may be simultaneously determined with hydrogen. Finally, another attractive feature of the combustion method is that it does not require high-vacuum equipment or elaborate furnaces.

Combustion is generally carried out in a horizontal-tube resistance furnace, although the use of a vertical-tube furnace with chunk uranium samples has been mentioned [91]. The combustion temperature is dependent on the metal being analyzed and the type of flux used, if any. For example, hydrogen has been determined in titanium using a lead flux at 900°C [89] and an iron–copper–tin flux at 1200°C [90]. A copper flux has been used with zirconium at 1300°C [94]. Other metals that have been analyzed for hydrogen by this method include steel [86, 87, 96], uranium [91], and magnesium [93]. Sodium, in an aluminum capsule, has been reacted with a flux of silica sand at 1000°C in an atmosphere of argon, followed by combustion in oxygen and weighing of the absorbed water [98].

The oxygen that is used for combustion must be purified to remove water prior to its entering the furnace. This is most easily accomplished by passing the oxygen through a water-absorber such as magnesium perchlorate. If desired, carbon dioxide can be removed using soda asbestos. Also, if trace amounts of organic contaminants are suspected, the gas should first be passed over copper oxide heated to about 850°C (see Chapter 4, Section I.B.3.f). An elaborate purification system for oxygen has been described in detail [89].

Porcelain [93] or alumina [90, 93] boats may be used to contain the sample. Boats should be preignited at a temperature sufficiently high to effect removal of water; they should then be stored in a muffle furnace at an elevated temperature, for example, at 500°C [93], until they are used. New combustion tubes should be conditioned prior to analysis. This has been accomplished [90] by heating and cooling the tube for 15-min periods and also by igniting 3 or 4 samples until a blank determination of the proper minimum amount (in this case 0.2 mg or less) was obtained. Quartz [89, 93] or 96% silica [89] combustion tubes have been used.

The sample weight will depend on the hydrogen content and the method for determining hydrogen. Combustion time will depend on oxygen flow rate, particle size of sample, temperature, flux, and type of sample. Argon may be used to reduce the violence of the combustion when analyzing magnesium [93] and to prevent suckback of gases from the laboratory atmosphere. The exit end, however, should contain a water-absorbent and a carbon dioxide–absorbent to protect the material used to absorb the water produced by combustion of the hydrogen in the sample.

As little as 5 μg hydrogen [90] has been measured by gravimetric determination of the water. Analysis times of 0.5 h [89] and 1 h [90] per determination have been reported for this method. To obtain increased sensitivity, Holt [91] used a manometric method by which he was able to determine as little as 0.1 μg hydrogen. Boenisch [95] determined less than 0.5 μg hydrogen by treating the water produced with naphthyloxychlorophosphine to form hydrochloric acid, which was then absorbed in water and measured titrimetrically. Rothmann et al. [96] determined the water formed, by absorbing it in pyridine–methanol and titrating it with Karl Fischer reagent, using the dead-stop endpoint technique [99]. One milliliter of reagent was equivalent to 1 mg of water, and the readability of the buret was 0.01 ml. The detection limit was about 1 ppm hydrogen, with an error of the same magnitude, the apparatus blank being 0.30 ppm.

Because of the difficulties encountered with this method, which were mentioned earlier, it is preferable to use more reliable techniques for determining hydrogen if they are applicable. However, for metals, such as magnesium, that cannot be analyzed by vacuum fusion or vacuum extraction [93] the combustion method is useful.

IX. MISCELLANEOUS METHODS

Other methods have been used with varying success for determining hydrogen in metals. Potter and Huber [100], using Debye-Scherrer X-ray-diffraction techniques in their study of electrolytic manganese, found that the average lattice parameter varied approximately linearly with the hydrogen content, the enlargement being 0.0003 %/ml of gas/100 g of metal. However, the parameters are not affected unless the metal is heated over 300°C. A similar study of titanium, zirconium, and niobium has been made by Kuznetsov et al. [101]. Barton and Lewis [102] have reviewed the determination of the hydrogen content of palladium and palladium alloys from measurement of the relative electrical resistance of the sample in an electrolyte solution. Data have been obtained for the range 0–0.9 H/Pd (atomic ratio), which is approximately equal to 0–0.75 wt % hydrogen. For convenience in measuring, the sample should be in the form of a wire. Sauerwald [103] determined hydrogen in magnesium by chlorination of the molten sample to form hydrogen chloride, which was then passed, along with liberated oxygen, over copper oxide, the water formed being determined gravimetrically. Bobalek and Shrader [104] determined hydrogen in magnesium alloys by bombarding the sample (as cathode) with mercury ions. Hydrogen was evacuated continuously during this ion bombardment. The method was inconvenient and inaccurate in comparison to vacuum extraction.

Ganiev [105] determined hydrogen in electroplated iron by anodic dissolution of the iron. Only a trace of hydrogen from the decomposition of water was found by Klyachko and Larina [106, 107], who used an electrolyte containing deuterium oxide in the anodic dissolution and analyzed the evolved gas by mass spectrometry. Also, apparently no hydrogen from the metal was lost to the solution. The residue remaining after anodic solution was analyzed by vacuum fusion; the sum of the hydrogen content determined by both techniques was the same as that determined on the original sample by vacuum fusion. To find the sample weight, a separate determination for the dissolved metal was usually performed, which added to the length and cumbersomeness of the method. Hillen and Thackray [108] determined helium, hydrogen, deuterium, and tritium in neutron-bombarded beryllium by extracting the gases from the molten sample and determining them by a combination gas-chromatographic–radiometric counting method. Yoshimori and Ishiwari [109] extracted hydrogen from various metals by heating them in a stream of purified argon. The hydrogen was oxidized to water over hot copper oxide, the water was converted to ammonia with sodium amide, and the ammonia was titrated with coulometrically generated hypobromite ion. Finally, Yavoiskii et al. [110] determined hydrogen in various metals by use of a laser-beam extraction system coupled to a mass spectrometer.

REFERENCES

1. R. Eborall, *Spec. Rep. 68*, Iron and Steel Institute, London, 1960, p. 192.
2. P. Kappes and A. Schiffers, *Chem.-Ztg.*, **86,** 255 (1962).
3. Z. M. Turovsteva and L. L. Kunin, *Analysis of Gases in Metals*, Academy of Sciences of the USSR, Moscow, 1959; trans. by J. Thompson, Consultants Bureau, New York, 1959.
4. T. D. McKinley, *Trans. AIME*, **212,** 563 (1958).
5. L. I. Sokol'skaya, *Gases in Light Metals*, Moscow, Metallurgizdat, 1959; trans. by G. F. Modlen and H. T. Protheroe, Pergamon, New York, 1961.
6. R. Suarez Acosta, *Inst. Hierro Acero*, **16,** 173 (1963); through *Chem. Abstr.*, **59,** 12171d (1963).
7. A. Wojcik, *Hutnik*, **30,** 281 (1963); through *Chem. Abstr.*, **60,** 6191h (1964).
8. J. F. Martin, R. C. Takacs, R. Rapp, and L. M. Melnick, *Trans. AIME*, **230,** 107 (1964).
9. G. M. Grigorenko, V. I. Lakomskii, and D. M. Rabkin, in B. B. Gulyaev, Ed., *Gases in Cast Metals*, Nauk, Moscow, 1964; trans. by Consultants Bureau, New York, 1965, pp. 247–50.
10. G. A. Smith, W. C. Lenahan, and D. S. MacLeod, *Metallurgia*, **79,** 121 (1969).
11. F. Dugain and J. Bril, *Rev. Met.*, **65,** 525 (7–8, 1968).
12. H. Bennek and G. Klotzbach, *Stahl Eisen*, **61,** 597 (1941).
13. J. J. Naughton, *Trans. AIME*, **162,** 385 (1945).
14. H. Wentrup, H. Fucke, and O. Reif, *Stahl Eisen*, **69,** 117 (1949).
15. D. J. Carney, J. Chipman, and N. J. Grant, *Trans. AIME*, **188,** 397 (1950).
16. B. M. Shields, J. Chipman, and N. J. Grant, *Trans. AIME*, **197,** 180 (1953).
17. H. Epstein, J. Chipman, and N. J. Grant, *Trans. AIME*, **209,** 597 (1957).
18. C. C. Carson, *Anal. Chem.*, **32,** 936 (1960).
19. M. W. Mallett, A. F. Gerds, and C. B. Griffith, *Anal. Chem.*, **25,** 116 (1953).
20. C. B. Griffith and M. W. Mallett, *Anal. Chem.*, **25,** 1085 (1953).
21. J. L. Brandt and C. N. Cochran, *J. Metals*, **8,** 1672 (1956).
22. D. T. Peterson and V. G. Fattore, *Anal. Chem.*, **34,** 579 (1962).
23. J. Parry, *J. Iron Steel Inst.*, **6,** No. 2, 238 (1872).
24. L. P. Pepkowitz and E. R. Proud, *Anal. Chem.*, **21,** 1000 (1949).
25. J. A. J. Walker and H. Seed, *Analyst*, **90,** 19 (1965).
26. F. R. Coe, N. Jenkins, and D. H. Parker, "Encapsulation as a Technique for the Determination of Hydrogen in Volatile Metals," *Proc. Soc. Anal. Chem. Conf.*, Nottingham, 1965.
27. R. C. Perriton and F. R. Coe, *Metallurgia*, **78,** 43 (1968).
28. F. R. Coe, N. Jenkins, and D. H. Parker, *Anal. Chem.*, **39,** 982 (1967).
29. J. H. Andrew, A. K. Bose, H. Lee, and A. G. Quarrell, *J. Iron Steel Inst.*, **146,** 203 (1942).
30. F. R. Coe, *Spec. Rep. 73*, Iron and Steel Institute, London, 1962, p. 111.
31. C. C. Carson and B. J. Alperin, *Trans. Amer. Foundrymen's Soc.*, **67,** 70 (1959).
32. American Society for Testing and Materials, *ASTM Standards, Part 32, Chemical Analysis of Metals*, Philadelphia, 1969, pp. 700–702.

33. American Society for Testing and Materials, unpublished Committee E-3 Task Force Report, March 1969.
34. N. Lounamaa and J. U. Aass, *Jernkont. Ann.*, **148,** 873 (1964).
35. Z. Falecki, *Hutnik*, **32,** 143 (1965).
36. J. Sannier and J. Leroy, *Comm. Energie At.* (France) *Rappt. CEA-R-2957*, 1966.
37. C. E. Ransley and D. E. J. Talbot, *J. Inst. Metals*, **84,** 445 (1955–56).
38. V. A. Danilkii, K. M. Konstantinov, and G. L. Bulatova, *Zavodsk. Lab.*, **27,** 259 (1961).
39. E. Marianeschi and R. Coletti, *Met. Ital*, **49,** 673 (1957); through *Chem. Abstr.*, **52,** 6853 (1958).
40. J. Calmettes, C. Dubois, and P. Bastien, *Mem. Sci. Rev. Met.*, **56,** 641 (1959).
41. C. E. A. Shanahan and F. Cooke, *J. Iron Steel Inst.*, **198,** 257 (1961).
42. A. Wojcik, *Hutnik*, **29,** 458 (1962) through *Chem. Abstr.*, **59,** 1059 (1963).
43. (a) S. Bergenfelt and C. A. Akerblom, *Jernkontor. Ann.*, **146,** 461 (1962); (b) H. Goto and M. Hosoya, *Nippon Kinzoku Gakkaishi*, **35,** 16 (1971); and (c) G. L. Hargrove, R. C. Shepard, and H. Farrar, IV, *Anal. Chem.*, **43,** 439 (1971).
44. E. Heyn, *Metallographist*, **6,** 39 (1903); through Ref. 45.
45. C. E. A. Shanahan and F. Cooke, *J. Iron Steel Inst.*, **190,** 381 (1958).
46. C. E. A. Shanahan, *Spec. Rep. 68*, Iron and Steel Institute, London, 1960, p. 250.
47. F. R. Coe, *Research* (London), **13,** 323 (1960).
48. F. R. Coe, British Patent 922,410, Apr. 3, 1963, assigned to British Welding Research Association.
49. F. R. Coe and N. Jenkins, *Spec. Rep. 68*, Iron and Steel Institute, London, 1960, p. 229.
50. F. R. Coe, *Spec. Rep. 68*, Iron and Steel Institute, London, 1960, p. 239.
51. A. T. J. Hayward, *Laboratory Practice*, Feb. 1960; through reference 50.
52. (a) K. Abresch, W. Dobner, and H. Lemm. *Arch. Eisenhuttenw.*, **31,** 351 (1960); and (b) D. I. Walter and G. A. Picklo, Jr., Pittsburgh Conference on Analytical Chemistry and Applied Spectroscopy, Pittsburgh, 1959.
53. L. A. Fergason, D. E. Seizinger, and C. H. McBride, *Nucl. Sci. Eng.*, **10,** 53 (1961).
54. R. J. Walter and H. G. Offner, *Anal. Chem.*, **36,** 1779 (1964).
55. E. Giegerl, *Arch. Eisenhuttenw.*, **33,** 453 (1962).
56. (a) J. Kashima and T. Yamazaki, *Bull. Chem. Soc. Jap.*, **41,** 2382 (1968); (b) G. Amati, S. Maneschi, and N. Vantini, U.S. Patent 3,520,171, July 14, 1970; assigned to Centro Sperimentale Metallurgico S.p.A., Rome, Italy.
57. H. Goto, S. Ikeda, and M. Hosoya, *Sci. Rep. Res. Inst., Tohoku Univ.*, **A17** (5), 259 (1965).
58. H. Goto, M. Hosoya, and Y. Otaka, *Sci. Rep. Res. Inst., Tohoku Univ.*, **A19** (4), 234 (1967).
59. A. D. McQuillan, *J. Inst. Metals*, **79,** 73 (1951).
60. T. D. McKinley, *J. Electrochem. Soc.*, **102,** 117 (1955).
61. E. W. Gill, *J. Iron Steel Inst.*, **201,** 960 (1963).

62. Cambridge Instrument Co., Ltd., British Patent 931,605, July 17, 1963.
63. Y. Dardel, *Metals Technol.*, **15** (8), Tech. Paper 2484 (1948).
64. D. J. Neil and A. C. Burr, *Rev. Met.*, **57,** 735 (1960).
65. D. J. Neil and A. C. Burr, *Mod. Cast.*, **39** (6), 86 (1961).
66. Y. Dardel, *Metal Ind.*, **76,** 203 (1950).
67. C. E. Ransley and H. Neufeld, *J. Inst. Metals*, **74,** 599 (1948).
68. D. P. Lovtsov, *Liteinoe Proizvodstvo*, **9,** 15 (1055); through *Chem. Abstr.*, **50,** 1559d (1956).
69. (a) R. Eborall, M. B. Bever, C. E. Ransley, N. D. G. Mountford, H. Udin, F. N. Rhines and E. A. Gulbransen, Discussion of Reference 63, *Trans. AIME*, **185,** 868 (1949); (b) W. R. Opie and N. J. Grant, *J. Metals*, **188,** 1237 (1950); and (c) W. Hofmann and J. Maatsch, *Z. Metallk.*, **47,** 89 (1956).
70. H. Plate, *Giesserei*, **52,** 556 (1965).
71. Anonymous, *Giesserei-Proxis*, **56,** 412 (1935); through Ref. 1.
72. N. F. Budgen, *J. Inst. Metals*, **42,** 119 (1929).
73. E. M. Brenner, *Metallen*, **21,** 81 (1965); through *Chem. Abstr.*, **63,** 14491e (1965).
74. C. E. Ransley, D. E. J. Talbot, and H. C. Barlow, *J. Inst. Metals*, **86,** 212 (1957–58).
75. M. F. Jordan, *J. Inst. Metals*, **86,** 524 (1957–58).
76. A. S. Gillespie, Jr., *Anal. Chem.*, **32,** 1624 (1960).
77. A. E. Vygodskii, V. G. Nesterenko, and D. G. Sherman, *Zavodsk. Lab.*, **29,** 1474 (1963).
78. A. Sieverts, G. Zapf, and H. Moritz, *Z. Phys. Chem.* (Leipzig), **A183,** 19 (1938).
79. A. Sieverts and W. Danz, *Z. Anorg. Chem.*, **247,** 131 (1941).
80. N. J. Hawkins, *U.S. At. Energy Comm. Rep.*, *KAPL-868*, 1953.
81. R. C. Frank, R. W. Lee, and R. L. Williams, *J. Appl. Phys.*, **29,** 898 (1958).
82. B. D. Holt, *Anal. Chem.*, **31,** 51 (1959).
83. C. Cercy, S. Titschenko, and F. Botter, *Bull. Soc. Chim. Fr.*, **1962,** 2315.
84. C. Evans and J. Herrington, *Anal. Chem.*, **35,** 1907 (1963).
85. C. Evans and J. Herrington, *Radioisotopes Phys. Sci. Ind., Proc. Conf. Use, Copenhagen, 1960*, **2,** 309 (1962).
86. F. K. Gerke and N. V. Zolotareva, *Zavodsk. Lab.*, **4,** 19 (1935); through *Chem. Abstr.*, **29,** 5381 (1935).
87. G. A. Moore, *Trans. AIME*, **162,** 404 (1945).
88. C. Kenty and F. W. Reuter, *Rev. Sci. Instrum.*, **18,** 918 (1947).
89. M. Codell and G. Norwitz, *Anal. Chem.*, **28,** 106 (1956).
90. R. B. Nunemaker and S. A. Shrader, *Anal. Chem.*, **28,** 1040 (1956).
91. B. D. Holt, *Anal. Chem.*, **28,** 1153 (1956).
92. N. Oda and K. Norishima, *Denki Kagaku*, **25,** 365 (1957); through *Chem. Abstr.*, **52,** 1853f (1958).
93. M. Codell and G. Norwitz, *Anal. Chim. Acta*, **18,** 265 (1958).
94. N. Oda, M. Kubo, and T. Yamagishi, *Bunseki Kagaku*, **11,** 214 (1962); through *Chem. Abstr.*, **57,** 25a (1962).

95. E. Boenisch, *Materialprufung*, **4** (7), 247 (1962); through *Chem. Abstr.*, **57**, 13175e (1962).
96. H. Rothmann, W. Keil, and H. Richter, *Z. Erzbergbau Metallhuttenw.*, **16**, 239 (1963).
97. G. Ya. Veinberg and S. I. Proshutinskii, *Zavodsk. Lab.*, **6**, 422 (1937); through *Chem. Abstr.*, **31**, 7793 (1937).
98. R. D. Gardner and W. H. Ashley, *U.S. At. Energy Comm., Rept. LA-3049*, March 31, 1964.
99. I. M. Kolthoff, R. Belcher, V. A. Stenger, and G. Matsuyama, *Volumetric Analysis*, Vol. III, Interscience, New York, 1957, p. 419.
100. E. V. Potter and R. W. Huber, *Phys. Rev.*, **68**, 24 (1945).
101. L. M. Kuznetsov, E. S. Makarov, and Z. M. Turovsteva, *Trudy Komissii Anal. Khim., Akad. Nauk S.S.S.R., Inst. Geokhim. i Anal. Khim.*, **10**, 122 (1960); through *Chem. Abstr.*, **55**, 6249 (1961).
102. J. C. Barton and F. A. Lewis, *Talanta*, **10**, 237 (1963).
103. F. Sauerwald, *Z. Anorg. Chem.*, **256**, 217 (1948).
104. E. G. Bobalek and S. A. Shrader, *Ind. Eng. Chem., Anal. Ed.*, **17**, 544 (1945).
105. Kh. V. Ganiev, *Doklady Akad. Nauk Tadzhik SSR*, **No. 20**, 63 (1957); through *Chem. Abstr.*, **53**, 11099i (1959).
106. Yu. A. Klyachko and O. D. Larina, *Zavodsk. Lab.*, **26**, 1047 (1960).
107. Yu. A. Klyachko and O. D. Larina, *Sb. Tr. Tsentr. Nauchm.-Issled. Inst. Chernoi Met.*, **No. 24**, 5 (1962); through *Chem. Abstr.*, **58**, 11947f (1963).
108. L. W. Hillen and M. Thackray, *J. Chromatogr.*, **10**, 309 (1963).
109. T. Yoshimori and S. Ishiwari, *Talanta*, **17**, 349 (1970).
110. V. I. Yavoiskii, L. B. Kosterev, V. L. Safonov, and E. V. Afanas, *Sb., Mosk. Inst. Stali Splavov*, **No. 62**, 57–71 (1970); through *Chem. Abstr.*, **74**, 150867e (1971).

CHAPTER

8

NITROGEN

HUGH F. BEEGHLY

Bruceton Mills, West Virginia

CONTENTS

I. INTRODUCTION

Nitrogen is a valuable alloying constituent for some metals and may be added intentionally. In other systems, nitrogen enters adventitiously, mainly from the atmosphere. Certain metal nitrides and/or carbonitrides, when present even in small concentrations, can distinctly alter the mechanical properties of the matrix metal. All of the metals, with the exception of the platinum-group metals and gold, form binary compounds with nitrogen either directly or indirectly, but only those compounds in which the nitrogen is the more electronegative constituent are called nitrides. Reliable analytical methods are required for determining the content of nitrogen and its compounds in metals and for relating these contents to specific properties of the alloy system. Methods are needed that can be used to analyze for nitrogen at any concentration level from approximately 0.0005 to 30 wt % or more.

The methods discussed in this chapter are mainly chemical in nature. Vacuum- and inert-gas-fusion techniques are discussed in Chapter 4. In addition to the methods presented in chapters on specific metals, a detailed discussion is given in Chapter 15 for the separation of nitrogen as ammonia followed by its determination photometrically, using Nessler reagent, or titrimetrically. Of all the metals analyzed for nitrogen, perhaps steel has received the most attention. The analysis of steel is of particular interest in that many metal nitrides (or carbonitrides) that are present in other metals treated in this book may be found in steel. For example, the caustic fusion procedure [1] described in Chapter 15 has been used in steel analysis for decomposing α-Si_3N_4 and β-Si_3N_4. It has also been used in the analysis of niobium [2]. Similar procedures have been used for decomposing a number of refractory metal nitrides and carbides. These have been discussed by Kriege [3] and are summarized in Table 8.1. The determination of nitrogen in steel [4] and the use of the ester-halogen method for isolating metal nitrides from steel [5–8] have been extensively investigated. The metal nitrides studied were

TABLE 8.1. Determination of Nitrogen in Refractory Nitrides and Carbides [3]

Nitride or Carbide	Dissolution Technique	Method of Determination
BN, Si_3N_4	Fusion with NaOH	Titrate NH_3 with HCl
NbN, TaN, TiN, UN	Dumas (CO_2), 900–1000°C	Measure N_2 volume
HfN, ZrN	Dumas (CO_2), 1050°C	Measure N_2 volume
Nb_2C, NbC, Ta_2C, TaC, TiC, W_2C, WC, UC, U_2C_3, UC_2, VC, Nb–U–C, Ta–U–C	Dumas (CO_2), 900–1000°C	Measure N_2 volume
HfC, ZrC, Hf–U–C, Zr–U–C	Dumas (CO_2), 1050°C	Measure N_2 volume

those of Al, Be, B, Nb, Si, Ti, V, and Zr. The ester-halogen method is treated further in this chapter and also in Chapter 15. Federov and Krischevskaya [9] in reviewing methods for determining nitrogen in steel have discussed the following: AlN, BN, CrN, iron nitrides, NbN, Nb_2N_3, Si_3N_4, TaN, TiN, VN, and V_2N_3. Jaudon [10] in a review covered the isolation from steel of the following: AlN, BN, Cr_2N, CrN, NbN, Si_3N_4, TaN, TiN, VN, and ZrN. In addition, fusion and chemical techniques in general were discussed.

Rottman and Nickel [11], in a lengthy study of high-temperature-resistant metal nitrides and carbonitrides, reviewed 15 chemical and physical methods and their applications. The methods are (a) vacuum fusion, (b) carrier-gas fusion, (c) Kjeldahl, (d) alkali fusion, (e) Dumas, (f) combustion in oxygen, (g) BrF_3, (h) chlorination, (i) isotope dilution, (j) emission spectroscopy, (k) activation, (l) x-ray emission, (m) residual resistance, (n) damping, and (o) solid-state mass spectroscopy. Of the chemical methods, the Dumas method was deemed universally applicable with only CO_2 (and not CuO) needed. As little as 0.5 ppm nitrogen in 1-g samples was determined volumetrically. The combustion was carried out between 900 and 1300°C for nitrides of Cr, Fe, Mn, Mo, Nb, Ta, Ti, U, V, W, and Zr.

A number of the metal nitrides react with alkali to release ammonia. The nitrogen compounds of the alkali and alkaline earth metals and of certain other elements (e.g., Al) that are partially ionic in character hydrolyze on contact with water. Other nitrides, boron nitride for example, have covalent structures and are less reactive chemically. In a third type of nitride, the nitrogen is located at the interstices between the metallic atoms. Examples are the nitrides of metals of groups IVB, VB, and VIB in the periodic table. These generally have a metallic appearance and conduct electricity. The nitride phases differ from metals in that they lack ductility and malleability and are extremely hard. Some of the interstitial nitrides—for example, those of zirconium and niobium—are superconductors up to about 10°K. The interstitial nitrides vary greatly in thermal stability, in reactivity at high temperatures, and in their resistance to attack by common laboratory reagents. Knowledge of the properties of metal intrides is presently inadequate; work with modern techniques on materials of known purity can be expected to yield data of greater accuracy.

Nitrides for synthetic standards may be prepared by such procedures as (a) direct reaction of a metal with elemental nitrogen, generally at elevated temperatures; (b) reaction of the metal with active nitrogen liberated from dissociated ammonia; (c) reaction of a metal hydride with elemental nitrogen or nitrogen from dissociated ammonia; (d) reaction of a metal halide with ammonia; (e) deposition of a metal on a hot filament in an atmosphere containing active nitrogen; and (f) reaction of a metal oxide by heating in ammonia.

For analytical purposes, the nitrogen can be considered to be in metal–nitrogen compounds of the matrix metal or its alloying constituents. It may be liberated quantitatively either in elemental form or as NH_3, by heating under carefully prescribed conditions, or as an ammonium salt by dissolving the compound in an acid. Details of these procedures vary considerably for different metals and with the objectives to be attained by the analysis. The principal steps, common to all procedures, are (a) sampling (b) decomposition of the sample, (c) removal of nitrogen from the sample, and (d) measurement of the nitrogen. Each will be considered separately prior to discussion of specific analytical procedures.

II. SAMPLING

The number of alloys in which nitrogen is a specified alloying constituent is relatively very small compared to the total number of alloy systems in which nitrogen is an adventitious component. The sampling procedures differ considerably depending on the alloy system and the purpose for which the analysis is to be made.

Sampling procedures such as drilling, milling, sawing, filing, and crushing are usually appropriate for sampling for nitrogen analysis. However, nitrogen often segregates in the matrix metal. Also, with reactive metals, it is easy to contaminate a sample with atmospheric nitrogen during sampling, as noted below.

In ferroalloys, nitrogen can be segregated between large and small lumps of the material. Nitrogen can also be nonuniformly distributed within a big lump of ferroalloy and between the material from the beginning and end of the casting or pouring of a ferroalloy. Obtaining a representative sample of this segregated material can present a real statistical problem. Often it is solved by agreement between producer and manufacturer that a specific sampling and riffling procedure will be used to produce a "representative" sample [12]. Titanium, zirconium, and alloys of these metals and alkali metals have a great affinity for nitrogen, and segregation is easily produced by a change in the partial pressure of nitrogen during production of sponge alloy or metal. Therefore, it is important to sample a heat, casting, or extruded shape at several locations and combine these samples as a representation of the metal. Errors can be easily introduced in the sampling process and care must be taken to prevent the metal from becoming hot and absorbing atmospheric nitrogen during milling, drilling, and cutting of the sample.

Sometimes solid nitrides are present in liquid metals used as heat exchangers. Because distribution of a solid in a liquid is nearly always nonuniform, obtaining a representative sample is not easy. Other nitrides may be soluble in the liquid metal, for which techniques described in Chapter 10 for sampling liquid metals should be employed.

Crushing refractory metal nitrides can lead to contamination of the nitrides with iron or metal carbide from the crusher or mortar. Only the hardest steel or carbides should be used for crushing operations. Sieving is recommended when crushing is employed in the sampling procedure, because crushing without intermittent sieving may result in too much fine material, which may contain a disproportionately high concentration of the nitrides in the analytical sample. Therefore, proportionate amounts of each of the size fractions should be included when the sample is weighed. In sampling pig iron, fine particles of nitride or carbonitride tend to adhere to the graphite [13a]. Often it is better to use coarse drillings or small chunk pieces of pig and cast iron to prevent segregation of nitrogen during sample preparation. Nitrogen content has been reported to vary with particle size in steel analysis [13b].

In most instances, surface metal is rejected in sampling. Cleaning the surface before sampling is usually accomplished by abrading or pickling.

III. STANDARD SAMPLES

For steel there are a number of standard samples that have certified values for nitrogen. Many are available from the U.S. National Bureau of Standards and the British Bureau of Analyzed Samples. Other steel standards or samples with stated nitrogen contents are available in France and Russia and through instrument manufacturers in the United States and Germany. Standard samples of titanium and zirconium alloys are also available from the National Bureau of Standards.

IV. DECOMPOSITION

A. GENERAL CONSIDERATIONS

Although gaseous nitrogen may be trapped in certain metals in voids or blow holes, occurrence of significant amounts in the elemental gaseous form is rare in metal of good integrity. For sampling and analytical purposes, the nitrogen can be considered to be present in the form of metal–nitrogen compounds. These metal–nitrogen compounds may be present as a compound of the matrix metal or as a compound of an alloy constituent of the matrix metal. Therefore, a variety of decomposition methods are used.

Both the metal (matrix or constituent) and the metal–nitrogen compounds may be extremely difficult to decompose, or one or both may be so reactive that it will combine with the atmosphere at room temperature. In many instances, this difference in chemical reactivity can be used to determine how the nitrogen is present in the matrix.

B. ACID DISSOLUTION

1. Kjeldahl

The method of Allen [14, 15], often referred to as the solution–distillation or Kjeldahl [16] method, probably is the most generally used method for separating and determining nitrogen in metals and in metal–nitrogen compounds. The distilled ammonia is usually determined by Nessler reagent photometry [17] or titrimetrically; the method of measurement is most often dictated by the amount of nitrogen in the sample. Usually the dissolution of the matrix is followed by a Kjeldahl-type digestion to liberate nitrogen bound in chemically stable nitrides.

The choice of solvent is generally dictated by the properties of the matrix metal and the nitrogen compounds present. Table 8.2 can be used as a guide in choosing a dissolving reagent. Nitric acid solutions cannot be used.

TABLE 8.2. Commonly Used Solvents for Metals and Alloys

Alloy	Solvents
Steels	HCl, HBr, HI, HF, HBF_4, H_2SO_4, H_3PO_4, $HClO_4$. See also Chapter 15.
Titanium, tungsten, and zirconium alloys	HCl, H_2SO_4, HF, HF–H_2SO_4, HBF_4–H_2SO_4. See also Chapters 17 and 18.
Niobium and tantalum alloys	HCl–H_2SO_4, HCl–HF–H_2SO_4, HCl–HF–H_2O_2. See also Chapter 18.
Alkali nitrogen compounds	H_2O, CH_3OH. See also Chapter 10.
Alkaline earth nitrides	H_2O, HCl. See also Chapter 11.
Miscellaneous metal nitrides	See Chapter 13.

2. HF–H_3PO_4–$K_2Cr_2O_7$

A method for determining low concentrations (1–50 ppm) of nitrogen in refractory metals has recently been developed by Kallman et al. [18]. The problems involved in such an analysis include minimizing the surface area of samples that may react with atmospheric nitrogen on crushing or milling, preparation of a fresh surface just prior to analysis, and minimizing the blank due to contamination of reagent chemicals. The method is applicable not only to niobium, molybdenum, tantalum, and tungsten, and their alloys but also to alloys of chromium, iron, cobalt, and nickel.

Massive, or "chunk," samples are abraded or pickled prior to analysis where necessary. The sample is then placed in a bottle made of fluorinated ethylene polymer and dissolved in K_2CrO_4 (converts to $K_2Cr_2O_7$ in acid), 5 ml H_3PO_4, and about 60 ml HF (HF + HCl for alloys of chromium). The amount of chromate added depends on the material being dissolved. The weight ratio of K_2CrO_4 to sample is 2 for Ta and W, 4 for Nb and Mo, and 5 for Co, Ni, and presumably Cr. (Oxidant is not needed for steel samples.) The bottle is loosely covered with platinum or a piece of polymer so that most of the acid will reflux. After complete dissolution of the sample, most of the HF is evaporated. The solution is diluted with hot water and transferred to a distilling flask, where conventional techniques are used to separate nitrogen as ammonia. Final determination is by indophenol photometry.

This method was used to determine 1–300 ppm nitrogen. Blanks were less than 0.5 μg nitrogen per gram of sample. One technique used to minimize the blank was the addition of KOH (3 g) to a solution of the K_2CrO_4 (700 g per 1000 ml water), followed by evaporation of the water under infrared heat until the solid appeared dry. The solid mixture was then transferred to a porcelain or Vycor crucible, gradually heated to 500°C, cooled, and stored in sealed bottles to prevent ingress of ammonia from the atmosphere prior to use.

3. Pressure Bomb

In the analysis of refractory metals, it is sometimes desirable to dissolve large solid cross sections to minimize inhomogeneity problems or because the production of chips by milling may result in reaction of the sample with nitrogen in the atmosphere as noted earlier. This problem is especially serious when the sample is low in nitrogen. Further, the long dissolution time required by large samples may result in contamination due to ammonia in the environment. To overcome these problems, Menis and co-workers [19] adapted a pressure-bomb dissolution technique used by Ito [20] and Bernas [21]. The sample is dissolved at about 150°C in a few milliliters of HF–HCl in a stainless-steel bomb lined with a cup made of fluorinated polyethylene.

C. CHEMICAL FUSION

Certain nitrides do not dissolve readily by acid attack, and a chemical fusion is often employed. This method, as noted earlier, has been used with various refractory nitrides (see Table 8.1). Such reagents as PbO, PbO + PbO_2, PbO + $PbCrO_4$, and Na_2O_2 are used with heat to decompose the nitride and liberate nitrogen in a flowing stream of carbon dioxide or helium. Chemical absorbers are used to remove carbon dioxide, sulfur dioxide, and other gases prior to the volumetric measurement of the nitrogen. Alternatively, nitrogen may be determined by gas chromatography. These are modified Dumas methods and have been recently discussed by various investigators [3, 9–11]. Attack of the sample with solid alkali hydroxides [1, 22, 23] has also been used with argon or hydrogen as the carrier gas; in such cases, ammonia is released for measurement.

D. SELECTIVE METHODS

Where the object of the analysis is to distinguish between the total nitrogen content and the nitrogen combined in the form of stable compounds, the problem becomes one of selecting a method for dissolving the metal matrix without decomposing the metal–nitrogen compounds. In simple alloy systems this will usually permit positive identification of the metal–nitrogen compound and quantitative measurement of the amount of nitrogen combined with each metallic element. In most useful engineering alloys, the system seldom is simple enough to permit analysis in this straightforward way. In the case of stable metal–nitrogen compounds in an alloy that is easy to dissolve, the problem of distinguishing between total nitrogen and nitrogen in stable compounds involves solution of the sample in acid, separation of the stable compound(s) by filtering or centrifuging, and measurement of the nitrogen content of the insoluble constituent(s). For the case of metal–nitrogen compounds that are easier to decompose with inorganic aqueous acids than is the metal matrix in which they are contained, other methods must be used to isolate the compounds. Displacement reactions, reaction of the metal matrix with anhydrous halogens at elevated temperatures, and various electrolytic procedures have been evaluated for this purpose. The methods are summarized in Table 8.3 [7]. In each case, the object has been to separate the matrix metal from the compounds, principally oxides, carbides, and nitrides or carbonitrides. Most of these procedures have been successfully used for isolating oxides and carbides and the more stable, or "refractory," nitrides.

Solutions of halogens in anhydrous aliphatic esters of organic acids have provided a means for separating aluminum nitride from a steel matrix [5]; bromine dissolved in methyl acetate has been used most extensively. The

TABLE 8.3. Types of Reactions Used for Isolating Nitrogen Compounds from Metals[a]

I. Displacement (neutral salts) using aqueous solutions
II. Halogens
    1. Aqueous solutions
    2. Organic solutions
    3. Anhydrous elements at elevated temperatures
III. Inorganic Acids
    1. Aqueous solutions
    2. Anhydrous halogen acids at elevated temperatures
IV. Electrolytic with aqueous electrolytes

[a] Reprinted by permission from *Analytical Chemistry* [7].

TABLE 8.4. Metals Soluble in Ester–Halogen Reagents

Aluminum	Niobium
Beryllium	Tin
Chromium	Titanium
Cobalt	Uranium
Iron	Vanadium
Manganese	Zinc
Nickel	Zirconium

reagent will dissolve a number of metals and alloys in addition to iron and many iron–base alloys. Table 8.4 lists metals known to be soluble in this reagent. The procedure used for solution of steels and some precautions to be observed in use of the reagents are given later.

Bandi and co-workers [24, 25] have chemically isolated nitrides or carbonitrides from steel and have analyzed various mixtures of these compounds along with metal carbides and oxides by differential thermal analysis–evolved gas analysis (see also Chapter 15). The evolved nitrogen was determined by gas-chromatographic techniques. Meyer et al. [26a] analyzed the nitrides of Mo, Nb, Ti, V, and Zr by heating at 1100°C in excess oxygen in a tube furnace, followed by gas-chromatographic measurement of the evolved nitrogen.

V. MEASUREMENT OF NITROGEN

A. TITRATION OF AMMONIA

Ammonia separated by distillation after a Kjeldahl digestion is often trapped in a standard-acid solution, and the excess acid is titrated with a standard base. A large number of acid–base indicators have been used but the choice is usually a matter of personal preference. More care must be taken in the distillation of ammonia for an acid–base titration than for a photometric measurement, because of the possible physical carry-over of caustic solution into the receiving flask by too rapid a distillation (see also Section VI.A). Direct distillation has been used [26b–c], but most workers use steam distillation.

B. AMMONIA: NESSLER REAGENT PHOTOMETRY

The most common method of measuring low concentrations of nitrogen as ammonia employs the use of Nessler reagent to form a collodial suspension of Hg_2ONH_3. This yellow orange compound is then measured photometrically. The reaction is thought to be

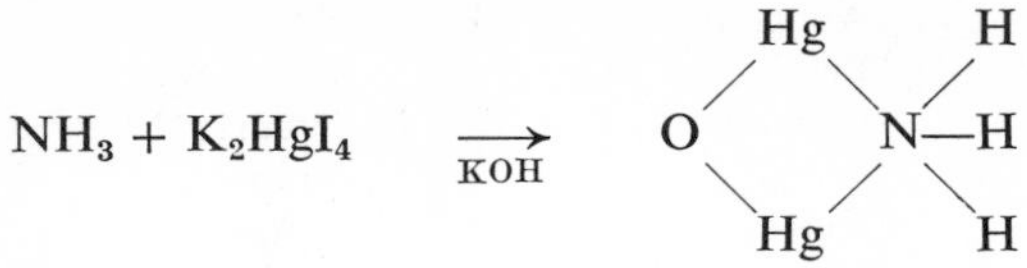

$$NH_3 + K_2HgI_4 \xrightarrow[KOH]{} O\langle^{Hg}_{Hg}\rangle N\langle^{H}_{H}\!-\!H$$

It is simple to carry out (see also Section VI.A).

C. AMMONIA: INDOPHENOL PHOTOMETRY

The indophenol method for nitrogen as ammonia has been extensively investigated by Kammori et al. [27] and more recently by Jenkins [28] in the analysis of steel. The latter was a group study involving a number of representatives of British steel companies; this work has been reviewed by Scholes [29]. An advantage of indophenol over Nessler reagent is that on reaction of the former with ammonia, a true solution is obtained.

In the work of Kammori et al. [27], phenol and sodium hypochlorite were added to the neutralized distillate containing ammonia. After a period of 10 min, the solution was saturated with sodium chloride, and the blue compound was extracted with isobutanol. The absorbance was measured at 660 nm.

In the British study [28], initial work showed that the sodium phenoxide reagent was unstable. When sodium nitroprusside was added to the reagent, there was rapid deterioration. When separate solutions of phenol (50 g/l.), sodium nitroprusside (0.1 % w/v), and sodium hypochlorite–sodium hydroxide (2.1 g NaClO and 25 g NaOH per liter) were used, stable and accurate results were obtained. The sodium nitroprusside solution must be prepared freshly as required but should not be used until 1–2 h after preparation. All three of these solutions should be stored in amber glass bottles. The use of a separate sodium nitroprusside solution is based on European Standard 50 [30].

An interesting observation was made concerning the blank. It was noted that the blank was higher when iron was present and reacted with the sulfuric acid, the additional blank being caused not only by the iron but presumably by the reaction of the generated nascent hydrogen with nitrates and other nitrogen forms in the sulfuric acid to form ammonium salts. Because high-quality sulfuric acid could not always be obtained and because of the mechanical difficulty of fuming sulfuric acid in the presence of salts, the procedure finally developed included:

1. Solution of the sample (1 g) plus 0.25 g of high-purity iron in hydrochloric acid in a centrifuge tube, with a second centrifuge tube containing 0.25 g of the high-purity iron treated likewise (for determination of the blank).

2. Addition of 0.25 ml H_2SO_4 (20% v/v) and 0.50 ml $BaCl_2$ solution (10% w/v) followed by separation of insolubles by centrifuging and decanting of the supernatant liquid.

3. Addition of 5.0 ml H_2SO_4 to each centrifugate and fuming at 320°C for 60 min.

4. Separation of nitrogen as ammonia by steam distillation into a 100-ml flask.

5. Addition of 8 ml phenol solution, 8 ml NaClO–NaOH solution, and 2 ml sodium nitroprusside solution.

6. Measurement of absorbance at 635 nm in a 1-cm cell after 40 min.

The use of the high-purity iron with and without the sample obviates the need to use denitrogenized iron in determining the blank. Only one batch of iron should be used for any given series of determinations.

Over the range 9–43 ppm nitrogen, the reproducibility at the 95% confidence level varied from 2 to 3 ppm. The detection limit was 2 ppm nitrogen.

D. AMMONIA: CHLORAMINE T (WITH THYMOL OR BIS-PYRAZOLONE) PHOTOMETRY

Goto et al. [31] have studied the photometric determination of ammonia after its reaction with thymol and chloramine T to form the blue indothymol which was extracted into isoamyl alcohol. The following parameters were investigated: (1) volumes of 3 *N* NaOH, 10% thymol–ethanol and 5% chloramine solutions used; (2) stability of the chloramine T solution; (3) heating time at 100°C to develop the colored product; and (4) effect of Fe, Mo, V, Cu, W, Cr, Ni, and Co.

In the analysis of steel, the sample was dissolved and the bulk of the iron was separated by precipitation with 6 *N* NaOH solution when the nickel, cobalt, and copper contents were low. (The distillation technique is to be used for high amounts of these elements.) To the separated ammonia contained in a 50-ml flask, 2 ml of 3 *N* NaOH solution, 2 ml of 10% thymol–ethanol solution, and 8 ml of 5% chloramine-T solution were added. The solution was diluted to 50 ml, heated at approximately 100°C for 13–15 min, cooled, and extracted with isoamyl alcohol. The absorbance was measured at 680 nm. Limited data showed good agreement with certified values for the range 0.005–0.068% nitrogen.

Kawamura et al. [32–34] investigated the reaction of ammonia with bis-pyrazolone and chloramine T to form rubazoic acid. In aqueous solution, this method is five times as sensitive as that with Nessler reagent. The effects of varying volumes of chloramine T, bis-pyrazolone and pyrazolone were studied along with the pH and standing times used at two different points in the procedure.

In the analysis of steel, a 1-g sample was dissolved in 20 ml of (1 + 1) HCl. The insolubles were removed by filtration (insoluble nitrides presumably were not encountered or were ignored), the filtrate going directly to a distillation flask; 50 ml of distillate were collected in a 100-ml flask. Five ml of pH-4.8 buffer (to achieve a pH of 5.3 ± 0.2) and 2 ml of 1% chloramine-*T* solution were added, and the solution was allowed to stand exactly 2 min at room temperature. Then 5 ml of 0.2% bis-pyrazolone solution (0.100 g of reagent in 50 ml of 0.5 *N* Na_2CO_3 solution) was added, and the solution was diluted with 10 ml water to obtain a pH of 9.9 ± 0.2. Following a period of 5 min, 10 ml of 0.25% (weight per volume) pyrazolone solution was added. After the blue purple color changed to red purple, the absorbance was measured (absorption maximum at 540 nm). For less than 5 μg nitrogen, the colored compound was extracted into 20 ml CCl_4, 1 ml of 0.5 *N* HCl was added, the mixture was again agitated, and the organic layer was separated. The absorbance was measured at 450 nm. The method is mainly applicable to less than 70 μg nitrogen. Excellent precision was shown at both the 3- and 21-μg-nitrogen levels.

E. MISCELLANEOUS

The determination of ammonia by electrical-conductivity measurements has been studied by Meunier and Demarez [35], who absorbed the distillate in dilute hydrochloric acid. Kawamura et al. [34] made direct conductometric measurements on the distillate and suggested that this method could be automated. Oelsen and Sauer [36] determined ammonia by absorbing it in barium perchlorate solution, coulometrically generating acid, and potentiometrically measuring the endpoint. A suggested technique for measuring ammonia in the distillate is that utilizing an ammonia gas-sensing electrode, such as is available from Orion Research, Inc., Cambridge, Massachusetts. For references to the gas-chromatographic determination of nitrogen, see Section IV.D.

VI. SELECTED METHODS FOR DETERMINING NITROGEN

A. TOTAL NITROGEN: KJELDAHL

1. Scope

The Kjeldahl method covers the determination of total nitrogen in concentrations greater than 0.0005%. It utilizes the solution–distillation technique and is based on the conversion of nitrogen to ammonia by treatment with an acid that will dissolve the matrix and ideally also dissolve the nitride phase [4]. The method is suitable for the determination of nitrogen accurately and rapidly in steels, niobium, titanium, and zirconium and other alloy systems that will liberate their nitrogen content as ammonia by acid treatment. Sometimes it is necessary to use a further fusion or digestion step to dissolve the nitride phase. This method includes a Kjeldahl treatment by which many, but not all, metal nitrides can be dissolved; other digestions have to be used for metal nitrides that cannot be dissolved by this treatment.

2. Apparatus

1. Kjeldahl digestion apparatus
2. Steam distillation apparatus (Fig. 8.1*a* and 8.1*b*)
3. Spectrophotometer.

3. Reagents

1. Dissolving reagents: Refer to Table 8.3 or to another chapter to determine what reagent is needed.

2. Ammonia-free water: Prepare by passing distilled water through a column containing a mixture of ion-exchange resins such as two parts by volume of Amberlite IRA 400 with one part of Amberlite IR 120. Use wherever water is specified.

3. Standard sulfuric acid, 0.01 *N* or 0.1 *N*.

4. Standard sodium hydroxide solution, 0.01 *N* or 0.1 *N*.

5. Sodium hydroxide solution, 33%, ammonia-free. Transfer 3000 ml distilled water to a 4000-ml or larger stainless-steel beaker, and mark the liquid level on the beaker. Add 1000 g NaOH and dissolve. Add 100 ml $BaCl_2$ solution (10 g $BaCl_2$/100 ml water) and 5 g Devarda's alloy and 700 ml water. Digest for a few hours below the boiling point, and then boil down to the 3000-ml mark. Cool and dilute to the mark with water. After the precipitate has settled, decant the clear liquid into a plastic bottle with a screw cap.

6. Nessler reagent: Dissolve 50 g KI in 35 ml water. Add a saturated solution of $HgCl_2$ slowly with constant stirring until a slight red precipitate of mercuric iodide persists. Add 400 ml 9 *N* NaOH solution, mix, dilute to 1 liter and allow to stand at least 24 h. Decant the clear supernatant liquid for use in the color development. The reagent is stable indefinitely.

7. Methyl purple acid–base indicator.

8. Nitrogen, standard solution (1 ml = 0.01 mg *N*): Dissolve 0.3832 dry NH_4Cl in 500 ml water, transfer to a 1-liter volumetric flask, dilute to volume and mix. Using a pipet, transfer 100 ml of this solution to a 1-liter volumetric flask, dilute to volume, and mix.

4. Procedure

Transfer an appropriate sample-weight, depending on the nitrogen content and whether the final measurement will be photometric or titrimetric, to a beaker or digestion flask. (Run a blank simultaneously.) Be sure that proper precautions are taken for those samples that may hydrolyze. Add an appropriate dissolving reagent and digest. Connect the digestion flask to the steam distillation unit, Fig. 8.1*a* for titrimetric determination or Figure 8.1*b* for photometric determination. The nitrogen distilling head (G) in Fig. 8.1*a* serves to trap any sodium hydroxide solution entrained with the ammonia.

a. Titrimetric Determination

Place 10 ml water, 4 drops methyl purple indicator, and sufficient standardized H_2SO_4 in a receiving flask (K in Fig. 8.1*a*) to ensure that an excess of the H_2SO_4 will be present after distillation of ammonia. To the Kjeldahl or Florence flask (F) containing the sample solution, add 125 ml 35% NaOH solution (or enough to provide an excess of NaOH) through the delivery tube and immediately close off the delivery tube. Heat both the water-flask B and sample-flask F so as to rapidly distill 100 to 110 ml of solution, as indicated by a mark on the receiving flask. After the distillate is collected, remove the receiver and heaters. Titrate the excess H_2SO_4 in the receiver with standardized NaOH solution until the color in the receiving flask changes from purple to green.

To determine the nitrogen in any undissolved material, remove the distillation flask, neutralize the excess NaOH, and acidify the solution with HCl. Filter through an ammonia-free asbestos mat. Wash with hot ammonia-free water. Transfer the mat to the sample Kjeldahl flask (F in Fig. 8.1*a*). Add 3 g $K_2S_2O_7$, 5 ml H_2SO_4, and digest initially at low heat. Then increase the heat so that the H_2SO_4 will reflux for 15 min after a clear solution is observed. Distill and titrate the ammonia as directed previously.

Determine the percent nitrogen from the volumes of standard H_2SO_4 and standard NaOH solution used. Calculate the nitrogen content as follows:

$$\text{Sulfuric acid equivalent } R = \frac{\text{Volume } H_2SO_4}{\text{Volume NaOH}} = \frac{V_A}{V_B}$$

$$\text{Blank, } K = V_{AK} - V_{BK}R$$

$$\%\text{N} = \frac{(V_{AS} - V_{BS}R - K)N_A \times 0.01401 \times 100}{S},$$

where N_A is the normality of H_2SO_4 and S is the sample-weight in grams.

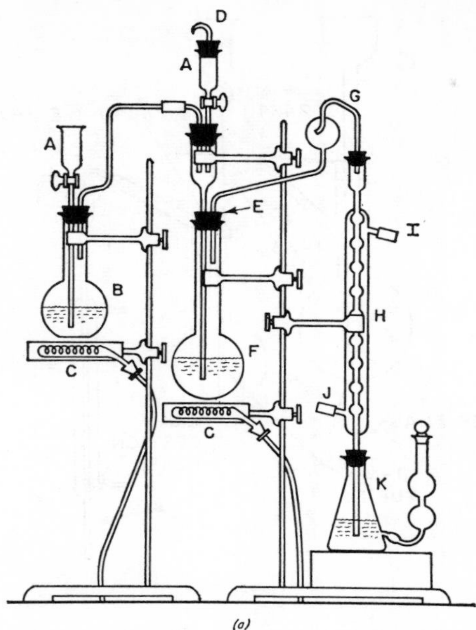

Fig. 8.1(a). Macro-steam-distillation apparatus. (Reproduced by permission from American Society for Testing and Materials [12].) A, 125-ml funnel. B, 1-liter Florence flask. C, Electric heater. D, Glass hook for hanging up stopper. E, Special Kjeldahl rubber stopper. F, 800-ml long-neck Kjeldahl flask. G, Nitrogen distilling head. H, Condenser. I, Water outlet. J, Water inlet. K, Modified Volhard nitrogen receiving flask.

. *Photometric Determination*

Prepare a calibration curve as follows: Using pipets, transfer 1, 2, 5, and 10 ml of standard nitrogen solution (1 ml = 0.01 mg N) to 50-ml volumetric flasks, and dilute each solution to approximately 45 ml with water. Transfer 45 ml water to a 50-ml volumetric flask as a blank. Add 1.0 ml Nessler reagent to each volumetric flask, dilute to volume, and mix. Allow the solution to stand for at least 5 min for full color development. Use water in the reference cell. Measure the reagent blank (which includes the cell correction) using absorption cells with a 1-cm light path and a light band centered at approximately 430 nm. Using the test cell, take the photometric readings of the calibration solutions. Plot the net photometric readings of the calibration solutions against milligrams nitrogen per 50 ml of solution.

To determine the nitrogen content of the sample, place a 50-ml volumetric flask under the condenser shown in Fig. 8.1*b*. Connect the sample or digestion flask to the micro-steam-distillation unit, and add 125 ml 33% sodium

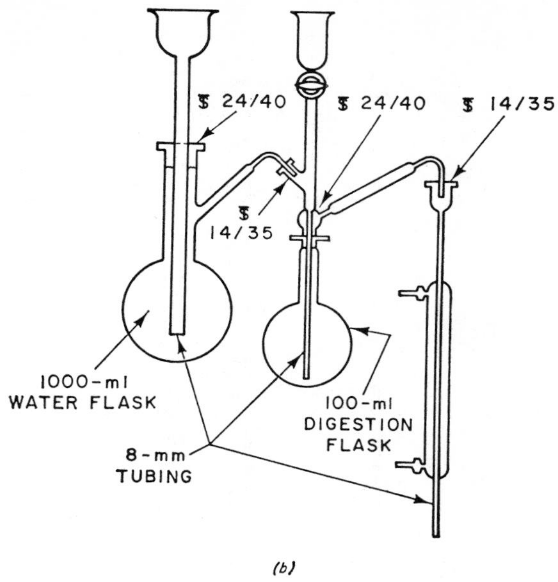

Fig. 8.1(b). Micro-steam-distillation apparatus (for photometric determination only). (Reproduced by permission from American Society for Testing and Materials [12].)

hydroxide solution to it. Steam distill 45 ml of the solution into the 50-m flask. Add 1 ml of Nessler reagent, and develop and measure the color a previously described. If there are any undissolved nitrides, treat the solutio in the sample flask as described in the titrimetric procedure. Convert th photometric readings to milligrams nitrogen by reference to the calibratio chart. Correct the nitrogen value for that of the blank.

Calculate the nitrogen content as follows:

$$\%\text{N} = \frac{A}{10B}$$

where A is milligrams of nitrogen from the calibration curve and B i grams of sample.

B. SELECTIVE METAL NITRIDES: BROMINE–METHYL ACETATE

1. Scope

This method is applicable to systems where the matrix metal is soluble and the metal nitride of interest is not.

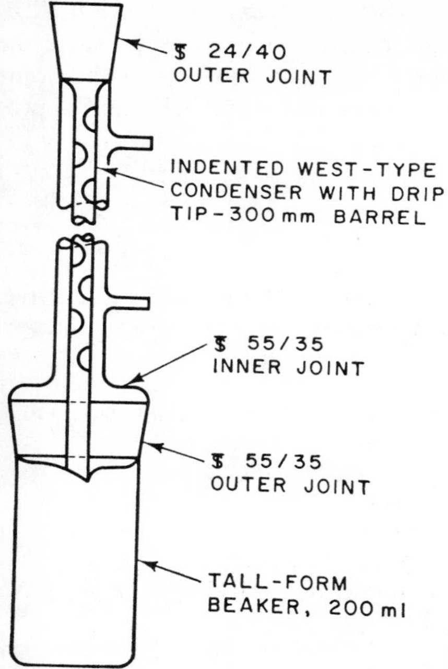

Fig. 8.2. Ester-halogen extraction apparatus. (Reproduced by permission from *Analytical Chemistry* [5].)

2. Apparatus

1. Dissolving apparatus: See Fig. 8.2.
2. Distillation apparatus: See Fig. 8.1*a* or 8.1*b*.

3. Reagents

1. Reagent-grade bromine.
2. Water-free methyl acetate.

4. Procedure

Transfer to the beaker in Fig. 8.2 a sample of 0.1–3.0 g, depending on the nitrogen content and the method of determination. Attach the condenser, support it as described [5], and turn on the cooling water. For each gram of sample or fraction thereof, add 3 ml bromine through the condenser. Cautiously add 2 ml methyl acetate through the condenser to initiate the reaction. (*Caution:* This reaction is vigorous and exothermic and is controlled by the rate of methyl acetate addition.) When the initial violent reaction subsides, continue the addition of methyl acetate in 2 ml portions until 15 ml has been added for each gram of sample. Heat the solution at reflux conditions until the metal matrix dissolves and only the nitride phase is left. Cool and filter the nitride using an asbestos [*Caution:* see Chapter 15, Section VI.B. 5.b.(4)] mat, filter paper, or appropriate organic membrane. Dissolve the nitride by applying an appropriate analytical procedure. (*Caution:* Some nitrides may decompose on contact with water vapor in the air, and appropriate precautions should be taken.)

REFERENCES

1. E. W. Beiter, R. H. Wynne, W. F. Harris, M. I. Mistrik, and F. A. Byrne, Pittsburgh Conference on Analytical Chemistry and Applied Spectroscopy, Pittsburgh, 1957.
2. A. Ul'yanov, *Zavod. Lab.*, **34,** 1442 (1968).
3. O. H. Kriege, Los Alamos Report 2306, Aug. 27, 1959; available from Office of Technical Services, U.S. Dept. Commerce, Washington, D.C.
4. H. F. Beeghly, *Ind. Eng. Chem., Anal. Ed.*, **14,** 137 (1942).
5. H. F. Beeghly, *Anal. Chem.*, **21,** 1513 (1949).
6. H. F. Beeghly, *Anal. Chem.*, **24,** 1095 (1952).
7. H. F. Beeghly, *Anal. Chem.*, **24,** 1713 (1952).
8. H. F. Beeghly, "The Determination of Nitrides in Metals," in *Spec. Rept. No. 68*, Iron and Steel Institute, London, 1960, pp. 183–91.
9. A. A. Federov and A. M. Krischevskaya, *Zavod. Lab.*, **34,** 1425 (1968).
10. E. Jaudon, *Chim. Anal.*, **51,** 59 (1969).
11. J. Rottman and H. Nickel, *Z. Anal. Chem.*, **247,** 208 (1969).
12. ASTM Standards, Pt. 32, American Society for Testing and Materials, Philadelphia, 1972.
13. (a) L. I. Levi and O. M. Borisova, *Zavod. Lab.*, **32,** 414 (1966); and (b) T.-K. Willmer and K. Zimmerman, *Arch. Eisenhuttenw.*, **42** (12), 8 (1971).
14. A. H. Allen, *J. Iron Steel Inst.*, **7,** 480 (1879).
15. A. H. Allen, *J. Iron Steel Inst.*, **8,** 181 (1880).
16. J. G. C. T. Kjeldahl, *Z. Anal. Chem.*, **22,** 366 (1883).
17. M. J. Taras, "Nitrogen," in D. F. Boltz, Ed., *Colorimetric Determination of Nonmetals*, Interscience, New York, 1958, pp. 75–160.
18. S. Kallman, E. W. Hobart, H. K. Oberthin, and W. C. Brienza, Jr., *Anal. Chem.*, **40,** 332 (1968).
19. O. Menis, *NBS Technical Note 454*, July 1968, pp. 72–74.
20. J. Ito, *Bull. Chem. Soc. Jap.*, **35,** 225 (1932).
21. B. Bernas, private communication, 1967, in Ref. 19.
22. P. Klinger, *Zavod. Lab.*, **3,** 20 (1933).
23. H. S. Karp, L. L. Lewis, and L. M. Melnick, *J. Iron Steel Inst.*, **200,** 1032 (1962).
24. W. R. Bandi, W. A. Straub, E. G. Buyok, and L. M. Melnick, *Anal. Chem.*, **38,** 1336 (1966).
25. G. Krapf, W. R. Bandi, E. G. Buyok, and L. M. Melnick, Pittsburgh Conference on Analytical Chemistry and Applied Spectroscopy, Cleveland, Ohio, 1972.

(a) R. A. Meyer, E. P. Parry, and J. H. Davis, *Anal. Chem.*, **39,** 1321 (1967); (b) *Spec. Rept. No. 62*, Iron and Steel Institute, London, 1958, pp. 3, 4; (c) E. Jaudon, *Rev. Met., Mem.*, **53,** 688 (1956); discussion of paper by J. Calmette and J. Drain, *ibid.*, p. 682; through Iron and Steel Inst. Translation No. 523.

O. Kammori, Y. Hiyama, and W. Hotta, *Trans. Jap. Inst. Metals*, **8,** 56 (1967); through *Chem. Abstr.*, **67,** 50095 (1967).

R. H. Jenkins, *The Determination of Small Amounts of Nitrogen in Steel*, Brit. Steel Corp. Rept. MG/CC/520/72 and Addendum, 1972.

P. H. Scholes, "Special Analytical Techniques and Their Application in Quality Control," in *The Determination of Chemical Composition; Its Application in Process Control*, Iron and Steel Institute, Preprint 131, London, 1971, pp. 129–140.

M. Jaudon, European Standard 50, Document CECA GT 20/128, 1964; through Ref. 28.

H. Goto, Y. Kakita, and I. Atsuya, *Sci. Repts. Res. Inst. Tohoku Univ.*, **A-Vol 19,** no. 1, 50–58 (1967).

K. Kawamura, S. Watanabe, T. Otsubo, and S. Goto, *Fuji Seitetsu Giho*, **17,** 37 (1968); through *Chem. Abstr.*, **70,** 83995 (1969).

K. Kawamura, S. Watanabe, T. Otsubo, and S. Goto, *Bunseki Kagaku*, **17,** 637 (1968); through *Chem. Abstr.*, **69,** 56780 (1968).

K. Kawamura et al., *Fuji Iron Steel Tech. Rept.*, **17,** 37 (1968); through Iron and Steel Inst. Translation No. 7049.

F. Meunier and A. Demarez, *Rev. Met.*, **9,** 647 (1952).

W. Oelsen and K. H. Sauer, *Arch. Eisenhuttenw.*, **38,** 141 (1967); through Iron and Steel Inst. Translation No. 5613.

CHAPTER

9

OXYGEN

LOUIS SILVERMAN

Formerly Atomics International
Canoga Park, California

AND

BERNARD D. LA MONT,*

Formerly Westinghouse Electric Corp.
Pittsburgh, Pennsylvania

CONTENTS

* Deceased.

I. INTRODUCTION

The purpose of this chapter is to survey, describe and discuss techniqu and equipment that are used in chemical determinations of oxygen in meta Fusion methods, activation methods and others, including the procedur details, for the determination of oxygen in specific metals appear elsewhe in this book.

Chemical methods are especially useful where there is need for dete mining the concentrations of different forms of oxygen (e.g., surface oxyg and internal oxygen) in a metal; such determinations usually require chemic separations. Chemical methods may also be employed to advantage for t determination of total oxide in certain metals. Examples are brominati methods for the determination of oxygen in aluminum or bismuth, ar distillation methods for the determination of oxygen in zinc or magnesiur

Other reasons for choosing chemical methods, in preference to fusio spectral, or activation methods, may be because of unavailability and cost special equipment or the infrequency with which the analysis is required.

II. OXYGEN IN METALS

A. FORMS OF OXYGEN

Oxygen is often present in a metal as an oxide of the base metal or of so minor constituent in the metal. Oxygen may also be present in solid soluti or as uncombined molecules trapped in voids. Zirconium [1] and sodium [2 for example, when rapidly cooled from the molten state may contain u combined and unreacted oxygen that will evade determination by mo methods.

If the oxygen is present as a stoichiometric oxide distributed through t metal or concentrated at the surface, most of the methods in this book shou eld reliable data. If the oxygen is present not as a true oxide but as a hy-'ated oxide, hydrated hydroxide, or carbonate, some methods may produce accurate results. For analysis of specific metals the reader is referred to the apters covering these metals.

THERMAL STABILITY OF THE METAL OXIDES

Many tables and graphs are available for thermodynamic calculations that oint out the likelihood of chemical reactions. Turovtseva and Kunin [3] ave included a table containing standard free energies for the formation of ides and their reduction by carbon and hydrogen. This table indicates the ermal stability of oxides and the optimum temperature ranges for carbon d hydrogen reduction of the oxides. Honig [4] has published melting and oiling points and extensive vapor pressure data for the more common ements (see Chapter 1). Such data are useful in the determination of oxide vacuum or reduced-pressure distillation of the metal. Graphs in Glassner's port [5] are helpful in the determination of oxides in sodium, magnesium, lcium, and zinc. This report also shows that above 750°C carbon monoxide more stable than carbon dioxide. Such information is pertinent to under-anding the carbon–bromine and carbon–chlorine methods (Section III.D).

. SAMPLE PREPARATION

The problems of sample-preparation for oxygen determination are detailed the following chapters of this volume. There are some points, however, that eserve general consideration.

The amount of surface oxide, which may be appreciable, varies with the etal and its treatment and must be considered before analysis. In some ses determination of only the surface oxide may be desired. Alternatively, ternal oxygen (exclusive of surface oxide) or total oxygen (including surface ide) may be wanted. For the latter, each replicate sample must be of entical dimensions; this is especially important where there is an oxide oncentration gradient varying with depth. Instructions for preparing specific mples are given in the chapters devoted to the respective metals.

III. CHEMICAL METHODS

. AQUEOUS SOLUTION

There are limitations to methods using aqueous solutions (acids, bases, d other inorganic compounds) for the separation of oxides. For example, such methods are not applicable to metals that contain uncombined oxyg[1]. Also, the aqueous solvent that is used must dissolve the metallic matr but not attack the oxide.

Beeghly [6] listed several reagents that have been used for isolation nonmetallic compounds from metals by displacement in aqueous solutior For steels, these aqueous reagents do not selectively separate oxides fro each other nor from other compounds. Furthermore, oxides such as aluminu oxide, silica, and tantalum oxide, are separated from the metal, but so a certain nitrides and carbides, some sulfides [7], and graphitic carbon. Aft the preliminary separation, individual elements are determined phot metrically and the respective oxides are calculated (see Chapter 15 for procedure on isolating the compounds in a solution of bromine and meth acetate or methanol).

Measurement of the hydrogen evolved after treatment of aluminum wi alkali, zinc with acid (plus a platinum catalyst), or sodium with water h been used as an indirect method for determining oxygen, assuming that t specific metal oxide is the only (or major) contaminant in the metal [8 The solubility of hydrogen in the solution is a large source of error for whi correction must be made.

During an investigation of the effect of oxygen on the properties of electr deposited chromium metal, the Adcock chemical method [9] for the dete mination of oxygen was compared with the vacuum-fusion method. Briefl in the Adcock procedure [10], the sample is first heated to 950°C in vacuur cooled, and dissolved in 10% hydrochloric acid at 80–90°C. The solution cooled and filtered; the insoluble residue is dried, ignited at 1000°C, cool and weighed as chromic oxide. Controlling factors appeared to be the tin and temperature of heating in vacuum, and the temperature and ac concentration of the dissolving solution. Although the purpose of the he treatment was to aggregate oxide particles, the heating also dehydrat partly hydrated chromic oxide. This was verified by Makariewa and Biryu off [11], who determined that the gas given off during the heat treatment w 5% water vapor and the balance almost completely hydrogen. Thus, when nonheated sample was analyzed by vacuum fusion, the results were high than those obtained by the Adcock method, because of reduction of water t graphite.

A procedure of acid dissolution, filtration, and ignition can be used determine alumina in sintered aluminum. Samples of about 1–3 g m produce oxide residues of about 50 mg. The solvent is usually (1 + hydrochloric acid containing an oxidant such as nitric acid or bromine. stronger acid solution may dissolve some of the alumina [12]. The temper ture and the time of contact should be minimized. Salt solutions such as tho of cupric chloride or ferric chloride have been used, instead of the stro cid, to reduce the solubility effects on the oxides. Filtration through a ne-textured paper or asbestos pad (pore size about 5 μ) is usually inadequate, ut duplicate results that agree within 0.05% can be obtained using 0.3-μ millipore or equivalent) filter disks. After filtration the residue is washed vith (1 + 20) nitric acid and ignited to constant weight. Silica may appear n the residue for which correction must be made.

A similar procedure of alkaline dissolution, filtration, and ignition can be sed to determine alumina in sintered aluminum. The solvent is usually 0–75 ml of 20% sodium hydroxide solution containing a small amount of romine. Dissolution time is about 45 min. The filtration problem is the ame as for that in the acid-solution procedure. The solution must be diluted vith water to reduce the sodium hydroxide concentration to less than 2.5%, ooled below 65°C, and filtered through a 0.3-μ disk. Considerable washing vith (1 + 10) nitric acid is required to remove adhering foreign salts and odium ions. The residue, after ignition and weighing, should be checked for mpurities. Results tend to be 0.2–0.3% higher than those obtained by the cid-solution methods.

3. ELECTROLYTIC SEPARATION

The matrix metal (as anode) is usually isolated in a porous cup or bag, vhich is immersed in an electrolyte called the *anolyte*. The anolyte, in turn, s usually contained in a porous cup or cylinder that contacts a liquid called he catholyte. The catholyte either surrounds the porous cup or contacts nly one wall of the porous cylinder. The cathode may be carbon or a metal hat is inert under the conditions of electrolysis. Solution of the matrix is ffected anodically at constant potential, and the insoluble oxide inclusions re collected by appropriate means.

Electrolytic solution methods have two main advantages over chemical olution methods. The sample need not be dissolved completely; the use of arge samples facilitates the isolation of small amounts of inclusions. Further, vhereas chemical methods require solvents that may dissolve certain oxides, lectrochemical methods utilize weaker solvents with the electric potential roviding part of the driving force for dissolution of the matrix. A disadvantage of the electrochemical method, especially in the case of steels, is that ther compounds, such as metal carbides and nitrides, are isolated with the xides, making determination of the oxides difficult or impossible. Beeghly 6] listed a number of electrolytes for isolating compounds from steel. Koch 13] described the technique in detail and gave numerous examples of its pplication in steel analysis.

C. DRY CHLORINATION AND HYDROCHLORINATION

This method, mainly applicable to titanium, requires a controlled atmos phere furnace. The maximum temperature required depends on the matri and the oxides to be isolated. It must not be exceeded, otherwise decomposi tion of the oxides will occur. After the sample is loaded into a ceramic boat the atmosphere is vented with dried argon or helium, and oxygen-fre chlorine is introduced. At a controlled temperature, the base metal (titaniun or zirconium) is completely volatilized as the metal chloride, and the residu is analyzed for the metallic portions of the respective oxides. Carbon usually present in titanium and zirconium, causes low results by reactin with the oxide residue to form carbon monoxide.

Colbeck, Craven, and Murray [14–17] applied the method to steels Ferrous and manganous oxides were attacked and low results were obtained (It is not known whether moisture and/or hydrogen chloride were present. Stable carbides and nitrides contaminated the residues and required corresponding correction of the results. Basic chlorides of molybdenum tungsten, and vanadium, if present, may volatilize and cause low results.

The dry hydrogen chloride (hydrochlorination) method has been applie to beryllium, iron, aluminum, titanium, and zirconium. In the procedure fo aluminum, hot dry hydrogen chloride reacts with aluminum metal to forn anhydrous aluminum trichloride (and also the dimer), which volatilize above 183°C. Ignited alumina is not affected. Lowenstein [18] obtaine consistent results with the aluminum metal then available (0.1% oxide) whereas with the dry chlorine procedure [19] he obtained high results. H noted that the boats were scored and had lost weight. Use of preignite alumina boats for the dry chlorine procedure gave results in agreement wit the hydrogen chloride procedure.

Brook and Waddington [20] used hydrogen chloride to volatilize aluminun and at first obtained slightly high results due to oxygen impurity in th hydrogen chloride. Eventually they used a (10 + 1) mixture of hydroge chloride and hydrogen, which they passed over platinized quartz (at brigh red heat), through drying agents, and finally over the sample at 250°C Urech et al. [21] also used the hydrogen chloride–hydrogen method fo refined, cast, and remelted aluminum samples. Because of the small amoun of insoluble residues obtained on analysis, they fused the residues an determined aluminum colorimetrically. Comparison with Werner's [22, 23 bromine–alcohol and iodine–alcohol procedures showed that the result obtained by the latter procedures were higher.

Finally, Federov and Linkova [24] compared the dry hydrogen chlorid method for the determination of Al_2O_3 in aluminum metal with technique using solutions of copper(I) chloride, oxalic acid, citric acid, and aceti acid, and obtained better results by the volatilization method. They preferred 0.5-g samples, quartz boats, and quartz tubes and allowed four hours at [illegible]00°C for complete volatilization.

D. CARBON–BROMINE AND CARBON–CHLORINE

A disadvantage of the dry chlorination (or bromination) method is the noticeable loss of residual oxygen as carbon monoxide when carbon is present. The addition of excess carbon, as graphite, results in the complete removal of oxide oxygen as carbon monoxide with either chlorination or bromination. Using argon as a carrier gas, volatile metal halides and the halogen gas are removed in a dry-ice trap and/or Ascarite. Carbon monoxide is then oxidized to carbon dioxide with hot copper oxide and appropriately measured (conductimetrically, gravimetrically, manometrically, or gas-chromatographically). Molybdenum and tungsten may interfere by forming oxy-salts.

Codell and Norwitz [25–27] and Codell et al. [28] used the carbon-bromination method for the determination of oxygen in titanium, zirconium, chromium, vanadium, iron and steel. For titanium the results agreed to within 0.02% with those obtained by vacuum fusion. The method is not applicable to steels containing alumina, silica, and possibly aluminates or silicates. Both the bromine and argon must be purified. Blanks may be attributable to the graphite, to the platinum, gold, or graphite boats, and to the ceramic tubes. Miyamoto et al. [29] used a micro-apparatus, and determined oxygen in zirconium in the range of 0.02–0.17%; the results were checked by vacuum fusion.

Elwell and Peake [30] described the carbon–chlorine procedure for the determination of oxygen in titanium and titanium alloys. Tabulated results by this method and by the vacuum-fusion method showed good agreement. The reaction with chlorine is faster than with bromine, and larger samples can be used in the determination of small amounts of oxygen.

E. BROMINE OR IODINE IN ORGANIC MEDIUM

A solution of bromine in an organic solvent such as anhydrous methyl acetate or methanol easily dissolves aluminum, beryllium, bismuth, uranium, zirconium, and iron. Insoluble residues consist of metal oxides, carbides, borides, silica, and other compounds. Iodine is less reactive than bromine and therefore may be used to isolate certain less stable oxides.

The apparatus is usually an Erlenmeyer flask with a ground glass connection to an air condenser. In the initial stage of dissolution of certain metals (for example, aluminum) the action of bromine is controlled by an ice bath to keep the reagents and sample in the flask. The insoluble residue is filtered on a medium of appropriate pore size and analyzed for the metallic constituents, from which the corresponding metal oxide contents are calculated.

Steinhauser [31] and Werner [22, 23] successfully used bromine instead of iodine. A 5-g sample of a 99.9% grade of aluminum was analyzed with 200 ml of methyl alcohol and 20 ml of bromine. The rate of reaction was rapid, requiring cooling and controlled additions of bromine. Werner claimed improved results over Hahn's [19] dry hydrogen chloride method. Agreement of results between the iodine and bromine methods was obtained at the 200 ppm level. Interlaboratory checks (four laboratories, five samples, 0.3–7% oxide) were quite good. Alloying elements such as iron (1%), copper (4%), magnesium (5%), manganese (0.7%), and silicon (1%) did not cause serious errors in the 100–2000-ppm range; the aluminum was determined colorimetrically.

Special care must be used if the sample contains more than the 30 mg of oxide or if the sample is in relatively large compact chunks. In these events, it is likely that the oxide may protect the metal such that the reaction is not completed, and it may be necessary to reheat the flask for an 8-h period with frequent shaking.

Solvents other than alcohols have been investigated [32] No reaction was obtained when benzene or chloroform was used. The reaction was slow and perhaps not quite complete in 16 h with ethyl acetate or acetonitrile. Dissolution was always complete with anhydrous 99% methyl alcohol.

F. FLUORINE AND FLUORINE COMPOUNDS

Fluorine attacks many metal oxides to form the metal fluoride and oxygen. Fluorination methods require great care in their application. The portion of the system in which the reaction occurs is usually made of nickel and the portion in which corrosive gases are retained can be polychlorotrifluoroethylene. The collection and measuring apparatus for the liberated oxygen is glass.

Emeleus and Woolf [33] discovered that bromine trifluoride reacts quantitatively with various metal oxides to produce metal fluorides and oxygen. A quartz apparatus was used in their study. Hoekstra and Katz [34] noted that the relatively high boiling point (126°C) of BrF_3 permits a liquid-phase reaction over a convenient temperature range. Further, BrF_3 can be prepared free of oxygen-containing impurities, whereas tank fluorine usually contains small amounts of oxygen, which are difficult to remove. Using a reaction system constructed of nickel, these investigators heated samples at 75°C for 2 h and obtained quantitative evolution of oxygen from TiO_2, V_2O_5, Cr_2O_3, $K_2Cr_2O_7$, $KMnO_4$, CuO, GeO_2, Nb_2O_5, TeO_2, Ta_2O_5, WO_3, PbO, PbO_2,

$BiPO_4$, and certain uranium oxides. At 200°C the oxides of aluminum, lanthanum, and calcium were completely fluorinated. Oxygen was measured manometrically. It was postulated that at 300°C and high pressure other metal oxides could be similarly decomposed. Moisture must be excluded from the reagent and the system.

Dupraw and O'Neill [35] used bromine trifluoride at 75°C in determining oxygen (and nitrogen) in titanium and its alloys. After measuring the combined pressure of oxygen and nitrogen, the oxygen was removed over hot copper and the nitrogen was determined manometrically. In all samples, except Ti–10Mo, the results agreed with those obtained by vacuum fusion. The molybdenum reportedly formed a stable intermediate, $MoOF_4$. The bromine trifluoride method was also recommended for determining oxygen in sulfur, selenium, and tellurium. These elements cannot be analyzed by a vacuum-fusion technique using a cupric oxide–ceric oxide catalyst, because they poison the catalyst. Monel metal and copper were suggested as replacements for the nickel reaction system because they are less expensive and more easily tooled. It was noted that traces of water caused high results even when the pumping time was extended.

Sheft et al. [36] discovered that the salt-like addition product $KBrF_4$, which is formed from the reaction between potassium fluoride and bromine trifluoride, could be substituted as the fluorination agent in the reaction with metal oxides. Potassium bromotetrafluoride can be used at temperatures in the vicinity of 500°C without accompanying high pressures [37]. This compound can be rather easily prepared; it is stable in dry air and melts at about 330°C. Goldberg et al. [37] analyzed BeO, MgO, Y_2O_3, ZrO_2, MoO_3, Mg, Y, Zr, and U by this method, heating the samples at 450°C for 2 h. Oxygen was measured manometrically. The coefficient of variation of results was usually less than 5%, the most suitable amount of oxygen to be determined being 0.02–0.10 mg. Between 25 and 30 samples could be analyzed with a 30-g charge of $KBrF_4$. Goldberg [38] used the same method for determining oxygen in lithium. Kirtchik and Lajcik [39] extended the use of the bromine trifluoride technique to the determination of oxygen in alkali metals by using the Brady method in which the molecular oxygen is measured spectrophotometrically by reaction with reduced sodium anthraquinone sulfonate. They obtained results in the range of 5–150 ppm oxygen.

G. HYDROGEN REDUCTION

Hydrogen reduction was one of the earliest methods used for determining oxygen in metals, dating back to 1882 when Ledebur used it on steel samples [40, 41]. Briefly, in a closed system that has been evacuated or purged with a dry inert gas the sample is melted and the oxide is reacted with dry hydrogen at elevated temperatures. The resulting water may be determined gravimetrically or by Karl Fischer titration [42]. The method is relatively rapid and the apparatus is inexpensive. The reaction has been carried out in both dynamic and static systems. In the latter case, the difference in hydrogen pressure (before and after reduction) is a measure of the water formed. The water is removed either by absorption or by freezing in a cold trap.

This technique is preferable to the vacuum-fusion method if the metal and its oxide are volatile at the temperature required for fusion. Another advantage is that standard samples can be prepared by first reducing the oxide in the melted sample and then adding a known amount of the specific metal oxide to the melt. Oxygen in the low-parts-per-million range or in the percent range may be determined by hydrogen reduction. A list of the elements that can be analyzed by this method may be made from a comparison of the free energies of formation of the metal oxides with that of water. If the free energy of formation of the metal oxide is less negative than that of water the reduction will occur. Thus, hydrogen will reduce the oxides of iron copper, zinc, tin, and lead but not those of aluminum, silicon, titanium zirconium, or thorium.

Interferences by other elements are possible. Carbon may react with the metal oxide to form carbon monoxide or carbon dioxide; phosphorus may be oxidized to the pentoxide, which may condense in a cooler portion of the system and react with the moisture. Lundell et al. [41] have discussed the use of this method for total oxygen in steel and have noted the numerous difficulties encountered. Samples must be in the form of fine chips to achieve complete reduction of oxides [43–46] and should preferably be prepared in the absence of oxygen to avoid surface oxidation. For steel, this method should only be used to determine specific oxides in certain compositions or grades. For example, iron oxide can be completely reduced at 1000°C in hydrogen, but iron silicate can be only partially reduced even when using a tin or antimony flux [47–49].

The hydrogen reduction method for oxygen in parts-per-million concentrations was applied by Baker [50] to copper, lead, and tin, and by Funston and Reed [51] to bismuth with an average deviation of less than 1 ppm. When analyzing copper, Funston and Reed encountered positive errors that were attributable to carbon and made appropriate corrections, assuming the carbon was oxidized to carbon dioxide during analysis. They also noted that samples should be thoroughly cleaned and dried to remove any contaminant that might reduce the metal oxide in steel. Thompson and Holm [52] used a nickel-on-thoria catalyst bed to reduce carbon dioxide to methane and water and then measured the water.

Fischer and Mehlhorn [42] reduced certain oxides of molybdenum and tungsten with hydrogen and determined the water produced by Karl Fischer titration.

H. MERCURY EXTRACTION

In this procedure the metal is dissolved in mercury, and the metal oxide, which does not react with mercury, separates from the amalgam. The oxygen in the oxide is determined by an appropriate method. The mercury-extraction method has been used mainly for sodium, sodium–potassium alloy, and tin. The mercury must be of triple distilled quality [53]. Because the amalgams are usually active reductants, amalgamation must be performed in an air-free system.

Perhaps the most widely studied analysis by amalgamation is that for sodium oxide in sodium, initiated by Pepkowitz and Judd [54]. The procedure has undergone many changes, but amalgamation is now generally performed in a stoppered funnel in a dry box [55]. Discussion of the problems encountered in this method are noted in a review of determinations of impurities in sodium metal [56]. Because many of these problems are common to the determination of oxygen in certain other metals by the mercury-extraction method, the solution of the problems for the analysis of sodium is described. In a study of the atmosphere in the dry box Holt [57] noted that in the presence of 15 ppm moisture and 5 ppm oxygen, sodium did not pick up more than 5 ppm oxygen in 60 min.

Although Hobart [2] reported uncombined oxygen in sodium, Goldberg [58] detected no free molecular oxygen, nitrogen or hydrogen; however, the history of the sample (long prior heating, if any) was not stated in the latter study. Hobart also noted that sufficient time must be allowed for the sodium oxide to rise to the top of the reaction mixture.

Impurities in the metal can cause error in the oxygen results. The final phase of the determination is usually performed by acid titration or by flame photometry. In order to improve accuracy it may be helpful to obtain results by both methods and then attempt to rationalize the combined information. Specifically, sodium carbonate and sodium oxide would give the same acid titration and the same sodium content by flame photometry; sodium hydride would yield half the apparent value for oxygen as sodium oxide by titration and the same apparent value as the oxide by flame photometry; lithium nitride would titrate as sodium oxide but would give a zero (sodium) value by flame photometry. Calcium and magnesium oxides would titrate as sodium oxide but also would give zero values for sodium by flame photometry. Silica would not be determined as sodium by either method.

The amalgamation method has also been applied to potassium [58]. Except for tin [59], no results by this procedure have been noted for the metals of the periodic groups 1B, 2B, 3A, 4A, and 5A, which amalgamate.

Oxygen in mercury (as mercurous or mercuric oxide) could be determined if oxygen-free sodium were heated with the mercury in a sealed tube [53] and the sodium oxide separated as described above. Another technique would be to heat the mercury to decompose the oxide and measure the oxygen by gas chromatography, by the Winkler process, or by the Brady method [60].

I. SULFUR

Three sulfur methods have been used for the determination of oxygen in metals, using elemental sulfur, sulfur monochloride and hydrogen sulfide.

1. Elemental Sulfur

Von Wartenburg [61], noting the occasional interference by sulfur in the determination of oxygen in metals by hydrogen reduction, suggested the use of sulfur for the determination. In this procedure, elemental sulfur is heated with the sample in a quartz tube through which is passed a stream of purified nitrogen. Sulfur replaces the oxygen in the oxide, and the oxygen is converted to sulfur dioxide, which is absorbed in acid solution and titrated iodimetrically. For example,

$$2ZnO + 3S = 2ZnS + SO_2$$

$$SO_2 + I_2 + 2H_2O = H_2SO_4 + 2HI.$$

In the case of zinc oxide Von Wartenburg found that the reaction could be made to occur quantitatively at 900–1000°C.

Eggertsen and Roberts [62] determined oxide in nickel, molybdenum, and tungsten, working at levels upward from 0.1% oxygen; analysis time was about 2 h. Babko and co-workers [63, 64] extended the method to the oxides of iron, manganese, cobalt, nickel, lead, cadmium, copper, arsenic, and antimony. They listed the lowest Celsius temperatures at which 100% reaction occurred as follows: Fe_2O_3, 800°; Mn_2O_3, 650°; Co_3O_4, 600°; As_2O_3, 700°; Sb_2O_4, 500°; CdO, 600°; CuO, 400°; and PbO, 600°. The following did not react at 1000°: SiO_2, B_2O_3, Al_2O_3, MgO, BeO, and TiO_2. Elemental oxygen in lead and 0.02% oxygen in copper were detected. They found that nitrogen reacted with sulfur to form a compound (decomposable in water to form NH_3 and SO_2) that caused high results. For this reason they carried out the reaction in vacuum. They also noted that water reacts with hot sulfur to produce sulfur dioxide and hydrogen sulfide, making it essential to use dried reagents. They made the final SO_2 measurement by the sensitive fuchsin-formaldehyde photometric determination.

2. Sulfur Monochloride

Kleiner [65–66] used sulfur monochloride (melting point, 80°C; boiling point, 135.6°C) to react with metal oxides to produce SO_2 for measurement of oxygen content. The reagent was carried by a stream of nitrogen over weighed samples of CuO at 480°; Fe_2O_3 at 700°; Al_2O_3, MgO, B_2O_3 and ZrO_2 at 900°; and SiO_2 at 950°C. The carrier gas stream was passed through mercuric sulfide to destroy the excess sulfur monochloride and then through an alkaline potassium iodide solution containing iodine. The SO_2 formed in reactions, such as

$$2Al_2O_3 + 6S_2Cl_2 = 4AlCl_3 + 3SO_2 + 9S \tag{1}$$

and

$$SiO_2 + 2S_2Cl_2 = SiCl_4 + SO_2 + 3S, \tag{2}$$

was determined iodometrically. Metals analyzed for oxide content by this method were copper, cadmium, nickel, chromium, titanium, zirconium, ferrosilicon, nichrome, and alloys of Fe–Zr–Cr, Ni–Mo and Ni–W [66–67].

Graphite boats and reaction tubes were used at the higher temperatures to avoid reaction of the sulfur monochloride with quartzware by the reaction in Eq. 2. Furthermore, to avoid clogging the apparatus with volatile chlorides at the higher temperatures (by reaction of S_2Cl_2 with metal) the metal samples were previously exposed to a stream of bromine-in-nitrogen in a quartz tube at 600–800°C before placing the boat and oxide residue in the apparatus for the sulfur monochloride reaction. The oxides did not react with bromine under the conditions of this pretreatment, and no oxygen was lost.

3. Hydrogen Sulfide

Hartmann and co-workers [68–71] studied the reduction of group-IIA metals and zinc with hydrogen sulfide. A stream of purified nitrogen was used to carry dry hydrogen sulfide over the molten metal at 550°C. The symbolic equation is

$$MeO + H_2S_{(g)} = MeS + H_2O_{(g)},$$

where Me is Mg, Sr, Ba, or Zn. (Reduction of BeO under these conditions is not thermodynamically possible.) The water is collected and measured gravimetrically.

J. DISTILLATION

Distillation under high vacuum has been applied to the separation of metal from its oxide in the cases of magnesium, calcium, strontium, barium, zinc, cadmium, tin, aluminum, sodium, and potassium.

Allsopp [72] and Holt and Goodspeed [73] studied the determination of oxygen in magnesium. In Allsopp's procedure a 3–4-g sample was weighed into a covered crucible, placed in a molybdenum boat and inserted into a silica tube. The pressure was reduced to 10^{-4} torr, and the furnace was heated to 900–1000°C. The magnesium distilled from the covered crucible and condensed in the cooler portions of the tube. After the furnace cooled and air was admitted, the boat and covered crucible were removed; the crude oxide was weighed; the sample was dissolved; and silica, iron, and aluminum were separated and determined photometrically. Calcium and magnesium were determined by titration with EDTA. Most of Allsopp's samples contained 1–2% magnesium oxide, and replicate values agreed within 0.02% MgO.

Rodyakin et al. [74] determined oxygen in magnesium by sublimation of the metal at 10^{-4}–10^{-6} torr and at temperatures increased stepwise up to 1000°C over a period of 2 h. They reported losses of 5–9% oxygen by entrainment of oxide particles in the magnesium vapors.

The determination of oxide oxygen in zinc by distillation of the metal has been investigated. Fischer and Bechtel [75] analyzed the distillation residue and reported results in the range of 10 ppm oxygen. See Chapters 11 and 12 for descriptions of the procedures used by Holt and Goodspeed [73] for the determination of oxygen in magnesium, zinc, and cadmium by distillation of the respective metals and subsequent carbon reduction of the residual oxides to carbon monoxide, which was oxidized to carbon dioxide and measured manometrically.

The determination of oxide oxygen in sodium has been extensively investigated because of the technological importance of this metal as a high temperature coolant. White [76], Walker et al. [77], and Bergstresser et al. [78] investigated the determination of combined oxides in the lower parts-per-million range. The residue after distillation contains many contaminants [56], for which corrections are suggested.

K. ALUMINUM REDUCTION

1. Steels

In the steel industry, aluminum is used to deoxidize certain types of steels. The same reaction (i.e., conversion of oxides to Al_2O_3) has been applied in the determination of oxygen in steel, especially where not all of the oxygen is present as alumina. A sample (5–10 g) is layered with an equal weight of aluminum in a graphite boat (covered with a graphite lid), heated to about 1200°C under vacuum [79–83] (or in a stream of hydrogen [81–85] or purified argon) for about 1 h, and cooled in a nonoxidizing atmosphere. The steel is dissolved in a solvent in which the impure alumina is insoluble. The solvent should be carefully chosen, and the solution time and temperature should be carefully controlled to prevent attack of the alumina. The impure residue is weighed and then analyzed for impurities, for which the residue weight is corrected. Carbon can be removed by ignition, but such compounds as silicates and titanium nitride may remain in the residue.

The sample can be in the form of disks (approximately 2.5 mm thick) cut from a 12.5-mm-diameter bar that has been ground to remove surface oxide [83]. The aluminum can be prepared by cutting strips of suitable size from sheets of 3-mm thickness. The surface of the aluminum reagent (high-purity grade) should be cleaned by lightly filing and should be thoroughly dried before use. The cleaning of the aluminum and the loading into the crucible should be performed in a dry box. Blank determinations should always be made.

The fusion atmosphere must be free of oxygen and nitrogen; the latter may react with certain elements to form nitrides insoluble in the dissolving medium and cause high results. Oxygen and nitrogen may be removed from argon or helium as described in Chapter 4. The temperature for the fusion of the aluminum and sample should be maintained at 1200°C or higher for at least an hour; otherwise, the alumina that is formed will be less resistant to hydration and to attack by the reagents used to dissolve the matrix metal and excess aluminum.

Wells [83] and Gray and Sanders [81, 84] found that their oxygen results compared well with those obtained by vacuum fusion. The latter applied their procedure to cast iron, ferrosilicon, and ferromanganese.

L. TOTAL OXIDATION

The dispersion of metal oxide particles (within certain ranges of size) in the parent or foreign metal is of metallurgical interest because of improved properties, such as tensile strength and corrosion resistance, of the resultant product. For example, sintered aluminum products, made by dispersion of alumina in aluminum metal, have corrosion properties that are superior to those of pure aluminum and certain aluminum alloys.

Because sintered aluminum products are usually prepared by mixing ignited alumina with powdered aluminum metal, two types of oxides are present in the final material: the ignited alumina and the surface oxide formed by atmospheric corrosion.

An oxidation procedure by which the total oxide (both types) in sintered aluminum may be determined follows: A 0.1-g sample is transferred to a tared 500-ml porcelain crucible, and the sample is slowly and carefully dissolved with aqua regia. The solution is evaporated to near dryness, and nitric acid is added. Repeated evaporation and nitric acid treatments are needed to remove chloride. The dried sample is ignited to constant weight at 1400°C. The ignition is the time-consuming portion of the procedure; as much as 20 h or more may be required.

Results for the sum of the internal alumina and the surface oxides [$Al_2O_3 \cdot 3H_2O$ and $AlO(OH)$] are calculated as follows: Let A = original sample-weight, B = final weight of ignited material, x = weight of oxide in original sample, and y = weight of metal in original sample. Then

$$A = x + y$$
$$B = x + 1.889y$$
$$x = 2.125A - 1.126B$$
$$\text{wt \% aluminum oxide} = \frac{x}{A} 100.$$

Duplicate results that agree within 0.05% Al_2O_3 are possible. However, the method should be limited to systems in which the nonoxygen component is at least 99.8% aluminum, lest the corrections for silicon dioxide, cuprous oxide, ferrous oxide, manganous oxide, and magnesium oxide introduce errors.

The alumina in sintered aluminum (exclusive of surface oxides) may be determined by acidic solution or alkaline solution methods described in Section III.A.

M. MISCELLANEOUS METHODS

1. Alkyl Halide Method for Sodium and Potassium Oxides

White et al. [86] showed that sodium metal could be completely reacted with butyl bromide (Würtz synthesis) and that the sodium oxide present was unaffected. (Pentyl halides may also be used.) The alkaline oxide was then dissolved in water and titrated with standard acid. Silverman and Shideler [87] reduced the detection limit by improving the purification of reagents. Smythe and de Bruin [88] made correction for a "salt error" in the titration for the alkali oxide and thereby improved the precision of the method.

2. Coulometric Reduction

A technique was developed by Lambert and Trevoy [89] for coulometric reduction of films on copper and other electropositive metals. The sensitivity of the method was increased by utilizing a granular electrode for rapid and thorough preelectrolysis of electrolyte to remove dissolved oxygen and traces of plateable cations. Results were reported for the reduction of films of cuprous oxide. As little as a quarter of a monolayer of a reducible film was detected. These workers were able to demonstrate the usefulness of the method in studying the topography of reducible surface films.

3. Carbon Monoxide Reduction

Carbon monoxide can be used to reduce copper oxide (750°C), producing carbon dioxide which can be conveniently measured. Weber [90] has studied the reaction of carbon monoxide with several metal oxides and concluded that it may be possible to distinguish between MnO and Mn_2O_3 and between VO and V_2O_5. According to Darken and Gurry [91], carbon monoxide is nearly as good a reductant as hydrogen. Thus, at 1000°C FeO, SnO_2, CoO, CuO, NiO, and PbO can be reduced with $CO:CO_2 > 100:1$.

4. Solution-Titration

Eberle et al. [92] reacted calcium metal with anhydrous methyl alcohol to produce calcium methylate. Calcium oxide does not react with the alcohol. A solution of anhydrous salicyclic acid–pyridine is added, and the water formed by the reaction of the calcium oxide with salicyclic acid is determined by titration with Karl Fischer reagent to calculate the oxygen present. Eberle et al. [92] usually used 2-g samples in determining about 0.8% oxygen in calcium. The lower limit of detection was about 0.4% oxygen. Sample inhomogeneity was considered a prime problem; therefore, meaningful precision data could not be developed. However, anhydrous calcium oxide was analyzed and a precision of 0.1 mg was obtained for 2.5–13 mg oxygen.

Sax and Steinmetz [93] applied this method to the determination of oxide in lithium metal. Jaworoski et al. [94] found the procedure difficult to handle. They suggested butyl cellosolve as solvent, and coulometric generation of the Karl Fischer reagent. Another proposal was to dissolve the lithium metal in anhydrous liquid ammonia, filter insoluble oxide (and carbonate), dissolve it in water, and titrate the alkali.

5. Infrared Spectrometry

Vasko [95], using an infrared spectrometer, determined as little as 9 ppm oxygen in amorphous selenium and established that at levels above 9 ppm, oxygen exists in amorphous selenium as SeO_2. In the determination of sodium monoxide in sodium [96] amyl chloride was reacted with the sodium to form sodium chloride; the sodium oxide was converted to sodium carbonate, and the latter was measured by infrared spectrometry at 11.38 μm. The lower limit of detection was 20 ppm oxygen, but modification of the procedure could extend this well below 10 ppm.

ACKNOWLEDGMENT

Ben Holt's additions to and editing of this chapter are sincerely appreciated.

REFERENCES

1. L. Silverman and W. Bradshaw, *Anal. Chim. Acta.*, **18,** 235 (1958).
2. E. W. Hobart, *USAEC Report TIM-900*, 1965.
3. Z. M. Turovtseva and L. L. Kunin, *Analysis of Gases in Metals*, trans. from the Russian, by James Thompson, Consultants Bureau, New York, 1959. Originally published by the Publishing House of The Academy of Sciences of the U.S.S.R., for the V. I. Vernadskii Institute of Geochemistry and Analytical Chemistry, Moscow, 1959.
4. R. E. Honig, *RCA Review*, **23,** 574 (1962).
5. A. Glassner, *USAEC Report ANL-5750*, 1957.
6. H. F. Beeghly, *Anal. Chem.*, **24,** 1713 (1952).
7. L. Silverman, *Ind. Eng. Chem., Anal. Ed.*, **7,** 205 (1935).
8. W. W. Scott, *Standard Methods of Chemical Analysis*, Vol. I, 5th ed., Van Nostrand, New York, 1938, p. 51.
9. F. J. Adcock, *J. Iron Steel Inst.*, **115,** (1), 369 (1927).
10. A. Lench, G. S. Martin, and G. Cumming, *Anal. Chem.*, **36,** 337 (1964).
11. S. P. Makariewa and N. D. Biryukoff, *Z. Elektrochem.*, **41,** 623 (1935).
12. F. O. Kichline, *Ind. Eng. Chem.*, **7,** 806 (1915).
13. W. Koch, *Metallkundliche Analyse*, Verlag Stahleisen m.b.H., Dusseldorf, und Verlag Chemie, G.m.b.H., Weinheim/Bergstrasser, 1965.
14. E. W. Colbeck, S. W. Craven, and W. J. Murray, *J. Iron Steel Inst.*, **134,** 251 (1936).
15. E. W. Colbeck, S. W. Craven, and W. J. Murray, *Spec. Rep. No. 16*, Iron and Steel Institute, London, 1937, p. 124.
16. E. W. Colbeck, S. W. Craven, and W. J. Murray, *Spec. Rep. No. 25*, Iron and Steel Institute, London, 1939, p. 177.
17. E. W. Colbeck, S. W. Craven, and W. J. Murray, *J. Iron Steel Inst.*, **143,** 332 (1941).
18. V. H. Lowenstein, *Z. Anorg. Chem.*, **199,** 48 (1931).
19. F. L. Hahn, *Z. Anal. Chem.*, **80,** 192 (1930).
20. G. B. Brook and A. G. Waddington, *J. Inst. Metals*, **61,** 309 (1937).
21. P. Urech, R. Sulzberger and E. Schaad, *Chimia*, **4,** 233 (1950); through *Chem. Abstr.*, **45,** 1914i (1951).
22. O. Werner, *Metall.*, **4,** 9 (1950).
23. O. Werner, *Z. Anal. Chem.*, **121,** 385 (1941).
24. A. A. Federov and F. V. Linkova, *Zh. Anal. Khim.*, **17,** 53 (1962); through *Chem. Abstr.* **57,** 2851f (1962).
25. M. Codell and G. Norwitz, *Anal. Chem.*, **27,** 1083 (1955).
26. M. Codell and G. Norwitz, *Anal. Chem.*, **28,** 2006 (1956).

27. M. Codell and G. Norwitz, *Anal. Chem.*, **30,** 524 (1958).
28. M. Codell, G. Norwitz, and S. Kallman, *ASTM Spec. Tech. Publ.*, **222,** 33 (1957).
29. O. Miyamoto, R. Nakashima, Y. Ishiguro, and K. Okumuro, *Nagoya Kogyo Gijutsu Shikensho Hokoku*, **7,** 322 (1958); through *Chem. Abstr.*, **58,** 16e (1963).
30. W. T. Elwell and D. M. Peake, *Analyst*, **82,** 734 (1957).
31. K. Steinhauser, *Aluminum*, **24,** 176 (1942).
32. L. Silverman, unpublished results.
33. H. J. Emeleus and A. A. Woolf, *J. Chem. Soc.* (London), **1950,** 164.
34. H. R. Hoekstra and J. J. Katz, *Anal. Chem.*, **25,** 1608 (1953).
35. W. A. Dupraw and H. J. O'Neill, *Anal. Chem.*, **31,** 1104 (1959).
36. I. Sheft, A. F. Martin, J. J. Katz, *J. Amer. Chem. Soc.*, **78,** 1557 (1956).
37. G. Goldberg, A. S. Meyer, Jr., J. C. White, *Anal. Chem.*, **32,** 314 (1960).
38. G. Goldberg, *Anal. Chem.*, **34,** 1343 (1962).
39. H. Kirtchik and T. Lajcik, *NASA Report CR-52184;* through Chem. Abstr. **61,** 15343b (1964).
40. A. Ledebur, *Leitfaden fur Eisenhutten Laboratorien*, 9th ed., p. 154; through Ref. 42.
41. G. E. F. Lundell, J. I. Hoffman, and H. A. Bright, *Chemical Analysis of Iron and Steel*, Wiley, New York, 1931.
42. W. Fischer and R. Mehlhorn, Reinststoffe Wiss. Tech. Intern. Symp., 1., Dresden 1961, p. 525, 1963; through *Chem. Abstr.*, **60,** 15132f (1964).
43. J. R. Cain and E. Pettyjohn, *Tech. Paper No. 118*, National Bureau of Standards, 1919.
44. P. Oberhoffer, *Stahl Eisen*, **38,** 105 (1918).
45. P. Oberhoffer, *Stahl Eisen*, **40,** 812 (1920).
46. T. E. Rooney, *J. Iron Steel Inst.*, **110,** 122 (1924).
47. P. Oberhoffer and O. Kiel, *Stahl Eisen*, **41,** 1949 (1921).
48. J. Keutman and P. Oberhoffer, *Stahl Eisen*, **45,** 1557 (1925).
49. P. Oberhoffer and J. Keutman, *Stahl Eisen*, **46,** 1045 (1926).
50. W. A. Baker, *Metallurgia*, **40,** 188 (1949).
51. E. S. Funston and S. A. Reed, *Anal. Chem.*, **23,** 190 (1951).
52. J. G. Thompson and V. C. F. Holm, *J. Res. Nat. Bur. Stds.*, **21,** 79 (1938).
53. L. Silverman, *Rev. Sci. Instum.*, **24,** 80, (1953).
54. L. P. Pepkowitz and W. C. Judd, *Anal. Chem.*, **22,** 1283, (1950).
55. D. E. Kuivinen, *NASA Report*, *Lewis Technical Preprint 6-63*, 1963.
56. L. Silverman, *Determination of Impurities in Nuclear Grade Sodium Metal*, Pergamon, New York, 1967.
57. B. D. Holt, *USAEC Report ANL-7123*, 1965.
58. G. Goldberg, *USAEC Report ORNL-3537*, 1964, p. 55.
59. L. Silverman and W. Gossen, *Anal. Chim. Acta*, **8,** 436 (1953).
60. L. Silverman, W. Bradshaw, *Anal. Chim. Acta*, **14,** 514 (1956).
61. H. von Wartenburg, *Z. Anorg. Allgem. Chem.*, **251,** 161 (1943).
62. F. T. Eggertsen and R. M. Roberts, *Anal. Chem.*, **22,** 924 (1950).
63. A. K. Babko, K. E. Kleiner, and L. V. Markova, *Zavod. Lab.*, **22,** 640 (1956); through *Chem. Abstr.*, **50,** 15339g (1956).

64. A. K. Babko, A. I. Volkova, and C. E. Drako, *Zavod. Lab.*, **23,** 136 (1957); through *Chem. Abstr.*, **51,** 17591c (1957).
65. K. E. Kleiner, *Ukr. Khim. Zh.*, **22,** 809 (1956); through *Chem. Abstr.*, **51,** 7233a (1957).
66. K. E. Kleiner and L. V. Markova, *Ukr. Khim. Zh.*, **23,** 236 (1957), through *Chem. Abstr.*, **51,** 12748e (1957).
67. K. E. Kleiner and N. V. Obolonchik, *Ukr. Khim. Zh.*, **25,** 370 (1959); through *Chem. Abstr.*, **53,** 21397c (1959).
68. H. Hartmann, W. Hofmann, and K. Schulte-Schrepping, *Z. Metallk.*, **43,** 350 (1952); through *Chem. Abstr.*, **47,** 1537i (1953).
69. H. Hartmann and G. Strohl, *Z. Anal. Chem.*, **144,** 332 (1955).
70. H. Hartmann and G. Strohl, *Z. Anal. Chem.*, **175,** 84 (1960); through *Chem. Abstr.*, **55,** 1284b (1961).
71. H. Hartmann, W. Hofmann, and G. Strohl, *Z. Metallk.*, **49,** 461 (1958); through *Chem. Abstr.*, **52,** 19687e (1958).
72. H. J. Allsopp, *Analyst*, **81,** 469 (1956).
73. B. D. Holt and H. T. Goodspeed, *Anal. Chem.*, **34,** 374 (1962).
74. V. V. Rodyakin, A. E. Andreev, A. M. Bragin, A. I. Boiko, and A. V. Riganelovich, *Zavod. Lab.*, **30,** 1203 (1964); through *Chem. Abstr.*, **62,** 3397c (1956).
75. J. Fischer and H. Bechtel, *Z. Erzberg. Metallhut*, **5,** 14 (1952).
76. J. C. White, *USAEC Report CF-56-4-31*, 1956.
77. J. A. J. Walker, E. D. France, W. T. Edward, *Analyst*, **90,** 727 (1965).
78. K. S. Bergstresser, G. R. Waterbury, C. F. Metz, *USAEC Report LA-3343*, 1965.
79. W. W. Stevenson and G. E. Speight, *J. Iron Steel Inst.*, **143,** 326 (1941).
80. T. Swinden and W. W. Stevenson, *J. Iron Steel Inst.*, **148,** 397 (1943).
81. N. Gray and M. C. Sanders, *J. Iron Steel Inst.*, **143,** 321 (1941).
82. N. Gray and M. C. Sanders, *J. Iron Steel Inst.*, **148,** 242 (1943).
83. J. E. Wells, *J. Iron Steel Inst.*, **184,** 185 (1956).
84. N. Gray and M. C. Sanders, *J. Iron Steel Inst.*, **137,** 348 (1938).
85. N. Gray and M. C. Sanders, *Spec. Report*, *No. 25*, Iron and Steel Institute, London, 1939, p. 103.
86. J. C. White, W. J. Ross, and R. Rowan, Jr., *Anal. Chem.*, **26,** 210 (1954).
87. L. Silverman and M. Shideler, *Anal. Chem.*, **27,** 1660 (1955).
88. L. E. Smythe and H. J. de Bruin, *Analyst*, **83,** 242 (1958).
89. R. H. Lambert and D. J. Trevoy, *J. Electrochem. Soc.*, **105,** 18 (1958).
90. J. Weber, *Kohaszati Lapok*, **91,** 122 (1958); through *Chem. Abstr.*, **52,** 18127g (1958).
91. L. S. Darken and R. W. Gurry, *Physical Chemistry of Metals*, McGraw-Hill, New York, 1953, p. 349.
92. A. R. Eberle, M. W. Lerner, and G. J. Petretic, *Anal. Chem.*, **27,** 1431 (1955).
93. N. I. Sax and H. Steinmetz, *USAEC Report ORNL-2570*, 1958.
94. R. J. Jaworoski, J. R. Potts, and E. W. Hobart, *Anal. Chem.*, **35,** 1275 (1963).
95. A. Vasko, *Phys. Status Solids*, **8,** K41 (1965); through *Chem. Abstr.*, **62,** 9783f (1965).
96. H. J. de Bruin, *Anal. Chem.*, **32,** 360 (1960).

CHAPTER

10

ALKALI METALS

SILVE KALLMANN

Ledoux and Company
Teaneck, New Jersey

CONTENTS

I. INTRODUCTION

Prior to 1950, sodium was the only one of the five alkali metals that was in commercial demand in any substantial amount. Its chief application was in the production of the sodium–lead alloy used in the manufacture of the anti-knock agent tetraethyl lead. During the last 15 years, applications for liquid alkali metals for various terrestrial and space uses have been extensively developed.

Lithium, which has the lowest atomic weight of all the alkali metals, has been investigated as a heat-transfer fluid in various solar and nuclear space power plants. Sodium and sodium–potassium alloys have been widely used as coolants or heat-transfer media in power plants, particularly in those employing nuclear reactors. In addition, sodium–potassium alloys are being considered as hydraulic fluids and for advanced flight vehicles [1]. Sodium, potassium, rubidium, and cesium are being investigated as thermodynamic working fluids in various Rankine cycle space power plants. Cesium, because it has the lowest ionization potential and largest atomic volume of all the liquid metals, is under investigation as an ion engine propellant.

Liquid alkali metals possess unique physical and nuclear properties that recommend them for various terrestrial and space power-plant applications [1]. A few such properties are good stability, good thermal conductivity, low vapor pressure and correspondingly high boiling points, and comparatively low melting points. This is especially true in situations where these metals are used as working fluids in thermoelectric plants, particularly in the space environments where heat rejection from the cycle for condensation of the working fluid must take place by radiation. In modern nuclear reactors the use of liquid alkali metals as coolants makes it possible to transport heat from the reactor at comparatively high temperature levels. This unique characteristic exerts a substantial influence on the efficiency attainable in nuclear power plants. Moreover, some of these heat-transfer media—sodium and potassium, for example—possess small neutron-absorption cross sections that are of considerable importance for such applications.

II. EFFECT OF CONTAMINANTS

While liquid alkali metals exhibit these many outstanding physical characteristics, they possess, on the other hand, chemical characteristics that create problems with their handling and end use.

The presence of contaminants (particularly oxygen and carbon) in liquid alkali–metals systems is of concern because of the resulting acceleration of corrosion and the impairment of mechanical and thermal performance of equipment. Soluble contaminants tend to increase corrosion rates drastically, and mechanical difficulties are caused by insoluble contaminants in the system circuitry. For instance, it has been shown that the presence of even 5 ppm oxygen may cause excessive embrittlement and corrosion of zirconium or niobium in contact with sodium at 1000°F [2, 3]. The reaction of sodium oxide and certain other alkali oxides with zirconium is so energetic that it has been used not only for the purification of the alkali metals [4] by so-called *hot-trapping processes*, but also in a method designed to measure the amount of oxygen in liquid alkali metals [5]:

$$Zr + 2Na_2O = ZrO_2 + 4Na \tag{1}$$

$$Na_2O + ZrO_2 \rightarrow Na_2ZrO_3. \tag{2}$$

Nitrogen in lithium may accelerate solution of iron, chromium, nickel, niobium, titanium, and molybdenum. Calcium in sodium increases the nitrogen solubility and often causes nitrogen transfer problems [6].

Hydrogen embrittlement of the fuel elements occurred at less than 300°C in the reactor at Dounreay, Scotland, from water vapor in the cover gas. Hydrogen, together with oxygen, rapidly increases the attack on iron [6] through the formation of NaOH.

The physical properties of a number of metals in contact with liquid alkali metals may be affected by the carbon present in the alkali metal. Austenitic stainless steel is carburized, while ferritic alloys are decarburized [7]. The analytical chemistry of carbon in the alkali metals has been discussed [8].

Because of the decreased solubility of contaminants, such as oxides, at lower temperatures, these contaminants may deposit as solids. In heat-transfer systems, for example, the entrance to the heat source represents the coldest region. Plugging in this region will result in poor flow-distribution and may lead to excessive temperatures and burn-out.

On the other hand, it is possible that low concentrations of gases may be beneficial at low temperatures and very deleterious at higher temperatures. For instance, it has been shown that low concentrations of oxygen depress the melting point of rubidium and thereby may reduce room-temperature handling problems. On the other hand, even low concentrations of oxygen in rubidium are known to increase corrosion problems in high-temperature applications [1].

III. REACTIONS OF ALKALI METALS WITH OXYGEN AND NITROGEN

A. INTRODUCTION

Atmospheric-source contamination is an ever-present problem in alkali-metal storage, handling, and application. Accidental occurrences, equipment failures, or improper techniques can rapidly and effectively poison an alkali metal for its intended use. The reactions of the alkali metals with atmospheric gases are of great importance, since they lead to the formation of various products that the analytical chemist may be asked to identify or analyze.

B. LITHIUM

Lithium metal when exposed to flowing dry air or to flowing dry nitrogen at room temperature for periods up to a week showed no significant weight change [9]. These results are consistent with the observed high ignition temperature (630–870°C) of lithium in oxygen [10] and the reported relative stability of lithium metal in dry air even at fusion temperatures (melting point, 180.5°C) [10]. However, the reaction between lithium and pure dry nitrogen is rapid and exothermic at about 170°C [12], leading to the formation of Li_3N. As a matter of fact, of all the alkali metals, lithium is the only one to form a stable nitride. It was shown that the rate of absorption of nitrogen by liquid lithium is independent of the nitrogen pressure [13]. The nitrides of the other alkali metals are poorly characterized, extremely unstable substances and are formed only as transients in the thermal decompositions of the alkali metal azides or by reaction of the free alkali metal with activated or atomic nitrogen in electric discharges [14].

The corrosion products formed by reaction of lithium with moist warm air are primarily anhydrous lithium hydroxide and lithium nitride, followed subsequently by anhydrous and hydrated lithium hydroxide with a corresponding diminution of the lithium nitride content due to the loss of ammonia [11]. Lithium oxide may be formed in a moist atmosphere by the following reactions:

$$2Li\ (solid) + H_2O\ (gas) \rightarrow Li_2O\ (solid) + H_2 \tag{3}$$

$$LiOH\ (solid) + Li\ (solid) \rightarrow Li_2O\ (solid) + \tfrac{1}{2}H_2. \tag{4}$$

The formation of lithium hydride is discussed later, together with that of the other alkali metal hydrides.

C. SODIUM

Solid sodium tarnishes almost immediately when exposed to air, owing to the formation of a film of oxide. Molten sodium burns readily in air to form dense fumes of sodium monoxide. With pure oxygen, molten sodium burns with a yellow flame forming a mixture of sodium monoxide and sodium peroxide. Kinetics of the reaction of sodium vapor with oxygen at low pressures and at 250°C indicate initial formation of the superoxide, NaO_2, followed by reaction with more sodium to form the peroxide, Na_2O_2 [11]. At room temperature, oxygen can exist in sodium as $NaOH$, Na_2O, Na_2O_2, or Na_2CO_3 [15]. However, at higher temperatures, $NaOH$ and Na_2O_2 are converted to Na_2O in the presence of excess sodium. Similarly, it is assumed that Na_2CO_3 is decomposed to Na_2O when sodium is heated to 400°C. Accordingly, if a sample of sodium is heated to 400°C in an inert environment Na_2O should be the only compound that need be considered. Sodium does not react with nitrogen under normal conditions, although nitrogen activated by an electrical discharge at low pressure readily reacts with sodium to form sodium nitride, Na_3N, which eventually is converted to sodium azide, NaN_3 [16]. Nitrogen may, however, be found in sodium, because of contamination with calcium [17].

D. POTASSIUM

When potassium is exposed to oxygen, potassium superoxide, KO_2, a very strong oxidizing agent, forms. It is known that KO_2 is transformed to

K_2O above 300°C. Some work with sodium–potassium alloy (NaK) at Oak Ridge National Laboratory (ORNL) shows KO_2 to be the stable oxide at room temperature in the presence of normal pressures of oxygen [18]. Explosions, which occasionally occur in handling potassium, have been attributed to KO_2 forming on the surface of the metal [19]. The mechanism that causes the stable superoxide to react violently with elemental potassium is not completely understood.

E. RUBIDIUM

It has been stated that rubidium ignites spontaneously in air [20]; other investigators, however, while reporting extreme chemical reactivity, noted no visible evidence of ignition [21]. The reaction of rubidium with oxygen is so rapid that in experiments involving additions of small amounts of oxygen to rubidium in a previously evacuated system, it was impossible to achieve a significant pressure [20]. There are conflicting reports on the composition of the most stable rubidium oxide, with both the superoxide [20] and the monoxide [1] being mentioned.

F. CESIUM

In terms of free energy of formation of the oxides, cesium is the least reactive of the alkali metals and is also less reactive than aluminum, magnesium, and silicon [4]. However, in terms of reaction kinetics, it is the most reactive of all the metals.

Contrary to statements found in the literature that cesium ignites spontaneously in air, it was observed that even at 260°C, cesium does not ignite [21]. However, as in the case of rubidium, the rate of oxidation is extremely rapid and is aided by the high solubility of the oxide in the parent metal.

When cesium reacts at ambient temperatures with air or oxygen, several oxides, including a peroxide, may form. The oxides that have been identified are Cs_2O, Cs_2O_3, Cs_2O_2, and CsO_2. Other oxides, such as Cs_7O, Cs_4O, Cs_7O_2, and Cs_3O, are mentioned in the literature, but they have not been completely characterized [22].

In contrast to the oxides of the lighter alkali metals, the higher cesium oxides are quite stable. The superoxide, rather than the monoxide, should be regarded as the most stable oxide of cesium. While cesium under ordinary conditions probably does not react with nitrogen, the possibility of nitride formation in the temperature range 870–1370°C cannot be excluded. Cesium nitride is, in fact, known [4].

IV. REACTIONS OF ALKALI METALS WITH HYDROGEN

The hydrides of the alkali metals are rather easily synthesized by direct union with hydrogen at temperatures ranging from 300–400°C for sodium to 580–620°C for cesium. Lithium hydride formed by the reaction of lithium with hydrogen at 500–800°C is well known. Hydrides may also form by reaction with traces of moisture in the inert gas covering the metal. The alkali metal hydrides are stable over different temperature ranges, with hydride decomposition and hydrogen liberation occurring at higher temperatures.

When hydrogen is present in sodium metal as the hydroxide, the following reactions occur when the metal is heated [20]:

$$2NaOH + 2Na \rightarrow 2Na_2O + H_2 \tag{5}$$

$$2Na + H_2 \rightarrow 2NaH. \tag{6}$$

These reactions occur at temperatures used during distillation or hot trapping. These reactions may also be expected with the other alkali metals. It has been stated [23] that in the purification of sodium by vacuum distillation, sodium hydride is completely decomposed at 5 μ and 573°K. However, it should be noted that recombination may occur at lower temperatures during condensation of the alkali metal unless the evolved hydrogen is efficiently removed.

V. SAMPLE STORAGE

Because of their extreme reactivity, the alkali metals are usually stored in sealed containers under a dry inert gas or under a liquid, saturated hydrocarbon. However, in the case of rubidium it has been observed that even traces of moisture in the hydrocarbon may lead to the formation of oxygen-containing rubidium contaminants, and there exists some evidence of contamination with carbonaceous material [20].

VI. SAMPLING

A. INTRODUCTION

The procurement or preparation of a representative and uncontaminated alkali metal sample is undoubtedly the most difficult part of the analytical scheme. Sampling of the alkali metals is extremely difficult because of their rapid reactions with oxygen, nitrogen, and moisture even at room temperature and with glass and quartz at elevated temperatures. One difficulty in obtaining a representative sample results from the fact that the solubilities of most impurities, including the gases, are temperature dependent. If in taking a sample of molten alkali metal for analysis the temperature is lowered, loss of impurities may occur, owing to precipitation and segregation. The problem of contamination during sampling has been demonstrated at Battelle [24] in the case of potassium. Three different and accepted sampling techniques were applied to gettered material (i.e., alkali metal depleted of oxygen by contact with another metal while hot) with an oxygen content of less than 15 ppm. Results for each of the three sampling techniques were satisfactorily reproducible, but they varied from 15–65 ppm of oxygen, depending on the sampling techniques employed. Therefore, in sampling alkali metals it is imperative that the following objectives be achieved:

1. There is no reaction of the alkali metal with the glass or metal container.
2. There is no exposure of the alkali metal to oxygen, nitrogen, hydrogen, carbon dioxide, or water vapor.
3. There is no differential freeze-out of insoluble compounds. For example, the solubilities of some impurities in liquid alkali metals are limited, and on cooling, these elements tend to freeze out before the alkali metal solidifies.

B. LIQUID METALS IN DYNAMIC SYSTEMS

Most of the reported techniques involve the sampling of sodium, NaK, and lithium. In the case of sodium, the greatest concern has been the sampling from loops at elevated temperatures. By different valving arrangements it is possible to direct small amounts of the liquid metal into analytical systems from flowing sources [25–27]. In a typical procedure [28], a vacuum-activated dip-tube sampler is employed. It is evacuated and purged with inert gas, then inserted through a ball valve into the liquid-metal system. A valve to an evacuated chamber at the upper end of the tube is opened, causing the molten metal to rise in the tube; the tube is then crimped to retain the sample. The use of this sampler is restricted to systems with internal pressures of less than 2 psig. It has been used successfully at temperatures in the range of 1100°C. Another sampling device that has been tested consists of a bypass circuit operating across a pressure drop produced by a component such as a pump, valve, or orifice. The bypass circuit, which contains four valves, is heated to the same temperature as the main stream fluid before the valves are opened. After 30 min or more, the valves are closed and the removable section is crimped, allowed to cool, and removed for analysis.

C. LIQUID METALS IN STATIC SYSTEMS

Figure 10.1 illustrates a typical setup [20] for transferring liquid alkali metals into a Pyrex bulb. To transfer the alkali metal from the container, which is under 20 psi pressure of argon, into the weighed sample bulb, the

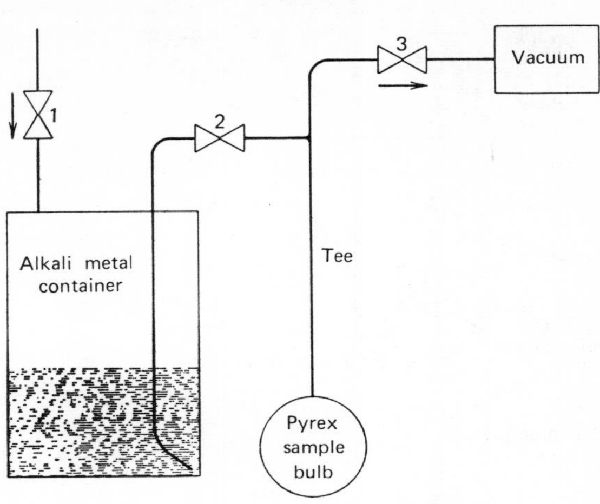

Fig. 10.1. Sampling of alkali metal from transfer container [20].

system is several times alternately evacuated to less than 1 torr and filled with argon (20 psi). During this operation valves 1 and 2 are closed and valve 3 is open while the alkali metal container, valves, and tee are heated. With vacuum to 2, and valve 3 closed, valve 2 is opened, whereupon the sample enters the sample bulb. Valve 2 is now closed, and vacuum is applied at valve 3. Valve 3 is subsequently closed and the sample-bulb stem is sealed with a torch.

The above technique is impractical for sampling lithium, because of the incompatibility of glass and liquid lithium. In this case the sample is transferred to high-density polyethylene tubing. A detailed method for transferring liquid cesium under oil and an argon atmosphere for glass-vial filling has been described by Brotherton et al. [29].

Both ORNL [30] and the Industrial Groups of the United Kingdom Atomic Energy Authority (UKAEA) [31] have designed metal-to-glass apparatus for the removal of alkali metal from static containers. The ORNL procedure involves a nickel-capsule sample-system, with improved sliding seals and ball valves. The UKAEA method was successfully applied to the syphoning of NaK to glass bottles using nitrogen as a carrier gas.

An efficient transfer system for cesium has been suggested by the U.S. Industrial Chemicals Company and is illustrated in Fig. 10.2 [32]. The sampling and transfer systems are purged with purified argon, as is indicated by the arrows, first with valve 3 closed and finally with valve 3 open. The cesium supply container is connected in a dry box to the transfer apparatus. With valve 3 closed and the whole system slightly warmed, purified argon is

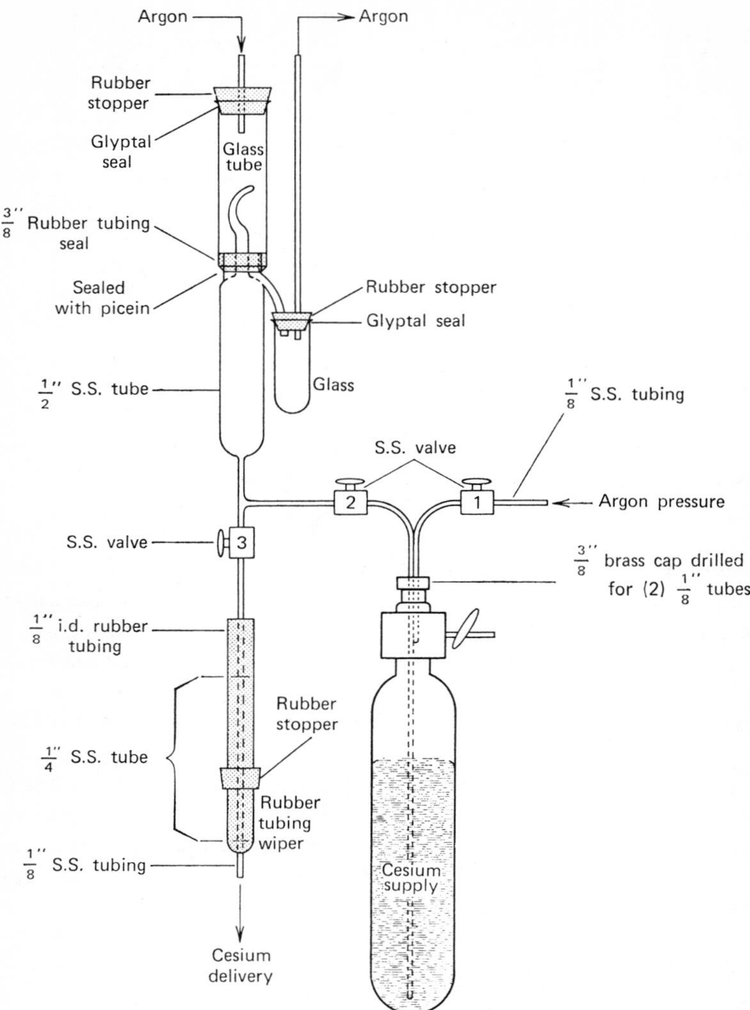

Fig. 10.2. Cesium constant-quantity transfer system [32].

introduced through valve 1. The cesium is subsequently forced, by manipulating carefully valve 2, into the $\frac{1}{2}$-in. stainless-steel tube, which is of known volume. Valve 2 is closed when a few drops of cesium overflow into the upper glass tube. Valve 3 is now opened to deliver the known volume of cesium metal. The rubber stopper indicated in Fig. 10.2 can be seated in the neck end of a flask. Obviously, the delivery tube can be so arranged to transfer the cesium metal into any desired analytical system.

D. SOLID METALS

Although there is always a danger that samples taken from a solid alkali metal may not be representative, a number of such procedures have been given. The Pepkowitz procedure [33] is to freeze NaK in an open glass container placed on a bed of solid carbon dioxide. Small pieces of metal can be cut off and transferred to an appropriate analytical system, thus avoiding dry-box techniques. Moisture from the surrounding air reacts too slowly with the binary alloy to have appreciable analytical consequences.

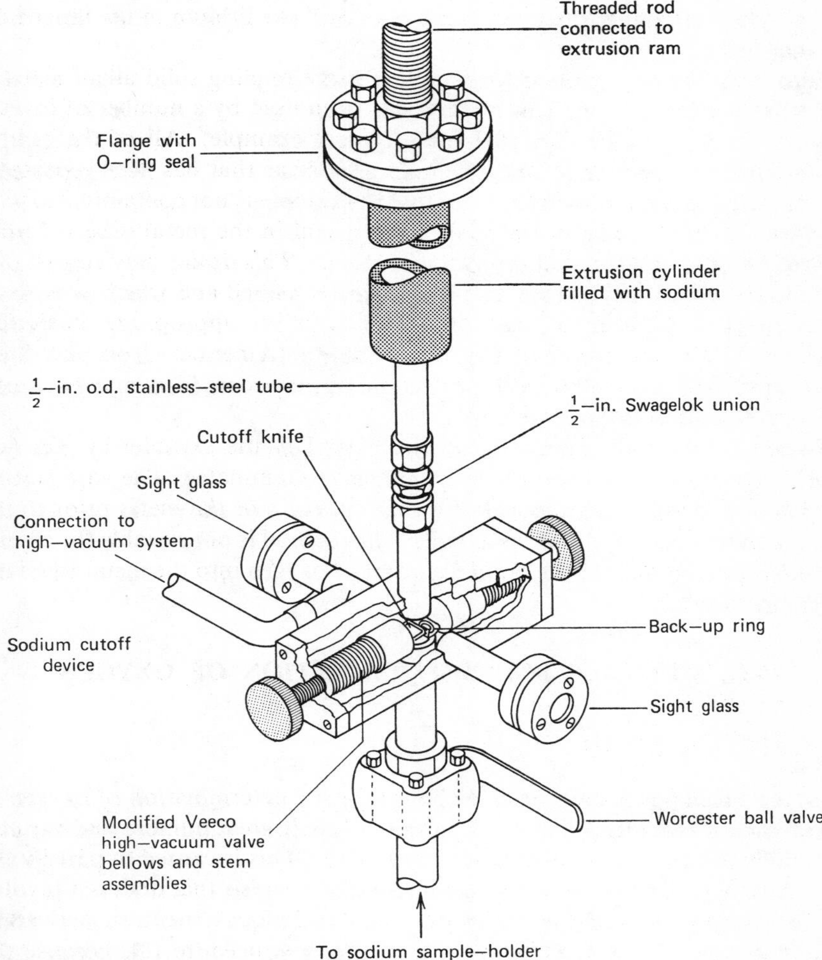

Fig. 10.3. Sodium extruder and cutoff device [38].

While this procedure has application for the lighter alkali metals, cesium reacts with carbon dioxide even at 0°C and with ice as low as −116°C. On the other hand, lithium forms an oxide film on the surface, which effectively allows one to handle the exposed lithium metal in air. A sampling technique common in the lithium industry [34] involves casting the molten metal in a $\frac{3}{4}$–1-in. internal-diameter stainless-steel tube and then taking transverse slices for analysis. The exposed ends may be dipped momentarily in water and then benzene before analysis. Sometimes the lithium is cast in pie-plate form, and then "drillings" are taken with a cork-borer. Pratt and Whitney [35] developed a modified stainless-steel pipet containing a hidden internal tube. After sampling, the pipet is cut open and the lithium in the inner tube is analyzed.

Probably the best method for handling and sampling solid alkali metal is that based on extrusion. This method has been used by a number of investigators [3, 36–39] (see Fig. 10.3 for a typical example). All of the sample manipulations are carried out within an apparatus that has been repeatedly evacuated and purged with argon so that the sample is not contaminated with oxygen. In brief, some of the solid alkali metal in the metal tube is forced out of one end and then cut off by a wire device. This device may consist of a thin knife, which in some systems is electrically heated and which provides a convenient 1–3-g sample that drops off into an appropriate analytical system. The alkali metal in the metal tubing (American Iron and Steel Institute type 316 stainless steel, or zirconium) may come from many sources, but typically it is from a hot trap.

When tubing with sample is cut and placed in the extruder system (see Fig. 10.3), the exposed end of the sample is contaminated. The wire-cutting device provides for the removal of a small portion of the metal prior to the extrusion and cutting-off of the sample. The method is not suitable for cesium or rubidium, because of the rapid diffusion of oxygen into the metal when the end is exposed.

VII. METHODS FOR DETERMINATION OF OXYGEN

A. INTRODUCTION

Several techniques have been proposed for the determination of oxygen in alkali metals (see also Chapter 9). In spite of their great number and variety, it is difficult to get agreement between methods. This is caused in part by the fact that there is no proven absolute method of analysis that does not involve some assumptions. Of the many proposed techniques, neutron activation seems to hold the greatest promise as a referee procedure [8], because the results are independent of the chemical form of oxygen in the sample. The fluorination method may yet be considered as a referee method because of its applicability to the determination of all oxygen contained in the sample, regardless of form. This evaluation of methods, which in part relies on projection into the future, eliminates the popular amalgamation and alkyl halide procedures, as well as the liquid ammonia technique as is evident from the ensuing discussions.

B. AMALGAMATION

1. Introduction

Pepkowitz and Judd [15] were the first to report on the use of the amalgamation method for the determination of oxygen in sodium. This procedure is based on the extraction of the alkali metal into mercury. This amalgam is soluble in an excess of mercury and is removed by successive mercury washes; the alkali metal oxide is insoluble and floats on the surface of the pool. When it has been determined that the sodium has been removed (see below) and the insoluble oxide is floating on the pure mercury, water is added to form the hydroxide, which is subsequently titrated with standard acid. Williams and Miller [40] improved the original procedure by simplifying the glass apparatus and by employing a vacuum technique. A similar apparatus is reproduced in Fig. 10.4 [41]. Vacuum stopcocks are used. The amalgamation chamber has a height of 30 cm and a diameter of 2.5–3 cm. The mercury reservoir has a capacity of 100–150 ml. From the reservoir, the mercury passes through a coarse fritted-glass disk into the amalgamation chamber. In this way, the mercury is introduced in the form of a "shower" that effectively washes the sides of the reaction chamber.

2. Procedure

The apparatus is assembled, as is indicated in Fig. 10.4, and connected to a high-vacuum–inert-gas manifold. While the whole apparatus is being flamed with a Bunsen burner to remove adsorbed gases and moisture, it is evacuated to below 10 μ and then flushed with purified argon (a very efficient purification train consists of hot—350°C—NaK, cold NaK, and zirconium chips at 650–760°C [42]). Then the apparatus is re-evacuated to below 10 μ, the lower stopcock is closed, and the sample bulb is broken by allowing the iron slug to fall upon it from a height of several centimeters. (The slug is also used as a stirrer and is manipulated externally by the use of a hand magnet). Approximately 20 ml mercury is added from the mercury reservoir while stirring vigorously with the iron slug. After the amalgam has cooled to room temperature, the bottom stopcock is opened and all but 1–2 ml of the amalgam is drained into the receiving flask. The surface of the mercury should not

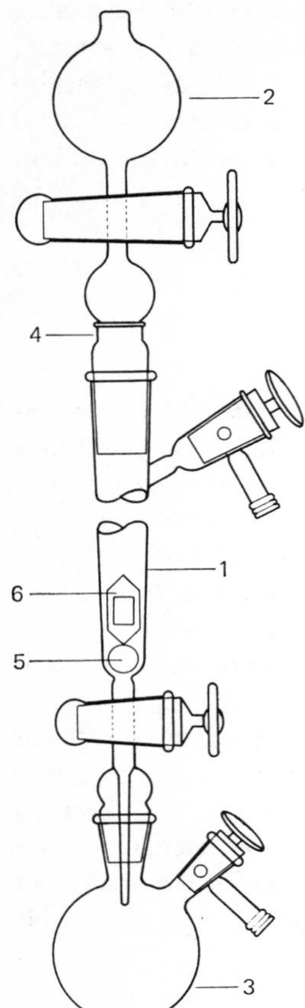

Fig. 10.4. Amalgamation apparatus [41]. 1, Amalgamation chamber. 2, Mercury reservoir. 3, Amalgam receiver. 4, Coarse fritted-glass disk. 5, Sample-ampul. 6, Glass-enclosed iron slug.

be allowed to fall to the stopcock opening; otherwise, some of the powdery oxide floating on the surface of the metal will be lost. More mercury is added, stirred, and drained, and this process is continued until an addition of water to the effluent mercury and a subsequent phenolphthalein test indicates complete removal of the sodium metal.

The amalgamation chamber is brought to atmospheric pressure, and both the mercury reservoir and the amalgam receiver are removed. About 20 ml water, previously boiled and adjusted if necessary to pH 7.0, is added to the amalgamation chamber. Then this solution is poured into a 150-ml beaker (be careful not to pour out any mercury). The mercury is transferred to another beaker and washed several times with water, the washings being combined with the main solution. This solution is then titrated with 0.005N hydrochloric or sulfuric acid to a phenolphthalein endpoint. The calculation is as follows:

$$\text{mg oxygen} = \text{ml acid} \times 0.005 \times 8.00. \tag{7}$$

The alkali sample weight, if it is unknown, is determined by adding a measured excess of 2.5*N* HCl to the combined amalgam with vigorous stirring and back titrating with standard sodium hydroxide to a phenolphthalein endpoint. It has been shown [36] that at least in the case of potassium, the alkali metal can be extracted from the amalgam with boiled distilled water, followed by titration with standard acid.

3. Modifications

To avoid contamination with oxygen and/or moisture, a number of improvements in equipment and manipulations have been introduced. Typical is the NASA apparatus [36] illustrated in Fig. 10.5. It features a sample container consisting of a stainless-steel tube $6\frac{3}{4}$ in. long and $\frac{5}{16}$ in. in diameter. It is vacuum-filled with liquid metal, capped, and stored under vacuum until used. For the determination of oxygen, the sample tube is mounted in the extruder-section apparatus. After a vacuum of about 10^{-5} torr has been obtained, a small portion of the sample is extruded from the tube, cut with a hot wire, and discarded into the waste tray. The analytical sample is then extruded, cut off, and dropped into the glass extraction system. The extruder is valved off, and the ball check valve is seated in place to confine the amalgamation reaction. After the amalgamation and separations the oxide is determined.

Instead of determining the oxide content of the alkali metal by titration, as described above, some analysts [15, 43, 44] applied flame photometry. A unique finish was described by Steinmetz and Minushkin [3]. In their procedure it is necessary only to remove the bulk of the sodium by amalgamation. A purified alcohol, such as ethylene glycol monoethyl ether, is added to dissolve the residual sodium oxide. The alcoholic solution is combined with an organic acid (such as salicyclic acid) to form an amount of water that is equivalent to the sodium oxide. The water is then determined by titration with Karl Fischer reagent. The pertinent chemical equations are as follows:

$$Na + Hg \rightarrow Na(Hg) \tag{8}$$

$$2Na(Hg) + 2ROH \rightarrow 2NaOR + H_2 + 2Hg \tag{9}$$

$$Na_2O + ROH \rightarrow NaOH + NaOR \tag{10}$$

$$NaOH + HA \rightarrow NaA + H_2O \tag{11}$$

$$NaOR + HA \rightarrow NaA + ROH \tag{12}$$

$$SO_2 + I_2 + 3C_5H_5N + CH_3OH + H_2O \rightarrow C_5H_5NHSO_4CH_3 + 2C_5H_5NHI. \tag{13}$$

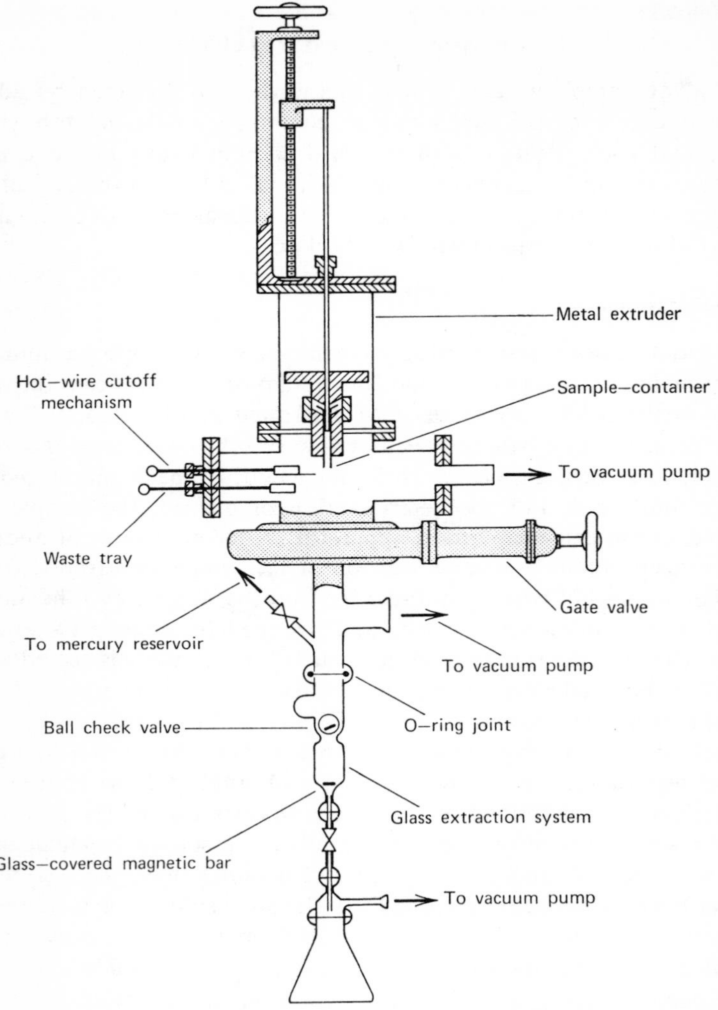

Fig. 10.5. Alkali metal analytical apparatus. (Reproduced by permission from *Analytical Chemistry* [36].)

In these equations ROH represents the alcohol and HA represents the organic acid. Equation 13 gives the reaction of Karl Fischer reagent with water.

4. Discussion

The applicability of the amalgamation procedure down to a level of 5–10 ppm oxygen in sodium, NaK, and potassium has been verified by the addition of known amounts of mercuric oxide to liquid sodium, and K_2O and KO_2 to liquid potassium at the time of the amalgamation. Although the reduction of KO_2 occurs above 300°C, the temperature rises enough during the amalgamation of potassium to cause the mercury to boil, and it is assumed that any KO_2 present is converted to K_2O.

One difficulty with the amalgamation procedure (and with the alkyl halide method discussed next) is that other alkali products, such as carbide, carbonate, hydride, hydroxide, and nitride, all contribute to the alkalinity after hydrolysis. The exact behavior of these compounds is in doubt, and a large error may result if all of these impurities are considered to react like an oxide.

There have been many unsuccessful attempts to apply the amalgamation method to the determination of oxygen in lithium [42]. Room-temperature amalgamation takes several days. Slight heating produces only incomplete amalgamation, while higher temperatures initiate an exothermic reaction which severely etches the glass chamber and causes high oxygen results. Attempts have been made to prevent the contact of hot lithium metal with glass by either introducing the sample (taken as a dip sample from molten lithium metal with a nickel bucket) below the surface of the mercury during the initial amalgamation or by using a protective metal cylinder or an all-metal amalgamation chamber. These modifications have been only partially successful. Lithium forms nitrides more readily than sodium or potassium, and the nitride content of metallic lithium is usually comparable to the oxide content. In addition, the hydride of lithium, and most probably the carbide and phosphide, are considerably more stable than the corresponding sodium and potassium compounds. Thus, while the nonspecificity of the amalgamation procedure is usually not too serious a disadvantage in the case of sodium and potassium, it may be serious in the case of lithium.

The amalgamation procedure is applicable neither to rubidium nor cesium [42, 45]. Standard additions of oxygen to cesium or rubidium are not quantitatively recovered because of the solubility of the oxide in the alkali amalgam and/or mercury. A modified method in which sodium amalgam is reacted with cesium oxide to form sodium oxide and cesium amalgam shows some promise but is fraught with operational difficulties.

To summarize, the amalgamation procedure is applicable only to sodium, potassium, and NaK.

C. ALKYL HALIDE

1. Introduction

The alkyl halide method introduced by White et al. [46] involves a modified Wurtz reaction between sodium and an alkyl halide using a hydrocarbon as a diluent:

$$2Na + RX \rightarrow RNa + NaX \tag{14}$$

$$RNa + RX \rightarrow NaX + R{-}R. \tag{15}$$

No reaction takes place between sodium oxide and the alkyl halide. After completion of the reaction, water is added and the sodium hydroxide is determined by titration with standard acid solution.

In the original method, a 1–2 g sample of sodium, sealed in glass, is placed in a 4-cm-by-20-cm tube with a reinforced bottom and containing 100 ml of a 50% (v/v) solution of purified *n*-butyl bromide in purified *n*-hexane. When handling different sample sizes, the reagent volume and tube size are adjusted accordingly. The glass capsule containing the sample is crushed by means of a heavy glass rod flattened at one end. As the reaction proceeds, the glass rod is used to expose continually fresh sodium to the reagent. The tube is heated at 60–70°C to complete the reaction, which takes 1–2 h. Care must be taken to prevent excessive boiling of the solution and subsequent loss of hexane.

The residue is cooled and pulverized to ensure the complete reaction of small particles of sodium. Subsequently, 100–200 ml of water from which carbon dioxide has been removed are added to dissolve the residue and to form a two-phase system consisting of an aqueous solution containing sodium bromide and sodium hydroxide and an organic phase containing *n*-hexane, excess reagent, and the newly formed octane. The phases are separated in a separatory funnel, and the water phase is potentiometrically titrated to pH 7 with 0.001*N* nitric acid. The volume of titrant is used to calculate the oxygen content:

$$\text{g oxygen} = \text{meq. acid} \times 0.008. \tag{16}$$

The sample weight is determined by titrating an aliquot of the neutralized solution with silver nitrate solution:

$$\text{g sodium} = \text{meq. silver nitrate} \times 0.023. \tag{17}$$

While the original work was carried out with a solution of *n*-butyl bromide in hexane [46], some investigators for various reasons preferred solutions of *n*-amyl chloride in hexane [38, 47]. Butyl bromide was purified by White et al. [46], by treatment with sulfuric acid to remove unsaturated hydrocarbons. The butyl bromide was subsequently neutralized with sodium carbonate, dried with calcium chloride, separated by distillation, and stored in an argon atmosphere in contact with activated alumina. A more efficient purification can be achieved by passing butyl bromide or amyl chloride through columns containing a 5:1 mixture of silica gel and Celite [38, 48]. The alkyl halide thus prepared contains less than 0.5 ppm of water. Traces of amyl alcohol in

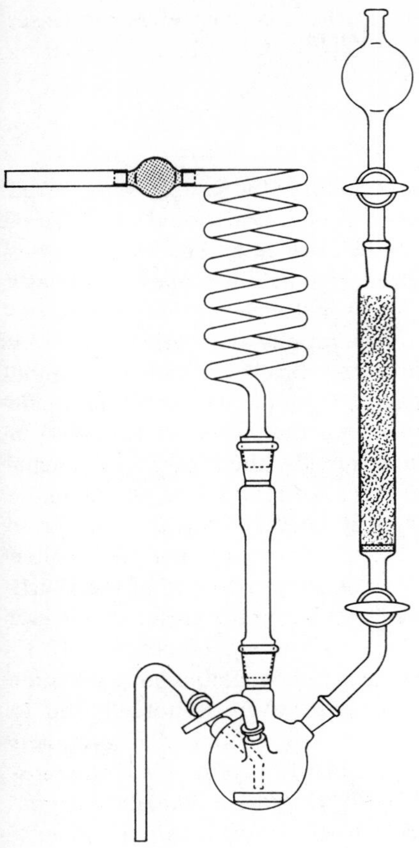

Fig. 10.6. Wurtz reaction apparatus.

amyl chloride can be removed by extraction with water, followed by passage of the amyl chloride through a silica-gel–Celite bed [38].

Hexane was purified by drying and then distilling over sodium ribbon. The resulting product was stored in a dark bottle containing sodium under oxygen-free argon [46]. Hexane can also be purified by passing it through a silica-gel–Celite column, followed by storage over anhydrous phosphorus pentoxide [38, 48]. The Wurtz reagent used for the determination of oxygen in sodium and consisting of a mixture of 60% *n*-amyl chloride in *n*-hexane (12% amyl chloride for potassium [49], 10% butyl bromide for NaK [48]) can be purified conveniently just prior to use by adding it to the reaction flask through a chromatographic column containing silica gel and/or freshly activated alumina [47, 49] (see Fig. 10.6). Tubing to the reaction vessel delivers purified argon, helium, or nitrogen as a blanket gas. Joined to one of the ground-glass openings of the reaction flask is either a copper reflux condenser [47] or a conventional water condenser [49] to prevent loss of hexane during the vigorous reaction.

2. Sample Introduction

In the original method the glass capsule containing the sample was crushed inside the reaction flask with a heavy glass rod [46]. Occasional breakage of the flask and subsequent loss of sample can be avoided by breaking the glass capsule in the reaction flask with modified channel-lock pliers with jaws bent at right angles to the handle [48]. Another device consists of a screw-actuated jaw, similar to a vise [20]. These operations are carried out under an inert-gas blanket, while the reaction flask contains dry hexane to a height of about 1 in. above the glass capsule. In another technique a scratch is made around the capsule with a diamond pencil and the container is cooled in liquid nitrogen to well below −100°C. Subsequently, the capsule is momentarily withdrawn from the liquid nitrogen, and a hot glass rod is placed on the scratch to crack the capsule, while leaving it intact. When the capsule is heated in the sampling apparatus to 110°C, it breaks and the molten sodium is transferred to the reaction flask prior to the addition of the Wurtz reagent [47]. Samples of NaK are solidified by placing the glass capsule in a reaction vessel containing hexane, followed by cooling to below −10°C. The capsule is then broken, the excess hexane is removed, and butyl bromide is added to initiate the reaction [48]. The temperature is not allowed to exceed 35°C until the bulk of the NaK has reacted. Sealed metallic containers holding sodium can be opened under mineral oil. The sodium is then transferred to a flask and melted in *n*-nonane at 100°C under a blanket of argon. The *n*-nonane is replaced by hexane prior to the addition of the *n*-butyl bromide [46]. A superior method of introducing the sample into the reaction flask is based on extrusion of the alkali metal in a stainless-steel tube [38].

3. Measurement of Oxygen Content

The actual measurement of the oxygen content of the alkali metal is carried out by titration of the alkali hydroxide with a standard acid. While the original method describes the potentiometric titration of the alkali hydroxide in the aqueous extract with $0.001N$ nitric acid [46], modifications of the procedure involve various acid–indicator combinations [31, 50]. Since the detection of the endpoint in aqueous media containing considerable amounts of alkali halides is difficult, it has been suggested that the reaction mixture be treated with a 90–95% (v/v) mixture of ethanol and water [49, 51]. The solubility of the alkali halide in this mixture is relatively small, and the titration of the alkali hydroxide can be carried out with improved precision.

A unique method for determining the oxygen content consists of converting the original sodium monoxide to carbonate by adding carbon dioxide and measuring the carbonate absorption band at 11.38 μ by an infrared-spectroscopic technique. After the Wurtz reaction, the sodium chloride–sodium oxide mixture is filtered and washed with *n*-hexane. The dry powder is dissolved in demineralized water saturated with carbon dioxide. The solution is filtered to remove any impurities that might interfere in the final measurement, evaporated to dryness, and heated at 200°C to convert any bicarbonate to carbonate. The powder is ground and pressed into a transparent disk, and its absorbance is measured [47].

4. Application of Method

The suitability of the method for sodium, potassium, and NaK appears well established. White et al. [46], realizing that the reactions

$$Na_2O + C_4H_9Br \rightarrow C_4H_9ONa + NaBr \tag{18}$$

$$C_4H_9ONa + C_4H_9Br \rightarrow C_4H_9OC_4H_9 + NaBr \tag{19}$$

might take place, conclusively showed that no ether is formed if the reaction temperature does not exceed 65°C. Upon hydrolysis, C_4H_9ONa yields an equivalent amount of hydroxide and therefore does not affect the results. On the other hand, it was shown that the oxides of rubidium and cesium may react with butyl bromide to give the metal halide and dibutyl ether [20].

The presence of oxides other than the monoxide will lead to erroneous results. Samples that are suspected of containing higher oxides may be heated in the presence of excess metal to effect reduction of these oxides to the monoxide. However, in the case of cesium, Cs_3O remains unreacted after the Wurtz reaction [22] and may react as follows:

$$2Cs_3O + 4H_2O \rightarrow 6CsOH + H_2, \tag{20}$$

giving a more alkaline solution, which results in an apparently higher oxygen content. The alkyl halide method is therefore not suited for the determination of oxygen in rubidium and cesium [45].

In addition, the alkyl halide method cannot be applied to lithium because some of the C_4H_9Li formed as an intermediate in the reaction

$$2Li + C_4H_9Br \rightarrow LiBr + C_4H_9Li \tag{21}$$

vigorously reacts with the water added later, yielding a large amount of LiOH.

One major disadvantage of the alkyl halide method is the uncertainty about the effect of other interstitial impurities on the Wurtz reaction and the final titration. There are only scanty data on the reaction between an alkyl halide and an alkali metal carbide, carbonate, hydride, hydroxide, or nitride [49]. If metal alkyls are formed, then a "false" hydroxide would result on hydrolysis. In addition, if these compounds do not react with the alkyl halide in the same manner as Na_2O and K_2O, they will hydrolyze as follows:

$$NaH + H_2O \rightarrow NaOH + H_2 \qquad (1 \text{ ppm H} = 8 \text{ ppm O}) \tag{22}$$

$$Na_3N + 4H_2O \rightarrow 3NaOH + NH_4OH \qquad (1 \text{ ppm N} = 2.3 \text{ ppm O}). \tag{23}$$

Carbonate would react with the acid titrant and would also be counted as oxygen.

The alkyl halide method is therefore applicable only to sodium, potassium, and NaK and is subject to a number of uncertainties.

D. FLUORINATION

Methods involving fluorination are based on the reaction between bromine trifluoride and metal oxides resulting in the liberation of elemental oxygen. The reaction

$$6MO + 4BrF_3 \rightarrow 6MF_2 + 2Br_2 + 3O_2 \tag{24}$$

was first mentioned by Emeleus and Woolf [52] and later applied by several investigators to various systems [53–55]. Because of the specificity of the reaction, it looks very attractive for the determination of oxygen in alkali metals. Goldberg [56] used the less volatile potassium bromotetrafluoride, the product of the reaction between potassium fluoride and bromine trifluoride. After determining the oxygen-plus-nitrogen content, the gases were passed over heated copper to remove the oxygen. The unreacted nitrogen was then measured manometrically. Kirtchik et al. [57], after many unsuccessful attempts to use oxygen-free fluorine or anhydrous hydrogen fluoride, finally concentrated their efforts on the application of bromine trifluoride to the determination of oxygen in potassium. Preliminary tests indicated that under specific conditions bromine trifluoride can be made to react in a controllable manner with potassium metal without difficulties. The method for measuring oxygen involves principles and apparatus recommended by Brady [58]. In essence, oxygen is absorbed in an alkaline solution of sodium anthraquinone β-sulfonate, the solution changing from red to colorless. The measurement for oxygen is made photometrically.

It is important to note that because of the corrosiveness of both bromine trifluoride and potassium bromotrifluoride, the reaction cell and other parts of the apparatus must be constructed of Monel, nickel, or inert fluorinated plastic.

E. METHANOL SOLUTION AND KARL FISCHER TITRATION

It was pointed out before that neither the amalgamation nor the alkyl halide method is applicable to lithium. Several different approaches have been used to determine the oxygen content of this metal. One method involves the conversion of the oxygen in the lithium to water and a subsequent dead-stop titration of this water with Karl Fischer reagent [59]. The following chemical reactions occur:

$$2Li + 2CH_3OH \rightarrow 2LiOCH_3 + H_2 \tag{25}$$

$$Li_2O + CH_3OH \rightarrow LiOH + LiOCH_3 \tag{26}$$

$$LiOCH_3 + C_6H_4OHCOOH \rightarrow C_6H_4OHCOOLi + CH_3OH \tag{27}$$

$$LiOH + C_6H_4OHCOOH \rightarrow C_6H_4OHCOOLi + H_2O \tag{28}$$

$$SO_2 + I_2 + 3C_5H_5N + CH_3OH + H_2O \rightarrow C_5H_5NHSO_4CH_3 + 2C_5H_5NHI \tag{29}$$

Nitride, carbide, and hydride do not interfere. Special precautions are necessary to prevent the introduction of contaminants, such as air or moisture, into the analytical apparatus. Various refinements in the titration procedure are necessary to obtain the desired accuracy and sensitivity. The reproducibility of this method is 10–20 μg oxygen. One serious disadvantage of this method is that a large blank correction is necessary for the water content of the methanol even though the methanol can be carefully dried to a water content of 5–10 ppm. The blank can be reduced by substituting distilled and molecular sieve-dried butyl cellosolve for the anhydrous methanol originally proposed [60]. The method can be further improved by coulometric generation of the Karl Fischer reagent. The resulting method is, however, so tedious and complex as to make it of little value.

F. LIQUID AMMONIA FOR OXYGEN IN LITHIUM

In this method (60), the lithium metal sample is transferred to a fritted glass filter and dissolved in purified liquid ammonia (distilled from a flask containing sodium) in an evacuated system. Lithium carbonate and lithium oxide are insoluble and are retained by the fritted glass filter. Nitrogen dissolved in lithium samples is not retained by the filter as indicated by analysis of the residue. After the reaction is completed and the system is equilibrated with purified argon, the residue is dissolved in water and the titration of lithium hydroxide and lithium carbonate is carried out by coulometric generation of hydronium ions in a 0.01 *M* sodium sulfate medium. It is stated that the titration curves almost always reveal the presence of carbonate in the solution of the residues in an amount approximately equivalent to the total carbon content of the lithium.

Only a limited amount of work has been done thus far to apply the liquid ammonia procedure to other alkali metals. There are indications that this technique may be applied to K, Na, and Cs.

G. DISTILLATION

This method has been applied to sodium, potassium, and NaK at several installations [61–64] and involves the removal of the alkali metal from the alkali oxide by distillation at 425°C at a pressure of less than 5 μ. In the case of sodium, a 400°C difference between the boiling point of sodium and the sublimation point of sodium oxide allows the determination of oxygen in highly purified sodium. The residue is dissolved in water and either titrated to determine the hydroxide or analyzed flame-photometrically to determine the sodium content. An evaluation of the procedure by Griesser et al. [65] indicates that the vacuum-distillation technique can be made to give reproducible results only with the most exacting care and diligence. The method is subject to errors because other impurities may remain in the residue and contribute to the amount of sodium determined.

H. PLUGGING METER

The plugging meter operates as a saturable-impurity monitor in which the plugging temperature of an in-system multiorifice plate is determined. The plugging temperature is then interpreted in terms of the solubility–temperature relation of the saturable impurity [66]. Unfortunately, it lacks the ability to determine the identity of the impurities that are separated, and therefore, some speculation is involved in interpreting the plugging temperature as an indication of oxygen level in any but an ideal sodium–sodium monoxide system [65]. It has been noted that the plugging temperature generally is different from true saturation temperature. It appears to be influenced by meter design and geometry, operating variables, and the separation characteristics of the specific impurities [67]. A fully automatic plugging meter has been introduced that gives a continuous indication of incipient plugging temperature [68]. Cooling is adjusted to the flow of sodium so as to maintain the flow at a constant value corresponding to incipient plugging of the orifice. The flow-meter output controls the cooling of the sodium. A rise in impurity content tends to plug the orifice and decrease the flow. The controls then act to increase the temperature at the orifice and maintain a constant flow. Thus, the measured orifice temperature shows continuously the plugging temperature of the sodium. While this device cannot differentiate between carbide, oxide, and nitride impurities, it undoubtedly represents a valuable tool in "on-line" monitoring of high-purity sodium. It must be emphasized that this technique is not applicable to rubidium and cesium, because of the great solubility of the various oxides in the parent metal.

I. ELECTROCHEMICAL OXYGEN METER

This device [67, 69, 70] is specific for oxygen in sodium. The sensing element consists of a galvanic cell formed by a closed-end ceramic tube that dips into the sodium stream. The ceramic is a solid electrolyte that, under proper conditions, conducts electricity at elevated temperatures solely by the movement of oxygen (hence, other impurities cannot interfere). The inside of the cell at the closed end contains a reference electrode that provides electrical contact with the inner ceramic surface and a fixed oxygen concentration. The sodium provides electrical contact with the outer ceramic surface and an unknown, but lower, oxygen concentration. An emf develops across the cell wall because of the difference in oxygen activity at its two faces. Under the proper conditions the cell emf is proportional to the logarithm of the oxygen concentration in the sodium and hence can be used as a sensitive measure of that concentration. The cell voltage is approximately 1 V. A twofold change in oxygen content (i.e., 10 to 20 ppm, 100 to 200 ppm, 1000 to 2000 ppm) changes the cell voltage by about 17 mV. A 10% change in oxygen concentration produces a 2.5 mV. change in cell voltage. Meters using ThO_2–Y_2O_3 electrolyte and a Cu–Cu_2O powder reference electrode have been most successful. The device described above is intended as a continuous on-line oxygen meter.

J. FREEZING-POINT DEPRESSION

This procedure appears to be a proven approach to determine oxygen in rubidium and cesium [42, 45]. It was observed that hot trapped (low-oxygen) rubidium exhibits a freezing point of 38.98°C. Oxygen additions, varying from 20 to 900 ppm, clearly indicate that this freezing point is depressed 0.0017°C/ppm oxygen. The curve appears to be linear to at least 900 ppm. It has been stated that the cesium–oxygen system has a eutectic at 10,000 ppm oxygen with a freezing point at −2°C [22]. Freezing points of cesium containing measured oxygen additions give a linear relationship up to 500 ppm oxygen with a depression of 0.0024°C/ppm oxygen. The apparatus is fitted with a thermometer well and a thermistor well and provisions are made for the quantitative addition of gaseous oxygen. The change in temperature during cooling is monitored with the aid of a thermistor bridge, the unbalanced potential of which is fed to a recorder. The recorded cooling curves show thermal arrests upon change of phase. Temperatures are read to $\pm 0.002°C$. There apparently is no statement in the literature on the effect of other interstitials on the depression of the freezing point.

K. RESISTIVITY

A 1 ppm increase in oxygen content increases resistivity in liquid alkali metals by about 1 part in 10^4. This effect is exploited in an electrical-resistivity meter (rhometer) [71, 72], the use of which, however, is not specific for oxygen. In one method of analysis the liquid alkali metals are brought in contact with an electrode of a metal (e.g., Zr, Hf, Al) that forms an oxide more stable than that of the alkali metal; the rectilinear rate of increase in electrical resistance of the circuit is caused by the oxidation of the metal electrode [73].

L. ACTIVATION ANALYSIS

Potentially the most satisfactory approach for determining oxygen in alkali metals is based on neutron activation (see also Chapter 3). This technique has received considerable attention [74a]. Activation analysis has the advantage over most of the methods previously discussed, because it is specific, rapid, and nondestructive. Further, the results obtained are practically independent of the matrix. There remain some problems of sample handling or preparation. As soon as they are solved, activation analysis, and more specifically fast-neutron activation analysis, will probably be the standard method for determining oxygen in alkali metals.

There exist a number of nuclear reactions applicable to the determination of oxygen in alkali metals, some examples of which are listed below:

$^{18}O(p, n)^{18}F$	[75]
$^{16}O(d, n)^{17}F$	[76]
$^{16}O(t, n)^{18}F$	[77]
$^{16}O(\gamma, n)^{15}O$	[78]
$^{18}O(n, \gamma)^{19}O$	[79]
$^{17}O(n, \alpha)^{14}C$	[80]
$^{16}O(n, p)^{16}N$	[81–83].

Of the above approaches, the (t, n) reaction has been used to some extent [84]. In this method [85] the alkali is surrounded by lithium fluoride during the activation. The thermal-neutron reaction $^{6}Li(n, \alpha)^{3}H$ produces a source of tritons, which in turn activate the oxygen nuclei in the sample to produce ^{18}F by the secondary reaction, $^{16}O(H_3, n)^{18}F$. Oxygen has an almost zero cross-section for thermal neutrons. Fluorine-18 decays by positron emission with a half-life of 112 min. The positrons (0.694 MeV) combine with electrons to produce annihilation γ-radiation (0.51 MeV). This γ-radiation is used to complete the analysis by γ-scintillation spectrometry. One of the complications that is limiting the use of this method is the relatively high oxygen content of the lithium compounds that are employed.

Of the four fast-neutron-induced reactions of the oxygen isotopes that result in the formation of a radioisotope [74],

$$^{16}O(n, p)^{16}N$$
$$^{16}O(n, 2n)^{15}O$$
$$^{17}O(n, \alpha)^{14}C$$
$$^{18}O(n, \gamma)^{19}O,$$

the neutron–proton reaction with ^{16}O is the most suitable for the determination of small amounts of oxygen in alkali metal. The $^{16}O(n, 2n)^{15}O$ reaction requires neutron energies of 16.65 MeV, which are produced in quantity only by large accelerators. The $^{17}O(n, \alpha)^{14}C$ reaction requires excessively long irradiation times owing to the 5700-yr half-life of ^{14}C. The $^{18}O(n, \gamma)^{19}O$ reaction has a very small cross-section, and the isotopic abundance of ^{18}O is too small. The $^{16}O(n, p)^{16}N$ reaction produces a short-lived radioisotope (half-life = 7.14 sec). Moreover, it requires only neutrons of energies above about 10 MeV.

Neutrons in the 10–15 MeV range are available from nuclear reactors and from accelerators. In the case of reactors, the 10–15 MeV neutron flux is several orders of magnitude lower than the flux of very low energy neutrons because of the primary fission neutron spectrum and the great degree of moderation required to sustain the fission process. Accelerators, on the other hand, produce only fast neutrons, with energies depending on the nature and energy of the projectile and on the target material. Small inexpensive accelerators can produce these very energetic neutrons, but they are limited to the thermonuclear reaction $^{3}H(d, n)^{4}He + 17.6$ MeV (of which about 14 MeV go to the neutron produced and about 3.6MeV to the α-particle). Accelerators provide 14.7-MeV neutron-output rates (4π) of about 10^{11} neutrons/sec, which enables one to expose small samples (1–10 g) to 14-MeV neutron fluxes of about 10^{8}–10^{9} neutrons/(cm^2)(sec) [86]. The analytical technique for applying the (n, p) reaction to alkali metals [74a] and to cesium, in particular [74b], has been fully developed. Automatic transfer of the sample to the neutron field and back to the detector, gamma-counting, and monitoring of the flux has been programmed in a step sequence that requires approximately 1 min. Cesium has been sealed in aluminum capsules and analyzed for oxygen over a two-month period with excellent results. Metal capsules are not suitable for reactor irradiations, because of the amount of heat produced, but they are fine for accelerator irradiations. In its present form, the method has a potential limit of detection of about 10 μg oxygen (e.g., 10 ppm in a 1-g sample or 1 ppm in a 10-g sample). Since alkali metals must be kept in sealed containers to prevent oxidation, the current practical limit of detection is approximately 100 μg oxygen or 10 ppm in a 10-g sample (due to the blank correction needed for the oxygen content of the container itself). Certain problems in sample handling must yet be solved.

It is feasible to determine oxygen in nonradioactive sodium loops for its oxygen content in situ by using intermittent activation and counting periods. For example, in an illustrative case (2-in.-diameter pipe, sodium flow rate of 1 ft/sec, 100-sec activation and counting period) oxygen levels of 10 ppm and 100 ppm could be determined to ± 2 ppm and ± 7 ppm, respectively. At higher flow rates, the sensitivity and precision would be poorer.

If, instead, an in situ sample chamber is filled, activated for 18 sec, transferred to the counter chamber, and then counted for 30 sec, the same sensitivities and precision can be obtained (e.g., 10 $\pm$ 2 ppm, 100 $\pm$ 7 ppm) independent of flow rate. After activation and counting, the sample can be pushed back into the flow. It would never be taken out of the system; hence, no problems of exposure to air, controlled-atmosphere handling, or sample-container would be encountered [86].

The only element that can interfere with the radioactivation determination of oxygen is fluorine:

$$^{19}F + {}^{1}n \rightarrow {}^{16}N + {}^{4}He.$$

In the above reaction, the product nucleus, ^{16}N, is identical to that formed from oxygen, so that there is no way to discriminate between these two reactions by product-nucleus examination. The presence of fluorine in a concentration 2.44 times that of oxygen would cause a 100% error in the oxygen results.

Fortunately, ^{19}F also undergoes a neutron–proton reaction, $^{19}F(n, p)^{19}O$, to produce 29-sec ^{19}O. By establishing calibration curves for both reactions, the interference of fluorine with the oxygen determination can be eliminated. Fluorine–oxygen ratios as large as 10 can be tolerated before the signal–noise ratio becomes too small for accurate analytical measurements.

Other irradiation procedures can only be touched on here. Samples of 1–2 g of sodium were irradiated with 35 MeV bremsstrahlung from an electron accelerator with a beam current of about 30 μA [87] and ^{15}O was produced by the (γ, n) reaction on ^{16}O. The irradiated sodium was first etched in a mixture of ethanol and water (1:1) to remove surface contamination and was then dissolved in dilute NaOH under N. The ^{15}O present as $Na_2{}^{15}O$ was converted into a NaOH solution, a portion of which was distilled rapidly from the mixture and counted by coincidence spectrometry between two NaI(Tl) detectors. The activity of ^{23}Na in the distillation residue was used as an internal standard. The determination of oxygen by proton activation (10 MeV, 4 μA) has also been described [88].

VIII. METHODS FOR DETERMINATION OF HYDROGEN

A. DIFFUSION GASOMETRIC PROCEDURE

In this unique method for the determination of hydrogen in sodium and sodium–potassium alloys, the sample is sealed in an iron [89, 90] or nickel [63, 91] capsule previously dehydrogenated. The capsule is placed inside the vacuum system, which is then evacuated. The capsule is heated to 700°C, whereupon the hydrogen in the sample evolves and diffuses through the walls of the capsule. From a determination of the hydrogen pressure inside a static system, the hydrogen content of the sample can be computed. Hydrogen present as sodium hydroxide is recovered because at temperatures in excess of 450°C the following reaction takes place:

$$2Na + 2NaOH \rightarrow 2Na_2O + H_2. \tag{30}$$

An average deviation of 0.0010 wt % hydrogen has been reported. Claims have been made that diffusion of hydrogen is not complete and is controlled by an apparent equilibrium between the liberated hydrogen both inside and outside the capsule [92]. This method is most probably not satisfactory for hydrogen in lithium, because of the exceptional stability of lithium hydride.

B. VACUUM FUSION AND SIMILAR METHODS

Some preliminary work on the determination of hydrogen in lithium using a vacuum-fusion apparatus with a tin bath has been reported [93]. This approach seems promising. A similar method has been used for the determination of hydrogen in lithium hydride [94, 95]. It is probably applicable also to the determination of hydrogen in lithium and other alkali metals. In this method, the sample is placed in a tin or mercury bath, which is then heated. The hydrogen evolved is determined by a gasometric method.

C. AMALGAMATION

When analyzing NaK by this procedure [61], the sample in a metal bucket is placed in a glass vial, which is then sealed under vacuum. For analysis the vial is broken in an inert-atmosphere dry box, and the bucket that contains the NaK is transferred to the reaction vessel of an amalgamation apparatus.

The alkali metal is then removed by the amalgamation procedure (see Section VII.B), leaving only the unreacted hydrides, oxides, and carbon compounds in the reaction vessel. The reaction vessel is next connected to the gas-transfer apparatus. Deaerated water is added under vacuum to the residue and the liberated hydrogen is determined by gas chromatography.

To differentiate between hydrogen in sodium in the free state and hydrogen present as hydride, the sample is first treated with mercury. The heat of amalgamation is sufficient to liberate the hydrogen and 50–80% of the hydrogen present as NaH. Undecomposed hydride hydrogen is determined after decomposition of the residual sample with H_2O [96].

D. ISOTOPE DILUTION

The isotopic-dilution procedure, introduced in 1959 [92], represents at the present time the most favored procedure for determining hydrogen in alkali metals and is therefore discussed here in some detail. In this procedure (see also Chapter 7), deuterium gas is heated with the alkali metal sample to 460°C. Under these conditions there is a rapid exchange of hydrogen isotopes. The isotopic composition of the gas, in equilibrium with the alkali metal, is analyzed with a mass spectrometer. From this analysis the hydrogen content of the alkali metal can be calculated. The standard deviation is about 3 μg hydrogen. The method has the advantage that it does not depend upon the quantitative extraction of hydrogen from the sample. The apparatus for carrying out the exchange is shown in Fig. 10.7.

Prior to the analysis, borosilicate sample-tubes (one or two are used for blank determinations) are heated for half an hour at 520°C. While the tube is still warm, the desiccator-type stopcock is greased and attached. The tubes are weighed, evacuated, and transferred to the sampling site. An approximately 2-g sample is taken at the operating temperature of the facility and put in the tube, which is in a glove box filled with inert gas. After being cooled and weighed, the sampling tube is attached to the reaction apparatus and evacuated through S_T, S_2, and S_1. S_2 is now turned to evacuate the deuterium line. Subsequently, the capillary U-tube is immersed in liquid nitrogen, S_1 is closed, and S_4 is opened to admit deuterium to the pipet bulb in slight excess of 200 torr. With S_4 closed, the pressure is adjusted to exactly 200 torr by means of S_1. S_2 is rotated so that the deuterium, confined in the bulb at the observed pressure, volume, and temperature is expanded into the sample tube. Barometric pressure is used to push the mercury from the reservoir below, up through the bulb to stopcock S_T. S_T is closed, the mercury is returned to the reservoir, and the sampling tube is removed.

To equilibrate the deuterium with sample hydrogen, whether present as hydroxide, hydride, or dissolved hydrogen, the tube is heated in a tube

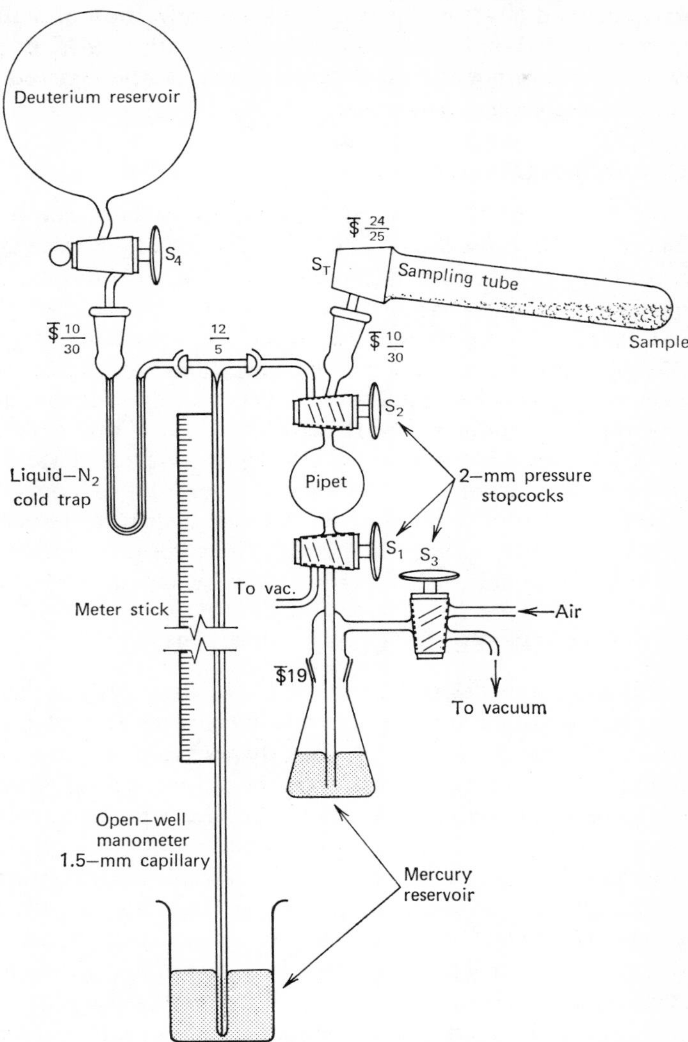

Fig. 10.7. Deuterium pipetter. (Reproduced by permission from *Analytical Chemistry* [92].)

furnace at 460°C for 5 min. It is then removed, cooled to room temperature, and flamed at its midsection for about half a minute with a burner to decompose the hydrides and deuterides deposited during the brief equilibration period. After being cooled again, the tube is attached to the mass spectrometer and the output of the mass numbers of H_2, HD, and D_2 are recorded.

The above method has been applied to the determination of hydrogen in sodium and NaK [92] and also to lithium and rubidium [42]. It has been shown [97] that deuterium and mass spectrometry can be replaced successfully by tritium and Geiger-counting.

E. MISCELLANEOUS

In a modification of the classical combustion method, sodium in an aluminum capsule was caused to react at 1000°C with a flux of silica sand in an atmosphere of argon. Combustion was completed in an atmosphere of oxygen. The resulting water was absorbed in $Mg(ClO_4)_2$ and weighed [98].

Application of the electrochemical meter for measuring oxygen (see Section VII.I) to the determination of hydrogen was investigated using 10 different electrochemical hydrogen sensors [99a]. For the determination of hydride hydrogen in sodium, the residue from the Wurtz reaction (see Section VII.C) was dissolved in water and the hydrogen evolved was measured manometrically. It is claimed that the method is sensitive to 1 ppm [99b].

During the distillation of sodium the sodium hydroxide present reacts with sodium metal to form Na_2O and NaH. The latter decomposes to yield hydrogen, which is withdrawn for analysis at a pressure of 0.01 torr [100].

IX. DETERMINATION OF NITROGEN

Nitrogen is commonly determined in lithium by hydrolysis of the sample, followed by steam distillation and determination of the resulting ammonia [20, 42, 101, 102] (see also Chapter 8). It is always assumed that all nitrogen present in lithium will form ammonia when the sample is dissolved in water. This seems to be a reasonable assumption and its acceptance is encouraged by results that correlate well with sample treatment.

In the case of lithium, which is known to form stable nitrides, the assumption that nitrogen is converted completely to ammonia on hydrolysis is relatively safe. However, when we consider sodium and the other alkali metals, which do not form stable nitrides, the assumption becomes questionable [8]. One report states [103], "Some modification of the Kjeldahl method for the determination of nitrogen seemed ideal for use with sodium metal. The sample containing nitride nitrogen is readily soluble, and the ammonia is easily removed by distillation. Is nitride the only form in which nitrogen is present in the sample? Does all nitrogen in the sample hydrolyze to ammonia? Could some of the nitrogen be present as cyanide through interaction with carbon? The proposed methods may well be perfectly valid, but until answers to these questions are forthcoming, we must bear in mind that it is based on unverified assumption and that it cannot be considered as an established method."

In a typical procedure, the sample is transferred to a reaction flask, using standard alkali-metal-sampling procedures involving vacuum or highly purified helium or argon. While maintaining an inert-gas flow, ammonia-free water or dilute boric acid is transferred into a receiving flask and the sample is decomposed, depending on the reactivity of the alkali metal, by either the slow or rapid addition of water [9, 42, 102, 104] or methanol [45].

In one procedure [16], potassium under hexane is dissolved by the dropwise addition of phosphoric acid. It has been postulated that in this medium all nitrogen compounds present would be reduced to ammonium ions by the reaction of the nascent hydrogen generated. The hexane must be expelled before the distillation of the ammonia from an alkaline medium. In a similar procedure, a hexane–butyl alcohol medium was used for dissolving sodium metal with phosphoric acid [105].

The rest of the analysis in all cases follows the usual pattern. Ammonia is distilled from an alkaline medium and is absorbed in dilute boric acid. Subsequently, the nitrogen content is determined by the usual Nessler technique with an aliquot or the total distillate. With absorption cells having a -cm path length and a sample-volume of 50 ml, the calibration range covers –40 μg nitrogen at a wavelength of 390 nm.

ACKNOWLEDGMENT

The assistance of Mr. H. Kirtchik of the General Electric Company in assembling some of the literature on this subject is gratefully acknowledged.

REFERENCES

1. R. H. Herald and W. D. Weatherford, Jr., paper presented before the International Conference on Aerospace Support & Exhibit, Washington, D.C., 1963.
2. R. L. Carter, R. L. Eichelberger, and S. Siegel, *Proceedings of the Second International Conference on the Peaceful Uses of Atomic, Energy* Volume VII, United Nations, Geneva, 1958, p. 72.
3. H. Steinmetz and B. Minushkin, *U.S. At. Energy Comm., NDA 2154-6*, 1961.
4. R. H. Herald, *Techn. Doc. Report No. ASD-TDR-63-703*, 1963.
5. J. W. Manstellar and S. J. Rodgers, British Patent 1,153,628; date applied 11/1/66.
6. J. R. Weeks, in *Rep. NASA SP-41*, NASA, 1963.
7. L. F. Epstein, in *NASA-AEC Proceedings of the Liquid Metals Corrosion Meeting, October, 1963, NASA SP-41*, NASA, 1964.
8. E. W. Hobart, Jr., "Remarks on the Analysis of Alkali Metals" before the Sodium Components Development Program, Information Meeting, Hollywood Calif., April, 1964.

9. M. M. Markowitz and D. E. Boryta, *J. Chem. Eng. Data*, **7,** 586 (1962).
10. C. Tyzak and P. B. Loughton, *U.K. At. Energy Authority Techn. Note 131* 1955.
11. C. B. Jackson and R. M. Adams, *Liquid Metals Handbook*, R. N. Lyon, Ed. 2nd ed., rev., U.S. Atomic Energy Commission and Dept. of the Navy Washington, D.C., 1954, p. 105.
12. F. D. Rossini, D. D. Wagman, W. H. Evans, S. Levine, and J. Jaffi, *Circ. 500* National Bureau of Standards, 1952.
13. E. F. McFarlane and F. C. Tompkins, *Trans. Faraday Soc.*, **58,** 1177 (1962)
14. N. V. Sidgwick, *The Chemical Elements and Their Compounds*, Vol. I, Oxford University Press, London, 1950.
15. L. P. Pepkowitz and W. C. Judd, *Anal. Chem.*, **22,** 1283 (1950).
16. H. Kirtchik and S. Melzer, *General Electric Report*, *R63FPD302*, 1963.
17. L. F. Epstein, paper presented before the Sodium Components Development Program, Information Meeting, Hollywood, Calif., April, 1964.
18. J. C. White, C. K. Talbott, and L. J. Brady, *Anal. Chem.*, **26,** 942 (1954).
19. T. W. Gilbert, *Chem. Eng. News*, **26,** 2605 (1948).
20. W. D. Weatherford, Jr., R. K. Johnston, M. L. Valtierra, and J. W. Rhoades *Air Force Report ASD-TDR-63-413*, 1963.
21. S. J. Rogers and W. A. Everson, *MSAR Report*, *Third Quarterly Progress Report*, 1963, under Air Force Contract AF 33(657)-8310.
22. G. Brauer, *Z. Anorg. Allg. Chem.*, **255,** 101 (1947).
23. G. W. Horsley, *U.K. At. Energy Res. Est. AERE M/R 1152*, (1953).
24. E. M. Simons, paper presented before the Sodium Components Development Program, Information Meeting, Hollywood, Calif., April, 1964.
25. H. Chevilliard, *Comm. Energie At. Rep. No. 834*, 1959.
26. W. J. Clabaugh, J. H. Germer, J. Jacobson, and M. L. Weiss, *General Electric Report GEAP-3230* 1959.
27. K. E. Stoffer and J. H. Phillips, *Anal. Chem.*, **27,** 773 (1955).
28. G. F. Schenck and C. E. Woodward, paper presented before the Sodium Components Development Program Meeting, Hollywood, Calif., April 1964
29. T. D. Brotherton, O. N. Cole, and R. E. Davis, *Trans. AIME*, **224,** 287 (1962)
30. G. Goldberg, A. S. Meyer, Jr., and J. C. White, *U.S. At. Energy Comm. Rept ORNL 2147*, 1956.
31. Anonymous, *U.K. At. Energy Auth.*, *IGO-AM/CA110 UKAEA*, 1958.
32. U.S. Industrial Chemical Co., "Cesium Constant Quantity Transfer Procedure," Cincinnati, 1962.
33. L. Pepkowitz, *Anal. Chem.*, **26,** 574 (1954).
34. Lithium Corporation of America, *Analysis and Handling of Lithium Metal* Technical Bulletin, 1958.
35. E. W. Hobart, Pratt and Whitney Corporation, Middletown, Conn., private communication, 1962.
36. W. A. Dupraw, J. W. Graab, and R. F. Gahn, *Anal. Chem.*, **36,** 430 (1964)
37. V. L. Hansley and R. A. Kolbeson, *Advan. Chem. Series*, **No. 9,** 163 (1957)
38. H. Kirtchik and G. Riechman, *General Electric Progress Report #2*, *Contract NaSr-12*, *DM-61-241*, 1961.

9. V. T. Kharlamoo, *Anal. Gazov V. Metallakh, Akad. Nauk SSSR; Komissiya Po Anal. Khim.*, **10,** 117 (1960); through *FTD-TT-62-1338/1 + 2* (1962).
0. D. D. Williams and R. R. Miller, *Anal. Chem.*, **23,** 1865 (1951).
1. Ye D. Malikowa and Z. M. Turovtseva, *Anal. Gazov V. Metallakh, Akad. Nauk S.S.S.R.; Komissiya Po Anal. Khim.*, **10,** 91 (1960); through *FTD-TT-62-1338/1 & 2.*
2. S. J. Rodgers, F. Tepper, W. J. Carter, and C. A. Palladino, *Air Force Report APL-TDR-64-34*, 1964.
3. G. Goldberg, *Paper No. 52a*, Pittsburgh Conference on Analytical Chemistry and Applied Spectroscopy, Pittsburgh, 1964.
4. H. Kirtchik, G. Riechman, and T. Lajcik, unpublished data, 1962.
5. F. Tepper and J. Greer, *Fifth Quarterly Report*, *MSAR 63-172*, 1963.
6. J. C. White, W. J. Ross, and R. Rowan, *Anal. Chem.*, **26,** 210 (1954).
7. H. J. de Bruin, *Anal. Chem.*, **32,** 361 (1960).
8. L. Silverman and M. Shideler, *Anal. Chem.*, **27,** 1660 (1955).
9. H. Kirtchik and G. Riechman, *General Electric Progress Report #1, Contract NASr-12, DM61-100*, 1961.
0. L. Champeix, R. P. Darras, and J. Duflo, *J. Nucl. Mater.* **2,** 113 (1959).
1. L. E. Smythe and H. J. de Bruin, *Analyst*, **83,** 242 (1958).
2. H. J. Emeleus and A. A. Woolf, *J. Chem. Soc.* (London), **1950,** 164.
3. W. A. Dupraw and H. J. O'Neill, *Anal. Chem.*, **31,** 1104 (1959).
4. G. Goldberg, A. S. Meyer, Jr., and J. C. White, *Anal. Chem.*, **32,** 315 (1960).
5. H. R. Hoekstra and J. M. Katz, *Anal. Chem.*, **25,** 1609 (1953).
6. G. Goldberg, *Anal. Chem.*, **34,** 1343 (1962).
7. H. Kirtchik and T. Lajcik, *NASA Report DM 63-233*, 1963.
8. L. J. Brady, *Anal. Chem.*, **20,** 1033 (1948).
9. H. Steinmetz and N. I. Sax, Division of Industrial and Engineering Chemistry, 133rd Meeting, American Chemical Society, San Francisco, Calif., 1958.
0. R. J. Jaworowski, J. R. Potts, and E. W. Hobart, *Anal. Chem.*, **35,** 1275 (1963).
1. G. Goldberg, *U.S. At. Energy Comm. Rept. ORNL 3537*, 1963.
2. J. C. White, *U.S. At. Energy Comm.*, *ORNL Rept. CF-56-4-31*, 1956; through *Nucl. Sci. Abstr.*, **11,** 8290 (1956).
3. V. I. Subbotin, F. A. Kozlov, E. K. Kuznetsov and N. N. Ivanovskiy, *Symposium on Alkali Metal Coolants*, Vienna, 1966.
4. K. W. Bergstresser, G. R. Waterbury and C. F. Metz, *U.S. At. Energy Comm. Report LA-3343*, 1965.
5. D. R. Grieser, G. G. Cocks, E. H. Hall, W. M. Henry, and J. McCallum, *Battelle Memorial Institute Rept. No. 1538*, *TID-4500*, 1961.
6. I. L. Gray, R. N. Neal, and B. G. Voorhees, *Nucleonics*, **14** (10), 34 (1956).
7. B. Minushkin, paper presented before the Sodium Components Development Program, Information Meeting, Hollywood, Calif., April 1964.
8. L. Vautrey, *Nucl. Eng.* **8,** 321 (1963).
9. M. Kolodney, B. Minushkin and H. Steinmetz, *U.S. At. Energy Comm. Report CONF. 641, 018-4*, 1964.

70. C. C. McPhaeters and J. H. Williams, *U.S. At. Energy Comm.* prepri *LA-DC 7743.*
71. L. R. Blake and A. R. Eames, *Nucleonics*, **19** (5), 66 (1961).
72. Anonymous, Atomic Power Development Associates, Inc., *Operation of t Sodium Technology Loop for Contamination Meter Evaluations, Rep APDA-163*, 1963.
73. J. W. Mansteller and S. J. Rodgers, *British Patent* 1,153,628; date ap 11/1/66.
74. (a) E. L. Steele and H. R. Lukens, Jr., *First and Second Quarterly Rep for NASA, GA-4855*, 1964; and (b) O. U. Anders and D. W. Briden, *An Chem.*, **37,** 530 (1965).
75. I. Fogelstrom-Fineman, O. Hohn-Hamsere, B. M. Tolbert, and M. Calv *Intern. J. Appl. Radiation and Isotopes*, **2,** 280 (1957).
76. P. Sue, *Compt. Rend.*, **242,** 770 (1956).
77. R. G. Osmund and A. A. Smales, *Anal. Chim. Acta*, **10,** 117 (1954).
78. P. Leveque, *First Geneva Conference Proceedings*, Vol. XV, United Natio Geneva, 1955, pp. 78–80.
79. N. Faull, *Report AERE-R/R, 1919*, 1957.
80. F. Ajzenberg and T. Lauritsen, *Rev. Mod. Phys.*, **27,** 77 (1955).
81. R. F. Coleman and J. L. Perkin, *Analyst*, **84,** 233 (1959).
82. E. L. Steele, W. W. Meinke, and W. Wayne, *Anal. Chem.*, **34,** 185 (1962).
83. D. J. Veal and C. F. Cook, *Anal. Chem.*, **34,** 178 (1962).
84. J. C. White, paper presented before the 133rd meeting, American Chemi Society, San Francisco, Calif., 1958.
85. L. C. Bate and G. W. Leddicotte, *Paper No. 40*, Pittsburgh Conference Analytical Chemistry and Applied Spectroscopy, Pittsburgh, 1958.
86. V. P. Guinn, paper presented before the Sodium Components Developm Program, Information Meeting, Hollywood, Calif., April 1964.
87. C. J. Lutz, *Anal. Chem.*, **42,** 531 (1970).
88. C. Engelmann, *Radioanal. Chem.*, **6** (1), 227 (1970).
89. L. P. Pepkowitz and E. R. Proud, *Anal. Chem.*, **21,** 1000 (1949).
90. J. A. J. Walker and H. Seed, *Analyst*, **90,** 19–23 (1965).
91. M. Hissink, *Atoomenergie Toepass*, **12,** 165 (1970); through *Anal. Abstr.* 2410 (1971).
92. B. D. Holt, *Anal. Chem.*, **31,** 51 (1959).
93. M. H. Mallet, F. Gerds, and C. B. Griffith, *Anal. Chem.*, **25,** 116 (1953).
94. K. S. Bergstresser and G. R. Waterbury, *U.S. At. Energy Comm. Re LAMS 1698*, 1954.
95. J. W. Frazer, C. W. Schoenfelder, and R. L. Tromp, *U.S. At. Energy Com Rept. UCRL 4944*, 1957.
96. G. Naud and J. Sanmier, *Bull. Soc. Chim. Fr.*, **12,** 27 (1963).
97. C. Evans and J. Herrington, *Anal. Chem.*, **35,** 1907 (1963).
98. R. D. Gardner and W. H. Ashley, *U.S. At. Energy Comm. Report LA-30* 1964.
99a. N. Furman, W. Capplinger, G. Lubitch, R. B. Holden, L. Ligget, and Lukasik, *U.S. At. Energy Comm. Report UNC 5232*, 1968.

b. V. S. Kopylov, M. N. Korotaevo and E. E. Komovalov, *Atomn. Energ.* **28,** 241 (1970); through *Anal. Abstr.*, **20,** 2310 (1971).

0. M. N. Ivanooski, G. O. Pablova, B. A. Shumatco, A. V. Milovidova, E. E. Konovalov, M. N. Arnoldov and A. D. Pleshivtsev, *At. Energ.* **24,** 227 (1968); through *Anal. Abstr.*, **17,** 45 (1969).

1. Foote Mineral Company, *Official Methods of Analysis of Lithium Metal and Lithium Hydride*, 1962.

2. T. W. Gilbert, Jr., A. S. Meyer, Jr., and J. C. White, *Anal. Chem.*, **29,** 1627 (1957).

3. C. H. Ward, R. D. Gardner, W. H. Ashley, and A. Zerwek, *U.S. At. Energy Comm. Rep. LA-2892*, 1963.

4. N. I. Sax, N. Y. Chu, R. H. Miles, and R. W. Miles, *U.S. At. Energy Comm. Rep. NDA-38*, 1957.

5. S. Kallmann, unpublished laboratory experiments, 1964.

CHAPTER

11

ALKALINE EARTH METALS

BEN D. HOLT

Chemistry Division, Argonne National Laboratory
Argonne, Illinois

CONTENTS

2. Nitrogen: Kjeldahl Method
- **a.** *Apparatus*
- **b.** *Procedure*

3. Oxygen
- **a.** *Beryllium*
- **b.** *Magnesium: Metal-Evaporation–Inert-Gas-Fusion Method*
- **c.** *Calcium: Methanol–Salicylic Acid Method*

I. METALLURGICAL ASPECTS

A. HYDROGEN

The metal–hydrogen systems of the alkaline earth elements can be divided into two groups. Beryllium and magnesium, in one group, do not react directly with hydrogen to form the hydrides; hydrogen is only slightly, endothermically soluble in the solid phase of these metals. Calcium, strontium, and barium, in the other group, react vigorously and exothermically with hydrogen at 200–300°C [1, 2].

The hydrides produced by the direct reaction of hydrogen with calcium, strontium, and barium are brittle, crystalline solids. These saltlike solids have lower specific volumes than the corresponding metals, and they are thermally stable at temperatures below about 600°C. In contrast, the hydrides of beryllium and magnesium are unstable compounds and are prepared only by indirect methods [1, 3]. They are white polymeric solids that decompose rapidly on heating (BeH_2 above 125°C and MgH_2 above 280°C).

The metallurgical properties of the alkaline earth metals can be somewhat anticipated and interpreted on the basis of the above-cited chemical nature of the metal–hydrogen systems [4]. In endothermic solutions of hydrogen in metal, solubility increases with temperature. Hence, hydrogen, dissolved at an elevated temperature, will on cooling tend to separate into the gas phase in various defects of the metal matrix [4–6]. In some systems the released gas tends to create new cavities that range in size from micropores to macroblisters. The solution of hydrogen in beryllium is apparently only marginally endothermic, because of entropic constraints [7], and therefore is not as adequately described by these general statements as is magnesium [8].

When the solubility of hydrogen in beryllium or magnesium is exceeded, no further reaction occurs. On the other hand, when the solubility of hydrogen in calcium, strontium, or barium is exceeded, as by lowering the temperature of the molten metal saturated with hydrogen, the corresponding hydride tends to precipitate within the solidified metal matrix as a separate solid phase [9].

The solubility of hydrogen in beryllium has been formulated by Jones anc Gibson [7] as

$$\log_{10} S(\text{std. ml/g atm}^{1/2}) = -(2.12 \pm 0.27) + (0.095 \pm 0.211)/T°\text{K}$$

and the diffusion rate as

$$\log_{10} D(\text{cm}^2/\text{sec}) = -(6.53 \pm 0.15) - (965 \pm 106)/T°\text{K}.$$

Jones and Gibson soaked the metal in tritium for this study, while earlie investigators [5, 6] introduced known quantities of hydrogen by proto bombardment. The solubility of hydrogen in magnesium just below th melting point has been estimated to be about 0.15–0.25 cc/g [10, 11]. A concentrations above this level adverse effects have been observed in th tensile strength and ductility of magnesium [8].

In the analysis of alkaline earth metals, it may be important to distinguis surface hydrogen from internal hydrogen. Some investigators of berylliun [12, 13] have concluded that essentially all of the hydrogen measured in thei work was on the surface and virtually none in the solid metal. Most of th surface hydrogen on beryllium that has been exposed to moist air is believe to be in the form of hydrated oxides, $BeO \cdot (H_2O)_x$ [14].

In studying the oxidation of magnesium and magnesium–beryllium alloy exposed to moist air and water vapor at 520–580°C, Zelenskii et al. [15 observed that intensive hydrogenation occurred during the formation o protective oxidation films on the metal. They found that molecular hydrogen segregated in the internal defects of the alloy, induced porosity, which led t disintegration of the specimens.

B. NITROGEN AND OXYGEN

The alkaline earth metals react readily with nitrogen and oxygen to forn stable compounds. The solubilities of nitrogen and oxygen in these metal have not been as extensively investigated as in other metals but are though to be generally low. Pemsler et al. [5] found the solubility of nitrogen i beryllium at 1000°C to be about 60 ppm. Nitrogen in excess of this concentra tion separated into a Be_3N_2 phase within the metal matrix. Solubility studie of oxygen in the alkaline earth metals have been hampered by the lack o suitable analytical methods for the low-parts-per-million range.

Beryllium and magnesium, both characteristically hard, rigid lightweigh metals, are used in several structural alloys. It is of interest, therefore, tc know the effect of nitrogen and oxygen impurities on the mechanical proper ties, particularly brittleness, of these alloys. Progress has been slow, however in the correlation of mechanical properties with low concentrations of ga impurities. Besides the lack of entirely adequate analytical methods, some o the metallurgical processes involved in the purification of these highly reactive metals have tended to limit the ultimately attainable purity. For example, the oxygen concentration in beryllium that is cast in beryllium oxide or magnesium oxide crucibles is not likely to be much lower than the solubility of oxygen in the metal at the temperature of the melt. Even the oxygen content of redistilled beryllium is sensitively dependent upon the quality of the vacuum system in which the high-temperature evaporation is performed [16]. Because of the smallness of the metal lattice parameters of beryllium, it has been conjectured that a given low concentration of oxygen may influence the physical properties of beryllium considerably more than does the same concentration in most other metals [17].

In modern metallurgical processes, both magnesium and beryllium are often used in powder form in which the large surface areas are characteristically accompanied by considerable oxygen contamination. To be sure, in some powder metallurgy processes, the presence of a controlled amount of oxide film on powder particles may be advantageous. For example, in extrusion of the heated material, a product of greater strength can result from retarded grain growth that is attributable to included oxide-film at powdered-particle boundaries.

All of the alkaline earth metals are used as minor components in various special-property alloys. Measurement and control of the oxygen and nitrogen concentrations of these additive metals are important because these gases may affect the properties of the final products.

Magnesium and calcium are used as reducing agents in metallurgical processes. Magnesium in particular is used in the production of the sponge metal forms of titanium and zirconium. As excess magnesium is distilled from the sponge product in the absence of air, the nonvolatile oxides and nitrides of magnesium remain behind as contaminants. Hence, the measurement and control of oxygen and nitrogen in the magnesium is important to the quality control of the end product.

Since all of the alkaline earth metals react readily with air, especially in the presence of moisture, it is desirable that the analysis of these metals be performed such that surface contaminants are distinguishable from internal oxygen and nitrogen.

II. STANDARD SAMPLES

A. BERYLLIUM

1. Hydrogen

There appears to be no generally suitable method for the preparation of standard hydrogen–beryllium samples by which methods of analysis can be tested. Only the specialized technique of proton bombardment [5, 6] has been reported for the addition of known quantities of hydrogen to beryllium.

2. Nitrogen

Nitrogen standards (11 and 55 ppm) have been prepared [16]. Vacuum melting and distillation techniques were used.

3. Oxygen

The New Brunswick, New Jersey, laboratory of the AEC has supplied standard beryllium samples for oxygen analysis [18, 19].

B. MAGNESIUM

1. Hydrogen

Attempts have been made to hydrogenate magnesium [11] by heating in the temperature range of 350–550°C for 1 h in hydrogen at low (3.5–7 torr) and at high pressures (2 atm). No detectable absorption of hydrogen was observed at the low pressures. At the high pressure, insufficient absorption of hydrogen, as well as appreciable volatilization of magnesium, prevented the preparation of reliable standards.

2. Oxygen

Apparently no oxygen–magnesium standards are commercially available. Evaporation methods of analysis can be tested by analyzing specimens of magnesium in which weighed quantities of magnesium oxide are encapsulated [20].

C. CALCIUM, STRONTIUM, AND BARIUM

1. Hydrogen

Standard samples of calcium, strontium, and barium containing known concentrations of hydrogen are not generally available. However, since hydrogen reacts readily with these metals, standards can be prepared by adding known quantities of hydrogen to reference material.

2. Oxygen

Known quantities of oxygen can be added to individual samples of these metals by heating in oxygen confined to a closed system. The decrease in oxygen pressure is a measure of the amount added.

III. SAMPLING AND SAMPLE PREPARATION

A. BERYLLIUM AND MAGNESIUM

Samples of beryllium and magnesium, preferably cut into cubes, should be abraded just prior to insertion into the analytical apparatus, to remove hydrated oxides that tend to accumulate on exposure to room atmosphere. *Caution:* Abrasion of beryllium metal must be confined to a well-ventilated hood to prevent the ingestion of metal or oxide dusts, which are highly toxic.

B. CALCIUM, STRONTIUM, AND BARIUM

Samples of calcium, strontium, or barium should be cut, cleaned, and weighed in a glove box filled with dry inert gas. If the weighing cannot be performed inside the box, each cleaned sample should be placed in a gas-tight, preweighed container (inside the glove box), reweighed (outside the glove box), and then transferred rapidly to the analytical apparatus. Although these precautions are advisable where practical, conditions may be such that contamination by atmospheric moisture is negligible. Peterson and Fattore [21] observed that calcium samples that were exposed to the air for 20 min showed the same hydrogen content as samples that were exposed for less than 1 min. They concluded that contamination, if any, occurred during the initial exposure to air.

IV. ANALYSIS

A. DETERMINATION OF HYDROGEN

1. Beryllium

Four methods that have been reported for the determination of hydrogen in beryllium are vacuum fusion [22–24], isotope dilution [12], hot extraction [5, 12] and spark mass spectrometry [25].

a. Vacuum Fusion

In the vacuum-fusion method the sample is vacuum-melted in a platinum bath contained in a graphite crucible. Hydrogen, carbon monoxide, and nitrogen, which are liberated from the sample, are pumped into a gas reservoir for subsequent separation and measurement. Although hydrogen is readily extracted from beryllium by this method, the method is unnecessarily complicated if hydrogen is the only gaseous element to be determined.

b. Isotope Dilution

The beryllium sample is heated in the presence of a known quantity of a mixture of hydrogen and tritium (or deuterium) until the isotopes are equilibrated throughout the two-phase system (metal and gas). The total amount of hydrogen originally associated with the beryllium sample (internal or on the surface) is calculated from the change in ratio of the isotopes before and after equilibration with the metal. Measurement of the isotope ratio is made by radioactivity counting if tritium is used [12] and by mass spectrometry if deuterium is used [26, 27].

c. Hot Extraction

Hydrogen can be determined in beryllium by heating a sample of the metal in vacuum to a temperature (below the melting point) at which the hydrogen quantitatively diffuses out of the metal. It is pumped into a reservoir for *PVT* measurements. Pemsler et al [5] observed that hydrogen was quantitatively removed when beryllium was vacuum-heated for one hour at 1050°C; that is, all bubbles disappeared from the grain boundaries of specimens that had been previously spiked with hydrogen by proton bombardment.

d. Spark-Source Mass Spectrometry

Bourguillot et al. [25] applied spark-source mass spectrometry to the analysis of beryllium for hydrogen (4.5 ppm), nitrogen (0.1 ppm), oxygen (11 ppm), and other impurities in the parts-per-million range.

e. Discussion of Methods

Of the first three methods cited above for the determination of hydrogen in beryllium, hot extraction appears to be the simplest. The equipment and procedure required for a vacuum-fusion analysis are considerably more complex than for hot extraction. The latter requires no fusion bath and no facilities for analyzing a mixture of hydrogen, carbon monoxide, and nitrogen normally obtained by vacuum fusion. It is desirable to test the purity of the extracted hydrogen occasionally by mass spectrometry, but such is not usually necessary for every hot-extraction analysis.

The isotope-dilution method is more complex in equipment and procedure than the hot-extraction method. Under the operating conditions of both methods, internal hydrogen diffuses through the solid-metal matrix to enter the gas phase at the surface, and surface hydrogen (present as hydrated oxides) is reduced by the hot metal and released as the gas. Both procedures entail precise gas measurements, yet, in addition, the isotope-dilution procedure requires isotopic analyses for each determination. Essential to the isotopic method are the reproducible establishment of isotopic equilibrium in the metal and gas phases, and the prevention of appreciable fractionation of the isotopes during subsequent purification and analysis.

The time required to reach equilibrium by the isotopic method is reported to be about 20 min when the metal is heated to 850°C. Presumably no more time than this is required for the collection of hydrogen by vacuum fusion or hot extraction. The measurement of hydrogen, subsequent to the sample reaction, is inherently less time-consuming by the hot-extraction method than by either vacuum fusion or isotope dilution.

A feeling for the range and precision of analytical data for hydrogen in beryllium may be gained by reference to some of the reported results. Still [23], using vacuum fusion, found an average of 0.044% hydrogen in 10 analyses of beryllium powder. Turovtseva and Kunin [28], also employing vacuum fusion, reported mean values of $(0.015 \pm 0.006)\%$ on 34 samples of powdered beryllium, $(0.022 \pm 0.010)\%$ on 18 samples of flake beryllium, and $(0.042 \pm 0.020)\%$ on 8 samples of another type of beryllium. Evans and Herrington [12] compared hydrogen data obtained by isotope dilution, $(7.5 \pm 0.6) \times 10^{-3}$ ml/cm^2, with hot-extraction data, $(7.0 \pm 0.7) \times 10^{-3}$ ml/cm^2, on the same material. These results were reported in terms of surface concentration, because prior experimental evidence had indicated to the analysts that essentially all of the measurable hydrogen was on the surface of the samples and not in the metal matrix.

2. Magnesium

At least nine different methods have been applied to the determination of hydrogen in magnesium and magnesium alloys. These include combustion [10, 29], hot extraction from solid samples [11], hot extraction from encapsulated samples [30], extraction into inert gas [29], vacuum evaporation [11, 29, 31, 32], tin fusion [11, 30], ion bombardment [29], chlorination [33] and bubble appearance [34].

a. Combustion

In the combustion method the magnesium is burned in a stream of purified oxygen that is appropriately mixed with purified argon to control the rate of reaction. The water produced from oxidation of the hydrogen in the sample is separated from the carrier-gas stream and measured either gravimetrically [10, 29] or manometrically [35].

Because of the low concentrations of hydrogen ordinarily found in magnesium (about 20 ppm in commercial grade) [10] large samples are required by the gravimetric procedure to produce conveniently weighable quantities of water. Bobalek and Shrader [29] and Codell and Norwitz [10] used sample weights of the order of 2.5 g and 15 g, respectively. Large samples necessitate large combustion boats that are free of appreciable quantities of moisture. Codell and Norwitz found it necessary to preignite their combustion boats at 1100°C and to store them in a furnace at 500°C.

In the combustion-manometric procedure, which was originally developed for the determination of microgram quantities of hydrogen in uranium [35] but is equally applicable to magnesium, the sample-size is 0.1–0.5 g. No combustion boats are used. The samples are dropped onto a bed of copper oxide at 900°C in a vertical furnace tube. The water is separated from the oxygen stream by freezing and then measured manometrically at 110°C.

b. Hot Extraction

Mallet et al. [30] and Berry et al. [11] found that the hot-extraction method (see Section IV.A.1.c) tends to yield low results for hydrogen in magnesium and is generally too time-consuming for convenient use. Presumably, only the interstitial hydrogen in solid solution is extracted, while molecular hydrogen, trapped in the pores and voids of the metal, is essentially completely retained. The temperature of extraction can not exceed about 500°C without excessive volatilization of magnesium.

c. Extraction from Encapsulated Samples

Mallett et al. [30] sealed magnesium samples in steel capsules of 0.038-in. wall thickness and vacuum-extracted at 800°C. Although the volatile magnesium was retained in the capsules while the hydrogen diffused through the thin walls for collection and measurement, the time required was impractically long (about 11 h).

d. Extraction into Inert Gas

Bobalek and Shrader [29] bubbled helium through a molten 30-g sample of magnesium contained in a graphite crucible that was heated inductively to 750°C. Hydrogen liberated from the melt was swept by the helium stream through a bed of copper oxide, where it was converted to water, and then through an absorption tube of magnesium perchlorate, where it was collected for weighing. The large-sample requirement appears to limit the usefulness of this method.

e. Vacuum Evaporation

Hydrogen can be rapidly and quantitatively extracted from magnesium during distillation of the metal from a heated container to a cooled surface within a vacuum system [29, 36]. The extracted hydrogen is pumped into a reservoir for *PVT* measurement. Heating of the sample is continued until the rate of evolution of gas equals the blank rate. Berry et al. [11, 31] at first heated the sample to 750°C in a nichrome-wound tube furnace but later found that by inductively heating at 1000°C, the rate of hydrogen removal was increased, the blanks were lowered, the precision was improved, and the life of the silica reaction tube was prolonged 20-fold. They preconditioned their apparatus by vaporizing 0.5 g of magnesium from the crucible to the walls of the silica tube before admitting samples for analysis. The magnesium that was condensed from this first vaporization apparently reacted with the water absorbed on the inner walls of the reaction tube, thus preventing subsequent production of spurious quantities of excess hydrogen. Tests were made on the reactivity of hydrogen with freshly condensed magnesium films. Known quantities of hydrogen were added to the reaction chamber before and during the evaporation procedure. No significant absorption of hydrogen by the freshly condensed magnesium was detected.

f. Tin Fusion

In the tin-fusion procedure [11, 30], the magnesium sample is dissolved in 100 g of molten tin contained in a borosilicate glass tube that is heated to 400–450°C by a resistance furnace. The tin bath remains fluid after addition of the specimen, and hydrogen is rapidly liberated with no appreciable evaporation of the metals. About 95% of the total gas is evolved within 15 min and about 100% within 1 h.

g. Ion Bombardment

Bobalek and Shrader [29] tried to quantitatively remove hydrogen and other gases from magnesium by making a rectangular piece of the solid sample (2 × 20 × 40 mm) the cathode in a vacuum-discharge tube. The liberated gases were pumped away from the tube and collected for analysis in a suitable apparatus. Comparative data [29] obtained by this procedure and by the combustion-gravimetric method indicated that an appreciable amount of residual hydrogen remained in the magnesium after degassing by ion bombardment, because ion bombardment probably had the greatest effect on hydrogen near the surface.

h. Chlorination

A chlorination method has been used [33] in which the hydrogen in magnesium is converted to hydrogen chloride and swept with pure oxygen through hot copper oxide for conversion to water. The water is collected and weighed in an absorption tube containing magnesium chloride and phosphorus pentoxide. Excess chlorine in the gas stream is absorbed in a potassium hydroxide solution. Two objectionable features of this method are the obnoxious reagent (chlorine) and the large sample-size necessary to produce a conveniently weighable quantity of water.

i. Bubble Appearance

A semiquantitative method was used for estimating the hydrogen content of magnesium–aluminum alloys [34]. The gas pressure over the molten sample and the temperature of the sample were measured at the appearance of the first gas bubbles on the surface. This technique is applicable only to magnesium–aluminum alloys that contain enough aluminum to maintain an oxide film on the surface, under which the bubbles of hydrogen are retained long enough for observation.

j. Comparison of Methods

Of the nine methods cited above for the determination of hydrogen in magnesium and magnesium alloys, the three that are likely to give the best results with the most practical apparatus and procedures are tin fusion, vacuum evaporation, and combustion manometric. The chief objections to the other six methods are indicated in the discussions above. Choice of one of the three more desirable methods should be based on the availability of equipment and on the applicability to the material to be analyzed. For example, with regard to the latter, the tin-fusion method is not applicable to a magnesium alloy that is not completely soluble in tin at about 450°C.

A comparison of the vacuum-evaporation and combustion-gravimetric methods for the determination of hydrogen in magnesium–aluminum alloys led Bobalek and Shrader [29] to conclude that such alloys are degassed in vacuum with much greater difficulty than is pure magnesium. By increasing the extraction time from 2–3 min to 15 min or more in the vacuum-evaporation method, the results of the two methods were brought into agreement. Some comparative data obtained by use of the two methods are shown in Table 11.1. Each reported value is the mean of two or more determinations. After making technique improvements in the combustion-gravimetric method that resulted in lower blanks and better precision, Codell and Norwitz [10] obtained the results listed in Table 11.2 on commercial-grade and sublimed magnesium.

A comparison of the evaporation and tin fusion methods was made by Berry et al. [11]. Their results are shown in Table 11.3. The evaporation method requires about 10–20 min of extraction time as compared to 1 h for tin fusion. The rate of extraction in tin fusion is governed by the rate of

TABLE 11.1. Determination of Hydrogen in Magnesium Alloys by Vacuum-Evaporation and Combustion Methods [29]

Alloying Metals, %	Hydrogen, ppm	
	Vacuum Evaporation	Combustion
—	23	24
2Mn	17	16
2Mn	23	30
3Al, 1Zn, 0.4Mn, 0.2Ca	19	26
9Al, 0.7Zn, 0.2Mn	24	27
8.5Al, 0.5Zn, 0.2Mn	11	10
6Al, 3Zn, 0.3Mn	10	13
9Al, 2Zn, 0.2Mn	<2	3
45Al ($Mg_{17}Al_{12}$)	<2	4
45Al ($Mg_{17}Al_{12}$)	10	13
42Al	11	14
6Al	<2	5

dissolution of the sample in the tin, which in turn is dependent upon temperature, concentration of magnesium in the bath, surface area of the sample, and mechanical agitation. The number and size of samples per bath are limited by the amount of magnesium that can be dissolved in the tin before saturation occurs. On the other hand, an inconvenience of the evaporation procedure is that the reaction chamber must be dismantled fairly often to remove magnesium that is condensed on the inner walls. Considerable time is then required to bake out the cleaned reaction tube sufficiently to restore a suitable blank.

The range of hydrogen concentration in magnesium and magnesium alloys, as indicated by the data of Tables 11.1–3, is about 3–20 ppm. With regard to precision, Bobalek and Shrader (Table 11.1) reported that the average deviations from mean values were $\leq$ 2 ppm and 5 ppm hydrogen, respectively, for the vacuum-evaporation and combustion-gravimetric methods. Berry et al. [11] reported a mean value for 12 samples of magnesium-rich alloy of

TABLE 11.2. Determination of Hydrogen (in ppm) in Magnesium by Combustion [10]

Commercial Magnesium	Sublimed Magnesium
20	6
17	6
18	7
20	7
21	8

TABLE 11.3. Determination of Hydrogen in a Magnesium Alloy by Vacuum Evaporation and by Tin Fusion [11]

	Hydrogen, ppm	
Material	Vacuum Evaporation	Tin Fusion
Magnox A12 sheet	4.5	4.7
	4.8	4.9
	5.9	4.9
	5.9	6.3
	6.6	6.7
		7.5
		7.6
Magnox A12 bar	15	18
	15	19
	15	20
	16	
Magnox A12 bar	11	11
	12	11
	12	12
	12	

8.8 ppm hydrogen with a standard deviation of 1.8 ppm. The author has tested the precision of results obtained by the combustion-manometric method [35]. Seven cube-shaped, 1-g samples of "chemically pure" magnesium rod, supplied by Merck and Company, Inc., Rahway, New Jersey, yielded a mean value of 8.1 ppm hydrogen with a standard deviation of 1.3 ppm. The surfaces of these specimens were abraded just before analysis. Another group of 5 specimens, prepared in the same way, were vacuum heated at about 400°C for 5 min to observe the effect of such treatment on the hydrogen content. The resulting average was 3.6 ppm hydrogen (compared to 8.1 ppm) with a standard deviation of 0.6 ppm indicating large effects that can be produced by vacuum heating prior to analysis.

3. Calcium, Strontium, and Barium

Very little has been published on methods of determining hydrogen in calcium, strontium, and barium. Two methods that appear to be applicable are combustion-manometric and tin fusion.

a. Combustion-Manometric

The author has used the combustion-manometric method (see Section IV.A.2.a) to determine hydrogen in magnesium. Inasmuch as the alkaline earth metals burn readily and their hydroxides are unstable in oxygen at a furnace temperature of 900°C, the combustion-manometric method appears to be as applicable to these metals as to magnesium.

b. Tin Fusion

Peterson and Fattore [21] successfully applied the tin-fusion method to samples of calcium containing from 200 ppm of hydrogen up to almost pure calcium hydride. They also attempted to use the hot-extraction method (see Section IV.A.1.c), but at temperatures that were sufficiently high to allow practical rates of extraction, calcium volatilization interfered with accurate analyses. With precision better than 4%, these investigators tested the accuracy of the tin-fusion method. A known quantity of hydrogen was added to a specimen of calcium, increasing its concentration from 0.022 to 0.177%. The sample was melted in a stainless steel capsule to ensure homogeneous composition and then was analyzed for hydrogen. The results are shown in Table 11.4.

TABLE 11.4. Analysis of Calcium Charged to 0.177% Hydrogen: Tin-Fusion Method [21]

Sample-Weight, Mg	Hydrogen Found, %
42.7	0.175
36.1	0.179
44.1	0.182
28.2	0.179
38.5	0.181

The possibility of losing hydrogen by adsorption on a metal film deposited on the inner walls of the apparatus was also investigated. A known quantity of hydrogen was introduced into the furnace tube when the bath was at 670°C and a tin mirror was present. The hydrogen was quantitatively recovered within experimental error. The investigators did not state whether or not calcium was present with the tin in the mirror film.

The evolution of hydrogen was 90% complete within 10 min. After half an hour the rate of pressure increase was equal to the blank rate. Up to 6 samples (2–200 mg each) were analyzed in one tin bath (40–50 g reagent grade tin) before changing the tin.

B. DETERMINATION OF NITROGEN

1. Kjeldahl

The Kjeldahl method for the determination of nitrogen is applicable to all the alkaline earth metals. It is generally regarded as a standard method by which other methods can be tested. No significant difficulty has been reported by investigators who have applied variations of the Kjeldahl procedure to the analysis of beryllium [16, 37–39], magnesium [29, 40], and calcium [41]. Bundy and Goode [42] eliminated the steam-distillation step in the analysis of zone-refined, high-purity beryllium by making a direct spectrophotometric determination of nitrogen, as ammonia, in the sample solution by the indophenol blue method [43].

2. Activation

Various methods of activation analysis have been applied to the determination of nitrogen in alkaline earth metals, particularly beryllium. Stalnaker et al. [44a] listed 130 references in a bibliography on the determination of nitrogen by activation analysis, 11 of which concern a beryllium matrix. Three nuclear reactions that have been applied in the measurement of nitrogen in beryllium are $^{14}N(d, n)^{15}O$ [44b, 44c], $^{14}N(p, \alpha)^{11}C$ [44d], and $^{14}N(\gamma, n)^{13}N$ [44e]. The chief advantages of activation methods are high sensitivity, speed of analysis, and nondestructiveness of the sample. A disadvantage for many laboratories is the unavailability of the specialized equipment that is required for activation.

3. Other Methods

The vacuum-fusion method has been used for the determination of nitrogen in beryllium [45] and the inert-gas-fusion method for nitrogen in magnesium [46], both without satisfactory results. A vacuum-evaporation technique [29, 47] yielded low results and poor precision. Spark-source mass spectrometry [25] and electron-probe analysis [48] have also been used for the determination of nitrogen and other elements in beryllium.

C. DETERMINATION OF OXYGEN

It must be noted that certain chemical methods will yield only the oxygen present as the oxide of the matrix metal.

1. Beryllium

Most of the analytical methods that have been used for oxygen in beryllium may be classified in 3 main categories: chemical separation methods by which the beryllium is dissolved in a reagent and the beryllium oxide residue is analyzed, fusion methods, and activation methods. Spark-source mass spectrometry [25] and microprobe analysis [48] have also been used.

a. Chemical Separation

The chemical methods are generally applicable to quality-control analyses of commercial grades of beryllium. The lower limit of the working range is about 0.1 % oxygen.

(1) HYDROGEN EVOLUTION [49]. A rough estimate of oxide concentration in beryllium is made by dissolving the metal in potassium hydroxide solution and measuring the amount of hydrogen liberated. The hydrogen is oxidized by hot copper oxide and weighed as water. The difference between the sample-weight and the weight of beryllium metal (calculated from amount of the liberated hydrogen) is assumed to be the weight of beryllium oxide in the sample.

(2) VOLATILIZATION WITH HYDROGEN CHLORIDE [39, 50]. A stream of dry hydrogen chloride gas is passed over a sample of beryllium at 600°C. The metal is converted to volatile beryllium chloride and is removed by the gas stream, leaving the beryllium oxide in the residue. Beryllium carbide and nitride impurities are also volatilized and removed as beryllium chloride. The residue is dissolved and analyzed for beryllium by the photometric 4-(*p*-nitrophenylazo)orcinol method, from which the beryllium oxide content is calculated.

This method is suitable for an oxygen range of 0.1–2.0 %, but it is not easily adaptable to high-purity beryllium. Two inherent sources of error make high sensitivity very difficult. There is a tendency for finely divided, lightweight beryllium oxide to be lost by entrainment in the gas stream, and some beryllium oxide may be produced by interaction of the hot metal with oxygen-bearing impurities in the hydrogen chloride.

(3) VOLATILIZATION WITH CHLORINE [51]. This method is carried out in about the same way as the preceding one, using dry chlorine instead of hydrogen chloride.

(4) HYDROCHLORIC ACID DISSOLUTION [51]. Dilute hydrochloric acid selectively dissolves beryllium metal, leaving the oxide virtually unattacked. The insoluble residue is separated by filtration and analyzed photometrically by the 4-(*p*-nitrophenylazo)orcinol method. Alternatively, the residue may be ignited and weighed as beryllia.

This rather simple method may be conveniently applied to the analysis of commercial grades of beryllium. Two disadvantages are the slight solubility of beryllium oxide in cool 10 % hydrochloric acid solution and the tendency toward colloidal suspension of some of the finely divided oxide in the acid filtrate. Besides these two features that contribute to low results, acid-insoluble carbides and silicates, if present, may cause high results. The tendency toward loss of beryllia by solution and by colloidal suspension can be largely minimized by heating the sample prior to analysis in an inert atmosphere, to convert the beryllia to a more inert form of larger particle size. Such treatment often requires long heating periods, however, and makes the method somewhat time-consuming.

(5) COPPER SULFATE DISSOLUTION [52]. The sample is treated with a buffered solution of copper sulfate containing mercuric chloride. The beryllium metal goes into solution, copper metal precipitates, and beryllium oxide remains undissolved. After filtration, beryllium is extracted from the residue as the fluoride complex, converted to the sulfate, and then precipitated as the hydroxide. To a neutralized suspension of this precipitate an excess of alkali fluoride is added, and standard acid is used to titrate the alkalinity produced by the reaction

$$Be(OH)_2 + 4F^- \rightarrow (BeF_4)^= + 2(OH)^-. \quad (1)$$

Alternatively, the beryllium may be determined photometrically or gravimetrically.

The method is applicable in the range of 0.1–2.0% oxygen. Beryllium carbide, if present, may lead to high results. The relative standard deviation is about 15% for powders and between 5 and 10% for other forms of the metal.

(6) BROMINE–METHANOL DISSOLUTION [53]. The metal is dissolved by treatment first with a bromine–methanol solution and finally with a solution of hydrochloric acid in methanol. The oxide residue is filtered off, washed with methanol until free of bromine, and dissolved for photometric determination with 4-(*p*-nitrophenylazo)orcinol. Electrolytic metal samples must be given a preliminary heat treatment to make the oxide resistant to the attack of the reagents. In the range of 0.1–2.0% oxygen, to which the method is best suited, moisture at concentrations less than 0.1% in the methanol does not interfere. This method is rapid and is easily adapted to routine analyses.

b. Fusion

In attempts to determine oxygen at concentrations lower than is possible with chemical dissolution methods, a great deal of work has been devoted to the application of vacuum-fusion [22, 54–60] and inert-gas-fusion [61] methods to the analysis of beryllium. In either procedure it is important to keep the partial pressure of carbon monoxide as low as possible in the reaction chamber (by fast pumping in the one case and by inert-gas dilution in the other) to minimize chemisorption by volatilized beryllium metal.

The vacuum-fusion method offers the possibility of a simultaneous determination of nitrogen and hydrogen with the oxygen, although previous attempts to apply it to the determination of nitrogen in beryllium have not been entirely successful [45]. Although the inert-gas-fusion technique has been used to determine microquantities of all three elements in other metals [62] no record of an attempt to apply it to the simultaneous determination of nitrogen, hydrogen, and oxygen in beryllium is known to have been published.

The applicability of fusion methods is not dependent upon the form (powder, flake, or chunk) nor upon the history (high-fired or electrolytic) of the sample. Total oxygen is obtained, whether it be present as dissolved oxygen, oxide coating, or refractory oxide inclusions.

The operating range most conveniently handled by vacuum fusion is 0.01–0.1% with a lower limit of detection of about 0.002%. Apparently there are no published results on the application of inert-gas fusion to the determination of oxygen in beryllium in concentrations below 0.05%, although for many other metals the two techniques are about equal in range and precision. Analysis by inert-gas fusion is inherently more rapid and more easily performed.

Two disadvantages of the fusion methods, in contrast with chemical methods, are that the apparatus is generally more expensive and that the relatively high specific volume of beryllium sharply limits the number of samples of appropriate weight that can be analyzed in a single graphite crucible.

(1) VACUUM FUSION [54, 56–60,63–65]. A 50-mg sample of beryllium is added with 70 mg of high-purity tin to a crucible containing 10 g of platinum [56]. The crucible is covered and heated in vacuum at 1400°C for 1 min and then at 1960°C for 2 min. The carbon monoxide, nitrogen, and hydrogen that are evolved from the sample are rapidly pumped away from the reaction chamber by a high-speed mercury-diffusion pump to a gas-analysis system for measurement.

The analysis is somewhat complicated by the low density of beryllium (causing slow mixing with the heavier platinum melt), by the high volatility of beryllium at temperatures necessary for the reduction of BeO by carbon, and by the gettering action of vaporized beryllium on carbon monoxide released from the bath. The addition of tin to the platinum bath with each sample tends to minimize the gettering effect [66].

Everett and Thompson [57] made a thorough, systematic study of various analytical conditions, including sample-size, crucible temperature, and addition (or omission) of tin with each sample (see also Chapter 4). Their data indicated that encapsulation of the samples in platinum and an increased pumping speed were the two most significant modifications to the previous procedure. Confinement of the sample within a platinum capsule while the temperature was raised to the melting point of the platinum bath apparently reduced the quantity of evaporated beryllium and thereby reduced the subsequent chemisorption of released carbon monoxide. Rapid, high-vacuum pumping was necessary to shorten the exposure of the gas molecules to the evaporated metal in the reaction chamber.

(2) INERT-GAS FUSION. Kallman and Collier [61] reported application of the inert-gas fusion method to oxygen in beryllium, using a commercial instrument. The sample was heated 1.5 min in molten nickel. The carbon monoxide, swept from the crucible by the argon stream, was oxidized to carbon dioxide and measured conductimetrically. Difficulty in extending the method to levels lower than the 0.05–1.0% range was reported. Analytical associates of the author currently determine oxygen in beryllium by inert-gas fusion, using an uninsulated graphite crucible (energized by a 20-kW induction heater) for the carbon reduction and manometric equipment for CO_2 measurement [62].

c. Activation

Activation methods for the determination of oxygen in beryllium are particularly applicable to the analysis of high-purity metal, because they are potentially extremely sensitive.

(1) $^{16}O(n, p)^{16}N$ REACTION [67–73]. The beryllium sample is irradiated in a 14-MeV neutron flux, which by a neutron–proton reaction converts normal oxygen to radioactive ^{16}N with a 7.4-sec activity of β-rays and 6–7-MeV γ-rays. By means of a special transfer apparatus the sample is rapidly moved from the neutron beam to a standard position in the counting apparatus. The ^{16}N activity is compared to that of a standard sample of similar size and shape containing a known amount of oxygen, by measuring either the β-activity using a proportional counter or the 6–7-MeV γ-ray activity using a scintillation counter. The latter method is considered to be less efficient in detection but more selective with respect to interference by radioactivity from other impurities in the beryllium [69]. The 14-MeV neutrons are produced by bombarding a tritiated target with deuterons from a Cockcroft-Walton or a Van de Graaff accelerator. The procedure of the method has been automated by modern users [71, 72].

Oxygen contents in the range of 0.0011–2.47% have been measured by this method [72]. The lower limits of measurement are made possible by precise control in handling cylindrically shaped samples during irradiation and counting and by the absence of other interfering impurities in the metal.

(2) $^{16}O(\gamma, n)^{15}O$ REACTION [18, 74–77]. When beryllium is energized by γ-rays, such as bremsstrahlung, the ^{16}O undergoes a (γ, n) reaction to form ^{15}O, which has a 126-sec half-life. In this method, an irradiated sample is

sandwiched between two large scintillation crystals of thallium-activated sodium iodide, and the induced positron emission is detected by coincidence counting of the annihilation quanta. The bremsstrahlung may be produced by a lead radiator, 1 mm thick, in the electron beam of a linear accelerator. The beryllium sample is placed in contact with the radiator.

The range claimed for this method is 10–5000 ppm oxygen with a precision of 20%. In the absence of interfering activities, the minimum amount of oxygen detectable is estimated to be 1 ppm. Since the counting technique detects positron annihilation, only those impurities that produce radioactive isotopes that decay by positron emission can interfere.

In addition to the high sensitivity, another advantage of this method is that carbon and possibly nitrogen may be measured simultaneously with oxygen.

(3) $^{16}O(^{3}He, p)^{18}F$ REACTION [78]. The sample is bombarded with ^{3}He ions accelerated to comparatively low kinetic energies (maximum about 31.2 MeV). Fluorine-18 of 110-min half-life is formed from oxygen by the reactions: $^{16}O(^{3}He, p)^{18}F$ and $^{16}O(^{3}He, n)^{18}Ne \rightarrow {}^{18}F$. The positron activity of the ^{18}F is measured with an end-window gas-flow proportional counter. The method is claimed to be rapid and absolute for the detection of oxygen and certain other elements over the concentration range of 1 ppm to 100%. Milligram amounts of sample may be used. An inherent difficulty with irradiation reactions of charged particles is the very limited penetration range of these particles in metals. This situation might be of advantage where the primary interest is in surface oxygen rather than internal oxygen. Accurate estimation of internal oxygen is dependent upon the thinness of the sample (<0.002 in.) and upon the absence of surface oxidation (a state that may not be known with certainty).

(4) OTHER NUCLEAR REACTIONS. Among other activation methods for oxygen in beryllium are those based upon the nuclear reactions resulting from the irradiation of oxygen-16 by 44-MeV α particles (79), 10-MeV protons [80], and 2.7-MeV deuterons [81].

d. Comparative Results

Investigators who have developed and used some of the methods discussed above have published data obtained in the analysis of beryllium by 2 or more methods. Most of these comparative results apply to oxygen concentrations in excess of 0.1%.

Eberle and Lerner [53] found good agreement of results by the bromine–methanol dissolution and hydrogen chloride-volatilization methods. Their data are shown in Table 11.5. Table 11.6 gives values that Kallmann and

TABLE 11.5. Comparative Results Obtained by Hydrogen Chloride-Volatilization and Bromine–Methanol Dissolution Methods [53]

	Oxygen, %	
Type of Beryllium	Hydrogen Chloride-Volatilization	Bromine–Methanol Dissolution
High-temperature reduction	0.75	0.75, 0.74
Vacuum cast	0.61, 0.68	0.67, 0.67
Vacuum cast	0.37, 0.38	0.38, 0.39
Vacuum cast	0.41, 0.33	0.34, 0.34
Vacuum cast	0.36, 0.31	0.33, 0.32
Electrolytic	2.79, 2.51	2.74, 2.76

TABLE 11.6. Comparative Results Obtained by Bromine–Methanol, Hydrogen Chloride-Volatilization, and Inert-Gas-Fusion Methods [61], in % Oxygen

Bromine–Methanol	Hydrogen Chloride-Volatilization	Inert-Gas Fusion
1.07	1.03	0.97, 0.99, 1.02, 1.10

TABLE 11.7. Comparative Results Obtained by Vacuum Fusion and $^{16}O(n, p)^{16}N$ Activation [57- 58]

	Oxygen, mean, wt %	
Type of Beryllium	Vacuum Fusion	Activation
Sintered rod	0.41	0.42
Sintered rod	0.43	0.42
Sintered rod	0.43	0.42
Sintered rod	0.41	0.42
Powder, 300–400 mesh	0.50	0.51

TABLE 11.8. Comparative Results Obtained by Gamma-Activation; Hydrogen Chloride-Volatilization, and Bromine–Methanol-Dissolution Methods [75]

	Oxygen, %		
Beryllium Identification	$^{16}O(\gamma, n)^{15}O$	Hydrogen Chloride-Volatilization	Bromine–Methanol Dissolution
Vacuum-melted flake	0.014	0.13, 0.08	0.06
		0.13, 0.26	0.06
"Temescal"	0.08	0.06, 0.14	0.24
		0.09, 0.27	0.34

Collier [61] obtained by the inert-gas-fusion method on beryllium that had been analyzed by the bromine–methanol and hydrogen chloride-volatilization methods.

Everett and Thompson [57] cooperated with Coleman et al. [58] to obtain the closely agreeing results shown in Table 11.7, by vacuum fusion and fast-neutron activation. Harris [59] compared the advantages and disadvantages of the two methods as they apply to other metals.

Gilman and Isserow [75] tested three methods: γ-activation, hydrogen chloride-volatilization, and bromine–methanol dissolution. Poor agreement of results (Table 11.8) led to the conclusion that all three methods should be investigated further. Anderson et al. [73] later found good agreement of results between the neutron-activation and bromine–methanol methods. Three laboratories cooperated in the comparative test. The data are listed in Table 11.9.

TABLE 11.9. Comparative Results Obtained by Bromine–Methanol and $^{16}O(n\text{-}p)^{16}N$ Activation [73], in % Oxygen

	Bromine–Methanol		Neutron Activation
Sample	Lab A[a]	Lab B[b]	Lab C[a]
1	1.03 ± 0.01	0.99 ± 0.01	1.07 ± 0.01
2	1.19 ± 0.02	1.17 ± 0.02	1.22 ± 0.02
3	0.50 ± 0.02	0.43 ± 0.02	0.483 ± 0.008
4	0.51 ± 0.02	0.43 ± 0.02	0.484 ± 0.007
5	0.17 ± 0.02	0.12 ± 0.02	0.120 ± 0.002
6	1.33 ± 0.02	1.34 ± 0.00	1.55 ± 0.02

[a] The ± values are standard deviations of the mean value based upon 6–8 determinations.
[b] The standard deviations are based upon 2–4 determinations.

e. Selection of a Method

Popularity of the neutron-activation method for oxygen in beryllium (as well as in other metals) has increased considerably during recent years [82]. Of the various methods cited above, neutron activation may be the most attractive. This method is rapid, applicable to both percent and parts-per-million ranges, and amenable to a relative error of 1% or lower [83]. The most common interfering elements, fluorine and fission products, are usually insignificant in high-purity beryllium.

If neutron activation is not practical, one of the fusion methods should be applied for oxygen concentrations below about 0.1%, and the simpler bromine–methanol method for higher levels of oxygen. Of the two fusion methods, the author prefers inert-gas fusion, because of its greater speed and simplicity. However, in the absence of published data to support the applicability of this technique to the low parts-per-million range of oxygen in beryllium, the vacuum-fusion procedure [56] is recommended.

2. Magnesium, Calcium, Strontium, and Barium

Methods that have been used for the determination of oxygen in alkaline earth metals other than beryllium may be grouped according to whether or not a preliminary separation of metal from metal oxide is made.

a. Without Preliminary Separation of Metal from Metal Oxide

(1) NEUTRON ACTIVATION. The neutron-activation method, discussed above for oxygen in beryllium, has also been applied to magnesium in the range of 0.03–0.72% oxygen [84]. The magnesium matrix was used as an internal standard for the $^{25}Mg(n, p)^{25}Na$ reaction.

(2) SULFUR MONOCHLORIDE. Sulfur monochloride is carried by a stream of nitrogen over the sample at 900°C [85]. The gas stream continues through mercuric sulfide (to destroy the excess sulfur monochloride), through a solution of iodine dissolved in potassium iodide solution, and finally through an alkaline potassium iodide solution. The reaction of the sulfur monochloride with the sample produces sulfur dioxide that is absorbed and determined iodometrically.

(3) HYDROGEN SULFIDE. Success has been reported in the determination of oxygen in oxides of magnesium, calcium, and strontium by passing purified hydrogen sulfide in a stream of purified nitrogen over the sample at 600–700°C [86, 87]. The metal oxide is converted to the corresponding sulfide and water. The water is absorbed in a weighing tube for gravimetric measurement.

(4) CARBON AND HYDROGEN CHLORIDE. In a method applied to magnesium metal [88], the sample is placed on powdered carbon in a graphite boat and heated to 900°C in the gas stream. The oxide is converted to carbon monoxide by reaction with hot carbon in a stream of hydrogen chloride. The excess hydrogen chloride is removed in a cold trap, and the carbon monoxide is oxidized to carbon dioxide for collection and gravimetric measurement. The lower limit of detection is 0.01% of oxygen, and the time required for an analysis is about 100 min.

(5) METHANOL AND SALICYLIC ACID. According to a procedure used for the determination of oxygen in calcium metal [89], the sample is treated with anhydrous methanol and the reaction products are dissolved by the addition of an anhydrous solution of salicylic acid in pyridine. The water formed by the reaction of the oxide with salicylic acid is determined by titration with Karl Fischer reagent. The investigators [89] reported results on the analysis

of anhydrous calcium oxide equivalent to 2.5 to 13 mg of oxygen with a precision of about 0.1 mg. They used the method to analyze calcium containing 0.4–2% of oxygen. The method is not applicable to magnesium or other alkaline earth metals.

b. With Preliminary Separation of Metal from Metal Oxide

(1) METAL EVAPORATION—CHEMICAL [90–91]. Magnesium metal is removed from the metal oxides of the sample by sublimation in vacuum. The residue is dissolved in dilute hydrochloric acid. Any iron or aluminum present is removed as the hydroxide, and the magnesium content of the residue is determined volumetrically by (ethylenedinitrilo) tetra-acetic acid (EDTA) titration or by flame photometry.

(2) GRIGNARD SOLUTION—CHEMICAL [92]. Ground magnesium metal is dissolved in an alkyl halide (Grignard reaction), and the residue containing magnesium oxide is filtered, dissolved in hydrochloric acid, and titrated with EDTA.

(3) METAL EVAPORATION—VACUUM FUSION [93]. The metal is separated from metal oxides in the sample by vaporization from a thin-walled steel capsule in vacuum. The capsule containing the oxide residue is transferred to a vacuum fusion apparatus (see Chapter 4) in which the total oxygen is determined.

(4) METAL EVAPORATION—INERT-GAS FUSION [20]. The metal is separated from oxides by evaporation from a graphite crucible (in a stream of argon at 100 torr) and condensed on a cold surface (removable condenser) inside the reaction chamber of an inert-gas-fusion apparatus. After the separation is completed, this condenser is removed, and the residue is analyzed for oxygen, in situ, by carbon reduction in the absence of metal flux.

c. Discussion of Methods

All of the methods cited in this section are applicable to oxygen contents above about 0.1%. Only the fusion methods and possibly the neutron-activation method are applicable in the lower parts-per-million range.

Twitty and Fritz [84] analyzed magnesium samples both by neutron activation and by Grignard solution-titration. The comparative results shown in Table 11.10 indicate good agreement over the range of study.

Both fusion methods, which are based on carbon reduction of oxide residues to carbon monoxide, tend to give low results when applied to magnesium. Using vacuum fusion, Berry et al. [93] obtained an average recovery of 91% (relative standard deviation of 15%) on a group of 12

TABLE 11.10. Comparative Results Obtained by Grignard Solution-Titration and Neutron Activation [84]

Sample	Oxygen, %	
	Grignard Solution-Titration[a]	Neutron Activation
1	0.028	0.026
2	0.144	0.148
3	0.239	0.240
4	0.271	0.264
5	0.392	0.392
6	0.719	0.716

[a] Average of four aliquots.

mixtures of CaO, BeO, MgO, and Al_2O_3 containing 10–200 μg oxygen per sample, with values of 80%, 89%, and 54% on mixtures in which MgO was the predominant oxide. These data were obtained by using only the vacuum-fusion part of the total procedure and therefore did not take into account possible errors connected with matrix-metal evaporation (such as loss of oxide by entrainment and variation of iron-capsule blanks).

The inert-gas-fusion procedure [20] yielded 84% recovery (standard deviation of 9%) of oxygen added as magnesium oxide to magnesium capsules in 9 samples. These low results (compared to 100 $\pm$ 2% for oxygen in zinc and 94 $\pm$ 4% for oxygen in cadmium) were probably caused by loss of carbon monoxide by gettering during carbon reduction and to mechanical loss of the low-density oxide during sublimation of the matrix metal.

The time required for the analysis of a group of 10 samples and 2 blanks by the vacuum-fusion procedure was 8 h for metal evaporation, 8 h for degassing the vacuum-fusion apparatus, and 5 h for the carbon-reduction runs. By the inert-gas-fusion procedure, the total time per sample was about 1.5 h. By either fusion method the loss of carbon monoxide by gettering during the carbon-reduction process is expected to be a greater problem in the analysis of calcium, strontium, or barium than in the analysis of magnesium. The best method for these more reactive metals is probably the methanol–salicyclic acid method [89].

D. RECOMMENDED METHODS

1. Hydrogen

a. Beryllium: Hot-Extraction Method

When hydrogen is the only gas to be determined in beryllium, the hot-extraction method is most conveniently used. The sample is dropped into a

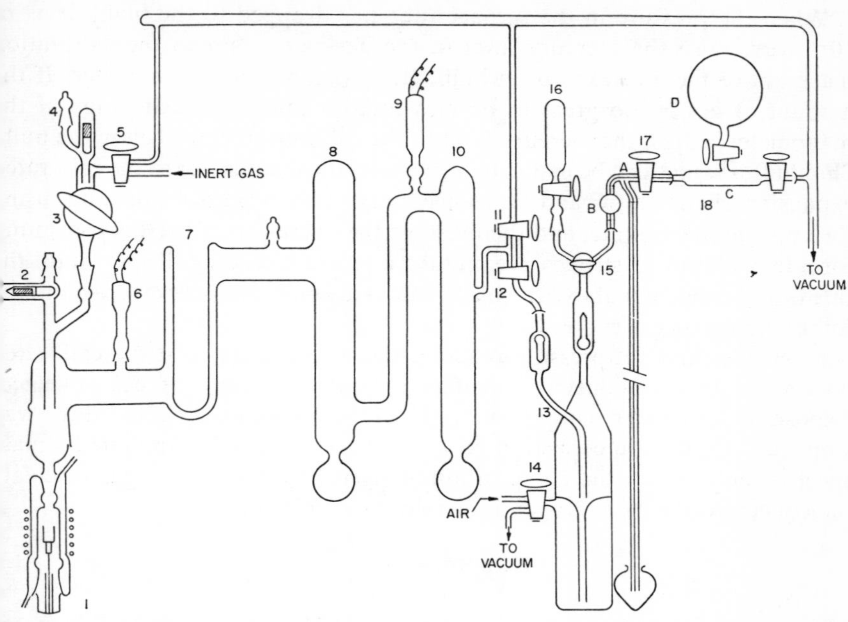

ig. 11.1. Hot-extraction apparatus. 1, Water-cooled quartz furnace tube; 2, Optical indow and metal-vapor shield; 3, Sample-stopcock; 4, Sample-dumper; 5, 11, 12, 14, and 5, Three-way stopcock; 6, Cold-cathode gauge tube; 7, Cold trap; 8, Single-stage mercury-ffusion pump; 9, Pirani gauge tube; 10, Two-stage, mercury-diffusion pump; 13, Auto-atic Toepler pump; 16, Gas-sample bulb; 17, Capillary monometer; 18, Calibrated pansion chambers. A, B, C, and D, of the capillary manometer.

egassed graphite crucible supported in a vacuum chamber. The crucible mperature is raised to 1050°C by induction-heating, and the hydrogen that volves from the sample is pumped into a calibrated vessel for measurement.

l) APPARATUS. Figure 11.1 shows the diagram of an apparatus for hot xtraction analyses.

2) PROCEDURE. With a stream of tank helium (grade A) flowing out of the ample-dumper, introduce a cleaned and weighed specimen of metal. Evacu-te the sample-dumper to about 10^{-3} torr and open the 14-mm-bore sample-opcock to the degassed, evacuated furnace tube. When the pressure in the xtraction line is reduced to 10^{-5} torr, drop the sample into the graphite rucible, supported on a tungsten rod in the water-cooled quartz furnace ibe. Raise the temperature of the crucible by induction heating to about 050°C. The mercury-diffusion pumps discharge the evolved hydrogen into ie automatic Toepler pump. The Toepler pump discharges the gas into the onnecting tube leading to the capillary manometer.

When the pressure in the furnace tube has dropped to the blank level o 10^{-5} torr, raise the mercury level in the Toepler pump to the calibratio mark above the 5-cc expansion bulb, B, of the capillary manometer. If th amount of gas is too great to be measured in this calibrated range of th manometer, adjust the mercury level to the calibration mark below the bulb Should this range also be inadequate, connect the manometer to the calibrate expansion chambers, C and D, as necessary. Take a pressure reading, usin the appropriate volume, and then resume the operation of the Toepler pum for a few additional strokes. Again take a pressure reading. Repeat until th rate of gas collection above the blank rate is insignificant in comparison to th total amount of gas evolved.

From the final net-pressure reading, the temperature, and the calibrate volume of the manometer, calculate the total quantity of gas collected Expose the evacuated sample bulb, 16, to the manometer via the three-wa stopcock, 15, for the collection of a sample of the gas for analysis by mas spectrometry. From the composition analysis of the collected gas calculat the concentration of hydrogen in beryllium as follows:

$$\text{H, ppm} = \frac{2.018\, nf}{w}, \qquad (2$$

where n is μmoles of gas, f is mole fraction of H_2 in the gas, and w is gram of sample.

(3) DISCUSSION. Some analysts use a heated palladium diaphragm in th hot-extraction vacuum line to separate hydrogen from other gases that ma be evolved from the heat sample and crucible. With mass-spectrometri analysis conveniently available, the author prefers to measure the total ga evolved and then to analyze for composition. By such a procedure concer about the quantitative transfer of hydrogen through the palladium is avoide (see Chapter 4 for a discussion of the palladium diaphragm). A mass spectro metric analysis is not usually required for every sample of a series. Th conditions of the procedure can usually be regulated such that the amount o gas other than hydrogen that is pumped from the furnace tube is minimize to a low reproducible value. The quantity of gas required for a mass-spectro metric analysis is about 0.3 cc-atm.

The open-well capillary manometer is used because of its simplicity, it ease of cleaning and maintenance, and its adequacy in precision of abou ± 1–2 relative percent. The manometer is calibrated by inverting and fillin the 1.5-mm capillary volume (bounded by the stopcock 17, the calibratio mark on the column, and the calibration mark above the expansion bulb with mercury. The mercury is drained out and weighed as a measure of th fixed volume A, the most sensitive range. The procedure is repeated, thi time filling to the mark below the expansion bulb, for a calibration of th

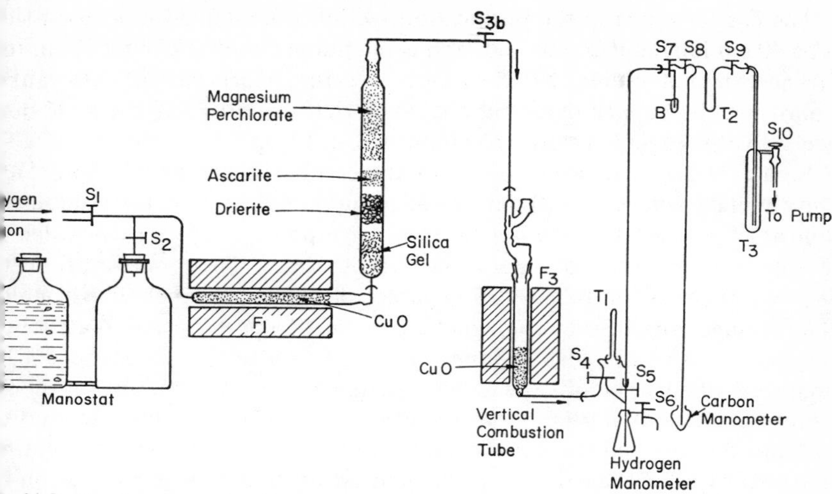

. 11.2. Combustion train. (Courtesy of *Analytical Chemistry* [35].)

xt most sensitive range (volume B). Volumes C and D are calibrated arately in a similar manner. The manometer leg extending to the mercury ll is made of precision-bore capillary; hence, the variable volume over the rcury column is proportional to the depression of the mercury below the o mark. A calibration curve, an equation, or a conversion table relating nometer readings to micrograms of hydrogen is derived from the measured ed volumes and the cross-sectional area of the capillary bore.

Magnesium: Combustion-Manometric Method

APPARATUS. Figure 11.2 is a schematic diagram of the apparatus used the simultaneous determination of microgram quantities of hydrogen d carbon in easily combustible metals. On combustion in oxygen, hydrogen d carbon are quantitatively released as H_2O and CO_2, respectively, at 0°C. The apparatus is composed of a carrier-gas purification section, a rtical combustion tube, and 2 capillary manometers, one for measuring H_2O at 110°C and the other for CO_2 at room temperature (Note 1).

Oxygen and argon, each regulated at a line pressure of 3 lb, are supplied the purification section through a three-way stopcock. The selected gas is mitted in sufficient quantity to allow a slow stream, in excess of that quired for the combustion train, to pass through the manostat (Note 2). e rate of gas flow through the train is governed by a grooved stopcock, S_8, the carbon manometer. The line pressure up to stopcock S_8 is equal the head established by the manostat. The pressure beyond stopcock S_8 pends upon the extent to which the stopcock is opened to the vacuum line.

The copper oxide in the purification section is contained in a fused sil tube 40 cm long and 26 mm o.d. and is maintained at 900°C by a 12-in. tu furnace, manufactured by Hevi-Duty Heating Corporation, Milwauk Following the copper oxide tube is an upright cylinder (50 cm × 46 mn packed with reagents (Note 3) as indicated in Fig. 11.2.

Figure 11.3 is a detailed sketch of the combustion tube. An 8-in. He Duty replacement unit, Model 123-8, is supported around the combusti tube at a level such that the hottest zone corresponds to the area at which sample burns when it falls upon the bed of oxides. The temperature of 584-W furnace is controlled by a variable voltage transformer. The iron-c glass plunger in the electromagnetic sample dumper is manipulated by variable transformer that supplies current to the electromagnetic coil. small tuft of silver wool is included near the exit of the combustion tube remove chloride impurities from the gas stream. The glass line between t combustion tube and stopcock S_4 on the hydrogen manometer is maintain at 100°C by a 2-ft heater tape to prevent adsorption of water vapor on t glass walls while en route to the manometer. The ball joint in this heat section is sealed with Apiezon-H grease.

Detailed drawings of the manometers used for measurements of hydrog

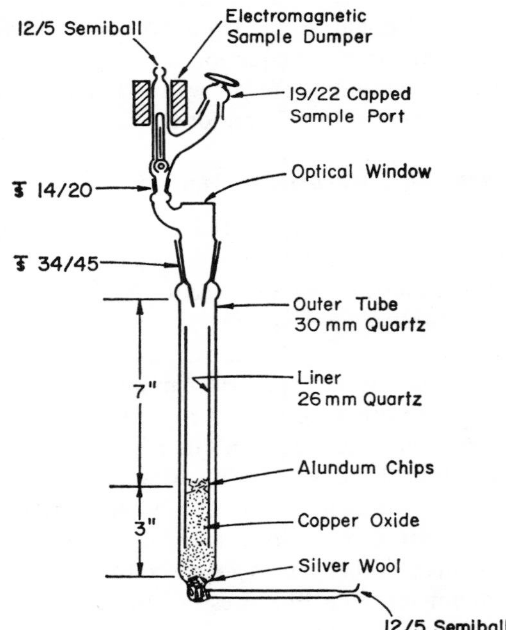

Fig. 11.3. Combustion tube. (Courtesy of *Analytical Chemistry* [35].)

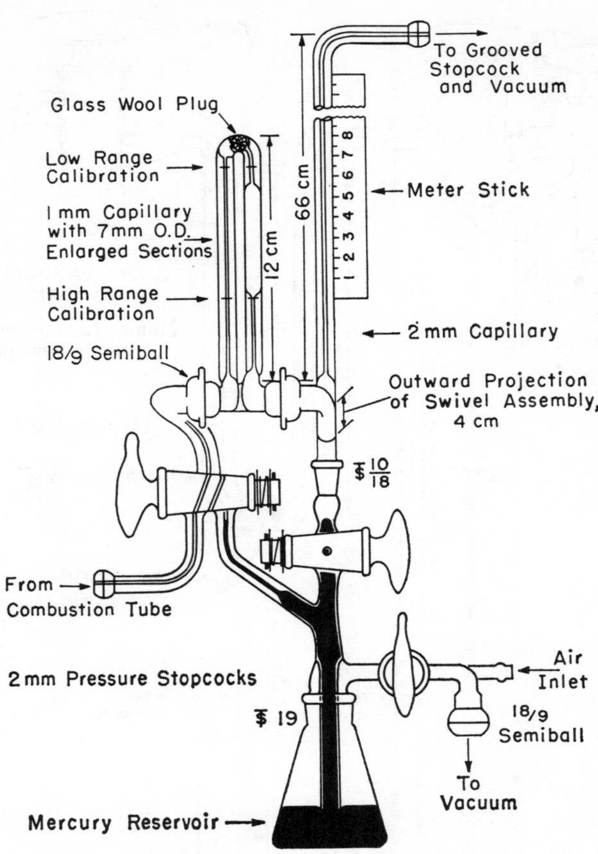

g. 11.4. Hydrogen manometer. (Courtesy of *Analytical Chemistry* [35].)

ıd carbon are shown in Figs. 11.4 and 11.5, respectively. The hydrogen anometer has a capillary U-tube that is swiveled upward and downward ı 2 pairs of 18/9 semiball joints, held firmly together by a wire coil spring. ı the downward position the U-tube is immersed in dry-ice–acetone slush to eeze out water vapor from the carrier gas stream. In the upward position becomes one leg of a constant-volume, closed manometer. Mercury is ıised to the calibration marks, indicated in Fig. 11.4, while the U-tube is ıclosed in a transparent glass chimney from which emerges a blast of air :gulated at 110°C. This hot air is supplied by a turbotype, electric heat gun, ıanufactured by the Master Appliance Company, Racine, Wisconsin. The ıimney is part of a 32-mm glass T-joint, indicated in Fig. 11.6, which couples) the barrel of the heat gun. An asbestos disk affixed to the U-tube near the

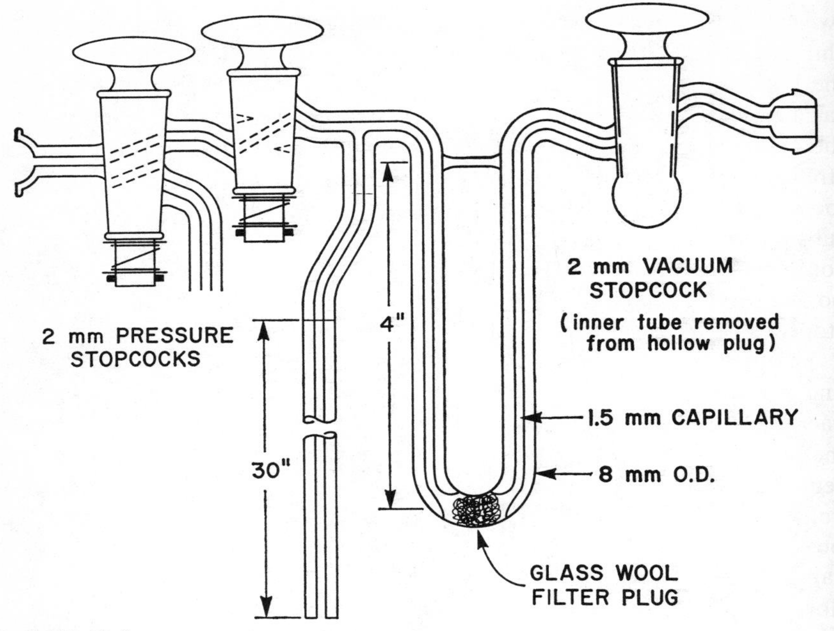

Fig. 11.5. Carbon manometer.

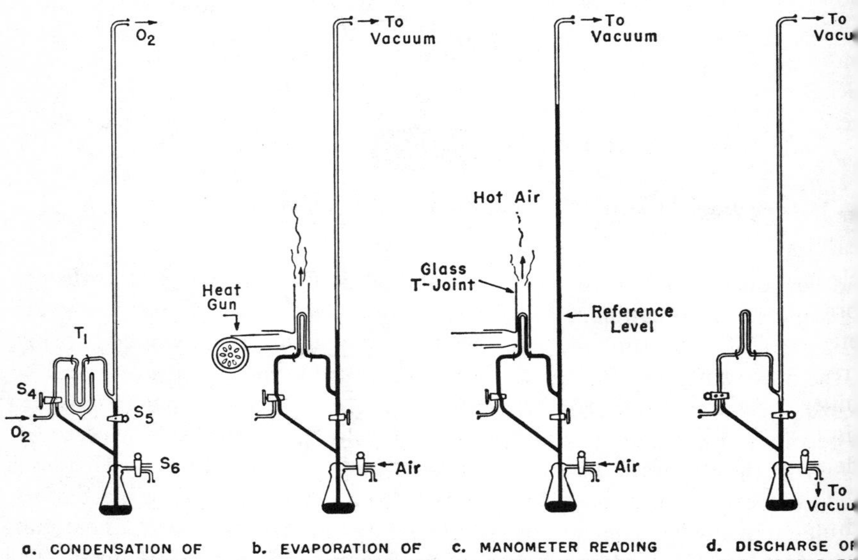

Fig. 11.6. Manipulations of hydrogen manometer. (Courtesy of *Analytical Chemistry* [35

wivel joints prevents the hot air from emerging at the lower end of the himney. The enlarged section in the capillary U-tube (Fig. 11.4) is part of he high-range fixed volume of the manometer. Larger quantities of water an be measured (at lower sensitivity) by using the calibration marks below he enlarged section. The glass-wool plug included in the path of the gas tream serves as a mechanical filter to retain ice crystals that tend to break ose from the walls of the capillary and pass through the cold trap to escape ubsequent measurement. Pressure-type stopcocks are used. Needle-valve ontrol is provided by tapered grooves in the plug of stopcock S_4 for the osition connecting the manometer to the combustion tube and in the plug of topcock S_6 for both air and vacuum positions.

The carbon manometer described by Smiley [94] has been modified [95] nd is shown in Fig. 11.5. Besides its grooved stopcock, it has a hollow acuum stopcock, which can serve as a reservoir to provide a dual range of neasurement. With the oblique inner tube of the hollow stopcock plug emoved, the volume enclosed by the plug and barrel becomes part of the olume of the manometer when the plug is rotated 180° from the open osition. Carbon dioxide is condensed from the carrier gas stream in the apillary U-tube at liquid-nitrogen temperature. The pressure of the gas lowing through the U-tube is regulated at about 50 torr by manipulation f the grooved stopcock. Neither oxygen nor argon is condensed under these onditions and the rate of gas flow is about 150 cc/min.

The liquid nitrogen trap, T_3 (Fig. 11.2), prevents back-diffusion of oil apors and other condensable gases from the vacuum pump. A Welch Duo-Seal pump provides vacuum for both the analytical train and the mercury eservoir of the hydrogen manometer.

2) PROCEDURE. At stopcock S_8, adjust the flow of argon through the nalytical train to produce a 5-cm depression on the mercury column in the arbon manometer. Open the capped port on the sample dumper, insert a leaned, weighed sample of magnesium, and close the port after flushing riefly with excess gas stored in the manostat. Arrange the hydrogen manom-eter as illustrated in Fig. 11.6a, with dry ice-acetone slush on the cold rap T_1, and immerse trap T_2 (Fig. 11.2) of the carbon manometer in liquid itrogen. Switch the carrier gas from argon to oxygen. After T_2 has cooled to iquid-nitrogen temperature, readjust the gas flow to maintain the 5-cm lepression of the mercury column in the carbon manometer.

Drop the sample into the furnace by manipulation of the dumper. (The lisplacement of argon by oxygen in the analytical train occurs at a rate onducive to even, nonviolent combustion of the metal.) Continue the run bout 5 min after the sample has ceased to glow, as observed through the ptical window at the top of the furnace tube.

Close stopcock S_4, and evacuate the two manometers. By turning th appropriate stopcocks, isolate the manometers from the vacuum line an from each other.

Manipulate the 3 stopcocks on the hydrogen manometer (Fig. 11.6b) t confine the condensed water with mercury in the up-positioned U-tube, an begin heating with the heat gun. As the heat expands the entrapped wate vapor, tending to drive the mercury level in the U-tube down below th heated zone, apply sufficient air pressure (room atmosphere) to the mercur well to maintain the level within the heated zone (otherwise, some of th water condenses on cooler surfaces and becomes lost to measurement) When approximate equilibrium is established between the column pressur of the confined, heated water vapor and the static pressure of the mercur column in the long leg of the manometer, allow 2–3 min for the temperatur of the chimney air to reach 110°C (indicated by a 6-in. thermometer suspende inside the chimney of the heat gun).

In the meantime, read the carbon dioxide pressure in the carbon manom eter, the U-tube of which is warmed to room temperature. Evacuate th manometer, and note whether or not the mercury level returns to the zer mark. Divergence from the zero position, set at the beginning of the analysi by adjustment of the elevation of the mercury well, may indicate a barometri pressure change during the analysis. Such an effect is usually negligible; i not, subtract this reading from the total reading.

Adjust the mercury level in the hydrogen manometer to the calibratio marks that are appropriate for the amount of water being measured (Fig 11.6c), and read the height of the mercury column above a predetermine zero level (Note 4). This represents the pressure of the water vapor confine to a fixed volume at 110°C.

Retract the mercury level in the hydrogen manometer to the positio shown in Fig. 11.6d by carefully recondensing the water in the U-tube whil simultaneously lowering the mercury column in the long leg of the manom eter. To keep the rate of recondensation under control, proceed as follows Lower the mercury level in the U-tube (by applying vacuum to the mercur well) to a point near the ball joints. Switch from hot to cold air in the hea gun and cool the U-tube to room temperatures. Remove the heat gun, an cautiously paint the tip of the U-tube with dry-ice–acetone slush. After al the water is recondensed, lower the mercury to the level shown in Fig. 11.6d and evacuate the manometer. Switch back to argon at stopcock S_1 (in prepara tion for the next sample), and with the grooved stopcock S_4 open enough t produce a depression of the mercury column in the carbon manometer o about 10 cm, heat the U-tube of the hydrogen manometer briefly with a flam to drive out absorbed moisture.

Follow the same procedure of analysis, without actually dropping ample, to obtain the respective blanks for hydrogen and carbon determinaions (if both are desired).

To leave the apparatus in standby condition, shut off the manostat and lirect a slow stream of oxygen through the bubbler on the carbon manometer.

(3) CALCULATIONS. When the fixed-volume manometer is operated at a constant temperature (110° ± 1°C), the quantity of measured hydrogen is proportional to the net-pressure reading. Then,

$$\text{H, ppm} = \frac{(R_s - R_b)F}{\text{wt of sample, g}} \tag{3}$$

where R_s is the pressure reading (torr) for the sample, R_b is the pressure reading (torr) for the blank, and F is the proportionality factor

$$(\mu\text{g/torr}) = \frac{2.016(\text{volume in cc})}{6.236 \times 10^4 \times 383 \times 10^6}.$$

The factor, F, can be determined empirically by injecting known quantities of hydrogen into the train, using a dosing stopcock described by Brown [96] (see also Chapter 4). A typical hydrogen manometer [35] had calibration factors of 0.0195 ± 0.0002 μgH/torr standard deviation for the low range and 0.127 ± 0.001 μgH/torr for the high range.

The quantity of carbon in the sample, if desired, may be calculated from a calibration equation of the form

$$\mu\text{g carbon} = Ax + Bx^2,$$

for the variable-volume manometer, where x is the manometer reading, A is a constant proportional to the fixed volume above the zero mark, and B is a constant proportional to the inside diameter of the precision-bore capillary of the manometer stem.

NOTES

1. Laboratory space can be conserved by placing the component parts in the left half of the diagram behind those in the right half.
2. The manostat consists of two 4-liter bottles containing silicone oil. An appropriate, low-pressure, high-flow gas regulator may be used in place of the manostat.
3. Thorough dehydration of the magnesium perchlorate is assured by heating the reagent (in 1-lb batches) for 6 h at 220°C in a stream of dry air at a pressure of 5 torr.
4. The zero level is attained when the U-tube is similarly heated but with no water confined.

c. Calcium: Tin-Fusion Method

(1) APPARATUS. The apparatus of the tin-fusion analysis consists of a silic crucible charged with 50 g tin inside a vertical, silica, furnace tube, a tw stage mercury-diffusion pump, and a gas-collection system. A sample-loadin tree with a capacity of about 6 samples is attached to the furnace tube. Th furnace tube is heated by a nichrome-wire resistance furnace. The diffusio pump is arranged to discharge the extracted hydrogen either into a 0.5– liter, multiple-volume collection system (connected to a McLeod gauge b which the total pressure is measured in the micron range) or into a Toepl pump for transfer to a capillary manometer by which the pressure is measure in the torr range. The latter arrangement is more conveniently used when th collected gas is to be sampled for mass-spectrometric analysis.

(2) PROCEDURE. Quickly transfer six weighed samples (2–200 mg, dependin on the approximate hydrogen concentration and on the capacity of the ga measuring system) from a desiccated storage container to the sample-loadin tree. Seal the tree to the furnace tube by a standard taper joint, using melte Apiezon W wax. Evacuate the apparatus and heat the tin bath to 670°C Continue heating until the blank rate of gas extraction from the furnace tub is reduced to 38 cc-torr/h. With the entire evacuated system closed off fro the mechanical vacuum pump, measure the residual pressure in the ga collection system and drop a sample into the tin bath. Collect the gas evolve from the sample until the rate of increase of pressure is equal to the blan rate. Correct the final pressure reading for the blank and for the initi pressure in the collection volume. Calculate the total quantity of gas from th known *PVT* data. Transfer a sample of the collected gas to a mass spectrom eter for analysis. From these data and the sample-weight, calculate th concentration of hydrogen in the calcium. Analysis of the extracted gas is no required for every sample if occasional analysis indicates that it is essentiall all hydrogen.

2. Nitrogen: Kjeldahl Method (see also Chapter 8)

a. Apparatus

Figure 11.7 shows the micro-Kjeldahl distillation apparatus [97].

b. Procedure

Place a 1-g sample of metal in a 50-ml flask, connect to a reflux condense and add through the condenser 5 ml ammonia-free water. If the metal i calcium, strontium or barium, add the water slowly and keep the flask coo

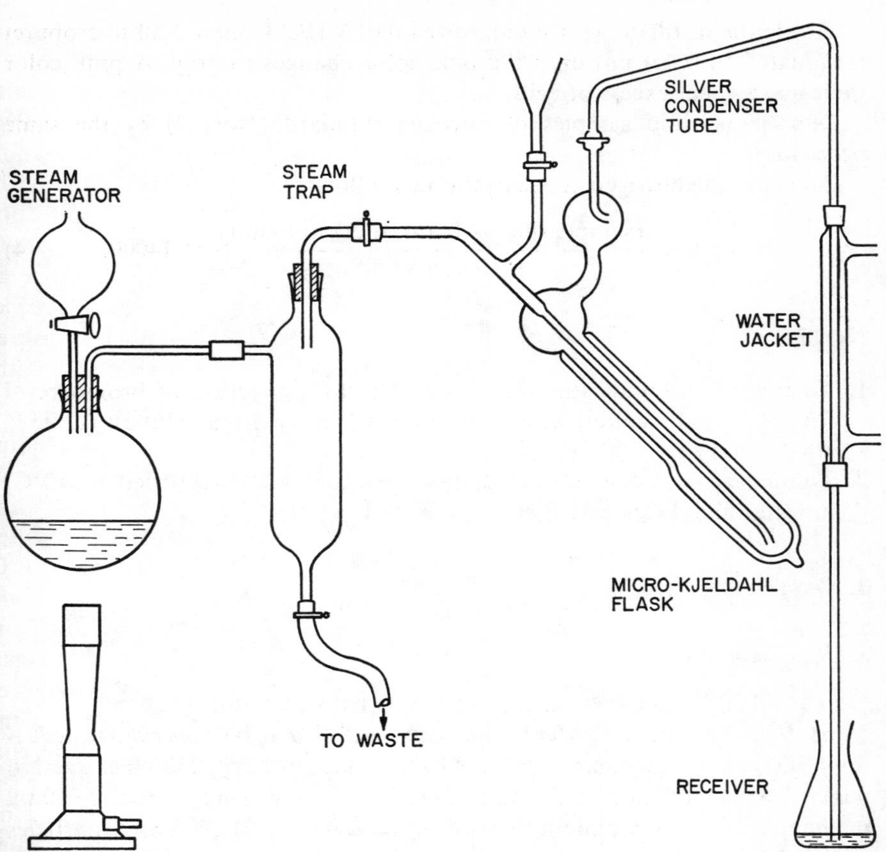

Fig. 11.7. Micro-Kjeldahl distillation apparatus. (Courtesy of *Analytical Chemistry* [98].)

to prevent the loss of ammonia from the basic solution. Add concentrated HCl dropwise through the condenser until the sample is completely dissolved.

Rinse the condenser into the flask. Transfer the solution to the distilling flask of a Parnas and Wagner micro-Kjeldahl distillation apparatus that has been freshly steamed out and used for establishing a constant blank. Add 15 ml 50% NaOH solution. The total volume of liquid, including rinse water, should not exceed $\frac{3}{4}$ of the jacketed volume of the distilling flask.

With the tip of the distillation condenser submerged in 5 ml 2% H_3BO_3 solution (containing 2 drops of mixed indicator; see Note 1), start the steam distillation. When about 25 ml of distillate has been collected, remove the receiver, rinse the condenser tip, and turn off the flame. Place a beaker containing 10 ml water under the condenser tip so that as the system cools, the water will be drawn into the apparatus to rinse out the distilling flask.

Titrate the distillate with standardized $0.01N$ HCl from a 2-ml microburet (graduated to 0.001 ml) until the blue color changes to gray. A pink color indicates a slight excess of acid.

Run blanks and samples of nitrogen standard (Note 2) by the same procedure.

Calculate the nitrogen concentration as follows:

$$\text{N, ppm} = \frac{(\text{sample titer} - \text{blank titer})\ \text{normality}}{\text{grams of sample}} \times 14000. \tag{4}$$

NOTES

1. To prepare the indicator, mix 5 ml of a 0.1% solution of bromocresol (in 95% ethyl alcohol) with 1 ml of a 0.1% methyl red solution (in 95% ethyl alcohol).
2. To prepare a nitrogen standard, dissolve 0.0382 g NH_4Cl (dried at 110°C) in water and dilute to 1 liter. 1 ml = 0.01 mg N.

3. Oxygen

a. Beryllium

(1) (> 0.1% O) BROMINE–METHANOL DISSOLUTION METHOD [53]

(*a*) *High-Temperature Metal* (*Vacuum-Cast or High-Temperature Reduction*). Place a 0.5-g sample of powder or fine chips in a dry, 250-ml extraction flask. Add 10 ml methanol. Add *slowly* in 2- and 3-ml portions, 50 ml methanol (Note 1) containing 3.0 ml bromine (Note 2). When all particles of the metal appear to be dissolved, add 50 ml of the methanol–hydrochloric acid solution (Note 3). Mix the contents by swirling every few minutes over a period of 30 min. Transfer the solution to a 150-ml beaker and wash any residue from the flask into the beaker with methanol. Filter the solution with vacuum through a 15-ml glass, fritted-disk crucible (Note 4). Wash the residue and crucible with methanol until the wash is free of bromine color. Remove the crucible from the filtering flask, and wash the bottom and outside of the crucible with water.

Transfer the crucible to the 150-ml beaker and add 15 ml of concentrated H_2SO_4 and 10 ml of concentrated HNO_3. Cover the beaker with a watch glass and dissolve the residue by heating on a hot plate until fumes of H_2SO_4 evolve. Cool the solution and dilute with 100 ml water. Transfer the solution to a 500-ml volumetric flask. Wash the crucible and beaker twice with 100-ml portions of hot water, adding the washings to the volumetric flask. Dilute the solution to 500 ml.

Into four of five 50-ml centrifuge tubes add, respectively, 0, 1.00, 3.00, and 5.00 ml of a standard beryllium solution (Note 5). Add 20 ml 3% H_2SO_4 to each tube. To the fifth tube add 20.0 ml of the sample solution. To all tubes add 1.0 ml aluminum nitrate solution (Note 6) and about 2 g solid NH_4Cl. Dilute each solution to about 35 ml with water. Stir the sample and standard solutions with glass rods, and precipitate the aluminum and beryllium by the addition of 5 ml concentrated NH_4OH. Stir thoroughly, and remove the glass rods. Centrifuge the suspensions, and carefully decant the supernatant liquids. Drain the tubes by standing them inverted on a dry towel for about 5 min. To each tube add 20.0 ml of the 4-(*p*-nitrophenylazo)-orcinol reagent (Note 7) and disperse the hydroxides in the solution by stirring with the glass rods. Warm the solutions in a water bath at 80–90°C for 5 min. Cool to room temperature. Transfer the solutions to 50-ml volumetric flasks and dilute to volume.

Measure the absorbance of each solution at 520 nm in a spectrophotometer using 2-cm light path cells. Use the solution containing no beryllium as a reference. Because Beer's law is not obeyed, draw a smooth curve through the plot of absorbance versus concentration of the standards. Calculate the beryllium oxide content by the following equation:

$$\mathrm{O}, \% = \frac{\mu\mathrm{g\ Be} \times 1.78 \times 10^{-4} \times 25}{\mathrm{wt.\ of\ sample,\ g}}. \qquad (5)$$

(*b*) *Low-Temperature Metal* (*Electrolytic*). Place a 1-g sample in a platinum boat. Insert the boat into a small combustion tube and sweep out the tube with helium from which oxygen is removed by a hot copper train. Heat the tube for 1 h at 200°C to remove moisture. Over a period of 30 min, raise the temperature to 600°C and retain at this temperature for 30 min. Cool the tube to room temperature under helium. Weigh out a 0.5 g sample of the heat-treated material and proceed as described above for high-temperature metal.

NOTES

1. Use methanol with a water content not exceeding 0.1%.
2. The reaction is very vigorous and should be carried out in a well-ventilated hood.
3. Prepare just before using by diluting 50 ml concentrated hydrochloric acid to 500 ml with methanol.
4. Use crucibles of fine porosity, which provide fast filtration.
5. Dissolve 1.00 g beryllium metal in HCl and dilute to 1 liter. Dilute 10.0 ml of this stock solution to 500 ml to obtain a standard solution of 20 μg Be/ml. Establish the calibration curve with each set of samples analyzed.
6. Dissolve 34.8 g aluminum nitrate nonahydrate in water and dilute to 500 ml.

7. To prepare the reagent, dissolve 24.0 g NaOH and 18.0 g H_3BO_3 in water and dilute to 1 liter. Transfer the solution to a 2-liter beaker and add 75 mg 4-(*p*-nitrophenylazo)-orcinol. Stir mechanically for 2 h and filter. Store in the dark. The 4-(*p*-nitrophenylazo)-orcinol reagent slowly precipitates upon standing. Filter the solution if it appears turbid when examined in a strong light.

(2) (< 0.1% O) VACUUM-FUSION METHOD. The reader is referred to publications of Booth and Parker [22, 55, 56] in regard to techniques that contribute to lower limits of detection (larger samples, lower blanks, etc.; see also Chapter 4).

(*a*) *Apparatus.* A diagram of the vacuum-fusion furnace tube, as used by Everett and Thompson [45], is given in Chapter 4 and the crucible assembly is described. Figure 11.8 shows the entire apparatus, including the low-pressure gas-analysis system. The gas-analysis system is not described in this section inasmuch as comparable systems are described elsewhere in this book. Furthermore, the analyst who applies the method might choose to analyze the evolved gas mixture in some alternative way, such as by mass spectrometry or by gas chromatography.

The furnace tube is exhausted by two mercury-diffusion pumps connected in series. The first, with a liquid-nitrogen-cooled cold trap in its intake throat and with a conductance from the crucible area of 3 liters/sec, is a 2-stage pump and has an effective pumping speed at the crucible of 2.6 liters/sec for pressures up to 3×10^{-2} torr. The second, 1-in., 2-stage pump discharges into the calibrated volume in which the total evolved gas is measured. The discharge end of this pump forms a boundary of the calibrated volume (about 500 ml). Additional calibrated volumes (500 and 1000 ml) can be added when needed by lowering the appropriate mercury cutoffs.

(*b*) *Procedure.* Weight out 5–30-mg freshly abraded chunks of beryllium into precleaned platinum capsules made by crimping the ends of $\frac{3}{8}$-in. lengths of 0.1-in. i.d. platinum tubing of 0.002-in. wall thickness. Enclose with the sample in each capsule 5 mg of degassed tin. After a room-temperature blank rate on the apparatus has been established at less than 0.4 μ-l./10 min, isolate the furnace tube from the gas collecting system. Slowly bring the pressure of the furnace tube up to 1 atmosphere with dry air, and place the weighed samples and 3.5 g of platinum pellets in the loading arms. Restore vacuum and heat the crucible to 2100°C for 2 hours. After the first hour fill the cold trap on the first exhaust pump with liquid nitrogen. Measure the degassing rate at intervals until a rate of <2.3 μ-l. S.T.P./10 min is obtained. Lower the temperature to 1850°C, and, leaving the power regulator of the induction heater at this setting, shut off the power.

Raise the lid of the crucible and lower the silica funnel until it just contacts the crucible. By means of a magnetically operated slug, push the platinum

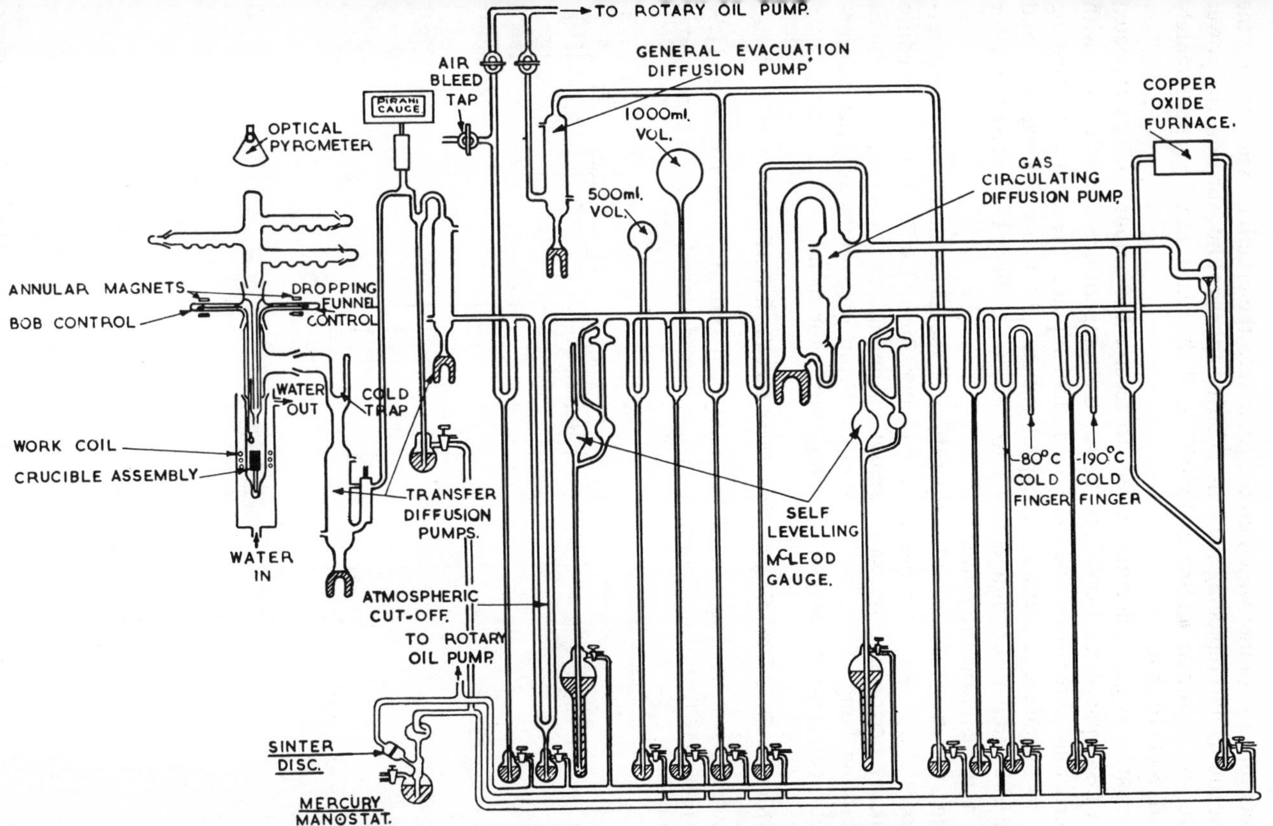

Fig. 11.8. Vacuum-fusion apparatus. (Courtesy of United Kingdom Atomic Energy Authority [45].)

pellets into the funnel which guides them into the crucible. After raising the silica funnel and replacing the self-centering spherical lid, turn on the power again and continue heating at 1850°C until the blank rate is approximately 2.3 μl. STP/10 min.

Prepare the gas-collecting system to receive the amount of gas expected from a sample of the beryllium. Lower the temperature to 800°C, and with the crucible lid removed and the silica funnel in place, use a magnetic pushing slug to introduce a platinum-encapsulated sample. Close the crucible, and raise the temperature immediately to 1850°C. Collect the evolved gas for 3 min and shut off the power.

Measure the pressure and temperature of the gas collected in a predetermined volume. Analyze the gas for composition. Establish the procedure blank by performing the same analytical operations on platinum capsules containing 5 mg of tin but no sample. From these data and the sample-weight, calculate the oxygen concentration in the metal.

b. Magnesium: Metal-Evaporation–Inert-Gas-Fusion Method

(1) APPARATUS. A diagram of the analytical train is shown in Fig. 11.9. The inert-gas fusion line contains a specially designed reaction tube and a

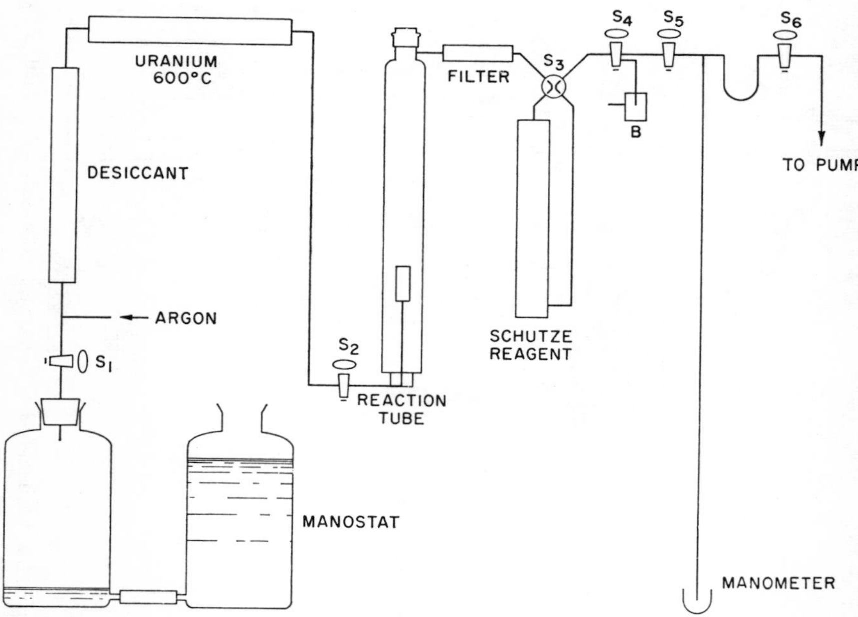

Fig. 11.9. Metal-evaporation–inert-gas-fusion train. (Courtesy of *Analytical Chemistry* [20].)

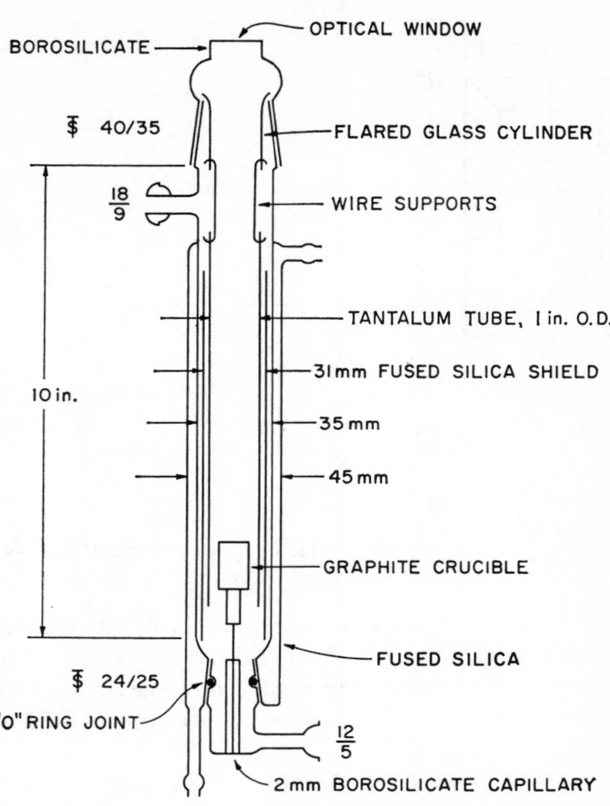

Fig. 11.10. Reaction chamber. (Courtesy of *Analytical Chemistry* [20].)

grooved stopcock, S_2, for adjusting low pressures in the reaction tube. A glass-wool filter is included to collect smokelike particles of metal produced by the condensation of metal vapor in the argon stream.

Figure 11.10 shows the details of the reaction chamber. The tantalum tube (8 in. in length 1 in. in diameter, and 0.02 in. in thickness) is suspended by three platinum wires from a glass cylinder, the flared edge of which rests on the male section of the 40/35 standard-taper joint. A slit, cut in the tube from the lower end upward (about $\frac{1}{8}$ in. wide and 7 in. long) minimizes the energy pickup from the induction coil. The fused silica liner, supported on the lower concave surface of the reaction tube, envelops the tantalum tube to prevent metal vapors, escaping through the slit, from condensing on the inner wall of the reaction chamber. It is necessary to keep this wall clear of magnesium to avoid interaction with the silica surface when the crucible is subsequently heated to high temperatures.

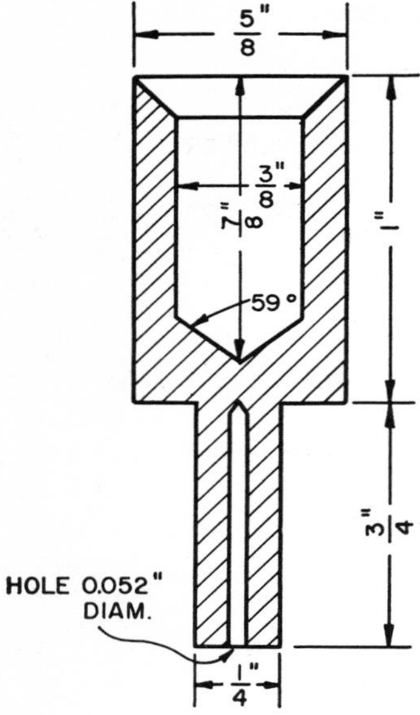

Fig. 11.11. Graphite crucible. (Courtesy of *Analytical Chemistry* [95].)

The tantalum tube and the fused silica tube can be removed from the reaction chamber (for cleaning in dilute acid) by a pair of 10-in. tweezers, the tips of which are bent outward to engage into holes in the tubes. Another pair of extra-long tweezers, with suitably contoured tips, is used to handle the crucible.

The graphite crucible (Fig. 11.11) and 20-mesh graphite chips are made from extruded rod, grade AUC, supplied by the National Carbon Company, New York. Induction heating is supplied by a 10-kW (output) generator. The Schutze reagent (silica gel impregnated with iodine pentoxide) is prepared as described by Smiley [98] (Note 1).

The capillary manometer in which carbon dioxide is collected and measured is illustrated in Fig. 11.5 and described in Section IV.D.1.b(1).

(2) PROCEDURE. Place a crucible, filled to a depth of $\frac{1}{2}$ in. with graphite chips, in the reaction tube and heat to 2100°C for 10 min in a stream of argon flowing out through the greased ST 40/35 joint. To obtain a 10-min blank, push the ST 40/35 joint together, and again heat the crucible to 2100°C with the argon stream (throttled to 150 ml/min at the grooved stopcock S_5) passing through the Schutze reagent and through the U-tube cold trap at liquid-nitrogen temperature. Isolate the carbon dioxide collected in the trap, and read its pressure on the capillary manometer at room temperature. Using the calibration data of the manometer, convert the pressure to micrograms of oxygen. Repeat the blank runs until the blank drops to 1 or 2 μg/10 min.

Insert the silica liner and the tantalum condenser tube. Reach in from the top of the reaction chamber with appropriate tweezers and remove the crucible. Pour the graphite chips out on clean aluminum foil. Place a weighed sample of magnesium in the crucible, cover it with the graphite chips and return the crucible to its position in the reaction chamber.

Bypass the Schutze reagent, and evacuate the line back to the stopcock S_2. Then, by throttling the flow at this stopcock, admit enough argon to produce a 100-mm depression of the mercury column in the manometer. Energize the heating coil at low power and increase the power gradually to raise the temperature of the crucible to 800°C (Note 2). Maintain this temperature until the sublimation of the magnesium from the crucible is complete (usually about 1 h), as evidenced by an optical field of uniform brightness. Cut off the power, and allow 5 min for cooling.

Increase the argon flow to an excessive amount through the manostat. Establish atmospheric pressure again in the system by closing the exhaust stopcock on the manometer and by slowly opening the grooved stopcock, S_2. Quickly remove the tantalum tube and quartz liner, and the allow argon to flow out through the greased ST joint for 1 min to flush the reaction tube of air and moisture. Push the ST 40/35 joint back together.

Turn S_3 to pass the gas through the Schutze reagent, immerse the U-tube in liquid nitrogen, adjust the argon flow to 150 ml/min at stopcock S_5, and raise the temperature of the crucible to 2100°C. Shut off the power after the crucible has been heated for 10 min, and allow argon to flush through 3 min more. Measure the CO_2 as described above for the blank and convert the manometer reading to micrograms of oxygen.

Calculate the concentration of oxygen by

$$\text{O, ppm} = \frac{\mu\text{g O in sample} - \mu\text{g O in blank}}{\text{g of sample}} \tag{6}$$

NOTES

1. Use the following procedure to prepare a suitable form of Schutze reagent. Mix a solution of 5 g iodine pentoxide in 25 ml water with 50 g coarse (6-16 mesh) silical gel. Dry the mixture at 145°C for 1.5 h. Stir in 10 ml of concentrated H_2SO_4, and after allowing it to stand overnight to ensure uniform mixing, sift on a 40-mesh screen and discard the fines. Place the remaining material in a glass combustion tube and heat to 200°C in a slow stream of air at 1–2 cm pressure. Both ends of the tube should be packed with silica gel to dry the entering air and to retain the H_2SO_4 and I_2 evolved from the mixture. Heat for 4 h. Cool to room temperature, and then transfer the product, which resembles a yellow or white sand, to a container in which it can be stored with a minimum exposure to air and moisture.
2. No reduction of magnesium oxide by graphite is detectable at 800°C.

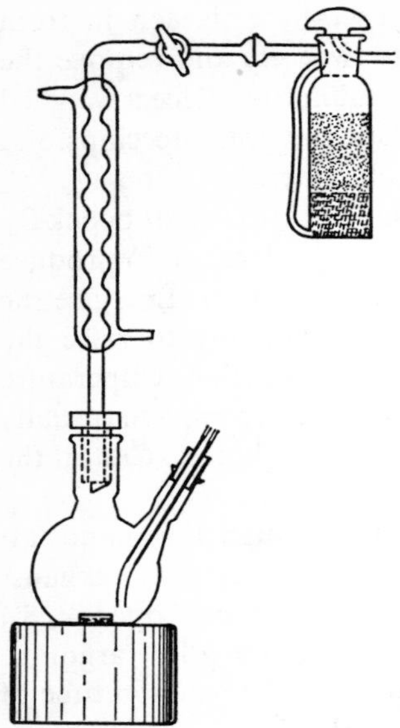

Fig. 11.12. Methanol–salicyclic acid reaction and titration apparatus. (Courtesy of *Analytical Chemistry* [89].)

c. Calcium: Methanol–Salicylic Acid Method [*89*]

(1) APPARATUS. The apparatus used for the reaction of calcium with methanol is shown in Fig. 11.12. A 250-ml borosilicate glass reaction and titrating flask, a 30-cm reflux condenser (with stopcock) and a Nesbitt absorption bulb (containing layers of Drierite and magnesium perchlorate) are connected by standard taper joints as indicated in the diagram. A magnetic stirrer is used to agitate the solution.

A Beckman Model K-F Aquameter, or equivalent, may be used for the Karl Fischer titration. The standard 125-ml titrating flask and dual electrodes supplied with the Aquameter are used in determining the water content of an aliquot of the methanol drawn from the reaction flask prior to the addition of calcium. The smaller flask is also used for titrating the salicylic acid–pyridine solution before addition of the acid to the reaction flask. In this procedure no correction for water in the salicylic acid–pyridine solution is necessary, since the acid is pretitrated. A special set of dual platinum electrodes is constructed for direct use in the 250-ml reaction flask.

(2) PROCEDURE. Dry the reaction flask, the standard-taper joint adapter, the side-arm stopper, the magnetic stirring bar, and the condenser in an oven at 120°C. Assemble the glass stopper in the side-arm. With the stopcock on the condenser closed, rinse the apparatus with two 50-ml portions of absolute methanol and discard.

Place 100 ml absolute methanol (Note 1) in the reaction flask through the side-arm and rapidly replace the stopper. Rinse the flask and condenser, but do not discard the alcohol. Attach the condenser through the ball-and-socket joint to the Nesbitt absorption bulb and open the stopcock.

Rinse a 25-ml pipet twice with absolute methanol, and then quickly withdraw through the side-arm a 25-ml aliquot of the alcohol from the reaction flask. Transfer the aliquot rapidly to the 125-ml titrating flask, which contains 25 ml pretitrated methanol. Titrate the aliquot with Karl Fischer reagent (Note 2) to a dead-stop endpoint and record the titration. Multiply by a factor of 3 to give the blank titration. Discard the solution.

Place a 2-g sample of calcium metal chips (Note 3) in the reaction flask through the side-arm, and tightly replace the stopper. Water-cool the condenser, and heat the flask gently with a small flame until the calcium reacts rapidly with the alcohol. Discontinue the heating (Note 4). Stir constantly, and occasionally warm the alcohol to dissolve the residual small particles of calcium metal, which are coated with calcium methylate. Allow 1 h for complete reaction.

Titrate 50 ml salicylic acid–pyridine solution (Note 5) in the previously used 125-ml flask with the Karl Fischer reagent; then transfer the acid rapidly to the reaction flask, and stir the contents with the magnetic stirrer for 10 min. Close the stopcock, and disconnect the condenser from the drying bulb. Rinse the flask and the condenser with the solution. Allow the condenser to drain for a few minutes, and then remove the condenser and stopper from the reaction flask. Insert the dual platinum electrodes made for the reaction flask. Connect the reaction flask to the buret of the Karl Fischer apparatus and titrate the sample solution with Karl Fischer reagent. Record the volume, and subtract the blank volume to find the net titration volume.

With a Kirk ultramicro transfer pipet add 25 mg water to the titrated solution. Titrate with Karl Fischer reagent, and determine the oxygen equivalent in milligrams per milliliter of reagent. Calculate the percent oxygen in the calcium metal according to the following equation:

$$\text{O, \%} = \frac{(\text{net titration volume, ml})\ \text{mg O/ml}}{10\ (\text{sample-wt, g})}. \tag{7}$$

NOTES

1. Reduce the water content of the alcohol to less than 0.005% by treatment of the alcohol with magnesium ribbon, followed by distillation.
2. Prepare the Karl Fischer reagent in the usual manner with 269 ml pyridine, 84.7 g iodine, 667 ml absolute methanol, and 45 ml liquid sulfur dioxide.
3. The weight of the sample should not exceed 5 g. Although a 2-g sample reacts completely with the methanol within 30 min with stirring, the dissolution of a 5-g sample may take an hour.

4. The reaction is very rapid at the start and goes nearly to completion within 15 min. As would be expected, large pieces of metal dissolve very slowly. Convenient particle size results if the sample is passed through a 6-mesh/in. sieve.
5. To prepare the solution, dissolve 100 g salicylic acid in 100 ml pyridine. Pyridine can be dried to less than 0.01 % water by adding a calculated 10 % excess of benzene over that required to remove the water initially present and fractionating out the benzene–water azeotrope and excess benzene.

REFERENCES

1. D. T. Hurd, *An Introduction to the Chemistry of the Hydrides*, Wiley, New York, 1952, p. 50.
2. C. L. Mantell and C. Hardy, *Calcium Metallurgy and Technology*, Reinhold, New York, 1945, p. 57.
3. J. C. Powers, D. W. Vose, and E. A. Sullivan, *Rept. WADD-TR-60-543*, Metal Hydrides, Inc., Beverly, Mass., 1960.
4. H. C. Rogers, *Science*, **159**, 1057 (1968).
5. J. P. Pemsler, R. W. Anderson, and E. J. Rapperport, *Rept. NMI-9815*, Nuclear Metals, Inc., Concord, Mass., 1962.
6. C. E. Ells and W. Evans, *Rept. CRGM-1041* or *AECL-1347*, Atomic Energy of Canada, Ltd., 1961.
7. P. M. S. Jones and R. Gibson, "Hydrogen in Beryllium," *AWRE-0-2/67*, Atomic Weapons Research Establishment, 1967.
8. M. V. Sharov, B. S. Morozov, and V. M. Pletenev, *Liteinoe Proizvodstvo*, **No. 6,** 16 (1953); through *Chem. Abstr.*, **48,** 8156 (1954).
9. D. T. Peterson and M. Indig, *J. Amer. Chem. Soc.*, **82,** 5645 (1960).
10. M. Codell and G. Norwitz, *Anal. Chim. Acta.*, **18,** 265 (1958).
11. R. Berry, J. A. J. Walker and A. E. Clarke, *DEG Report 107 (C)*, U.K. At. Energy Auth., 1961.
12. C. Evans and J. Herrington, *Radioisotopes in the Physical Sciences and Industry*, Vol. II, International Atomic Energy Agency, Vienna, 1962, p. 309.
13. P. Cotterill, R. E. Goosey, and A. J. Martin, Monograph and Report Series No. 28, Institute of Metals, Chapman and Hall, London, 1963.
14. J. L. English, "Corrosion," in D. W. White, Jr. and J. E. Burke, Eds., *The Metal Beryllium*, American Society for Metals, Cleveland, 1955, p. 530.
15. V. F. Zelenskii, I. A. Petel'guzov, and L. D. Kolomiets, *Zh. Prikl. Khim.*, **40,** 1013 (1967).
16. L. R. Aronin, F. A. Bauman, and A. K. Wolff, *Rept. AD-626987*, Nuclear Metals, Inc., Concord, Mass., 1965.
17. G. E. Darwin and J. H. Buddery, *Metallurgy of the Rarer Metals, Beryllium*, Academic, New York, 1960, p. 377.
18. W. Bradshaw, R. Johnson, and D. Beard, *Rept. LMSD-288231*, Lockheed Aircraft Corp., Sunnyvale, Calif., 1960.

19. W. G. Bradshaw, private communication, June 4, 1964.
20. B. D. Holt and H. T. Goodspeed, *Anal. Chem.*, **34,** 374 (1962).
21. D. T. Peterson and V. G. Fattore, *Anal. Chem.*, **34,** 579 (1962).
22. E. Booth, F. J. Bryant, and A. Parker, *Analyst*, **82,** 50 (1957).
23. J. E. Still, *Spec. Rept. No. 68*, Iron and Steel Institute, London, 43, 1960.
24. J. Sannier and J. Leroy, *Com. Energie At.* (*France*) *Rappt. CEA-R-2957* (1966); through *Chem. Abstr.*, **65,** 4650c (1966).
25. R. Bourgillot, A. Cavard, A. Cornu, R. Stefani, *Chim. Anal.* (Paris), **49,** 315 (1967).
26. B. D. Holt, *Anal. Chem.*, **31,** 51 (1959).
27. A. N. Zaidel and A. A. Petrov, *Zh. Tekh. Fiz.*, **25,** 2571 (1955).
28. Z. M. Turovtseva and L. L. Kunin, *Analysis of Gases in Metals*, Academy of Sciences of the USSR, Moscow, 1959; trans. by Consultants Bureau, New York, 1959, p. 297.
29. E. G. Bobalek and S. A. Shrader, *Ind. Eng. Chem.*, *Anal. Ed.*, **17,** 544 (1945).
30. M. W. Mallett, A. F. Gerds, C. B. Griffith, *Anal. Chem.*, **25,** 116 (1953).
31. R. Berry and J. A. J. Walker, *Analyst*, **88,** 280 (1963).
32. A. Colombo, *Energia Nucl.*, **13,** 251 (1966); through *Chem. Abstr.*, **66,** 16308e (1967).
33. F. Sauerwald, *Z. Anorg. Chem.*, **256,** 217 (1948); *Chem. Abstr.*, **43,** 2119 (1949).
34. A. P. Gudchenko and V. V. Serebryakov, *Tr. Moskov. Aviatsion. Tekhnol. Inst.*, No. 49, 160 (1961); through *Chem. Abstr.*, **55,** 23168 (1961).
35. B. D. Holt, *Anal. Chem.*, **28,** 1153 (1956).
36. H. Winterhager, *Aluminum-Arch.*, **12,** 7 (1938); through Ref. 29.
37. J. M. Googin, *Rept. Y-1324*, U.S. AEC, 1960.
38. U.K. Atomic Energy Auth., *Rept. IGO-AM/S-190; AERE-AM-43*, 1959.
39. C. J. Rodden and F. A. Vinci, "Analytical Chemistry of Beryllium," in D. W. White, Jr., and J. E. Burke, Eds., *The Metal Beryllium*, American Society for Metals, Cleveland, 1955, p. 669.
40. A. P. Tikhonova, A. A. Morozova, and V. V. Rodyakin, *Sb. Tr. Vses. Nauchn.-Issled. Proekt. Inst. Titana*, **1,** 293 (1967); through *Chem. Abstr.*, **68,** 26642q (1968).
41. C. V. Banks and B. D. La Mont, *Rept. ISC-584*, U.S. AEC, 1955.
42. J. K. Bundy and G. C. Goode, *Anal. Chim. Acta.*, **37,** 394 (1967).
43. J. A. Tetlow and A. L. Wilson, *Analyst*, **89,** 453 (1964).
44. (a) N. D. Stalnaker, M. Kahn, and B. T. Kenna, *Rept. SC-R-68-1723*, Sandia Corp., Albuquerque, N. Mex., 1968; (b) K. R. Blake, C. V. Parker, L. D. England, and I. L. Morgan, *Rept. ORO-2980-14*, Texas Nuclear Corp., Austin, Tex., 1966; (c) K. R. Blake, T. C. Martin, I. L. Morgan, and C. R. Houston, Proceedings, Intern. Conf. Modern Trends Activation Anal., College Station, Tex., 1965; (d) C. Engelmann and G. Cabane, *ibid.*; and (e) C. Engelmann and M. Loeuillet, *Bull. Soc. Chim. Fr.*, **No. 2,** 544 (1965).
45. Production Group, *PG-Report-381*, U.K. At. Energy Auth., 1962.
46. J. J. Schmidt-Collerus and A. J. Frank, *AD-260, 980; WADD-TR-60-482* U.S. Dept. Com., Office Tech. Serv., 1961.

47. V. V. Rodyakin, A. E. Andreev, A. M. Bragin, A. I. Boiko, and A. V. Riganelovich, *Zavod. Lab.*, **30,** 1203 (1964); through *Chem. Abstr.*, **62,** 3397c (1965).
48. G. V. T. Ranzetta and V. D. Scott, *Brit. J. Appl. Phys.*, **15,** 263 (1964).
49. W. A. Bergholz, *J. Res. Nat. Bur. Stand.* **48,** 201 (1952).
50. E. B. Read, *Rept. NMI-4911*, Nuclear Metals, Inc., Concord, Mass., 1958.
51. J. Rynasiewicz, *Rept. AECD-3710*, U.S. AEC, 1951.
52. C. G. Wallace, *Rept. AERE-AM-16*, U.K. At. Energy Auth., 1959.
53. A. R. Eberle and M. W. Lerner, *Metallurgia*, **59,** 49 (1959).
54. J. N. Gregory and D. Mapper, *Analyst*, **80,** 230 (1955).
55. E. Booth and A. Parker, *Analyst*, **83,** 241 (1958).
56. E. Booth and A. Parker, *Analyst*, **84,** 546 (1959).
57. M. R. Everett and G. E. Thompson, *Analyst*, **87,** 515 (1962).
58. R. F. Coleman, R. S. Shaw, and R. Todd, private communication in Ref. 57.
59. W. F. Harris, *Talanta*, **11,** 1376 (1964).
60. N. F. Litvinova and Z. M. Turovtseva, *Geokhim. Anal. Khim.*, **12,** 341 (1960); through *Chem. Abstr.*, **54,** 20645 (1960).
61. S. Kallmann and F. Collier, *Anal. Chem.*, **32,** 1616 (1960).
62. B. D. Holt and H. T. Goodspeed, *Anal. Chem.*, **35,** 1510 (1963).
63. J. Brill and F. Dugain, *Bull. Soc. Chim. Fr.*, **No. 2,** 562 (1965); through *Nucl. Sci. Abstr.* **19,** 24419 (1965).
64. J. Brill and F. Dugain, *J. Nucl. Mater.*, **16,** 162 (1965); through *Nucl. Sci. Abstr.*, **19,** 26259 (1965).
65. E. Sudo, *Bunseki Kagaku*, **17,** 1369 (1968); through *Nucl. Sci. Abstr.*, **23,** 6081 (1969).
66. A. Parker, *PG-Rept.-171*, U.K. At. Energy Auth. Prod. Group, 1960, p. 81.
67. R. F. Coleman and J. L. Perkin, *Analyst*, **84,** 233 (1959).
68. R. F. Coleman and J. L. Perkin, *Analyst*, **85,** 154 (1960).
69. R. F. Coleman, *Analyst*, **87,** 590 (1962).
70. J. H. McCrary, I. L. Morgan, and L. L. Baggerly, Proceedings, Intern. Conf. Modern Trends Activation Anal., College Station, Tex., 1961.
71. J. T. Byrne, C. T. Illsley, and H. J. Price, *RFP-522*, Kaman Aircraft Corp. Colorado Springs, Colo., 1965.
72. R. W. Benjamin, K. R. Blake, and I. L. Morgan, *Anal. Chem.*, **38,** 947 (1966).
73. G. H. Anderson, P. C. Kempchinsky, and A. Laverty, *Trans. Amer. Nucl. Soc.*, **9,** 107 (1966).
74. D. B. Beard, R. G. Johnson, and W. G. Bradshaw, *Nucleonics*, **17** (7), 90 (1959).
75. A. R. Gilman and S. Isserow, *Rept. NMI-1234*, Nuclear Metals, Inc., Concord, Mass., 1960.
76. P. Albert, C. Englemann, S. May, and J. Petit, *Compt. Rend.*, **254,** 119 (1962).
77. R. Basil, J. Hure, P. Leveque, and C. Schul, *Compt. Rend.*, **239,** 422 (1954).
78. S. S. Markowitz and J. D. Mahony, *Anal. Chem.*, **34,** 329 (1962).
79. C. Engelmann, *Compt. Rend.*, **258,** 4279 (1964).
80. C. Engelmann and G. Cabane, Intern. Conf. Modern Trends Activation Anal., College Station, Rex., 1965.

81. K. R. Blake, C. V. Parker, L. D. England, and I. L. Morgan, *ORO-2980-14*, Texas Nuclear Corp., Austin, Tex., 1966.
82. G. J. Lutz, R. J. Boreni, R. S. Maddock, and W. W. Meinke, Eds., "Activation Analysis, A Bibliography, Parts 1 and 2," *Tech. Note 467*, U.S. Dept of Commerce, 1968.
83. S. S. Nargolwalla, M. R. Crambs, and J. R. DeVoe, *Anal. Chem.*, **40,** 666 (1968).
84. B. L. Twitty and K. M. Fritz, *Anal. Chem.*, **39,** 527 (1967).
85. K. E. Kleiner, *Ukr. Khim. Zh.*, **22,** 809 (1956); through *Chem. Abstr.*, **51,** 7232 (1957).
86. H. Hartmann and G. Ströhl, *Z. Anal. Chem.*, **175,** 84 (1960); through *Chem. Abstr.*, **55,** 1284 (1961).
87. H. Hartmann, W. Hofmann, and G. Ströhl, *Z. Metallk.*, **49,** 461 (1958); *Chem. Abstr.*, **52,** 19687 (1958).
88. N. Oda, K. Norishima, and M. Kubo, *Bunseki Kagaku*, **11,** 526 (1962); through *Chem. Abstr.*, **57,** 2846 (1962).
89. A. R. Eberle, M. W. Lerner, and G. L. Petretic, *Anal. Chem.*, **27,** 1431 (1955).
90. H. J. Allsopp, *Analyst*, **81,** 469 (1956).
91. E. D. Malikova and Z. M. Turovtseva, *Inst. Geokhim. Anal. Khim.*, **10,** 103 (1960).
92. D. J. Anderson and H. W. Thieman, *Rept. MCW-1475*, U.S. AEC 1962.
93. R. Berry, J. A. J. Walker, and R. E. Johnson, *DEG Rept. 111* (*C*), U.K. At. Energy Auth., 1960.
94. W. G. Smiley, *Anal. Chem.*, **27,** 1098 (1955).
95. B. D. Holt, *Anal. Chem.*, **27,** 1500 (1955).
96. E. H. Brown, *Ind. Eng. Chem., Anal. Ed.*, **14,** 551 (1942).
97. Committee for the Standardization of Microchemical Apparatus, *Anal. Chem.*, **23,** 523 (1951).
98. W. G. Smiley, *Nucl. Sci. Abstr.*, **3,** 391 (1949).

CHAPTER
12

ALUMINUM, ZINC, CADMIUM, TIN, LEAD, ANTIMONY, AND BISMUTH

BEN D. HOLT

Chemistry Division, Argonne National Laboratory
Argonne, Illinois

CONTENTS

2. Fusion Methods
   a. *Vacuum Fusion (Sn, Bi, Al)*
   b. *Metal Evaporation—Inert-Gas Fusion (Zn, Cd)*
3. Activation Methods
   a. *Neutron Activation (Zn, Cd, Sn, Bi, Al)*
   b. *Other Activation Methods (Al)*
4. Comparison of Methods

D. Recommended Methods
1. Hydrogen
   a. *Aluminum: Hot Extraction*
   b. *Zinc, Cadmium, Tin, Lead, Antimony, Bismuth: Combustion—Manometric*
2. Nitrogen: Kjeldahl
3. Oxygen
   a. *Aluminum: Neutron Activation*
   b. *Zinc and Cadmium: Metal Evaporation—Inert-Gas Fusion*
   c. *Tin, Lead, Antimony, and Bismuth: Hydrogen Reduction*

I. METALLURGICAL ASPECTS

Most of the low-melting metals of this chapter contain only traces of dissolved hydrogen, nitrogen, and oxygen. The effects of these gaseous impurities on physical and chemical properties, however, are greater than that of comparable concentrations of dissolved metallic elements [1]. Dissolved gases, especially oxygen, are corrosive to the surfaces of other metals that are used to contain melts of these metals [2].

Hydrogen is the only gas that has an appreciable solubility in aluminum [3]. Its solubility increases with temperature, being considerably higher in molten aluminum (about 0.6 ppm at 660°C). Molten aluminum containing hydrogen near saturation reaches supersaturation on freezing, and the excess hydrogen tends to separate into gas bubbles, causing porosity in the solid metal.

All of the metals of this chapter are oxidizable in air to stable oxides, and each can contain considerably more oxygen as surface oxides than as internal oxygen. The extremely thin, yet remarkably tight and adherent, transparent film of oxide that forms on solid aluminum serves as a protective coating against further oxidation and is thereby self-limiting [3].

Although the solubility of oxygen in aluminum is very low, aluminum–aluminum oxide alloys may contain 6–14% oxide. These alloys, fabricated from aluminum powders by combining heat and pressure, are the strongest and most stable aluminum alloys above about 300°C [4].

II. ANALYSIS

A. HYDROGEN

The hydrides of aluminum, zinc, cadmium, tin, lead, antimony, and bismuth are all unstable compounds. Melting points and boiling points are listed in Table 12.1 [5]. The hydrides of tin, antimony, and bismuth are gases

TABLE 12.1. Hydrides of Aluminum, Zinc, Cadmium, Tin, Lead, Antimony, and Bismuth [5]

Hydride	Melting Point, °C	Boiling Point, °C
$(AlH_3)_x$	Decomposes	—
ZnH_2	Decomposes	—
CdH_2	Decomposes	—
SnH_4	−150	−52
PbH_4	−135	−13
SbH_3	−89	−17
BiH_3	Unknown	22

at room temperature, and the hydrides of aluminum, zinc, and cadmium are unstable, white, polymerized solids. All except aluminum hydride decompose at room temperature into the elements. Aluminum hydride hydrolyzes rapidly with water and oxidizes spontaneously upon exposure to air or oxygen. Because of the instability of these compounds, the hydrogen concentration in each metal is governed by its solubility, which is generally in the low parts-per-million range.

1. Aluminum Methods

Several analytical methods have been used for the determination of hydrogen in aluminum, some for solid metal and some for liquid. Laboratory methods include vacuum fusion, tin fusion, vacuum sublimation, hot extraction, first bubble, and isotope dilution. On-the-spot field tests include a method of equilibration with recirculating inert gas and a method of hot extraction from ladles of molten sample.

a. Vacuum Fusion

The vacuum-fusion method (described in Chapter 4) was tested by Sloman [6] for the determination of oxygen, nitrogen, and hydrogen in aluminum and aluminum–base alloys. Hydrogen was completely evolved from the sample at temperatures considerably below that necessary for rapid reduction of oxides or nitrides. A rapid evolution of gas, which occurred as the metal melted, ceased after about 3 min. This gas was essentially hydrogen but contained small quantities of carbon monoxide and nitrogen. Sloman observed excellent reproducibility in hydrogen results on specimens taken from the same sample of metal; the results were the same whether the analysis was carried out with no molten bath below 1100°C or with a steel bath at about 1500°C.

Klyachko et al. [7] determined hydrogen in aluminum by vacuum-melting at 1000°C without an iron bath and at 1700°C with an iron bath. In a later application with a modified, vacuum-fusion procedure, Pfundt [8] enclosed a surface-cleaned sample of the metal in a glass ampul of high softening point and negligible permeability to hydrogen. The metal was heated above its melting point for 15 h, after which the ampul was broken in an enclosed apparatus and the released gas was measured and analyzed by gas chromatography or mass spectrometry.

b. Tin Fusion

Griffith and Mallett [9] determined hydrogen in wrought aluminum alloys and in porous aluminum welds by dissolving 6-g samples in molten tin and measuring the released gas. This method, which was a modification of a procedure reported by Carney, et al. [10], was capable of determining 0.1 ppm or more hydrogen. The apparatus consisted of a furnace section where the hydrogen was extracted, a section for the collection and measurement of the evolved gas, and a section for the analysis of the collected gas. Depending on the type of alloy, complete solution of the sample required 1–1.5 h. The gas was analyzed for hydrogen by measuring its pressure in a calibrated volume before and after diffusion of the hydrogen through a heated palladium tube. Internal hydrogen was differentiated from hydrogen associated with oxide film on the surface. The internal hydrogen for 6 wrought alloys was about 0.3 ppm and for porous welds about 1.1 ppm.

c. Vacuum Sublimation

Colombo [11] measured hydrogen in surface-cleaned samples of aluminum alloys (50–100 mg) by subliming at 1000°C for 15 min at 10^{-6} torr and collecting the evolved gases for analysis by gas chromatography. Interference by sources of hydrogen within the apparatus was minimized by subliming several milligrams of aluminum prior to the analysis of samples. The method was sensitive to about 1 ppm hydrogen.

d. Hot Extraction

Hydrogen is quantitatively extracted from aluminum and aluminum alloys by heating in vacuum at temperatures slightly below the solidus.

Ransley and Talbot [12] cited a procedure, described by Eborall and Ransley [13], that had given satisfactory results in routine use over a long period. The gas extracted from the solid sample was continuously removed by a mercury-diffusion pump and was stored in a low-pressure chamber of calibrated volume for pressure measurements before and after the selective removal of hydrogen through a heated palladium membrane. The extraction of gas from a cylindrical sample 1 cm in diameter was completed within 2 h of heating by a resistance furnace. Ransley and Talbot reported results on duplicate determinations for each of 43 samples of pure metal and alloys (containing up to 1.0% magnesium) with a standard deviation of 0.014 ppm hydrogen, the range of concentration being 0.06–0.40 ppm.

Ransley and Talbot [12] discussed surface-adsorbed moisture and sublimation of volatile components as sources of error in the hot-extraction method. Nikiforov and Silant'eva [14] suppressed these interferences by pretreating the sample with a chromate solution before washing, drying, and calcining.

Danilin [15] used a reaction tube which permitted the removal of the sample from the heating zone without disturbing the vacuum. Moisture, adsorbed on the walls of the reaction tube, was removed by reaction with sublimed magnesium and zinc from the first of a series of samples placed in the apparatus.

Aschehoug et al. [16] analyzed aluminum and aluminum alloys by hot extraction, using a copper oxide furnace so that the hydrogen released was converted to water, which was then condensed in a cold trap for measurement in a semiautomated McLeod gauge. The oxygen pressure, maintained in the extraction chamber by the hot copper oxide, was shown to suppress the formation of reactive films of volatile components (magnesium, zinc, etc.) on the inner walls of the apparatus.

e. First Bubble

An empirical method of estimating the hydrogen content of aluminum alloys was introduced by Dardel [17] and later adopted by other investigators [18–20]. A small sample of the metal is maintained in the molten state while the pressure of the surrounding atmosphere is progressively reduced. A relation is established for the pressure at which the first bubble appears, the sample temperature, and the hydrogen content of the metal. A detailed discussion of this method is given in Chapter 7.

f. Isotope Dilution

Hydrogen in aluminum and other metals (Zn, Zr, Ta, Fe, steels) has been determined by equilibrating the hydrogen with a known quantity of deuterium in a closed system and spectroscopically measuring the H/D ratio in the gas phase before and after the equilibration [21–23]. The hydrogen content of the metal was calculated from the change in ratio, assuming the coefficient of distribution of H and D between the metal and gas to be unity. The mean square deviation for a single determination was 3–10%, and the detection limit, which depended upon the weight of the sample and the degree of degassification, was 1 ppm in a 5-g sample. Evans and Herrington [24] used tritium instead of deuterium and determined the isotope ratios radiochemically. Gillespie [25] used tritium as a tracer for tagging each of several possible sources of hydrogen contamination in aluminum in certain metallurgical processes. The radioactive isotope was introduced under controlled conditions and then quantitatively removed from the solid sample by hot extraction. The radiochemical measurement was capable of detecting hydrogen (as tritium) to a level of 3×10^{-10} ml at standard temperature and pressure.

g. Recirculating Inert Gas (Telegas)

Ransley and co-workers [26] designed and developed an instrument to give direct readings of the hydrogen content of molten aluminum during melting and casting. By their method a small volume of inert gas is circulated through a probe that contains a hydrogen-permeable membrane submerged in the liquid metal. The hydrogen concentration in the inert gas reaches equilibrium with that of the molten metal within 2–10 minutes. The instrument is standardized by relating measurements of hydrogen in the gas mixture (by a thermal-conductivity detector) to hydrogen concentrations in the metal (measured by hot extraction). (See also Chapter 7 for a discussion of this method.) A similar procedure was reported by Kashima and Yamazaki [27]. They used gas chromatography to measure the hydrogen in a nitrogen stream that was exposed to the molten metal.

Another field method, designed to be used at the site of industrial melting and casting, was described by Shiota [28]. A sample of the molten metal, contained in a small deep ladle, is placed under a bell jar, which is then evacuated. The thermal conductivity of the released gas is measured, giving results for hydrogen content within 10 min.

h. Discussion of Methods

The choice of a method for determining hydrogen in aluminum depends upon the availability of analytical equipment, the condition of the sample (state, shape, size, percentage of alloyed metals), the location of the analysis (in the laboratory or in the plant), or the need for the determination of other gaseous elements, such as oxygen and nitrogen.

Hydrogen should generally be determined by vacuum fusion if analysis for xygen on the same material is also needed; hydrogen can be simultaneously ieasured with little extra effort, the rapid, quantitative release of hydrogen t temperatures above the melting point being the chief advantage of this iethod. By the usual vacuum-fusion procedure, total hydrogen is measured, icluding internal hydrogen and hydrogen released from interaction between ydrated surface oxide and the metal substrate. If the method is applied to luminum alloys containing volatile components, such as zinc or magnesium, ie latter may interfere with the transfer of energy from the work coil to the rucible when heating inductively, by condensing out as a conducting film n the walls of the reaction chamber surrounding the crucible.

The hot-extraction method has been extensively used for hydrogen in olid aluminum, and it is regarded as a reference method by which other iethods are tested. No crucible is required to contain the sample while eating; the evaporation of volatile, alloying elements is not extensive nough to be troublesome, and the contribution of hydrogen by hydrated urface oxides is smaller and easier to ascertain than by the vacuum-fusion iethod. The chief disadvantage of the hot-extraction method is the lengthy eating time (about 2 h) required to quantitatively extract the gas. The tin-ision and isotope-dilution methods do not appear to offer any distinct dvantage over the hot-extraction procedure.

The first-bubble method is a relatively easy test that yields hydrogen data f sufficient accuracy for many metallurgical applications. The precision is bout 0.05 ppm in the range of 0.08–0.50 ppm, the analysis can be performed i about 2 min, and the rugged apparatus is simple to use and can be set up ither in the laboratory or in the foundry.

The chief advantage of the Telegas method over the others is the provision or continuous monitoring (2–10-min response time) of the hydrogen oncentration in molten aluminum during casting and other industrial lant operations.

. Zinc, Cadmium, Tin, Lead, Antimony, and Bismuth

Four methods by which hydrogen has been determined in these metals are otope dilution, vacuum fusion, combustion-manometric, and photoneutron ctivation [29].

Because molten tin is used as a bath in the extraction of hydrogen from ther metals by vacuum fusion [30], this method can obviously be applied) the analysis of tin itself. The other metals of this grouping (Zn, Cd, Pb, b, and Bi) can also be melted in vacuum to release hydrogen for collection nd measurement. The vapor pressure of each metal limits the temperature) which it can be practically heated for the release of hydrogen. In most ases it is necessary to provide a condenser in the system for collection of aporized metal.

The author of this chapter has used the combustion-manometric meth([31], described in Section IV.D.1.b of Chapter 11, for the determination hydrogen in zinc, cadmium, and tin. Presumably, the method is also appli able to lead, antimony, and bismuth, inasmuch as they are all combustib in oxygen at 900°C and the corresponding hydroxides dehydrate to the oxid at temperatures below 900°C. By using a sensitive capillary manometer, which the water-vapor pressure is measured at 110°C, hydrogen concentr tions as low as 0.1 ppm can be determined. (Carbon can be determine simultaneously.) The analysis can be completed within a half-hour.

The isotope-dilution method (see Section II.A.1.f. above) offers litt advantage over vacuum fusion. Both techniques depend upon the diffusic of hydrogen through the metal to the surface, yet the fusion method inherently more sensitive. The combustion-manometric method is as sensiti as the vacuum-fusion method and does not involve high vacuum; comple removal of the hydrogen from the metal is ensured by combustion of the tw to produce water and metal oxide, respectively.

B. NITROGEN

The micro-Kjeldahl procedure is the method most generally applied to tl determination of nitrogen in metallic aluminum, zinc, cadmium, tin, lea antimony, and bismuth. The nitrogen is converted to ammonium chloric and steam-distilled as ammonia into boric acid solution for measurement l titration (see Section IV.D.2 of Chapter 11).

Namiki et al. [32] determined microamounts of nitrogen in aluminum l spectrophotometric measurements of the blue compound formed from an monia, phenol, and hypochlorite (chloramine-T). The compound is extracte with isobutyl alcohol or isoamyl alcohol by using a salting-out reagent.

Other methods that have been investigated for the determination of nitr gen (and other elements) in aluminum and other metals are activation anal sis using γ-photons and charged particles [33, 34], and microprobe analys [35].

C. OXYGEN

Several methods have been employed in the determination of oxygen in tl metals of this chapter, none of which is generally applicable to all. Tl methods are of three general types: chemical, fusion, and neutron activatio (see also Chapters 3 and 9). Certain chemical methods will yield only tl oxygen present as the oxide of the matrix metal.

. Direct Chemical Methods

In direct chemical methods the oxygen is converted by an appropriate eagent to a form in which it can be isolated and measured. Reagents that ave been used in these analyses are hydrogen, ammonium acetate solution, cetic acid containing hydrogen chloride, hydrogen sulfide, sulfur vapor, ılfur monochloride, and bromine trifluoride.

. *Hydrogen Reduction (Zn, Sn, Pb, Bi)*

Zinc [36], tin [37], lead [36–39], and bismuth [40] have been analyzed for xygen by exposing the sample to hydrogen at elevated temperatures and easuring either the water that is produced or the hydrogen that is consumed. n analyzing lead and lead–tin alloys, Baker [37] heated the sample for 30-ıin periods in hydrogen at 600°C until the water vapor, measured mano-ıetrically, reached a constant value. The precision was about 4% of the mount of oxygen present in the sample. Funston and Reed [40] determined xygen in bismuth at the 2-ppm level (with an average deviation of less than ppm) by heating a 10-g sample with hydrogen for 30 min at 850–900°C ı a closed system containing magnesium perchlorate. The calculation was ased on the decrease in hydrogen pressure during the reaction. The analysis f solid zinc was less satisfactory by this method [36], the relative experi-ıental error being about 20%. Generally, the best results are obtained when he reaction is carried out above the melting point of the metal sample.

. *Ammonium Acetate (Zn)*

Since zinc oxide is soluble in ammonium acetate solution, while the metal s relatively insoluble, the oxide can be estimated in zinc powder by digesting he sample for 2 h in a 30% ammonium acetate solution, filtering, and etermining the zinc either in the filtrate or in the residue [41]. This method s not suitable for the determination of oxygen in the parts-per-million range.

. *Hydrogen Chloride (Zn, Al)*

Eberius and Kowalski [42] dissolved samples of zinc dust in water-free cetic acid, saturated with hydrogen chloride; the zinc oxide, thereby onverted to water, was measured by a Karl Fischer titration. The average elative precision for several analyses of samples containing 3–7% zinc oxide vas 0.6%. Serrini and co-workers [43] described a similar method of separat-ng aluminum oxide from aluminum through volatilization of the matrix s aluminum chloride.

d. Hydrogen Sulfide (Zn, Cd, Pb)

Hartmann et al. [44–47] used purified hydrogen sulfide to convert oxid impurities in zinc, cadmium, and lead to water for collection and gravimetri measurement. The hydrogen sulfide is carried by a stream of purified nitroge through a reaction tube containing the sample at 500°C. The water produce from the reaction

$$MO + H_2S \rightarrow MS + H_2O \qquad (1$$

is adsorbed on phosphorus pentoxide and weighed. Oxygen concentration as low as 1.2 ppm were determined in remelted zinc. Typical results for oxyge in zinc were 1–3 ppm and in cadmium and lead, 4 ppm and 8 ppm, respec tively. Good precision and accuracy in the gravimetric measurement of wate require rather large samples (about 90 g).

e. Sulfur Vapor (Zn, Cd, Pb, Sb)

Zinc, cadmium, lead, and antimony have been analyzed for oxygen b heating the sample (600–800°C) in a stream of purified nitrogen that wa charged with sulfur vapor from a sulfur boiler maintained at 500–600°C [48–50]. The metal oxide was converted to the corresponding sulfide, and a equivalent amount of sulfur dioxide was carried by the nitrogen stream into 10% potassium iodide–5% sodium bicarbonate solution containing a know concentration of iodine (0.01*N*). The oxygen concentration in the meta sample was calculated from the amount of iodine consumed by the absorbe sulfur dioxide. The detection limit of the method, limited by the blank, wa about 5–20 ppm oxygen.

f. Sulfur Monochloride (Cd, Al)

Oxygen has been determined in cadmium [51] and aluminum [52] b passing sulfur monochloride in a stream of nitrogen over the sample heate in a graphite boat at 600–900°C. The sulfur dioxide formed by oxidation o the sulfur monochloride was collected in an absorbing solution and deter mined iodimetrically.

g. Bromine Trifluoride (Pb, Sb, Bi)

Hoekstra and Katz [53] showed that the oxygen in the oxides of lead antimony, and bismuth could be determined by treating the sample with th powerful fluorinating agent bromine trifluoride. Presumably, the techniqu would be applicable to the corresponding metals containing oxide. The meta oxide was converted to metal fluoride and oxygen; the latter was pumpe through cold traps to a gas measuring system. After the gas was measured s composition was determined (by mass spectrometry) because nitrogen 'as also released from the sample during fluorination. Nickel valves and ther nickel hardware which resist attack by fluorinating agents were used ı fabrication of the analytical apparatus. The normal blank was such that ıe method is not applicable to oxygen concentrations before about 0.01%.

. *Mercury Extraction (Sn, Pb)*

When mercury is added to tin or lead, an amalgam forms leaving the ısoluble oxide to separate and appear on the surface of the liquid alloy. ilverman and Gossen [54] extracted the metal from tin samples until the ıercury "washings" were free of tin; the tin oxide residue was then dissolved ı hydrochloric acid, and tin was determined by a photometric procedure ıvolving reduction of phosphomolybdate with stannous ion to the molyb- enum blue complex. The procedure was reported to be applicable to 1 ppm, r less, of oxygen in 2.5-mg samples.

Black [55] has estimated the lead oxide in lead dross by dissolving a 1-g ample in 5 ml of mercury and treating the resulting surface scum with an queous solution of ammonium chloride. Ammonia, produced by the eaction

$$PbO + 2NH_4Cl \rightarrow PbCl_2 + H_2O + 2NH_3,$$

'as steam-distilled into boric acid solution, in which it was titrated with a andard acid. The method is not suitable for measuring the oxide in quantities ss than about 10 mg.

. **Fusion Methods**

In order for vacuum or inert-gas fusion methods to be applicable, the ıetal oxide must be of sufficient thermal stability that it does not decompose 'hile the crucible is heated rapidly to a temperature suitable for the oxide– arbon reaction; at such a temperature, the vapor pressure of the metal must e such that it does not evaporate too rapidly. Table 12.2 gives the melting oints and boiling points of the metals discussed in this chapter.

TABLE 12.2. Liquidus Ranges of Aluminum, Zinc, Cadmium, Tin, Lead, Antimony, and Bismuth

Metal	Melting Point, °C	Boiling Point, °C
Al	660	2450
Zn	420	906
Cd	321	765
Sn	232	2270
Pb	327	1725
Sb	631	1380
Bi	271	1560

a. Vacuum Fusion (Sn, Bi, Al)

Determination of oxygen by vacuum fusion has been reported for tin [56] bismuth [57], and aluminum [6, 58, 59]. An examination of the boilin points in Table 12.2 suggests that tin (often used as a flux for other metals i vacuum fusion analyses [60]), bismuth, lead, and possibly antimony could b analyzed by this method. Consideration should be given to the therma stability of the oxides involved. According to Griffith and Mallett [57], th oxide of bismuth is apparently reduced by carbon at 750°C, thus permitting relatively low temperature for the vacuum-fusion analysis of this metal.

Sloman [6] investigated the vacuum-fusion method for the determinatio of oxygen in aluminum and aluminum–base alloys. The sample was dissolve in a steel bath at 1550°C in which the concentration of aluminum was les than 10–15% by weight. Under these conditions the oxide of the sample wa completely reduced, and the volatilization of aluminum from the covere graphite crucible did not seriously interfere with recovery of the evolve carbon monoxide. When the aluminum concentration in the bath exceede about 30%, the melt tended to solidify by carbide formation and to damag the crucible.

More than 20 years later, Colombo and Rodari [58, 59] worked out vacuum-fusion procedure for aluminum, aluminum–aluminum oxid composites, and Al–Mg–Al_2O_3–MgO composites in which a sample (30–6 mg) was sealed in a graphite capsule with cleaned copper filings (weigh ratio, 1:4) and dropped into the hot crucible. The crucible was held a 1100°C for 15 min and the extracted gas was discarded; the temperature wa then held at 1600–1700°C for 15 min, and the released carbon monoxide wa collected for measurement. The procedure gave good results for specimens c aluminum and aluminum–aluminum oxide composites with oxygen content ranging from some tens of parts-per-million to some weight percent. Th detection limit was about 20 ppm for 100-mg samples.

b. Metal Evaporation–Inert-Gas Fusion (Zn, Cd)

Zinc and cadmium have been analyzed for oxygen by evaporating 1- samples of the metal in a low-pressure stream of inert gas and determining th oxygen in the oxide residue by the inert-gas fusion technique [61]. Th lower limits of detection were about 10 ppm for zinc and 20 ppm for cadmium About 1 h was required for the analysis. Recoveries of oxygen added to th metals as oxides were 100 ± 2% for zinc and 94 ± 4% for cadmium.

A direct application of the inert-gas-fusion method, without prior separa tion of the metal from the oxide, has been reported for aluminum [62], zinc lead, antimony, and bismuth [63]. The relative error was 10–15%, and th time required was about 50 min.

. Activation Methods

. *Neutron Activation (Zn, Cd, Sn, Bi, Al)*

The neutron activation method—14 MeV neutrons for the $^{16}O(n, p)^{16}N$:action—has been used to determine oxygen in tin [64], aluminum, bismuth, ıdmium, zinc, and lead [65]. Hoste and co-workers [65] investigated ıatrix effects of several nonferrous metals on the ^{16}N measurement of the O(n, p)^{16}N reaction and found that lead can interfere seriously and ıuminum to a lesser degree.

Several other investigators [66–70] have applied the method specifically to .uminum and to aluminum–aluminum oxide alloys. Considerations were ;ven to preparation of the sample before irradiation, standardization of ıe neutron flux, correction for attenuation of neutrons inside the sample by :attering and deceleration, correction for fluorine interference by the F(n, α)^{16}N reaction, and the sample-transfer system. Dugain et al. [70] ·ported that surface cleaning of the aluminum sample (by etching with a ›lution of sodium hydroxide) after irradiation is extremely important, ·sulting in a significant decrease of the apparent content of oxygen, regard-ss of the method of preparation of the surface prior to irradiation and ·gardless of the nature of the gas around the sample during irradiation. alues as low as 2 ppm were obtained; the interference at this level because of ıorine may be appreciable. The method has been used on large (12–17 g) ʼlindrical samples of aluminum. At levels of 20–50 ppm the method has a andard deviation of about 2–3 ppm.

Other Activation Methods (Al)

Other activation methods that have been proposed and tested for the deter-ination of oxygen in aluminum include irradiation by γ-photons to give the O(γ, n)^{15}O reaction [34, 71] and bombardment by α-particles and helium ions • produce the reactions $^{16}O(\alpha, pn)^{18}F$, $^{16}O(\alpha, d)^{18}F$, $^{16}O(\alpha, 2n)^{18}Ne \xrightarrow{B^+} {}^{18}F$, O(^{3}He, p)$^{18}F$; and $^{16}O(^3He,n)^{18}Ne \xrightarrow{B^+} {}^{18}F$ [72–74]. Irradiation of aluminum ırfaces by 2-MeV tritons has been used by recent investigators [75–77] for ıe determination of surface oxygen by the $^{16}O(t, n)^{18}F$ reaction.

The determination of oxygen and nitrogen in aluminum has been performed . the surface of solid samples by electron microprobe analysis [35].

Comparison of Methods

A comparison of methods for each specific metal shows that the more nsitive ones are the hydrogen-reduction method for tin, lead, and bismuth;

the hydrogen sulfide method for zinc, cadmium, and lead; the vacuun fusion method for tin, bismuth, and possibly antimony; and the neutro activation method for aluminum. Inasmuch as the hydrogen sulfide metho requires very large samples (90 g) and a highly toxic gas, the evaporation inert-gas-fusion method is preferred for zinc and cadmium. The othe methods that require the use of obnoxious reagents such as sulfur vapo sulfur monochloride, and bromine trifluoride are also less attractive than th hydrogen reduction and the carbon reduction methods.

D. RECOMMENDED METHODS

1. Hydrogen

a. Aluminum: Hot Extraction

The apparatus and procedure described in Chapter 11, Section IV.D.1. for the determination of hydrogen in beryllium are applicable to the analysi of aluminum.

b. Zinc, Cadmium, Tin, Lead, Antimony, Bismuth: Combustion-Manometric

Follow the procedure, presented in Chapter 11, Section IV.D.1.b. Th sample is burned in purified oxygen to produce water, which is condense from the oxygen stream for subsequent manometric measurement at 110°C

2. Nitrogen: Kjeldahl

Analyze the sample for nitrogen by applying the Kjeldahl procedure a given in Chapter 11, Section IV.D.2.

3. Oxygen

a. Aluminum: Neutron Activation

See Chapter 3 for descriptions of apparatus and procedures.

b. Zinc and Cadmium: Metal Evaporation–Inert-Gas Fusion

A description of the apparatus and procedure for the analysis of magnesiun is presented in Chapter 11, Section IV.D.3.b. The procedures for zinc an cadmium are somewhat different because of variations in metal vapor pres sures and in oxide stabilities.

(1) ZINC. Place in the reaction chamber (Fig. 11.10) a graphite crucible (Fig. 11.11) filled to a depth of $\frac{1}{4}$ in. with 20–40-mesh graphite chips. With argon flowing through the chamber and out through the loosened 40/35 joint, condition the graphite by heating to 1900°C for 10 min. To make a blank run, tighten the 40/35 joint and direct the argon stream through the analytical train at a rate of 150 cc/min (throttled at stopcock S_5, Fig. 11.9). Place liquid nitrogen on the U-tube cold trap, and heat the crucible for 10 min at 1850°C; shut off the power to the induction heating coil (Note 1).

Open the reaction chamber at the top, and with argon flowing out, insert first the cleaned and dried quartz liner and then the slitted tantalum tube. Insert an 18-mm glass tubing funnel, extending to the crucible, and drop in a weighed sample of zinc.

Close the grooved stopcock S_2; open the other stopcocks downstream toward the pump, bypassing the Schutze reagent (Note 2). When the pressure in the reaction chamber is diminished to about 5 torr, as indicated by the manometer of the capillary trap, partially open S_2 to give a manometer reading of about 100 torr. Energize the induction heating coil at low power and increase the power to raise the temperature of the crucible rim to 820°C (Note 3). When the evaporation is complete (usually about 25 min for a 1-g sample of zinc), as evidenced by an optical field of uniform brightness, shut off the power; allow 5 min for cooling.

Follow the same procedure as outlined for magnesium for the removal of the tantalum tube and the quartz shield and for the subsequent carbon reduction of the oxide residue. Convert the manometer readings for the sample and the prevailing blank to micrograms of oxygen, and calculate the oxygen content as follows:

$$\text{O, ppm} = \frac{(\mu\text{g O in sample}) - (\mu\text{g O in blank})}{\text{g of sample}}. \qquad (3)$$

(2) CADMIUM. Instead of dropping the sample through a funnel to the bed of graphite chips in the crucible as in the zinc procedure, remove the crucible, as in the magnesium procedure, and bury the sample in the graphite chips before the metal is evaporated (Note 4). To control the rate of evaporation, maintain the crucible at dull redness (about 570°C) until the cadmium is volatilized. This temperature is too low to be measured by optical pyrometer (Model 8622-C, Leeds and Northrup Company, Philadelphia, or equivalent) (Note 5).

For quantitative carbon-reduction of the cadmium oxide, raise the temperature as rapidly as possible to 1900°C. By this procedure, any O_2 that may be produced by the thermal decomposition of the relatively unstable oxide will have a maximum exposure to hot carbon (for conversion to carbon monoxide) before it escapes into the carrier-gas stream.

NOTES

1. The blank should decrease to about 3 torr (capillary-manometer reading) or about 2 μg of oxygen within two or three 10-min runs. If not, make a low-pressure bakeout run by throttling at S_2 instead of at S_5. This enhances removal of moisture contamination in the reaction tube. Repeat the blank runs if necessary.
2. Refer to Note 1 in Section IV.D.3.b of Chapter 11 for preparation of the Schutze reagent.
3. The contents of the crucible remain at a lower temperature (hence, a darker optical field) than the graphite until all the zinc has evaporated.
4. By this procedure the more thermally unstable cadmium oxide is in better contact with carbon during the carbon-reduction step after evaporation of the metal. Without this precaution, some of the oxygen may be lost by thermal decomposition of the oxide to give molecular oxygen, which escapes from the crucible before reaction with the hot carbon.
5. The low boiling point of cadmium (765°C) necessitates the low evaporation temperature.

c. Tin, Lead, Antimony (Note 1), and Bismuth: Hydrogen Reduction (38, 41)

(1) APPARATUS. The apparatus shown in Fig. 12.1 consists of a reduction chamber, a desiccant tube, a manometer, and a gas microburet.

The 5-cc microburet, B, graduated in 0.01-cc divisions, is sealed to a 50-cc expansion bulb at the lower end and to 2-mm capillary tubing at the upper end, which leads to the reduction unit and to the three-way stopcock shown in the diagram.

The 22-mm quartz reduction chamber, J, is fitted with a greased ST 35/20 ball-joint end cap that is held firmly in place by a spring clamp. The vertical section of the quartz tube, H, filled with 30-mesh anhydrous magnesium perchlorate, is joined to the rest of the unit by a greased ball joint at G. The reaction chamber is heated by a small nichrome-wire resistance furnace, the temperature of which is controlled by a variable transformer.

(2) PROCEDURE. Enclose in the reduction chamber a clean, dry porcelain boat containing a 1–10-g sample, the size depending upon the approximate oxygen content. Fill the buret with mercury, and close off at the pinch clamp, M. Evacuate the system for 5 min and check for leaks. Alternately fill with hydrogen and evacuate 5 times to remove extraneous gases. Fill the system with hydrogen at a slight positive pressure, open the pinch clamp, and lower the mercury to fill the buret with hydrogen. When the system has come to room temperature, adjust the height of the mercury leveling bulb until atmospheric pressure is attained, as indicated by the manometer. Record the

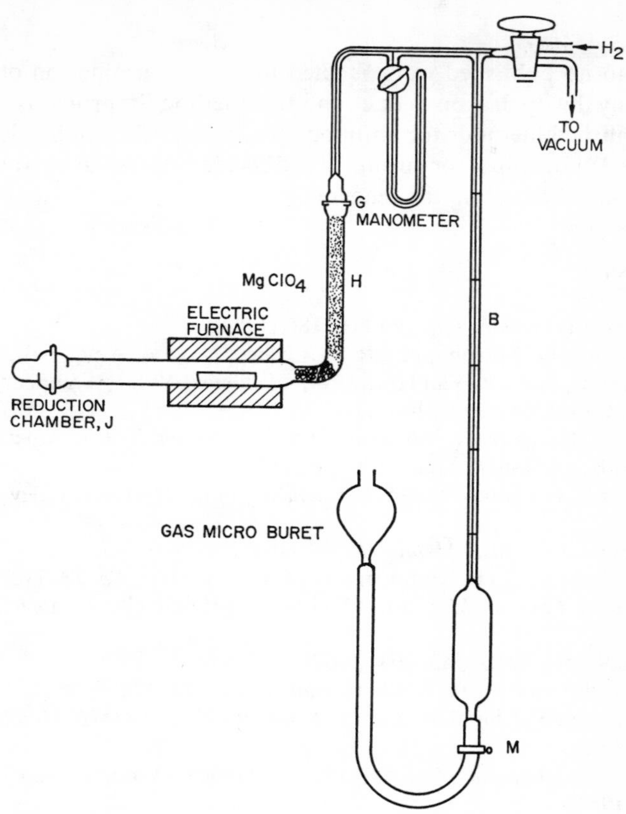

Fig. 12.1. Hydrogen-reduction apparatus.

buret reading (cc), the room temperature, and the barometric pressure (torr). Lower the leveling bulb, and heat the sample for 30 min at 850–900°C. Shut off the furnace, and cool the system to room temperature by directing a stream of air on the furnace. Adjust the leveling bulb to bring the gas to atmospheric pressure, and again record the buret reading, the room temperature, and the barometric pressure.

To calculate the oxygen content, add each buret reading to the predetermined volume of the apparatus, and correct the two total volumes to standard temperature and pressure; then,

$$\text{O, \%} = \frac{16 \times 760 \times \Delta V}{6.236 \times 10^4 \times 273} \times \frac{100}{\text{g of sample}} \tag{4}$$

$$\text{O, \%} = \frac{0.0714\ \Delta V}{\text{g of sample}}. \tag{5}$$

NOTE

1. Although no published work is cited for the determination of oxygen in antimony by hydrogen reduction, the method is probably applicable. The published methods for antimony oxide (described in Section II.C.1.a) that use sulfur vapor or bromine trifluoride are less attractive from the standpoint of handling the reagents.

REFERENCES

1. D. Maykuth, *Product Eng.*, **24** (11) 186 (1953).
2. D. L. Smith, G. Murphy, and R. W. Fisher, *IS-1350*, Ames Lab, Iowa, 1966.
3. J. L. Brandt, in K. R. Van Horn, Ed., *Aluminum*, Vol. I, American Society for Metals, Cleveland, Ohio, 1967, p. 1.
4. J. P. Lyle, Jr., in K. R. Van Horn, Ed., *Aluminum*, Vol. I, American Society for Metals, Cleveland, Ohio, 1967, p. 337.
5. D. T. Hurd, *An Introduction to the Chemistry of the Hydrides*, Wiley, New York, 1952.
6. H. A. Sloman, *J. Inst. Metals* **71,** 391 (1945).
7. Yu. A. Klyachko, L. L. Kunin, and E. M. Chistyakova, *Sb. Tr. Tsentr. Nauchn.-Issled. Inst. Chernoi Met.* **24,** 42 (1962); through *Chem. Abstr.*, **58,** 11947 (1963).
8. H. Pfundt, *Aluminium*, **43,** 363 (1967).
9. C. B. Griffith and M. W. Mallett, *Anal. Chem.*, **25,** 1085 (1953).
10. D. J. Carney, J. Chipman, and N. J. Grant, *Trans. AIME Metals Div.*, **188,** 397 (1950).
11. A. Colombo, *Energia Nucl.*, **13** (5) 251 (1966); through *Chem. Abstr.*, **66,** 16308 (1967).
12. C. E. Ransley and D. E. J. Talbot, *J. Inst. Metals*, **84,** 445 (1956).
13. R. Eborall and C. E. Ransley, *J. Inst. Metals*, **71,** 525 (1945).
14. G. D. Nikiforov and S. A. Silant'eva, *USSR*, *161,960* April 1964; through *Anal. Chem.*, **39,** 105 R (1967).
15. V. A. Danilin, *Metody Opred. Issled. Sostoyaniya Gazov. Metal.*, **1968,** 24; through *Chem. Abstr.*, **71,** 9376 (1969).
16. H. Aschehoug, B. F. Haegland, P. K. Kofstad, J. L. Lauritzen, and P. A. Simensen, *Z. Metallk.*, **60,** 864 (1969).
17. Y. Dardel, *Trans. AIME*, **180,** 273 (1949).
18. D. J. Neil and A. C. Burr, *Rev. Met.*, **57,** 735 (1960).
19. A. P. Gudchenko and A. I. Leont'ev, *Tr. Moskov. Aviatsion. Tekhnol. Inst.*, **1961** (49), 137; through *Chem. Abstr.*, **55,** 23168 (1961).
20. M. Heckler, *Aluminium*, **43** (4), 239 (1967).
21. A. N. Zaidel and A. A. Petrov, *Zh. Tekh. Fiz.*, **25,** 2571 (1955); *U.S. Atomic Energy Comm.*, *Translated Report AEC-tr-3162*.
22. A. N. Zaidel and A. A. Petrov, *Ser. Fiz. Khim.*, **No. 2,** 40 (1957); through *Chem. Abstr.*, **52,** 4394 (1958).

23. A. N. Zaidel, A. A. Petrov, and K. I. Petrov, *Fiz. Sbornik. L'vov. Univ.*, **No. 4,** 206 (1958); through *Chem. Abstr.*, **55,** 5236 (1961).
24. C. Evans and J. Herrington, *Radioisotopes in the Physical Sciences and Industry*, Vol. II, International Atomic Energy Agency, Vienna, 1962, p. 309.
25. A. S. Gillespie, Jr., *Anal. Chem.*, **32,** 1624 (1960).
26. C. E. Ransley, D. E. J. Talbot, and H. C. Barlow, *J. Inst. Metals*, **86,** 212 (1958).
27. J. Kashima and T. Yamazaki, *Rept. Castings Research Lab.*, **10,** 89 (1959); through *Chem. Abstr.*, **55,** 7149 (1961).
28. N. Shiota, *Trans. Jap. Inst. Metals*, **5** (3) 142 (1964).
29. N. P. Mazyukevich and V. A. Shkoda-Ul'yanov, *At. Energ.*, **21,** 388 (1966); through *Nucl. Sci. Abstr.*, **21,** 19540 (1967).
30. C. C. Carson, *Anal. Chem.*, **32,** 936 (1960).
31. B. D. Holt, *Anal. Chem.*, **28,** 1153 (1956).
32. M. Namiki, Y. Kakita, and H. Goto, *Talanta*, **11** (5), 813 (1964).
33. P. Albert, C. Englemann, S. May, and J. Petit, *Compt. Rend.*, **254,** 119 (1962).
34. P. Albert, *Chimia*, **21,** 116 (1967); *Nucl. Sci. Abstr.*, **21,** 21896 (1967).
35. P. Lanusse, C. Le Pennec, F. Pichoir, and G. Roy, *Rech. Aerospatiale*, **132,** 33 (1969); through *Chem. Abstr.*, **72,** 62408 (1970).
36. J. Maatsch, *Metall*, **7,** 873 (1953); through *Chem. Abstr.*, **51,** 13651 (1957).
37. W. A. Baker, *Metallurgia*, **40,** 188 (1949).
38. K. Barteld and W. Hofmann, *Z. Erzbergbau Metallhuttenw*, **5,** 102 (1952); through *Chem. Abstr.*, **46,** 6033 (1952).
39. G. Serrini and W. Leyendecker, *European Atomic Energy Community, EUR-3072i*, Ispra, Italy, 1966; through *Nucl. Sci. Abstr.*, **21,** 5 (1967).
40. E. S. Funston and S. A. Reed, *Anal. Chem.*, **23,** 190 (1951).
41. G. H. Osborn, *Analyst*, **76,** 114 (1951).
42. E. Eberius and W. Kowalski, *Z. Erzbergbau u. Metallhuttenw*, **7,** 229 (1954); through *Chem. Abstr.*, **51,** 13650 (1957).
43. G. Serrini, J. Collin, and W. Leyendecker, *Rapp. EUR-2207i*, 1965.
44. H. Hartmann, W. Hofman, K. Schulte-Schrepping, *Z. Metallk.*, **43,** 350 (1952); through *Chem. Abstr.*, **47,** 1537 (1953).
45. H. Hartmann, W. Hofman, and G. Strohl, *Z. Erzbergbau Metallhuttenw*, **8,** 18 (1955); through *Chem. Abstr.*, **49,** 4451 (1951).
46. H. Hartmann, W. Hofman, and G. Strohl, *Z. Metallk.*, **49,** 461 (1958); through *Chem. Abstr.*, **52,** 19687 (1958).
47. H. Hartmann and G. Strohl, *Z. Anal. Chem.*, **175,** 84 (1960); through *Chem. Abstr.*, **55,** 1284 (1961).
48. H. V. Wartenberg, *Z. Anorg. Allg. Chem.*, **251,** 161 (1943); through *Chem. Abstr.*, **37,** 4326 (1943).
49. A. K. Babko, K. E. Kleiner, and L. V. Markova, *Zavod. Lab.*, **22,** 640 (1956); through *Chem. Abstr.*, **50,** 15339 (1956).
50. K. E. Kleiner, *Geokhim. Anal. Khim.*, **10,** 150 (1960); through *Chem. Abstr.*, **55,** 7149 (1961).
51. K. E. Kleiner, and N. V. Obolonchik, *Ukr. Khim. Zh.*, **25,** 370 (1959); through *Chem. Abstr.*, **53,** 21397 (1959).

52. K. E. Kleiner, *Ukr. Khim. Zh.*, **22,** 9 (1961); through *Anal. Chem.*, **33,** 81R (1961).
53. H. R. Hoekstra and J. J. Katz, *Anal. Chem.*, **25,** 1608 (1953).
54. L. Silverman and W. Gossen, *Anal. Chim. Acta*, **8,** 436 (1953).
55. R. M. Black, *Analyst*, **76,** 208 (1951).
56. U.K. Atomic Energy Auth., Production Group, *Pg-Report-381*, 1962.
57. C. B. Griffith and M. W. Mallett, *J. Amer. Chem. Soc.*, **75,** 1832 (1953).
58. A. Colombo and E. Rodari, *Anal. Chim. Acta*, **42,** 133 (1968).
59. A. Colombo and E. Rodari, *Anal. Chim. Acta*, **49,** 584 (1970).
60. C. Venkateswarlu and M. W. Mallett, *Anal. Chem.*, **32,** 1888 (1960).
61. B. D. Holt and H. T. Goodspeed, *Anal. Chem.*, **34,** 374 (1962).
62. W. G. Smiley, *Anal. Chem.*, **27,** 1098 (1955).
63. N. Yoneda, *Nippon Kinzoku Gakkaishi*, **21,** 392 (1957); through *Chem. Abstr.*, **56,** 7993 (1962).
64. K. G. Broadhead, D. E. Shanks, and H. H. Heady, *Proceedings*, Intern. Conf. Modern Trends Activation Anal., College Station, Tex., 1965.
65. J. Hoste, D. DeSoete, and A. Speecke, *Ghent Rijksuniversiteit*, *EUR-3565e*, Belgium, 1967; through *Nucl. Sci. Abstr.*, **22,** 1948 (1968).
66. I. Fujii, K. Takada, and H. Muto, *Bunseki Kagaku*, **15,** 1239 (1966); through *Chem. Abstr.*, **67,** 39864 (1967).
67. E. T. Bramlitt, *Rept. AI-CE-74*, Atomics International, Canoga Park, Calif., 1967.
68. D. Gibbons, G. Olive, P. Sevier, and J. E. Deutschman, *J. Inst. Metals*, **95,** 280 (1967).
69. D. Brune and K. Jirlow, *J. Radioanal. Chem.*, **2,** 49 (1969).
70. F. Dugain, M. Andre, and A. Speecke, *Radiochem. Radioanal. Lett.*, **4,** 35 (1970).
71. M. Kobayashi, T. Sawai, S. Nagatsuka, and S. Maeda, *Proc. Japan Conf. Radioisotopes*, *5th*, **No. 3,** 179 (1963); through *Nucl. Sci. Abstr.*, **17,** 30119 (1963).
72. M. Deyris and P. Albert, *Compt. Rend.*, *Ser. C*, **262,** 1675 (1966); through *Nucl. Sci. Abstr.*, **20,** 35425 (1966).
73. P. Albert, M. Deyris, and G. Revel, *Compt. Rend.*, *Ser. C*, **262,** 1774 (1966); through *Nucl. Sci. Abstr.*, **20,** 36741 (1966).
74. T. Sawai and P. Albert, *Radioisotopes* (Tokyo), **16,** 509 (1967); through *Nucl. Sci. Abstr.*, **22,** 53 (1968).
75. J. N. Barrandon and P. Albert, *Modern Trends in Activation Analysis*, Vol. II, National Bureau of Standards, Washington, D.C., 1969, p. 794.
76. L. Quaglia, G. Robaye, M. Cuypers, and J. N. Barrandon, *Nucl. Instrum. Methods*, **68,** 315 (1969).
77. J. N. Barrandon, L. Quaglia, J. L. Debrun, M. Cuypers, and G. Robaye, *J. Radioanal. Chem.*, **4,** 115 (1970).

CHAPTER

13

CHROMIUM, MANGANESE, AND VANADIUM

LAWRENCE A. DE'AETH

Union Carbide Corporation, Mining and Metals Division
Niagara Falls, New York

CONTENTS

I. CHROMIUM

A. METALLURGICAL ASPECTS

1. Introduction

Metallurgical uses of chromium have been limited because of poor mechanical properties, thought to be caused in part by gaseous impurities. An example is the ductile-to-brittle transition at or above room temperature. Considerable work has been devoted to the development of techniques for producing high-purity chromium of low-temperature ductility and impact resistance.

High-purity chromium cannot be obtained by direct reduction of the ore. Pure metal is made by precipitation of chromium hydroxide from sodium chromate solution, ignition, and conversion of the oxide. Chromium is also produced by Goldschmidt's aluminothermic process or by reduction with silicon in an electric furnace. High-purity metal, suitable for basic material in metallurgical studies, is made by thermal dissociation of the iodide, the van Arkel-deBoer process. Another method of production is fused-salt electrolysis. Melts of K_3CrF_6 in NaCl and K_3CrF_6 in $CaCl_2$ are electrolyzed to produce dendritic crystals of chromium, which are then consolidated by arc melting.

Most commerically available high-purity chromium is in the form of electrolytic plate, deposited from chromic sulfate solution. The oxygen content of electrolytic chromium can be reduced by hydrogen treatment or vacuum techniques. Electrolytic chromium is used in the production of cobalt–chromium and nickel–chromium "superalloys" and of chromium carbide and cermets. High-purity metal is desired for quality end products and particularly for vacuum-melted materials.

2. Hydrogen

The effects of oxygen, hydrogen, and nitrogen on the mechanical and physical properties of chromium have been of considerable interest over the past several years. There is still some controversy regarding the effects of interstitial elements on the ductility of chromium.

The solubility of hydrogen in chromium is 0.3 ppm at 500°C, increasing to 2.3 ppm at 1000°C [1]. It is reported that hydrogen can exist in chromium as the free element and combined with oxygen in basic compounds. Thermal expulsion of hydrogen from electrolytic chromium involves release of water vapor from inclusions of hydrous chromic oxide and reduction of the water vapor by the chromium to yield hydrogen [2].

Hydrogen is perhaps of the least concern of the gaseous impurities because it can be removed successfully from chromium by heating in air at 500°C. In a study by Potter and Lukens [3], the bulk of the gas (97% hydrogen) that was liberated from the metal upon heating was absorbed, entrapped, or present as a supersaturated interstitial solution. Early investigators attributed hardness to hydrogen content; however, this was later refuted [2].

3. Oxygen and Nitrogen

The solid solubility of oxygen in chromium is 0.03% at 1350°C and less at low temperatures [4]. Oxygen in electrolytic chromium is present as chromic oxides, very likely hydrous chromic oxide $Cr_2O_3 \cdot nH_2O$. Anhydrous chromic oxide is not easily soluble in acids, whereas the deposits in chromium metal dissolve completely in dilute hydrochloric acid. $Cr(OH)_3$ does not exist [5]. Oxygen is present as Cr_2O_3, SiO_2, and Al_2O_3 in iodide-, silicon-, and aluminum-reduced chromium, respectively.

Oxygen, long suspected as the cause of room-temperature brittleness of electrolytic chromium, appears to have little effect on the ductile–brittle transition temperature, at least in quantities greater than 0.001% [6]. Although oxygen has little influence on the transition temperature, it may reduce the ductility above this temperature. Concentrations of 0.006–0.37% O do not significantly effect the bend-transition temperature of wrought chromium [7].

In a study of the influence of oxygen on the hardness of electrolytic chromium [8], no definite relationship was established, although there was a tendency for hardness to increase with the oxygen content. In a report on the effect of oxygen on the electrical resistivity of electrodeposited chromium [8], a wide range of values was attributed to variations in the oxygen content and the soundness of the deposits.

The solid solubility of nitrogen in chromium is 0.08% at 1200°C and less at lower temperatures. The brittleness of chromium below the ductile–brittle transition temperature, 190–195°C, is produced by nitrogen contents of less than 0.04% [6]. Embrittlement of ductile chromium occurs on heating in nitrogen at 700°C [9]. A nitrogen content of 0.02% produces brittleness in cold-worked chromium while 0.001% nitrogen appears to embrittle recrystallized material. Concentrations of 0.001–0.032% N apparently do not affect bend ductility [7]. Oxygen and nitrogen impurities can increase the temperature of transition (from a brittle to a ductile state) from 175 ± 10°C to 325 ± 50°C [10]. Allen et al. [11] have discussed quite well the effect of impurities on the tensile-transition temperature of chromium.

B. HISTORICAL DEVELOPMENT

Chromium has been analyzed for gaseous impurities by a variety of methods. Although fusion techniques are now generally used for the determination of gases in chromium, early chemical methods for oxygen and nitrogen proved to be quite useful.

The acid-solution method proposed by Adcock has been used by many workers [12]. The specimen is heated in vacuum for one hour at 850°C to convert the oxygen to Cr_2O_3. A pulverized 1-g sample is dissolved in hot 10% hydrochloric acid, and the solution is cooled and filtered. The residue is washed with cold 4% hydrochloric acid, ignited, and weighed, and the oxygen content is calculated. The advantages of this method are its simplicity and its applicability to large numbers of samples; however, it is only suitable for high-purity material that does not contain acid-insoluble carbides, silicides, or other similar nonmetallic compounds. In a modification of the method for oxygen in high-purity chromium [13], the sample is calcined in a quartz tube (evacuated at room temperature) at 805–900°C for 2 h, 10% hydrochloric acid is added, and the solution is cooled and filtered. The washed residue is calcined at 800–900°C for 1.5 h. Oxygen is determined by weighing and calculating. Accuracy is reported to be 10%. Again, the method is not applicable in the presence of SiO_2, Al_2O_3, and other oxides.

The bromination–carbon-reduction method has been applied to chromium at 925°C, using helium as a carrier gas [14]. The method is long, and purification of reagents is a problem.

By a sulfur method, adapted to the determination of oxygen in chromium [15], the metal is treated with sulfur vapors in a vacuum with the resulting sulfur dioxide determined colorimetrically by the fuchsin-formaldehyde reaction. The analysis takes considerable time (5–6 h), and purification of reagents can be a problem. The sensitivity of the method is good (10^{-4}%), and 0.1–0.2-g samples can be analyzed.

Although spectrographic and isotope-dilution methods have yielded accurate results when applied to chromium, they are not widely employed, chiefly because of expensive instrumentation. In a spectrographic method for oxygen and nitrogen in chromium used by Alpatou et al. [16], oxygen is determined by a low-voltage pulse discharge in a helium atmosphere (700 torr) with the sample as the cathode and a graphite pencil as the anode. Nitrogen is determined by a low-voltage spark discharge in a helium atmosphere (700 torr) with the sample as the cathode and a tungsten pencil as the anode. The mean error is reported to be 25–55% for 0.005–0.2% oxygen and 0.009–0.15% nitrogen.

The isotopic method for the determination of oxygen in chromium was proposed by Kirshenbaum [17]. The sample and "master alloy" containing ^{18}O are fused with graphite and part of the oxygen is liberated as carbon monoxide. The ^{18}O–^{16}O ratio in the carbon monoxide is determined with a mass spectrometer, and the oxygen content is calculated. An isotope-dilution procedure has also been applied to the determination of nitrogen in chromium and other metals with a relative sensitivity of 1×10^{-4}% [18].

Lassner [19] determined oxygen in chromium by a chemical-dissolution-gravimetric method. A 20-g sample was dissolved in 400 ml methyl alcohol and 50 ml bromine. The solution was heated to boiling and filtered through a dry filter. The residue was thoroughly washed, ignited, and weighed as metal oxides. An accurate oxygen determination required an analysis of the metal oxide mixture.

Debrun et al. [20–22] have studied activation analysis of high-purity chromium, iron, and nickel, using ^{4}He and ^{3}He ions to determine trace oxygen. Since ^{18}F was the radioisotope measured in the oxygen determination, the production of ^{18}F by nuclear reactions of elements other than oxygen was examined.

C. STANDARDS AND SAMPLING

In the absence of commercially available standards for the determination of gases in chromium metal, the usual approach is to "spike" the samples with known amounts of chromium oxide and check for quantitative recovery of oxygen. Synthetic mixtures of oxide and metal powders have been analyzed with a reasonable degree of accuracy [23]. In some cases the error in recovery of oxygen from the encapsulated Cr_2O_3 was attributed to the use of an open graphite crucible resulting in ejection of some of the metal powder [24].

Chromium is not usually analyzed in the massive state but rather as a powder of some particular mesh size. As the surface-area–volume ratio is increased there is a considerable increase in gas content. It is advisable, therefore, to avoid preparing the material any finer than is necessary to get a representative sample for analysis. The producer and consumer of chromium should analyze samples of the same particle size.

The effect of sample preparation on the nitrogen content of electrolytic chromium has been demonstrated by Martin and Wilms [25]. Nitrogen pickup was noticed in all turning and drilling operations; however, contamination was kept to a minimum by machining under a suitable liquid such as alcohol.

The deposits from chromic–chromous plating baths used for the commercial electrowinning process for chromium lose about half their hydrogen and gain about 0.06% oxygen during about 100 days of aging [26]. The hydrogen loss is explained as caused by oxidation of the elementary dissolved hydrogen, and the oxygen gain is attributed to oxidation of the skeletonlike chromous oxide inclusions that are characteristically present in the metal deposits.

D. ANALYSIS

1. Hydrogen

As mentioned previously, hydrogen in chromium is readily removed by heating in air at 500°C; hence, the simplest method for determining hydrogen is vacuum extraction. The sample is heated in vacuum at 500–1000°C, and the evolved gas is then analyzed.

The combustion method is applicable for the determination of microquantities of hydrogen [27]. The sample is burned in a stream of oxygen, and the resulting water is measured in a constant-volume manometer.

A spectrographic method for hydrogen in chromium has been proposed by Shubina [28].

Although these methods are adequate for hydrogen, the usual procedure is to determine hydrogen by vacuum-fusion techniques along with oxygen.

2. Oxygen

A vacuum-fusion method for the determination of oxygen in chromium was proposed by Vacher and Jordan [29] as early as 1931. An iron bath and a temperature of 1600°C were used. The iron bath was also used by Sloman [30], and a 40–50% maximum chromium content was recommended. A graphite cover on the fusion crucible was used with a high-speed diffusion pump (80 liters/sec) to obtain quantitative results [31]. An iron bath with a 25% tin addition at 1600°C was reported by Horton and Brady [23]. Aluminum-foil and iron-wire wrappings were used as an encapsulation technique, and this was reported to be superior to that employing nickel capsules [32]. Several other investigators have cited advantages of the iron-bath procedure. A 75% iron–25% tin bath at 1600°C was used by Turovtseva and Kunin [33]. A sensitivity of 1 ppm oxygen in high-purity chromium with the iron-bath technique was reported [34], and good agreement between the iron-bath–vacuum-fusion method and the acid-solution method was shown by Fowler et al. [35].

Lench et al. [36] have critically evaluated the Adcock and vacuum-fusion methods for the determination of oxygen in electrodeposited chromium. They postulated that the higher oxygen contents of non-heat-treated chromium result from the decomposition of the water of hydration by carbon in the vacuum-fusion furnace to form carbon monoxide and hydrogen. Results of the two methods are most correctly compared when material heat-treated at 950°C for 1 h is used. The Adcock method gives lower results than the vacuum-fusion method amounting to 3% relative at the 0.26% oxygen level and increasing to 21% relative at the 0.014% level. It should be noted, however, that the oxygen present as water of hydration in electrodeposited chromium is contributable to the steel or alloy to which it is added.

The determination of oxygen in chromium was reported as being satisfactory by dissolving the sample in a molten solvent, such as the low-vapor-pressure, metallic and metalloid elements of groups I–IV and VII, in which carbon has a finite solubility [37]. A platinum–tin bath gave results that were reported to be in agreement with those obtained with the iron–tin bath in the 80–250-ppm range [38]. A nickel bath with a maximum concentration of chromium in the melt of 15–20% has been recommended [39]. Dilution with nickel was said to result in greatly reduced volatilization of chromium at the operating temperature of 1600°C. A nickel–iron bath (80:20) has also been used with nickel- or tin-flux additions with each sample of chromium [19].

Dallmann and Fassel [40] have shown the applicability of the inert-gas-fusion method, which is considerably simpler in procedure and apparatus than vacuum fusion, to the determination of oxygen and nitrogen in chromium. The procedure of Holt and Goodspeed [41], by which hydrogen as well as oxygen and nitrogen can be determined, is also applicable to chromium.

3. Nitrogen

Nitrogen in chromium metal has usually been determined by vacuum fusion [30] and by chemical methods. Lumley [42] recommended solution of the sample in sulfuric acid and collection of the undecomposed nitrides in a precipitate of $BaSO_4$ formed by the addition of $BaCl_2$ to the solution. The nitrides are decomposed by boiling with a sulfuric acid–potassium sulfate mixture. Distillation from a basic solution is followed by nesslerization of the distillate. An error of 0.0004% is reported at the 0.001% nitrogen level on a 0.5-g sample.

Decomposition of the insoluble nitrides is accomplished also either by the addition of $K_2S_2O_8$ and fusion of the residue with $KHSO_4$; by heating with H_2SO_4, K_2SO_4, and $CuSO_4$; or by use of perchloric acid.

Nitrogen in chromium has also been determined using isotope dilution [18].

E. RECOMMENDED METHODS

1. Hydrogen and Oxygen

A well-established procedure is to determine oxygen and hydrogen simultaneously using vacuum fusion. A recommended procedure for the determination of oxygen involves fusion of the sample in a bath of molten iron at a temperature of 1600–1700°C. The concentration of chromium should not exceed 40–50%, and tin additions are recommended. Total time for a complete extraction of oxygen is 10–15 min. The sample size and the sensitivity of the analysis depend upon the type of equipment employed, the skill of the operator, and the amount of gas in the metal.

Inert-gas fusion has largely taken the place of vacuum fusion for the determination of oxygen in many metals including chromium. It is therefore recommended as the best method unless all three gases (oxygen, nitrogen, and hydrogen) are to be determined and vacuum-fusion equipment is available for the analysis (see Chapter 4).

2. Nitrogen

The acid digestion (Kjeldahl) method is the simplest and most rapid method and the one usually employed for the analysis for nitrogen. Solution of the sample with HCl is followed by decomposition of the carbides with H_3PO_4–H_2SO_4 mixture and addition of $K_2S_2O_8$. Any residue after this treatment is fused with $KHSO_4$. After dissolution of the residue in weak acid, ammonia is distilled from a basic solution and titrated with a weak standard acid. Results on low nitrogens have been obtained that agree well with those obtained by the photometric method [33, 43].

II. MANGANESE

A. METALLURGICAL ASPECTS

1. Introduction

Manganese is widely used in the steel industry. It is essential in the manufacture of cast and wrought steel. It is a deoxidizer and also combines with sulfur in steel, which results in improved rolling qualities. Manganese is added to steel during melting in the electrolytic form or as a ferroalloy (see also Chapter 16).

Up to 1943 most manganese metal was produced by reduction of oxides with silicon or aluminum. The metal produced by these processes usually contained substantial amounts of oxides and other impurities. In 1936 a successful commercial process for the electrolytic production of manganese was developed by the U.S. Bureau of Mines. This product is deposited from an aqueous electrolyte containing manganese and ammonium sulfate and is marketed as cathode fragments about $\frac{1}{8}$-in. thick.

Manganese has also been produced on a small scale by distillation, reduction by carbon or hydrogen, and electrolysis of fused salts. Distillation of manganese produces very pure metal.

The most important nonferrous alloys of this metal contain more than 50% manganese. Manganese–copper alloys (60–80% manganese) have high electrical resistivity and are good standard resistance alloys. Substitution of nickel for some of the copper in these alloys increases thermal stability. Manganese–iron–aluminum alloys (50% manganese) have a low coefficient of expansion and high electrical resistance. The largest use of manganese as a nonferrous alloying metal is as a constituent of alpha brass.

High-manganese steel containing 10–14% manganese (Hadfield steel) has remarkable properties. It has high tensile strength, excellent elongation, and resistance to abrasion and distortion. Hadfield steel is used for parts subjected to severe mechanical conditions in service.

2. Hydrogen

Manganese is similar to iron, and it is likely that the effect of hydrogen in manganese imparts brittleness and distortion of the lattice parameter much as it does in iron. Hydrogenated manganese can be crushed and ground much easier than dehydrogenated material.

There is a minimum in the solubility of hydrogen at about 500°C [44, 45], which is accounted for by the fact that manganese has four crystal modifications. The solubility in the alpha form decreases with increase in temperature; in the beta and gamma forms, solubility increases with increase in temperature. The most probable temperature values for the three transformations in solid manganese are as follows:

$$\alpha \rightleftharpoons \beta \ 700°\text{C}$$
$$\beta \rightleftharpoons \gamma \ 1079°\text{C}$$
$$\gamma \rightleftharpoons \delta \ 1143°\text{C}.$$

The gamma modification produced by electrodeposition is unstable and at room temperature transforms to the alpha form.

The hydrogen content of electrolytic manganese and its removal have been investigated [46], and two points are of particular interest. One is that approximately 250 cc of hydrogen are usually absorbed or entrapped in 100 g of electrolytic manganese; the other is that hydrogen can be effectively removed by heating the manganese to 400–500°C for at least 1 h in either air or an oxygen-free atmosphere, at atmospheric pressure or in a vacuum.

3. Oxygen and Nitrogen

Manganese has been nitrided by heating for 3 h in a stream of nitrogen at 850–1100°C [47]; the highest nitrogen concentration observed in the metal after 3 h at 900°C was 6.15%. The different kinetic effect at lower temperatures was attributable in part to the lower solubility of nitrogen in the alpha and beta forms. Precipitation of nitrides retards diffusion of nitrogen through the metal. The solubility of nitrogen in liquid manganese at 1400°C is 1.98% [48].

The effect of oxygen, nitrogen, and hydrogen on the mechanical properties of manganese metal has received relatively little attention through the years because of its brittle nature. The physical properties of manganese are of little consequence to the user.

B. STANDARDS AND SAMPLING

No standards are commercially available for the determination of gaseous impurities in manganese metal. Heating the sample in a stream of oxygen at 1000°C and recording the gain in weight as oxygen is the best procedure for providing a "standard."

The same sampling procedures outlined in Section I.C for chromium should be followed for manganese. Excessive crushing and grinding should be avoided to prevent oxygen and hydrogen pickup; the choice of a representative sample requires extensive work.

C. ANALYSIS

Little has been done toward developing analytical techniques for the analysis for gases in manganese metal. The main concern of the analyst has been the effect of the metal on the determination of gases in manganese alloys by conventional fusion methods. Because of the very high vapor pressure of the metal, even below its melting point, the determination of oxygen has been given relatively little attention, and many analysts avoid completely the determination of oxygen in high-manganese alloys.

The "gettering" effect of manganese on the determination of oxygen by the vacuum-fusion method was investigated by Thanheiser [49] (see also Chapter

4, Section I.A.4). It was noticed that the interference of manganese on oxygen recovery was much more pronounced in analyses performed in a furnace with a relatively cool wall than in a furnace with a warm wall. The "gettering" effect of the deposited film was greater when the furnace wall was cooled with a water jacket. Vacher and Jordan [29] also observed poor recovery of oxygen in samples analyzed in a furnace that had a deposited manganese film.

One of the earliest investigations [30] of the determination of oxygen in manganese made use of the iron bath, with a maximum manganese concentration of 10% at 1550°C. The oxides and nitrides readily react with carbon at this temperature. The iron bath lowers the partial pressure of the volatile manganese and thereby decreases its interference. The iron-bath method has also been used with periodic additions of tin [50]. The latter vaporizes and then condenses on cooler portions of the system, thus covering freshly deposited reactive manganese. An iron–tin bath was also recommended by Yeaton [51]. Tin does not react with evolved carbon monoxide.

In analyzing titanium–manganese alloys by the micro-vacuum-fusion procedure, Wood [52] advocated a sample–tin-bath ratio of 3:1 at a crucible temperature of 1500°C. These conditions allow the manganese to volatilize with the tin over a period of 10 min before the conversion of oxygen to carbon monoxide is effected at 1800°C. In baths containing up to 1.6% manganese, the interference of this element on oxygen recovery in the analysis of steels disappeared with tin addition [24].

Hitoshi and Furuya [53] used an isotope-dilution technique to study the determination of oxygen in high-manganese steel without a bath in the crucible and with iron, platinum, and tin baths. Their data indicated that platinum and iron baths can promote reaction between carbon and oxygen in a melt and that tin does not promote reaction but is effective in preventing adsorption of carbon monoxide in the furnace. They concluded that a combination of iron (or platinum) and tin should be used as a bath.

Shul'te et al. [54] used a nickel bath containing a small amount of tin for the analysis of manganese and iron–manganese alloys. The manganese concentration in the nickel was 2–3%, and the temperature of the crucible was 1550°C.

In the vacuum-fusion analysis of manganese an iron bath or coarse graphite powder with tin has been proposed [33]. With the graphite method, the sample-weight is about 4 g, the fusion temperature is 1250–1300°C, and the extraction time is 15 min. The graphite method might be disputed in view of the findings of Beach and Guldner [55] in their study of the effect of evaporated films on recovery of gases evolved during vacuum fusion. Because of the lack of repeatability, the graphite method appears questionable.

Klyachko and Chistyakova [56] proposed the use of a copper bath for the determination of oxygen in manganese by vacuum fusion. This was based on thermodynamic data and calculation of the energy of mixing and the molar heat of mixing of various systems. The optimum conditions were a copper bath, with a manganese concentration not to exceed 30%, and a 10–15 min extraction time at a temperature of not more than 1100°C. Insufficient data were presented to establish this technique as superior to the iron–tin bath method.

An inert-gas-fusion procedure was used by Goto and co-workers [57] to determine oxygen in metallic manganese. A tin bath [15–20 times the weight of the manganese sample] was operated at 1650 ± 30°C with an argon flow of 250 cc/min for 10 min. The range of recovery of oxygen was 95–110%. Engelsmann et al. [58] determined oxygen in manganese–antimony by inert-gas fusion, using a tin bath at 1500°C. The limit of detection was 0.1 μg oxygen.

In the inert-gas-fusion method, the evaporation of manganese and deposition of active metal films are considerably reduced, and the sorption error is therefore decreased. Regardless of whether vacuum or inert-gas fusion is utilized, the gases evolved during fusion must be rapidly removed from the crucible and furnace area. This is especially necessary when working with metals of high vapor pressure. In vacuum fusion it has been shown that with a 4% manganese concentration in the melt at a pump speed of 15 liters/sec, satisfactory results for oxygen are obtained; however, when the speed is reduced to 6 liters/sec, low results occur [59]. The use of a rapid transfer pump to reduce "gettering" has been discussed [48, 60].

In a chlorination method used for the determination of oxygen in titanium–base alloys containing manganese [61], the sample was mixed with graphite and chlorinated in an atmosphere of argon; the purified carbon monoxide was oxidized to carbon dioxide and weighed.

D. RECOMMENDED METHODS

1. Hydrogen and Oxygen

As in the case of chromium, hydrogen is readily removed from manganese by heating in vacuum at low temperatures; however, since simultaneous oxygen analyses are usually required, there is little incentive to use vacuum extraction instead of vacuum fusion.

Based on literature findings and the author's own experience, the iron–tin bath employed in vacuum fusion seems the most satisfactory. The manganese concentration in the bath should not exceed 10%, and periodic additions of tin are necessary. The temperature should be approximately 1500–1600°C. A rapid pumping system to transfer the gases from the furnace to the analytical section is essential.

2. Nitrogen

Nitrogen is readily determined in manganese by the standard Kjeldahl or micro-Kjeldahl method as described by Beeghly [62] (see also Chapter 8). Manganese nitrides are easily decomposed in acids and a chemical determination is preferred to the vacuum-fusion method.

III. VANADIUM

A. METALLURGICAL ASPECTS

Most of the vanadium produced is used in the steel industry as ferrovanadium [63]. However, this element is also used as a master alloy with aluminum to stabilize the beta-phase in Ti–V–Al alloys, thereby obtaining improved tensile properties [64]. Increased corrosion resistance at elevated temperatures has similarly been achieved in the aluminum industry. A 3% vanadium–copper alloy is used in making copper–base alloys [63]. High-purity (99.7%) vanadium is available [63, 64].

Oxygen affects the mechanical properties of vanadium more than nitrogen or hydrogen [65]. Studies of the effect of oxygen on X-ray, hardness, and bend-test data [65] and on strain aging have been conducted [66]; the effect of oxygen on the physical and mechanical properties of vanadium has been broadly covered [67]. Hydrogen causes embrittlement of vanadium, while nitrogen has a deleterious effect on the ductility of the metal [64].

B. STANDARDS

Standards for gases in vanadium are presently not commercially available. Fassel and Altpeter [68] prepared synthetic calibration standards for oxygen by dissolving pure vanadium pentoxide in low-oxygen content vanadium. Several small holes were drilled in 40-g specimens of vanadium. The V_2O_5 was then packed in the cavities to yield about 0.2% oxygen in a button prepared by arc melting. Tapered cylindrical caps of the base material were forced into the holes with a hydraulic press. A low-current arc in purified helium was then deployed around the circumference of the caps to seal them. Arc melting was then carried out in the same gas at atmospheric pressure. The buttons were flipped over at least six times to insure homogeneity of the oxygen. Standards of lower oxygen content were prepared by dilution with the base metal. Vacuum-fusion analysis in a platinum bath was used to check the accuracy of the standards. For standards calculated to have 0.026, 0.056, 0.106, and 0.206% oxygen, the vacuum-fusion results were lower, respectively, by only 0.001, 0.002, 0.001, and 0.002%.

C. SAMPLING

Vanadium (99.9% V) has been analyzed both in lump and powdered form (presumably from the same source) for all three gases by Fuchs et al. [69], using vacuum fusion–gas chromatography. Fusion temperatures for the two forms, respectively, were 1730°C and 1800°C. Powdered or granular samples were pressed into cylinders before charging into the vacuum-fusion furnace. Disregarding the small temperature difference, the difference in results, especially for oxygen, is presumably caused by surface oxidation and occlusion of air during compacting the sample (see Table 13.1).

TABLE 13.1. Analysis of Vanadium [69]

Form	H(ml/100 g)	N(%)	O(%)
Lump	20	0.015	0.06
Powder	30	0.02	0.17

In addition to the reactivity of vanadium, its ductility and tendency to exhibit segregation militate against obtaining a good sample [64]. If the sample is remelted or subjected to working, better homogeneity and improved precision are obtained but at the expense of accuracy, because gases are usually lost by these techniques. Alternatively, during remelting, gases may be adsorbed by the vanadium because of inadequate purging of the furnace or because of contaminants in the inert atmosphere used.

D. ANALYSIS

1. Hydrogen

Little has been reported on the determination of hydrogen in vanadium. In the vanadium industry [64], hydrogen is usually determined by vacuum fusion, because generally the oxygen content is desired. However, hot extraction at 1000–1200°C may also be used.

Holt and Goodspeed [70] used inert-gas (helium) fusion in a platinum bath at about 1950°C, oxidizing the hydrogen to water over hot copper oxide. The water was isolated in a capillary-manometer cold trap immersed in dry-ice–acetone, and then vaporized at 110°C for subsequent pressure measurement. Values of 13 ppm H were reported as compared to 12 ppm by hot extraction. It was noted that the sensitivity and precision were about the same as those for vacuum fusion. Fuchs et al. [69], as noted above, used vacuum fusion–gas chromatography.

Orlova and Petrov [71] used the spectral isotopic method (see Chapter 6) and obtained a relative standard deviation of 20%. Sample-weights of 0.05–1.0 g for vanadium (and other metals) were equilibrated at 1100–2000°C for 10–25 min. The systematic error of the determination, caused by the different solubility of the hydrogen isotopes in vanadium, decreases when the temperature of analysis is increased. Further work using this technique has been reported by Zaidel et al. [72].

2. Nitrogen

Nitrogen has been determined in vanadium by dissolution of the metal in $9M$ H_2SO_4 followed by addition of 30% H_2O_2 to decompose any residual nitride [73]. Ammonia was then steam distilled from alkaline solution and determined photometrically with either a sodium phenoxide–sodium hypochlorite reagent (for 5–90 ppm combined N) or Nessler reagent (for 90 ppm–2% combined N). For the lower range, the method had a negative bias of 5% and a coefficient of variation of the same magnitude. For the 300–600-ppm range, recoveries were quantitative and the coefficient of variation was 3%.

Winge and Fassel [74] determined both nitrogen and oxygen in vanadium using the dc-carbon-arc–gas-chromatographic technique (see Chapter 5). The arc discharge must not continue beyond about 20 sec, as nitrogen and carbon monoxide will react with metal vapor and will be lost; in addition nitrogen will react with carbon vapor to form cyanogen. With a sample-weight of 0.30 ± 0.03 g a platinum flux of 1.0 ± 0.1 g was used. Helium at 680 torr was the supporting static atmosphere. The sample was contained in the graphite anode during a 20 sec excitation at 15 A. The extracted nitrogen and carbon monoxide were allowed to mix with the helium for 60 sec, after which a 5-ml aliquot was transferred to the gas chromatograph. Average analysis time was 6 min per sample in a chamber containing 8 electrodes.

Gottwald et al. [75] studied the behavior of oxygen and nitrogen in vanadium during melting or annealing under high vacuum. The best results for nitrogen were obtained with the platinum-capsule or platinum-capsule–platinum-bath method.

In the vanadium industry [64], nitrogen is usually determined by acid dissolution, separation of nitrogen by distillation as NH_3, and photometric determination with Nessler reagent (see Section E.2).

It is quite possible that valid nitrogen determinations could be obtained by fusion, using the single-sample-per-crucible technique (see Chapter 4, Sections I.L.8 and II.C.2.d, e, and g).

3. Oxygen

In general, vacuum or inert-gas fusion has been used for determining oxygen in vanadium. Other techniques utilized include bromination–carbon-reduction, isotope dilution, activation, and emission spectrometry. With bromination–carbon-reduction [76], the crushed sample was mixed with high-purity graphite powder, placed in a platinum boat, and brominated at 925°C in a stream of helium. The evolved carbon monoxide was oxidized to carbon dioxide, which was absorbed in Ascarite and weighed. In vanadium alloys containing aluminum or silicon, the oxygen present as alumina or silica may not be recovered [77].

Vitol [78] used ^{18}O in an isotopic-exchange study of vanadium and other metals, while Perdijon [79] used a small neutron generator with a flux of 10^{11} neutrons/sec to obtain oxygen results comparable to those from vacuum fusion. In addition to the method [74] cited earlier, Fassel and Altpeter [68] determined oxygen in vanadium by emission spectrometry.

Smith and Waterbury [80] determined oxygen in vanadium and refractory metals by the inert-gas-fusion–platinum-bath method and generally obtained higher values than by vacuum extraction. Kamin et al. [81] used both platinum and platinum–75% nickel in a 1:1 ratio with vanadium to obtain oxygen results over the range of 102–2200 ppm that were in agreement with those provided by vacuum fusion. Care must be taken to outgas the nickel prior to analysis. The relative standard deviation, as a measure of precision, was 3%. Although elemental nickel volatilizes above 1800°C, no signs of nickel volatilization from the Ni–Pt bath were noted until 2200°C.

Fassel et al. [82] compared oxygen results obtained by vacuum fusion, inert-gas fusion, and dc-carbon-arc–emission spectrometry. In the vacuum-fusion analysis, the sample, weighing 0.1–0.4 g, was added along with 0.5 g of 12-gauge platinum wire to a platinum bath of 20–60 g. Extraction was performed at 1900°C for 15–30 min. The concentration of vanadium in the bath should not exceed 15%. With fusion in argon at 2100°C and 2200°C, respectively, in a platinum bath and dry crucible, quantitative recovery was achieved in 8 min. A 3:1 platinum–vanadium ratio was employed with the sample and flux being added together. It was concluded that although there were isolated disagreements, all three methods gave concordant results.

E. RECOMMENDED METHODS

1. Hydrogen

As noted earlier, a recommended method [64] for determining hydrogen alone is that involving hot extraction (vacuum or inert gas) at 1000–1200°C. Where oxygen is also to be determined, vacuum fusion or inert-gas fusion should be used.

2. Nitrogen

The Kjeldahl procedure is commonly used [64]. Dissolve the sample, consisting of 0.1 g of filings, in 10 ml (1 + 1) H_2SO_4, 5 ml $HClO_4$, and 5 drops

HF. Heat gently until light fumes of $HClO_4$ appear. Cool and transfer the solution to a Kjeldahl distillation flask. Add 30 ml 60% aqueous NaOH solution, distill, and collect 20 ml distillate in 2 ml H_2O and 1 ml Nessler reagent. Dilute the cooled distillate to 25 ml in a volumetric flask, mix thoroughly, and measure the absorbance at 425 nm.

3. Oxygen

Grady [64] states that the platinum-bath and platinum-flux techniques are applicable to the determination of oxygen in vanadium. This was subsequently confirmed by Fassel et al. [82] as cited earlier. The former author [64] recommends taking a 0.1–0.25 g lump of vanadium and degreasing it with an organic solvent such as acetone or ether. For the platinum-flux technique, weigh the cleaned sample, wrap it in at least eight times its weight of platinum and compact it (use low-oxygen platinum as described in Chapter 4, Section I.A.4.b). Charge the sample and platinum into the degassed and blanked crucible at 1850°C, and fuse in vacuum for 30 min. A platinum bath (about 20 g) may be used, but as noted earlier [82], the weight of samples in the crucible should not exceed 15% of the bath weight. Also, shorter analysis times can be obtained with inert-gas fusion at higher temperatures [64, 82].

REFERENCES

1. L. Luckemeyer-Hasse and H. Shenck, *Arch. Eisenhuttenw.*, **6,** 209 (1932).
2. S. P. Markariewa and N. D. Biriikoff, *Z. Electrochem.*, **41,** 838 (1935).
3. E. V. Potter and H. C. Lukens, *Trans. Met. Soc. AIME*, **175,** 699 (1948).
4. D. Caplan and A. A. Burr, *J. Metals*, **7,** 1052 (1955).
5. J. B. Cohen *J. Electrochem. Soc.*, **86,** 441 (1944).
6. A. H. Sully, E. A. Brandes, and K. W. Mitchell, *J. Inst. Metals*, **81,** 585 (1953).
7. D. J. Maykuth, W. D. Klopp, R. I. Jaffee, and H. B. Goodwin, *J. Electrochem. Soc.*, **102,** 316 (1955).
8. A. Brenner, P. Burkhead, and C. Jennings., *J. Res. Nat. Bur. Std.*, **40,** 31 (1948).
9. H. L. Wain, I. Henderson, and S. T. M. Johnstone, *J. Inst. Metals*, **83,** 133 (1954).
10. N. V. Ageev and V. A. Trapeznikov, *Nekotorye Probl. Prochnosti Tverd. Tela, Sb. Statei*, **1959,** 172.
11. B. C. Allen, D. J. Maykuth, and R. I. Jaffee, *Trans. Met. Soc. AIME*, **227,** 724 (1963).
12. F. Adcock, *J. Iron Steel Inst.* (London), **115,** 369 (1927).
13. N. V. Ageev, A. I. Ponomarev and V. A. Trapeznikov, *Zh. Prikl. Khim.*, **32,** 2479 (1959); through *Chem. Abstr.*, **54,** 7423f (1960).
14. M. Codell and G. Norwitz, *Anal. Chem.*, **28,** 2006 (1956).

15. A. K. Babko, A. I. Volkova, and O. F. Drako, *Zavod. Lab.*, **23,** 136 (1957).
16. M. S. Alpatou, P. N. Galonov, K. A. Sukhenko, and M. S. Fal'kova, *Anal. Khim. Akad. Nauk. SSSR*, **12,** 288 (1960).
17. A. D. Kirshenbaum, *Anal. Chem.*, **29,** 980 (1957).
18. K. B. Orlova and E. V. Vitol, *Zh. Anal. Khim.*, **20,** 694 (1965); through *Nucl. Sci. Abstr.*, **19,** 40444 (1965).
19. E. Lassner, *Z. Erzbergbau Metallhuttenw.*, **20,** 466 (1967).
20. J. L. Debrun, J. N. Barrandon, and Ph. Albert, in J. R. DeVoe, Ed., *Modern Trends in Activation Analysis*, Vol. II, National Bureau of Standards, Washington, D.C., 1969, p. 774.
21. J. L. Debrun, J. N. Barrandon, and Ph. Albert, *Bull. Soc. Chim. Fr.*, **3,** 1011 (1969).
22. J. N. Barrandon, D. L. Debrun, and Ph. Albert, in H. G. Ebert, Ed., *Proceedings of the 2nd Conference on Practical Aspects of Activation Analysis with Charged Particles*, European Atomic Energy Community, Brussels, 1968, p. 277.
23. W. S. Horton and J. Brady, *Anal. Chem.* **25,** 1891 (1953).
24. N. A. Gokcen, *Trans. Met. Soc. AIME*, **212,** 93 (1958).
25. G. S. Martin and G. R. Wilms, in *Ductile Chromium*, American Society for Metals, Cleveland, 1955, p. 194.
26. K. A. Moon and C. Levy, *J. Electrochem. Soc.*, **109,** 301 (1962).
27. B. D. Holt, *Anal. Chem.*, **28,** 1153 (1956).
28. S. B. Shubina, *Inst. Chernoi Met.*, **1,** 244–253 (1961).
29. H. C. Vacher and L. Jordan, *J. Res. Nat. Bur. Std.*, **7,** 375 (1931).
30. H. A. Sloman, *J. Inst. Metals*, **71,** 391 (1945).
31. A. Lench and G. S. Martin, *Trans. Nat. Vac. Symp.*, **8,** 776 (1962).
32. G. S. Martin, *Australian Engin.*, **46,** 58 (1955).
33. Z. M. Turovtseva and L. L. Kunin, *Analysis of Gases in Metals*, 1st ed., Academy of Sciences of the USSR, Moscow and Leningrad, 1959; trans. by J. Thompson, Consultants Bureau, New York., 1961.
34. J. T. Sterling, in *Ductile Chromium*, American Society for Metals, Cleveland, 1955, p. 188.
35. R. M. Fowler, T. C. Lancaster, and G. Porter, in *Ductile Chromium*, American Society for Metals, Cleveland, 1955, p. 121.
36. A. Lench, G. S. Martin, and G. Cumming, *Anal. Chem.*, **36,** 337 (1964).
37. H. S. Spacil, U.S. Pat. 3,091,689, May 28, 1963.
38. T. D. McKinley, Pittsburgh Conference on Analytical Chemistry and Applied Spectroscopy, March, 1959.
39. Yu. A. Klyachko and E. M. Chistyakova, *Inst. Chernoi Met.*, **1963,** 31.
40. W. E. Dallmann and V. A. Fassel, *Anal. Chem.*, **39,** 133R (1967).
41. B. D. Holt and H. G. Goodspeed, *Anal. Chem.*, **35,** 1510 (1963).
42. E. J. Lumley, noted in Ref. 9.
43. H. F. Beeghly, *Ind. Eng. Chem.*, *Anal. Ed.*, **14,** 137 (1942).
44. A. Sieverts and H. Moritz, *Z. Phys. Chem.*, **180,** 249 (1937).
45. E. V. Potter and H. C. Lukens, *Metals Tech.*, **13,** No. 6, *Trans. AIME Inst. Met. Div., Tech. Publ. No. 2032* (1946).

6. E. V. Potter, E. T. Hayes, and H. C. Lukens, *Trans. Met. Soc. AIME*, **161,** 373 (1945).
7. I. B. Baratashvili and V. M. Berezhiani, *Soobshch. Akad. Nauk Gruz. SSR*, **27,** 169 (1961).
8. N. A. Gokcen, *Trans. Met. Soc. AIME*, **221,** 200 (1961).
9. G. Thanheiser, *J. Iron Steel Inst.* (London), **134,** 359 (1936).
0. Z. M. Turovtseva and N. F. Litvinova, *Geneva Conference on Peaceful Uses of Atomic Energy*, Moscow Acad. Sci., 1959.
1. R. A. Yeaton, *Vacuum*, **2,** 115 (1952).
2. D. F. Wood and S. A. Oliver, *Analyst*, **84,** 436 (1959).
3. K. Hitoshi and K. Furuya, *Bull. Chem. Soc. Jap.*, **41,** 1256 (1968); through *Nucl. Sci. Abstr.*, **22,** 42623 (1968).
4. Yu. A. Shul'te, N. A. Gederevich, and V. S. Shitikov, *Ind. Lab.*, **34,** (trans.) 1434 (1968).
5. A. L. Beach and W. G. Guldner, *Anal. Chem.*, **31,** 1772 (1959).
6. Yu. A. Klyachko and E. M. Chistyakova, *Zavod. Lab.*, **26,** 1335 (1960).
7. H. Goto, S. Ikeda, and A. Onuma, *Sci. Rept. Res. Inst.*, **17,** 237 (1965).
8. J. J. Engelsmann, A. Meyer, and J. Visser, *Talanta*, **13,** 409 (1966).
9. T. Raine, and J. B. Vickers, "7th Report on Heterogeneity of Steel Ingots," *J. Iron Steel Inst.* (London), 100 (1937).
0. W. Hessenbruch, and D. Oberhoffer, *Arch. Eisenhuttenw.*, **1,** 583 (1927).
1. W. T. Elwell and D. M. Peake, *Analyst*, **82,** 734 (1957).
2. H. F. Beeghly, *Anal. Chem.*, **23,** 523 (1951).
3. J. Strauss, "Vanadium and Vanadium Alloys," in R. E. Kirk and D. F. Othmer, Eds., *Encyclopedia of Chemical Technology*, Vol. II, Interscience, New York., 1955, pp. 583–593.
4. H. R. Grady, "Vanadium," in I. M. Kolthoff and P. J. Elving, Eds., *Treatise on Analytical Chemistry*, Pt. 2, Vol. VIII, Wiley, New York., 1963, pp. 177–272.
5. S. A. Bradford and O. N. Carlson, *Amer. Soc. Metals, Trans. Quart.*, **55,** 1969 (1962); through *Chem. Abstr.*, **56,** 12633C (1962).
6. S. A. Bradford and O. N. Carlson, *Trans. AIME*, **224,** 738 (1962); through *Chem. Abstr.*, **57,** 8306f (1962).
7. S. A. Bradford, Univ. Microfilms Order No. 61-6155 and *Diss. Abstr.*, **22,** 2348 (1962); through *Chem. Abstr.*, **56,** 1130h (1962).
8. V. A. Fassel and L. L. Altpeter, *Spectrochim. Acta.*, **16,** 443 (1960).
9. A. Fuchs, H. Reinhard, J. Niebuhr, and R. Beck, *Arch. Eisenhuttenw.*, **34,** 361 (1963).
0. B. D. Holt and H. T. Goodspeed, *Anal. Chem.*, **35,** 1510 (1963).
1. N. M. Orlova and A. A. Petrov, *Vestn. Leningrad Univ., Ser. Fiz. Khim.*, **22** (10), no. 2, 55–63 (1967); through *Chem. Abstr.*, **68,** 9051d (1968).
2. A. N. Zaidel, T. F. Ivanova, A. A. Petrov, V. V. Fedorov, and N. M. Chumakova, *Zavod. Lab.*, **29,** 693 (1963); through *Chem. Abstr.*, **59,** 10756b (1963).
3. U.K.A.E.A. Industrial Group, Chem. Services Dept. (Operations Branch, Springfield, Lancs.), *Report IGO-AM/S-9*, U.K. At. Energy Agency, 1958; through *Anal. Abstr.*, **6,** 1271 (1959).
4. R. K. Winge and V. A. Fassel, *Anal. Chem.*, **37,** 67 (1965).

75. H. J. Gottwald, K. Krone, and J. Krueger, *Metall*, **23,** 1284 (1969); throug *Chem. Abstr.*, **72,** 103092x (1970).
76. M. Codell and G. Norwitz, *Anal. Chem.*, **28,** 2006 (1956).
77. S. Kallman, private communication in Ref. 76.
78. E. N. Vitol, *Zavod. Lab.*, **29,** 649 (1963); through *Anal. Abstr.*, **11,** 2492 (1964
79. J. Perdijon, *Rev. Met.* (Paris), **63,** 27 (1966); through *Chem. Abstr.*, **68,** 26655 (1960),
80. M. E. Smith and G. R. Waterbury, *TID-11794*, U.S. AEC, 1960; throug *Chem. Abstr.*, **59,** 10755h (1963).
81. G. J. Kamin, J. W. O'Laughlin, and C. V. Banks, *Anal. Chem.*, **35,** 1053 (1963
82. V. A. Fassel, W. E. Dallman, R. Skogerboe, and V. M. Horrigan, *Anal. Chem* **34,** 1364 (1962).

CHAPTER

14

FERROALLOYS

LAWRENCE A. DE'AETH

Union Carbide Corporation, Mining and Metals Division
Niagara Falls, New York

CONTENTS

I. METALLURGICAL ASPECTS

The inclusion of a chapter on ferroalloys in this book points to the need for information on analytical techniques for the determination of gases in these materials. Iron and steel manufacturers are looking more closely at the purity of the alloys used in making steel and thus need to know the gas content of the alloys used.

A ferroalloy can best be defined as "irons so rich in some particular element, other than carbon, that they are used as vehicles for introducing that element in the manufacture of iron and steel" [1]. The term *alloy additives* refers to alloys of low iron specification, such as aluminum-vanadium and nickel-boron. These can generally be analyzed by the methods used for ferroalloys.

A detailed discussion on the metallurgy and uses of the various alloys is beyond the scope of this chapter. A list of some common ferroalloys and their nominal compositions is given in Table 14.1.

TABLE 14.1. Nominal Composition of Some Common Ferroalloys [2]

Ferroalloy	Nominal Composition
Standard ferromanganese	74–82% Mn and not more than 1.25% Si and 7.50% C
Low carbon ferromanganese	80–85% Mn
Silicomanganese	65–68% Mn, 18–20% Si, 1.5% C max.
Spiegeleisen	16–28% Mn, 6.5% C max.
Ferrosilicon	50% Si
Ferrochromium	65–72% Cr, 2% Si max.
Ferrovanadium	35–55% V
Ferromolybdenum	55–75% Mo
Ferrotitanium	40% Ti
Zirconium	12–15% Zr, 39–43% Si; 35–40% Zr, 37–52% Si
Ferroniobium	50–60% Nb, up to 8% Si
Ferrotantalum–niobium	20% Ta, 40% Nb

There is disagreement as to the effects of impurities in alloy additives on the steel product. However, the degree of contamination by interstitials in a ferroalloy is usually regarded as important by the consumer.

Two types of inclusions in ferroalloys are those that are natural or indigenous and are formed in molten metal during cooling and solidification and those that are exogenous and originate from slag and refractories entrapped during solidification. Producers of ferroalloys try to minimize exogeneous contaminants resulting from the mechanics of smelting and pouring molten metal.

The formation of nonmetallic inclusions in ferroalloys is discussed by Saunders et al. [3]. A correlation is seen between the oxygen content, as determined by vacuum fusion, in systems of low oxygen solubility, and the amount of inclusions measured metallographically. In alloys where the solid solubility of oxygen is high, e.g., ferrotitanium, metallographic ratings can lead to false estimates of oxygen content. In experimental steel heats, it has been found that there was no relationship between the inclusions in the alloy additions and those present in the steel even when using materials of abnormally high non-metallic matter.

In a discussion of hydrogen in ferroalloys (4), it was estimated that less than 3% of the total hydrogen available in the production of a commercial heat of stainless steel is introduced by the ferroalloys.

II. STANDARDS AND SAMPLING

There are no standards available for gases in ferroalloys. Perhaps the best approach to this problem has been the cooperative analysis of a specified omogeneous alloy by a number of laboratories, followed by use of this lloy as a reference material.

The sampling of ferroalloys for gas analysis is one of the most important teps of the analytical procedure. The problems involved in obtaining a truly epresentative sample of the particular alloy under investigation should be ecognized. Also, two considerations in preparation of a ferroalloy sample for as analysis are (1) the danger of contaminating the analytical sample with xygen, hydrogen, and nitrogen during the crushing or grinding procedures nd (2) a risk of picking up atmospheric gases during storage of the alloy. The merican Society for Testing and Materials (ASTM) [5] has established a ampling procedure for ferroalloys for chemical analysis; however, the rushing and grinding methods used can seriously contaminate many alloys uch that the gas content will no longer be representative of the original roduct.

The ideal particle size for an analytical sample is that equivalent to what is sed in making steel. Thus, if minus 2-in. pieces of ferrosilicon are added to he steel heat, the analytical sample should consist of minus 2-in. material. t is readily seen that this method of sampling is impossible in most cases. n order to minimize the effects of crushing and grinding and still obtain a epresentative sample, the Ferroalloy Division of Union Carbide Corporation as followed the ASTM procedure for sampling and stopped at the minus -mesh stage (8 mesh = 0.0926 in. = 2380 μm). This fraction is then used or the analytical sample for gas analysis.

Molten samples, such as are taken by steelmakers for determination of ydrogen in steel, are impractical for analysis of ferroalloys. The gas content f the solid ferroalloy, as it is used in steelmaking, is of much more signifi- ance.

Saunders et al. [4] have shown the effects of crushing on several ferroalloys. slight increase in hydrogen is noticed in ferrochromium in crushing from .5-in. lumps to minus 40 mesh. This increase is more pronounced in alloys f the more active metals such as titanium and niobium. Manganese products re also prone to hydrogen contamination, but the susceptibility tends to ecrease with higher nitrogen and carbon contents of the product. Limpach nd Marincek [6] showed that there was an increase in gas content with a ecrease in particle size in all the alloys they tested. Also, many of the ferro- lloys were contaminated during storage through the adsorption of gases rom the atmosphere, particularly water vapor. An appreciable pickup of xygen was noted in the storage in air of 75% ferrosilicon. Ferrochromium nd 90% ferrosilicon are quite stable with respect to hydrogen content even hen exposed to extreme conditions of moisture [4].

In general, the gas content of a ferroalloy is related to the affinity of the rimary constituent for oxygen, hydrogen, and nitrogen. There are many types, grades, and sizes of alloy additives, and to outline a single sampling technique applicable to all would be impractical. One should begin the procedure with the ASTM method for sampling bulk lots and then proceed to the choice of an analytical sample, being careful to minimize contamination during crushing and storage.

III. ANALYSIS

A. INTRODUCTION

With increased emphasis on the production of clean steel by vacuum-induction melting, vacuum-arc melting, and vacuum-degassing practices, iron and steel producers are obligated to investigate all possible sources of contaminants. While there are few actual specifications for gas content of ferroalloys, many users want a gas analysis of materials that are added to steel.

The analysis of ferroalloys for gaseous impurities differs from most other gas-in-metal analyses, in that there is a very wide range of gas contents in these alloys. The analytical chemist may have to determine oxygen contents as low as 100 ppm or as high as several percent. Hydrogen and nitrogen contents also vary widely depending upon the particular alloy.

B. HYDROGEN

Hydrogen can be determined in ferroalloys by methods of hot extraction [7–12], vacuum fusion [6, 9, 10, 12, 13], or inert-gas extraction [14]. In hot extraction procedures, the sample is heated in vacuum to a temperature (usually 900–1200°C) below the melting point, which is sufficient for quantitative diffusion of the hydrogen out of the metal into a collection chamber for measurement. At the higher temperatures hydrogen extraction is more rapid, but it is accompanied by volatilization of high-vapor-pressure components of the alloy and by expulsion of carbon monoxide and nitrogen from the solid sample. Lower extraction temperatures minimize these interferences but tend to give low results for hydrogen because of incomplete extraction. Lassner obtained quantitative yields of hydrogen from ferrochromium alloys at 1100–1200°C [10], from ferromanganese at 1000–1100°C [11a], and from undried ferrosilicon powder at 1850°C [11b]. Etterich et al. [12] reported that hydrogen removal from ferromanganese, ferrochromium, and ferrotitanium at 600–1050°C was not always successful, and Christensen and Gjermundsen [9] showed that significant amounts of hydrogen remained in ferrosilicon alloys after vacuum heating at 900°C. Hot extraction at 600–700°C has yielded reproducible results for hydrogen in ferromanganese, ferrotungsten, and ferrosilicon [12–75% Si) [8], indicating that only hydrogen was evolved; however, at higher temperatures (800–900°C), carbon monoxide and nitrogen were also obtained. Even though carbon monoxide and nitrogen are not expected to be produced in a given hot-extraction procedure, the collected gas should be at least occasionally analyzed to verify hydrogen purity.

In vacuum fusion procedures, hydrogen is quantitatively released with oxygen and nitrogen. The gas mixture is collected, measured, and analyzed by gas chromatography, mass spectrometry, or chemical separation methods. The determination of hydrogen by vacuum fusion has been reported for iron–manganese alloys at 1550°C [13], iron–chromium alloys at 1400°C [10], and other ferroalloys [6, 9, 12].

An inert-gas-extraction–gas-chromatographic technique was used by Kashima and Yamazaki [14] to determine hydrogen in ferrous metals. The hydrogen diffused from the heated sample into a stream of argon, which flowed through a cold precolumn, where the hydrogen was adsorbed and temporarily stored for subsequent desorption and measurement by gas chromatography.

C. OXYGEN

Union Carbide Corporation, Ferroalloy Division, has long been interested in the gas content of ferroalloys, and the analytical techniques employed follow closely those proposed by Sloman [15] and Hatfield and Newell [16]. Fusion of the sample in a carbon-saturated iron bath at 1800–2000°C for approximately 15 min works very satisfactorily and is applicable to a variety of alloys. The sample is usually placed in a high-carbon steel capsule, and tin is added for the analysis of manganese-containing alloys.

Donovan et al. [17] used an iron bath at 1750°C and wrapped the samples in nickel foil for the determination of oxygen in ferrosilicon, ferromolybdenum, and ferroaluminum. Fuchs et al. [18] also applied the vacuum-fusion method to the determination of gases in ferroboron, ferromolybdenum, ferroniobium, ferrotantalum, ferrotitanium, and ferrovanadium. Hydrogen and oxygen were determined simultaneously and, in some instances, also nitrogen. Temperatures were in the 1500–2100°C range, and the analysis of the evolved gases was accomplished by gas chromatography. There was considerable evidence in such investigations [17–19] of the effects of particle size and sample preparation on the concentration of gas in a particular sample.

Fassel and co-workers [20] analyzed standard steel samples for oxygen by vacuum fusion, using an iron bath at 1650°C and 2150°C, and a platinum bath at 1850°C. They then studied the applicability of their platinum-bath procedure to the determination of oxygen, and particularly nitrogen, in several other steels and ferroalloys. A nickel–iron bath at 1600–1700°C was used by Lassner [10] to determine oxygen in ferrochromium. Reproducibility was improved by washing the samples with carbon tetrachloride (note toxicity of solvent) and drying at 200–300°C. Shitikov and co-workers determined oxygen in ferrochromium [21a] and ferromanganese [13] by dissolving the sample in a nickel bath at 1650°C containing a small amount of tin. Nitrogen and hydrogen were also determined in ferrochromium [21b].

A carbon-chip, tin–nickel-foil (vacuum-fusion) technique was used by Sudo et al. [22, 23] to determine oxygen in ferrosilicon alloys. Carbon chips (10–16 mesh) and tinfoil were placed in a covered graphite crucible. After degassing, a 0.1-g sample, wrapped in nickel foil, was placed in the crucible at 1100–1200°C and the temperature was raised to 1750°C. The results agreed well with those obtained by fast-neutron radioactivation analyses performed on the same alloys.

Little has been reported in the literature on the determination of oxygen in ferroalloys by inert-gas fusion, although the technique can generally be substituted where vacuum-fusion procedures are applicable.

D. NITROGEN

Nitrogen is almost universally determined in ferroalloys by a suitable chemical method. Other less commonly used methods include vacuum fusion, inert-gas fusion, arc extraction, and isotope dilution.

The procedure followed in the usual chemical method is to dissolve the sample in a suitable acid and decompose the insoluble nitrides and carbonitrides by addition of another agent or agents. Such decompositions have been carried out in a number of ways; heating with H_2SO_4, K_2SO_4, and $CuSO_4$ [24]; $HClO_4$ [25]; or with H_3PO_4, H_2SO_4, and $K_2S_2O_8$, followed by fusion of the insolubles with K_2SO_4. After separation of ammonia by distillation, the ammonia is determined photometrically or titrimetrically.

The use of a trap containing HCl or H_2SO_4 is recommended to prevent loss of ammonia when the sample is treated with H_3PO_4 or H_2SO_4 [26]. In manganese alloys the nitrogen compounds are not converted completely to NH_4Cl when using HCl, and dilute H_2SO_4 is recommended to dissolve the sample [27].

A direct nesslerization method was used by Newell [28] for nitrogen in ferrochromium. Karp and co-workers [29] determined nitrogen in silicon steels by the caustic-fusion method, and this method should be applicable to many silicon-containing ferroalloys.

Fassel et al. [20] made a critical comparison of the nitrogen results obtained by a modified platinum-bath vacuum-fusion procedure and by the Kjeldahl, sotope-dilution, caustic-fusion, and arc-extraction techniques. Their results ndicated quantitative measurement of nitrogen in several steels and ferro-lloys by the vacuum-fusion procedure, the average analysis time of which vas about 20 min per sample.

An isotope-dilution procedure, using NH_4^+ enriched with ^{15}N, was used or determination of nitrogen in titanium, manganese, and chromium ferro-lloys [30]. NH_4^+ in the distillate was oxidized to N_2 by hypobromite and $^5N^1$ was assayed by determining the $29^+/28^+$ ion current ratio in a dual-ollector, isotope-ratio mass spectrometer. Pearce [31] compared the methods f isotope dilution, vacuum fusion, and chemical (Kjeldahl) for the determina-ion of nitrogen in a silicon–iron alloy. The results indicated superiority of he isotope-dilution method for precision and accuracy.

IV. RECOMMENDED METHODS

. HYDROGEN AND OXYGEN

In view of the findings of other workers and our own experience at Union Carbide Corporation, Ferroalloy Division, it is not possible to suggest a nethod for the determination of gaseous impurities in ferroalloys that would ncompass the entire field. It is reasonable to state that vacuum-fusion tech-iiques are generally applicable to the determination of oxygen and hydrogen, nd chemical methods are generally employed for nitrogen determination.

The vacuum-heating method for hydrogen in chromium and manganese lloys can be used at 800–1000°C, followed by analysis of the evolved gas. Oxygen and hydrogen can be determined simultaneously by vacuum fusion, nd this is the usual approach to the analysis. Sampling of the various alloys s a major problem; to ensure accurate results a large (5–10 g) sample should e used when possible. The vacuum-fusion apparatus must be capable of andling these samples as well as large and small volumes of gas. Encapsula-ion of the alloy in iron or steel capsules promotes solution in the bath and essens the chance of reaction of the evolved gases with metal vapors.

An iron or iron–tin bath for manganese, chromium, and titanium alloys s recommended at approximately 1500–2000°C, depending upon the alloy. The choice of sample-size, metal bath, temperature, extraction time, and nethod of gas analysis is governed by the type of ferroalloy and by its pproximate gas content.

. NITROGEN

Nitrogen in ferroalloys may be determined by vacuum-fusion techniques, lthough the usual approach is to use the chemical method as described in

Section III.D and Chapter 8. The latter method consists of solution of th sample, decomposition of the insoluble portion, distillation, and finally deter mination of ammonia by photometric or titrimetric procedures. The caustic fusion method [29] seems particularly applicable to silicon-containing alloys

REFERENCES

1. W. H. Dennis, *Metallurgy of the Non-ferrous Metals*, 1st ed., Pitman, Londo 1954, p. 379.
2. U.S. Steel Corp., *The Making, Shaping and Treating of Steel*, 9th ed., H. E McGannon, Ed., U.S. Steel Corp., Pittsburgh, 1971.
3. E. R. Saunders, W. D. Forgeng, and J. W. Farrell, *Trans. AIME Elec. Furn Proc.*, **18,** 93 (1960).
4. E. R. Saunders, J. L. Lamont, and G. Porter, *Trans. AIME Elec. Furn. Proc.* **18,** 76 (1960).
5. American Society for Testing and Materials, *Annual Book of Standards*, Pt. 32 *Chemical Analysis of Metals; Sampling and Analysis of Metal-Bearing Ores* Specification E32, Philadelphia, 1970, p. 123.
6. R. Limpach and B. Marincek, *Archiv. Eisenhuttenw*, **31,** 639 (1960).
7. E. Piper, *Stahl Eisen*, **73,** 817 (1953).
8. K. T. Kurochkin, and L. M. Melnikov, *Tr. Ural Politekh. Inst. in S. M. Kirova* **52,** 105 (1955).
9. N. Christensen and K. Gjermundsen, *Iron and Steel*, **22,** 243 (1959).
10. E. Lassner, *Z. Erzbergbau Metallhuttenw.*, **19,** 526 (1966); through *Chem Abstr.*, **66,** 121758 (1967).
11. (a) E. Lassner, *Z. Erzbergbau Metallhuttenw.*, **20,** 420 (1967); through *Chem Abstr.*, **68,** 26623 (1968); and (b) E. Lassner, *Erzmetall*, **23,** 281 (1970); throug *Chem. Abstr.*, **73,** 47809 (1970).
12. O. Etterich, H. Taxhet, and W. Thomich, *Arch. Eisenhuttenw.*, **35,** 613 (1964) through *Chem. Abstr.*, **61,** 10369g (1964).
13. Yu. A. Shul'te, N. A. Gederevich, and V. S. Shitikov, *Zavod. Lab.*, **34,** 119 (1968); through *Chem. Abstr.*, **70,** 63960d (1969).
14. J. Kashima and T. Yamazaki, *Bull. Chem. Soc. Jap.*, **41,** 2382 (1968); throug *Chem. Abstr.*, **70,** 8623w (1969).
15. H. A. Sloman, *J. Iron Steel Inst.* (London), **148,** 235 (1943).
16. W. H. Hatfield and W. C. Newell, *J. Iron Steel Inst.* (London), **148,** 351 (1943)
17. P. D. Donovan, J. L. Evans, and G. H. Bush, *Analyst*, **88,** 771 (1963).
18. A. Fuchs, H. Reinhard, J. Niebuhr, and R. Beck, *Arch. Eisenhuttenw.*, **34** 361 (1963).
19. J. Calmettes, L. Glenat, and J. Bertin, *Centre Doc. Sider. Circ.*, **3,** 749 (1963)
20. V. A. Fassel, F. M. Evens, and C. C. Hill, *Anal. Chem.*, **36,** 2115 (1964).
21. (a) V. S. Shitikov and N. A. Gederevich, *Zavod. Lab.*, **33,** 1503 (1967); throug *Chem. Abstr.*, **69,** 8174e (1968); and (b) V. S. Shitikov, N. A. Gederevich, an Yu. A. Shulte, *Zavod. Lab.*, **36,** 149 (1970).

22. E. Sudo, M. Saito, and H. Inoue, *Jap. Analyst*, **17,** 1364 (1968); through *Chem. Abstr.*, **70,** 73940m (1969).
23. E. Sudo and M. Saito, *Jap. Analyst*, **17,** 1482 (1968); through *Chem. Abstr.*, **70,** 63964h (1969).
24. T. K. Cunningham and H. L. Hammer, *Ind. Eng. Chem., Anal. Ed.*, **11,** 30314 (1939).
25. C. M. Johnson, *Iron Age*, **134,** 688 (1944).
26. B. A. Generozov, *Zavod. Lab.*, **13,** 314 (1947).
27. H. Kempf and K. Abresch, *Arch. Eisenhuttenw*, **19,** 17 (1948).
28. W. C. Newell, *J. Iron Steel Inst.* (London), **152,** 333 (1945).
29. H. S. Karp, L. L. Lewis, and L. M. Melnick, *J. Iron Steel Inst.* (London), **200,** 1032 (1962).
30. H. G. Staley and H. J. Svec, *Anal. Chim. Acta*, **21,** 289 (1959).
31. M. L. Pearce, *Trans. Met. Soc. AIME*, **227,** 1393 (1963).

CHAPTER

15

IRON AND STEEL

WILLIAM R. BANDI, JOHN F. MARTIN, AND LABEN M. MELNICK

United States Steel Corporation
Research Laboratory
Monroeville, Pennsylvania

CONTENTS

I. INTRODUCTION

An excellent review on gases in steel has been written by Bruch [1]. The article covers the influence of hydrogen, oxygen, and nitrogen on steel; sampling and analyses for these gases; and application of the results.

II. METALLURGICAL ASPECTS

A. HYDROGEN

Hydrogen in steel comes mainly from moisture in iron ore, from ferro-alloys, from rust on scrap, and from the combustion of fuel if the steel is made by a gas-fired process [2]. This element can have a deleterious effect on the mechanical properties of steel. Steels generally have hydrogen concentrations ranging from 1–10 ppm [3]. In concentrations greater than 2 ppm hydrogen plays an important role in a phenomenon known as *flaking*. This is manifested as internal cracks or bursts usually occurring during cooling from rolling or forging and is more pronounced in heavy sections and in higher carbon steels [4]. Such defects have resulted in the disintegration of large rotors during use in power plants because of the internal stresses that occur. Modern steel-making practices (e.g., vacuum casting) produce steels of lower hydrogen contents so that this need no longer be a problem. Reduction of hydrogen concentration can also be effected by slow cooling after rolling and forging.

The amount of hydrogen removed from steel by diffusion is a function mainly of temperature, time, and size or geometry of the piece of steel. Further, hydrogen diffuses much more rapidly from ferritic steel than from austenitic steel. In the production of sound iron castings, the hydrogen content should be kept below 2 ppm to prevent pinholes and general porosity [5].

The ability of hydrogen to diffuse at an appreciable rate from ferritic steel lessens the incidence of embrittlement of steel. During heat treatment under certain conditions of temperature and atmosphere, hydrogen is absorbed by steel, more particularly in high-carbon steel. Tension tests in such steels performed shortly after rolling will show low ductility, which will increase on aging at room temperature or after shorter times at slightly elevated temperature [4].

Hydrogen embrittlement may arise when cold working follows too soon after pickling, during which nascent hydrogen penetrates the steel [6]. As implied above, this effect is not permanent and may be eliminated by aging. Hydrogen embrittlement occurs mainly in martensitic grades of steel; it is less acute in ferritic grades and virtually unknown in austenitic grades [7]. Also, it generally increases with hardness and carbon content. Hydrogen, at sufficiently high partial pressures, has been known to attack steel in service at elevated temperatures, causing decarburization and fissuring [8].

In cast iron, any gas pockets or blowholes caused by hydrogen will lead to weakness [4]. The effect of hydrogen on cast iron has been studied by the British Cast Iron Research Association [9]. Further, a review on the effect of hydrogen on iron and steel has been written by Buzzard and Cleaves [10].

B. NITROGEN

All steels contain nitrogen. The amount present depends on the method of steel production; type, amount, and manner of addition of alloying elements; how the steel was cast or poured; and finally whether nitrogen was purposely added. Table 15.1 classifies most commercial steels according to production method and alloy content and may be used as a rough guide for predicting the expected nitrogen concentration range. It is necessary, however, to emphasize that steels may be found that vary widely from this expected range. Steels melted in vacuum or cast in vacuum can be expected to contain less nitrogen than air-melted or air-cast steels. Cast irons can have nitrogen contents as high as 500 ppm.

Some of the factors affecting the solubility of nitrogen in a molten-steel bath are the bath temperature, partial pressure of nitrogen over the bath, and composition of the steel. Nearly pure liquid iron in equilibrium with nitrogen at a pressure of 760 torr and a temperature of 1600°C will dissolve about 0.045% nitrogen. Pelke and Elliott [11] have shown that a liquid-steel

TABLE 15.1. Nitrogen Contents of Commercial Steels

Type of Steel	Nitrogen (ppm)
Plain-carbon–basic-oxygen process, open-hearth, or electric furnace	30–100
Alloy–basic-oxygen process, open-hearth, or electric furnace	30–200
Stainless	100–500
High-alloy, vacuum-melted	$<$10–100
Austenitic Mn–Cr–N (N added)	2000–4000
Plain-carbon–basic converter, air-blown Bessemer	150–250

bath containing 20% chromium will dissolve six times as much nitrogen as a pure-iron bath; this demonstrates the marked effect of steel composition. Langenberg [12] has developed a formula for predicting the concentration of nitrogen in molten steel based on its composition.

The rate of solution of nitrogen in the liquid bath is dependent on bath composition, surface-area–volume ratio of steel, turbulence of the bath, and temperature of the molten steel. Oxygen concentration has a pronounced effect on the rate of nitrogen solution: the lower the oxygen content, the higher the rate of nitrogen solution.

The concentration of nitrogen soluble in the solid state is only a fraction of that soluble in the liquid state. When the nitrogen content of molten steel exceeds that for the solid material, porous ingots may be formed in the casting process. Adams et al. [13] have discussed this problem.

Nitrogen may be present in steel interstitially, as a metal nitride, or as nitrogen dissolved in a second phase, such as a metal carbide. Nitrogen soluble in solid steel exists in the iron lattice and at dislocations [14, 15]. Interstitial nitrogen imparts many of the same properties, such as hardening, to iron that interstitial carbon does. In alloyed iron the effect of interstitial nitrogen is not as clearly defined, because of the presence of the alloying metal. Most of the nitrogen present in steel exists as what may be defined generally as a metal nitride. Some examples of the metallurgical significance of nitrides are briefly cited.

A steel is said to have undergone strain aging when after having been stored for some time, the steel cannot be deep drawn (e.g., into automobile fenders), because it tears and does not draw uniformally in all directions [7]. This is caused by large grain size and precipitation of Fe_4N at the grain boundaries. Strain aging is also accompanied by an increase in the yield strength and a loss of ductility and notch toughness as the steel ages [16]. The addition of a strong nitride-former such as aluminum [14], boron [17], titanium [18], or vanadium [19–21] can eliminate strain aging.

The formation of chromium nitride (Cr_2N) at the grain boundaries in stainless steel depletes the chromium content of the steel at the boundary and causes a phenomenon known as intergranular corrosion. This detrimental effect can be prevented by the addition of titanium, which forms titanium nitride preferentially [22].

Additions of boron are sometimes used to increase the hardenability of steel [23]. However, when boron and nitrogen are present in the same steel, they can form boron nitride and the hardenability effect is no longer achieved [17]. Therefore, titanium or zirconium is added to the steel to combine with the nitrogen before the boron addition is made.

The formation of Si_3N_4 in transformer-grade silicon steel affects the grain orientation and the magnetic permeability of the steel [24, 25a]. The formula

Si_3N_4 is used throughout this chapter for both the alpha and beta forms of this compound, which reportedly are $Si_{11.5}N_{15}O_{0.5}$ [25b] and Si_6N_8 [25c], respectively.

One example where nitrogen is an alloying element [26] rather than part of a second phase is in the manganese austenitic stainless steels, which vary in composition according to the producer but always contain high amounts of chromium and manganese, and up to 0.5% nitrogen. It is known that nitrogen lowers the upper critical temperature at which transformation from the austenitic structure to ferritic structure occurs [23]. The austenitic structure is therefore stable at ordinary ambient temperatures. The properties of these austenitic steels have been described by Langenberg [26].

TABLE 15.2. Nitrides Found in Steel

Simple	
Fe_4N	α-Si_3N_4
ZrN	β-Si_3N_4
SiN	TaN
Cr_2N	hexagonal NbN
CrN	cubic NbN
TiN	Nb_2N
VN	BN
	AlN
Complex	
$(Cr, Mo)N_x$	NbC_xN_y
Si_2ON	VC_xN_y
TiC_xN_y	$NbO_{0.1}N_{0.9}$
	$NbO_{0.6}N_{0.3}$

Most of the elements that are now added to special grades of alloy steel can form nitrides under the proper conditions. These elements include boron, aluminum, manganese, chromium, vanadium, molybdenum, titanium, tungsten, niobium, tantalum, zirconium, silicon, and the rare earths. When one considers that there are several simple and complex nitrides for many of the nitride formers, it appears possible that as many as 75 nitrogen compounds can form in steels. Table 15.2 lists some typical nitride compounds that have been found in steel. The complex compounds may consist of two metals with nitrogen or one metal with nitrogen and carbon or oxygen.

Knowledge of the complex nitrides is very limited, and definite formulas are not really descriptive of the phase that exists in the steel. The use of an indefinite formula to describe an interchangeable range of carbon and nitrogen with the designation carbonitride has also led to confusion. An example of this was reported by Lund [27], who said that the term *carbonitride* is sometimes used to describe two different compounds. Titanium carbide with nitrogen dissolved in the lattice responds to heat treatment as normal titanium carbide does, while titanium nitride with carbon dissolved in the lattice is like normal titanium nitride and does not respond to heat treatment. The same sort of confusion may exist with other carbonitrides. Even the simple nitrides are not always definite stoichiometric compounds in steel. Turkdogan and Fenn [28] noted the nonstoichiometric nature of vanadium nitride, while Hume-Rothery et al. [29] described titanium nitride as a phase containing from 10–25% nitrogen. A similar situation with niobium and nitrogen and niobium with oxygen and nitrogen is also known to exist.

Some nitrides, formed in the bath or ingot, are thermally stable at the temperature of molten iron. Once formed, these nitrides are not decomposed during subsequent rolling or heat-treating operations and are therefore described as non-heat-treatable nitrides. Examples are titanium nitride [26, 27, 30] and zirconium nitride [27]. Other nitrides, such as those of vanadium and aluminum [31], chromium [32], and silicon [33], are designated as heat-treatable because they may be decomposed or formed during heat treatment. Some nitrides do not decompose during heat treatment but undergo a change in crystal structure [34–36]. From the analytical viewpoint this is important, since it means that the nitrogen in a bath may exist in a form different from that in the heat-treated steel. The size of the nitride particles formed is also of great metallurgical and analytical significance because only fine nitride particles that have not agglomerated prevent grain growth in the steel and thus give it superior yield strength. Dahl et al. [37] have reported aluminum nitride particles as small as 100 A, as did Hsiao [38]. Particles of niobium and vanadium carbonitrides smaller than 50 A have been studied by Gray and Yeo [39]. Very fine particles of silicon nitride caused analytical difficulty to Karp et al. [33].

Several attempts [31, 40, 41] have been made to predict the existence and concentration of nitrides present, by developing thermodynamic expressions based on the free energies of formation of the various metal nitrides. However, such predictions are somewhat meaningless unless the rate of formation of the particular nitride is also known. Konig et al. [42] have demonstrated how vanadium nitride and aluminum nitride contents depend on the heat-treating times and cooling rates for the steel and have further shown that vanadium nitride can exist even though thermodynamic calculations predict that only aluminum nitride should be present. Beeghly [31, 40] has also shown this. Several workers have studied the rate of formation of aluminum nitride [40, 43] and vanadium nitride [21, 31, 44]. Turkdogan and Ignatowicz [44] have shown that one week is required to reach equilibrium in formation of chromium nitride at moderate temperatures.

C. OXYGEN

Oxygen in steel can originate from a variety of sources such as air, moisture in the air or in the addition agents, and oxide in scrap [5]. Lesser concentrations come from broken or eroded refractories; usually the particles from these sources are so relatively large they will rapidly rise to the surface of the molten metal in the ladle or mold and can be separated [3]. Oxygen, like hydrogen, can affect the mechanical properties of steel deleteriously. Not only is the oxygen concentration important in this regard but also the number, type, size, and distribution of the oxygen-bearing inclusions [45]—namely, metal oxides, silicates, aluminates, oxysulfides, and similar second-phase inclusion compounds. Steels and cast irons generally have oxygen concentrations varying from about 10 to 800 ppm, the value depending upon whether the steel is of the rimmed, semikilled, or killed type and the method by which the steel is made (e.g., basic oxygen process, vacuum induction, consumable electrode remelting). It is desirable to remove oxygen from steel, because during solidification, carbon and oxygen in solution react to give carbon monoxide, which can cause blowholes [4]. Also, on cooling, oxygen can come out of solution as FeO, MnO, and other oxide inclusions that may impair the hot and cold workability and the ductility, toughness, fatigue resistance, and machinability of steel [5]. Oxygen, along with nitrogen and carbon, can cause aging or a spontaneous increase in hardness at room temperature, which is accelerated by raising the temperature.

In cast iron, oxides may react with carbon while the casting is solidifying, thus causing porosity and weakness in the product [4].

III. HISTORICAL DEVELOPMENT

A. HYDROGEN

The development of methods for determining hydrogen in steel is covered mainly in Chapter 7. After years of evolution, the preferred method for hydrogen in steel appears to be extraction from the sample in the solid state, in vacuum, or in an inert gas, for which several commercially available analyzers have been used. The single-sample-per-crucible technique involving fusion in inert gas seems to be gaining acceptance.

B. NITROGEN

The evolution of methods for determining nitrogen in steel is treated, in part, in Chapters 4 and 8. The first attempts to determine the nitrogen content of steel were made over a hundred years ago. Baker and Stuart [46] in 1864 passed hydrogen over heated steel to form NH_3, which was then determined titrimetrically. Results and blanks were scattered and inaccurate, resulting in an erroneous conclusion that nitrogen was rarely present and had no effect on the properties of steel. About 1872, Allen [47] used a modification of a method by Boussingault [48] to analyze for nitrogen in steel. Allen's disclosure that steel contained nitrogen was not completely accepted, and the concentration of nitrogen found was not considered to be metallurgically significant. His method consisted of solution of the steel in hydrochloric acid, followed by distillation of ammonia from an alkaline solution and treatment of the distillate with Nessler reagent [49]. While many analysts have used iodometric [50–52] or acidimetric [53] titration of the distilled ammonia and others have eliminated the distillation step [54–57], the original method, with the use of Nessler photometry, is commonly used to determine nitrogen in unalloyed carbon steels. There has, however, been a tendency to replace this method with more sensitive photometric procedures. One of these utilizing indophenol [58, 59] has been known for many years, but because of problems with the stability of reagents, it has been accepted only recently [60]. Extraction of the indophenol was studied by Kammori and co-workers [61]. Others have studied the photometric determination of nitrogen with bis-pyrazolone–chloramine T [62–64].

With the production of stainless steels and the alloying of carbon steels, a voluminous amount of data appeared in the literature that indicated that metal nitrides insoluble in hydrochloric acid were present in many steels. In addition, many workers stated that alloy cast irons had carbonitrides that could not be dissolved in the presence of graphite. Most of the literature concluded that these metal nitrides could be dissolved by a modified Kjeldahl method that included fuming the sample or the insoluble portion of the sample in sulfuric acid with the addition of sulfate salts [65–71]. Other workers [62, 72, 73] recommended modification of the Dumas oxidative fusion method. The use of perchloric acid [22] or oxidizing agents in the presence of mineral acid [67] was recommended some time ago to decompose stable nitrides. Other investigators [74–79] attempted to heat-treat the steel in order to make all the nitrogen soluble in hydrochloric acid. Karp and co-workers [33] used a caustic-fusion procedure [80] to decompose α- and β-Si_3N_4. Recently there has been a renewed interest in the use of perchloric acid for the dissolution of steels for determination of nitrogen. The basis of this renewed interest is the realization that the formation of hypochlorous acid, which interferes with the nitrogen determination [81], can be prevented by proper choice of dissolution vessel [82, 83] and control of dissolution temperature [84]. By dissolving the sample and fuming lightly in an open beaker to allow the evolution of chlorine, the formation of hypochlorous acid is prevented.

As the metallurgical understanding and complexity of steels grew, the determination of individual nitrides in the steel became increasingly important.

Significant contributions to the identification and determination of nitrides were made by Beeghly [31, 40, 85], who showed that nitrides in low-alloy steels, with the exception of iron nitride, could be isolated from the matrix by use of iodine or bromine dissolved in an aliphatic alcohol or ester (so-called *ester-halogen method*).

The use of different dissolving acids to decompose nitrides in steel has often been successfully employed to help in identification and determination of certain metal nitrides [21, 31, 69–71, 86, 87]. These determinations are metallurgically significant because the results can be used to explain the development of certain mechanical properties of the steel so that proper heat-treating methods can be used. However, the identification and determination of the nitrides is limited to simple alloys where not many nitride formers are present. With more complex alloys, where several nitrides might be present in conjunction with metal carbides, sulfides, and oxides, the determination of a particular nitride is somewhat questionable. Nevertheless, quite often the determination of nitrogen, soluble in dilute hydrochloric acid or in the subsequently added sodium hydroxide solution used for the distillation of ammonia, is made and compared with the total nitrogen result so that some indication as to how the nitrogen is present may be discerned.

The use of heat treatment, as mentioned earlier, to convert the nitride insoluble in hydrochloric acid to a soluble form is unreliable because non-heat-treatable nitrides are not affected. Nevertheless, the determination of heat-treatable nitride nitrogen is of metallurgical significance [31, 32]. To determine the concentration of specific nitrides and carbonitrides in steel, Bandi and co-workers [88–90] have been applying differential thermal analysis–evolved gas analysis to the residues isolated by chemical or electrochemical methods.

The determination of metallurgically dissolved nitrogen in steel is of interest in strain-aging studies. This determination has been performed by heating fine millings of the sample in a stream of hydrogen [86, 91–94] and determining the evolved ammonia coulometrically or photometrically. The magnitude of the results is dependent on the reduction temperature and time, the size of the sample particles, and the composition of the steel with regard to nitride compounds. Dulski and Raybeck [92] obtained best results by heating. -80 mesh (less than 177 μm) millings for 10 min at 550–600°C; unannealed aluminum-killed steels containing less than 0.4% Si and less than 0.03% Cr were analyzed. The results agreed with those obtained by the Beeghly ester-halogen method [40, 85]. Willmer and Zimmerman [86], however, warned that nitrogen pickup could occur in preparing fine particles of sample in air. This, however, has not been our experience with steels of standard composition. They [86] further noted that at reduction temperatures approaching 750°C, any metallurgically dissolved aluminum in the steel can react with the free nitrogen to form AlN. In addition, they noted that ZrN decomposes around 600°C.

Koch et al. [93] recommended reduction at 500°C for 20 min. Fine millings were obtained with a carbide cutting tool. It was noted that AlN particles less than 80 A could not be determined by the Beeghly method but that if the particles exceeded approximately 100 A, results were quantitative. The pore size of their filter medium was not specified. Regardless of the technique used (hydrogen reduction or ester halogen) special care must be exercised in obtaining the results. With the latter method, the metallurgically dissolved nitrogen is determined by subtraction of the combined nitrogen from the total nitrogen value. Although the AlN invariably found in steel is the hexagonal form, which can be dissolved in dilute acid or alkaline solution, Koch [94] has reported a cubic (Al, Fe)N that is insoluble in these solutions.

The determination of nitrogen in steel by classical vacuum fusion or by inert-gas fusion as initially developed has been used by comparatively few workers. Up to about 1965, many laboratories preferred to analyze for nitrogen by chemical techniques that involved less capital expense and were not as complex as vacuum fusion. In addition, chemical and fusion methods often gave discordant values on the same material. Explanations for the poor recovery of nitrogen by fusion were put forth by Goward [95], Gerhardt et al. [96, 97], and Koch and Lemm [98]. This led to the development of the single-sample-per-crucible technique by Gerhardt et al. (spinning crucible) [97, 99, 100], Koch and Lemm (fresh crucible) [98, 101, 102], and Lemm [103]. At about this time Vasserman and Turvotseva [104] developed the pulse-heating method for oxygen in metals using a fresh crucible for each determination. Their work was extended by Goldbeck et al. [105, 106]. One of these techniques is embodied in at least one analyzer manufactured by the following companies: Balzers (spinning and fresh crucible), LECO (pulse heating and fresh crucible), and Strohlein (fresh crucible). The analyzers are described in Chapter 4.

For a number of years investigators in both the United States and the Soviet Union have worked on the development of spectroscopic procedures (see Chapters 5 and 6) for the determination of nitrogen [107–114]. These techniques are not commonly used as yet. Electrical extraction has also been used followed by a gas-chromatographic measurement [115] (see Chapter 5).

Several investigators [116–118] have determined nitrogen in steel by isotope dilution, but this method has not become popular, because it is costly and time-consuming.

C. OXYGEN

After the fundamental work of Walker and Patrick [119], Schenck and co-workers [120, 121], and Jordan and Eckman [122], the classical work of

Sloman and associates [123–132] and many others (see Chapter 4) firmly established vacuum fusion as a standard technique for determining oxygen in steel. This was followed by the application of inert-gas fusion (Chapter 4), activation (Chapter 3), and emission spectrometry (Chapter 5). The use of the single-sample-per-crucible technique came about mainly because of the work on nitrogen (see above and Chapter 4). The isolation of oxide inclusions from steel was studied by many workers; this is reviewed in Chapter 9. Koch and co-workers [27] have provided much of the information concerning isolation, identification, and quantification of many specific oxides in steel. The electrochemical determination of dissolved oxygen in molten steel is described in Section V.C.3.

In comparison to steel, much less has been done with regard to developing methods for oxygen (and also nitrogen) in cast iron, and interestingly almost all of the work on this material has been carried out in Europe and Japan. Specific reference to this work is given in Section V.

IV. SAMPLING, SAMPLE PREPARATION, AND STANDARDS

A. INTRODUCTION

Because of the distinct problems associated with sampling liquid and solid iron and steel for analysis for hydrogen, nitrogen, and oxygen the two phases are treated separately. The sampling of liquid steel has been extensively reviewed by King [133].

B. SAMPLING

1. Introduction

A fundamental difference exists between the method of sampling liquid iron and steel for hydrogen analysis as opposed to oxygen and nitrogen analysis [1]. Most investigators recommend taking a spoon of metal from the bath from which samples are obtained for hydrogen, whereas for oxygen and nitrogen the bath is usually sampled directly. Cassler and Fitterer [134], in comparing bath and spoon tests, have shown that considerably more oxygen is present in the latter, apparently because of surface oxidation.

2. Liquid Steel—Hydrogen

The spoon used in sampling is first lined with slag from the melt. Then the liquid sample is obtained, slag is skimmed off the surface, and sufficient

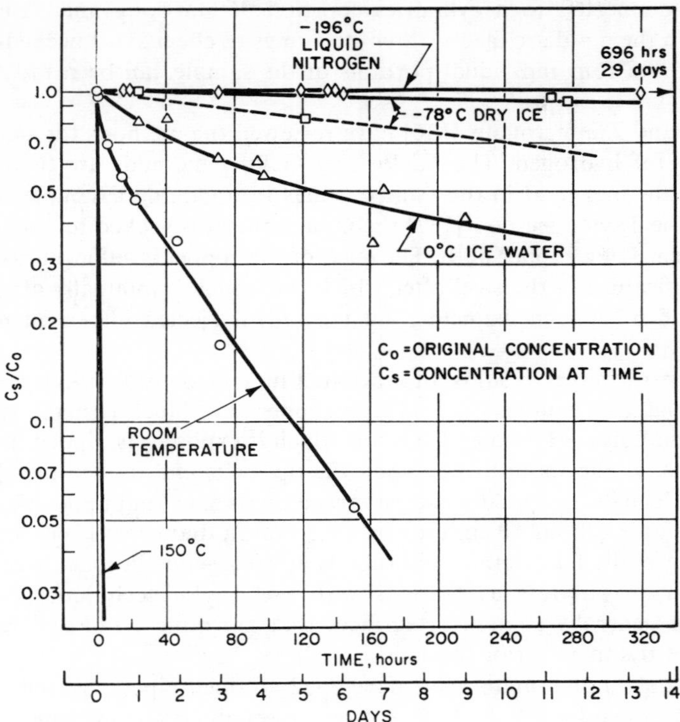

Fig. 15.1. Variation of hydrogen content with time in Armco iron samples (6.3 × 6.3 × 9.5 mm) at different temperatures (Reproduced by permission from the author [135].)

aluminum (usually in the form of wire) is added quickly to halt the evolution of carbon monoxide; this also reduces the rate of evolution of hydrogen. The aluminum addition should weigh between 0.2 and 0.4% of the iron or steel; it is not necessary for fully killed steels. Analytical samples are then taken from the spoon as described below. In each case the solidified sample should be quenched as quickly as possible by moving it about in cold water, dried, tagged for identification, and then placed in a Dewar containing dry ice, liquid nitrogen, or similar medium. The importance of drastically cooling the specimen is shown in Fig. 15.1 [135], which shows the rate of diffusion of hydrogen from a sample of specific geometry and composition under different conditions. The International Committee for the Study and Rationalization of Methods for the Determination of Gases in Steel and Pig Iron [1] has developed a copper mold for receiving samples poured from a spoon or directly from the casting stream. This mold consists of two halves of a cylinder, cut along the axis and held together by steel tongs. A funnel-shaped portion is connected to a cylindrical portion 96 mm long and 12 mm in diameter. In the mold are placed three steel rings to effect short necked-down sections so that four individual portions of the sample can be readily separated.

Willmer and Zimmermann [136] have reviewed the methods for sampling liquid steel for hydrogen. The Taylor and Geffner methods are two of the more popular ones used in the United States in obtaining a sample from a spoon. In the Taylor method [137] the molten metal is sucked into a copper mold by means of a rubber bulb. The mass of the copper is sufficient to cause rapid solidification of the steel after which the sample (nominally about 7.6 cm long and 6.3 mm in diameter, but tapered) is ejected into a quenching bath.

The Geffner sampler [138] is an evacuated tube made of borosilicate glass (Pyrex, Kimax, or equivalent) held in a copper chill block (usually split to facilitate handling). The sealed glass tip, which is exposed, is dipped into the molten metal in the spoon, whereupon the tip melts and molten metal rises in the tube. The sample-holder is then turned on its side, and upon solidification, the sample (about 10 cm long and 6.3 mm in diameter) is ejected into a quenching bath. The Geffner pin tube is advantageous in that there is no space for hydrogen loss as there is with the Taylor technique. Geffner samples are generally solid with surfaces that are quite clean and require little surface treatment prior to analysis.

Willmer and Zimmermann have developed an open pipette sampler. This has been described by Bruch [1] as a Pyrex vessel about 40 mm long with an inside diameter of 10 mm. The pipet is connected at an angle of 45° to a glass tube about 200 mm long, by means of a capillary about 2 mm in diameter. This device can be used to receive a sample from the casting stream or from a spoon.

3. Liquid Cast Iron—Hydrogen

The sampling of cast iron for hydrogen analysis has been reviewed by Dawson and Smith [139]. They noted that considerable amounts of hydrogen may be picked up when the sample is poured into a greensand mold. During such sampling the hydrogen content was markedly increased by increasing manganese (up to at least 5 %) and magnesium (up to at least 0.1 %) contents, and by 0.01–0.05 % aluminum. Although a thin sample sucked into a copper mold apparently gave the most accurate results, the standard technique finally adopted involved a graphite chill mold to produce a pin about 1.43 mm in diameter. Levi and Aleksandrova [140] recommended dipping a 10-mm-diameter quartz tube, containing two 5-mm-diameter holes located 100 mm from a sealed end, into the melt, followed by quenching of the sample.

4. Liquid Steel—Oxygen and Nitrogen

Brower and Larsen [141] reported in 1945 on a study of the sampling of molten steel for oxygen. They concluded that the main source of uncertainty in the measurement of the oxygen content was the sampling, not the analysis. They stated, "The difficulty arises from the fact that any method of sampling is of itself apt to alter the amount of oxygen in solution in the liquid steel because this oxygen is in a readily reactive condition." It is no small wonder that there have since been many investigations on sampling molten steel for oxygen analysis. One of the more frequently used methods of sampling has been the bomb technique described by Huff et al. [142], modified by Gilbert and Bailey [143], used by Langenberg and Snook [144], and Jackson [145], and investigated for comparison purposes by Padget [146]. The mold is generally a two-piece cast-iron shell driven into a ring to provide the equivalent of a monolithic mold. The ring is welded on the end of a long (about 3.5 m) handle. In the bomb, which should have a clean, smooth interior, is placed enough balled-up fine aluminum wire to yield about 0.4% Al in the steel sample weighing about 1000 g. The entrance to the bomb is covered with an aluminum or steel foil cap, which in turn is covered with a block of wood wired to studs welded on the bomb ring. The wood protects the cap during immersion through the slag. The sampler and holder are covered with a mixture of borax and feldspar to prevent welding of bath metal to them.

Samples from partially filled bombs or from bombs with a slaggy appearance should not be used. In addition, samples indicating a local boil and loss of oxygen in the vicinity of the mold mouth should not be used. Heterogeneity of the bomb sample can be reduced by forging it to a round or rectangular bar and sampling the entire cross section.

Bomb samples do not provide for rapid analysis. Also, samples obtained from a spoon have been shown [134] to give high values. Therefore, a variety of immersion samplers have been developed to obtain simultaneously a disk sample for spectrographic analysis and a pin sample for oxygen analysis. One of these is the LECO Diskpin [147] contained in a heavy cardboard protection tube. Aluminum is contained in the mixing chamber at the end of the tube that is covered by a thin metal cap. The pin, which is cast in heat-resistant glass, is 57 mm long and 7 mm in diameter. The thermally insulating character of the protection tube prevents any discernible chilling of the bath in the vicinity of the sampling device. This pin sample may obviously also be used for nitrogen analysis where desirable.

5. Solid Steel—Hydrogen

Because hydrogen diffuses at an appreciable rate from ferritic steels at ambient temperatures, few solid steels are sampled for hydrogen analysis

However, rotor forgings that are deleteriously affected by hydrogen have received much attention (see Section II.A); they are large enough, so that sometimes relatively large concentrations of hydrogen have been retained in this material cast in air. Sampling is accomplished with a long trepan tool, tubular in nature so that a core about 2.5 cm in diameter is obtained; during this operation the steel is cooled intensively with water. Analyses have generally been performed along the entire length of the core.

Coated or plated steels are usually analyzed without removing the coating, because it may actually prevent or impede the loss of hydrogen before analysis.

Methods for sampling solid steel for hydrogen have been reviewed by Zitter and Schwarz [148].

6. Solid Steel and Cast Iron—Oxygen and Nitrogen

In cutting samples for oxygen, care should be taken to avoid overheating and oxidizing the specimen. The edge of the gross sample should be avoided. For round bars, an entire cross section should be taken if possible. Otherwise, a wedge-shaped sample should be taken. For plates and sheets, cross-sectional samples should be taken. If the plate is too thick, samples should be taken from the quarter point and center. Porous materials should not be analyzed because of possible entrapment of air in the voids and the impossibility of proper surface preparation.

Blanc and Blondel [149] have noted that cast iron samples for oxygen analysis should be stored in a vacuum desiccator to protect them from moisture, which can be adsorbed on the graphite in the sample and therefore affect the results.

C. SAMPLE PREPARATION

1. Hydrogen

For steel, cut a 1–5 g sample from the gross specimen, and while minimizing sample-preparation time, remove all evidence of oxidation, smeared metal, and cracks by gentle filing. While holding the sample in a pair of tongs, dip it in a grease-free solvent, rinse in reagent-grade acetone, and dry in a stream of warm air. Weigh the sample to the nearest 0.1 mg, and transfer it to the analytical apparatus.

For cast iron [139], lightly machine the sample to remove surface oxide, which will react with hydrogen on analysis. Care must be taken to prevent the sample from overheating during preparation to prevent loss of hydrogen. Many cast-iron samples, however, are extremely hard and brittle and difficult to cut or machine.

2. Oxygen and Nitrogen

Remove the oxide film on a slow-speed grinding wheel by using 180-grit silicon carbide paper or a clean, fine file. Wash the sample successively in acetone, benzene, and acetone both of reagent-grade purity. Remove all traces of solvents by holding the sample in a stream of warm, dry, oil-free air. Weigh the sample to the nearest milligram, and insert it in the analytical apparatus.

D. STANDARD SAMPLES

The following information is not meant to be exhaustive but to be indicative of the standards for gases in steel available.

The preparation of hydrogen-in-steel standards is discussed in Chapter 7, Section III.A. Nitrogen and oxygen in steel standards are available from the U.S. National Bureau of Standards, Washington, D.C. Nitrogen standards may be obtained from the Bureau of Analyzed Samples, Ltd., Newham Hall, Middlesbrough, England; the Soviet government also has available nitrogen standards that presently are issued only in the Soviet Union. Standards for oxygen in steel have been prepared by the Chemikerausschuss des Vereins Deutscher Eisenhuttenleute in West Germany (Berlin, Dortmund, and Dusseldorf). To obtain oxygen standards of good homogeneity and convenient geometry, Cooke et al. [150] mixed weighed amounts of aluminum isopropoxide and iron powder, which were then heated, compacted, and sintered. While each batch of samples was rather homogeneous with respect to oxygen content, the discrepancies between the calculated and determined oxygen contents made the samples unsuitable for use as primary oxygen standards.

V. ANALYSIS

A. HYDROGEN

Methods for determining hydrogen in steel are generally covered in Chapter 7. A standard ASTM method for determining hydrogen by vacuum extraction from the solid sample has been developed by Martin et al. [151]. A study by an international committee [152] showed no significant difference in the standard deviation of results obtained by carrier-gas and vacuum extraction.

In a study of failed steel boiler tubes, Walter and Picklo [153] determined by mass spectrometry that considerable methane was present in the steel. This compound was released by decomposition of the sample in copper(II) chloride solution and measured by Orsat or mass-spectrometric methods. It has since been shown that vacuum extraction from the solid sample at 1000°C results in conversion of any methane to hydrogen; this will not occur at the 600–650°C used by many analysts. Any form of hydrogen not extracted at the lower temperatures has commonly been called nondiffusible hydrogen as opposed to the diffusible form. Methane is decomposed to yield hydrogen during classical vacuum-fusion analysis [153].

The determination of hydrogen in cast and pig iron has been studied by a number of investigators [139, 140, 154–160]. Dawson and Smith [139] determined hydrogen in white iron by vacuum extraction at 1070°C, followed by a pressure measurement after diffusion of the hydrogen through palladium. Samples were chill-cast in a graphite mold and the surface was removed by machining. A determination required 15–20 min. Equivalent amounts of hydrogen (with different extraction times) were obtained at both 650° and 1000°C. Extraction at 500° and 800°C gave lower results, the latter temperature coinciding with the α- to γ-transition in iron. It has been reported [158–160] that vacuum fusion yields higher values for cast iron than does vacuum extraction; in one case [159] the former were higher by a factor of 2–3 than those obtained by the latter technique at 1000°C. Kato [160] recommended the use of a fused alumina crucible for vacuum fusion determinations at 1300°C. At this temperature, however, the sample will be liquid only if it contains 3% carbon or greater. Further, increased hydrogen values at higher temperatures may reflect the manner in which the hydrogen is present in the sample, only the atomic hydrogen diffusing from the solid sample at the lower temperatures.

B. NITROGEN

1. Chemical Dissolution of Nitrides

The use of acid-dissolution procedures for the determination of nitrogen in steel has been criticized in the past because the results do not include nitrogen present in molecular form. Pigott [32] and Peifer [161] have discounted this criticism by claiming that only very minute amounts of molecular nitrogen have been found when appropriate methods have been employed. Therefore, if complete decomposition of the nitrides can be attained, results obtained by acid dissolution represent the total nitrogen. The biggest problem with chemical methods has always been the decomposition of acid-resistant nitrides that may be present in steel.

The determinations of both total nitrogen and individual nitrides have been shown to be dependent on the solubility and/or rate of solution of the various nitrides in chemical reagents. It therefore is important to summarize some of the published studies on the solubility of nitrides under certain conditions. Beeghly [40] showed that nitrides of niobium, tantalum, titanium, and vanadium are insoluble in hydrochloric acid. Hague and co-workers [67] showed that they were also insoluble in sulfuric acid, as are certain nitrides in SAE 3140 chromium–nickel steel.

Pigott [32] published an excellent summary of the difficulties encountered in chemically dissolving some nitrides. The Nitrogen Group of the British Iron and Steel Research Association studied this problem in some detail, and their report [66], although a notable accomplishment, deals with only a few of the simpler nitrides in relatively uncomplicated steels.

Solution rates for the nitrides in steel cannot be definitely established, because of the wide range of nitride particle sizes and the great variation in purity of the nitride phase. It is possible, however, to make generalized statements regarding solubility that may be helpful to the analyst in identifying some of the common nitrides in steel.

As noted in Section III.B, the aluminum nitride, commonly found in steel, is somewhat soluble in acid or water and readily soluble in caustic solution [85]. It can be quantitatively isolated by the Beeghly ester-halogen technique, as noted earlier, but the electrolytic dissolution of the steel matrix in an aqueous medium yields low results [76]. Only recently have 100 A particles of aluminum nitride been isolated by a modified ester-halogen technique [37, 38]. The crystallographic identification of fine particles of such compounds is dependent on a sufficient concentration of the particles and on the use of a retentive filter medium; pore sizes down to 100 A are commonly available.

Iron nitride (Fe_4N), manganese nitride (Mn_4N), Cr_2N, and some molybdenum nitride phases can be easily dissolved in dilute mineral acids. Certain chromium nitride phases, such as CrN, are insoluble in acid except after lengthy fuming in sulfuric acid or after acid fusion of the insoluble phase in bisulfate [65, 67].

Popova and Kabanik [163] have recently published a thorough study of the solubility of niobium and tantalum nitrides. These nitrides resist most acids except a vigorously oxidizing one or hydrofluoric acid. Zirconium nitride hydrolyzes to some extent in hot water or steam [85, 162] and is at least partially soluble in most acid media. Several authors [33, 66, 164–166] have noted that α- and β-Si_3N_4 are almost completely acid-resistant. As a result certain chemical methods have been proposed for steels known to contain these nitrides. In addition, after investigations of the solubility of metal nitrides, three recommended dissolution procedures [66, 67, 167] designed to decompose the chemically stable nitrides have been advanced. All three methods specify a lengthy Kjeldahl-type digestion of the sample or of the insoluble residue after dissolution of the matrix. However, the absence of nitrogen in the insoluble residue or the appearance of an apparently clear solution after digestion is not sufficient proof of total recovery. Very fine nitride particles can resist this long chemical digestion, and in addition, nitrogen recovery from the insoluble residue may be low because the fine particles containing nitrogen may not be present in the insoluble residue obtained by filtration or centrifuging.

As an alternative dissolution procedure, the authors recommend initial dissolution with perchloric acid containing several drops of hydrofluoric acid. The sample is then fumed at a controlled temperature (190–220°C) in an open beaker and transferred to a steam-distillation apparatus similar to that used by Beeghly [168]. This treatment, however, will not dissolve α- or β-Si_3N_4 nor CrN. The dissolution time is much less than that required in the above-mentioned methods [66, 67, 167]. Forty nitrogen determinations can be performed in 8 h by a single analyst with Nessler reagent used to measure the distilled ammonia.

2. Determination by Chemical and Fusion Techniques

Final measurement of the ammonia distilled from the dissolved steel is usually made either by acid titration or by Nessler photometry (see Chapter 8). Titration of very small amounts of ammonia is more difficult than photometric measurement. A brief discussion of this problem and the merits of several indicators has been published by Sher [169]. The choice of the best indicator is largely one of personal preference.

While Nessler photometry may be preferable for the determination of small amounts of ammonia, higher amounts, such as may be present in manganese–nitrogen austenitic steels are better determined by titration. The use of Nessler reagent to measure high amounts of ammonia necessitates undesirable modifications of the method, such as selection of a small aliquot, development of color in a large volume, or use of a very small sample weight. When Nessler reagent is used to measure ammonia, the physical transfer of caustic solution to the receiving flask through bumping or entrainment during the distillation is not important as in titration; therefore, a less complicated distillation apparatus may be employed [170].

Indophenol [58, 59] has been used for the photometric determination of nitrogen as ammonia, as noted earlier. The bis-pyrazolone–chloramine-T procedure has been used after solvent extraction of the ammonia from the distillate [62–64]. For greater than 5 μg nitrogen, no solvent extraction was found necessary [64]. All of these procedures yielded distinctly better detection limits than that obtainable with Nessler reagent. The use of photometric methods without distillation, involving either Nessler reagent or indophenol, is not recommended, because highly colored ions and compounds such as hydrocarbons can cause errors.

For the determination of total nitrogen in high-silicon electrical steels where α- or β-Si_3N_4 may be present, either a reduction fusion (vacuum or inert gas at sufficiently high temperature, using a clean crucible for each determination) or a caustic fusion [33] may be used. In the latter, the measurement of nitrogen as ammonia may be made either photometrically or coulometrically. The caustic-fusion method is not recommended for routine application. Best results are obtained with very fine steel millings, which permit a good attack by the sodium hydroxide. The method is described in detail in Section VI.B.1. Evens and Fassel [115] have shown that a dc carbon arc can be used to liberate nitrogen for chromatographic measurement from a wide range of steel compositions.

The vacuum- and inert-gas-fusion techniques for determining nitrogen are described in Chapter 4, Sections I.A.4.a, I.L., and II.C. Much of this information is directly applicable to steel analysis, especially that concerning means to effect evolution of nitrogen from the bath. It is generally recognized that samples should be analyzed either by using a fresh crucible for each sample (Balzers, LECO, Strohlein systems) or by emptying the crucible after each determination (Balzers system).

Nitrogen has been determined in cast iron and steel by Jecko and Touvenin [171], using a LECO TN-14 Automatic Nitrogen Analyzer. Good fusions were obtained with white iron poured as a thin plate (about 2 mm thick) and then crushed. At the 66 ppm level, the coefficient of variation was 1.7%. For steels at about the 100 ppm level, the coefficient of variation was 1.9%. The same analyzer was used by Cline et al. [172], who reported a repeatability at the 95% confidence level of 1 ppm at a 40-ppm nitrogen concentration. It was further noted that when more than 4% titanium was present in a sample, nitrogen recovery was poor.

Nitrogen has been determined in cast iron by both chemical and inert-gas-fusion techniques by Kashima and Yamazaki [173]. For both methods the sample was obtained as a thin plate chilled in a copper mold. In the chemical method the sample was dissolved in dilute HCl. After addition of NaOH solution and steam distillation, the ammonia was determined titrimetrically. The accuracy of both methods was 0.0004% N for 0.0030–0.10% N.

Levi and Kitaev [174] used a more drastic chemical treatment because earlier work [175] showed that some of the nitrogen was associated with the undecomposed graphite after dissolution of the matrix; presumably this nitrogen was present as a carbonitride. This was especially true for cast irons to which nitrogen was deliberately added. After initial treatment of fine shavings of the sample with HF and H_2SO_4, $BaCl_2$ solution was added to precipitate sulfate and drag down the insoluble matter. After separation by centrifuging and decanting the liquid, the residue was treated with H_2SO_4 and heated to fumes of SO_3 for 1.5 h. The two portions of the sample were then combined for a conventional distillation, followed by Nessler photon etry. Nitrogen over the range 0.009–0.018% was determined with an erro of 0.0005% or less.

Berg and Freiberg [176] have evaluated the Strohlein Dinometer, Heraeu Evolograph, LECO, and Argonex instruments for the determination of bot nitrogen and oxygen in steel, and the Balzers Exhalograph for oxygen. The reported reproducibilities of 5–10% at the 50 ppm level and 2–5% at th 100-ppm level. Similar studies have been made by Strohlein [177] an Effenkammer [178].

C. OXYGEN

1. Neutron Activation

The determination of oxygen in steel by this technique is covered in Chapte 3. At the authors' laboratory, neutron activation is used, where sampl geometry permits, for 20 ppm oxygen and higher. Agreement with vacuu fusion is generally quite good. Hanin [179] has reported on an extensiv comparison of results for oxygen in steel obtained by vacuum fusion, inert gas fusion, and neutron activation. Results over the range of 40–120 ppr for three steels showed the inert-gas-fusion method to be higher by abou 20 ppm. For a number of samples containing 500–800 ppm oxygen, th agreement among all three methods was generally quite good, especiall between the neutron activation and inert–gas-fusion methods. Simila studies reported by Bruch [1] showed good agreement between the two fusio methods for both unkilled and aluminum-killed steels for 100–1200 ppr oxygen, good agreement for unkilled steels analyzed by neutron activatio and both fusion methods over the range 150–1200 ppm oxygen, but somewha higher results (about 50–200 ppm higher) for the range 400–830 ppm oxyge by activation as compared to inert-gas-fusion analysis of aluminum-kille steels. No reason for this deviation was found.

The advantages and disadvantages of neutron activation are discussed i Chapter 3 and in the introduction to Chapter 4.

2. Oxide Inclusions

A number of methods have been developed for isolating oxide inclusion chemically or electrochemically from steel. The isolated residue can then b analyzed for the various cations or examined microscopically. Much of th literature in this area has been summarized by Koch [180]. In the authors laboratory, a method has been developed for low-carbon, low-alloy steel that consists of (1) neutron activation analysis of the sample for total oxygen (2) dissolution of the same sample with bromine–methyl acetate, and (3 pectrographic analysis of the separated residue after fusion and dissolution. Details of the procedure are given in Section VI.C.3. The following elements re determined and the oxygen or oxide equivalents are calculated: Fe, Mn, i, Al, Ca, and Mg. In the range of 20–300 ppm oxygen, the sum of the ndividual oxygen values generally agrees with the neutron-activation result o within 10–15% of the amount present.

. Electrochemical Determination in Furnace Bath

The direct determination of oxygen dissolved in liquid metal was suggested y Fischer and Hoffman [181], who utilized the potential developed between n air–ZrO_2 electrode and liquid metal. This was based on the work of Kiukkola and Wagner [182], who studied high-temperature oxidation eactions. After about seven years a number of investigators [183–190] emonstrated the possible applicability of galvanic cells for determining xygen in liquid steel. Gatellier et al. [191] have written a lengthy review overing iron and steel. The technique has also been used to determine oxygen n lead [192], sodium [193], copper [194], and copper and silver [195]. The ollowing discussion is limited to the application of the electrode to the etermination of oxygen in steel.

The measurement of oxygen dissolved in iron (but not as SiO_2, Al_2O_3, etc.) y the galvanic cell or sensor technique is based on the flow of oxygen ions n the direction of the fall in pressure from I to III [195]:

I	II	III
Air	Solid	Liquid Metal
p'_{O_2}	Electrolyte	p_{O_2}

If the cell is reversible and electronic conduction is negligible, the emf leveloped is given by

$$E = \frac{RT}{nF} \ln \frac{p_{O_2}}{p'_{O_2}}, \tag{1}$$

where R is the gas constant, T is the temperature of the molten metal, n is the number of electrons involved, and F is the Faraday constant.

Detailed theoretical discussions of the technique have been published [182, 185, 186, 188, 197, 198]. An oxygen sensor using chromic oxide for supply of oxygen rather than air, as described later, is shown in Figs. 15.2 and 15.3 [199, 200]. The oxygen potential is measured using the Pt/10% Rh half of the thermocouple and a molybdenum wire, the latter in contact with the molten steel. The type of response curve obtained (Figs. 15.4 and 15.5) will depend on the rapidity with which the sensor is inserted through the slag layer into the melt, the mass of the sensor, and the speed of response of

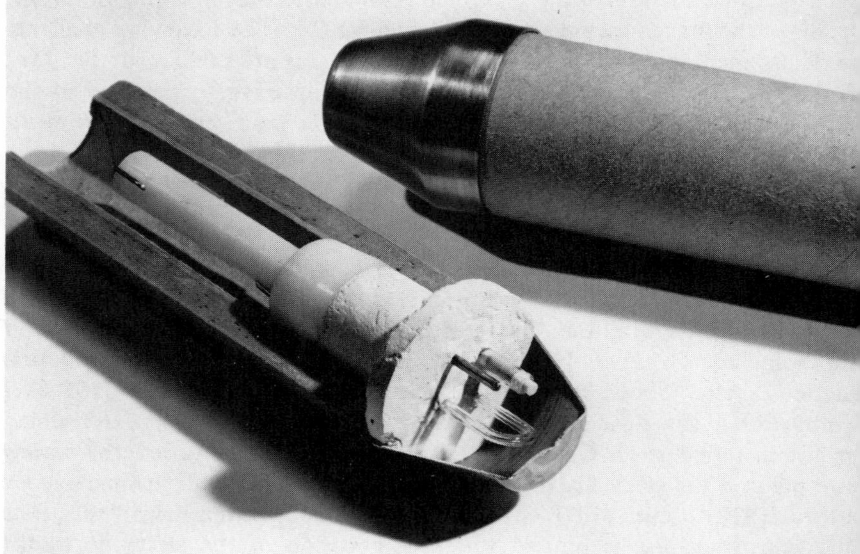

Fig. 15.2. Oxygen sensor. (Reproduced by permission from *Journal of Metals* and *The Metallurgical Society of AIME* [199].)

the voltmeter [201]. Temperature measurements must be accurate; a variation of 10°F (5.6°C) yields a 4% relative error in the oxygen potential. A series of calibration curves is shown in Fig. 15.6 [200]. It should also be kept in mind that gross changes in bath composition can affect the response of the sensor. Most applications to date have been in essentially iron–carbon baths where measurements have been made during refining.

Many investigators have contributed to the development of the sensor, and detailed discussion of the evolution of it is beyond the scope of this book. However, certain specific problems of earlier probes will be mentioned prior to discussion of a specific probe for steel used at the time of this writing.

Lime stabilized zirconia is often used as the electrolyte for several reasons. Among these is the fact that this mixture (15 mole % CaO) expands uniformly with increasing temperature and melts at about 2600°C; pure zirconia, on the other hand, undergoes a structural transformation with a reversible inversion and a volume change near 1000°C. The lime, in addition to stabilizing the cubic structure of zirconia, produces a deficiency of oxygen ions in the anion lattice, since bivalent calcium is substituted for tetravalent zirconium. Thus, oxygen ions can be conducted through the electrolyte with ease, but at the same time it is impervious to gases, including oxygen, and to liquid metals. Only the tip of the sensor contains the electrolyte; an all-electrolyte tube could be susceptible to oxygen gradients in the bath [202].

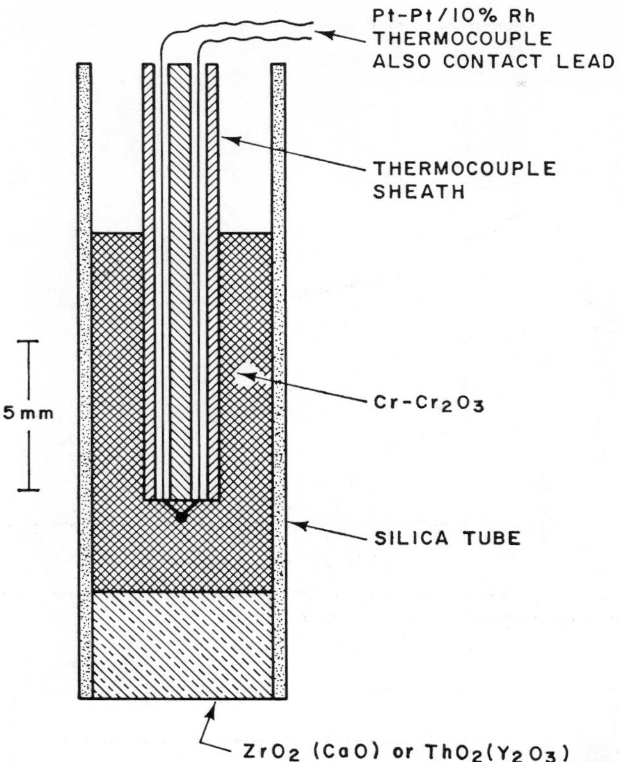

Fig. 15.3. Schematic diagram of oxygen sensor. (Reproduced by permission from *Transactions of the Metallurgical Society of AIME* and *The Metallurgical Society of AIME* [200].)

Fitterer [189] has considered some of the problems encountered with the design of the sensor and the materials used in it:

1. Thermal shock resistance
2. Temperature and oxygen pressure limits for each electrolyte where electronic instead of ionic conduction takes place
3. Preparation of the electrolyte so as to prevent formation of capillary leaks through the electrolyte barrier with resultant failure of the cell
4. Selection of the standard state for oxygen
5. Consideration of side-reactions
6. Kinetics of the reaction that establishes equilibrium and/or a stable voltage shortly after immersion
7. Sensitivity of the calibration curve relating oxygen content and voltage
8. Thermoelectric effects involving the electrolyte and the elements it contacts

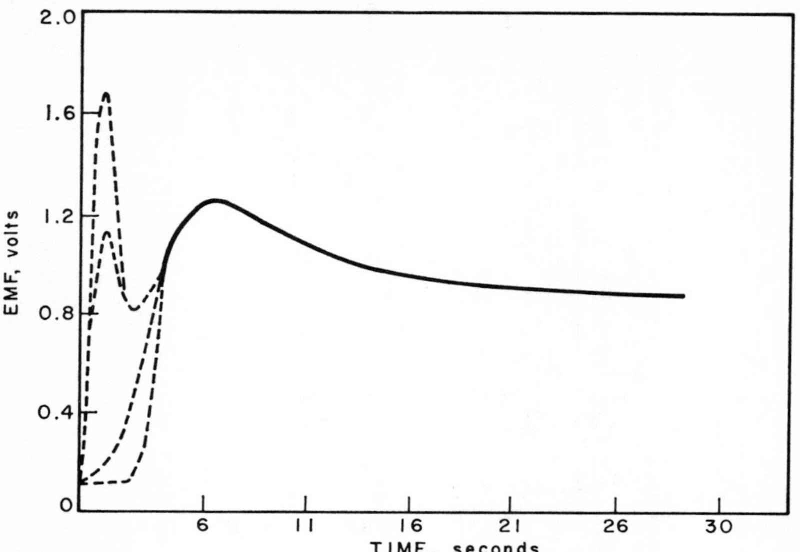

Fig. 15.4. Wave forms of sensor responses. (Reproduced by permission from *Journal* *Metals* and *The Metallurgical Society of AIME* [203].)

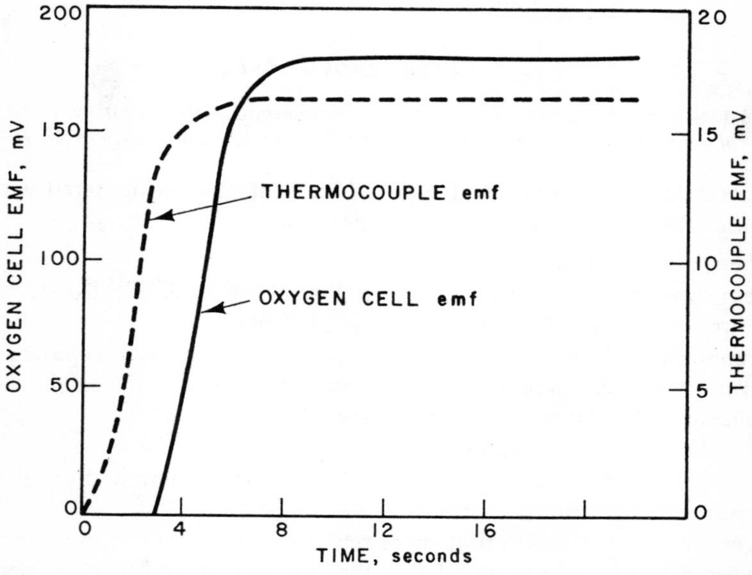

Fig. 15.5. Representation of emf traces for oxygen sensor. (Reproduced by permissio from *Journal of Metals* and *The Metallurgical Society of AIME* [199].)

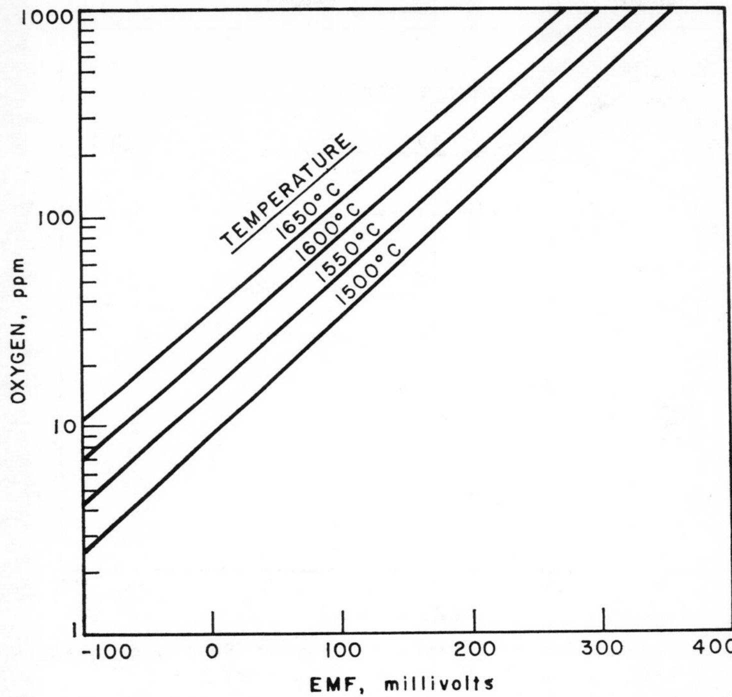

Fig. 15.6. Calibration curves for oxygen sensor using $Cr-Cr_2O_3$ reference electrode. (Reproduced by permission from *Transactions of the Metallurgical Society of AIME* and *The Metallurgical Society of AIME* [200].)

5. Development of a device that can be plunged through a slag into molten steel without any contamination of the device.

Such problems have been investigated by various workers [188–190, 199, 200, 202–208].

To obtain good thermal shock resistance in the sensor, Fitterer [188, 189] suggested the use of a quartz tube with a $CaO-ZrO_2$ disk fused on the end. Because quartz and zirconia do not form a true fused joint and because of other problems encountered in the use of this system, Pargeter [196] made a sensor of a quartz tube with one constricted end into which was rammed a mixture of zirconia powder and a suitable binder that was subsequently dried. The sensor tip was then sintered before use. The quartz tube was mounted in a ceramic plug placed in the end of a ceramic-coated steel tube. Air, the source of oxygen, was fed into the sensor by a tube held in a lance. The sensor was protected from slag on entry into the bath by a steel cap, which melted off in less than 1 sec. Such sensors were reportedly used in excess of 10 times each to give what was considered acceptably accurate

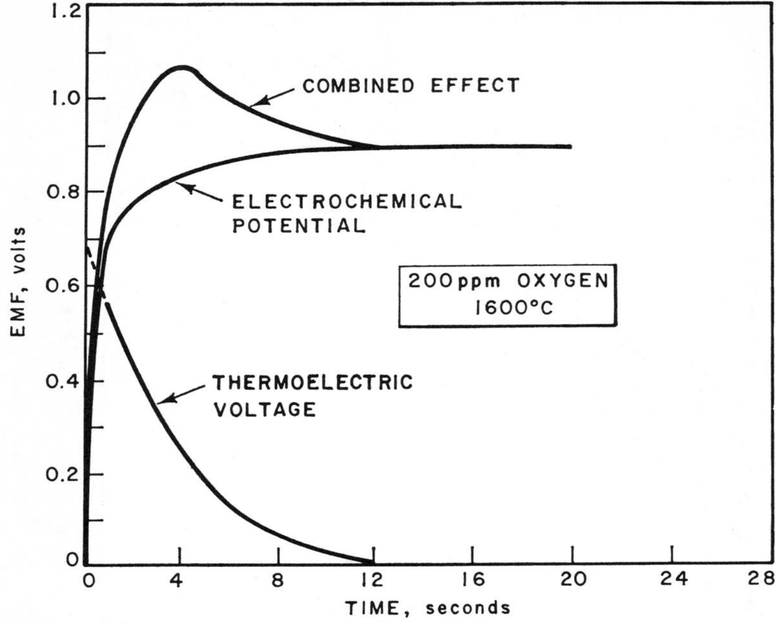

Fig. 15.7. Calculated wave-form for oxygen sensor response. (Reproduced by permissio from *Journal of Metals* and *The Metallurgical Society of AIME* [203].)

readings; for example, a deviation of 50 mV caused a difference of 9 ppn oxygen at about the 15-ppm level at 1538°C.

Pargeter and Faurschou [203] noted that a temperature gradient acros the cell caused a thermoelectric voltage to be superimposed on the electro chemical potential, and it required at least 12 sec for the former to drop t zero (Fig. 15.7); after reaching this point, a stable electrochemical potentia reading for oxygen could be obtained. The optimum time to reach thi stable oxygen potential was 20–25 sec.

Air has been used as the source of oxygen by most investigators. Flov rates must not be too high, otherwise, the inside of the probe will be at a temperature below that of the liquid steel and an additional voltage effec will occur yielding high oxygen values [185]. Pure oxygen will destroy th electrolyte [189]. (Dilution of air with an inert gas yields a curve parallel t that obtained with air, so that there is no increase in sensitivity.) Because th use of carbon dioxide gas as a source of oxygen presented a problem, a smal amount of calcium carbonate was added to each sensor. This compounc decomposes spontaneously at about 800°C. Further, the small amount o calcium carbonate gives far more carbon dioxide than needed for the 10–15 sec immersion; therefore, a vent was installed in the probe to permit the gas to escape while maintaining a constant pressure of one atmosphere. Too much calcium carbonate can affect the reading, because of the endothermic nature of the reaction. Fruehan et al. [200] chose to use a mixture of chromium and chromic oxide because the oxygen dissociation pressure of Cr_2O_3 is in the range of that for oxygen in liquid steel. (The Cr–Cr_2O_3 mixture also does not melt at liquid-steel temperature nor does it react with $CaO{\cdot}ZrO_2$, quartz, or the thermocouple junction.)

Comparison of the calibration data of Fitterer et al. [202] with those of Fruehan et al. [200] shows that the use of carbon dioxide at 1650°C is not applicable to 30 ppm oxygen or less, whereas Cr–Cr_2O_3 at this temperature can be used down to about 11 ppm oxygen. Below the 10-ppm level and at certain temperatures and oxygen pressure, the zirconia stabilized with 4% lime will exhibit electronic conductivity. Under these conditions, thoria stabilized with 7% yttria is suitable down to 1 ppm oxygen [200].

With the Cr–Cr_2O_3 system and the zirconia–4% lime electrolyte, a Pt–Pt, 10% Rh thermocouple was used, with the Pt, 10% Rh half also used as one conductor for the oxygen measuring cell. A molybdenum wire in contact with liquid iron served as the other conductor. Molybdenum does not deoxidize iron and only very slightly dissolves in it.

Equation 1 can also be expressed as

$$E = \frac{RT}{nF} \ln \frac{a_0}{Kp_{O_2}^{1/2}}, \tag{2}$$

where P_{O_2} is the oxygen pressure in equilibrium here with Cr–Cr_2O_3, $n = 2$, and K is the equilibrium constant for the reaction

$$\tfrac{1}{2}O_2(1 \text{ atm}) = O(1 \text{ wt } \% \text{ in Fe}). \tag{3}$$

The temperature dependence of the equilibrium constant K and $p_{O_2}^{1/2}$ for the Cr–Cr_2O_3 equilibrium are given by

$$\log K = \frac{6120}{T} + 0.15 \tag{4}$$

$$\log p_{O_2}^{1/2} \, (\text{atm}^{1/2}) = -\frac{19{,}700}{T} + 4.47. \tag{5}$$

The activity of oxygen, taken equal to the concentration in weight percent in iron–oxygen binary melts, is

$$\log a_o = 4.62 - \frac{13{,}580 - 10.08E}{T} \tag{6}$$

where E is in millivolts and T is in degrees Kelvin. The activity coefficient of oxygen in iron alloys may be estimated by

$$\log f_o = \sum_i e_0^{(i)}(\%i) \tag{7}$$

where $(\%i)$ is the concentration of each alloy element and $e_0^{(i)}$ is the interaction coefficient, which is known reasonably well for the alloying elements usually found in steel. For low-carbon steels, the concentrations of these elements are small, Eq. 7 may be ignored, and concentration can be substituted for activity in Eq. 6. Conversely, Eq. 7 is an approximation that should be used only for moderate concentrations of alloying elements. Fischer and co-workers have studied the activity of oxygen in molten iron containing manganese, titanium, or boron [209] and molybdenum, tungsten, niobium, or tantalum [210]. The effect of both chromium and vanadium [211] and the oxygen permeability in stabilized zirconia, mullite, and alumina [212] have been studied. In addition, various solid-metal–metallic oxide mixtures were investigated [213] as references in oxygen measuring cells between 600° and 1700°C. Errors in potential readings (4 mV or greater) were attributed to uncontrollable thermovoltages, reactions between the electrolyte and the electrodes, and additional galvanic voltages. A temperature variation of 20°C caused only a 3% deviation with the air electrode, but with solid electrodes the error was 8% of the oxygen content with only a 10°C variation.

With the $Cr–Cr_2O_3$ electrode a steady value of emf ($\pm$ 2 mV) is achieved within 10–15 sec after immersion in the melt [199]. This value stays constant for 30 min or more if the bath temperature is constant. Failure of a sensor is indicated by a large fluctuating emf or by a rapid increase or decrease of about 25 mV/min. The sensor has been used to determine oxygen in iron containing silicon, chromium, aluminum, or vanadium with excellent results. Further, the sensor does not have to be preheated; after all experiments the sensors were free of cracks indicating good thermal shock resistance. The use of this sensor has been described by Dukelow et al. [214]. It is available from Leeds and Northrup Company, Philadelphia. Similar probes are available from Leigh Instruments, Ltd., Ontario, Canada, or Leigh-AARCO, Inc., Chagrin Falls, Ohio [215, 216]; ASEA Central Development Department, Vasteras, Sweden [217]; Fitterer Engineering Associates, Oakmont, Pennsylvania; the Broken Hill Proprietary Company, Ltd., Shortland, New South Wales, Australia; and Electro-Nite Company, Philadelphia. A review of developed sensors has appeared [218].

4. Vacuum and Inert-Gas Fusion

The determination of oxygen by vacuum or inert-gas fusion technique is treated in Chapter 4, Sections I.A.4.b, I.L., and II.C.2.a, b, c, e, f, g, and h; much of the information contained therein is applicable to iron and steel analysis. Schwarz and Zitter [219] reported on a round-robin analysis involving nine laboratories using vacuum fusion and/or inert-gas fusion with various means for determining the oxygen-bearing gas. Oxygen in the range of 0.002–0.003% was determined. For the range of 0.002–0.009%, marked differences (about 0.002%) between laboratories were found; these steels were killed by aluminum or silicon, and the differences may be caused by segregation of oxides.

Another factor that must be considered when performing multiple analyses in a crucible is the loss of oxygen by reaction with condensed metal films (gettering). This can be quite complex and should be investigated for each type of steel analyzed. Gettering is usually less of a problem with inert-gas fusion than it is with vacuum fusion and presents no apparent difficulty when each sample is presented to an empty crucible. Gettering is treated in detail in Chapter 4. In steel analysis both platinum and nickel have been commonly used as bath materials, the latter by many operators of the Balzers Exhalograph EA-1. Cobalt, although more expensive than nickel, has been found to be better in some cases. The use of the Balzers Exhalograph EA-1, with and without the spinning crucible (the latter for determining nitrogen), has been reported by Chistyakova et al. [220], who analyzed various grades of steel. Other instrument evaluations for steel analysis, as noted earlier, have been performed [176–178]. In addition, studies were performed by Frohberg and Leygraf [221]. Bruch [1] has stated that the vacuum- and inert-gas-fusion methods, as practiced by the best techniques, can be considered to be equivalent.

Bagshawe et al. [222] determined oxygen in zone-refined iron and in plain carbon, alloy, stainless, and tool steels by a vacuum-fusion–mass-spectrometric technique that they recommended for small samples or low oxygen contents. The detection limit was 0.15 μg oxygen. The use of a mass spectrometer is especially advantageous to examine the gas mixture evolved during fusion for oxygen-bearing compounds other than CO, such as COS and SO_2. The treatment of CS_2 and COS has been discussed in Chapter 4, Section I.B.3.f, and the effect of SO_2 has been mentioned. Abe and Takazawa [223] have used the graphite-capsule technique to convert SO_2, COS, and any CO_2 to CO during vacuum-fusion analysis of steel. Tsugane and Kamakura [224] found positive errors in oxygen determination in high-sulfur steels caused by evolution of CS and CS_2, which were oxidized to CO_2 and SO_2, thereby affecting the final pressure measurement. To prevent this error the sulfide gases were condensed in a liquid oxygen trap. Limited experience at the authors' laboratory with two free-machining steels (0.26–0.35% sulfur) showed no CS_2, COS, or SO_2 to be produced when vacuum-fusion analyses were performed on a modified Balzers Exhalograph EA-1. The two samples containing about 50 and 180 ppm oxygen, respectively, were analyzed first by neutron activation and then by vacuum fusion. Similar results were obtained by both methods. In addition, after fusion in vacuum, gas samples were taken from the apparatus and analyzed by gas chromatography; none of the above sulfur compounds was detected.

Goto et al. [225] determined oxygen in cast iron by vacuum fusion in a tin bath. Using the fractional technique, they decomposed the following oxides at the respective Celsius temperatures: FeO, 1050; MnO, 1150; SiO_2, 1450; and $Al_2O_3 + TiO_2$, 1750. The argon-fusion technique followed by absorption and measurement of CO_2 in a solution of pH 10.5–10.8 has been used by Berg et al. [226] for pig iron and steel.

VI. RECOMMENDED METHODS

A. HYDROGEN

A recommended method for determining hydrogen in steel is vacuum extraction from the solid sample. This method is described in Chapter 7, Section III.A, and a diagram of the apparatus is shown in Fig. 7.1. The sample weight will depend on the apparatus used and the preference of the analyst. Weigh the sample to the nearest 0.1 mg.

Great care must be taken in sampling molten or solid steel for hydrogen and in storing the samples prior to analysis as noted in Section IV. In addition, the precautions observed in preparing the samples must be closely followed.

B. NITROGEN

All chemical nitrogen determinations should be made in a laboratory that has an ammonia-free atmosphere.

1. Total Nitrogen—Caustic Fusion [33]

This method is applicable to any steel composition provided that the nitrogen content is 0.002% or higher. The apparatus is shown in Fig. 15.8.

a. Apparatus

(1) NICKEL COMBUSTION TUBE. Fabricate from high-purity nickel a tube 75 cm long, 3.1 cm o.d., and 2.5 cm i.d. Reduce one end of the tube, and add a nickel section that is about 5 cm long, 0.8 cm o.d., and 0.6 cm i.d. Braze a copper water-jacket onto each end of the tube to dissipate the heat generated from the furnace.

(2) NICKEL COMBUSTION BOATS.

(3) FURNACE. A 12-in. split-type resistance furnace that will operate continuously at 950°C should be used. Provide for temperature monitoring of the furnace by making room for a thermocouple.

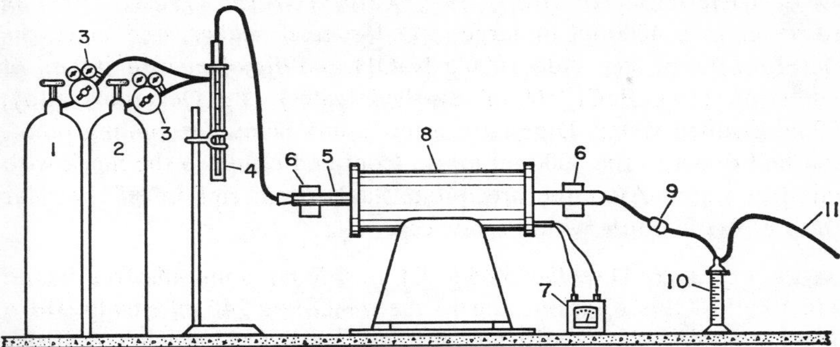

1. HYDROGEN
2. ARGON
3. GAS REGULATOR
4. FLOW-METER
5. NICKEL TUBE
6. WATER CONDENSER
7. PYROMETER
8. 12" SPLIT-TYPE RESISTANCE FURNACE, 950°C
9. BALL-SOCKET JOINT CONTAINING GLASS WOOL
10. AMMONIA RECEIVER 50 ml
11. EXHAUST TO OUTSIDE

Fig. 15.8. Caustic-fusion apparatus. (Reproduced by permission from the *Journal of the Iron and Steel Institute* [33].)

(4) GAS SUPPLY AND AMMONIA COLLECTOR. Set up a cylinder of hydrogen and a cylinder of argon (extra low oxygen content) with a flow meter and appropriate controls so that either gas can be passed through the apparatus at a controlled and known flow rate. Connect the gas cylinders to the flow meter and the large end of the combustion tube with Tygon tubing, the latter by means of a one-hole rubber stopper and a short length of glass tubing. Lay the tube in the furnace so that about 30 cm of the reduced end extends beyond the end of the furnace. Connect the reduced end of the combustion tube with Tygon tubing to a standard-taper ball-and-socket joint containing a glass-wool plug. Then connect the joint with Tygon tubing to a gas-scrubbing bottle to collect the ammonia (Note 1). Attach tubing to the outlet side of the scrubber and vent the gases to the atmosphere outside the building.

b. Reagents

(1) SODIUM HYDROXIDE, REAGENT-GRADE.

(2) SODIUM CARBONATE, REAGENT-GRADE.

(3) MERCURIC CHLORIDE, REAGENT-GRADE.

(4) AMMONIA-FREE WATER. See Chapter 8. This water is to be used wherever rinsing, washing, or dilution with water is specified in the procedure.

(5) SODIUM HYDROXIDE SOLUTION, 33% AMMONIA-FREE. Transfer 3000 ml distilled water to a 4000-ml or larger stainless-steel beaker, and mark the liquid level on the beaker. Add 1000 g NaOH and dissolve. Add 100 ml of $BaCl_2$ solution (10 g $BaCl_2$/100 ml distilled water), 5 g Devarda's alloy, and 700 ml distilled water. Digest for a few hours below the boiling point, and then boil down to the 3000-ml mark. Cool and dilute to the mark with ammonia-free water. After the precipitate has settled, siphon off the clear liquid into a plastic bottle with a screw cap.

(6) NESSLER REAGENT. Dissolve 62.5 g KI in 250 ml ammonia-free water. Set aside 10 ml of this solution, and to the remaining 240 ml slowly add a saturated solution of $HgCl_2$, stirring constantly until a red precipitate forms that does not redissolve. About 500 ml will be required. Now add the 10 ml KI solution that was set aside, to dissolve the precipitate. Continue adding $HgCl_2$ solution, drop by drop, until a trace of red precipitate forms that does not redissolve. Add 400 ml ammonia-free NaOH solution, and dilute to 2000 ml with ammonia-free water. Let stand one to two weeks, and decant the clear liquid into a borosilicate glass bottle.

c. *Preparation of Calibration Curve*

To prepare the calibration curve, transfer by means of a microburet or standardized buret exactly 0.5, 1, 2, 3, 4, 5, 10, 20, 30, and 40-ml portions of standard ammonium chloride solution (see Chapter 8; 1 ml contains 0.01 mg ammonia nitrogen) to 100-ml volumetric flasks. Dilute the solution in each flask to 95 ml with water, add 3 ml Nessler solution, and, finally, dilute to the mark with water. Mix thoroughly by shaking and determine the absorbance at 410 nm. Plot absorbance against milligrams of nitrogen per 100 ml of solution.

d. *Procedure*

Before assembly, clean the combustion tube and scrubber by rinsing them alternately with (1 + 1) hydrochloric acid and water. Remove water from the combustion tube by rinsing it with a small volume of methyl alcohol, and dry the tube. Fill the scrubber two-thirds full with water, and add 2 drops (1 + 1) hydrochloric acid. Pass argon through the apparatus at a flow rate of about 500 ml/min, and begin to heat the furnace. Do not heat the furnace beyond 300°C before adding the sample. Add about 10 g sodium hydroxide and 0.5 g sodium carbonate to a nickel combustion boat, melt the mixture over a gas burner, and allow the melt to solidify. (This prefusion is made to remove any ammonia from the reagents.) Scatter about 1 g of finely milled sample, weighed to the nearest 0.1 mg, on the fused mass, and push the boat into the combustion tube until the forward end of the boat is about 12 cm from the end of the furnace (Note 2). Continue passing argon through the apparatus for 5 min to remove any air present. Pass hydrogen through the apparatus for 5 min, and then raise the furnace temperature to 950°C. Continue heating for 20 min after that temperature is reached. Turn off the furnace, flush hydrogen from the apparatus by sweeping with argon for 5 minutes, and remove the scrubber. Carefully withdraw the boat to avoid spilling any molten sodium hydroxide (Note 3).

Transfer the scrubber solution to a volumetric flask, add 3 ml Nessler reagent, and dilute to the mark with ammonia-free water. Stopper the flask and mix by shaking. Measure the absorbance at 410 nm with a photometer. Determine the nitrogen content from the calibration curve.

e. Calculation

Where A is milligrams nitrogen from the calibration curve and B is grams of sample,

$$\% \text{ Nitrogen} = \frac{A}{10B}. \tag{8}$$

NOTES

1. Some water vapor and, on occasion, a white substance condense within the end of the combustion tube and the tubing leading to the scrubber. This condensate is basic and should not contain any ammonia. Replace the glass wool after each determination.
2. Run two reagent blanks, one before, and one after each set of determinations. The reagent blanks should be low; they are comparable to the blanks obtained by the usual chemical-Kjeldahl procedure.
3. If the fused sample is broken and examined visually, no shiny metallic surface will be visible following a successful fusion.

2. Total Nitrogen—Acid Dissolution and Treatment of Residue

This method is applicable to all types of steel except those containing α- and β-Si_3N_4. The titration procedure is preferred for determining more than 0.1% nitrogen.

a. Apparatus

(1) KJELDAHL DIGESTION APPARATUS.

(2) STEAM DISTILLATION APPARATUS (see Fig. 8.1a–b).

b. Reagents

(1) CONCENTRATED SULFURIC ACID.

(2) PERCHLORIC ACID, 35%. To 1 liter of ammonia-free water, slowly add 1 liter 70–72% $HClO_4$.

(3) AMMONIA-FREE WATER (see caustic-fusion method and also Chapter 8).

(4) STANDARD SULFURIC ACID, 0.01 *N*.

(5) STANDARD SODIUM HYDROXIDE SOLUTION, 0.01 *N*.

(6) SODIUM HYDROXIDE SOLUTION, 33%. Treat to remove nitrogen compounds (see caustic-fusion method).

(7) METHYL PURPLE, ACID–BASE INDICATOR.

(8) NESSLER REAGENT (see caustic-fusion method).

(9) STANDARD AMMONIUM CHLORIDE SOLUTION (see Chapter 8).

c. Procedure

Transfer 2 g of sample weighed to the nearest 0.1 mg to a 400-ml beaker (run a blank simultaneously). Add 50 ml 35% $HClO_4$ and several drops 48% HF. Dissolve the sample at low heat. When the sample is completely dissolved, add 50 ml water, and filter through an 11-cm Whatman No. 42 paper (or equivalent) with pulp, both the paper and pulp having been previously washed with water. Wash the paper and residue with water until free of $HClO_4$.

Evaporate the filtrate to 75 ml at low heat; then cool, and reserve. Place the paper containing the residue from the filtration in a 500-ml Kjeldahl flask. Add 10 g K_2SO_4, 1 g $CuSO_4$, and 20 ml H_2SO_4. Heat the flask moderately at first and then at a temperature below the boiling point of the acid until frothing ceases. Increase the heat so as to boil the contents of the flask for 30–60 min after the solution has become colorless. Cool to room temperature, add 40 ml water, and heat gently to dissolve the iron salts (chromium salts will not dissolve). Cool to room temperature, and transfer the reserved filtrate and the Kjeldahled solution to a 500-ml Florence flask.

(1) TITRIMETRIC DETERMINATION. Place 10 ml water, 4 drops methyl purple indicator, and sufficient standardized H_2SO_4 in the receiving flask of the distillation system to ensure that an excess of the H_2SO_4 will be present after distillation of the NH_3. The bottom tip of the condenser must be immersed in the standard H_2SO_4. To the 500-ml Florence flask that contains the sample solution, add 125 ml 33% NaOH solution through the delivery tube, and

immediately close off the delivery tube. Heat both the water flask and the Florence flask to distill as rapidly as possible 100–110 ml of solution as indicated by a mark on the receiver. After the distillate is collected, remove the receiver and heaters. Titrate the excess H_2SO_4 in the receiver with standardized NaOH solution until the color in the receiving flask changes from purple to green. Determine the percent nitrogen from the volumes of standard H_2SO_4 and standard NaOH solution used. The blank volume should not exceed 0.3 ml and should be reproducible.

Calculate the nitrogen content (with all volumes in milliliters) as follows: Sulfuric acid equivalent R,

$$R = \frac{V_A}{V_B} \tag{9}$$

where V_A is the volume of H_2SO_4 and V_B is volume of NaOH solution. Blank,

$$\text{Blk} = V_A - V_B R. \tag{10}$$

Nitrogen content of sample,

$$\%\text{N} = \frac{(V_A - VR_B - \text{Blk})N_A \times 1.401}{\text{weight of sample (g)}} \tag{11}$$

where N_A is the normality of H_2SO_4.

(2) PHOTOMETRIC DETERMINATION. Prepare the calibration curve as described in the caustic-fusion method.

Evaporate the portion of the sample soluble in $HClO_4$ to 75 ml and transfer to a 500-ml distillation flask. Transfer the Kjeldahled portion to the same flask, and add 125 ml 33% sodium hydroxide solution. Steam distill about 75 ml of the solution into a 100-ml volumetric flask, add 3-ml Nessler reagent, and dilute to the mark with water. Stopper the flask, and mix by shaking. After the sample has stood for 1 h, measure the absorbance at 410 nm. with a spectrophotometer. Convert the absorbance reading to milligrams nitrogen using the calibration curve. Correct for any blank. Calculate the nitrogen content according to Eq. 8.

3. Total Nitrogen—$HClO_4$ Dissolution

This method is applicable to all types of steel except those containing α- and β-Si_3N_4 or an insoluble chromium nitride phase (Note 1). It is preferred for samples where the nitrogen content is between 0.001 and 0.01%.

a. Apparatus

(1) STEAM DISTILLATION APPARATUS (see Chapter 8).

b. Reagents

(1) AMMONIA-FREE WATER (see caustic-fusion method and also Chapter 8).

(2) PERCHLORIC ACID, 35%. Slowly add 1000 ml 70–72% $HClO_4$ to 1000 ml ammonia-free water.

(3) SODIUM HYDROXIDE SOLUTION, 33%. Treat to remove nitrogen compounds (see caustic-fusion method).

(4) NESSLER REAGENT (see caustic-fusion method).

(5) STANDARD AMMONIUM CHLORIDE SOLUTION (see Chapter 8).

c. Preparation of Calibration Curve (see caustic-fusion method)

d. Procedure

Transfer 1 g (Note 2) of the sample, weighed to the nearest 0.1 mg, to a 400-ml beaker, add 40 ml of 35% $HClO_4$ (Note 3), and warm to dissolve the sample (run a blank simultaneously). Add several drops of 48% HF, and evaporate to fumes in an open beaker on a hot plate having a surface temperature of about 230°C. Fume for 5 min after the chromium has been oxidized or after a clear solution is observed. Cool, dilute to 100 ml with water, and warm to dissolve salts. Transfer the solution to the distillation flask. Place a 100-ml volumetric flask at the outlet of the water-cooled condenser. Introduce 40 ml of 33% sodium hydroxide solution to the sample flask. Heat the steam generator and sample flask so as to distill as rapidly as possible 80–90 ml. Remove the heat from both flasks. To the distillate, add 3 ml Nessler reagent, mix, and dilute to volume with water. Allow 0.5–1 h for color development, and determine the absorbance of the solution in a cell of appropriate path length. Correct for the blank and determine the nitrogen content by reference to the calibration curve. Calculate the percent nitrogen according to Eq. 8.

NOTES

1. For steels that contain this chromium nitride phase, treat as follows. After the initial fuming and dilution of the sample, filter the solution through a Whatman No. 42 (or equivalent) filter paper. Place the paper and residue in a 500-ml Kjeldahl flask, add 20 ml H_2SO_4, 10 g K_2SO_4, and 1 g $CuSO_4$, and digest as described in the previous method. Transfer this digested solution and the filtrate to the distillation apparatus, add 100 ml 33% NaOH solution, and continue as in the procedure.

2. For stainless steels containing 0.04% nitrogen, weigh 0.25 g. For high-manganese–high-nitrogen austenitic steels containing about 0.4% N, weigh 30 mg. The results with this small weight may not be sufficiently accurate; alternatively, a photometer cell of shorter light path can be used with a larger sample.
3. If the sample will dissolve more easily in HCl, use this acid and then proceed with the $HClO_4$ addition.

4. Acid-Soluble Nitrogen

Transfer 1–5 g of sample (the weight depending on the nitrogen concentration and whether the ammonia is to be determined titrimetrically or photometrically) to a beaker or Kjeldahl flask, and add 100 ml 10% HCl. Dissolve at a temperature that does not exceed 50°C, and proceed with the distillation by either method 2 or 3.

5. Ester-Halogen Nitrogen

This method is applicable to the isolation of the nitrides of aluminum, silicon, zirconium, titanium, vanadium, niobium, tantalum, and certain other elements from all types of steel.

a. Apparatus

(1) DISSOLUTION APPARATUS (see Fig. 8.2).

(2) STEAM-DISTILLATION APPARATUS (see Fig. 8.1).

b. Reagents

(1) AMMONIA-FREE WATER (see caustic-fusion method and also Chapter 8).

(2) SODIUM HYDROXIDE SOLUTION, 33% (see caustic-fusion method).

(3) NESSLER REAGENT (see caustic-fusion method).

(4) ASBESTOS. Use asbestos that has been treated with (1 + 1) HCl for preparing Gooch crucibles. Dry the crucibles and mat overnight at 105°C. The dilute HCl should be prepared with ammonia-free water. Use precaution when handling asbestos so that the material does not become airborne and ingested by personnel.

(5) METHYL ACETATE. Use anhydrous reagent-grade compound.

(6) BROMINE. Reagent-grade.

(7) ABSOLUTE ETHYL OR METHYL ALCOHOL.

c. Procedure

Remove all traces of water from the dissolving apparatus by drying in an oven. Cool and then wash with absolute alcohol. Transfer 2 g of sample (Note 1) to the flask of the dissolving apparatus. Add the reagents and dissolve as described in Chapter 8, Section VI.B. After dissolution, cool and filter through the Gooch crucible, using a slight suction, and wash the residue free of bromine and bromides with methyl acetate. Dry the Gooch crucible and contents at 105°C for 15 min.

Transfer the contents of the crucible to the distillation flask, add 20 ml (1 + 1) HCl (Note 2), and swirl to disintegrate the asbestos mat. Add 30 ml water, and swirl to mix the water with the acid. Connect the flask to the distillation apparatus, and finish the determination according to method 3, starting with the distillation of the ammonia.

NOTES

1. The dissolution of certain high-temperature alloys causes the formation of bromide salts that are less soluble than $FeBr_3$ and that present problems in filtering and washing. In these instances, use a 1-g sample.
2. This procedure is written for the determination of nitrogen present as aluminum nitride. For other nitrides other acid treatments may be necessary.

6. Total Nitrogen—Vacuum or Inert-Gas Fusion

File or machine the block or pin sample. Then degrease by dipping first in reagent-grade benzene and then in reagent-grade acetone. Dry in a stream of warm air. Weigh a pin or milled sample to the nearest 0.1 mg. Select the dimensions of the pin and the weight of the sample according to the instrument being used. Determine the nitrogen content with an analyzer that incorporates the single-sample-per-crucible method. Such analyzers are described in Chapter 4. Calibrate the instrument using standards of accurately measured total nitrogen content and degassed crucibles.

C. OXYGEN

1. Activation (see Chapter 3, Section IV)

2. Vacuum or Inert-Gas Fusion

After preparation of the sample as described in Section IV, it may be analyzed with any of the instruments described in Chapter 4 that are used for this purpose. The single-sample-per-crucible technique described there generally provides for faster analyses with fewer gettering problems. The older analyzers, utilizing the multiple-samples-per-crucible technique, can be of advantage in that they usually will accept larger samples, so that segregation effects can be reduced. Bath and flux materials used with the latter analyzers should be examined carefully for their oxygen contents. The fusion technique is preferred over activation analyses where less than about 20 ppm oxygen is to be determined. Standard samples (see Section IV) must be analyzed to ascertain the precision and accuracy of the particular method used.

3. Oxide Inclusions

For a typical 4-g sample, this method can detect as little, respectively, as 5, 3, 2.5, 10, 2, and 0.6 ppm of the following oxides in steels: FeO, MnO, SiO_2, Al_2O_3, CaO, and MgO. The upper limit can be extended by aliquoting. Zr, Ti, V, and Cr may be present as an oxide, carbide, nitride, or sulfide but are reported as the element.

a. Reagents

(1) BROMINE. Reagent-grade.

(2) ANHYDROUS METHYL ALCOHOL.

(3) METHYL ACETATE. Water-free reagent-grade.

(4) WASH SOLUTION, 6% by weight K_2CO_3.

(5) FUSION MIXTURE. Mix three parts Na_2CO_3, three parts K_2CO_3, and four parts $Na_2B_4O_7 \cdot 10H_2O$ by weight.

(6) INTERNAL STANDARD COBALT SOLUTION. Dissolve 300 mg high-purity cobalt sponge in a minimum amount of HCl and HNO_3, and dilute to 1 liter with distilled H_2O. One ml contains 0.30 mg cobalt.

(7) MASTER SOLUTION. Dissolve 50 mg fusion mixture and 0.30 mg cobalt per 10 ml of 10 volume % HCl.

b. Sample Preparation

Machine or hand-file the sample to remove any oxide film or scale from the surface. Where possible, use a sample that has previously been analyzed by neutron activation for oxygen. Immediately store in a labeled polyethylene vial under absolute methyl alcohol until ready for analysis. When this sample is also used for neutron-activation analysis, it is immediately replaced in the vial after this analysis.

c. Chemical Procedure

Weigh a 4–6-g sample to the nearest 0.1 mg, and quickly transfer to a 250-ml ground-glass, stoppered Erlenmeyer flask previously dried in an oven and containing the appropriate amount of water-free methyl acetate. Use

15-ml methyl acetate per gram of sample. Add 5 ml bromine per gram of sample, and insert the stopper. Place on a shaker at slow speed, and dissolve overnight at room temperature.

After dissolution, treat one sample at a time. Filter with suction, using two 55-mm-diameter Whatman No. 44 (or equivalent) papers. Wash the walls of the filter funnel rapidly with methyl acetate. Police the flask, and rinse into the funnel with additional methyl acetate. Maintain a small quantity of methyl acetate on the residue in the filter at all times during transfer. After the residue has been completely transferred to the filter, wash six additional times with methyl acetate. Next, wash with a minimum of six portions of anhydrous methyl alcohol or until all brown coloration is removed from the paper. Then wash six times with hot (near boiling) 6% K_2CO_3 solution. Allow about 1 min contact with the residue for each washing. Finally, wash with at least three portions of water, and finish with two washings of methyl alcohol.

Remove the top paper from the filter funnel, and place it in a clean platinum crucible of about 25-ml capacity. Swab any remaining residue with the second (lower) paper, and add to the first paper in the crucible. With a tweezer carefully remove any undissolved sample from the filter, dry, and weigh. Subtract this weight from the original sample weight to determine the weight of sample dissolved.

Burn off the paper slowly in a low-temperature muffle furnace until charred, and then complete the ignition in a high-temperature muffle furnace (about 930°C) for about 10 min. (The usual procedure is to filter all the samples first and then ignite them as a group.) Before starting filtrations for other samples, wash the filter funnel and other pertinent portions of the apparatus with methyl acetate to remove all traces of moisture.

Add 50 mg fusion mixture to the ignited residue in the platinum crucible, fuse over a Meeker burner, cool, and dissolve in 5 ml 10% HCl. Heat to dissolve only if necessary. Quantitatively transfer the solution to a 10-ml volumetric flask, add 1 ml cobalt internal standard solution, and dilute to the mark with additional 10% HCl. Mix well.

For a sample weight less than 5 g, transfer the solution to a clean, dry cylindrical plastic vial; cap if desired. For a sample heavier than 5 g, aliquot 5 ml of solution to a plastic vial, and add 5 ml master solution. The sample is now ready for solution spectrographic analysis. Following is a summary of the equipment and procedure used.

d. Optical-Emission Spectrometric Procedure

(1) EQUIPMENT OR PROCEDURE.

(*a*) *Spectrometer*. Applied Research Laboratories, Inc., Quantometer or equivalent.

(*b*) *Grating*. 2-m grating, 961 grooves/mm.
(*c*) *Analytical Gap*. 3 mm.
(*d*) *Width of Primary Slit*. 50 μm.
(*e*) *Width of Exit Slits*. 75 μm.
(*f*) *Exposure time*.
Prespark—35 sec.
Integration—40–60 sec., initially established by adjusting the sensitivity potentiometer in the internal standard channel.
(*g*) *Source Parameter*.
Discharge current—1.95 A.
Primary current—4.0 A.
Primary voltage—120 V.
(*h*) *Rotating-Disk Assembly*. Applied Research Laboratories, Inc., tantalum mandrel or equivalent.
(*i*) *Upper Electrode*. High-purity graphite electrode, $\frac{1}{4}$-in. diameter, cone-shaped with an included angle of 30° and a $\frac{1}{16}$-in.-radius tip; ASTM Designation C-5 [227].
(*j*) *Lower Electrode*. High-purity, graphite rotating-disk electrode, $\frac{1}{2}$-in. diameter, $\frac{1}{8}$-in. width, with a $\frac{1}{8}$-in.-diameter hole; ASTM Designation D-1 [227].
(*k*) *Rotational Speed of Electrode*. 5 rev/min.
(*l*) *Depth of Electrode in the Solution*. 2 mm.
(*m*) *Solution Container*. Teflon boat.
(*n*) *Volume of Solution*. 2 ml.

(2) SPECTRAL LINES USED FOR ANALYSIS.

Element	Wavelength (A)	Concentration Range[a] (mg/10 ml)
Mn	2576.1[b]	0.021–1.001
	2933.1	0.021–3.001
Si	2881.6[b]	0.010–1.010
	2516.1	0.010–3.01
Al	3944.0	0.066–3.016
Mg	2852.1[b]	0.0014–0.1246
Fe	2599.4[b]	0.034–2.002
Ca	4226.7	0.0032–0.3012
Zr	3438.2	0.0043–0.2264
Ti	3242.0	0.0036–0.600
V	3110.7	0.0011–0.168
Cr	2677.2	0.0030–0.2083
Co (Internal standard)	2286.2	0.3

[a] Concentrations expressed as the oxides specified earlier for the first six elements only.
[b] Second-order diffraction spectrum.

ACKNOWLEDGMENT

The help of Dr. J. K. Hurwitz and Dr. C. K. Russell of the U.S. Steel Research Laboratory, who, respectively, provided the optical-emission spectrometric procedure for oxide inclusions and reviewed the section on the electrochemical determination of oxygen in molten steel, is sincerely appreciated.

REFERENCES

1. J. P. Bruch, "Determination of Gases in Steel and Application of the Results," in *Determination of Chemical Composition—Its Application in Process Control Spec. Rept. 131*, Iron and Steel Institute, London, 1971, pp. 64–73.
2. W. O. Philbrook, M. Tenenbaum, and F. M. Washburn, "Refining Practice," in W. O. Philbrook, M. B. Bever, H. B. Emerick, and B. M. Larsen, Eds., *Basic Open Hearth Steelmaking*, 2nd ed., AIME, New York, 1951, pp. 294–338.
3. R. Eborall, "The Determination of Hydrogen in Metals: A Review," in *The Determination of Gases in Steel, Spec. Rept. No. 68*, Iron and Steel Institute, London, 1960, pp. 192–218.
4. H. E. McGannon, Ed., *The Making, Shaping, and Treating of Steel*, 9th ed., U.S. Steel Corp., Pittsburgh, 1971.
5. E. C. Pigott, *Ferrous Analysis*, 2nd ed., Wiley, New York, 1953.
6. E. W. Husemann et al., "Pickling of Iron and Steel," in T. Lyman, Ed., *Metals Handbook*, Vol. II, *Heat Treating, Cleaning, and Finishing*, American Society for Metals, Cleveland, Ohio, 1964, pp. 346–356.
7. C. H. Avery et al., "Heat Treating of Stainless Steel and Heat-Resisting Alloys," in T. Lyman, Ed., *Metals Handbook*, Vol. II, *Heat Treating, Cleaning, and Finishing*, American Society for Metals, Cleveland, Ohio, 1964, pp. 243–253.
8. C. H. Samans et al., "Selection of Steel for High-Temperature Service in Petroleum Refinery Applications," in T. Lyman, Ed., *Metals Handbook*, Vol. I, *Properties and Selection of Materials*, American Society for Metals, Cleveland, Ohio, 1961, pp. 585–603.
9. J. V. Dawson and L. W. L. Smith, "The Determination of Hydrogen in Cast Iron," in *The Determination of Gases in Metals, Spec. Rept. No. 68*, Iron and Steel Institute, London, 1960, pp. 219–228.
10. R. W. Buzzard and H. E. Cleaves, *N.B.S. Circular 511*, National Bureau of Standards, Washington, 1951.
11. R. D. Pehlke and J. F. Elliot, *Trans. AIME*, **218,** 1088 (1960).
12. F. C. Langenburg, *Trans. AIME*, **206,** 1099 (1956).
13. G. R. Adams, W. O. Binder, and J. Thompson, *Proc. AIME Electric Furn. Conf.*, **15,** 17 (1957).

14. A. H. Cottrell and A. T. Churchman, *J. Iron Steel Inst.* (London), **162,** 271 (1949).
15. W. C. Leslie and R. L. Rickett, *Trans. AIME*, **197,** 1021 (1953).
16. R. W. Fountain and J. Chipman, *Trans. AIME*, **212,** 737 (1958).
17. E. R. Morgan and J. C. Shyne, *Trans. AIME*, **209,** 65 (1957).
18. W. Wegner, *Archiv. Eisenhuttenw.*, **26,** 71 (1955).
19. S. Epstein, U.S. Patent 2,356,450, Aug. 22, 1944.
20. S. Epstein, H. S. Cutler, and J. W. Frame, *J. Metals*, **188,** 830 (1950).
21. J. W. Frame and F. B. S. Schunk, *AIME Special IMD Report Series No. 6*, 1958, p. 57.
22. C. M. Johnson, *Iron Age*, **134,** no. 4, 10 (1934).
23. B. Phrenosil, *Hutnicke Listy*, **12,** 289 (1957).
24. J. D. Fast, *Phillips Res. Report*, **11,** 490 (1956).
25. (a) E. E. Schumaker, *Trans. AIME*, **188,** 1097 (1950); (b) P. Grieveson, K. H. Jack, and S. Wild, *Special Ceramics*, **4,** 237 (1968); *ibid.*, **5,** 271 and 385 (1972); through W. Roberts, P. Grieveson, and K. H. Jack, *J. Iron Steel Inst.* (London), **210,** 931 (1972); and (c) D. Hardie and K. H. Jack, *Nature*, **180,** 332 (1957).
26. F. C. Langenberg and M. J. Day, *Proc. AIME Elec. Furn. Conf.*, **15,** 7 (1957).
27. C. H. Lund and H. J. Wagner, *Memorandum 160*, Defense Materials Information Center, Battelle Memorial Inst., Columbus, Ohio, Nov. 15, 1962.
28. E. T. Turkdogan and E. M. Fenn, *J. Iron Steel Inst.* (London), **181,** 343 (1955).
29. W. Hume-Rothery, G. V. Raynor, and A. T. Little, *J. Iron Steel Inst.* (London), **145,** 1261 (1942).
30. Yu. G. Gurevich, *Izv. Vyssh. Uchebn. Zaved. Chern. Met.*, **5** (7), 71 (1962); through *Chem. Abstr.*, **57,** 14809d (1962).
31. H. F. Beeghly, *Anal. Chem.*, **24,** 1713 (1952).
32. E. C. Pigott, *Metallurgia*, **37,** 129 (1948).
33. H. S. Karp, L. L. Lewis, and L. M. Melnick, *J. Iron Steel Inst.* (London), **200,** 1032 (1962).
34. T. Mori, K. Fujita, M. Tokizane, and K. Yamaguchi, *Tetsu To Hagane*, **50,** 911 (1964).
35. R. Grimaldi, S. Maneschi, and N. Vantini, *Archiv. Eisenhuttenw.*, **38,** 401 (1967).
36. T. Mori, M. Tokizane, Y. Nakazima, and T. Saheki, *Tetsu To Hagane*, **51,** 2031 (1965).
37. W. Dahl, P. Schwab, and H. Hergstenberg, *Archiv. Eisenhuttenw.*, **32,** 475 (1961).
38. C. C. Hsiao, *Nature*, **181,** 1527 (1958).
39. J. M. Gray and R. B. G. Yeo, *Trans. ASM*, **61,** 255 (1968).
40. H. F. Beeghly, *Anal. Chem.*, **24,** 1095 (1952).
41. J. Pearson and U. J. C. Ende, *J. Iron Steel Inst.* (London), **175,** 52 (1953).
42. P. Konig, W. Scholz, and H. Ulmer, *Archiv. Eisenhuttenw.*, **32,** 541 (1961).
43. P. Werthebach and H. Hoff, *Stahl Eisen*, **78,** 736 (1958).
44. E. T. Turkdogan and S. Ignatowicz, *J. Iron Steel Inst.* (London), **188,** 242 (1958).

45. W. C. Schulte et al., "The Selection of Steel for Fatigue Resistance," in T. Lyman, Ed., *Metals Handbook*, Vol. I, *Properties and Selection of Materials*, American Society for Metals, Cleveland, Ohio, 1961, pp. 217–224.
46. G. Baker and W. J. Stuart, *J. Chem. Soc.* (London), **17,** 390 (1864).
47. A. Allen, *J. Iron Steel Inst.* (London), **7,** 480 (1879).
48. Boussingault, *Compt. Rend.*, **53,** 77 (1861).
49. J. Nessler, *Inaugural Dissertation*, University of Freiburg, 1856.
50. P. Artman and A. Skrabal, *Z. Anal. Chem.*, **46,** 5 (1907).
51. J. Hurum and H. Fay, *Chem. Met. Eng.*, **26,** 218 (1922).
52. M. B. Donald, *Analyst*, **49,** 375 (1924).
53. L. W. Winkler, *Z. Angew. Chem.*, **26,** 231 (1913).
54. A. Gotta and H. Seehof, *Z. Anal. Chem.*, **124,** 216 (1942).
55. S. Acosta, *Inst. Hierro y Acero*, **11,** 258 (1958); through *Chem. Abstr.*, **53,** 8925a (1959).
56. N. B. Myakina, *Pochvovedenie*, **No. 1,** 106 (1958); through *Chem. Abstr.*, **52,** 8835c (1958).
57. W. C. Newell, *J. Iron Steel Inst.* (London), **152,** 333 (1945).
58. J. Calmettes and J. Drain, *Rev. Met.* (Paris), **53,** 682 (1956).
59. R. Boulin, J. Coulombeau, and E. Jaudon, *Rev. Met.* (Paris), **57,** 623 (1960).
60. P. H. Scholes, "Special Analytical Techniques and Their Applications in Quality Control," BISRA Conference, Sept. 28–30, 1970.
61. O. Kammori, Y. Hiyama, and W. Hotta, *Trans. Jap. Inst. Metals*, **8,** 56 (1967)
62. T. Mori, K. Fujita, and I. T. Kim, *Suiyokaishi*, **14,** 35 (1959); through *Chem. Abstr.*, **55,** 19591b (1961).
63. H. I. Kamada and K. Sato, *Bunseki Kagaku*, **6,** 150 (1957).
64. K. Kawamura et al., *Fuji Iron Steel Tech. Report*, **17,** 37 (1968); through *Brit. Iron Steel Inst. Transl. 7049*, Oct. 1971.
65. T. R. Cunningham and H. L. Hamner, *Ind. Eng. Chem., Anal. Ed.*, **11,** 303 (1939).
66. Iron and Steel Institute, *Spec. Rept. No. 62*, London, 1958.
67. J. L. Hague, R. A. Paulson, and H. A. Bright, *J. Res. Nat. Bur. Standards*, **43,** 201 (1949).
68. H. Kempf and K. Abresch, *Archiv. Eisenhuttenw.*, **13,** 419 (1940).
69. L. I. Levi and O. M. Borisova, *Zavodsk. Lab.*, **32,** 414 (1966).
70. L. I. Levi and Y. A. Kitaev, *Zavodsk. Lab.*, **35,** 1438 (1969).
71. A. A. Federov and A. M. Krichevskaya, *Zavodsk. Lab.*, **34,** 1425 (1968).
72. P. Klinger, *Archiv. Eisenhuttenw.*, **5,** 29 (1931).
73. J. Coulombeau and E. Jaudon, *Chim. Anal.* (Paris), **42,** 61 (1960); through *Anal. Abstr.*, **7,** 4282 (1960).
74. C. L. Shapiro, *Trans. ASM*, **27,** 666 (1939).
75. C. Holthaus and W. Holtmann, *Archiv. Eisenhuttenw.*, **17,** 247 (1944).
76. P. Tyou, J. Vanstiphout, and M. Lacomble, *Rev. Univ. Mines*, **12,** 641 (1956).
77. J. Marot, *Ind. Chim. Belge*, **22,** 1161–1174 (1957).
78. H. Kempf and K. Abresch, *Archiv. Eisenhuttenw.*, **17,** 261 (1944).
79. J. D. Hobson, "The Determination of Nitrogen in Metals," in *Spec. Rept. No. 68*, Iron and Steel Institute, London, 1960, pp. 151–182.

80. E. W. Beiter, R. H. Wynne, W. F. Harris, M. I. Mistrik, and F. P. Byrne, Pittsburgh Conference on Analytical Chemistry and Applied Spectroscopy, Pittsburgh, 1957.
81. F. B. Moore and H. Diehl, *Anal. Chem.*, **46,** 1638 (1962).
82. M. Sicha, *Hutnicke Listy*, **15,** 729 (1960).
83. H. Kempf and K. Abresch, *Archiv. Eisenhuttenw.*, **17** (5, 6), 119 (1943).
84. S. Takano, M. Iida, and Z. Tsutsumi, *Nippon Kinzoku Gakkaishi*, **18,** 494 (1954); through *Chem. Abstr.*, **52,** 10797e (1958).
85. H. F. Beeghly, *Anal. Chem.*, **21,** 1513 (1949).
86. V. T. Willmer and K. Zimmermann, *Archiv. Eisenhuttenw.*, **42,** 877 (1971).
87. Yu. A. Klyachko, M. M. Shapiro, and E. F. Yakovleva, Ti. *Seminara po Zharostoikim. Materialaiv, Akad. Nauk. Ukn. USSR, Inst. Metallokoram i Spets. Splavoo, Kiev*, **No. 6,** 59 (1960); through *Chem. Abstr.*, **55,** 10259c (1961).
88. W. R. Bandi, E. G. Buyok, G. Krapf, and L. M. Melnick, "The Use of Differential Thermal Analysis–Effluent Gas Analysis in Metallurgical Research Investigations," in R. W. Schwenker and P. D. Garn, Ed., *Thermal Analysis*, Vol. II, Academic, New York, 1969, pp. 1363–1376.
89. W. R. Bandi, W. A. Straub, E. G. Buyok, and L. M. Melnick, *Anal. Chem.*, **38,** 1336 (1966).
90. G. Krapf, W. R. Bandi, E. G. Buyok, and L. M. Melnick, Pittsburgh Conference on Analytical Chemistry and Applied Spectroscopy, Cleveland, Ohio, 1972.
91. W. Oelson and K. H. Sauer, *Archiv. Eisenhuttenw.*, **38,** 141 (1967).
92. T. R. Dulski and R. M. Raybeck, *Anal. Chem.*, **41,** 1025 (1969).
93. K. H. Koch, M. Kretchsmer, G. Langenscheid, and G. Schmolke, *Hoesch Ber.*, **6,** 40 (1971).
94. W. Koch, Report of Meeting of Working Group, *Determination of Gas Contents of Steel*, Chemical Comm. of VDEH, Dusseldorf, Feb. 9, 1966; through Ref. 83.
95. G. W. Goward, *Anal. Chem.*, **37,** 117R (1965).
96. A. Gerhardt, Doctoral thesis, Technische Universität, Berlin, 1965.
97. A. Gerhardt, T. Kraus, and M. G. Frohberg, *Giess. Techn. Wiss.*, **17,** 203 (1965).
98. W. Koch and H. Lemm, *Z. Anal. Chem.*, **215,** 377 (1966).
99. A. Gerhardt, T. Kraus, and M. G. Frohberg, Pittsburgh Conference on Analytical Chemistry and Applied Spectroscopy, Pittsburgh, 1966.
100. A. Gerhardt, T. Kraus, and M. G. Frohberg, *Neue Hutte*, **12,** 115 (1967).
101. W. Koch and H. Lemm, *Archiv. Eisenhuttenw.*, **38,** 881 (1967).
102. W. Koch and H. Lemm, Pittsburgh Conference on Analytical Chemistry and Applied Spectroscopy, Cleveland, Ohio, 1968.
103. H. Lemm, Pittsburgh Conference on Analytical Chemistry and Applied Spectroscopy, Pittsburgh, 1967.
104. A. M. Vasserman and Z. M. Turovtseva, *Zh. Anal. Khim.*, **20,** 1359 (1965).
105. C. G. Goldbeck, S. P. Turel, and C. J. Rodden, Eleventh Conference on Analytical Chemistry in Nuclear Technology, Oak Ridge, Tennessee, 1967.

106. C. G. Goldbeck, S. P. Turel, and C. J. Rodden, *Anal. Chem.*, **40,** 1393 (1968).
107. S. L. Mandel'shtam and O. B. Fal'kova, *Zavodsk. Lab.*, **16,** 430 (1950); through *Chem. Abstr.*, **46,** 1387e (1952).
108. K. A. Sukhenko, P. P. Galonov, and T. V. Barasheva, *Izv. Akad. Nauk. SSSR, Ser. Fiz.*, **23,** 1123 (1959); through *Chem. Abstr.*, **54,** 6397g (1960).
109. E. F. Runge and F. R. Bryan, *Appl. Spec.*, **10,** 68 (1956).
110. V. A. Fassel, W. A. Gordon, and R. W. Tabeling, *Determination of Gases in Metals, ASTM Spec. Tech. Publication No. 222*, Philadelphia, 1957.
111. V. A. Fassel, W. A. Gordon, and R. J. Jasinski, *Proceedings of the Second International Conference on the Peaceful Uses of Atomic Energy* (Geneva), **28,** 583, 1958, United Nations, New York.
112. E. F. Runge and F. R. Bryan, *Spectrochim. Acta*, **12,** 96 (1958).
113. V. A. Fassel and H. Kamada, *Spectrochim. Acta*, **17,** 121 (1961).
114. A. N. Zaidel and A. A. Petrov, *Zavodsk. Lab.*, **28,** 552 (1962).
115. F. M. Evens and V. A. Fassel, *Anal. Chem.*, **35,** 1444 (1963).
116. A. D. Kirshenbaum and A. V. Grosse, *Anal. Chem.*, **26,** 1955 (1954).
117. I. A. Maslov, *Zavod. Lab.*, **30,** 51 (1964).
118. M. L. Pearce and G. R. Masson, *Trans. AIME*, **224,** 1134 (1962).
119. W. H. Walker and W. A. Patrick, *Orig. Com. 8th Intern. Cong., Applied Chem.*, **21,** 139 (1912); through *Chem. Abstr.*, **6,** 3381 (1912).
120. P. Oberhoffer and H. Schenck, *Stahl Eisen*, **47,** 1526 (1927).
121. W. Hessenbruch and H. Schenck, *Archiv. Eisenhuttenw.*, **1,** 583 (1928).
122. L. Jordan and J. R. Eckman, *Sci. Pap. Nat. Bur. Stand.* **20,** 445 (1925).
123. H. A. Sloman, *Spec. Rept. No. 9*, Iron and Steel Institute, London, 1935, pp. 71–89.
124. H. A. Sloman, *Spec. Rept. No. 16*, Iron and Steel Institute, London, 1937, pp. 82–100.
125. H. A. Sloman, *Spec. Rept. No. 25*, Iron and Steel Institute, London, 1939, pp. 43–62.
126. H. A. Sloman, *J. Iron Steel Inst.* (London), **43,** 298 (1941).
127. H. A. Sloman, *J. Inst. Metals*, **71,** 391 (1945).
128. H. A. Sloman, *Engineering*, **160,** 385 (Nov. 9, 1945).
129. H. A. Sloman, *Engineering*, **160,** 404 (Nov. 16, 1945).
130. H. A. Sloman, *Engineering*, **160,** 419 (Nov. 23, 1945).
131. H. A. Sloman, *Report No. 5066*, National Physical Laboratory, Metallurgy Division, Oct. 9, 1947; through R. A. Yeaton, *Vacuum*, **2,** 115 (1952).
132. H. A. Sloman, C. A. Harvey, and O. Kubaschewski, *J. Inst. Metals*, **80,** 391 (1951–1952).
133. T. B. King, *Spec. Rept. No. 68*, Iron and Steel Institute, London, 1960, pp. 3–18.
134. C. D. Cassler and G. R. Fitterer, *J. Metals*, **17,** 655 (1965).
135. M. L. Hill, private communication, 1956.
136. T.-K. Willmer and K. Zimmermann, *Archiv. Eisenhuttenw.*, **41,** 1131 (1970).
137. C. Taylor and J. Chipman, *Trans. AIME*, **154,** 228 (1943).
138. J. G. Bassett, J. W. Dougherty, and G. R. Fitterer, *Proc. 39th Conf. Nat'l Open Hearth Comm.*, Iron Steel Div. AIME, **39,** 44 (1956).

139. J. V. Dawson and L. W. L. Smith, *Spec. Rept. No. 68*, Iron and Steel Institute, London, 1960, pp. 219–228.
140. L. I. Levi and A. N. Aleksandrova, *Zavodsk. Lab.*, **34,** 666 (1968).
141. T. E. Brower and B. M. Larsen, *Trans. AIME*, **162,** 712 (1945).
142. G. F. Huff, G. R. Bailey, and J. H. Richards, *Trans. AIME*, **194,** 1162 (1952).
143. S. Gilbert and G. R. Bailey, *Trans. AIME*, **200,** 1383 (1954).
144. F. C. Langenberg and J. M. Snook, *Trans. AIME*, **215,** 706 (1959).
145. P. L. Jackson, *J. Metals*, **18** (6), 725 (1966).
146. G. Padget, *Steel Coal*, **187** (4968), 648 (Oct. 4, 1963).
147. Anonymous, *Metallurgia*, **81,** (484), 79 (1970).
148. H. Zitter and W. Schwarz, *Stahl Eisen*, **88,** 862 (1968); through *Brit. Iron Steel Inst. Trans. No. 6685*, London.
149. G. Blanc and A. Blondel, *Brit. Foundryman*, **50,** 245 (1957); through *Chem. Abstr.*, **52,** 3630d (1958).
150. F. Cooke, S. Hall, and C. E. A. Shanahan, *J. Iron Steel Inst.* (London), **207,** 204 (1969).
151. J. F. Martin, R. C. Takacs, R. Rapp, and L. M. Melnick, *Trans. AIME*, **230,** 107 (1964).
152. P. Dickens, *Stahl Eisen*, **84,** 776 (1964); through Ref. 1.
153. D. I. Walter and G. A. Picklo, Jr., Pittsburgh Conference on Analytical Chemistry and Applied Spectroscopy, Pittsburgh, 1959.
154. B. B. Bach, J. V. Dawson, and L. W. L. Smith, *Brit. Cast Iron Res. Assoc. J. Res. Dev.*, **5,** 490 (1954).
155. K. Zimmermann and H. Uhlitzsch, *Neue Hutte*, **2,** 607 (1957); through *Chem. Abstr.*, **52,** 7013d (1958).
156. J. V. Dawson and L. W. L. Smith, *Modern Castings*, **33** (2), 39 (1958).
157. J. Vaclavinek, *Slevarenstvi*, **10,** 83 (1957); through *Chem. Abstr.*, **52,** 1879d (1958).
158. T. Shimomura and M. Yamamoto, *Tetsu To Hagane*, **43,** 569 (1957); through *Chem. Abstr.*, **54,** 6400b (1960).
159. L. I. Levi and A. N. Aleksandrova, *Liteinoe Proizvod.*, **No. 1,** 25 (1971); through *Chem. Abstr.*, **74,** 134549 (1971).
160. E. Kato, *Tetsu To Hagane*, **49,** 641 (1963); through *H. Brutcher Translation 6060*, Altadena, Calif.
161. W. A. Peifer, *Proc. AIME Elec. Furn. Steel Conf.*, **15,** 27 (1957).
162. I. E. Campbell, *High Temperature Technology*, Wiley, New York, 1956, p. 181.
163. O. I. Popova and G. T. Kabannik, *Russian J. Inorg. Chem.*, **5,** 447 (1960).
164. K. Narita, *Nippon Kagashu Zasshi*, **79,** 1564 (1958); through *Chem. Abstr.*, **53,** 18740b (1959).
165. H. Sawamura, T. Mori, K. Fujita, and H. Miura, *Suryokwai Shi*, **13,** 681 (1958); through *Chem. Abstr.*, **53,** 11096d (1959).
166. H. A. Sloman, *J. Iron Steel Inst.* (London), **182,** 307 (1956).
167. ASTM Methods for Chemical Analysis of Metals, American Society for Testing and Materials, Philadelphia, 1960, p. 156.
168. H. F. Beeghly, *Ind. Eng. Chem., Anal. Ed.*, **14,** 137 (1942).
169. I. H. Sher, *Anal. Chem.*, **27,** 831 (1955).

170. C. E. Sims and C. W. Briggs, *Proc. AIME Electric Furn. Conf.*, **17,** 104 (1959).
171. G. Jecko and R. Touvenin, *Rev. Met.* (Paris), **66,** 823 (1969).
172. R. W. Cline, H. S. Karp, and L. M. Melnick, Pittsburgh Conference on Analytical Chemistry and Applied Spectroscopy, Cleveland, Ohio, 1970.
173. J. Kashima and T. Yamazaki, *Bull. Chem. Soc. Jap.*, **39,** 1453 (1966); through *Chem. Abstr.*, **65,** 14420f (1966).
174. L. I. Levi and Yu. A. Kitaev, *Zavodsk. Lab.*, **35,** 1438 (1969).
175. L. I. Levi and O. M. Borisova, *Zavodsk. Lab.*, **32,** 414 (1966).
176. H. J. Berg and G. Freiberg, *Giessereitechnik*, **16** (12), 402 (1970); through *Chem. Abstr.*, **74,** 134468 (1971).
177. J. Strohlein, *Chim. Anal.* (Paris), **51** (12), 644 (1969); through *Chem. Abstr.*, **72,** 96370 (1970).
178. K. Effenkammer, *Hutnicke Listy*, **25,** 190 (1970); through *Chem. Abstr.*, **73,** 21080 (1970).
179. M. Hanin, *Chim. Anal.* (Paris), **51,** 9 (1969).
180. W. Koch, *Metallkundliche Analyse*, Verlag Stahleisen, M.B.H. Dusseldorf and Verlag Chemie, G.M.B.H. Weinheim/Bergstr., 1965.
181. W. A. Fischer and A. Hoffman, *Archiv. Eisenhuttenw.*, **28,** 739 (1957).
182. K. Kiukkola and C. Wagner, *J. Electrochem. Soc.*, **104,** 379 (1957).
183. K. Sanbongi, M. Ohtani, Y. Omori, and H. Inoue, *Tetsu To Hagane*, **50,** 1823 (1964); English transl. in *Trans. Iron Steel Inst. Jap.*, **6,** 76 (1966).
184. Y. Matsushita and K. Goto, *Tetsu to Hagane*, **52,** 827 (1966); English translation in *Trans. Iron Steel Inst. Jap.*, **6,** 21 (1966).
185. W. A. Fischer and W. Ackermann, *Archiv. Eisenhuttenw.*, **36,** 643 (1965).
186. W. A. Fischer and W. Ackermann, *Archiv. Eisenhuttenw.*, **37,** 43 (1966).
187. R. Baker and J. M. West, *J. Iron Steel Inst.*, **204,** 212 (1966).
188. G. R. Fitterer, *J. Metals*, **18,** 961 (1966).
189. G. R. Fitterer, *J. Metals*, **19,** 92 (1967).
190. K. Schwerdtfeger, *Trans. TMS-AIME*, **239,** 1276 (1967).
191. C. Gatellier, K. Torsell, M. Olette, N. Meysson, M. Chastant, A. Rist, and P. Vicens, *Rev. Met.* (Paris), **66,** 673 (1969).
192. C. B. Alcock and T. N. Belford, *Trans. Faraday Soc.*, **60,** 822 (1964).
193. M. Kolodny, B. Minushkin, and H. Steinmetz, *Electrochem. Tech.*, **3** (9–10), 244 (1965).
194. T. C. Wilder, *Trans. TMS-AIME*, **236,** 1035 (1966).
195. C. M. Diaz and F. D. Richardson, in C. B. Alcock, Ed., *Electromotive Force Measurements in High Temperature Systems*, Elsevier, New York, 1968.
196. J. K. Pargeter, *J. Metals*, **20,** 27 (1968).
197. R. Littlewood, *Can. Met. Quart.*, **5** (1), (1966); through *Chem. Abstr.*, **65,** 7999c (1966).
198. H. Peters and G. Mann, *J. Electrochem.*, **63,** 244 (1959).
199. C. K. Russell, R. J. Fruehan, and R. S. Rittiger, *J. Metals*, **23,** 44 (1971).
200. R. J. Fruehan, L. J. Martonik, and E. T. Turkdogan, *Trans. Met. Soc. AIME*, **245,** 1501 (1969).
201. C. K. Russell, private communication, June 1972.
202. G. R. Fitterer, C. D. Cassler, and V. L. Vierbicky, *J. Metals*, **21** (8), 46 (1969).

203. J. K. Pargeter and D. K. Faurschou, *J. Metals*, **21** (3), 46 (1969).
204. C. Wagner, *Proc. Int. Comm. Electrochem. Thermodynamics and Kinetics*, 7th Meeting, London, 1955, Butterworth, London, 1957.
205. H. Schmalzried, *Z. Elektrochem.*, **66,** 572 (1962).
206. B. C. H. Steele and C. B. Alcock, *Trans. TMS-AIME*, **233,** 1359 (1965).
207. J. W. Patterson, E. C. Bogren, and R. A. Rapp, *J. Electrochem. Soc.*, **114,** 752 (1967).
208. W. A. Fischer and D. Janke, *Archiv. Eisenhuttenw.*, **39,** 89 (1968).
209. W. A. Fischer and D. Janke, *Archiv. Eisenhuttenw.*, **42,** 691 (1971).
210. W. A. Fischer and D. Janke, *Archiv. Eisenhuttenw.*, **42,** 695 (1971).
211. P. A. Cerkasov and W. A. Fischer, *Archiv. Eisenhuttenw.*, **42,** 699 (1971).
212. P. A. Cerkasov, W. A. Fischer, and C. Pieper, *Archiv. Eisenhuttenw.*, **42,** 873 (1971).
213. W. A. Fischer and G. Pateisky, *Archiv. Eisenhuttenw.*, **41,** 661 (1970); through *Brit. Iron Steel Inst. Trans. 8798*, January 1971.
214. D. A. Dukelow, J. M. Steltzer, and G. F. Koons, *J. Metals*, **23** (12), 22 (1971).
215. R. Littlewood, "Application of Solid Electrolytes to Monitoring of Gases and Liquid Metals," Pittsburgh Conference on Analytical Chemistry and Applied Spectroscopy, Cleveland, Ohio, 1970.
216. S. L. Gertsman, D. K. Faurschou, and J. C. Pope, *Foundry*, **98** (9), 78 (1970).
217. Anonymous, *Iron Age*, **206,** 29 (December 10, 1970).
218. E. T. Turkdogan and R. J. Fruehan, *Can. Met. Quart.*, **11,** 371 (1972).
219. W. Schwarz and H. Zitter, *Berg. Hutten. Monatsh.*, **113,** 1 (1968); through *Brit. Iron Steel Inst. Trans. 6676*, February 1969.
220. E. M. Chistyakova, T. A. Izmanova, and N. N. Timoshenko, *Zavodsk. Lab.*, **37,** 889 (1971).
221. M. G. Frohberg and H. Leygraf, *Giesserei*, **57** (17), 531 (1970); through *Chem. Abstr.*, **73,** 126639 (1970).
222. B. Bagshawe, J. B. Headridge, R. Pemberton, E. D. Rawsthorne, and D. F. Wilson, *Metallurgia*, **82** (494), 237 (1970).
223. Y. Abe and S. Takazawa, *Tetsu To Hagane*, **57,** 1350 (1971); through *Chem. Abstr.*, **75,** 70978 (1971).
224. F. Tsugane and M. Kamakura, *Tetsu To Hagane*, **57,** 1543 (1971); through *Chem. Abstr.*, **75,** 94320 (1971).
225. H. Goto, S. Suzuki, J. Kimura, and A. Onuma, *Sci. Rep. Res. Inst., Tohoku Univ., Ser. A.*, **11,** 271 (1959); through *Chem. Abstr.*, **54,** 4253g (1960).
226. H. J. Berg, E. Hesse, W. Richling, and R. Zimmermann, *Neue Hutte*, **14,** 242 (1969); through *Chem. Abstr.*, **71,** 56328 (1969).
227. American Society for Testing and Materials, "Recommended Practice for Designation of Shapes and Sizes of Graphite Electrodes," E130 in *1970 Annual Book of A.S.T.M. Standards*, Pt. 32, Philadelphia, Pa., 1970, pp. 613–618.

CHAPTER

16

COBALT, NICKEL, AND COPPER

WILLIAM G. GULDNER

Bell Telephone Laboratories
Murray Hill, New Jersey

AND

DEAN I. WALTER

U.S. Naval Research Laboratory
Washington, D.C.

CONTENTS

- **2.** Oxygen and Nitrogen
  - **a.** *Vacuum Fusion: Cobalt*
  - **b.** *Vacuum Fusion: Nickel*
- **B.** Copper
  - **1.** Hydrogen
    - **a.** *Hot Extraction*
    - **b.** *Vacuum Fusion*
  - **2.** Oxygen
    - **a.** *Hydrogen-Reduction*
    - **b.** *Isotopic Dilution*
    - **c.** *Vacuum Fusion*
    - **d.** *Specialized Techniques for Surface Analysis*

VI. RECOMMENDED METHODS

- **A.** Cobalt and Nickel
  - **1.** Hydrogen
  - **2.** Oxygen (Hydrogen and Nitrogen)
- **B.** Copper
  - **1.** Hydrogen
  - **2.** Oxygen
  - **3.** Nitrogen

Cobalt and nickel are related to copper by virtue of their atomic-number sequence, a relationship that provides order and convenience of textural organization. The chemical, metallurgical, and electrical properties of cobalt and nickel, however, differ widely from those of copper and preclude a common treatment of the three metals in terms of their gas–metal relation-ships and their analytical requirements. Cobalt and nickel will be treated individually in areas where meaningful differences exist, but most of the analytical evidence suggests that the two metals can be treated as one within the analytical context of this chapter.

I. METALLURGICAL ASPECTS

A. COBALT

In modern metals technology cobalt is used primarily as an alloying element, and great interest has focused on the unique properties of this alloy family. Fewer applications have been found for the pure metal, and as a result, no major efforts have been expended to relate trace impurity compositions to particular physical properties of cobalt. Zone refinement of cobalt, by which overall purity was increased from 99.95 to 99.99%, with significant oxygen removal, resulted in lower hardness for the metal [1]. Other factors can influence hardness values, however, such as the relative amounts of α- and β-phases* present and the orientation of the hardness impression with a crystal. Cobalt has a higher Curie point than any other metal and is one of three elements that exhibit ferromagnetism at room temperature, thus assuring continued research interest in its comprehensive characterization.

Analysis of cobalt metal, with particular emphasis on gases, has more than academic interest. Producers of cobalt metal are concerned with the effectiveness of degassing processes in cobalt cathodes [2] and in meeting the ever-tightening compositional requirements of customers, who in turn are facing demands for alloys of greater integrity. Cobalt is used extensively in complex high-temperature alloys for aircraft-engine parts, and in the electrical and electronic fields, where metals having a low coefficient of linear expansion are required. One of the oldest and still most important uses of cobalt is in permanent magnet materials. Of the current U.S. annual consumption of metallic cobalt of more than 2 million pounds, almost half is used in making magnetic materials [3].

It is in the area of magnetic alloys that the influence of gases, particularly oxygen, is best known and is most dramatic. A wide range of soft, magnetic alloys respond to magnetic annealing (heat treatment in a magnetic field). Heidenreich, Nesbitt, et al. [4–6] have explained this on the basis of the oxygen present as an impurity. In determining the response of a series of iron–nickel–cobalt alloys to field heat treatment, their studies show that the oxygen added by conventional heat treatment (nonmagnetic) can be more important than variations within the basic alloy composition. An oxygen content of 0.001% is sufficient to produce magnetic annealing, with proper heat treatment, while an alloy containing 0.0001% oxygen fails to respond to field heat treatment.

The hysteresis loops shown in Fig. 16.1 [6] are typical of the squaring effect produced by heat treatment in a slightly oxidizing atmosphere for the family of soft magnetic alloys.

B. NICKEL

Nickel is used extensively as an alloying element in steels, high-temperature alloys, and magnetic materials and constitutes the matrix for a distinguished series of alloys. In addition, it has its own identity as a high-purity metal with special electronic applications [7]. Low-temperature electrical-resistance

* Based on the order of their discovery, the closed-packed hexagonal phase is designated α and the face-centered cubic β.

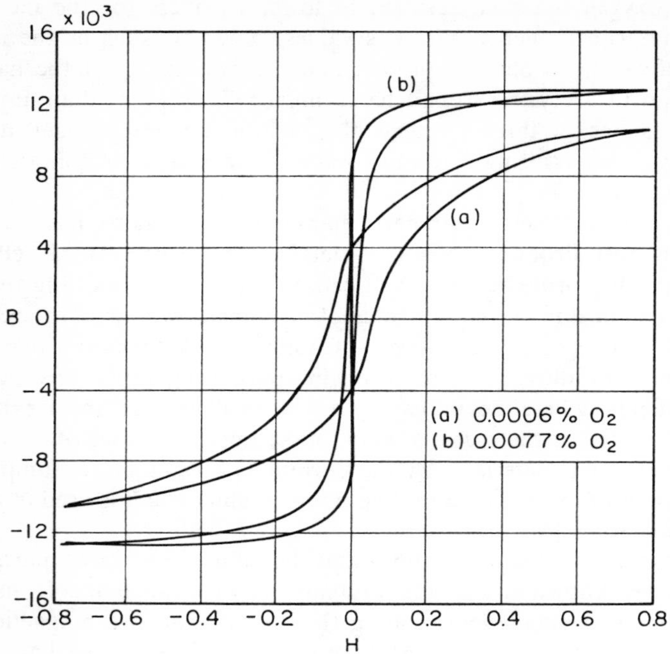

Fig. 16.1 Hysteresis loops on single-crystal ring of a soft magnetic alloy showing effects of annealing atmosphere: (a) annealed at 1100°C for 16 h, furnace cooled in H_2, followed by magnetic-field heat treatment; (b) same as (a) except for slightly oxidizing atmosphere. Flux density in Gauss (B); magnetizing force in Oersteds (H). (Reproduced by permission from *Journal of Applied Physics* [6].)

measurements are commonly used as a measure of total solid solution impurity in relatively pure materials. Working with high-purity nickel in which no single impurity was greater than 0.005% [8], Olsen and Larkin [9] showed that low-temperature resistance measurements are sensitive to as little as 0.0001% hydrogen (1 atom of hydrogen in 600,000 atoms of nickel). The fact that hydrogen is continuously evolved from nickel at room temperature, with an attendant change in resistivity, must be considered in interpretation of data obtained by low-temperature resistance measurements on high-purity nickel. For example, Fig. 16.2 shows that low-temperature resistance measurements on high-purity nickel change considerably as the nickel ages and that this is caused by the slow evolution of hydrogen from the nickel at room temperature.

Shahinian and Achter [10], in a comparison of the creep of nickel in a nitrogen atmosphere and in vacuum at 482–816°C, found that nitrogen

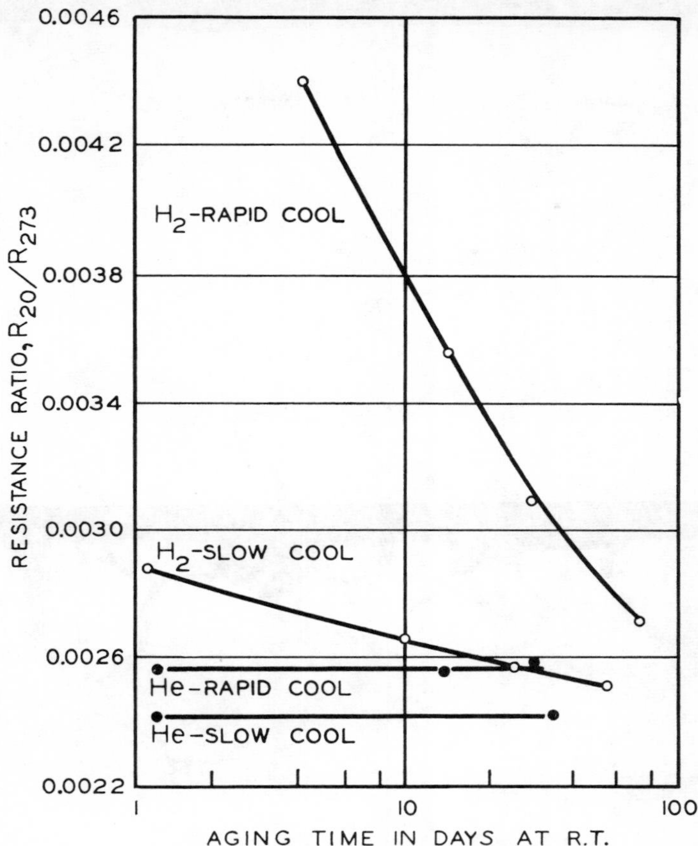

Fig 16.2. Aging characteristics of acid-cleaned high-purity nickel after heat treatment in indicated atmosphere. (Reproduced by permission from *Journal of the Electrochemical Society* [9].)

promotes intergranular cracking, increases creep rate, and reduces time to rupture (Figs. 16.3 and 16.4). The effects are much greater for a relatively impure (99.8%), than for a purer (99.99%), grade of nickel and apparently are caused by the reaction of nitrogen with foreign atoms that are concentrated at the grain boundaries rather than with the nickel itself. A possible mechanism is that a layer of gas is adsorbed on the impurities, forming a layer of nitrides, thus lowering the energy required to propagate a crack. The amount of nitrogen required to form two monolayers in the transverse grain boundaries is too small to be measured with any degree of certainty by present methods of analysis. These investigators also reported that oxygen

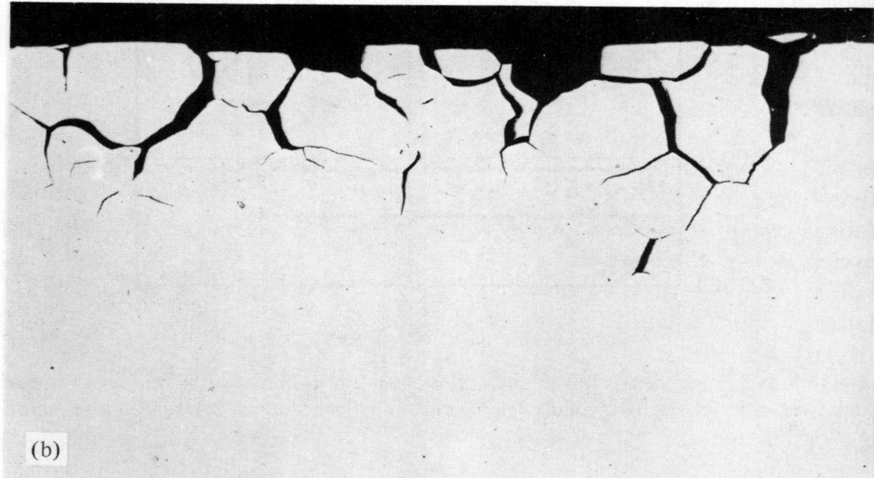

Fig. 16.3. Intergranular cracking produced in nickel during creep at 482°C and 30,000 psi: (a) in vacuum, 617 h to rupture; (b) in nitrogen, 84 h to rupture (100 ×). (Reproduced by permission from *Institute of Mechanical Engineers* (London) [10].)

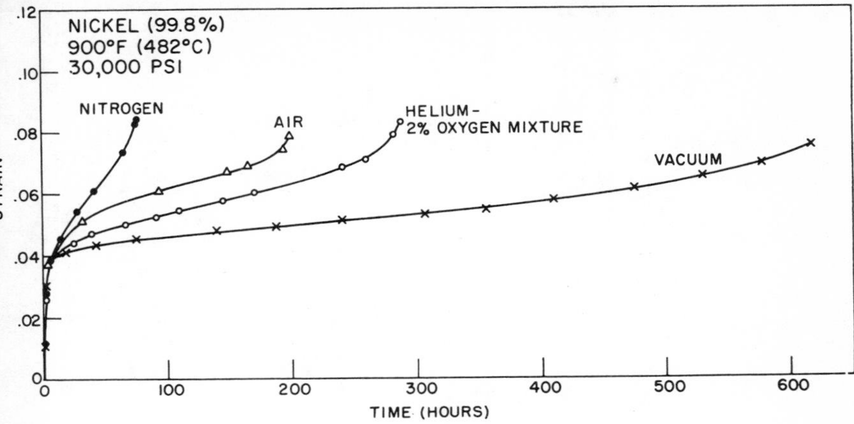

Fig. 16.4. Effect of atmosphere on creep rate of nickel. (Reproduced by permission from *Institute of Mechanical Engineers* (London) [10].)

may decrease creep strength at low temperatures and at fast strain rates but conversely under highly oxidized conditions can improve strength.

Again, Achter and his associates [11], studying the effects of metal fatigue in gaseous environments under varying temperature and pressure, showed that small amounts of impurities in inert gases may substantially reduce fatigue strength of nickel. At an oxygen pressure of 10^{-6} torr the number of cycles-to-failure was found to be 10 times the number required to produce failure when the oxygen pressure increased to 10^{-3} torr (Fig. 16.5). The latter pressure is equivalent to an inert-gas environment containing 4 ppm oxygen and represents the saturation point for adsorption of oxygen on crack surfaces in nickel at 300°C. Although not within the scope of this treatment, sulfur, even at the 0.001% level or less, should not be overlooked for its possible influence on the physical properties of nickel [12].

Residual gas phases in nickel for structural applications present no major problem or, as yet, are not clearly defined. Continuing concern will doubtlessly be in the areas of magnetic materials, high-purity nickel, and effects produced by environmental conditions.

Nickel is used extensively in the electronics field, either as a high-purity metal or with small amounts of selected additives such as tungsten, silicon, carbon, titanium, magnesium, aluminum, zirconium, and other reducing elements. These elements may volatilize or diffuse through cathode nickel at elevated temperatures to react with the active barium–strontium oxide coating. This type of reaction is of extreme importance in the initial activity level of the cathode and extension of tube life [7]. Other reactions, involving

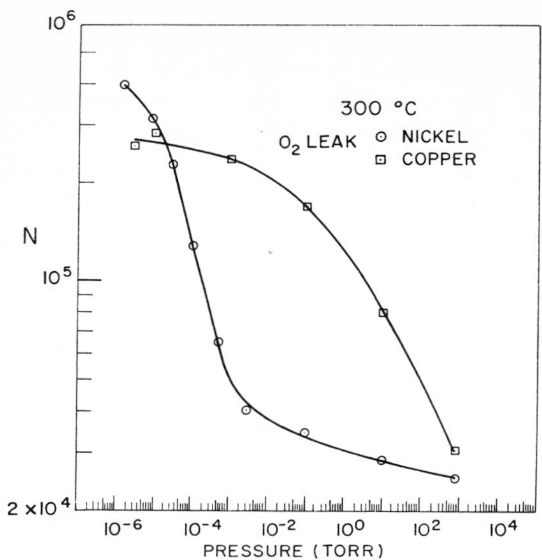

Fig. 16.5. Cycles-to-failure (N) at 300°C as a function of oxygen pressure (p). (Reproduced by permission from *Transactions of AIME* [11].)

the slow release of gas from the metallic components and the glass-to-metal seals, are of course, intolerable. Such slow outgassing is often attributed to the reaction of carbon and metal oxides to form carbon monoxide.

C. COPPER

The commercial production of copper involves the intentional retention of residual gases for desired casting characteristics. Hydrogen and sulfur are removed from the impure metal by flowing air into the melt until 5–8% of the copper is converted into cuprous oxide. The cuprous oxide oxidizes the sulfides of copper to sulfur dioxide, which is swept out of the melt by the atmospheric nitrogen. Excess oxygen is then removed by covering the surface of the melt with charcoal and forcing green wood poles beneath the surface of the melt. This process of "poling," now being replaced by gaseous reduction systems, lowers the oxygen content to a minimum of 0.02–0.04%. At the end of this process, the gas-forming compounds provide sufficient minute bubbles within the melt to offset the normal shrinkage of the casting. This product is known as *tough pitch copper* and has good electrical conductivity.

The structural usefulness of tough pitch copper is limited, however, to conditions in which reducing atmospheres will not be encountered. In the presence of hydrocarbons or hydrogen at elevated temperatures, reduction of the cuprous oxide takes place with the formation of water vapor, among other things, resulting in internal cracking. This reaction serves as the basis for a physical test, known as the hydrogen embrittlement test, for the presence of cuprous oxide in copper.

When further deoxidation is required, small amounts of phosphorus (0.02%) are added to the melt. While the addition of phosphorus lowers the oxygen content, it also, unfortunately, lowers the conductivity. A low-oxygen copper can be produced by controlled deoxidation with charcoal, followed by casting in the absence of air. This oxygen-free high-conductivity (OFHC) copper has as little as a few parts per million oxygen and excellent electrical conductivity. More recent improvements involving continuous casting, zone refining, heat treating, and other purifications have enabled the production of sound castings with a purity of 99.999%.

The slight solubility of nitrogen eliminates it from most practical considerations in terms of influencing copper metallurgy, although nitrogen in alloy systems involving copper may be of interest.

II. HISTORICAL DEVELOPMENT

A. COBALT

Early investigators either had little occasion to analyze cobalt for gases or treated it as nickel with much the same conditions for analysis. Sloman was the first to publish any details of the analysis [13]. He used the vacuum-fusion approach and reported that at temperatures just above the melting point of the metal, the evolution of carbon monoxide was very slow. He suggested that the slow diffusion rate of carbon into the metal at this temperature was responsible for the slow rate of gas evolution. However, an iron bath maintained at 1550°C gave rapid evolution of carbon monoxide when the concentration of iron in the bath was kept above 50%. In addition, hydrogen and nitrogen were evolved rapidly and presented no problem.

Analytical data on the concentrations of gases in cobalt have been reported by Morral [1] in a comparison of the composition and physical characteristics of various high-purity cobalt samples. The analytical work, although not described in Morral's publication, was done by Mallett [14] with conventional vacuum-fusion equipment and an iron bath at 1600°C. More recently, Smith, Lenahan, and Macleod have studied in great detail the various factors affecting the hot-extraction method for hydrogen [2], the inert-gas-fusion method for oxygen, and the Kjeldahl method for nitrogen [15].

B. NICKEL

The wider scope of applications for high-purity nickel, particularly in the electronics industry, has made it necessary to evaluate with great care the inherent variables affecting the determination of gases in the parts-per-million range. The fact that no standards are presently available to serve as references accentuates the need for understanding in depth the reactions and side-reactions involved in an analytical procedure.

1. Absorption of Gases by Evaporated Films of Nickel in Vacuum-Fusion Analyses

The first problem in any determination of gases in metals is that of providing the thermodynamics for evolution of the gases from the fusion, and the second problem is that of removing the gases from the furnace without side-reactions of a gettering nature.

All of the published data on the determination of gases in nickel point to the fact that no basic problem exists in reducing the oxides of nickel or in decomposing the hydrides and nitrides [13, 17, 32]. However, the work of Beach and Guldner on evaporated films [16] brings to light a problem in the quantitative removal of gases from a furnace in which nickel has been melted. The work was carried out as follows: A vacuum-fusion apparatus was modified to make it feasible to admit small known volumes of carbon monoxide, hydrogen, or nitrogen to a furnace chamber in which a specific fresh metal film (nickel, iron, etc.) had been deposited under conditions simulating an actual vacuum-fusion analysis. A bypass was inserted, connecting the mercury cutoff with the sample loading arm of the furnace. A high-vacuum stopcock was inserted in the bypass to control the admittance of gas into the furnace. Initially, about 5 g of metal were admitted to the previously degassed crucible. With the crucible at fusion temperature, metal vapor deposited continuously on the furnace walls. Small quantities of gas, ranging from 2 to 150 cc-torrs were admitted from a gas-storage bottle, collected, and accurately measured in the capillary pipette. Then, the mercury cutoffs were adjusted to permit the gas to flow from the capillary pipette, through the bypass, and into the hot preconditioned furnace containing the evaporated film. The gas was removed with a two-stage mercury-diffusion pump at a rate similar to that in an actual metal fusion and collected again in the capillary pipette for remeasurement. The recovery of gas was measured at 5-min intervals for a total of 30 min. In most cases the recovery was complete in about 15 min. In this way, the magnitude of the sorption of gas by metal films was determined. Each metal film was exposed to a specific gas, the

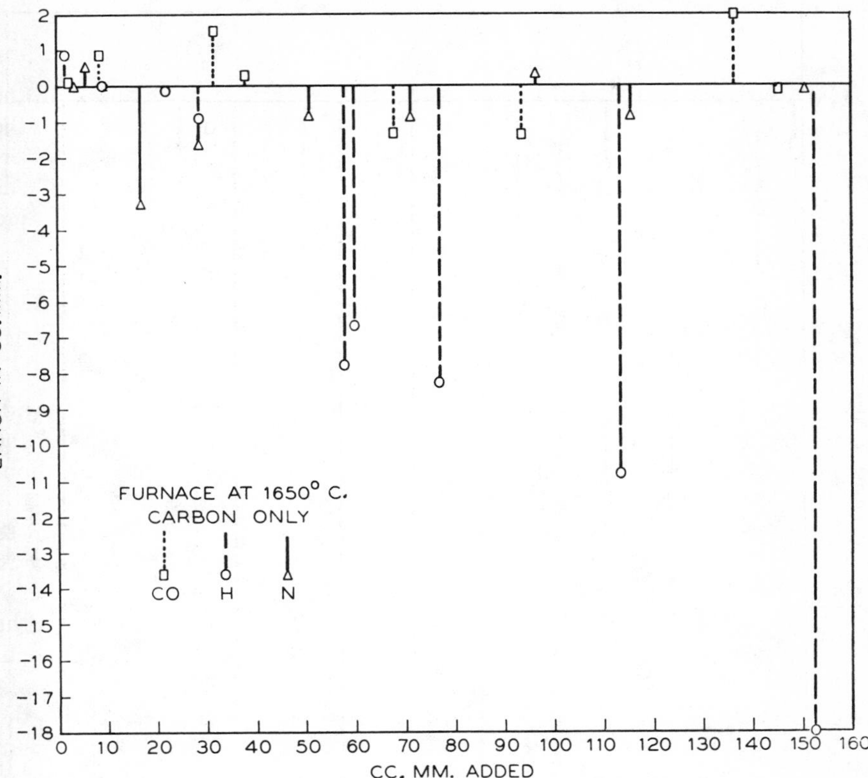

Fig. 16.6. Gas recoveries from furnace at 1600°C containing graphite components. (Reproduced by permission from *Analytical Chemistry* [16].)

amounts of which were admitted in random order. No tests were carried out using gas mixtures.

For comparison purposes, data are given here for carbon, iron, and nickel, as summarized in Figures 16.6–16.8. It will be noted that for carbon and iron, recoveries of carbon monoxide and nitrogen are excellent over the entire range studied. Small losses in hydrogen are observed, however, and in general indicate a linear relationship with amounts added, up to about 80 cc-torrs. Beyond this point, the error in recovery appears to be about constant.

The data obtained with nickel show excellent recovery for nitrogen over the entire range studied. The error in recovery of carbon monoxide is somewhat greater than for carbon and iron and is both positive and negative. However, from a practical standpoint, these errors are still small when an

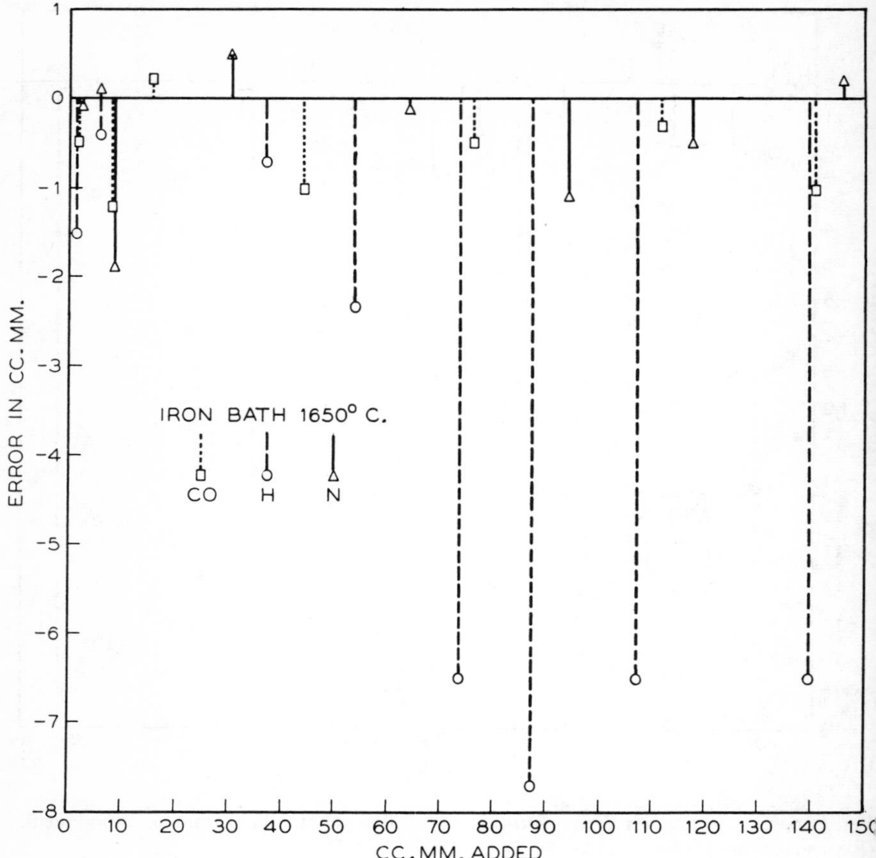

Fig. 16.7. Gas recoveries from furnace at 1650°C containing iron. (Reproduced by permission from *Analytical Chemistry* [16].)

analysis is reported on a weight-percent basis. The hydrogen errors in recovery are also positive and negative up to about 50-cc-torrs addition, and then follow a similar pattern of negative error as with carbon and iron.

Assuming a 5-g sample of nickel at the 3–4-ppm level of hydrogen, the maximum error due to sorption on the walls of the furnace would be 0.2 ppm. No studies were conducted in the presence of tin to assess its possible minimizing effect upon the gettering action of the freshly deposited films; however, it seems likely that its presence would make a lower and more consistent error.

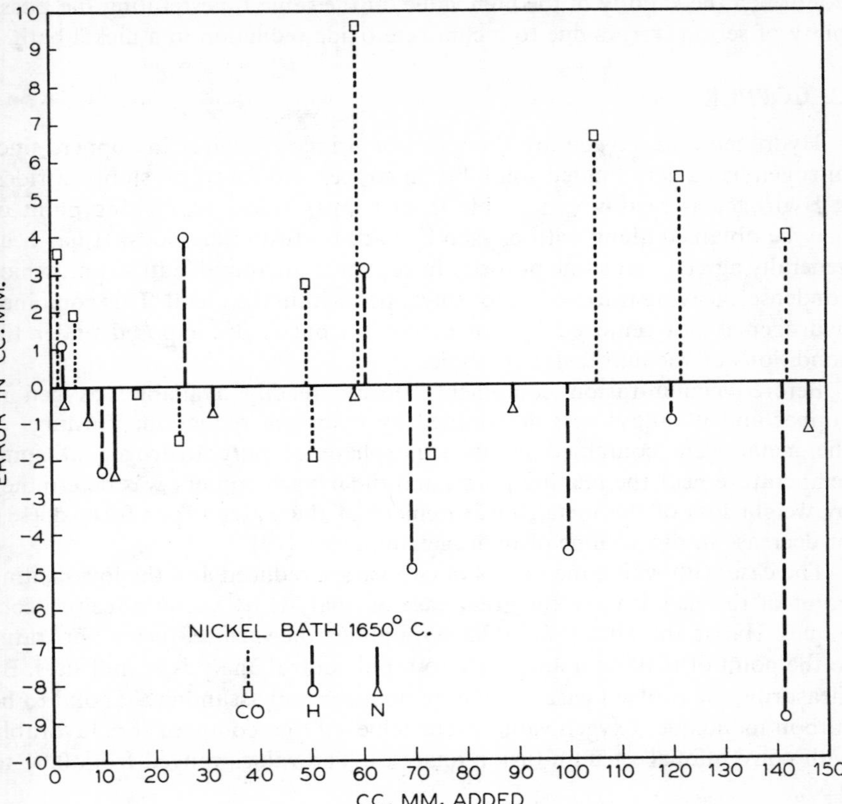

Fig. 16.8. Gas recoveries from furnace at 1650°C containing nickel. (Reproduced by permission from *Analytical Chemistry* [16].)

2. Nickel as a Bath Material for Vacuum-Fusion Analyses

Various properties of nickel recommend it as an attractive bath material (see also Chapter 4). It forms low-melting-point baths with many of the higher-melting-point metals; it remains more fluid than iron over the temperature ranges required for vacuum-fusion analysis; and, being magnetic, it can be used to encapsulate nonmagnetic materials for insertion into a vacuum system through a mercury cutoff. Further, high-purity nickel is available in many convenient geometries.

In the first cooperative analytical program on oxygen in molybdenum, sponsored by the office of Naval Research, Walter [17], with a nickel–tin bath at 1850°C, obtained the highest values (68 ppm) of any of the 10 cooperators. Later work by Guldner and Beach, with other bath conditions, confirmed the validity of the high value, at the same time refuting the possibility of serious errors due to incomplete oxide reduction in a nickel bath.

C. COPPER

Hydrogen and oxygen are the gases of primary interest in copper, since nitrogen has a very limited solubility in copper and forms no stable nitrides.

Hydrogen is readily extractable from copper below its melting point or may be obtained along with oxygen by vacuum-fusion methods. It has been generally agreed that some porosity in copper is attributable to steam, which condenses, leaving traces of water entrapped within the metal. This combined hydrogen is not removed by hot extraction but is decomposed under the conditions of vacuum-fusion analysis.

Before vacuum-fusion equipment became readily available, oxygen in copper and its alloys was determined by hydrogen reduction. Turnings of the metal were contained in an atmosphere of pure hydrogen at some temperature near the melting point, and the oxygen content was determined by weight loss of the metal, measurement of the water vapor formed [18] or decrease in the volume of hydrogen admitted [19].

The ease with which the oxides of copper are reduced and the low melting point of the metal make for great ease of analysis by vacuum-fusion techniques. Harris and Hickam [20] have simplified the vacuum-fusion apparatus to the point of making it adaptable to rapid control analysis at mill sites. By measuring the evolved gases with a manometer and assuming the total to be carbon monoxide, oxygen values were achieved that compare very favorably with conventional vacuum-fusion analyses. The values ranged from 0.01 to 0.5%.

The original work on the reaction between bromine trifluoride and oxides by Ruff and his co-workers [21] was further explored by Emeléus and Woolf [22], with the conclusion that at least a dozen oxides, including cupric oxide, react quantitatively to release molecular oxygen. The powerful oxidizing and corrosive characteristics of bromine trifluoride, however, make its analytical applications somewhat unattractive, especially when other methods are available.

Isotope-dilution methods have been used for the determination of oxygen in copper, from as little as 0.01% to very rich oxide mixtures [23]. The methods, although comparable in accuracy to other accepted methods, are somewhat more involved and less adaptable to routine applications.

III. STANDARD SAMPLES

No standard samples for gases in cobalt, nickel, and copper are currently available from either private or commercial sources. The problem of choosing a source material of composition suitable to the broad spectrum of technological needs and the even greater problem of finding laboratories with sufficient time to devote to cooperative standardization programs combine to form a task of Herculean proportions. Frequently the needs of an individual laboratory are sufficiently restricted or peculiar to warrant the local development of standards for the specific need. Exchange of such standard samples with a few laboratories of similar missions and interests is usually sufficient to provide analytical confidence.

IV. SAMPLING AND SAMPLE PREPARATION

A. COBALT AND NICKEL

A thorough study of cobalt sampling for the determination of hydrogen, oxygen, and nitrogen by hot-extraction, inert-gas-fusion, and Kjeldahl methods, respectively, was conducted by Smith et al. [2, 15]. Of particular interest was the fact that the hydrogen content of cobalt showed no decrease on sample-aging. In fact, a slight increase was indicated on samples that had been stored from two to seven months. The authors gave no explanation for this seeming anomaly. Samples for nitrogen determination followed a more predictable pattern. Because solid pieces of cobalt are slow to dissolve in acid media suitable for Kjeldahl nitrogen determination, the samples were pulverized. Higher nitrogen values than would be expected from cathode cobalt were obtained.

In many instances lack of correlation of chemical compositions with physical properties of nickel has been caused by improper sampling. Often the original stock material has been altered to such a large extent that it is necessary for the chemist to avail himself of the sample history. An excellent example is shown in Fig. 16.9 [24]. The oxygen content is plotted against the surface area per gram of nickel and shows the increase in oxygen content of a normal cathode nickel, Inco 220, with surface area. It also shows the relationship in terms of sample thickness. The two points at the 4-mil-thickness level represent typical analyses for high-purity nickel (passive) and nickel containing free reducing agents (active). They also show how the curve would be shifted for relative thicknesses of such nickels. It will be noted that raw stock or ingot material has an oxygen content of less than 0.001%.

Detailed work by Beach and Guldner [24a] revealed that the increased oxygen content with surface area cannot be accounted for in terms of room-temperature surface-adsorption phenomena involving a monolayer or two, even if a high roughness factor is assumed. However, it was found that the amount of oxygen picked up as a result of annealing is directly related to the ratio of surface to volume, as shown in Fig. 16.9. Further, it was found that the oxygen pickup is retained in the first 0.1-mil surface layer. Removal of this surface layer from a 4-mil active sheet with an acid etch dropped the oxygen content to the level of the raw stock, with no tendency for oxygen to build up to the pre-etch level. Fabrication of 2-mil sheet from ingot stock, without annealing, also showed no oxygen increase.

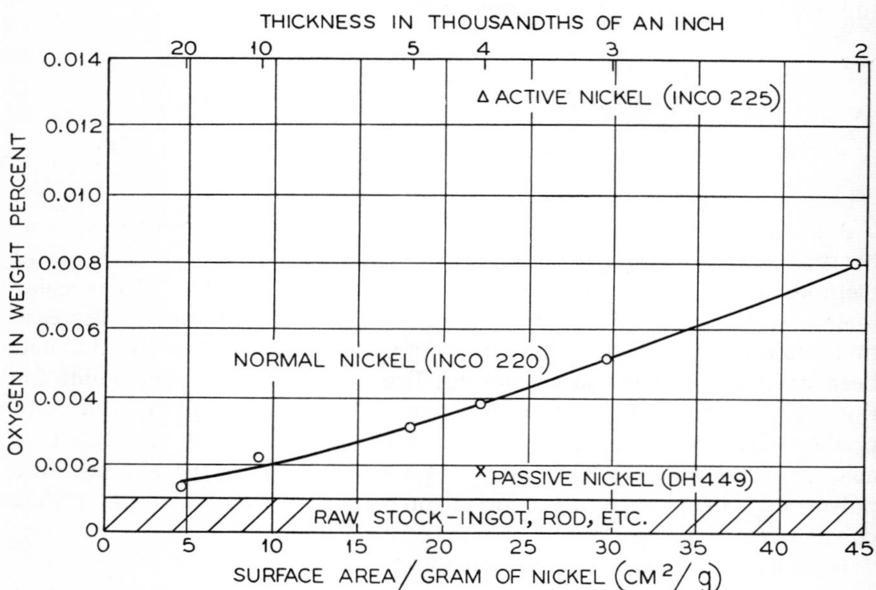

Fig. 16.9. Oxygen in cathode nickel. (Reproduced by permission from American Society for Testing and Materials [24].)

These data have been projected to show factors that must be carefully considered before making a comparison of the oxygen content of samples. Too often, comparisons of samples prepared from the same ingot material are made without due respect to their individual histories.

The following treatments are given the bath materials, just prior to use:

Iron, for bath material: (1) Degrease with solvent, (2) wash in dilute HCl (1 + 1) for 2–4 min, (3) rinse in distilled water, (4) rinse in acetone, and (5) dry.

Nickel (or Cobalt): (1) Degrease with solvent, (2) treat in bright acid dip for 1–2 min at 60–75°C, (50 acetic acid–30 nitric acid–3 hydrochloric acid), (3) rinse in distilled water, (4) rinse in acetone, and (5) dry.

Tin (sheet, capsules): (1) Clean in NaOH solution (20%) for 5 min, (2) rinse in distilled water, (3) dip in HCl (1 + 1) for 5 min, (4) rinse in distilled water, (5) rinse in acetone, and (6) dry.

B. COPPER

Sampling and sample preparation of copper become more important as the purity of the sample increases. While the variations in amounts of surface gases due to surface–weight ratio differences and various preparation treatments may not be significant if tough pitch copper is being analyzed, such variations may render the analytical data for vacuum-melted OFHC samples meaningless.

The work of Guardipee [24b], summarized in Tables 16.1 and 16.2, illustrates dramatically the need for careful consideration of the surface gases in the analysis of copper. The data in Table 16.1 show the reproducibility of analyses made on samples taken from the same stock, cut to approximately the same surface geometries, cleaned, handled, and analyzed in the same

TABLE 16.1. Total Oxygen and Hydrogen in Copper Samples (Varying Surface-Area–Weight Ratios)[a]

Materials	Portion	Surface Area (mm^2 per g)	Total O (ppm)	Total H (ppm)
OFHC, sample 1	1	83	0.69	0.28
OFHC, sample 1	2	83	0.79	0.28
OFHC, sample 1	3	83	0.65	0.25
OFHC, sample 1	4	301	1.37	0.31
OFHC, sample 1	5	303	1.15	0.31
OFHC, sample 1	6	301	1.06	0.33
OFHC, sample 2	1	100	1.52	0.54
OFHC, sample 2	2	91	1.49	0.54
Vacuum-Melted:				
OFHC, sample 1	1	84	0.27	0.0034
OFHC, sample 1	2	83	0.14	0.0029
OFHC, sample 1	3	83	0.11	0.0033
OFHC, sample 1	4	277	0.64	0.0175
OFHC, sample 1	5	275	0.69	0.0178
OFHC, sample 2	1	103	0.19	0.0040
OFHC, sample 2	2	222	0.37	0.0077
OFHC, sample 3	1	115	0.17	0.0024
OFHC, sample 3	2	189	0.39	0.0081
Tough Pitch, sample 1	1	141	416	0.32
Tough Pitch, sample 1	2	262	457	0.38

[a] Courtesy *Analytical Chemistry* [24b].

TABLE 16.2. Surface Versus Bulk[a] Oxygen and Hydrogen in Copper[b]

Material	Oxygen (ppm)			Hydrogen (ppm)		
	Surface	Bulk	Total	Surface	Bulk	Total
OFHC, sample 1	0.15	0.48	0.63	0.01	0.22	0.23
Vacuum-Melted:						
OFHC, sample 1	0.08	0.27	0.35	0.0051	<0.0005	0.0051
OFHC, sample 2	0.14	<0.06	0.14	0.0014	<0.0005	0.0014

[a] Gases were evolved in two stages from each sample: below melting point for "surface" and above melting point for "bulk."
[b] Courtesy *Analytical Chemistry* [24b].

manner. They also show the increased oxygen and hydrogen levels with an increase in the surface-area–weight ratio for the three most commonly encountered types of copper.

Table 16.2 shows separate surface and bulk gas levels for various OFHC samples. These data were obtained by a two-stage evolution process for the gases from each sample. This information can serve as a useful guide in the design of meaningful analytical programs for gases in copper.

Nitrogen is not chemically bonded to copper under the conditions of current gases-in-metals studies, and in all but the most specialized research efforts, is not of significant interest.

Surface preparation treatments must also be considered carefully when oxygen values in the low parts-per-million range are of interest. Results of a study [25], in which various surface preparations were tried, are summarized in Table 16.3. The samples were solid pieces of OFHC copper cut from the

TABLE 16.3. Effect of Sample-Preparation on Oxygen Content of Copper [25]

Sample-Preparation	Oxygen (ppm)
Filed, stored in vacuum	1.1, 1.3, 1.0, 1.2
Filed, overnight air storage	1.4, 1.4
Solvent degreased, bright acid etch	4.3, 4.0, 3.3
Solvent degreased, HNO_3 (1 + 1) etch	1.4, 2.0, 2.0, 1.9
Solvent degreased, NaCN solution (10%) etch	7.0, 5.6

same stock, and all analyses were made in the same laboratory by the vacuum-fusion method. From the data reported in Table 16.3 it is apparent that a simple abrading treatment of a solid copper sample, followed by vacuum storage or immediate insertion into the vacuum analysis system, is preferred. Cleaning with acid etches can be equally effective if precautions are observed to prevent oxidation of the activated surfaces. The samples cited in Tables 16.1 and 16.2 [24b] were etched briefly in nitric acid (1 + 1), followed immediately by rinses in hydrochloric acid (1 + 1), distilled water, acetone, and ether. This treatment was given the samples immediately before final weighing and insertion into the vacuum system.

Romanova et al. [26] have also reported that blank levels achieved after chemical etching with an acetic–phosphoric–nitric acid mixture were indistinguishable from those obtained after abrading or mechanical polishing. They reported 2 ppm as their oxygen sensitivity, however, which would tend to obscure small differences in blank measurements.

When samples are being prepared for hot extraction of hydrogen, even though oxygen values are not involved, the surface of the sample should be abraded to remove any copper oxide film. It is possible to obtain low hydrogen values as a result of its conversion to water vapor by reaction with the oxide film at the sample surface.

V. ANALYSIS

A. COBALT AND NICKEL

1. Hydrogen

a. Hot Extraction

If the need arises for a separate determination of hydrogen, hot-extraction techniques similar to those for iron and steel are applicable [2]. Hatfield and Newell [27], in the analysis of raw materials used in steelmaking, determined the hydrogen in both nickel and cobalt. With nickel in particular they reported the slow diffusion of hydrogen out of the metal at room temperature and a much greater rate at higher temperatures. The hydrogen content of a sample containing originally 76 ppm hydrogen dropped to 2 ppm after heating for 1 h at 850°C in air. These data show the ease with which hydrogen is removed from nickel and suggest the facility with which the extraction can be effected under the more amenable conditions of vacuum analysis. These investigators also reported a loss of oxygen, with an attendant brightening of the sample surface, when nickel was heated in air at 650°C. This would suggest caution in the interpretation of hot-extraction data for samples containing appreciable amounts of oxygen and the need for care in the preparation of sample surfaces prior to extraction.

More recent work on the aging of high-purity nickel [9], shows dramatic changes in the resistance ratio (R_{20}–R_{273}), because of the evolution of hydrogen (see Figs. 16.2 and 16.10).

Hot extraction has been used routinely at the Naval Research Laboratory with nickel and nickel–palladium alloys for the determination of hydrogen and deuterium additions [28]. For this work a traditional vacuum-fusion apparatus [29] is used, and the samples are dropped into a graphite crucible at 1000°C. The graphite of the crucible eliminates any possible error resulting from the presence of oxides by combining with the oxygen, and the temperature is low enough to permit rapid outgassing to a negligible blank. Samples are allowed to accumulate in the crucible, and determinations can be made sequentially as rapidly as the extracted gases can be analyzed. Extraction time is 10 min, and the usual analysis time is 20 min.

Deuterium provides no special problems in extraction, but mass-spectrometric techniques are required for obtaining the hydrogen–deuterium ratio.

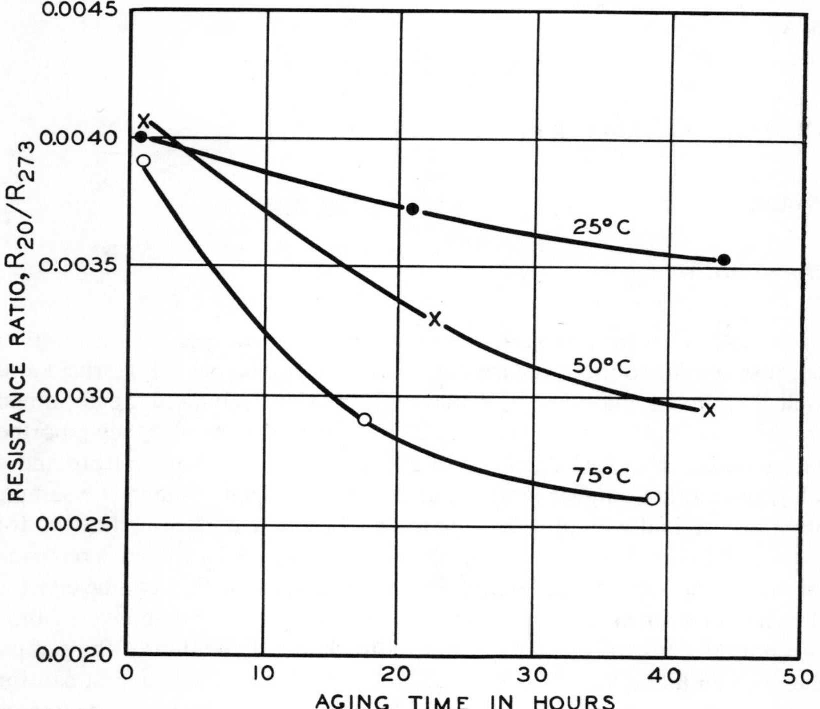

Fig. 16.10. Effect of aging temperature on resistance ratio of acid-cleaned high-purity nickel strips rapidly cooled after 1 h at 1000°C in H_2. (Reproduced by permission from *Journal of the Electrochemical Society* [9].)

b. Vacuum Fusion

Usually the determination of hydrogen in nickel and cobalt is made simultaneously with that of oxygen and nitrogen by vacuum-fusion techniques.

2. Oxygen and Nitrogen

a. Vacuum Fusion: Cobalt

Although inert-gas-fusion methods have been used for both oxygen and nitrogen in cobalt [15], most investigators have used the vacuum-fusion approach. It permits a wide variety of sample geometries and composition ranges and simultaneously yields the three major gases of interest. It also permits a selection of readout devices, including the versatile mass spectrometer [30].

Vacuum-fusion techniques described by Fraser utilized fractional freezing for the separation of water and carbon dioxide [31], although more-recent studies were done with a mass spectrometer [30]. In this work, analyses were conducted with and without an iron bath, both with tin, at temperatures in the neighborhood of 1800°C. One stock material of silicon-killed cobalt was used for the series of analyses, and all the samples were cut from a portion of the ingot that gave metallurgical evidence of being reasonably homogeneous. Determinations were made on different days and with different crucibles to provide statistical variations. Without an iron bath the average oxygen value was 0.022%, with a variation of ±0.001% (Table 16.4). With an iron

TABLE 16.4. Gases in Cobalt by Vacuum-Fusion Analysis [30]

Sample-Weight	Bath	Oxygen	Oxygen Blank	Hydrogen	Nitrogen
3.0 g	None	0.023%	0.0004%	<0.0001%	0.0003%
4.5	None	0.023	0.0006	0.0001	0.0002
5.3	None	0.021	0.0002	0.0001	0.0002
2.4	None	0.022	0.0004	<0.0001	0.0002
3.3	Iron	0.028	0.0013	<0.0001	0.0002
3.3	Iron	0.022	0.0007	0.0001	0.0002
3.7	Iron	0.028	0.0006	0.0001	0.0001

bath the average value was 0.026%, with a variation of ±0.003%. Fraser attributed the less-consistent values obtained with the iron bath to more-rapid solution of the crucible, which led to blank irregularities.

Mallett [14] used an iron bath at 1600°C for cobalt. A 5-g sample of iron was used to condition the crucible, and no upper limit was established for the cobalt–iron ratio. Reproducible analyses were obtained on material ranging from zone-refined ingots (approximately 0.0030% oxygen) to granular material (approximately 0.030% oxygen). Sloman's observations [13] relating to the ease with which nitrogen and hydrogen could be liberated were confirmed by Mallett. Hydrogen values ranged from less than 0.0001% for the zone-refined material to 0.0030% for electrolytically refined material, with intermediate values for the granular. Nitrogen in all materials was 0.002% or less.

b. Vacuum Fusion: Nickel

A detailed description of the vacuum-fusion method for nickel was jointly prepared by F. C. Benner* and W. G. Guldner for ASTM [32]. Equipment of the National Research Corporation type was specified with the Guldner-Beach furnace [33]. This equipment provides for outgassing at temperatures as high as 2400°C, so that suitable blanks are achieved in 2–4 h. Blanks equivalent to 0.0001% oxygen on a 2-g sample at a furnace temperature of 1650°C are desirable for samples in the 0.001% range. Samples are fused at 1650°C, and the evolved gases are collected over a 20–30 min period. Occasionally the collection period is extended to insure the complete collection of gases and to check the consistency of operational parameters. Analysis of the gases is conducted by catalytic oxidation of the hydrogen and carbon monoxide, followed by absorption of the water vapor with anhydrous magnesium perchlorate and freezing of the carbon dioxide with liquid nitrogen. The residual gas is measured as nitrogen.

The method outlined above and detailed in the *Annual Book of ASTM Standards* [32] does not provide for an iron bath nor for the use of tin additions; however, participants in comparative analysis programs have not encountered the problem of low yields originally reported by Sloman [13]. It would appear that the temperatures "just above the melting point" reported by Sloman for determining oxygen in both nickel and cobalt may have been too low and that the higher temperature of 1650°C, without the addition of iron, would have overcome the slow rate of gas evolution. Correspondence by the authors with a dozen or more organizations interested in nickel from the standpoint of manufacture, fabrication, application, and research reveals that no analytical problem relating to residual gases is currently present, except in the parts-per-billion range [10].

* Norton Vacuum Equipment Division, part of Varian Associates.

B. COPPER

1. Hydrogen

a. Hot Extraction

Hydrogen may be removed quantitatively from copper and its alloys in a vacuum-extraction system over a fairly wide temperature range. Hydrogen that exists as water vapor is not obtained by this method, and the possibility always exists, when much cuprous oxide is present, that additional reactions will take place with correspondingly low hydrogen results. A temperature between 800° and 900°C is usually recommended.

b. Vacuum Fusion

The hot reducing conditions provided by the furnace crucible in vacuum-fusion analysis decompose all hydrogen-bearing compounds to yield hydrogen. If necessary, a combination of hot-extraction and vacuum-fusion techniques can be combined in vacuum-fusion equipment by raising the temperature slowly to the melting point and observing the gas evolution during the heating cycle. Gas evolution curves have been interpreted to provide practical information for improving manufacturing processes.

2. Oxygen

a. Hydrogen-Reduction

The combustion method for the determination of oxygen in copper involves the hydrogen reduction of the oxides of copper to form water vapor. Usually, the sample is reduced to turnings and exposed to a pressure of purified hydrogen at 900°C, or if a slug-type sample is used, it may be melted to hasten reduction time. The water vapor formed may be absorbed and weighed if the amount of oxygen present is sufficiently large, or it may be collected in a dry-ice–acetone trap during the reduction process, followed by expansion of the water vapor into a calibrated volume equipped with a suitable pressure measuring device [18]. Sample-sizes are chosen in accordance with the parameters of the particular analytical system. If very small amounts of oxygen are to be determined by this method, measurement of the volume of hydrogen consumed yields the highest degree of measurement sensitivity. Funston and Reed [19] reported a sensitivity of 2 ppm oxygen with this method when a 10-g sample was employed.

b. Isotopic Dilution

The attractive feature of isotopic-dilution methods for the determination of gases in metals lies in the fact that it is not necessary to extract all of the gases from the reaction crucible if the labeled isotopic addition is completely equilibrated with its sister atoms in the metal.

In determining the oxygen content of copper samples, Kirshenbaum and Grosse [23] used a copper–oxygen master alloy containing a known amount of ^{18}O. Weighed amounts of the sample to be analyzed, the master alloy, and a small amount of preanalyzed graphite were intimately mixed and placed in a hydrogen-reduced molybdenum or platinum crucible. The crucible and its contents were placed in a furnace tube, which was then evacuated to 10^{-4} torr and sealed. Induction heating was used to maintain the crucible at a temperature of 1100–1200°C for 30–60 min, during which time isotopic equilibrium was achieved and a portion of the oxygen was liberated as carbon monoxide. The liberated gases were removed to a sample tube for analysis by mass spectrometry.

Corrections must be applied for the oxygen contributed by the crucible, the graphite powder, and the furnace system. Any air inleakage is readily detected mass-spectrometrically by the presence of the argon peak, and appropriate corrections can be applied.

When very small amounts of oxygen are present in the copper sample, it is conceivable that the blank may be higher than the amount of oxygen to be determined. In these instances, it is recommended that no graphite powder be used to provide reduction of the oxides. Instead, the oxygen for the mass-spectrometric analysis should be obtained by thermal decomposition of the oxides.

While the oxygen values obtained by isotopic-dilution techniques for the analysis of copper are compatible with those obtained by other methods, it would appear that for routine analytical work, especially for nonrefractory metals such as copper, the method is unnecessarily complex and lends itself to the introduction of errors by the uninitiated. It could be an invaluable tool, however, in the evaluation of oxidation or corrosion studies and in environmental studies where tagged oxygen could be used.

c. Vacuum Fusion

The low temperature range in which the oxides of copper may be reduced and the relatively small weight percentages of hydrogen and nitrogen usually encountered make it possible to effect great simplification of the traditional vacuum-fusion equipment for the determination of oxygen in copper. The equipment and procedure developed by Harris and Hickam [20] incorporate these simplifications to provide oxygen analyses on a schedule adaptable to mill control demands but with sufficient accuracy and reproducibility for most analytical requirements.

The Harris-Hickam apparatus, illustrated in Fig. 16.11, is constructed of borosilicate glass, with the exception of the 9-in. quartz furnace tube. Spectrographic-grade graphite is used for the sample-boat and is of sufficient

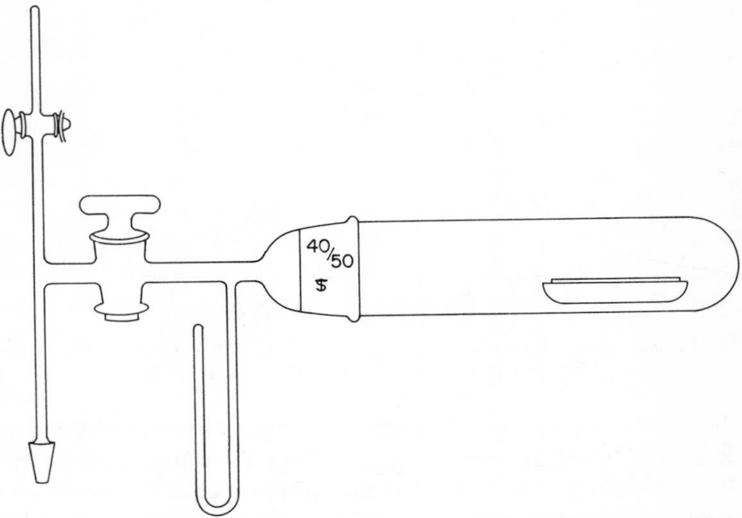

Fig. 16.11. Harris-Hickam apparatus for oxygen in copper. (Reproduced by permission from *Analytical Chemistry* [20].)

size to accommodate a 10-g sample of copper. Working calibrations for the mercury manometer and the attached furnace system are obtained by admitting measured volumes of carbon monoxide to the system under operating conditions and with the sample-boat heated to 1150°C. A 5-kW high-frequency generator is adequate for heating the sample boat.

Figure 16.12 represents a typical extraction cycle for a 3-g sample containing approximately 0.06% oxygen. The direct manometer reading, based on the appropriate calibration data for a particular pressure, is satisfactory for all except the most critical demands, in which case a mass-spectrometric analysis of the collected gases is made. Analytical studies of the gases evolved from typical copper samples have provided data indicating an average of over 97.5% carbon monoxide by volume, approximately 2% hydrogen, and 0.1% or less each of carbon dioxide and water vapor. Converted to weight percent of the gas evolved, the total carbon monoxide present averaged

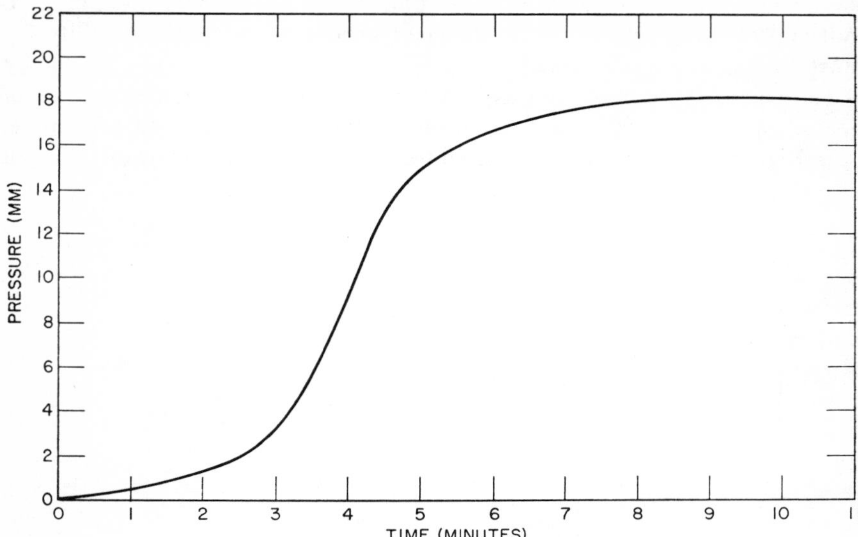

Fig. 16.12. Typical extraction cycle for oxygen (as CO) from copper by Harris-Hickam method. (Reproduced by permission from *Analytical Chemistry* [20].)

99.6%. No sulfur-containing gases were found, even when samples containing as much as 0.02% sulfur were analyzed. These data also indicate the low probability of finding a significant amount of nitrogen in unalloyed copper.

In practice an analysis can be completed in 15 min, and no additional degassing of the boat is required between determinations. This method has been used for the oxygen range of 0.01–0.5%, with accuracy and consistency comparable to that provided by traditional vacuum-fusion methods (Table 16.5).

TABLE 16.5. Comparison of Oxygen Values in Copper Obtained by Different Methods [20]

	Oxygen (%)	
Sample	Vacuum Fusion	Harris-Hickam
1	0.027	0.026
2	0.052	0.053
3	0.059	0.060
4	0.047	0.048
5	0.047	0.046
6	0.038	0.044
7	0.040	0.043
8	0.038	0.040

d. Specialized Techniques for Surface Analysis

Frequently the need arises in research studies for the determination of gases on the surface of a sample. Thin film work, surface-chemistry studies, and materials for outer-space applications have their own peculiar requirements in terms of surface characterization. Several electronic devices have been used effectively to measure surface gases for specialized needs.

(1) ION MICROPROBE MASS SPECTROMETER. The ion microprobe mass spectrometer was used by Evans [34] for the determination of oxygen on copper surfaces with a precision equal to that usually associated with bulk sample analysis. For this type of study the sample is located in the vacuum chamber of an ion probe so that the surface area of interest can be bombarded with a beam of ions, usually 5–55 keV Ar^+. The resultant sputtering causes neutral atoms, as well as positive and negative ions, to be ejected from the sample. It is the common practice to extract the positive ions into a mass spectrometer and to record the selected ions by means of suitable detection devices. Since the primary ion beam erodes the sample continuously at the rate of Angstrom thicknesses per second (depending on the beam energy), concentration variations with depth can be measured. Also, by movement of the primary beam or the sample, concentration gradients across the sample-surface can be effected. On surface studies, as described by Evans, extreme care must be observed to control the oxygen blank, which can reflect the integrity of the vacuum system, the condition of surfaces within the probe, and impurities in the duoplasmatron gas supply. With proper attendance to these sources of oxygen, analyses can be performed in the 10–20-ppm range.

(2) SCANNING ELECTRON PROBE MICROANALYZER. The scanning electron probe is similar in operating modes to the ion probe and is currently more generally available to the research scientist who is concerned with the resolution of surface composition problems. The newer and updated scanning electron probes have imaging capabilities approaching those of a scanning electron microscope, and when this capability is added to the electron probe's analytical versatility, compositional information can be obtained and related to topographical features on a sample-surface. A search for the source, form, and location of oxygen on a copper surface is used to illustrate the analytical capabilities of the electron probe [35].

A copper surface that gave metallurgical evidence of contamination was examined for embedded residues of abrasive compounds. The sample's metallurgical and service history narrowed the field of suspected compounds to the oxides of iron, aluminum, and magnesium, and to particles of metal dust. In searching the surface of the sample in the scanning electron probe, an area displaying an abrasion-type blemish was located and examined

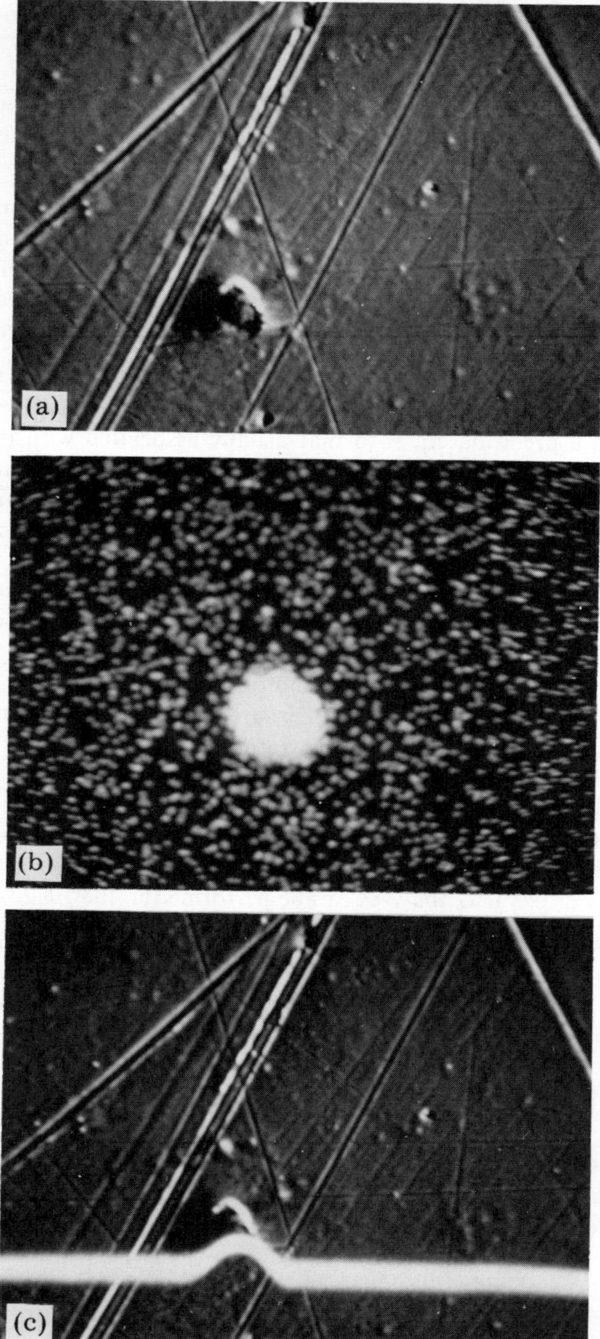

Fig. 16.13. Scanning electron probe displays of aluminum oxide identification data: (a) scanning electron micrograph of particle on copper surface; (b) aluminum X-ray energy responses superimposed on particle site; (c) oxygen X-ray trace superimposed on particle site. (Courtesy of E. J. Brooks, Naval Research Laboratory.)

analytically. Figure 16.13 shows the data confirming the presence of particulate matter of an aluminum oxide composition.

Picture (a) of Figure 16.13 is a scanning electron micrograph of the area of interest, showing the embedded particle at a magnification of 1000×.

Picture (b) shows the same area, with the probe operating in an analytical mode and the analyzing crystal set for the K-alpha wavelength of aluminum. The heavy concentration of X-ray energy responses superimposed on the site of the particle identifies it as an aluminum concentration site. (Similar scans for iron and magnesium gave no such concentration sites.)

When similar rastering was done at the oxygen wavelength, the X-ray energy responses were clustered on the particle site, but the density increase above background was not as definitive as for aluminum. This is to be expected with the lower-atomic-number elements. A different analytical display was chosen for oxygen. This is shown in picture (c).

In (c) the X-ray trace for oxygen is superimposed on the electron micrograph of the particle site. By comparing the height of the X-ray energy peaks with those of a standard aluminum oxide, the inclusion may be definitely identified.

VI. RECOMMENDED METHODS

A. COBALT AND NICKEL

1. Hydrogen

If hydrogen only is to be determined, any hot-extraction method is satisfactory that will provide outgassing temperatures in the 1000°C range and a suitable pressure measuring device capable of determining the completion of extraction and total hydrogen collected. A mercury lift or other suitable device for sample-loading after system evacuation is recommended for samples that lose hydrogen when stored under vacuum.

2. Oxygen (Hydrogen and Nitrogen)

Vacuum fusion is recommended for the determination of oxygen in cobalt and nickel and for the simultaneous determination of oxygen, hydrogen, and nitrogen. Any vacuum-fusion design that provides satisfactory results for iron and steel analysis may be used for cobalt and nickel; however, if equipment is to be purchased or developed primarily for cobalt and nickel analysis, the detailed directions as prepared by Guldner for ASTM are recommended [32].

The analytical conditions for general utility are summarized in Table 16.6.

TABLE 16.6. Experimental Conditions for Determination of Oxygen in Cobalt and Nickel

Analytical facility	Vacuum-fusion gas-analysis equipment capable of operating under conditions specified below
Furnace assembly	Guldner-Beach type assembly
Crucible and funnel	Fabricated from National Carbon AGR-grade graphite or equivalent, or Ultra Carbon custom-made; graphite powder of equivalent grade, −200 mesh
Furnace heater	Induction heater, capable of maintaining crucible temperature of 2400°C; 5-kW units adequate
Analytical system	Solid absorbents or fractional freezing for gas separation; catalysis by mixed oxides
Apparatus outgassing	Minimum of 2h at 2400°C
Blank	Collect for 30 min; should be equivalent to 0.0002% oxygen or less on basis of 1-g sample
Sample-size	2–5 g
Extraction temperature	1650°C for nickel 1750°C for cobalt
Extraction time	30 min
Bath	No bath necessary, although iron bath at 1650°C is common; tin additions between analyses
Remarks	Mercury lift should be used if hydrogen is to be determined

B. COPPER

1. Hydrogen

If hydrogen alone is to be determined, a hot-extraction procedure conducted in a graphite crucible is preferable. This eliminates any possibility of low hydrogen values due to side-reactions with cuprous oxide and also provides for the decomposition of water vapor that may have been formed in the final stages of metallurgical processing. If dissolved hydrogen or entrapped molecular hydrogen is the only concern, an extraction temperature less than the melting point of copper is satisfactory and will simplify the analytical problem; if, however, hydrogen combined either as water vapor or hydrocarbons is desired, the sample should be brought to the melting point. Any vacuum-fusion equipment with a variable furnace temperature and an attached analytical system is satisfactory.

For copper alloys containing elements suspected of gettering activities, liberal additions of tin to the crucible will provide a low-melting-point bath and thus reduce the gettering potential.

2. Oxygen

The Harris-Hickam [20] method for oxygen in copper is recommended when the oxygen level is 0.01% or higher and when hydrogen is not to be determined. It is a specialized approach for use where copper is the only metal of interest. The experimental conditions are tabulated in Table 16.7.

TABLE 16.7. Experimental Conditions for the Determination of Oxygen in Copper

Analytical facility	Modified vacuum-fusion apparatus (Fig. 16.11)
Furnace assembly	Quartz tube, 9-in. length, connected by 40/50 standard-taper ground-glass joint to system
Crucible (boat)	Spectrographic grade graphite boat, $1\frac{1}{2} \times \frac{1}{2} \times \frac{3}{8}$ in., fitted with $\frac{1}{8}$-in. thick cover
Furnace heater	Induction heating, 5-kW generator
Analytical system	No gas separations; pressure read with mercury manometer; total volume of system less than 200 cc
Apparatus outgassing	Approximately 20-min, at 1150°C, no additional outgassing between samples
Blank	Approximately 1 mm or less on manometer
Sample-size	2–8 g
Extraction temperature	1150°C
Extraction time	10 min or until manometer reading is constant
Bath	None
Remarks	If gases other than oxygen are to be determined, gas samples may be taken for mass-spectrometric analysis

3. Nitrogen

If the need arises for the determination of nitrogen in copper alloys, vacuum fusion is recommended, or in the absence of such capabilities, a Kjeldahl analysis as detailed by Elwell and Scholes [36] is satisfactory.

REFERENCES

1. F. R. Morral, *J. Metals*, **10,** 662 (1958).
2. G. A. Smith, W. C. Lenahan and D. S. Macleod, *Metallurgia*, **79,** 121 (1969).
3. E. A. Nesbitt, in Roland S. Young, Ed., *Cobalt—Its Chemistry, Metallurgy and Uses*, Reinhold, New York, 1960, pp. 184–209.
4. R. D. Heidenreich, E. A. Nesbitt, and R. D. Burbank, *J. Appl. Phys.*, **30,** 995 (1959).
5. E. A. Nesbitt and R. D. Heidenreich, *J. Appl. Phys.*, **30,** 1000 (1959).
6. E. A. Nesbitt, R. D. Heidenreich, and A. J. Williams, *J. Appl. Phys.*, **31,** 228S (1960).

7. H. E. Kern and E. T. Graney, *Advan. Electron Tube Techniques*, **2,** 235 (1962).
8. K. M. Olsen, *Met. Progr.*, **72,** 105 (1957).
9. K. M. Olsen and C. F. Larkin, *J. Electrochem. Soc.*, **110,** 86 (1963).
10. P. Shahinian and M. R. Achter, *Proceedings of the Joint International Conference on Creep*, Institute of Mechanical Engineers, London, 1963.
11. M. R. Achter, G. J. Danek, Jr., and H. H. Smith, *Trans. Met. Soc. AIME*, **227,** 1296 (1963).
12. K. M. Olsen, C. F. Larkin, and P. H. Schmitt, Jr., *ASM Preprint No. 203*, 1960.
13. H. A. Sloman, *J. Inst. Metals*, **71,** 391 (1945).
14. M. W. Mallett, Battelle Memorial Institute, private communication, 1965.
15. G. A. Smith, W. C. Lenahan and D. S. Macleod, *Metallurgia*, **79,** 171 (1969).
16. A. L. Beach and W. G. Guldner, *Anal. Chem.*, **31,** 1722 (1959).
17. M. W. Mallett, Minutes of Second Meeting, ONR Advisory Committee on Molybdenum, Subcommittee on Analysis of Molybdenum for Small Traces of Oxygen, Nitrogen, and Other Gases, Battelle Memorial Inst. Columbus, Ohio, March 1954.
18. W. A. Baker, *J. Inst. Metals*, **65,** 345 (1939).
19. E. S. Funston and S. A. Reed, *Anal. Chem.*, **23,** 190 (1951).
20. W. F. Harris and W. M. Hickam, *Anal. Chem.*, **31,** 281 (1959).
21. O. Ruff and W. Mengel, *Z. Anorg. Allg. Chem.*, **202,** 49 (1931).
22. H. J. Emeléus and A. A. Woolf, *J. Chem. Soc.* (London), **35,** 164 (1950).
23. A. D. Kirshenbaum and A. V. Grosse, *Anal. Chem.*, **26,** 1955 (1954).
24. (a) A. L. Beach and W. G. Guldner, *ASTM Symposium on Determination of Gases in Metals*, *Spec. Tech. Publ. 222*, 1958, pp. 1–9; and (b) K. W. Guardipee, *Anal. Chem.*, **42,** 469 (1970).
25. V. M. Horrigan, Minutes of Meeting, Division I, Committee E-3, American Society for Testing and Materials, St. Joseph, Mich., November 1963.
26. V. P. Romanova, N. F. Litvinova and V. I. Karyakina, *Zavod. Lab.* **36,** 7 (1970).
27. W. H. Hatfield and W. C. Newell, *J. Iron Steel Inst.* (London), **148,** 351 (1943).
28. G. A. Ferguson and A. I. Schindler, *Rept. of NRL Prog.*, **42,** (April 1962).
29. D. I. Walter, *Anal. Chem.*, **22,** 297 (1950).
30. M. L. Aspinal, *Analyst*, **91,** 33 (1966).
31. W. A. Fraser, *Rept. of NRL Prog.*, **22** (April 1964).
32. *Annual Book of ASTM Standards*, Pt. 32, American Society for Testing and Materials, Philadelphia, 1971, pp. 508–511.
33. W. G. Guldner and A. L. Beach, *Anal. Chem.*, **22,** 366 (1950).
34. C. A. Evans, Jr., *Anal. Chem.*, **42,** 1130 (1970).
35. E. J. Brooks and F. W. Fraser, Naval Research Laboratory, Internal Report, 1972.
36. W. T. Elwell and I. R. Scholes, *Analysis of Copper and Its Alloys*, Pergamon, Oxford, 1967, pp. 107–110.

CHAPTER

17

TITANIUM, ZIRCONIUM, AND HAFNIUM

LOREN C. COVINGTON AND MAURICE J. MILES

Titanium Metals Corporation of America
Henderson Technical Laboratories
Henderson, Nevada

CONTENTS

I. METALLURGICAL ASPECTS

Because of their position in the periodic table of the elements, titanium, zirconium, and hafnium are in general similar in their behavior toward dissolved gases, and they therefore offer similar problems in analysis. The principal gaseous impurities found in these metals are oxygen, nitrogen, and hydrogen. Like most impurities in metals, these may have favorable or unfavorable effects on the physical and mechanical properties of the metals.

A. HYDROGEN

Hydrogen [1, 2] is of concern mainly because in sufficient concentration it causes embrittlement. This is true of all three metals, but the susceptibility to embrittlement varies considerably from alloy to alloy. In-service failures of components made from these metals have been found to result from hydrogen contamination. For this reason, careful control of the hydrogen content is required at all steps in processing.

B. NITROGEN

Titanium, zirconium, and hafnium react with nitrogen to form nitrides at elevated temperatures. In general, nitrogen causes an increase in hardness and strength. In the case of zirconium [2] nitrogen has been found to be very detrimental to the corrosion resistance of the metal.

C. OXYGEN

Oxygen has a very pronounced strengthening effect on the metals and also increases the hardness. Low oxygen concentrations may be helpful in strengthening the metal, while high concentrations may be harmful with regard to ductility. Oxygen reacts very rapidly with these metals at elevated temperatures and therefore creates a problem in welding and rolling operations involving exposure of the metal to the air.

In general, oxygen and nitrogen have a substantial strengthening effect. However, they cause weld embrittlement and a decrease in machinability. Forming operations are made more difficult because of a decrease in uniform elongation and ductility. Their detrimental effects are magnified by the presence of hydrogen.

II. HISTORICAL DEVELOPMENT

For the reasons cited in the previous section, it was early recognized that a knowledge of the gas content of these metals was important if predictable physical and mechanical properties were to be obtained. This was especially true of titanium, which has been developed commercially to a much greater extent than zirconium or hafnium. The analytical methods for oxygen, nitrogen, and hydrogen presently used for zirconium and hafnium are essentially the same as those employed for titanium. Since they were originally developed for titanium, their historical development will be traced relative to this element.

In 1951 when the production of titanium on a commercial scale was first begun, there were no proven methods for determining oxygen, nitrogen and hydrogen in this metal. Some of the first attempts to determine the gaseous impurities in titanium were made using methods borrowed from analytical practices in the steel industry.

A. HYDROGEN

For many years practically all hydrogen results were obtained during vacuum-fusion analysis for oxygen. This method, however, was too costly and time-consuming for routine control analysis, and so, as the commercial production of titanium increased, a number of methods were soon developed for the direct estimation of hydrogen [3]. These methods can be classified into four general categories: (1) vacuum-extraction methods, (2) equilibrium-pressure methods, (3) combustion methods, and (4) isotope-dilution methods.

1. Vacuum Extraction

The vacuum-hot-extraction methods developed for steel [4] were found, with some modifications, to work well with titanium. Young and Cleaves [5] developed a method for determining hydrogen in titanium by vacuum extraction at about 950°C. The apparatus used was simple and easy to assemble and repair. The accuracy was adequate for analyses of samples of moderately high hydrogen content. At this temperature, however, there is a measurable hydrogen residue left that is not constant for all samples but varies with the composition and physical form of the sample. Other investigators [6–8] determined hydrogen in zirconium and zirconium alloys (about 100-mg samples) by vacuum heating at 1100–1200°C.

Vacuum extraction at 1400°C was found to yield an apparently complete and much faster extraction [3, 9]. This method is capable of high precision, but the equipment is more complicated and expensive. It can be performed in vacuum-fusion equipment. Commercial equipment designed specifically for this purpose is available.

The vacuum-fusion extraction is performed above the melting point of the metal, at about 1950°C. At this temperature oxygen and nitrogen are also evolved, so that the determination of hydrogen is more complicated. Hydrogen is usually determined by difference, so the accuracy may at times be questionable. It is possible, however, to make a direct determination of hydrogen by use of a palladium tube (see Chapters 4 and 7), in which case the accuracy would be as good as the method involving extraction at 1400°C. Other techniques for analyzing the extracted gas mixture for hydrogen are mass spectrometry [10] and gas chromotography [9, 11]. Usually vacuum-fusion analysis for hydrogen is employed only in those cases where the oxygen content is also of interest.

2. Equilibrium Pressure (see Chapter 7)

In 1955, McKinley [12] published a method for determining hydrogen in titanium by measuring the equilibrium pressure generated over a sample in a container of fixed volume at a controlled temperature. The apparatus employed was simple and inexpensive. The method was reasonably rapid (about 15 min per determination), and the results were in good agreement with vacuum extraction at 1400°C and vacuum fusion. Similarly, Jones [13] measured equilibrium pressures of hydrogen over zirconium in a closed system at temperatures ranging from 900° to 1000°, with collection times of 10–30 min.

The principal limitation to this method is that the hydrogen equilibrium relationship varies for each alloy system, so that an equilibrium pressure curve must be determined in advance for each alloy system under test.

3. Combustion

Combustion methods have been studied in many laboratories and several companies have apparatus available commercially. In general, the sample is burned in pure oxygen, and the resulting water is absorbed and determined gravimetrically. Various combustion accelerators are used such as lead, tin, or a mixture of tin, iron, and copper.

Combustion methods require simple equipment and are relatively fast, the time required varying from 5 to 30 min. However, these methods are not widely used, because accuracy is not as good as with vacuum extraction; in general, the combustion results are slightly higher. Also, the reproducibility is not as good.

4. Isotope Dilution

Ashley and Denovan [14] described an isotope-dilution method for the determination of hydrogen in zirconium alloys. The sample was equilibrated with deuterium gas at an elevated temperature, and the resulting isotopic dilution was determined by mass spectrometry. The precision and accuracy over a wide range of sample size and hydrogen concentration were approximately the same as obtainable by vacuum extraction or fusion methods.

B. NITROGEN

Early in the development of the titanium industry, the U.S. Army Ordnance Corps created a Metallurgical Advisory Committee on Titanium, which organized (among others) a Panel on Methods of Analysis; part of the panel consisted of a Task Force on Nitrogen. The Task Force selected and tested available methods with the result that the micro-Kjeldahl distillation–titration method was recommended by the panel [15] and accepted by ASTM [16]. In this method, the sample is dissolved in acid, converting all the nitrogen present to ammonium ion. Ammonia is separated by steam distillation and titrated with standard acid.

Variations of the Kjeldahl method have also been reported for the determination of nitrogen in zirconium and zirconium alloys [17–19].

Other methods have been investigated since the ASTM method was accepted. One [20] is based on the liberation of molecular nitrogen from the sample by vacuum fusion. The oxygen is usually converted to carbon dioxide and absorbed; hydrogen is converted to water and either absorbed by magnesium perchlorate or frozen out in a cold trap. The remaining pressure is measured and attributed to nitrogen.

Chromatographic methods have been used to determine nitrogen and other elements in the gases liberated by vacuum fusion [20] or arc fusion [21] of steel samples. The gaseous components were separated with the aid of molecular sieves and measured by their thermal conductivities.

Goward [22] reviewed several reports on nitrogen determination by vacuum fusion and inert-gas fusion. He cited a prevailing conclusion (neither strongly supported nor denied) that conventional vacuum-fusion analyses for nitrogen in titanium, zirconium, and hafnium were not reliable.

The Leco TN-14 Direct Reading Nitrogen Determinator [23] (forerunner of the Leco 589-900; see Chapter 4) has been demonstrated by Howe [24] to be applicable to the determination of 40–500 ppm nitrogen in unalloyed titanium and in Ti–6Al–4V and Ti–8Al–1Mo–1V alloys. Each sample, consisting of a single piece and weighing about 0.1 g, is wrapped with platinum foil weighing about 10 times as much as the sample. A pellet is then made on a press to exclude air and to effect close contact of the platinum and sample. Results obtained by this technique agree closely with those obtained by standard chemical methods.

Less commonly used methods that have been reported for nitrogen in titanium include proton activation—giving the $^{15}N(\rho, \alpha\gamma)^{12}C$ reaction—[25], isotope dilution [26, 27], and fluorination by bromine tetrafluoride [28].

C. OXYGEN

The determination of oxygen in titanium and its alloys has been a particularly difficult problem. The initial application of vacuum-fusion analysis to titanium proved disappointing. The results were erratic, and the method was time consuming. The Panel on Methods of Analysis of the Metallurgical Advisory Committee on Titanium, through the cooperative efforts of the government and private laboratories forming its membership, investigated and evaluated many techniques, each of which can be classified as either halogenation or fusion. Activation analytical methods have since been developed for oxygen in titanium and zirconium.

1. Halogenation Methods

a. Chlorination

In chlorination [29] the titanium was chlorinated at 400°C, to convert the metal to volatile titanium tetrachloride, leaving oxygen behind as titanium dioxide. The oxygen was calculated from the amount of the titanium dioxide residue. This method was found to be unsuitable because of a high blank and interference due to carbon, silicon, and chromium. Also, it was difficult to prevent mechanical loss of the titanium dioxide during chlorination.

Another chlorination procedure, basically equivalent to the bromination procedure described in the next section, was reported by Elwell and Peake [30]. It gave reproducible results and had the advantage of simpler equipment than that of vacuum fusion.

A hydrochlorination method was developed by Negina et al. [31] for the determination of oxygen in titanium hydride and titanium metal. The sample was heated at 500–670°C for 2 h in a stream of dry hydrogen chloride. The oxygen in the oxide residue was determined gravimetrically or by a photometric determination of the titanium in the residue.

b. Bromination

This method was developed by Codel and Norwitz [32, 33] and was used quite widely for a short time in the analysis of titanium and zirconium [34]. The titanium sample was mixed with graphite powder in a gold boat and exposed to bromine, with argon used as a carrier gas at 900°C. The oxygen of the sample combined with the graphite to form carbon monoxide, while the titanium was converted to titanium tetrabromide. This was removed in a cold trap and the carbon monoxide was converted to carbon dioxide by hot copper oxide. The carbon dioxide was absorbed in Ascarite and weighed.

This method was slow, and the blank was too high for accurate determinations.

c. Bromine Trifluoride

This reagent liberates total oxygen as molecular oxygen from all metals that form volatile fluorides or fluorides that are soluble in bromine trifluoride. If the fluorides are not volatile or soluble, they form a film on the surface of the sample, thus preventing its reaction with the bromine trifluoride. O'Neill [35] applied this reagent to the determination of oxygen and nitrogen in titanium. His results were comparable to those obtained by the bromination method. The bromine trifluoride method was never widely adopted.

2. Fusion Methods

The fusion methods consist of vacuum fusion, of which there are several variations, and inert-gas fusion. The reaction is the same in all cases (see Chapter 4, Sections I and II). The liberated gases are collected and analyzed by a variety of techniques.

a. Vacuum Fusion

Among early efforts to modify the vacuum-fusion method so that it could be applied to titanium were those of Derge [36], who proposed the use of an iron bath, and Walter [37], who used no molten bath but instead increased the active surface area of the carbon by the addition of carbon chips to the crucible. The iron-bath technique was improved [38] but did not win lasting favor because carbides tended to form in the bath and agglomerate, giving erratic results. The method of Walter was limited to about 4 or 5 samples in a crucible, because all available carbon surfaces then are covered with a metal film. Further, determination took an hour or more; however, a skilled operator could produce accurate, reliable results. This method was approved by the Panel on Methods of Analysis.

Because the method of Walter was slow, costly, and not suited to control analysis, a faster, cheaper method was required. The platinum-bath technique appeared to offer the most promise of filling this need. It was first used by Gregory and Mapper [39] in the analysis of beryllium. They also suggested using it for zirconium. In 1955, Smiley [40] used a platinum bath to analyze iron, steel, aluminum, and thorium samples by his inert-gas-fusion–capillary-trap method. His experiments with titanium, however, were disappointing.

In 1956, Wilkins [41] successfully used a platinum-bath technique on titanium. This technique was soon tried by a number of laboratories and was made the subject of an investigation by the Panel on Methods of Analysis. Difficulty was experienced in obtaining consistent results. An investigation by Hansen et al. [42] showed this to be caused by inadequate mixing of the bath, which resulted in an excessive ratio of titanium to platinum in portions of the bath. They proposed a platinum-flux technique in which the required amount of platinum was dropped into the crucible with each sample, thus assuring the proper ratio of titanium to platinum at all times.

The vacuum-fusion method using a platinum-flux technique was evaluated by the Panel on Methods of Analysis and recommended to ASTM as an acceptable method of determining oxygen in titanium. This method is widely used by industry today. Improvements in equipment [43] have reduced the time per determination to as little as 5 min.

b. Inert-Gas Fusion

Among the earliest efforts in inert-gas fusion are the capillary-trap method of Smiley [40] and the spectrographic method of Fassel [44, 45]. The capillary-trap method was not widely used. The spectrographic method was rapid (80 determinations in an 8-h day) and accurate. It was not an absolute method, however, and had to be calibrated by reference to another method, usually vacuum fusion.

The inert-gas-fusion method, which is widely used today (see Chapter 4, Section II), was also evaluated by the Panel on Methods of Analysis and recommended as an approved method for determining oxygen in titanium.

3. Activation Methods

Two activation methods that have been successfully used for the determination of oxygen in titanium and zirconium are 14-MeV neutron activation [46–48] and irradiation with ^3He nuclei of energy below 11 MeV [49].

Twitty and Fritz [47] reported an internal standard technique of neutron-activation analysis of titanium that eliminated the necessity of weighing the sample or of measuring the neutron flux. A summation of the three γ-rays of ^{48}Sc [$^{48}Ti(n, p)^{48}Sc$] was chosen as the internal standard to which the ^{16}N activity of the $^{16}O(n, p)^{16}N$ reaction was compared. The oxygen results for five samples of titanium containing 0.054–0.260% O were compared with the averages of multiple, vacuum-fusion results. The standard deviation for the internal standard technique was 0.005%.

Revel and Albert [49] reported the determination of oxygen in zirconium and other metals by irradiation with ^3He nuclei. The limit of sensitivity was 10^{-8} and no other impurity interfered with the oxygen analysis.

III. STANDARD SAMPLES

The National Bureau of Standards (NBS) has standard titanium samples available with certified oxygen, nitrogen, and hydrogen contents, as shown in Table 17.1. Oxygen standards are available in unalloyed titanium and in Ti–6Al–4V alloy. Nitrogen standards are available in Ti–4Al–4Mn, Ti–6Al–4V, and Ti–5Al–2.5Sn alloys. The hydrogen standards were developed with unalloyed titanium containing 32, 98, and 215 ppm hydrogen.

Many laboratories have developed their own standards for hydrogen [50] and oxygen by sectioning and analyzing a relatively large quantity of titanium sheet or wire. Portions yielding results outside of a selected tolerance are rejected. Numerous subsequent analyses on the remaining portions eventually lead to a stock of material having a well-established range and average that

TABLE 17.1 NBS Titanium Standards for Oxygen, Hydrogen, and Nitrogen

NBS Sample Designation	Alloy Grade	Certified %O	Certified ppm H	Certified %N
355	Unalloyed	0.303		
356	Ti–6Al–4V	0.133		
352	Unalloyed		32 ± 2	
353	Unalloyed		98 ± 5	
354	Unalloyed		215 ± 6	
173	Ti–6Al–4V			0.018
174	Ti–4Al–4Mn			0.012
173a	Ti–6Al–4V			0.018
176	Ti–5Al–2.5Sn			0.010

will yield results falling within statistically established control limits. These materials are regularly used along with samples certified by the NBS, where available, to monitor production and control laboratories and to assure results free from errors. The standards available are expected to be replaced by the NBS when exhausted, thus assuring an increasing supply of future standards for each of the three gases of interest to the producers and users of titanium.

The NBS has certified a Zircaloy sample for nitrogen; until more such certified samples become available, the titanium standards may be used to monitor laboratory performance when analyzing zirconium or hafnium for oxygen, hydrogen, or nitrogen.

IV. SAMPLING AND SAMPLE PREPARATION

When analyzing for oxygen and hydrogen, samples are preferred in chunk or cube form. They may be cut from the material to be tested by shearing or sawing in a way that will avoid excessive heating as evidenced by sparks or discoloration of the cut surface. Chip or powder samples should be avoided. Such materials can be analyzed for oxygen and nitrogen by compacting and melting them into button form under an inert atmosphere. When used for nitrogen determination by the Kjeldahl method, samples are preferred in the form of chips or turnings. Holson and Swinburn [51] identified a potential source of contamination as atmospheric nitrogen during sample preparation. Milling or drilling, either with dull tools or at excessive speeds, were shown to induce a contaminating surface reaction.

Sample preparation will vary somewhat, depending on the analysis desired. In all cases the surface should be cleaned thoroughly of any contamination. This may be done by abrading the surface with a clean file or, if hydrogen content is of no concern, by a brief pickle in 5% hydrofluoric acid solution. Pickling can be a source of hydrogen contamination and must be avoided when hydrogen analyses are desired. The final step in preparation should be a thorough rinse in an organic solvent, such as reagent-grade acetone, which will not leave a residue when it evaporates. Samples should be handled only with clean tweezers from this point on.

V. ANALYSIS

A. HYDROGEN

1. Hot Extraction

Hot extraction or vacuum extraction at 1400°C is probably the most widely used method for determining hydrogen. The equipment used by some laboratories is somewhat complicated and expensive, being a simplified version of a vacuum-fusion apparatus. Many laboratories use vacuum-fusion equipment to determine hydrogen by this method, by the simple expedient of lowering the furnace temperature to 1400°C. Since only hydrogen is evolved at this temperature, it is only necessary to have a known volume in which the hydrogen is collected and the pressure measured, usually with a McLeod gauge. This method is capable of high precision, a standard deviation of 1.3–2.3 ppm often being reported on NBS standards at the 100-ppm level. The method is also rapid, the time per determination ranging from 3–10 min depending on the apparatus used.

Other techniques that can be used to measure the extracted hydrogen include cryogenic collection in a liquid-helium cold trap for subsequent pressure–volume measurement by a sensitive manometer [7], and gas chromatography [9].

2. Vacuum Fusion

This method is primarily used for determining oxygen. Hydrogen determinations are usually regarded as a by-product. For hydrogen this method offers no apparent advantage over vacuum extraction at 1400°C.

Vacuum fusion is normally performed at 1900–1950°C. At this temperature, the oxygen, nitrogen, and hydrogen are all evolved. It is then necessary to separate and measure each gas. Hydrogen is usually determined by difference, in which case cumulative errors can be introduced. A palladium tube is sometimes used, which permits the direct measurement of the hydrogen and elimination of this source of error. With careful work, hydrogen determinations can be fairly precise, although usually the reproducibility is not as good as that obtained with vacuum extraction at 1400°C. Therefore, vacuum fusion is generally used for hydrogen only when oxygen is also desired.

B. OXYGEN

1. Inert-Gas Fusion

The inert-gas-fusion method is widely used today, largely because of the availability of the rapid oxygen analyzer manufactured by Laboratory Equipment Corporation. With this equipment, analysis times as short as 5 min have been reported. Accuracy and precision are comparable to those obtained with vacuum fusion. As mentioned earlier, the method as recommended for rapid control analysis by the Panel on Methods of Analysis of the Metallurgical Advisory Committee on Titanium specifies the use of a platinum flux.

2. Vacuum Fusion

As noted earlier, two fusion methods have been recommended by the Panel for the determination of oxygen in titanium, the Walter (dry-crucible) [37] and the platinum-flux techniques [42]. These two methods require essentially the same equipment. Because of the time-consuming nature of the Walter (dry-crucible) technique, it has been largely supplanted by the platinum-flux technique, which involves dropping each sample together with 10 times its weight of platinum into the crucible. This platinum–titanium ratio was recommended by the Panel on Methods of Analysis and probably requires a little more platinum than is essential, since Hansen et al. [42] reported good results at a ratio of 8:1. A platinum bath, which was giving low erratic results, was analyzed in the authors' laboratories and found to have a platinum–titanium ratio in the top portion of only 5:1. This indicates that the minimum platinum–titanium ratio essential for complete evolution of the gases is between 5 and 8.

The platinum acts as a fluxing agent to promote better contact between the sample and the carbon of the crucible. In addition, a highly exothermic alloying reaction takes place between the platinum and titanium, which is believed to contribute to extremely rapid gas evolution.

The number of samples that can be analyzed per crucible with the platinum-flux technique is limited only by the physical capacity of the crucible, since carbon (about 1%) dissolves in the bath and is readily available for reaction with the oxygen. Unpublished data from Titanium Metals Corporation show that as many as 95 samples, averaging 0.1 g in weight, can routinely be analyzed in a single crucible with an inside diameter of $\frac{3}{4}$ in. and a depth of $2\frac{3}{4}$ in.

Analysis times of 20 min to 1 h per determination were required using the platinum-flux technique with equipment available before 1960, but these have been reduced to as little as 5 min, with no sacrifice of accuracy or precision, by using the apparatus designed by Covington and Bennett [43]. A standard deviation of 0.007% at the 0.3% level is routinely obtained with the rapid technique.

C. NITROGEN

Most of the laboratories engaged in the analysis of titanium and zirconium use the titrimetric or photometric methods for nitrogen described by ASTM [16]. These procedures are similar to the recommended procedure presented below; most of the variations in the methods lie in the type of equipment used rather than the method itself.

Some laboratories use vacuum-fusion methods where nitrogen is determined by difference. This approach is attractive for the analysis of products having considerable nitrogen content, but the added errors involved in the oxygen and hydrogen removal make it less dependable for production control. Strong claims for the speed and accuracy of vacuum-fusion nitrogen determinations have been made by some manufacturers of this equipment. Extra platinum is required to attain the high temperature necessary to release all of the nitrogen. In analyzing titanium nitride or other high-nitrogen samples, titanium having a known nitrogen content may be added with platinum to create the necessary release temperature. The added nitrogen is deducted as a blank. However, the chemical methods recommended below are used for most of the nitrogen determinations in the laboratories of the producers and fabricators of these metals.

VI. RECOMMENDED METHODS

A. HYDROGEN

Determine hydrogen in titanium, zirconium, or hafnium by the vacuum-hot-extraction method at 1400°C. At this temperature, there are no interferences (Note 1).

Cut a 0.1–0.5-g piece from the gross sample (Note 2). Degrease the sample by dipping in an organic solvent that will leave no residue on evaporation, such as reagent-grade trichlorethylene or carbon tetrachloride, rinse in reagent-grade acetone, dry in air, and weigh to the nearest milligram, being careful to handle the sample with clean tweezers.

After a satisfactory blank has been obtained (Note 3), open the analytical volume to the evacuating pump until the system attains minimum pressure. Isolate the analytical volume from the evacuating pumps, and drop the sample into the furnace zone heated to 1400°C ±50°C. The hydrogen is usually evolved at this temperature in less than 1 min. Thicker samples will require more time. Constant pressure in the analytical volume or a drop to blank pressure in the furnace section indicates complete transfer of all hydrogen from the sample and furnace volume. This requires 2 min or more, depending on the apparatus used. Read the pressure due to hydrogen or determine hydrogen from the difference between the total pressure and the pressure after removing hydrogen.

Evacuate the analytical volume, and repeat the cycle for successive samples. Calculate the hydrogen content as shown in Chapter 4, Section I.E.

NOTES

1. Hydrogen may also be determined along with oxygen by the vacuum-fusion method described in Section VI.B. However, provision must then be made either to remove hydrogen by diffusion through palladium or by catalytic oxidation to water followed by removal of the water.
2. This method is applicable to all titanium, zirconium, and hafnium base materials containing more than 5 ppm of hydrogen. Materials containing more than 200 ppm are unusual and may require use of smaller samples, a larger calibrated collection volume, or both.
3. The blank pressure in the analytical volume should be less than 1% of the normal reading obtained with samples containing approximately 100 ppm. The average of several blank determinations should be used in calculating the results.

B. OXYGEN

Determine oxygen by the vacuum-fusion–platinum-flux method [16, 42, 52] (Note 1). This method is recommended for titanium, titanium alloys, zirconium, and hafnium containing 0.03–0.5% oxygen, which covers the range normally found in commercial alloys. These limits may be extended by adjustment of the sample size or varying the size of the gas-collection volume of the vacuum fusion apparatus. The lower limit, however, is dependent on the amount of gas present in the apparatus blank and the platinum flux. Under the conditions described, no element present in these commercially available metals or alloys is known to interfere.

Cut the sample in chunk or cube form, where possible, from the material to be analyzed in any manner that will avoid oxidation. Do not use chip samples. The sample size should be such that the expected gas from the sample will be at least 10 times the gas from the apparatus blank and platinum flux. Normally, a 0.15–0.25 g sample can be taken. Any surface oxides should be removed by filing. Degrease the sample by rinsing in reagent-grade acetone, air dry, and weigh to the nearest milligram. Do not touch the samples with fingers during and after the final stages of cleaning. Use clean tweezers for handling the specimens.

Outgas the vacuum-fusion furnace by heating at a temperature of about 2350°C until the gas evolution has dropped to the normal operating-blank level. Lower the temperature of the crucible to 1900°C, and maintain at this temperature during the blank determination.

Determine the blank by exactly following the same sequence of operations as is used in analyzing a sample. Measure the pressure. Then analyze the gas collected for carbon monoxide, hydrogen, and nitrogen according to the instrument manufacturer's instructions. Condition the furnace by adding a piece of platinum (the same size as used with samples) to the crucible, and determine the gas content of the platinum. This should be low and reproducible. The blank thus obtained will be a combined furnace and platinum blank. The magnitude of the normal operating blank will vary depending on the extraction time used.

After a suitable blank has been obtained, add the sample together with at least 10 times its weight of platinum to the crucible (Note 2). This is followed almost immediately by a brilliant flash as the platinum melts and alloys with the titanium. At this time, most of the gases in the sample are evolved. Successive readings of the pressure in the furnace or collection volume will indicate when all the evolved gases have been transferred to the analytical section.

The gases are now ready for analysis; the details of this procedure vary somewhat according to the type of equipment being used. Generally, the carbon monoxide is removed by oxidizing it to carbon dioxide and either freezing it out with liquid nitrogen or absorbing it with Ascarite. If Hopcalite at room temperature is used as the oxidizing agent, remove the hydrogen by diffusing through palladium prior to oxidation. If copper oxide heated to 325°C is used as the oxidizing agent, hydrogen will also be oxidized. In this latter case, circulate the gas through a $Mg(ClO_4)_2$ trap to remove the water before passing the gas into a liquid-nitrogen-cooled trap to remove the CO_2. The residual gas is assumed to be nitrogen; measure its pressure. Remove the liquid-nitrogen flask from the cold trap, release the carbon dioxide, and combine it with the residual gas. Measure the increase in pressure and calculate the amount of oxygen evolved from the sample (see Chapter 4, Section I.E).

NOTES

1. Vacuum-fusion systems are produced commercially, and some of these are discussed in Chapter 4. The platinum-flux technique can be employed

using any of these units. A means is provided for introducing the sample and platinum into the crucible when desired. This is an "air-lock" type device that permits charging the samples as the run progresses without disturbing the vacuum.

2. It is recommended that the titanium and platinum be pressed together before being introduced into the crucible. If platinum wire is used, it may be wrapped around the sample so that they enter the crucible as one unit.

C. NITROGEN

For titanium, zirconium, and hafnium samples containing more than 0.01% nitrogen, determine nitrogen by the distillation titration method [16]. For samples containing less nitrogen, determine it photometrically with Nessler's reagent. This method is discussed in Chapter 8.

1. Apparatus

The modified Kjeldahl apparatus consists of a 2-liter, single-neck (ST 29/42), round-bottom steam generator flask in a controllable heating mantle. This flask is connected by two consecutive sections (both inner parts of Corning 7729 vacuum trap), the latter delivering steam to the bottom of a 250-ml Erlenmeyer flask with a ST 29/42 mouth. The latter section has a side-arm that is connected to a condenser arranged to deliver distillate to the bottom of a 250-ml Erlenmeyer receiving flask.

2. Reagents

Use deionized water for all reagent solutions (see Chapter 8).

(1) AMMONIUM CHLORIDE, STANDARD SOLUTION (1 ml = 0.1 mg nitrogen). Transfer 0.3820 g pure dry NH_4Cl to a 1-liter volumetric flask, dissolve, adjust to ambient temperature if necessary, dilute to volume with deionized distilled water, and mix.

(2) BORIC ACID. Saturated solution.

(3) HYDROCHLORIC ACID. Standard solution, 0.007 N (1 ml = ~0.01% N on a 1-g sample). Dilute 100 ml to 1 liter for use. Check the standardization occasionally by treating an aliquot of the standard NH_4Cl solution according to the procedure and titrating the distillate.

(4) SODIUM HYDROXIDE. 500 g per liter. Dissolve 500 g reagent-grade NaOH pellets in 600 ml water with stirring and cooling, and dilute to 1 liter. If the blank determinations are not satisfactory, treat the solution with 1 g Devarda's alloy, cool, and decant.

(5) NESSLER'S REAGENT. See Chapter 8.

(6) INDICATOR SOLUTION. Dissolve 0.167 g bromocresol green in 100 ml ethanol. Dissolve 0.033 g methyl red in 100 ml ethanol. Add equal volumes of the two solutions to a dropping bottle.

3. Preparation of Calibration Curve

Dilute 10 ml ammonium chloride solution to 100 ml (1 ml = 0.01 mg of N). Transfer 3, 6, 9, and 12 ml dilute standard NH_4Cl solution to 50-ml volumetric flasks containing 25 ml water. Pipet 1 ml Nessler's reagent into each flask, dilute to the mark with water, stopper, and mix. To prepare a reference solution, pipet 1 ml Nessler's reagent into another 50-ml volumetric flask containing only 35 ml water, dilute to volume with water, stopper, and mix. Allow the solutions to stand 5–30 min (Note 1). Measure the absorbance of each solution at 430 nm, and plot absorbance against milligrams of N per 50 ml of solution.

4. Procedure

Dissolve a 5.000 g sample in 75 ml HCl or $(1 + 4)H_2SO_4$ in an atmosphere free of ammonia or ammonium salts. For zirconium or hafnium samples, use 20 ml HF and plastic containers or 40 ml $(1 + 1)H_2SO_4$ with dropwise additions of HF. Dissolve insoluble residues by adding a few small selenium crystals, and fume with about 20 ml H_2SO_4. Additions of drops of HF or H_2O_2 before fuming may aid the decomposition of residues. Cool, transfer to a 100-ml volumetric flask, adjust the temperature, dilute to volume with deionized water, and mix.

Pipet a 20-ml aliquot to an Erlenmeyer flask having a ST 29/42 mouth and containing 50–75 ml water. Add 30 ml NaOH solution and immediately attach to the steam-delivery section.

For samples containing 0.01% or more nitrogen, place a 250-ml Erlenmeyer receiving flask containing 25 ml saturated boric acid solution, 25 ml water, and 8 drops mixed indicator under the condenser with the liquid in contact with the condenser tip. Steam distill approximately 50 ml into the receiving flask. Remove the flask, rinsing the condenser tip with water. Titrate the distillate with standard HCl to the same endpoint as was obtained by titrating the blank. Determine the blank, using all the reagents and steps in the procedure except that no sample is added to the flask. Redetermine the blank each time water is added to the steam generator.

For samples containing less than 0.01% nitrogen, place a 50-ml volumetric flask under the condenser with a Tygon adapter extending into the flask. Steam distill 35–40 ml into the flask. Remove the flask, pipet 1 ml Nessler's reagent into the flask, adjust to room temperature, dilute to volume with water, stopper, and mix. Allow the solution to stand 5–30 min, depending on the nitrogen content (Note 1). Measure the absorbance of the solution against a reagent blank at 430 nm.

5. Calculations

See Chapter 8.

6. Repeatability

The volumetric method used in rapid routine work has a repeatability of 0.002% *N*. The repeatability of the photometric method is 0.0005% *N*.

7. Time of Analysis

One operator with three distillation units can easily produce three results per hour.

NOTE

1. The calibration curve should be prepared under exactly the same conditions that are used for the samples [53]. A reaction time of 5 min may be suitable for samples of high nitrogen content, while 30 min may be necessary for low nitrogen concentrations.

REFERENCES

1. D. N. Williams, *TML Report No. 100*, Battelle Memorial Institute, Columbus, Ohio, 1958.
2. G. L. Miller, Ed., *Metallurgy of the Rarer Metals*, Vol. II, *Zirconium*, Butterworth, London, 1954, p. 178.
3. T. D. McKinley, *The Determination of Hydrogen in Titanium and its Alloys, A Critical Review*, *Memorandum TML*, Battelle Memorial Institute, Columbus, Ohio, 1957; *Trans. Met. Soc. AIME*, **212,** 563 (1958).
4. C. E. Sims and G. A. Moore, *Trans. AIME*, **176,** 248 (1948).
5. R. K. Young and D. W. Cleaves, *Anal. Chem.*, **28,** 372 (1956).
6. P. Ebling, G. W. Goward and R. K. McGeary, *U.S. At. Energy Comm. WAPD-CTA* (*GLA*), **183** (1958).
7. E. A. Gulbransen and K. F. Andrew, *Electrochem. Technol.*, **5,** 471 (1967).
8. M. Robinson, *AERA-AM-107*, Atomic Energy Research Establishment, Harwell, England, 1968.
9. L. Brhacek and O. Kalandra, *Chem. Prum.*, **18,** 417 (1968); through *Nucl. Sci. Abstr.*, **23,** 11413 (1969).

10. R. E. Taylor, *Anal. Chim. Acta*, **21,** 549 (1959).
11. R. Lesser and H. Gruber, *Z. Metallk.*, **51,** 495 (1960).
12. T. D. McKinley, *J. Electrochem. Soc.*, **102,** 117 (1955).
13. W. H. Jones, *WADC-TN-57-294*, U.S. Atomic Energy Commission, 1957.
14. R. W. Ashley and A. S. Denovan, *At. Energy of Canada*, *CRDC-1212*, Chalk River, Ontario, 1965.
15. U.S. Army Materials Research Agency, *Information Bulletin No. T8*, *Part I*, Watertown Arsenal, Watertown, Mass., 1955, p. 57.
16. American Society for Testing and Materials, *Annual Book ofASTM Standards*, Part 32, Chemical Analysis of Metals, Philadelphia, 1973.
17. J. J. Harter, A. W. Perrine, and J. F. Rodgers, *WAPD-CTA* (*GLA*) *170*, U.S. Atomic Energy Commission, 1958.
18. C. F. Bush, *Anal. Method AM-78*, Atomic Energy Research Establishment, Harwell, England, 1961.
19. N. A. Selezneva and O. M. Plotnikova, *Isv. Akad. Nauk Kaz. SSR*, *Ser. Khim.* **18** (1968); through *Chem. Abstr.*, **70,** 73936q (1969).
20. L. L. Lewis and L. M. Melnick, *Anal. Chem.*, **34,** 868, (1962).
21. F. M. Evens and V. A. Fassel, *Anal. Chem.*, **35,** 1444 (1963).
22. G. W. Goward, *Anal. Chem.*, **37,** 117R (1965).
23. Form No. C258, Laboratory Equipment Corp., St. Joseph, Mich.
24. G. Howe, Private communication to L. M. Melnick, Feb., 17, 1970.
25. T. M. Sabine, J. R. Bird, and B. K. Mak, BNL-50083; through *Nucl. Sci. Abstr.*, **22,** 30427 (1968).
26. K. B. Orlova and E. N. Vitol, *Zh. Anal. Khim.*, **21,** 1263 (1966); through *Nucl. Sci. Abstr.*, **21,** 1297 (1967).
27. T. F. Ivanova and V. V. Fedorov, *Zh. Anal. Khim*, **23,** 1750 (1968); through *Nucl. Sci. Abstr.*, **23,** 2244 (1969).
28. W. A. Dupraw and H. J. O'Neill, *Anal. Chem.*, **31,** 1104 (1959).
29. T. A. Sullivan, B. J. Boyle, A. T. Mackie and R. A. Pott, *Determination of Oxygen in Titanium*, *Report 5834*, Bureau of Mines, 1961.
30. W. T. Elwell and D. M. Peake, *Analyst*, **82,** 734 (1957).
31. V. R. Negina, E. A. Kozyreva and Z. S. Bogatova, *Zh. Anal. Khim.*, **23,** 563 (1968); through *Chem. Abstr.*, **69,** 15860d (1968).
32. M. Codell and G. Norwitz, *Anal. Chem.*, **27,** 1083 (1955).
33. M. Codell and G. Norwitz, *Anal. Chem.*, **30,** 524 (1958).
34. Y. Otake, *Nippon Kinzoku Gakkaishi*, **22,** 341 (1958); through *Chem. Abstr.*, **55,** 26837b (1961).
35. H. J. O'Neill, *The Determination of Oxygen and Nitrogen in Titanium and Titanium Alloys Using Bromine Trifluoride*, *Armour Research Foundation*, *Contract No. DA-11-022-505-ORD-2189*, 1957; and *Anal. Chem.*, **31,** 1104 (1959).
36. G. Derge, *J. Metals*, **1,** 31 (1949).
37. D. I. Walter, *Anal. Chem.*, **22,** 297 (1950).
38. R. S. McDonald, J. E. Fagel and E. W. Balis, *Anal. Chem.*, **27,** 1632 (1955).
39. J. N. Gregory and D. Mapper, *Analyst*, **80,** 230 (1955).
40. W. G. Smiley, *Anal. Chem.*, **27,** 1098 (1955).

41. D. H. Wilkins and J. F. Fleischer, *Anal. Chim. Acta.*, **15,** 334 (1956).
42. W. E. Hansen, W. M. Mallett and M. I. Trzeciak, *Anal. Chem.*, **31,** 1237 (1959).
43. L. C. Covington and S. J. Bennett, *Anal. Chem.*, **32,** 1334 (1960).
44. V. A. Fassel and R. W. Tabeling, *Spectrochim. Acta*, **8,** 201 (1956).
45. V. A. Fassel and W. A. Gordon, *Anal. Chem.*, **30,** 179 (1958).
46. J. Hoste, D. DeSoete and A. Speecke, *EUR-3565e*; through *Nucl. Sci. Abstr.*, **22,** 1948 (1967).
47. B. L. Twitty and K. M. Fritz, *Anal. Chem.*, **39,** 527 (1967).
48. G. H. Andersen, in P. Polishuk, Ed., *Nucleonics in Aerospace*, Plenum, New York, 1968, p. 317.
49. G. Revel and P. Albert, *J. Nucl. Mater.*, **25,** 87 (1968).
50. M. J. Trezeciak, *Anal. Chem.*, **32,** 72 (1960).
51. J. D. Holson and D. Swinburn, *Analyst*, **83,** 376 (1958).
52. S. Vigo and others, *U.S. Army Materials Research Agency*, *Information Bulletin No. T8*, *Part III*, Watertown Arsenal, Watertown, Mass., June 8, 1962.
53. M. J. Taras, "Nitrogen," in D. F. Boltz, Ed., *Colorimetric Determination of Nonmetals*, Interscience, New York, 1958, p. 92.

CHAPTER

18

NIOBIUM, TANTALUM, MOLYBDENUM, AND TUNGSTEN

MANLEY W. MALLETT

Battelle Memorial Institute
Columbus, Ohio

CONTENTS

I. METALLURGICAL ASPECTS

The four refractory metals, niobium, tantalum, molybdenum, and tungsten, are characterized by their high melting points (about 2460°C). However, they fall in different positions in the periodic table of the elements, niobium and tantalum in group 5A and molybdenum and tungsten in group 6A. The chemical behavior of these metals toward gases differs widely according to group.

A. NIOBIUM AND TANTALUM

Niobium and tantalum show remarkably high gas solubilities. Both form monohydrides. At temperatures above about 200°C, these hydrides transform to solid solutions. Thus, solutions approaching concentrations of 50 at % are possible. Niobium forms a solid solution with oxygen that reaches 1 wt % at 1100°C [1], while tantalum dissolves 0.2 wt % at 900°C. In the same temperature region, the nitrogen solubility in niobium is about 0.3 wt % and in tantalum about 0.4 wt %.

The large solubilities at elevated temperature result in appreciable absorption of the gases, which then tend to precipitate as compounds (hydrides, oxides, and nitrides) upon cooling. In solution, the gases cause important changes in the lattice parameters of the metals. The gas–metal compounds precipitate and cause increased hardness of the metals. In fact, if the gas pickup is limited to a single gas, hardness may be used as a measure of gas content.

The increase in hardness with increasing nitrogen content in tantalum parallels an increase in tensile strength and yield point. Tantalum with a low nitrogen content shows an upper and lower yield point and a corresponding range of work hardening. Hydrogen in tantalum has a similar effect. With high nitrogen contents, tensile stress falls after the yield point is reached and brittle fracture follows. Nitrogen also increases the electrical resistivity of tantalum. The behavior of gases with niobium is analogous to that with tantalum.

B. MOLYBDENUM AND TUNGSTEN

Molybdenum and tungsten have rather meager solubilities for hydrogen, oxygen, and nitrogen. Neither forms a hydride. Data on the solubility of hydrogen in molybdenum are conflicting. Hill [2] reports that hydrogen solubility increases with temperature at least to 1700°C for which a value of 2.0 ppm is indicated. Hydrogen is said to embrittle all body-centered cubic metals including tungsten and molybdenum [3]. Molybdenum shows a maximum oxygen solubility of about 120 ppm at 2100°C; the nitrogen solubility at this temperature is about 100 ppm. These solubilities decrease to a few parts per million below 800°C, and the excess oxygen and nitrogen precipitate in the grain boundaries as oxide and nitride to cause severe reduction in fracture strength. The very low solubilities of oxygen and nitrogen in tungsten indicate that substantial solid solution hardening should not be expected [4]. Nitrogen [5] has been reported to have no detectable effect on fine tungsten wires. Oxygen and nitrogen decrease recrystallization temperature [6]. In view of the behavior of other refractory metals, some effect of gas content on ductility is expected. Quantitative data supporting these impressions are largely lacking for tungsten. This is in part because of the lack of analytical methods sensitive in the range less than 10 ppm.

II. HISTORICAL DEVELOPMENT

A. HYDROGEN

The analysis for hydrogen in metals is simpler than the analyses for oxygen and nitrogen. Hydrogen is extracted rapidly and quantitatively from nearly all metals by melting in vacuum. Thus, satisfactory hydrogen values often are obtained even though extraction of concurrently sought oxygen may be incomplete. Spurious hydrogen values are sometimes found for metals with high vapor pressures when the vaporized metal reacts with water molecules adsorbed on apparatus walls. This does not occur as much with refractory metals, because they are less volatile and because the analysis is generally made in an apparatus rather completely degassed at very high temperatures (2000–2500°C). Although hydrogen analysis by diffusion of hydrogen from the solid refractory metal sample at 1000°C or higher seems promising, there appear to be few investigations in this field. With the exception of the isotopic-dilution [7] methods, hydrogen methods for refractory metals are a by-product of methods devised for oxygen determination.

B. OXYGEN

The vacuum-fusion method probably was the earliest method used for precision analysis for oxygen in the refractory metals. As initially performed, samples of refractory metals were dissolved in a degassed bath of iron [8] at about 1600°C. Subsequently, iron–tin [9] platinum [10, 11] platinum-tin [10, 12] and other metals have been used to flux the samples. Gases were evolved, collected, and analyzed according to the procedures established for steels. In the early vacuum-fusion apparatus, the gas often was removed from the vacuum system with a Toepler pump and analyzed in a wet chemical (Orsat) apparatus [13]. This technique has been almost entirely replaced by low-pressure analysis through absorption by dry chemicals or through fractional freezing. These techniques are faster and more precise than the older technique. In some (usually special) cases, the extracted gas is analyzed mass-spectrometrically or gas-chromatographically.

The inert-gas-fusion method carried out in an apparatus typified by the LECO conductometric oxygen analyzer [14] (Chapter 4, Section II is reported to have been used for the determination of oxygen in all four refractory metals. However, the validity of the use of the method for oxygen levels less than about 20 ppm in molybdenum and tungsten has not been established.

A diffusion-extraction (or vacuum-extraction) method developed by Hansen and Mallett [15] for the determination of oxygen in niobium later was used by them for tantalum. Use of this method was extended to molybdenum and tungsten by Fagel et al. [16]. An emission spectrographic method for oxygen in niobium and tantalum was developed by Fassel [17, 18] and his associates.

The methods presently favored are vacuum fusion and diffusion extraction in that order. There also is a tendency in vacuum-fusion analysis to use the iron-bath techniques for molybdenum and tungsten and the platinum bath for niobium and tantalum [19].

A vacuum-fusion iron-bath technique for oxygen in molybdenum was developed and evaluated in a cooperative study sponsored by the Office of Naval Research Advisory Committee on Molybdenum. The work was performed through a subcommittee on the Analysis for Gases in Molybdenum and resulted in recommendation of details of technique. Final validation of the method continued under the guidance of the Task Group on Oxygen in Molybdenum, ASTM Committee E-3 on Chemical Analysis of Metals. The effort resulted in a Tentative Method of Test for Oxygen in Molybdenum. As ASTM Designation E174-60T the tentative method [20] was approved in 1960 for use pending adoption as standard.

C. NITROGEN

Historically the earliest successful methods for determination of nitrogen in metals were those based on the solution-distillation principle (see Chapter 8, Section III) where the sample is dissolved by acid attack and nitrogen as ammonia is distilled from alkaline solution and measured by titration or other means. This method is still the one most widely used for refractory metals.

Nitrogen in molybdenum, tungsten, and niobium has been determined by alkali (sodium hydroxide) fusion. The ammonia produced is determined by titration. This method may become more popular, particularly for analyses where digestion of the sample by the Kjeldahl technique is difficult.

The vacuum-fusion method has been used to determine nitrogen along with oxygen in refractory metals. The validity of the method at all nitrogen levels requires further substantiation. The method has good potential for application to nitrogen concentrations from 1 to 10 ppm if the necessary accuracy can be established. This would permit analysis at levels below the range of applicability of the Kjeldahl method.

A unique approach [21, 22] to the determination of nitrogen in thin films of tantalum was developed. Through application of a xenon flash lamp, the film was evaporated and stable compounds such as tantalum nitride were dissociated. The gas was collected by vacuum-analytical techniques and analyzed in a gas chromatograph. The detection limit appears to be 0.1 μg nitrogen or about 70 ppm.

Hot extraction in vacuum of nitrogen from solid samples appears attractive for molybdenum but not for tungsten or tantalum [16]. The inert-gas-fusion method has been used for nitrogen in niobium. The nitrogen was measured with a gas chromatograph. Precisions of 10 ppm are claimed, but both methods require further development before they could be considered satisfactory alternatives to the inexpensive Kjeldahl method.

III. STANDARD SAMPLES

There presently is no source of standard samples for gases in the refractory metals. However, Division I, ASTM Committee E-3 acquired materials that become well categorized as a result of cooperative analytical studies.

IV. SAMPLING AND SAMPLE PREPARATION

Refractory metal samples for vacuum fusion, vacuum extraction, and inert-gas fusion are prepared by the "standard" techniques used for these methods. Samples for isotopic equilibration (for hydrogen) and diffusion extraction (for oxygen) should be no more than about 1 mm thick to avoid long equilibration and extraction times. In general, no stringent storage conditions are required for the samples of refractory metals, although concentrations of surface gases (particularly oxygen) can be sufficiently high to influence decisively analytical results [23]; hence, the surfaces should be cleaned just prior to analysis. In the case of molybdenum charged with hydrogen at 1 atm pressure at 1280–1700°C, Hill [2] concluded that there was no significant loss of hydrogen upon quenching to room temperature. Lawley et al. [24] have presented data that appear to contradict this. However, in private conversation with Lawley it was learned that his value of 400 ppm given for hydrogen solubility at 2500°C is in terms of atoms. Lawley's analytical value is about 6 ppm hydrogen by weight, which is in reasonable agreement with a value of 4.2 ppm by weight obtained by extrapolation of Hill's equation to 2500°C. Thus, the data support Hill's conclusion as to absence of hydrogen loss upon quenching. Hydrogen in tungsten would be expected to behave in a similar manner. Concentrated solutions of hydrogen and hydrides in niobium and tantalum are quite stable and require no refrigeration. However, heating of samples during sectioning should be avoided to prevent loss of hydrogen. On the other hand, hydrogen pickup in samples has been attributed to use of a water-cooled cutoff wheel. Acid pickling of samples and chemical milling should be avoided. Recommended sample-handling procedures are given in Section V.D.

V. ANALYSIS

A. HYDROGEN

1. Vacuum Hot Extraction

The method of vacuum extraction from the solid metal would seem to be almost universally applicable to metals with melting points above 500°C. However, almost no reference is made to its use for the refractory metals.

Eborall [25] pointed out that equilibrium studies suggest that the method should be applicable to niobium and tantalum. Norton and Marshall [26] indicated that molybdenum samples (0.070 in. thick) heated in hydrogen at 1600–2000°C yielded no hydrogen at 1760°C after a preliminary vacuum bakeout of 30 min at about 450°C, followed by 30 min at 800°C. Tungsten with a similar bakeout evolved no significant amount of hydrogen at high temperature. This suggests that hot extraction at 800°C constitutes a satisfactory method for determining hydrogen in molybdenum and tungsten. Hill [2] apparently determined the total hydrogen content of molybdenum samples used in his solubility studies by hot extraction at 575–980°C. Analytical results in a "reasonable time" were obtained by vacuum extraction of niobium at 1000°C [27]. A lower limit of detection of 0.02 ppm for a 25-g sample was reported. Lakomskii and Kalinyuk [28] determined hydrogen in niobium by heating specimens weighing 0.15–0.30 g for 10 min at 1400°C. Fagel et al. [16] obtained hydrogen values for molybdenum, tungsten, and niobium in the course of vacuum-hot-extraction analysis at 2000°C for oxygen (see Section V.B.3). This high temperature, of course, would not be required to determine hydrogen only.

2. Carrier-Gas Hot Extraction

At least two methods have been reported in which hydrogen diffuses from the solid metal sample at about 1100°C into a stream of argon for removal to an analytical device. Walter and Offner [29] analyzed 10-g samples of niobium by heating in a Leco furnace for 20 min in a stream of argon, passing the gas over copper oxide to quantitatively convert the hydrogen to water vapor. The water vapor was collected in an absorption tube and weighed. The method was useful down to hydrogen contents of $22/A$ ppm, where A is the sample weight in grams. Yoshimori and Ishiwari [30] also converted the extracted hydrogen to water by conducting the argon carrier through a bed of hot copper oxide. They converted the water vapor to ammonia by reaction with sodium amide at 80°C and then titrated the ammonia with electrolytically generated hypobromite ion.

3. Vacuum Fusion

Several analysts have reported hydrogen values obtained concurrently with vacuum-fusion oxygen analysis of the refractory metals. Hansen and Mallett [10] and Sloman and Harvey [31] analyzed molybdenum; Sloman [32], tungsten; Mikhailov et al. [33], molybdenum, tungsten, and niobium; and Turovtseva and Litvinova [34], molybdenum, tungsten, niobium, and tantalum. In general, the analyses are reported without comment on the method or evaluation of the results. However, it is concluded that reproducible and apparently accurate values may be obtained by the vacuum-fusion method. The analyses of molybdenum, and sometimes niobium and tungsten, are made with an iron bath at 1650°C, while those of niobium and tantalum usually are made with a platinum bath at 1900°C. Extraction times are 15–30 min. The proportion of refractory metal in the iron bath should not exceed 30%. The weight ratio of platinum to niobium or tantalum may be as low as 4:1. The precision of analysis is the same as for hot extraction; a sample-weight of 5–10 g will usually permit a precision of ±0.1 ppm hydrogen. Wood and Wolfenden [35] developed a gas-chromatographic procedure for the analysis of gases evolved from tungsten and tungsten-base alloys by vacuum fusion and compared the results with those of conventional differential pressure analysis. Their method was shown to be the more sensitive, allowing hydrogen to be determined down to about 0.2 ppm on a 0.1-g sample. Another advantage of the gas-chromatographic finish was that small quantities of hydrogen evolved from the crucible as methane could be identified and measured. The sample-weight and form are usually dictated by the needs for the oxygen analysis but, in general, a sample of minimum surface–mass ratio is desired.

No determinations have been made on the accuracy of vacuum-fusion hydrogen methods for the refractory metals. The preparation of hydrogen standards for molybdenum and tungsten would be very difficult; however, the preparation of standards of niobium and tantalum would be relatively simple using a Sieverts apparatus. Statistical analysis of the hydrogen data of Turovtseva and Litvinova [34] show coefficients of variation of 10–15% in the 2–10-ppm range. These hydrogen results presently appear typical for the vacuum-fusion analysis of the refractory metals.

4. Isotopic Equilibration

In this method, the hydrogen content of the sample is equilibrated in a closed volume with an ambient atmosphere composed of a known quantity of hydrogen isotope mixture rich in deuterium. Equilibration occurs by diffusion of deuterium into the sample and hydrogen out of the sample. The method has been used by Zaidel and Petrov [7] for determining hydrogen at the 10–50-ppm level in tantalum and niobium. The equilibrium time was 15 min at 800°C for plate specimens 1 mm thick. There is no need to extract all of the gas from the metal, since the hydrogen content of the sample can be determined from the isotope mass ratios in the gas phase. The relative standard deviation was about 10%. Turovtseva and Kunin [36] state that despite the difference in mass numbers of hydrogen and deuterium, the method is absolute and promising. Eborall [25] takes exception stating that "the apparatus is complicated, the method indirect, and the assumption of equal solubilities for hydrogen and deuterium not justified." The latter point is substantiated by reference to studies showing that the solubilities differ by 20% or more. The equilibration and hot-extraction methods require comparable times for analysis, and both methods are affected to the same extent by hydrogen from spurious sources. Therefore, the simple hot-extraction method would appear to be the better choice of techniques.

5. Combustion

Hydrogen has been determined in molybdenum and tantalum by combustion procedures. Gardner et al. [37] burned molybdenum in oxygen at 1100°C. They used fused magnesium oxide sand to absorb the molybdic oxide that was produced, thus preventing its sublimation in the combustion tube. The water produced by combustion of the sample was absorbed in anhydrous magnesium perchlorate and weighed. The average recovery obtained on 62 samples, containing known amounts of hydrogen in the 10–500 ppm range, was 99.1%.

A combustion-manometric method [38] was applied by McDougall et al. [39] to the determination of small amounts of hydrogen and carbon in tantalum and other metals. The water formed from the hydrogen in the sample was collected in a swivel-mounted dry-ice cold trap and then measured manometrically at 100°C. The method permits small samples (100 mg) to be analyzed for hydrogen down to the 5-ppm level.

B. OXYGEN

1. Inert-Gas Fusion

As a result of early work on the method, Smiley [40] reported that the inert-gas fusion with a platinum flux was unsuitable for analysis of these four refractory metals. Analyses of tantalum with iron, platinum–cobalt, and platinum–iron baths gave low results.

In an ASTM Committee E-3, Division I [41] round-robin study of analysis for oxygen in niobium, difficulty with the inert-gas-fusion method was reported when no bath was used in the crucible. With a platinum or tin bath, results approaching agreement with those by vacuum fusion were obtained at the 50-ppm level, but results were 10–20% low at the 140-ppm level. Table 18.1 shows selected laboratory averages from the round robin. Each average represents five individual analyses. Several averages differing more than 15% from the "class" average for a particular material and method were discarded.

Further studies have since been reported, establishing the applicability of the method to various metals, including niobium, molybdenum, tantalum,

TABLE 18.1. Comparison of Vacuum-Fusion and Inert-Gas-Fusion Methods for Oxygen in Niobium

	Vacuum Fusion–Pt Bath[a] (ppm)		Inert-Gas Fusion (ppm)		
Laboratory	50[b]	140[b]	50[b]	140[b]	Method
1			55		Pt flux
1	45	137	49		Sn flux
2	51	153			
3	58	137	53	122	No flux or bath
4		166			
5	47	143			
6	47	144			
7	47	141			
8			48	138	Pt flux
9	53	149			
10	43	141			
11			47	119	Pt bath
12	51	144			
13	48	144	47		Pt bath
14	52	150			
15				137	No flux or bath
16				116	No flux or bath
17	48	129			
18	52	154			
Average	49	145	50	126	

[a] 1900°C ± 100°C, Pt–sample ratio = 4:1 to 10:1, 10–20-min extraction.
[b] Nominal.

and tungsten [42–47]. Kamin et al. [42] used a nickel–platinum bath with a platinum-flux technique for niobium and tantalum. Oxygen contents of 100–1000 ppm were determined on standard samples with a relative standard deviation of 5%. Winge and Fassel [44] analyzed niobium and other refractory metals for oxygen and nitrogen by heating the sample and a platinum flux with a dc carbon arc in a static helium atmosphere. An aliquot of the resulting gas mixture was passed through a gas chromatograph, and the respective peak heights were related to the oxygen and nitrogen contents of the metal. Kashima et al. [45] used a silicon-fusion technique instead of the platinum-bath method to decompose oxides of tantalum, niobium, molybdenum, and tungsten. Dallman and Fassel [47] made an extensive study of the inert-gas-fusion method for the determination of oxygen and nitrogen in several base metals, including molybdenum, tantalum, and niobium. They optimized extraction conditions with respect to platinum–tin bath composition, bath–sample ratio, and crucible temperature. The composition of the reaction medium ranged from 80 (Pt):20(Sn) at extraction temperatures in the range of 1725–1950°C to 95(Pt):5(Sn) for extraction temperatures greater than 1950°C. Good agreement was obtained between inert-gas-fusion results and vacuum-fusion results for these metals.

For determination of oxygen in tungsten, the LECO inert-gas-fusion gas-chromatographic oxygen analyzer (see Chapter 4, Section II.C.2.c), is recommended by the manufacturer. A sensitivity of 2 ppm for a 1-g sample is claimed. The extraction conditions are 2200°C for 5 min.

A group of laboratories cooperated in evaluating oxygen results obtained on tungsten and molybdenum using the inert-gas-fusion, vacuum-fusion, vacuum-extraction, and neutron-activation methods [48]. It was concluded that in competent hands, any one of the methods can apparently provide satisfactory results.

Vasserman and Turovtseva [49] showed that oxygen can be determined in some refractory materials (including tantalum and niobium metal powders) by an impulse heating method. The sample was heated rapidly to temperatures above 3000°C in a graphite crucible containing no bath. The liberated carbon monoxide was measured by an infrared-absorption analyzer.

2. Vacuum Fusion

a. Molybdenum and Tungsten

Identical conditions are used for the vacuum-fusion analysis of molybdenum and tungsten. The analysis is carried out in an iron bath at 1600–1650°C with a minimum extraction time of 20 min. The oxygen results are the same at 1800°C as at 1650°C, but results at 1650°C are more reproducible [33]. The refractory metal content of the bath is limited to 30 wt %. This method [10, 19, 50, 51] is described in detail in Section VI.B Although descriptions of the method make major reference to molybdenum, the method has had an extensive and satisfactory application to tungsten [32, 34, 52a]. An iron–nickel bath with a tin flux has been used for molybdenum, while a cobalt bath was used with tungsten (W–CO = 1:5) [52b]. Extractions on 1–2-g samples were performed at 1650°C for 7–8 min.

Use of about 3% tin in the iron bath with like additions of tin between samples has been reported [51]. Results appear to be the same as those obtained without tin. Other tin additions have been of the order of 25% [10]. Under the conditions for the analysis, tin is generally considered innocuous. In the case of slow pumping speeds, additions of tin may be useful in suppressing gettering by vaporized metals. However, vaporization of molybdenum and tungsten is negligible at 1650°C.

An 80-platinum–20-tin bath also has been used [10]. The composition approaches the eutectic in the platinum–tin system (72Pt:28Sn; melting point, 1080°C) and gives a fluid bath. Loss of tin at 1650°C is said to be quite slow.

Wilkins [10] made an earlier use of a platinum bath at 1850°C for molybdenum. An incidental addition of iron was made to the bath before introducing the sample.

Whether the platinum bath gives results as good as or better than those by the iron bath is yet to be resolved [19, 53]. Molybdenum and tungsten–platinum phase diagrams [19, 54] suggest than an extraction temperature of 2100–2200°C may be required to eliminate solid phases from the platinum bath. At lower temperatures, only part of the sample may dissolve, and yet oxygen recovery may not be impaired. That is, the oxygen may be removed by the vacuum-hot-extraction mechanism in which case the platinum would appear to serve no useful purpose. However, studies show that 30–40 wt % molybdenum is soluble in a 90-platinum–10-tin bath at 1900°C. These very favorable conditions of temperature and sample solubility merit further investigation.

It has been reported [55] that microsamples of molybdenum and tungsten as well as niobium and tantalum were successfully analyzed in a platinum bath at 1900°C. The precision was 5 relative % or ±1 ppm hydrogen, ±10 ppm oxygen, and ±5 ppm nitrogen. More recently, a gas-chromatographic method was developed for the analysis of gas evolved from tungsten and tungsten–base alloys in a platinum bath at 1900°C [35]. The method is suitable for determining 5–3000 ppm oxygen.

b. Niobium and Tantalum

Mikhailov et al. [33, 34] used the same iron-bath conditions for niobium as for molybdenum and tungsten with reproducible results. Benner [56] has used the iron bath at 1850°C for niobium. Care must be taken that the sample is completely immersed in the iron bath [10]. Mallett et al. [19] in a brief study obtained very low oxygen values when graphite precipitate appeared in the bath. Better, but still incomplete, oxygen extraction was obtained from a bath of vacuum-melted iron. The "trouble-free" platinum bath was recommended. Klaus et al. [56b] also used platinum but at 2300°C for powdered niobium or tantalum. Turovtseva and Kunin [36] found that oxygen results for tantalum by the iron-bath technique were low and erratic.

The platinum-bath vacuum-fusion method is the most versatile and satisfactory method for analyzing niobium and tantalum for oxygen [53, 57, 58]. The same operating conditions may be used for both metals. The method is given in detail in Section VI.B In essence, the method calls for dissolving the sample (1–2 g) in a degassed platinum bath at 1900°C. Extraction time is

20 min, and the niobium and tantalum content of the bath should not exceed about 25 wt %. At the 100 ppm level, results showing a coefficient of variation of less than 5% are readily obtained [19]. Typical oxygen results obtained under the prescribed operating condition are given in Table 18.2. Smaller

TABLE 18.2. Oxygen Analysis of Niobium and Tantalum Using a Platinum Bath

	Niobium (ppm)	Tantalum (ppm)
	168	105
	154	99
	165	103
	161	102
	168	106
	159	104
	169	100
	161	107
Average	163	103
Standard deviation	5.3 ppm	2.8 ppm
Coefficient of variation	3.3%	2.7%

sample-weights have been reported—for example, 0.25 g niobium [59] and 0.025–0.075 tantalum [34]. However, use of small samples is not essential to the success of the method as it may well be in the case of beryllium (see Chapter 11). The larger sample is recommended to permit greater precision of analysis. The precision with a 2-g sample is usually ±3 ppm oxygen at levels of about 10–100 ppm.

Lesser and Gruber [60] have used a platinum bath at 1900°C for determining oxygen, hydrogen, and nitrogen in niobium, tantalum, molybdenum, and tungsten. In their apparatus, the gases are transferred by a Toepler pump to a gas chromatograph for separation and measurement.

Harris et al. [61] made quantitative additions to niobium of oxygen rich in ^{17}O and ^{18}O and analyzed the product by the platinum bath technique. Mass spectrometric analysis of the extracted gas showed complete recovery of the isotopes. Thus, the accuracy of the technique is apparently limited only by the precision of the vacuum-fusion technique.

Karpov et al. [62] studied the sources of errors of the vacuum-fusion method for determining oxygen in niobium. Deviation of the blank value due to surface impurities and errors of analysis of the extracted gases due to gettering and incomplete extraction were cited as the most prominent errors. A more effective method of mixing the melt for enhancement of extraction, was cited as a needed improvement.

3. Vacuum Hot Extraction

This method consists of heating a solid sample in a graphite crucible in vacuum and collecting and analyzing the evolved gases. The mechanism is one of diffusion of oxygen atoms to the metal surface where they combine with carbon atoms vaporized from the crucible. The CO product is desorbed and pumped from the furnace and analyzed as in the vacuum-fusion method. Thin sample-sections shorten diffusion times. Certain disadvantages sometimes associated with the iron bath are eliminated. The method was first demonstrated by Hansen and Mallett [15] with niobium. Oxygen was quantitatively removed from a 0.3 g sample at 2000°C in 30 min. The analysis precision was about ±20 ppm. This can be improved with a larger sample as long as thin samples are used.

The vacuum-hot-extraction technique was applied to molybdenum, tungsten, and tantalum by Fagel et al. [16]. Friedrich [63] applied the method to molybdenum and found it superior to the bath technique. For tantalum, Mallett [19] obtained results lower by 10% at 2000°C and 7% at 2100°C than those by the platinum-bath technique. Klaus et al. [56b] extracted oxygen from niobium and tantalum at 2300°C.

4. Emission Spectrography

Preliminary results with the emission-spectrographic method [64, 65] (Chapter 5) using platinum to flux the sample show that this method is applicable to niobium and tantalum containing a few hundred ppm oxygen [66]. Although the method has been demonstrated [67] to be satisfactory for oxygen levels as low as 20 ppm for iron and 60 ppm for vanadium, it is doubtful that it can be extended to low oxygen containing (less than 10 ppm) niobium and tantalum. Part of the difficulty lies in the oxygen content of the platinum flux and in degassing from the walls of the arc chamber. There are as yet no reports of experimental application of the method to molybdenum or tungsten.

5. Neutron Activation

Activation of oxygen by fast (14 MeV) neutrons to produce the $^{16}O(n, p)^{16}N$ reaction is a suitable method for determining small amounts of oxygen in molybdenum, tungsten, and niobium [48, 68–70]. Harris [68] investigated niobium and other metals over the range 0.0008–0.1% and compared results obtained by fast-neutron activation and vacuum fusion. Kallmann et al. [48] reported comparative results obtained for oxygen in molybdenum and tungsten by fast-neutron activation, vacuum fusion, vacuum extraction, and inert-gas fusion. Clark and Stensland [70] reported that a sensitivity and precision of about 6 ± 3.3 μg oxygen in a single determination was obtained when no sample-container was used for solid (chunk) samples.

6. Other Methods

Revel and Albert [71] determined oxygen in molybdenum, tungsten and other metals by irradiation with ^3He nuclei of energies below 11 MeV. Cabane and Engelmann [72] detected oxygen at parts-per-million and parts-per-billion levels in niobium, iron, and silicon by photon and charged-particle activation.

Techniques of isotope equilibrium have been reported for determining oxygen in refractory metals. Lazeeva et al. [73] analyzed tungsten by establishing oxygen isotope equilibrium within 30 min at 1900°C, using induction heating. Ivanova and Fedorov [74] used a dc arc to melt niobium and titanium for isotopic equilibrium between sample oxygen and added stable oxygen isotopes. The sensitivity of these isotope-equilibrium techniques was about 10^{-2}–10^{-3}%.

Vidal et al. [75] reviewed results obtained by investigators using spark-source mass spectrometry to determine oxygen and nitrogen in niobium.

Hoekstra and Katz [76] have used bromine trifluoride in the determination of oxygen in the oxides of niobium, tantalum, and tungsten. The reagent converts combined oxygen to free oxygen as shown in the following reaction:

$$3MO_2 + 4BrF_3 \rightarrow 3MF_4 + 2Br_2 + 3O_2. \quad (1)$$

The method is not applicable to molybdenum oxide, since molybdenum forms a stable oxyfluoride [76], which is retained in the reaction medium.

Dupraw and O'Neill [77] extended the method to titanium containing more than 500 ppm oxygen with average deviations as low as 30 ppm. It appears that the method could be developed to apply to niobium, tantalum, and tungsten for oxygen levels greater than 100 ppm but not to the lower levels of oxygen these metals normally contain.

Babko et al. [78] determined oxygen in molybdenum by reacting the sample with sulfur vapor in vacuum and determining the sulfur dioxide product colorimetrically with a fuchsine-formaldehyde reagent.

Geyer and Friedrich [79] determined oxygen in metal powders of molybdenum, tungsten, and rhenium by heating in a stream of hydrogen and measuring the water product in a thermal conductivity cell. Ten ppm of oxygen was measured in a 1-g sample within 5–10 min.

C. NITROGEN

1. Vacuum Fusion

In general, nitrogen results are obtained incidental to the determination of oxygen and under conditions designed solely to give optimum analyses for oxygen. Often the nitrogen values are ignored or reported without comment, even in journal articles. This is because the validity of the method as applied to nitrogen in most metals has not been established. In cases where chemical nitrogen analyses are questioned, vacuum-fusion values often are sought to resolve the apparent problem. This applies to analysis of the refractory metals as well as other metals. Typical sets of nitrogen (and oxygen) vacuum-fusion data are presented by Turovtseva and Litvinova [34] and Mikhailov et al. [33]. Molybdenum, tungsten, and niobium were analyzed as 1–3-g samples in an iron bath at 1650°C and tantalum as 50–200-mg samples in a platinum bath at 1900°C. Analyses of molybdenum and niobium with known nitrogen contents also were reported. Reasonable recovery was obtained in the 0.3–1 $\times 10^{-2}$ wt % range for niobium and the 4.5–7.5 $\times 10^{-2}$ wt % range for molybdenum.

Everett [80] reported the use of 50-mg samples in a 3.5-g platinum bath at 1900°C for the determination of nitrogen in niobium, molybdenum, and tungsten. The precision of analysis appears to be ± 10 ppm nitrogen (corrected from original). Sloman and Harvey [31] obtained good recovery of nitrogen from the powder of a mixture of molybdenum metal and Mo_2N containing a total of 5.3% nitrogen. Additions of this powder contained in 20-g capsules of molybdenum gave equally good results. The iron-bath technique at 1560°C (maximum molybdenum content, 40 wt %) was used in both cases [33].

Fagel et al. [16] found the iron–tin bath at 1800°C [81] to give nitrogen results for molybdenum comparable to those obtained by vacuum hot extraction. However, for tungsten and tantalum, the iron-bath results were higher by a factor of 4–8.

Kallmann and co-workers [48] presented nitrogen results obtained on tungsten and molybdenum by several cooperating laboratories as a by-product of their vacuum-fusion oxygen analyses. The agreement was not good, although the nitrogen level was quite low (less than 5 ppm). Friedrich et al. [58] reported work showing the possibility of determining nitrogen in tantalum; they concluded that a suitable method for nitrogen determination in niobium had not been found.

Davis et al. [82] compared nitrogen results obtained on tantalum alloys by Kjeldahl and vacuum-fusion procedures. In the 5–25-ppm range the relative standard deviation was 3–9% by the Kjeldahl procedure and 4–8%

by vacuum fusion. This was regarded as a measure of the homogeneity of the material as well as of the reproducibility of results.

Mallett and Griffith [8, 10] analyzed surface-nitrided molybdenum powders and wires by the iron-bath technique at 1650°C and the Kjeldahl method. In general, the results showed good agreement in the range 30–1000 ppm. However, in cases where the nitrogen was dissolved or otherwise dispersed in the bulk of the sample, iron-bath results were lower than both Kjeldahl and diffusion-extraction results. Klaus et al. [56b] reported that neither vacuum fusion with a platinum bath or vacuum-hot-extraction both at 2300°C gave quantitative nitrogen recovery.

2. Vacuum Hot Extraction

This method was first used for the determination of nitrogen in molybdenum by Mallett and Griffith [8, 10]. Nitrogen was diffused from the solid sample in vacuum in a graphite crucible at 1800°C. Complete extraction required 30–50 min. Agreement was very good between the vacuum extraction and Kjeldahl methods for the range 10–300 ppm. Iron-bath vacuum-fusion results at 1650°C were quite low by comparison. Vacuum extraction at 2000°C for 20–30 min was used [16] for molybdenum, tungsten, and tantalum. Results for molybdenum agreed with those by vacuum fusion, but those for tungsten and tantalum were low by factors of 4 and 8, respectively.

3. Inert-Gas Fusion

An inert-gas-fusion method [43, 83] has been used for determining nitrogen in niobium. The nitrides (and supposedly solid solution nitrogen) are decomposed by fusion of the metal in a graphite crucible by high-frequency induction heating. The nitrogen (and carbon monoxide) is swept by a carrier gas into a liquid-nitrogen cold trap containing a molecular-sieve packing. Upon completion of the extraction, the gases are eluted, separated by means of a chromatographic column, and passed through thermal-conductivity cells. The electrical output of the cells goes to two integrators, one of which registers nitrogen, the other carbon monoxide. A nitrogen determination requires 6 min per sample. Apparent precision of the analysis is 10 ppm.

Dallmann and Fassel [47] tested the inert-gas-fusion method for nitrogen determination in 19 different base metals, including molybdenum, tantalum, and niobium. A metal bath of 80(Pt):20(Sn) at 1850–1950°C was used for molybdenum, 95(Pt):5(Sn) at 1950–2050°C for niobium and 95(Pt):5(Sn) at 2150–2250° for tantalum. Concordant results were obtained by helium fusion, vacuum fusion, and chemical dissolution for niobium and tantalum.

For molybdenum a comparison was not made because of high blanks encountered during the lengthy chemical-dissolution procedure.

Winge and Fassel [44] applied a dc carbon-arc, gas-chromatographic technique to the determination of nitrogen and oxygen in refractory metals. With a platinum flux the nitrogen was liberated from the metal into a static helium atmosphere. An aliquot of the resulting gas mixture was passed through a gas chromatograph for measurement of the nitrogen. Ivanova and Federov [74] used a dc arc to determine nitrogen and oxygen in niobium by introducing stable isotopes as internal references. A maximum sensitivity of 5×10^{-4} wt % was reported for nitrogen.

4. Alkali Fusion

Harris [59] analyzed niobium for nitrogen by an alkali-fusion method previously used by Beiter et al. [84] for analysis of molybdenum and tungsten. The fusion apparatus consisted of a Vycor tube, 30 in. long and $\frac{1}{3}$ in. in diameter, fitted with ground-glass joints at each end and tubes for passage of carrier gas. A nickel boat containing a fusion mixture was placed inside a nickel sleeve, which protected the combustion tube from spattered alkali. The mixture was heated by a split-tube furnace.

Niobium, in the form of fine powder or millings, was fused with sodium hydroxide at 600°C for $1\frac{1}{2}$ h in a slow stream of hydrogen. Ammonia formed during the fusion was absorbed in water or boric acid and determined by the usual titrimetric or colorimetric techniques. A blank was determined in the same way. Blanks as low as 1 μg were obtained by prefusing the sodium hydroxide before use. Otherwise, the blank amounted to 20–25 μg nitrogen, which may be tolerable at high nitrogen levels.

LECO made a nitrogen analyzer based on alkali fusion [85]. Although used principally for nitrides and highly nitrided (3–30 wt % nitrogen) materials, it has been applied to molybdenum metal. The weighed sample is placed in a nickel crucible with 6–10 g sodium hydroxide. The mixture is heated to 600°C in an induction furnace. The ensuing reaction results in the formation of ammonia. A carrier gas of hydrogen or argon sweeps the ammonia into the titration vessel of a special titrator. Here, the ammonia neutralizes a portion of a dilute solution of sulfuric acid. After a 20-min fusion period, the unreacted sulfuric acid is back titrated with sodium hydroxide automatically, and the percentage of nitrogen calculated.

Ul'yanov [86] dissolved niobium in potassium hydroxide solution and transferred the liberated ammonia by a stream of argon into 0.01 N H_2SO_4. Nessler's reagent was added, the absorbance was measured at 440 nm., and the amount of nitrogen was calculated by means of a calibration curve.

5. Chemical Methods

a. Solution Distillation

In determining nitrogen in the refractory metals, the major problem is in the digestion and complete solution of the samples. Various digestion procedures [51, 87–89a] must be used depending on the metal and type and quantity of alloying elements present. A detailed discussion of the use of HF–H_3PO_4–$K_2Cr_2O_7$ [89a] for sample-decomposition is given in Chapter 8. The pressure-bomb technique is also discussed [89b]. In general, acids that produce or leave an insoluble film or residue must be avoided. In the various solution-distillation methods both the dissolved and combined nitrogen contents of the metal are converted to ammonia, which is determined by titration or spectrophotometry (see detailed methods for molybdenum, tungsten, niobium, and tantalum in Section VI.C).

VI. RECOMMENDED METHODS

A. HYDROGEN

1. Vacuum Hot Extraction

Although no specific studies have been made to establish optimum extraction conditions for the refractory metals, it is recommended that the conditions proven satisfactory for titanium and zirconium be used (see Chapter 19).

2. Vacuum Fusion

Use the methods recommended (Section VI.B) for oxygen. Discard residual gases after determination of hydrogen to conserve time.

B. OXYGEN

1. Vacuum Fusion

(1) Molybdenum and Tungsten. Use the Tentative Method of Test for Oxygen in Molybdenum ASTM Designation E174-60T [21]. Also, see Chapter 4, Section I for description of vacuum fusion apparatus and techniques. The analytical details bearing specifically on molybdenum and tungsten are as follows: Perform the analysis by the iron-bath technique. Use a 1–4-g sample in cube or chunk form (do not use chips). Choose the sample-weight so that the expected amount of gas is at least 10 times the apparatus and iron–tin bath blank. The concentration of molybdenum and/or tungsten

in the bath shall not exceed a total of 30 wt %. File the surfaces of the samples and the iron with a clean fine-mill file. Handle samples with clean forceps during final stages of cleaning. Avoid so-called degreasing treatments. Carry out the extraction at 1650°C for 20 min. The samples may be intermixed, but where possible, analyze the low-oxygen-content samples first. Start with a bath of 15 g iron. Drop and degas 0.3 g tin before each sample, and add 10 g iron after every second sample.

The blank should be maintained at less than about 3 μl/20 min. The precision of the analysis is about 0.2 ppm for a 1-g sample at the 10-ppm level and 2.0 ppm at the 100-ppm level. The method has been established for the 50-ppm level but requires further affirmation at 10 ppm.

(2) NIOBIUM AND TANTALUM. Use the basic method described in Recommended Method for the Determination of Oxygen in Commercial Grade Niobium submitted for member approval by the Niobium Task Force, Oxygen Subgroup Division M, Committee E-3, ASTM, April 1960. This is the platinum-bath technique similar to that used by Booth et al. [90]. Use the same procedure for both niobium and tantalum. Intermix samples if necessary. In a series, analyze low-oxygen-content samples first (see Chapter 2 for vacuum-fusion theory and Chapter 4 on apparatus and techniques). Principal specific details for niobium and tantalum follow.

Use a 0.5–2-g sample in chunk or nearly cubic shape (do not use chips). Choose the smallest weight for oxygen levels above 100 ppm. Abrade samples with a clean fine-mill file. Degrease platinum with warm acetone. The concentration of niobium and/or tantalum in the bath shall not exceed 30 wt %. At the start, drop and degas 2 g of platinum for each gram of sample to be analyzed in the crucible. In the case of a long series of samples, one may wish to reserve half the platinum to be used as a fresh addition after half the total weight of the samples has been analyzed. Make the gas extraction at 1900°C for 20 min or until the gas evolution returns to blank rate. The precision of the analysis is about 2 ppm for a 1-g sample at the 100-ppm oxygen level. The method is satisfactory for the 100–500 ppm oxygen levels and with care appears applicable down to 10 ppm (with a 2-g sample).

2. Vacuum Extraction

(1) MOLYBDENUM, TUNGSTEN, NIOBIUM AND TANTALUM. Following is a brief description of the method. Take a sample weighing 1–5 g. For the heavier samples, cut so that the cross section is no thicker than $\frac{1}{8}$ in. Clean sample-surfaces by abrading with a clean fine-mill file. Drop the sample into a graphite crucible (no bath or flux) at 2000°C, and extract gases for 30 min. If it can be demonstrated that the geometry of the sample in use permits gas evolution to return to blank rate in 20 min, the shorter time may be used.

Complete the analysis of gases by the normal procedure for vacuum fusion. The precision of analysis is about 2 ppm for a 1-g sample at the 100-ppm level. At the 10-ppm level, the precision becomes about 0.2 ppm for a 1-g sample.

C. NITROGEN

1. Solution Distillation

The following procedures presented by Huber and Chase [87] are recommended.

(1) SOLUTION OF SAMPLES FOR NITROGEN DETERMINATION

(a) *Molybdenum*. Transfer a weighed sample (0.5–1.0 g) to a 30-ml Kjeldahl flask and add 6 ml 1:1 H_2SO_4 and 1 ml 60% H_3PO_4 solution. Heat on a digestion rack until excess water is driven off and the acid refluxes about midway up the flask. Continue the digestion until no residue remains. Cool, dilute with 10 ml water, and transfer to the Parnas-Wagner nitrogen distillation apparatus with 5-ml portions of water.

(b) *Niobium, Tantalum, and Tungsten*. Place a finely divided weighed sample (0.5 g) into a 40-ml platinum crucible, and add 5 ml 1:1 H_2SO_4 along with 2 ml HF and 1 ml 30% H_2O_2 solution. Cover with a platinum lid and heat on a hot plate until most of the HF is driven off. If a residue remains (which is likely), repeat the additions of HF and H_2O_2 until solution is complete. Decompose all H_2O_2 by boiling. Cool the solution and transfer to the distillation apparatus with three 3-ml portions of water.

(2) DISTILLATION AND DETERMINATION OF NITROGEN. Add 25 ml NaOH solution (prepared by dissolving 340 g NaOH in 1000 ml H_2O, adding 0.5 g Devarda's alloy, and boiling until the volume is about 900 ml). Distill and collect the condensate in a 50-ml conical flask containing 5 drops mixed indicator solution (0.0166 g methyl red and 0.0834 g bromocresol green in 100 ml neutral ethanol) [91] and 10 ml 0.16% H_3BO_3 solution (prepared by dilution of 4% stock solution). Adjust the heat input to the still so that a 10-ml volume of condensate is collected within 5 min. Titrate the condensate with 0.007*N* H_2SO_4 (prepared by dilution of standard 0.07*N* H_2SO_4). Alternatively, complete the determination colorimetrically by Nessler's method (see Chapter 8 for details of this method).

(3) REMARKS. The precision of the nitrogen determination is ±10 ppm nitrogen. Contamination by ammonia must be avoided and blanks must be run carefully. The titrimetric finish is usually used at all nitrogen levels. In general, the colorimetric method is not used for concentrations above 100

ppm. Its potentially greater sensitivity would give this method an advantage in the range below 10 ppm. However, the advantage is usually not realized because the uncertainty in the blank often amounts to ± 5 ppm.

REFERENCES

1. G. L. Miller, *Tantalum and Niobium*, Academic, New York, 1959, pp. 603, 617.
2. M. L. Hill, "Solubility and Diffusivity of Hydrogen in Molybdenum" Abstract of talk given at the Fall Meeting of the Metallurgical Society, Philadelphia, October 17, 1960. *J. Metals*, **12,** 725 (1960).
3. H. S. Spacil and J. Wulff, "Effects of Oxygen, Nitrogen, and Carbon on the Ductility of Wrought Molybdenum," in J. J. Harwood, Ed., *The Metal Molybdenum*, American Society for Metals, Cleveland, Ohio, 1958, p. 280.
4. J. W. Semmel, Jr., "Oxidation Behavior of Refractory Metals and Alloys," in M. Semchyshen and J. J. Harwood, Ed., *Refractory Metals and Alloys*, Interscience, New York, 1961, p. 154.
5. G. L. Davis, *Metallurgia*, **58,** 177 (1958).
6. J. W. Semmel, Jr., *Trans. ASM*, **52,** 1015 (1960).
7. A. N. Zaidel and K. I. Petrov, *Zavod. Lab.*, **24,** 1000 (1958).
8. M. W. Mallett and C. B. Griffith, *Trans. ASM*, **46,** 375 (1954).
9. A. L. Beach and W. G. Guldner, *Spec. Tech. Publ. No. 222*, American Society for Testing and Materials, Philadelphia, 1958, p. 15.
10. M. W. Mallett and W. R. Hansen, "Determination of Gases in Molybdenum," in J. J. Harwood, Ed., *The Metal Molybdenum*, American Society for Metals, Cleveland, Ohio, 1958, p. 365.
11. J. N. Gregory and D. Mapper, *Analyst*, **80,** 230 (1955).
12. T. D. McKinley, Pittsburgh Conference on Analytical Chemistry and Applied Spectroscopy, 1959.
13. M. W. Mallett, *Trans. ASM*, **41,** 870 (1959).
14. LECO, *Conductometric Oxygen Analyzer for Ferrous and Nonferrous Metals and Alloys*, Laboratory Equipment Corporation, St. Joseph, Mich., 1959.
15. W. R. Hansen and M. W. Mallett, *Anal. Chem.*, **29,** 1868 (1957).
16. J. E. Fagel, R. F. Witbeck, and H. A. Smith, *Anal. Chem.*, **31,** 1115 (1959),
17. *Semiannual Summary Research Report in Chemistry (Spectroscopy) ISC-1049.* Ames Laboratory at Iowa State College, U.S. At. Energy Comm. Res. Dev. Rep. (November 1, 1958).
18. V. A. Fassel, W. A. Gordon, and R. J. Jasinski, *Second United Nations International Conference on the Peaceful Uses of Atomic Energy*, A/Conf. 15/P/917, USA (June, 1958).
19. M. W. Mallett, D. F. Kohler, R. B. Iden, and B. G. Koehl, *Tech. Rep. WAL TR 823/5*, Battelle Memorial Institute, 1962.
20. American Society for Testing Materials, *Methods for Chemical Analysis of Metals*, Philadelphia, 1960, p. 558.
21. W. G. Guldner, *Anal. Chem.*, **35,** 1744 (1963).

22. W. G. Guldner, I.E.E.E.-E.I.A. Electronic Components Conference, May 5–7, 1964, Washington, D.C.
23. K. Friedrich, J. Barthel and J. Kunze, *J. Less-Common Metals*, **14,** 55 (1968).
24. A. Lawley, W. Liebmann, and R. Maddin, *Acta Met.*, **9,** 841 (1961).
25. R. Eborall, *Spec. Rep. 68*, Iron and Steel Institute, London, 1960, p. 192.
26. F. J. Norton and A. L. Marshall, *Trans. AIME*, **156,** 351 (1944).
27. U.K. Atomic Energy Authority, Production Group, *PG-Report-412(S)*, Springfields, Lancaster, England, 1962.
28. V. I. Lakomskii and N. N. Kalinyuk, *Avtomat. Svarka*, **No. 6,** 21 (1966); through *Nucl. Sci. Abstr.*, **21,** 45026 (1967).
29. R. J. Walter and H. G. Offner, *Anal. Chem.*, **36,** 1779 (1964).
30. T. Yoshimori and S. Ishiwari, *Talanta*, **17,** 349 (1970).
31. H. A. Sloman and C. A. Harvey, *J. Inst. Metals*, **80,** 391 (1951).
32. H. A. Sloman, *Metalluriga*, **32,** 223 (1945); *Engineering*, **160,** 385, 404 (1945).
33. G. V. Mikhailov, Z. M. Turovtseva and R. Sh. Khaltov, *Zh. Anal. Khim.*, **12,** 338 (1957).
34. Z. M. Turovtseva and N. F. Litvinova, *Proceedings of Second United Nations International Conference on Peaceful Uses of Atomic Energy*, Geneva, Switzerland, September 1–13, 1958, Paper p/2205, USSR, p. 593.
35. D. F. Wood and G. Wolfenden, *Anal. Chim. Acta*, **38,** 385 (1967).
36. Z. M. Turovsteva and L. L. Kunin, *Analysis of Gases in Metals*, Academy of Sciences of the USSR, Moscow, 1959; trans. by Consultants Bureau, New York, 1960, p. 27.
37. R. D. Gardner, W. H. Ashley and P. Bergstresser, *LA-3225* Los Alamos Scientific Laboratory, 1965.
38. B. D. Holt, *Anal. Chem.*, **28,** 1153 (1956).
39. C. S. MacDougall, M. E. Smith and G. R. Waterbury, *LA-4260* Los Alamos Scientific Laboratory, 1969.
40. W. G. Smiley, *Mater. Spec. Tech. Publ.*, **222,** 25 (1958).
41. Minutes of meeting, Division I, Committee E-3, Am. Soc. Testing Mater., at E. I. du Pont de Nemours & Co., Inc., Wilmington, Delaware, December 5–6, 1962.
42. G. J. Kamin, J. W. O'Laughlin, and C. V. Banks, *Anal. Chem.*, **35,** 1053 (1963).
43. F. T. Coyle and W. C. Gannon, 8th Conference on Analytical Chemistry in Nuclear Technology, Gatlinburg, *CONF-721-8*, 1964.
44. R. K. Winge and V. A. Fassel, *Anal. Chem.*, **37,** 67 (1965).
45. J. Kashima and T. Yamazaki, *Bull. Chem. Soc. Jap.*, **39,** 1448 (1966).
46. B. Migaud, *Rev. Met* (Paris), **63,** 137 (1966).
47. W. E. Dallmann and V. A. Fassel, *Anal. Chem.*, **39,** 133R (1967).
48. S. Kallmann, R. Liu, and H. Oberthin, *Tech. Report AFML-TR-65-194; AD-622275*, Ledoux and Company, Inc., 1965.
49. A. M. Vasserman and Z. M. Turovtseva, *Zh. Anal. Khim.*, **20,** 1390 (1965).
50. R. F. Coleman and J. L. Perkin, *Analyst*, **84,** 233 (1959).
51. M. W. Mallett, *Talanta*, **9,** 133 (1962).
52. (a) M. F. Qualey, private communication, March 5, 1963; and (b) L. B. Kuznetsov and V. T. Burstev, Monokrist. Tugloplavkikl Redk. Metal., Tr.

Soveshch. Poluch., Strukt. Fiz. Svoistvam Primen, Monokrist, Tugloplavkikh Redk, Metal., 1st 2nd 1966–1967 (pub. 1969) 43–49, E. M. Savitskii, Ed., Izd., "Nauka," Moscow, USSR; through *Chem. Abst.*, **73,** 94305K (1970).
53. E. S. Tankins, *Met. Soc. Conf.*, **34,** Pt. 1, 325 (1968).
54. J. J. English *DMIC Report 152*, Defense Metals Information Center, 1961.
55. U.K. Atomic Energy Authority, Production Group, *PG-Report-391*, Springfields, Lancaster, England, 1962.
56. (a) F. C. Benner, in R. F. Bunshah, Ed., *Vacuum Metallurgy*, Reinhold, New York, 1958, p. 407; and (b) F. Klaus, E. Lassner, T. Kraus, G. Paesold, and F. Schlaet, High Temp. Materials, Pap. Plansee Semin., 6th, 1968 (pub. 1969) 1016–1027; F. Benesovsky, Ed., Springer-Verlag, Vienna; through *Chem. Abstr.*, **72,** 74449a (1970).
57. E. Lassner and G. Kraft, *J. Less-Common Metals*, **22,** 83 (1970).
58. K. Friedrich, E. Lassner and G. Paesold, *J. Less-Common Metals*, **22,** 429 (1970).
59. W. F. Harris, in B. W. Gonser and E. M. Sherwood, Ed., *Technology of Columbium* (*Niobium*), Wiley, New York, 1958, p. 27.
60. R. Lesser and H. Gruber, *Z. Metall.*, **51,** 495 (1961).
61. W. F. Harris, W. M. Hickam, M. H. Loeffler, and D. H. Shaffer, *Trans. AIME*, **218,** 625 (1960).
62. Yu. A. Karpov, G. G. Clavin and L. L. Kunin, *Zh. Anal. Khim.*, **24,** 276 (1969); through *Nucl. Sci. Abstr.*, **23,** 15788 (1969).
63. K. Friedrich, *Acta Chim. Hung.*, **28,** 187 (1961).
64. V. A. Fassel and R. W. Tabeling, *Spectrochim. Acta*, **30,** 179 (1956).
65. V. A. Fassel and W. A. Gordon, *Anal. Chem.*, **30,** 201 (1956).
66. V. A. Fassel, W. A. Gordon, and R. W. Tabeling, *Spec. Tech. Publ. 222*, American Society for Testing and Materials, Philadelphia, 1958, p. 41.
67. V. A. Fassel, *Spec. Rep. 68*, Iron and Steel Institute, London, 1960, p. 103.
68. W. F. Harris, *Talanta*, **11,** 1376 (1964).
69. S. R. Abdurakhmanova, V. A. Kireev, L. V. Navalikhin and Yu. N. Talanin, *Zh. Anal. Khim.*, **23,** 1188 (1968); through *Nucl. Sci. Abstr.*, **22,** 45020 (1968).
70. R. G. Clark and W. A. Stensland, *J. Radioanal. Chem.*, **4,** 365 (1970).
71. G. Revel and Ph. Albert, *J. Nucl. Mater.*, **25,** 87 (1968).
72. G. Cabane and Ch. Engelmann, *Rev. Phys. Appl.*, **3,** 365 (1968).
73. G. S. Lazeeva, A. A. Petrov and G. A. Yusupova, *Vestn. Leningrad Univ., Ser. Fiz. i Khim.*, **20** (1), 141 (1965).
74. T. F. Ivanova and V. V. Fedorov, *Zh. Anal. Khim.*, **23,** 1750 (1968); through *Nucl. Sci. Abstr.*, **23,** 2244 (1969).
75. G. Vidal, P. Galmard, and P. Lanusse, *Recherche Aerospatiale* (France), **1967,** 33; through *Nucl. Sci. Abstr.*, **22,** 25273 (1968).
76. H. R. Hoekstra and J. J. Katz, *Anal. Chem.*, **25,** 1608 (1953).
77. W. A. Dupraw and H. J. O'Neill, *Anal. Chem.*, **31,** 1104 (1959).
78. A. K. Babko, A. I. Volkova, and C. F. Drako, *Zavod. Lab.*, **23,** 136 (1957).
79. R. Geyer and K. Friedrich, *Z. Anal. Chem.*, **213,** 259 (1965).
80. M. R. Everett, *Analyst*, **83,** 321 (summary) (1958).
81. R. S. McDonald, J. E. Fagel, Jr., and E. W. Balis, *Anal. Chem.*, **27,** 1632 (1955)

82. W. F. Davis, J. W. Graab and E. J. Merkle, *Talanta*, **18,** 263 (1971).
83. Laboratory Equipment Corporation, St. Joseph, Mich., Form 1102.
84. E. W. Beiter, F. P. Byrne, W. F. Harris, M. I. Mistrik, and R. H. Wynne, Pittsburgh Conference on Analytical Chemistry and Applied Spectroscopy, 1957.
85. Laboratory Equipment Corporation, St. Joseph, Mich., Form No. 171.
86. A. I. Ul'yanov, *Zavod. Lab.*, **34** (12), 1442 (1968); through *Chem. Abstr.*, **70,** 83994w (1969).
87. F. E. Huber and D. L. Chase, *Chemist-Analyst*, **50,** 71 (1961).
88. J. R. Ciaranello and K. E. Combs, *KAPL-M-JRC-3*, Knolls Atomic Power Laboratory, 1960.
89. (a) S. Kallmann, E. W. Hobart, H. K. Oberthin, and W. C. Brienza, *Anal. Chem.*, **40,** 332 (1968); and (b) O. Meris, *NBS Tech. Note 454*, July 1968, pp. 72–74.
90. E. Booth, F. J. Bryant and T. Parker, *Analyst*, **82,** 50 (1957).
91. T. S. Ma and G. Zuazaga, *Ind. Eng. Chem., Anal. Ed.*, **14,** 280 (1942).

CHAPTER

19

SCANDIUM, YTTRIUM, LANTHANIDE, AND ACTINIDE ELEMENTS

WAYNE E. DALLMANN AND VELMER A. FASSEL

Ames Laboratory-USAEC
and
Department of Chemistry
Iowa State University
Ames, Iowa

CONTENTS

I. INTRODUCTION

The elements included here constitute well over one-third of the known metals in the periodic system. Even with the exclusion of those elements that to date have not been prepared in the metallic state in amounts sufficient to make realistic oxygen, nitrogen, and hydrogen determinations, there remain 19 metals to be covered in this chapter. Although most of these elements possess somewhat unique similarities in their chemical properties, the differences in their physical properties actually make them less similar to each other than elements belonging to other groups. For example, the melting temperatures of the lanthanide elements vary by a factor of two (from 795°C for cerium to 1652°C for lutetium), and their vapor pressures at 1000°C vary by a factor of over 3 billion (from 4.9×10^2 torr for ytterbium down to 1.4×10^{-7} torr for lutetium).

Treatment of each of these elements individually would result in much repetition and scattering of information that in many respects may be logically related by appropriate groups. In the discussions that follow, scandium, yttrium, and lanthanum will be included in the treatment of the lanthanide elements (atomic numbers 58–71). This grouping is appropriate not only because these elements have been associated together in an historical sense but also because their compounds occur together in nature. To further expedite the discussions on the lanthanide series, these metals will be classified in subgroups based on certain similarities or differences in their physical properties. Discussions on the actinide series will of necessity be limited primarily to those members that have been prepared in quantity—namely, thorium, uranium, and plutonium.

II. METALLURGICAL ASPECTS

Of all the common contaminants, the nonmetallic impurities, oxygen, nitrogen, and hydrogen, generally exert the most pronounced effects on the properties of the host metal. The solubilities of these impurities in the lanthanide and actinide metals generally are believed to be quite low [1–5], and consequently would be expected to exist primarily as compound inclusions (oxides, nitrides, hydrides). If these impurities are present as compounds, their volume concentrations in the host metal may be significant, and the ultimate effects on the properties of the metal could be quite extensive. The state of occurrence of trace amounts of oxygen, nitrogen, and hydrogen in the lanthanide and actinide metals and the specific effects exerted by these impurities on the physical and mechanical properties of the metals have not been systematically evaluated.

Harrington and Ruehle [5] reported the solubility of oxygen in uranium to be approximately 0.05 at. % at the melting temperature of the metal and less at lower temperatures. Microscopic evidence of an oxide phase in uranium that contained only 0.0003% oxygen was reported by Sloman and co-workers [6]. These authors also presented evidence for a rather significant increase in the solubility of hydrogen in uranium as the temperature was increased from 600 to 1000°C. Tucker et al. [7] reported that the solid solubility of oxygen in yttrium is less than 0.06 wt % at 600°C, and that this increases to 3.3% at the eutectoid temperature of 1180°C. Gschneidner and Waber [8] reported the solubility of nitrogen in cerium to be about 0.001 wt % between 450°C and 550°C, and presented evidence indicating that the solubility of oxygen in cerium is less than 0.002 wt % at room temperature.

If these reported solubilities are taken as typical for the lanthanide and actinide elements in general, virtually all of the reported measurements of the physical and mechanical properties at ambient temperatures have been made on specimens for which the solubility limits of these impurities were exceeded.

The melting temperature of yttrium has been found [7] to be somewhat dependent upon the oxygen content of the metal. For thorium this dependence was attributed to the formation of a "refractory skin" on the specimen surface, but this behavior would not be expected for yttrium [9]. The oxygen content also was found to exert varying degrees of influence on the hardness of yttrium, the magnitude of which depended on the metallurgical history of the metal [10, 11]. Bohlander [12] indicated that the ability to "work" yttrium mechanically was quite dependent on the oxygen content. Carlson and co-workers [10] surveyed the effects of oxygen, nitrogen, and hydrogen on the mechanical properties of yttrium and found the ductility and strength of the metal to be influenced surprisingly little. The yttrium, however, contained 0.8 wt % of other impurities, including 0.5% titanium; hence, the observations should be evaluated with respect to possible effects of small amounts of such impurities in the metal.

In general, oxygen impurities in thorium appear to exert only minor influences on the hardness, strength, ductility, and other mechanical properties of the metal [2, 3, 13–16]. Nitrogen appears to have more pronounced effects on the mechanical properties of thorium and, above about 0.25 wt % concentration, may cause serious embrittlement of the metal [2]. The presence of an oxide phase was found to cause surface defects in uranium and to have detrimental effects on the ductility of the metal [5]. Hydrogen impurities cause surface defects in uranium [5], affect the ductility of both thorium [2] and uranium [17], and affect the tensile properties of uranium [17]. At equivalent concentrations (expressed as parts per million by weight) hydrogen normally would exert a far greater influence than oxygen or nitrogen on the properties of the lanthanide and actinide metals.

Relatively high purity is a requirement for good mechanical workability of a metal; also, high-purity specimens are required for unequivocal characterization of the metal through accurate determinations of its properties. This need for very high purity lanthanide and actinide metals has created a demand for more accurate and precise procedures for determining the oxygen, nitrogen, and hydrogen contents.

III. HISTORICAL DEVELOPMENT

No single analytical technique is considered to be the ultimate one for the determination of oxygen, nitrogen, and hydrogen in the subject metals. In general, each of these impurities may be best determined by some modification of different general techniques.

A. HOT EXTRACTION

The hot-extraction technique [18], in which the metal specimen is heated in a dynamic vacuum to a temperature somewhat below its melting temperature, is usually the most reliable for the hydrogen determination. Because the sample does not melt, the extraction process is entirely dependent upon diffusion of the gas through the metal. The diffusion coefficients of hydrogen in the metals discussed in this chapter are generally sufficiently large that quantitative extractions would be expected.

Hydrogen is usually vacuum-extracted from uranium at 500–900°C [19–22]. The anomalous solubility behavior of hydrogen in the metal [6] and the potential health hazards involved make this a desirable technique for uranium. For the determination of hydrogen in thorium, extraction temperatures of 1300–1400°C appear to be necessary [23]. The hydrogen contents of a number of the lanthanide elements have been determined by hot extraction at temperatures 100–150°C below their respective melting temperatures [24, 25].

B. CHEMICAL METHODS

Chemical dissolution–distillation techniques [26, 27] are generally the most accurate for the nitrogen determination. Difficulties in completely decomposing metal specimens in the simple acids has led to extensive use of various composite dissolution reagents. Recent experimental evidence [28–30] indicates that many of the reagents now commonly employed in the decomposition phase contain nitrogenous impurities that are not detected in operating blanks taken in the conventional manner. In the presence of a metal, such impurities may, however, be reduced to ammonium ion. It is important to recognize that this reduction can occur, and if it does occur and appropriate corrections are not applied, high nitrogen results will be obtained. The lanthanide and actinide metals are, however, quite readily soluble in the simple acids normally used in the chemical procedures; hence, sample decomposition and operational blanks present no unusual problems.

C. FURNACE FUSIONS

The high-temperature fusion of the metal sample in a graphite crucible, both in vacuum [31] and under an inert atmosphere [32, 33], is the most

TABLE 19.1. High-Temperature Furnace Fusion Conditions Applied to Lanthanide and Actinide Metal Systems

Reaction Medium	Temperature, °C	Metals Analyzed and Related References	Impurities Determined
Vacuum Fusion:			
Graph. cruc. only	1800	U[36]	O N H
Fe	1600, 1850	La[37]	H
Fe	1650	Pu[38]	O N
Fe	1800	U[6]	O N H
Fe	1860–1900	U, Th[23]	H
Fe	1900	Th(6)	O N H
Fe–Sn	1750	U[39]	O N
Fe–Sn	1900	U, Pu[40]	O N H
Pt	1800–1820	Y[41], U[42]	O N H
Pt	1850	U[21], rare earth metals[43]	O
Pt	1860–1900	U[23, 40, 41], Th[23, 42], Pu[40]	O N H
Pt	1900	U[44]	O
Pt	1950	Th[25, 45]	O
Pt	1950–2000	Th[35]	O N
Pt–Sn	1650–1700	Y, Th[25, 45]	O
Pt–Sn	1675–1725	Sc, Y, Gd, Tb, Dy, Ho, Er, Lu[35]	O N
Pt–Sn	1700	Sc, Y, Gd, Tb, Dy, Ho, Er, Lu[24, 25]	O N H
Ni	1600, 1900	La[37]	O N H
Carrier-Gas Fusion:			
Pt	1800	Th[33, 46], U, Pu, Y, La, Nd, Gd, Dy, Ho, Er, Yb[46]	O
Pt	1950	Er, Gd[47]	H
Pt	1950	Pu[48]	O N H
Pt	2100–2200	Y[49]	O
Pt	2500	Th[25, 45]	O
Pt–Sn	2300	Y[24, 25, 45], Th[25, 45], Sc, Gd, Tb, Dy, Ho, Er, Lu[24, 25]	O
Pt–Sn	2275–2325	Th, Sc, Y, Gd, Tb, Dy, Ho, Er, Lu[35]	O N

commonly used technique for the determination of oxygen in metals. Hydrogen is usually quantitatively evolved along with the oxygen, but the nitrogen may not be [34, 35].

A tabulated summary of furnace fusion conditions which have been applied to determinations of oxygen, nitrogen, and hydrogen in the lanthanide and actinide metals is presented in Table 19.1. Under the impurities-determined column are listed those impurities for which quantitative recovery was either assumed or validated with substantiating data. Notable exceptions to this are the stipulation that nitrogen recovery was always significantly low for uranium and thorium in Refs. 23, 39, and 42; the vacuum-fusion conditions appeared somewhat marginal for the nitrogen determinations in thorium and some of the lanthanide metals studied in Ref. 35; and, only approximately 90% recovery of oxygen was achieved for lanthanum under the optimal nickel bath conditions explored in Ref. 37.

D. ARC-DISCHARGE EXTRACTIONS

Another gas-extraction technique based on high-temperature fusion of the metal specimen in a graphite receptacle involves a high-current dc carbon-arc discharge in a static, inert atmosphere [50]. The gases may be separated chromatographically and determined [51].

The arc-extraction process has been shown to be suitable for determining oxygen in lanthanum [50], yttrium and thorium [52], and plutonium [53], the impurity levels of which were determined by spectrographic procedures. In conjunction with a gas-chromatographic measurement system, the basic extraction technique has been successfully applied to the determination of both oxygen and nitrogen in yttrium, terbium, and thorium [51].

E. ACTIVATION ANALYSIS

The nondestructive fast-neutron activation technique [54], which is based on the $^{16}O(n, p)^{16}N$ nuclear reaction, may be used for oxygen determinations. This technique is particularly useful when the sample cannot be sacrificed, when comparative analysis by another technique is desirable, or when the vapor pressure of the metal is excessive. The oxygen contents of the very volatile metals samarium and ytterbium were determined by scintillation counting after irradiation with 14-MeV neutrons [55]. Fast-neutron irradiation of fluorine also produces the ^{16}N reaction product, so this element may cause erroneous oxygen results unless its concentration in the sample is far less than that of the oxygen impurity [56].

The fissionable actinide elements cannot be analyzed by the fast-neutron technique, because the energies of the gamma emission produced by irradiation of the metal cause large counting interferences [56]. For fissionable materials, the ^{3}He-induced nuclear reaction of oxygen may be used to advantage [57]. By controlling the bombarding energy of the ^{3}He particles, a coulomb barrier restriction may be imposed on the heavy elements so that charged-particle fission in the actinide elements is excluded [58]. Thus, the $^{16}O(^{3}He, p)^{18}F$ nuclear reaction has been used to determine oxygen in thorium, plutonium, and americium by measurement of the gamma annihilation radiation [58].

IV. STANDARD SAMPLES

One of the most appropriate and widely accepted methods of evaluating the accuracy of analytical results is the analysis of standard samples. There are no commercial sources of oxygen, nitrogen, or hydrogen standard specimens for the lanthanide or actinide elements; consequently, standards for these metals must be prepared in the laboratory.

Two general methods may be used to prepare standard samples of known nonmetallic impurity content, the *gas-addition* procedure, as employed in the preparation of lanthanum–oxygen standards [37], and the *master-alloy* procedure, as used to prepare yttrium–oxygen and thorium–oxygen standards [45]. For either method, it is essential that the base metal be of high purity with respect to the element that is intentionally added. With respect to other impurities, the composition of the base metal should correspond as closely as possible to that of typical specimens of the metal to be analyzed.

A. GAS-ADDITION TECHNIQUE

In the gas-addition procedure, a weighed specimen of the high-purity metal is heated with a measured quantity of the gas in a container of constant volume. If it is assumed that all of the gas represented by the decrease in pressure during heating reacted only with the metallic mass, the weight percent of the added impurity can be calculated.

For certain lanthanide and actinide elements, notably the lighter members of both series, the gas tends to form compounds with the metal on and near the surface, thereby preventing uniform distribution of the added-impurity content. This is especially true with oxygen. Thus, the entire specimen usually is taken as a single standard sample. Furthermore, care must be exercised in handling the prepared standards to avoid loss of any of the impurity-containing surface layers. Another point of concern with this technique of preparing standards is that the added impurity may not respond to a method of analysis in the same manner as would the same impurity in a typical metal sample.

B. MASTER-ALLOY TECHNIQUE

Difficulties in obtaining metal nitrides that are entirely free of oxide and the relatively low thermal stabilities of most metal hydrides have limited the application of the master-alloy procedure to the preparation of metal–oxygen standards. Consequently, the gas-addition and master-alloy procedures can complement each other, depending on the impurity to be added.

The master alloy is prepared by drilling several small holes into a large mass of the high-purity base metal and packing these cavities with known amounts of previously degassed high-purity oxide of the metal. Tapered plugs of the base metal are fitted into the cavity openings, and the contained oxide is further degassed by application of a low-current arc in an argon atmosphere. The oxide–metal composite is then arc-melted in a purified argon atmosphere. The metallic button is "flipped" a number of times to insure homogeneous distribution of the added impurity, the concentration of which is calculated from the weights of the base metal and the oxide. Standards of lower oxygen content may be similarly prepared by successive dilutions of the master alloy with known amounts of the base metal.

V. SAMPLING AND SAMPLE PREPARATION

The geometrical form and the nonmetallic impurity content of some metals are largely dependent on the method of preparation and refining techniques employed in the production of the metal. Metals obtained by the metallothermic reduction of lanthanide [59, 60] or actinide [61] halides are in the form of ingots that contain appreciable amounts of the reductant metal. Vacuum distillation of lanthanide metal ingots results in bright, well developed crystals from which most of the reductant metal and part of the nonmetallic impurities are removed. The more volatile lanthanides, samarium, europium, and ytterbium, prepared by the metallothermic reduction of the oxides with lanthanum metal [59], are obtained originally as crystalline condensates. The intermediate alloy process, used to produce yttrium and thorium on a large scale, yields a sponge-metal product after vacuum heat-treatment of the crushed Y–Mg [60], Th–Zn [62], or Th–Mg [63]. Metal iodide decomposition [64] and fused-salt electrolysis [65] processes are used in refining thorium metal produced by the intermediate alloy methods; a crystalline product is obtained by these refining techniques. A method of recent interest for highly purifying small quantities of the lanthanide and actinide metals is one of electrotransport of the interstitial impurities [66, 67]. The typical electrotransport analytical specimen is a rod of several millimeters length and diameter and is analyzed as received.

It is normally desirable to have the final metal product in massive form; thus, the crystalline and sponge metals frequently are consolidated into ingots by arc-melting procedures. Although most analytical specimens are submitted as massive-metal samples, analysis of sponge and crystalline samples are frequently desired for production and fabrication control.

Proper sample preparation prior to the analysis is one of the most critical phases of the entire analytical process. Massive specimens are best prepared by carefully filing all the surfaces. The porous sponge-metal and crystalline specimens normally are given no preanalysis treatment, although washing in a non-residue-forming solvent, such as methanol, followed by thorough drying, may be advisable in some instances. Chemical etching of any kind is not recommended. Personal contact with the specimen should be avoided, particularly after completion of any preanalysis preparations.

The apparent reactivities of certain lanthanide metals with the atmosphere [12, 68, 69] require special precautions in their sampling and storage. Effects on the oxygen contents of lanthanum, cerium, praseodymium, and neodymium imposed by variations in the preparation and storage of the analytical sample were examined at the Ames Laboratory. Results of the study, which were obtained by carrier-gas-fusion procedures, are summarized in Table 19.2. All of the data reported in this table represent the average values of at least triplicate determinations.

The oxygen values shown in the last row of Table 19.2 were obtained under the following conditions of sample preparation, storage, and loading: samples of appropriate size were first cut and filed clean in air. Then, within a dry-argon filled vacuum dry-box, the specimens were filed, weighed, encased in platinum capsules, and placed in polyethylene screw-cap vials. The atmospheric exposure factor was almost completely eliminated by loading the prepared samples through an air-lock valve into the apparatus sample-arms, which were either under vacuum or filled with purified helium. The samples were prepared and loaded just prior to the carrier-gas-extraction process, thus minimizing storage effects also. The oxygen values obtained by following these precautions were used as the basis for determining the apparent increase in the oxygen content when such precautions were intentionally neglected.

All of these precautions were executed on the samples involved in the storage studies, except that the vials containing the prepared specimens were left open for specific periods of time within each particular storage environment. The results of the storage studies indicate that analytical samples of any of these four metals may be stored under a good vacuum for reasonable periods of time without causing significant increase in the oxygen contents of the specimens. Neodymium and praseodymium samples may also be similarly stored in well-sealed desiccators freshly charged with $CaSO_4$ or activated molecular sieves. The magnitudes of the oxygen-content increase caused by exposure of the samples to room atmosphere, especially for lanthanum and cerium, indicate the necessity for minimizing exposure by loading the samples into the extraction apparatus through an air-lock valve assembly.

Table 19.2 shows the effect of omitting the prefiling step. The samples used for this phase of the study were cut to appropriate sizes and stored sufficiently long in glass vials in room atmosphere to cause formation of heavy oxide layers. All of the previously stipulated precautions were then exercised in preparation of the analytical samples, except that the precleaning step in air was omitted. This study also revealed that all final preparations, including surface cleaning, weighing, and encapsulation in platinum, must be performed under a dry, inert atmosphere, rather than in air, in order to avoid intolerable oxygen "pickups."

TABLE 19.2. Sample Preparation and Storage Effects on the Oxygen Contents of the Most Reactive Lanthanide Metals

	La	Ce	Pr	Nd
Ppm oxygen content increase per hour of storage in:				
1. room atmosphere	12	44	4	1
2. argon atmosphere	4	7	0.5	0.1
3. dry air ($CaSO_4$ desiccant)	4	0.6	0.2	<0.1
4. dry air (molecular-sieve desiccant)	—[a]	0.9	0	0.1
5. vacuum (mechanical pump)	0	0.1	<0.1	0
Ppm oxygen content increase due to the condition of:				
1. not pre-filing the sample	55	133	69	14
2. routine preparation in air	28	31	15	20
Ppm oxygen content obtained under optimal preparation, storage and loading procedures	210	93	280	135

[a] Not analyzed.

The results presented in Table 19.2 may not always appear to be analytically significant, nor the conclusions based on them to be analytically valid. However, considering the relatively high oxygen-impurity levels in the four base specimens, the fact that certain effects, although seemingly small in some cases, were even distinguishable as oxygen-content increases should be cause for concern when analytical samples of these reactive metals are prepared or handled.

The filing of samples weighing 0.1–0.3 g in a dry box is a tedious and difficult task. As an alternative, anodic etching (electropolishing) in a perchloric acid electrolyte offers promise for cleaning the surfaces of the reactive lanthanides. Comparative data on the two surface cleaning procedures are compiled in Table 19.3. The two sets of data (i.e., for "special handling" and "electropolishing") were obtained at different periods in time; consequently, some of the observed deviations in results may be attributable to this factor. The data in Table 19.3 show that most of the lanthanides (samarium, europium, ytterbium, and any porous-type sample being exceptions) may be successfully handled by the electropolishing approach. Although the mechanism by which the sample-surfaces are deactivated has not definitely been proven, it appears from cathodoluminescence examinations that a protective oxide coating (probably a monolayer) is responsible for the marked decrease in reactivity to the atmosphere. Even though electropolishing renders the sample surfaces inactive, it is advisable to

TABLE 19.3. Comparison of Sample-Surface Cleaning by Electropolishing and Other Methods

	Vacuum Fusion			
Metal	ppm O	ppm N	ppm H	Cleaning Method
Sc	142	3	29	Routine preparation
	109	4	34	Electropolishing
Y	422	15	2	Routine preparation
	431	12	2	Electropolishing
La	267	87	19	Special handling
	296	77	22	Electropolishing
La	60	8	2	Special handling
	67	10	2	Electropolishing-1
	73	10	3	Electropolishing-2
	104	9	8	Electropolishing-3
La	60	8	2	Special handling
	73	11	3	Electropolishing
Ce	63	4	3	Special handling
	89	5	6	Electropolishing
Nd	69	21	3	Special handling
	58	19	3	Electropolishing
Nd	39	15	5	Special handling
	36	19	1	Electropolishing
Tb	241	2	27	Routine preparation
	231	2	27	Electropolishing
Tb	220	2	24	Routine preparation
	210	2	28	Electropolishing
Tb	194	2	16	Routine preparation
	200	2	19	Electropolishing
Er	40	1	8	Routine preparation
	44	4	10	Electropolishing

polish the very reactive lanthanides just prior to loading them into the storage portion of the analytical facility.

The rate of reactivity of europium metal with the atmosphere also was examined. The high volatility of this metal at temperatures necessary to effect carbon-reduction prevented determination of the oxygen content by either furnace fusion technique. Based on the conclusion of Spedding and coworkers [69] that the compound initially formed by reaction of the metal with atmospheric moisture is $Eu(OH)_2 \cdot H_2O$, determination of the degree of hydrogen "pickup" was made instead, by employing a tin bath at a vacuum-fusion temperature of 850°C. Two europium specimens of different hydrogen contents were employed in the study, the results of which are summarized

TABLE 19.4. Effect of Atmospheric Exposure on the Hydrogen Content of Europium Metal

	Hydrogen (ppm)	
Sampling and Handling Conditions	Eu-1	Eu-2
Optimal for reactive lanthanides	64	5
Prepared in air; loaded via air-lock valve into evacuated sample-arms	70	—[a]
Prepared in air; loaded into sample-arms in conventional manner	72	23
Prepared in air; exposed to atmosphere for 18 h	—[a]	1235

[a] Not analyzed.

in Table 19.4. All of the data reported in this table represent the average values of duplicate determinations. The dependency of the hydrogen content of europium on atmospheric exposure time is clearly demonstrated. The apparent hydrogen content of the low-hydrogen-content specimen was increased by a factor of nearly 5 simply by preparing and loading the samples in the conventional manner. Because the europium compound formed by reaction with the atmosphere is a hydrated hydroxide, a similar effect may be assumed for oxygen.

The potential health hazards related to the radioactivity of the actinide metals should alone be sufficient reason to carry out their sample-preparation in an enclosed atmosphere. The health-hazard potential with plutonium is sufficiently great that maximum precautionary measures should always be observed, even to the extent that the extraction portion of the analytical system be isolated [38, 70]. Also, surface deterioration of the actinide metals visually appears similar to those of some of the more reactive lanthanide metals. For these reasons, the special sampling and handling precautions previously discussed for the reactive lanthanide metals should also be observed with the actinide metals. Finally, because of commercial restrictions and requirements, fusion melts containing any of the radioactive elements should be held separate for reclaiming of any precious metals.

VI. ANALYSIS

A. HYDROGEN

1. Hot Extraction

The thermal stabilities of metal hydrides generally are sufficiently low that only moderate temperatures are needed to cause disruption of metal–hydrogen bonds and quantitative formation of molecular hydrogen. The diffusion coefficient of hydrogen generally is sufficiently large to permit quantitative evolution from the solid metal sample in a reasonable time. The extent to which oxygen and nitrogen impurities may be extracted at moderate temperature is generally insignificant because of the high thermal stabilities of metal–oxygen and metal–nitrogen bonds.

The hot-extraction technique has been applied to the determination of hydrogen in uranium at 500°C [19], 750°C [22], and 900°C [20, 21], thorium at 1300–1400°C [23], and a number of the lanthanide metals at 100–150°C below their respective melting temperatures [24, 25]. Typical extraction times vary from as little as 10 min to as much as 4 h, the efficiency being dependent not only on the extraction temperature but also on the conformation and size of the sample, which may range from several tenths of a gram to 25 g. Nickel capsules can be used [22] to facilitate handling of finely divided specimens.

Recent evidence obtained at the Ames Laboratory indicates that the vacuum-hot-extraction technique does not provide conditions that cause complete extraction of hydrogen from the lanthanide metals. The data summarized in Table 19.5 reveal a hot-extraction deficit of 1.5 ppm, which

TABLE 19.5. Determination of Hydrogen in Several Lanthanide Metals

Metal	Vacuum Fusion: Number of Samples	Vacuum Fusion: ppm	Vacuum Hot Extraction: Number of Samples	Vacuum Hot Extraction: ppm
Erbium	10	4.9 ± 0.8	6	3.6 ± 0.3
Yttrium	12	2.3 ± 0.7	13	0.68 ± 0.25
Gadolinium	10	13.2 ± 0.7	10	11.8 ± 0.3

does indeed appear real in that it remained essentially constant over a range of 2–13 ppm hydrogen. (The vacuum-hot-extraction data were obtained at temperatures 100–200°C below the melting temperature of the particular metal; the vacuum-fusion data were obtained under conditions previously described [35].) On samples containing more than 10–15 ppm hydrogen, this magnitude of difference is almost impossible to detect, and comparative data at these impurity levels generally are acceptable if the differences are on the order of 1–2 ppm; hence, it is not surprising that the hot-extraction deficit has not heretofore been mentioned.

2. Furnace Fusions

Hydrogen results obtained by vacuum- or carrier-gas-fusion procedures, concomitant to the desired oxygen determinations, are generally considered to be quantitative [18]. In a study [24] involving a number of the lanthanide metals, for which the analytical data are reproduced in Table 19.6, vacuum fusion hydrogen results were validated by comparison with hot extraction determinations. The various vacuum and carrier-gas fusion reaction media and temperatures for the hydrogen determinations on the lanthanide and actinide metals are compiled in Table 19.6.

Table 19.6. Analytical Comparative Data for Weight Percent Hydrogen in Lanthanide Metals

Metal	Hot Extraction	Vacuum Fusion
Scandium	0.0033	0.0026
Yttrium	0.0098	0.0095
Gadolinium	0.0021	0.0027
	0.0048	0.0046
Terbium	0.0075	0.0062
	0.018	0.015
	0.019	0.017
Dysprosium	0.0054	0.0047
Holmium	0.0011	0.0011
	0.0065	0.0064
Erbium	0.0009	0.0013
	0.0010	0.0012
Lutetium	0.0041	0.0043

B. OXYGEN

1. Furnace Fusions

The accuracy of the analysis depends on whether the environmental conditions provide (a) a suitable carbon-containing reaction medium in which the carbon reduction reaction occurs and in which the solubility of carbon monoxide is negligible; (b) temperatures sufficient to effect the carbon-reduction reaction and the thermal decomposition reactions; (c) some means for minimizing vaporization-gettering effects of the materials within the reaction zone; and (d) tolerable operating blanks. Usually these requirements can be met by providing a molten solvent bath in the graphite crucible into which the specimen is introduced or by simultaneously introducing the specimen and solvent metal into the graphite crucible. In the historical development and refinement of the furnace fusion techniques, a variety of solvent metals have been used as bath and flux materials for the lanthanide and actinide metals (see Table 19.7). Although iron is still used as a

TABLE 19.7. Recovery of Oxygen from Thorium by the Iron-Bath Vacuum-Fusion Procedure

Wt % Oxygen		
Absolute Synthetic Standards	Vacuum Fusion, Fe Bath	(%)
0.020	0.012	60
0.040	0.028	70
0.070	0.054	77
0.12	0.092	77

reaction medium, it fails to remain sufficiently fluid throughout extended fusion periods. According to Lassner and Wolfel [71], the amount of carbon dissolved is directly dependent on both the fusion time and temperature. The formation of graphite flakes near the top of the melt then leads to a high viscosity [72]. With reference to the lanthanide and actinide metals, very few attempts have been made to verify the accuracy of oxygen determinations when an iron bath was employed. In one such critical study of carbon-reduction techniques for the determination of oxygen in thorium and yttrium metals, Fassel and co-workers [45] evaluated the iron bath procedure of Sloman and Harvey [6] and found this reaction medium to be entirely inadequate for quantitative recovery of oxygen from thorium. This is evidenced by the analytical results summarized in Table 19.7 for several thorium–oxygen synthetic standards. The merit of employing platinum as the reaction medium for both the carrier-gas and vacuum-fusion techniques is clearly shown by the analytical data summarized in Table 19.8 [45]. The recovery of oxygen from yttrium metal by a carrier-gas-fusion platinum-flux procedure patterned after that described by Banks and co-workers [49] was found to be somewhat less than quantitative, even when the extraction time was extended considerably. A variation of this flux procedure, in which the fusion crucible was first preconditioned with a platinum bath, was adequate for quantitative recovery of oxygen from both yttrium and thorium under vacuum-fusion-extraction conditions and for oxygen in thorium under carrier-gas-fusion conditions. For thorium, carrier-gas extraction at 1800°C, previously used by Smiley [33, 46], always led to low oxygen results, and any temperature below 2500°C was found to be inadequate with the simple platinum reaction medium. For either furnace fusion technique, a platinum–tin reaction medium was found to be the most appropriate for both yttrium

TABLE 19.8. Recovery of Oxygen from Thorium and Yttrium with Various Platinum-Based Reaction Media

	Wt % Oxygen					
	Carrier-gas fusion			Vacuum fusion		
Absolute Synthetic Standards	Pt Flux, 2100–2200°C	Pt Bath, 2500°C	Pt–Sn Bath, 2300°C	Pt Bath, 1900°C	Pt Bath, 1950°C	Pt–Sn Bath, 1700°C
Thorium:						
0.020	—[a]	0.020	0.020	—	0.021	0.021
0.040	—	0.038	0.038	—	0.042	0.042
0.070	—	0.067	0.073	—	0.070	0.074
0.12	—	0.12	0.12	—	0.12	0.12
0.19	—	0.19	—	—	0.19	—
Yttrium:						
0.080	0.069	—	0.081	0.079	—	0.083
0.13	0.11	—	0.13	0.13	—	0.14
0.20	0.18	—	0.20	0.20	—	0.20
0.31	0.29	—	0.32	0.31	—	0.31
0.48	0.46	—	0.48	0.45	—	0.48

[a] Not analyzed.

and thorium because quantitative oxygen recoveries were achieved at fusion temperatures significantly lower than those required with the simple platinum medium. As a result of these studies, it was cautioned [45] that the absolute accuracy of analytical results must be viewed with reservations unless quantitative recovery can be demonstrated with the particular analytical apparatus employed.

Tin is not very suitable as a solvent bath when used alone, but it forms an excellent bath material when alloyed with platinum. The possible functions of the tin have been discussed in Chapter 4 and elsewhere [73–77]. One other possibility that deserves consideration is that the solubility of carbon monoxide may be significantly lower in fusion melts containing tin.

The platinum–tin-bath, platinum-flux, vacuum-fusion, and carrier-gas-fusion procedures [45] were subsequently applied to many of the lanthanide

TABLE 19.9. Study Groups for Determination of Oxygen in the Lanthanide Metals

Group I	Group II	Group III	Group IV
Scandium	Lanthanum	Samarium	Europium
Yttrium	Cerium	Thulium	Ytterbium
Gadolinium	Praseodymium		
Terbium	Neodymium		
Dysprosium			
Holmium			
Erbium			
Lutetium			

TABLE 19.10. Analytical Results on Determination of Oxygen in the Group-I Lanthanide Metals

	Oxygen (wt %)		
Metal	Carrier-Gas Fusion	Vacuum Fusion	Synthetic Standards
Scandium	0.038	0.037	—[a]
	0.070	0.071	—
	0.36	0.36	—
Yttrium	0.037	0.036	—
	0.20	0.20	0.20
	0.27	0.26	—
	0.29	0.30	—
	0.48	0.48	0.48
Gadolinium	0.081	0.081	—
	0.087	0.087	—
Terbium	0.22	0.22	—
	0.34	0.34	—
	0.43	0.43	—
Dysprosium	0.0042	0.0042	—
	0.0071	0.0078	—
	0.015	0.016	—
	0.11	0.11	—
Holmium	0.17	0.17	—
	0.23	0.21	—
	0.24	0.24	—
Erbium	0.17	0.017	—
	0.026	0.026	—
	0.12	0.12	—
Lutetium	0.22	0.22	—
	0.23	0.23	—
	0.46	0.43	—
	1.13	1.10	—

[a] Not analyzed.

metals, plus thorium [35]. Because of widely varying volatilities and reactivities [12, 68, 69, 78–83], the lanthanide metals were classified into the four study groups shown in Table 19.9. Group I consists of metals with low volatilities (thus, vaporization under high-temperature fusion conditions would not give rise to serious experimental difficulties) and low rates of reactivity with the atmosphere (so special sampling precautions are not required). The group-II metals have volatilities sufficiently low that vaporization is not a problem at typical fusion temperatures, but they require special sampling precautions. Such sampling precautions are not required for the metals in group III, but the rates of vaporization are significant at normal fusion temperatures. The excessive rates of vaporization of europium and ytterbium in group IV preclude direct application of high-temperature fusion techniques.

Group-I lanthanide metals have been the subject of several studies on the applicability of the platinum–tin-bath platinum-flux furnace fusion procedures to the determination of oxygen [24, 35]. Comparative oxygen results reproduced from these studies (Table 19.10) show that the basic procedures successfully applied earlier to the thorium and yttrium systems [45] are equally suitable for all the metals in group I. Under both vacuum- and carrier-gas-extraction environments, the platinum–tin binary reaction medium was adequate for oxygen determinations in the group-II and -III lanthanide metals, as shown by the authors' analytical results summarized in Table 19.11.

TABLE 19.11. Analytical Results on Determination of Oxygen in the Group-II and Group-III Lanthanide Metals

	Oxygen (wt %)		
Metal	Carrier-Gas Fusion	Vacuum Fusion	Fast-neutron Activation
Lanthanum	0.016	0.017	[a]
La(4% Gd)	0.021	0.020	—
La(2% Gd)	0.021	0.018	—
Cerium	0.0045	0.0046	—
Praseodymium	0.028	0.031	—
	0.070	0.067	—
	0.095	0.096	—
	0.097	0.10	—
	0.16	0.16	—
Neodymium	0.013	0.013	—
	0.063	0.063	—
	0.068	0.069	—
Samarium	—	0.016	0.016
	0.21	0.24	—

[a] Not Analyzed.

The only notable modifications to the general analytical procedures specified for the group-I lanthanides [35] were the following: (a) analytical samples of the group-II metals were prepared under a dry argon atmosphere to avoid exposure of the samples to air; and (b) the concentration of the group-III metals in the fusion melt was maintained at less than 5 wt %.

Nickel does not appear to be very satisfactory as a reaction medium for the extraction of oxygen from the lanthanide metals. Peterson and Beerntsen [37] reported maximal oxygen recovery on lanthanum to be only 90% complete with the nickel bath under a vacuum environment. Loss of oxygen was attributed primarily to absorption of some of the evolved carbon monoxide by a nickel-rich condensate that collected on the furnace-tube walls. Under similar extraction conditions, the nickel reaction medium was found unacceptable for several of the group-I lanthanide metals, as is shown by the authors' analytical data compiled in Table 19.12.

TABLE 19.12. Recovery of Oxygen from Group-I Lanthanide Metals by the Nickel-Bath Vacuum-Fusion Procedure

	Oxygen (wt %) Vacuum Fusion	
Metal	Pt–Sn Bath	Ni Bath
Gadolinium	0.78	0.40
Terbium	0.32	0.098
Dysprosium	0.013	0.002
	0.043	0.005
Lutetium	0.23	0.13

In an attempt to utilize a nickel reaction medium for the lanthanide metals in general, Dallmann [24] found that under the conditions of carrier-gas fusion the molten nickel dissolved carbon to the extent that the melt eroded through the graphite crucible.

2. Arc-Discharge Extraction

The extraction of oxygen from several of the lanthanide and actinide metals has been performed in the anode receptacle of a high-current dc carbon-arc discharge in a static, inert atmosphere [50–53]. Gas evolution occurs by the same mechanisms as in the vacuum- and carrier-gas-fusion techniques. In this extraction technique a precipitous thermal gradient established between the electrode–globule interface and the anode spot of the arc discharge caused vigorous stirring of the molten sample and very rapid evolution (less than 10 sec) of the carbon monoxide and nitrogen. The arc-extraction technique has been employed for determining the oxygen content in lanthanum [50], yttrium and thorium [52], and plutonium [53], for which the impurity levels were measured spectrographically, and in thorium, yttrium, and terbium, for which gas chromatography was employed for the analytical measurements [51].

The accuracy of the arc-extraction oxygen results on thorium and yttrium has been substantiated by comparative vacuum-fusion and carrier-gas-fusion data, as well as by metal–oxygen synthetic standards [25, 45]. The primary advantage of the arc-extraction technique is the short time period required for the gas-evolution process. When gas-chromatographic techniques are used for the analysis of the extracted gases, only a few minutes are required for analysis.

3. Activation Analysis

The fast-neutron-activation technique has been used to determine trace amounts of oxygen in samarium and ytterbium [55]. Regarding accuracy, only one comparative result with one independent technique has been given, as shown for samarium, in Table 19.11. The ^{16}N product is also produced by fast-neutron irradiation of fluorine. Analytical interferences due to fluorine could thus be especially crucial for the lanthanide elements, because metallothermic reduction of the fluorides is often used for preparing the metals.

For health safety purposes and to avoid radioactive contamination of the apparatus, it often is undesirable to utilize fusion-extraction facilities for determining the oxygen contents of the fissionable actinide metals. Activation analysis again may afford a suitable alternative technique. Because of the interfering gamma emission produced by irradiation of such specimens with fast neutrons, the ^{3}He-induced nuclear reaction of oxygen, $^{16}O(^{3}He, p)^{18}F$, has been utilized instead for oxygen determinations in thorium, plutonium, and americium [58]. Interference by activation of other elements may be largely controlled by restricting the ^{3}He bombarding energy.

Analytical specimens normally employed in the activation methods are significantly larger (up to 20 g or more) than those for fusion-extraction analysis (usually tenths of a gram). Thus, inhomogeneities in the distribution of the oxygen impurity in the metal are not as readily apparent when the activation techniques are employed.

C. NITROGEN

1. Chemical Methods

In the past, the chemical dissolution–distillation technique has gained the reputation of being the most accurate for the nitrogen determination and almost invariably is used as the "referee" method for evaluating other analytical techniques for nitrogen determinations. Nitrogen present in solid-solution and in the combined form is liberated in the atomic state when the sample is dissolved in an acid solution [26, 27] and forms the appropriate ammonium salt with the acid anion. The ammonia liberated by rendering the resulting solution alkaline is steam-distilled into an absorbing solution, and the absorbed ammonia may be determined volumetrically when suitable amounts of nitrogen are present or colorimetrically when the nitrogen contents are small. Free nitrogen gas, which may be present on the surface of the metal or within voids in the lattice structure, is liberated in the molecular state, does not form the ammonium salt and thus is not detected by the chemical technique.

The lanthanide and actinide metals do not show any great resistance to attack by the common laboratory acids and normally do not present any significant problems in the first phase of the chemical method. Dissolution of 1-g-samples in (1:1) hydrochloric acid usually is completed in less than 1 h. Complete conversion of the nitrogen content to the ammonium salt, which has been labeled as the "main problem in determining nitrogen by chemical methods" [26], remains a largely indeterminable factor for all metals, including the lanthanides and actinides. The distillation phase does not present any problems, nor does the final determination, for which a number of equally satisfactory alternatives are available (see Chapter 8).

2. Furnace Fusions

Incomplete recovery of nitrogen by vacuum-fusion and carrier-gas-fusion procedures is commonly experienced for nearly all metals when using the multiple-sample-per-crucible technique. (see Chapter 4, Section I.A.4.a for a discussion of the single-sample and multiple-sample-per-crucible techniques). In most studies on the lanthanide and actinide metals, for which a reference summary is provided in Table 19.1, no substantiating data were presented to confirm the furnace-fusion nitrogen results. In a few cases, vacuum fusion results were reported as being consistently lower than chemical values for uranium [23, 39, 42], for thorium [23, 42], and, when an iron reaction medium was employed, for lanthanum [37]. Quantitative nitrogen recovery was indicated, however, when the lanthanum samples were fused in a nickel medium [37].

A detailed study on the high-temperature extraction of nitrogen from lanthanide metals and from thorium established that the platinum–tin reaction medium led to maximal nitrogen recoveries for both extraction techniques [35]. If concordance of results with the chemical technique can be taken as an estimate of the accuracy of the nitrogen results obtained by other techniques, then the comparative nitrogen data reproduced in Table 19.13

TABLE 19.13. Analytical Results on Determination of Nitrogen in the Group-I Lanthanide Metals and Thorium

	Nitrogen (wt %)		
Metal	Carrier-Gas Fusion	Chemical	Vacuum Fusion
Scandium	0.0023	0.0021	0.0016
	0.0040	0.0038	0.0043
	0.0067	0.0078	0.0049
Yttrium	0.0034	0.0034	0.0028
	0.019	0.018	0.019
	0.034	0.037	0.032
Gadolinium	0.0072	0.0063	0.0055
	0.017	0.017	0.014
Terbium	0.044	0.045	0.042
	0.057	0.055	0.054
	0.072	0.074	0.072
Dysprosium	0.0033	0.0026	0.0027
	0.011	0.010	0.009
Holmium	0.0090	0.0090	0.0090
	0.044	0.040	0.037
	0.052	0.051	0.046
Erbium	0.0025	0.0028	0.0024
	0.016	0.014	0.014
	0.10	0.096	0.098
Lutetium	0.0027	0.0035	0.0031
	0.0092	0.0094	0.0060
	0.025	0.024	0.019
	0.084	0.083	0.083
Thorium	0.0054	0.0054	0.0053
	0.0096	0.0092	0.0062
	0.011	0.012	0.009
	0.016	0.016	0.013
	0.15	0.14	0.14
	0.36	0.35	0.34

adequately demonstrate that the Pt–Sn-bath carrier-gas-fusion procedure [35] provides accurate nitrogen determinations on the group-I lanthanide metals and thorium. The Pt–Sn-bath vacuum-fusion procedure [35] gave reasonable nitrogen recoveries on these same metals also, although there are scattered vacuum-fusion results in Table 19.13 that appear to be low. The low results cannot, however, be associated with any particular metal. The frequency of occurrence and the degree to which the results are low indicate that the vacuum-fusion environmental and extraction conditions were somewhat marginal [35].

Later studies on the extraction of nitrogen from the lanthanide metals showed that the Pt–Sn-bath carrier-gas-fusion procedure [35] was also satisfactory for the group-II members, as evidenced by the comparative data summarized in Table 19.14. The frequency of low nitrogen recoveries by the

TABLE 19.14. Analytical Results on Determination of Nitrogen in the Group-II Lanthanide Metals

	Nitrogen (wt %)		
Metal	Carrier-Gas Fusion	Chemical	Vacuum Fusion
Lanthanum	0.0015	0.0014	—[a]
	0.0030	0.0025	—
	0.0050	0.0053	—
	0.0082	0.0095	0.0067
La (6% Gd)	0.0094	0.0089	—
La (4% Gd)	0.0099	0.0085	0.0079
La (2% Gd)	0.010	0.011	0.007
Cerium	0.0020	0.0018	—
	0.0030	0.0031	—
	0.0053	0.0059	0.0048
	0.019	0.018	—
	0.022	0.025	—
Praseodymium	0.0046	0.0041	0.0030
	0.0069	0.0071	0.0071
	0.018	0.019	0.018
	0.018	0.019	0.017
	0.165	0.134	0.163
Neodymium	0.0013	0.0016	0.0016
	0.0039	0.0041	0.0029
	0.019	0.021	0.015

[a] Not analyzed.

vacuum-fusion procedure was, however, considerably higher for this group of metals than for the group-I members. For one praseodymium specimen, both furnace fusion results were considerably higher than the value reported by the chemical technique, suggesting the presence of significant amounts of molecular nitrogen within the lattice structure.

3. Arc-Discharge Extraction

The ability to extract quantitatively the nitrogen impurity from yttrium, terbium, and thorium, by means of a high-current dc carbon-arc discharge in a static helium atmosphere, has been demonstrated [51]. The metal sample is melted in the anode receptable of special design, where the molten globule undergoes vigorous stirring due to the temperature gradient that occurs between the anode spot of the discharge and the electrode–globule interface. The temperature of the anode spot has been estimated to be of the same order of magnitude as the boiling temperature of the globule [84], which would approach or exceed 3000°C for the three metals mentioned. Thus, the arc-extraction technique allows very rapid evolution of the nitrogen-impurity content. The general applicability of this technique to the determination of nitrogen in the lanthanide and actinide metals has not been established.

VII. RECOMMENDED PROCEDURES

A. HYDROGEN

1. Hot Extraction

Several commercial analytical systems have been designed specifically for extraction from the solid state (see Chapter 7). For some of the lanthanide metals and thorium the most efficient extraction temperature may exceed the maximum attainable temperature of the analyzers, and conventional vacuum-fusion systems are more suitable.

The following general analytical procedure is based on the use of a Varian Vacuum Division (formerly NRC Equipment Corporation) vacuum-fusion gas-analysis facility (see Chapter 4), with accompanying crucible assembly and furnace heater as described in Table 19.15. The procedure for a particular

TABLE 19.15. Vacuum-Extraction Apparatus for Determination of Gases in Lanthanide and Actinide Metals

Analysis facility	Vacuum-fusion gas-analysis unit, Varian Vacuum Division, Model 912 S [31].
Crucible assembly	Ultra Carbon Corp., C-625 graphite crucible and F-703 graphite funnel, or equivalents, floated on −200 mesh UCP-2 graphite powder, or equivalent, in a quartz thimble suspended by platinum-wire hooks within an air-cooled borosilicate glass jacket, patterned after the design of Guldner and Beach [85]
Furnace heater	Induction heater, Lepel High Frequency Corp., Model T-2.5B, 2.5-kW output, 450-kHz nominal frequency.
Analytical system	Conventional low-pressure oxidation and selective absorption and condensation separation [31].

lanthanide or actinide metal may, however, be adapted to any hot-extraction apparatus that provides the following: (a) furnace temperatures up to 1600°C; (b) a reaction crucible that does not react with the metal sample and that is thermally stable at the extraction temperature; (c) efficient removal of the evolved hydrogen from the reaction zone; (d) separation of the hydrogen from any nitrogen or carbon monoxide that may be evolved; and (e) collection and measurement of the hydrogen extracted. The analytical procedures and reaction conditions for hot extraction of hydrogen from the lanthanide and actinide metals are outlined in Table 19.16.

TABLE 19.16. Analytical Procedures and Reaction Conditions for Determination of Hydrogen in Lanthanide and Actinide Metals by Vacuum Hot Extraction[a]

Apparatus degassing	Crucible temperature raised to 1600–2000°C over ½-h period and maintained at the selected temperature until the pressure level is less than 10^{-6} torr. (The higher degassing temperature permits more rapid attainment of a lower, more stable hydrogen blank-rate).
Furnace blank	0.02 to 0.1 μg hydrogen, depending on temperature.
Extraction temperatures	900°C, U; 1200°C, Gd, Tb; 1300°C, Dy, Ho; 1400°C, Sc, Y, Er, Tm; 1500°C, Lu, Th. Probable extraction temperatures for remaining lanthanide metals: 900°C, Nd, Sm; 800°C, La, Pr; 700°C, Ce; Eu and Yb not considered.
Extraction time	Normally 15 min for sample-sizes stipulated. Larger samples require extended extraction periods.
Samples	Carefully filed and cleaned specimens weighing 0.1–0.4 g. Special sampling and handling precautions previously discussed to be practiced on specimens of U, Th, La, Ce, Pr, and Nd.

[a] It should be recognized that a deficit of about 1.5 ppm hydrogen is likely to be observed at hydrogen concentrations less than approximately 15 ppm. In these cases, the vacuum-fusion method would be a more suitable choice.

2. Tin Fusion

The hydrogen content of europium and ytterbium metals is best determined by vacuum fusion in a tin melt at 850°C. The same vacuum-extraction apparatus described in Table 19.15 may be used as may any other hot-extraction facility that fulfills the previously mentioned requirements. The apparatus degassing procedure and the extraction time would be similar to those indicated in Table 19.16. Fifteen to 20 g of tin is introduced to the degassed crucible, which is at dull-red heat. This melt is degassed as 850°C until the furnace pressure has been reduced to about 5×10^{-7} torr, which should result in an operating blank of less than 0.05 μg hydrogen. For europium it is imperative that the special sampling and handling precautions be observed.

3. Vacuum Fusion

The hydrogen contents of many of the lanthanide metals may also be determined under the vacuum-fusion experimental conditions discussed in the following section on oxygen. This should, in fact, be the method of choice whenever possible, because of the apparent 1.5-ppm deficiency of hot-extraction hydrogen values.

B. OXYGEN

1. Vacuum Fusion

The general analytical procedure discussed here is based on the use of the vacuum-extraction apparatus described in Table 19.15. Again, the particular analytical apparatus employed is not critical, providing the environmental criteria necessary for quantitative evolution and recovery of the oxygen content, as previously discussed, can be achieved. The specific procedures and reaction conditions for vacuum-fusion determinations of oxygen in the lanthanide and actinide metals are summarized in Table 19.17.

2. Carrier-Gas Fusion

The general analytical procedure discussed here is based on the use of the experimental carrier-gas extraction and analysis apparatus shown schematically in Fig. 19.1 and described in Table 19.18. The special crucible assembly is shown in Fig. 19.2. The operation of this facility has been thoroughly discussed elsewhere [35]. Again, the only factor governing the use of a particular analytical apparatus is that the criteria necessary for quantitative extraction and recovery of the oxygen be achieved. Operating conditions and experimental details pertinent to the determination of oxygen in the lanthanide and actinide metals are summarized in Table 19.19.

3. DC Carbon Arc

The carbon-arc extraction procedure discussed here is based on the experimental design shown in Figs. 19.3–19.5 and described in Table 19.20. The mode of operation of this apparatus has been discussed in detail elsewhere [87] and in Chapter 5. The analytical procedures and operating conditions that apply specifically for the determination of oxygen in yttrium, terbium, and thorium and that may prove suitable for other lanthanide and actinide metals are outlined in Table 19.21.

TABLE 19.17. Analytical Procedures and Reaction Conditions for Determination of Oxygen and Nitrogen in Lanthanide and Actinide Metals by the Vacuum-Fusion Technique

Apparatus degassing	Crucible temperature raised to 2400°C over 1-h period and maintained at this temperature until the pressure is less than 10^{-6} torr.
Bath degassing	Platinum-bath procedure for thorium and uranium: not less than 20 g or more than 60 g of 12-gauge platinum wire added to the degassed crucible at 1700°C and degassed at 1950°C until the pressure level is less than 10^{-6} torr. Platinum–tin-bath procedure for thorium and the lanthanide metals of groups I–III: not less than 20 g or more than 40 g of a a composite 80(Pt):20(Sn) bath added intermittently to the degassed crucible at 1400°C and degassed at 1700°C until the pressure level is less than 10^{-6} torr.
Bath conditioner	0.3–0.5 g sample metal (yttrium preferable) added to degassed bath.
Furnace blank	<5 μg oxygen and <2 μg nitrogen for extraction time and temperature stipulated.
Extraction temperature	Platinum-bath procedure, 1950°C. Platinum–tin bath procedure, 1700°C.
Extraction time	Normally 15 min for sample-sizes stipulated. Larger samples require extended extraction periods as may high-nitrogen content specimens.
Samples and flux	Carefully filed and cleaned specimens weighing 0.1–0.3 g, fluxed with 0.5-g 12-gauge platinum wire, all encased in a 0.35-g platinum capsule formed from a 1-in. square of 1-mil platinum foil. Special sampling and handling precautions previously discussed to be practiced on samples of U, Th, La, Ce, Pr, and Nd.
Platinum flux blank	Platinum wire: 6–8 ppm oxygen; <2 ppm nitrogen. Platinum foil: 10–15 ppm oxygen; <2 ppm nitrogen.
Maximum solute metal concentration	Not to exceed 10 wt % in bath in any case, and not over 5 wt % for Sm or Tm.

4. Activation Analysis

Activation techniques have not as yet been widely adopted for the determination of oxygen in the metals covered in this chapter. The activation techniques are most useful in instances where sample destruction is to be avoided, when materials possessing high health-hazard potentials need to be analyzed, and when sample volatility prohibits the use of more conventional and commonly accepted analytical techniques.

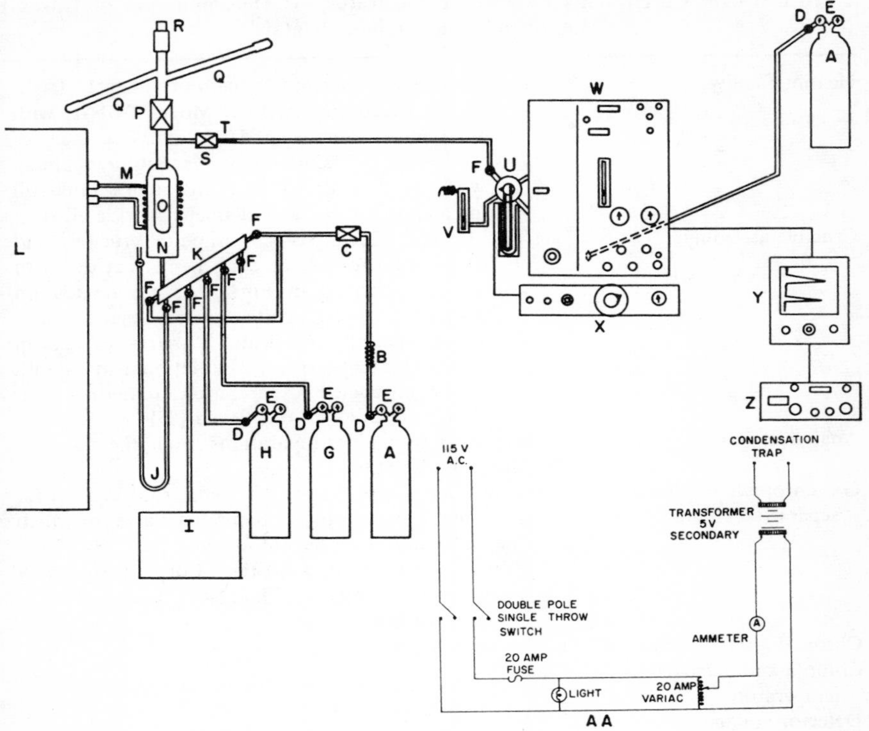

Fig. 19.1. Carrier-gas extraction and analysis experimental facility.

A, Helium supplies, Bueau of Mines, Grade A.
B, Helium purification column.
C, Hoke flow-control needle valve.
D, Matheson shut-off valves.
E, Matheson two-way pressure regulators.
F, Hoke toggle valves.
G, Calibration dry-nitrogen supply; Linde Air Products.
H, Calibration carbon monoxide supply; Matheson, CP.
I, Welch mechanical vacuum pump.
J, Mercury manometer.
K, Gas manifold.
L, Induction generator.
M, Work coil.
N, Vycor furnace jacket.
O, Crucible assembly.
P, 1-inch Jamesbury ball valve; NRC Equipment Corp.
Q, Sample storage arms.
R, Sighting prism and optical flat.
S, $\frac{1}{4}$-inch Jamesbury ball valve; NRC Equipment Corp.
T, Glass-wool filter plug.
U, Precision gas sampling valve-collection trap assembly.
V, Flow meter.
W, Gas chromatograph.
X, Collection trap heater unit.
Y, Recorder.
Z, Printing integrator.
AA, Electrical circuit for collection trap heater.

TABLE 19.18. Carrier-Gas Extraction Apparatus for Determination of Gases in Lanthanide and Actinide Metals

Heating facility	High-frequency induction generator, Lepel High-Frequency Laboratories, Inc., Model T-5N-1, with nominal power output of 5-kW and frequency of 400 ± 100 kHz. Maximum power output coupling to work load achieved with maximum tank-coil setting and an 8-turn, 4-in. diameter work coil.
Crucible assembly	Ultra Carbon Corp., C-625 graphite crucible and F-703 graphite funnel, or equivalents, cut down to dimensions specified in Fig. 19.2, are floated on −200-mesh UCP-2 graphite powder, or equivalent, within a specially fabricated "pyrolytic" boron nitride thimble [86] suspended by platinum-wire hooks within an air-cooled Vycor jacket, patterned after the design of Guldner and Beach [85]
Analytical system	Gas-chromatographic separations and thermal-conductivity detection.
Gas chromatograph and separation column	Perkin-Elmer Corp., Vapor Fractometer, Model 154 D, including a Precision Gas Sampling Valve, of which the collection trap is an integral part. Synthetic zeolite column I, Perkin-Elmer Corp., periodically reactivated in situ for 8–12 h at 190°C under a 100-cc/min flow of helium.
Chromatograph carrier gas	92 cc/min helium.
Column and detector temperature	80°C
Detector voltage	8 V
Recorder	Leeds & Northrup Co., Speedomax "H", AZAR.
Recorder sensitivity and response	0–2 mV, full-scale; 1 sec, full-scale.
Recorder chart speed	$\frac{1}{2}$-in./min
Integrator	Perkin-Elmer Corp., Printing Integrator, Model 194 B

a. Fast-Neutron Activation

The analytical procedures and conditions summarized in Table 19.22 were extracted from a report [55] on the determination of oxygen in samarium and ytterbium, in which the authors thoroughly discuss the operation of their particular fast-neutron-activation system. Schematic diagrams of the monitor and "sample-rabbits" used in this procedure are shown in Fig. 19.6. Although the technique is suitable for nearly all of the lanthanide elements (except yttrium, cerium, praseodymium, and neodymium), charged-particle fission in the actinide elements renders the method useless for this latter series of metals.

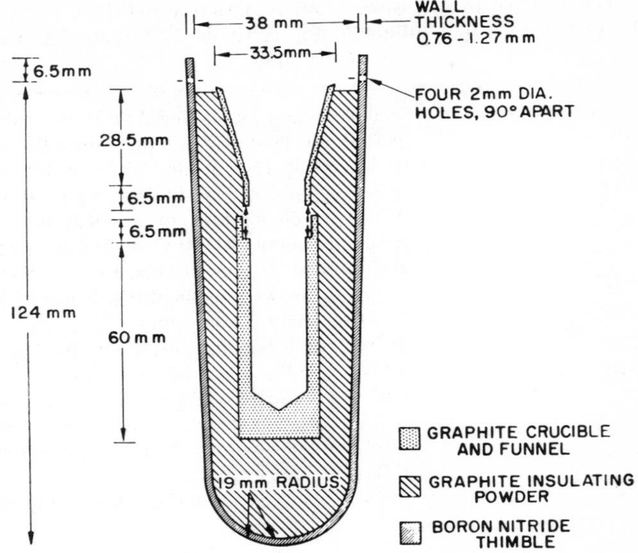

Fig. 19.2. Carrier-gas-fusion crucible assembly with pyrolytic boron nitride thimble.

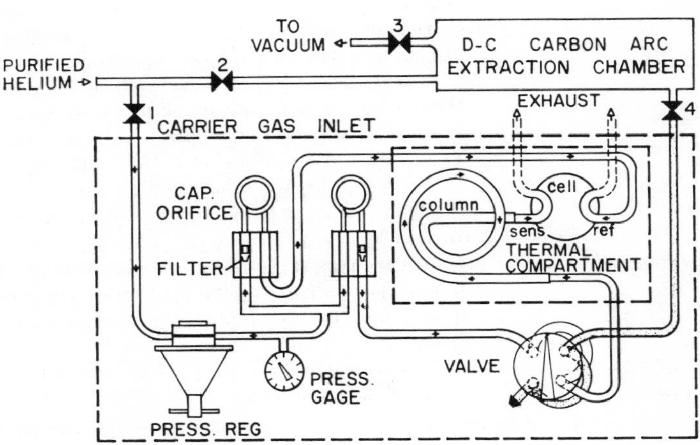

Fig. 19.3. Schematic diagram of arc-extraction–gas-chromatograph assembly.

TABLE 19.19. Analytical Procedures and Reaction Conditions for Determination of Oxygen and Nitrogen in Lanthanide and Actinide Metals by Carrier-Gas-Fusion Technique

Apparatus degassing	Crucible temperature is raised to 2450–2500°C over 1-h period and held at this temperature for 45 min.
Bath degassing and conditioning	Not less than 15 g of the 95(Pt):5(Sn) binary bath added intermittently to the degassed crucible at 1750°C. Temperature of crucible and bath then gradually raised to 1900°C and 0.3–0.5 g of sample or steel added as conditioner for reaction medium. Crucible and bath temperature then raised to 2350°C and degassed for 30–45 min.
Furnace carrier-gas	275–300 cc/min helium, purified by passing it through a 12-ft coiled length of $\frac{3}{8}$-in.-o.d. copper tubing packed with approximately 50 g of 20–50-mesh 13X Molecular Sieve maintained at liquid-nitrogen temperature. Sieve originally activated by heating in vacuo at 180–190°C for 24 h.
Furnace blank	0.4–0.6 μg oxygen/min and 0.5–0.8 μg nitrogen/min.
Extraction temperature	2350°C
Extraction time	8 min
Sample and flux	Carefully filed and cleaned specimens weighing 0.1–0.3 g, encased in 0.35-g platinum capsule formed from 1-in. square of 1-mil foil. Additional 12-gauge platinum-wire flux added when necessary to maintain 3:1 flux ratio. Applicable to thorium, uranium, and group I–III lanthanide metals. Special sampling and handling precautions previously discussed to be observed on samples of U, Th, La, Ce, Pr, and Nd.
Platinum flux blank	Platinum wire: 6–8 ppm oxygen; <2 ppm nitrogen. Platinum foil: 10–15 ppm oxygen; <2 ppm nitrogen.
Collection of evolved impurity gases	Impurity gases preconcentrated from the helium carrier stream by means of a stainless-steel trap packed with approximately 2 g of 20–50-mesh 13X Molecular Sieve, which is maintained at liquid-nitrogen temperature for the collection interval. Collected impurity gases released from the trap by 12–15-sec heating period with circuitry illustrated in Fig. 19.1. Sieve originally activated by heating in an oven at 180–190°C for 24 h.
Integrator calibration	Specified as integrator counts per microgram of carbon monoxide or nitrogen obtained by analysis of known quantities of nitrogen and carbon monoxide standard gases.
Maximum solute metal concentration	Not to exceed 10 wt % in bath in any case and not over 5 wt % for Sm or Tm.

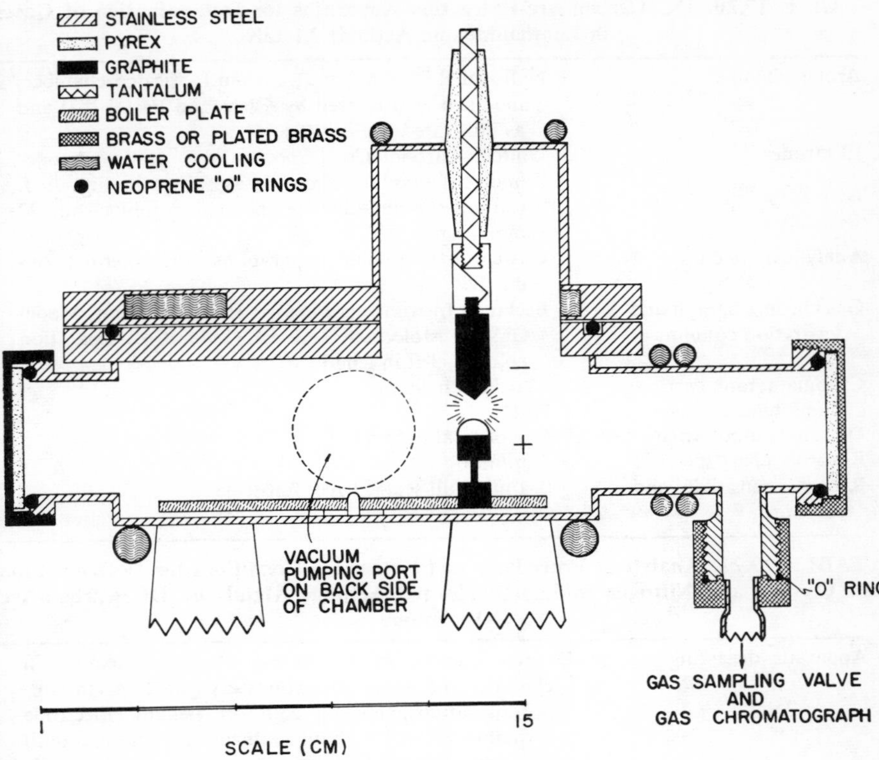

STAINLESS STEEL
PYREX
GRAPHITE
TANTALUM
BOILER PLATE
BRASS OR PLATED BRASS
WATER COOLING
NEOPRENE "O" RINGS
VACUUM
PUMPING PORT
ON BACK SIDE
OF CHAMBER
"O" RING
GAS SAMPLING VALVE
AND
GAS CHROMATOGRAPH
1
15
SCALE (CM)

Fig. 19.4. Schematic diagram of arc-extraction chamber.

6.4 mm
8.0
SAMPLE
PLATINUM
BATH
0.8
4.8
1.6
6.4
6.4
7.9 mm

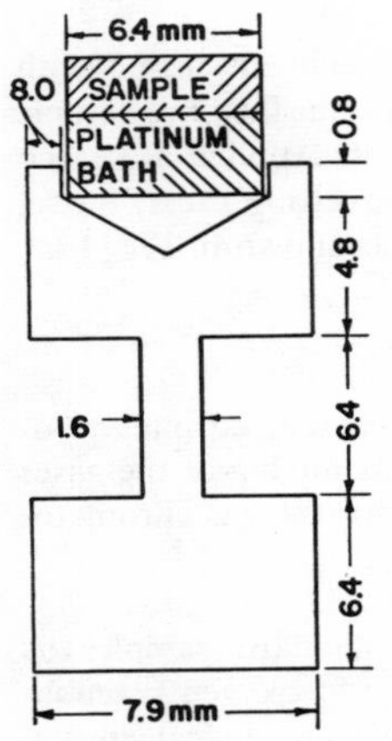

Fig. 19.5. Electrode assembly.

TABLE 19.20. DC Carbon-Arc-Extraction Apparatus for Determination of Gases in Lanthanide and Actinide Metals

Arcing chamber	National Research Corp., custom fabricated chamber, modified as discussed by Evens and Fassel [87] and as illustrated in Fig. 19.4.
Electrodes	United Carbon Co., Spectro-Tech grade. Anode: undercut graphite electrode as shown in Fig. 19.5. Cathode: 8-mm-diameter pointed graphite rod, 22 mm long.
Analytical system	Gas-chromatographic separations and thermal-conductivity detection.
Gas chromatograph and separation column	Beckman Instruments Inc., gas chromatograph, Model GC-2A. Molecular Sieve 5A (Linde Co.) separation column, 3-ft in length.
Chromatograph carrier-gas	72 cc/min helium.
Column temperature	70°C
Detector output attenuation	5 ($\frac{1}{5}$ of total output)
Recorder chart speed	$\frac{1}{2}$-in./min
Recorder sensitivity and response	0–1 mV, full-scale; 1 sec, full-scale.

TABLE 19.21. Analytical Procedures and Extraction Conditions for Determination of Oxygen and Nitrogen in Lanthanide and Actinide Metals by DC Carbon-Arc Technique

Apparatus degassing	Desired number of supporting electrodes loaded in chamber and arced 30 seconds each at 25 A, in supporting atmosphere of 250 torr helium. Electrode turntable rotated without extinguishing the arc until all electrodes are degassed. Chamber evacuated while still hot, then brought to atmospheric pressure with helium. Samples loaded as quickly as possible. Chamber then degassed by arcing to auxiliary electrode for two 2-min periods to remove atmospheric gases that may have been resorbed during the sample-loading procedures.
Samples and flux	Carefully filed and cleaned 0.3-g specimens fluxed with 1-g 12-gauge platinum wire. Platinum flux and sample pressed together at 8000 psi in an Applied Research Laboratory (Sunland, Calif.) briquetting press, model No. 52, to form a bilayer pellet as shown in Fig. 19.5.
Supporting atmosphere	Helium, 680 torr
Arcing current	15 A
Arcing period	20 sec
Gas sampling and measurement	After the sample extraction is completed, an interval of 60 sec is allowed for homogeneous mixing of the gases prior to sampling a 5-cc aliquot for the gas-chromatographic analysis.
Total blank	Oxygen: 8–15 μg; nitrogen: 1–2 μg
Calibration	Analytical curves established on standard samples as "peak height observed" versus wt % oxygen (vacuum fusion values) or versus wt % nitrogen (chemical values).

TABLE 19.22. Analytical Procedures and Conditions for Determination of Oxygen in Lanthanide Metals by Fast-Neutron Activation

Type of sample	Lanthanide metals, excepting yttrium, cerium, praseodymium, and neodymium
Nuclear reaction	$^{16}O(n, p)^{16}N$
Neutron-beam energy	14-MeV beam produced by deuterons accelerated to 150 kV energies with maximum current of 2.5 A impinging on a tritiated titanium target (3–5 Ci/in.^{2}).
Irradiation time	21 sec; sample rotated during irradiation.
Counting	Samples are counted in a well in a 4-in. NaI(Tl) scintillation crystal and the flux monitored with a solid 3-in. NaI(Tl) crystal, each in conjunction with a single-channel analyzer. A multichannel pulse height analyzer is used to observe discriminator settings on the single-channel analyzer to minimize counting of interfering activities formed in the sample-matrix. Samples are counted for 30 sec; flux monitored for 45 sec. Monitor count used to normalize sample-counts to the same relative neutron dose.
Sample system	Flux monitor: water-filled, permanently sealed polyethylene capsule, as shown in Fig. 19.6. Sample: Metal specimen contained in polyethylene capsule within polyethylene rabbit shown in Fig. 19.6. At the sample-counting station, the polyethylene rabbit is stopped in the receiver 2 ft from the NaI(Tl) crystal, and the encapsulated sample is blown from the rabbit into the crystal well.

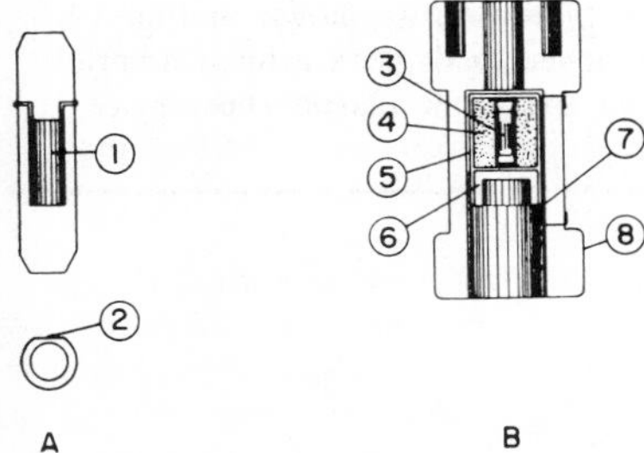

Fig. 19.6. (A) Monitor Rabbit. 1, water; 2, one side of the rabbit is flat to prevent it from spinning during transfer. (B) Sample Rabbit. 3, polyethylene encapsulated metal oxide sample; 4, metal sample; 5, polyethylene capsule; 6, polyethylene cap; 7, bronze leaf spring; 8, polyethylene rabbit.

b. ^{3}*He-Induced Reactions*

The ^{3}He-activation technique is most commonly employed for the determination of oxygen in the fissionable actinide elements. The operational functions of the technique are discussed in detail by Demildt [58], and the experimental parameters condensed in Table 19.23 were taken from his studies on thorium, plutonium, and americium. The platinum-seal mount for the monitor and sample is shown in Fig. 19.7.

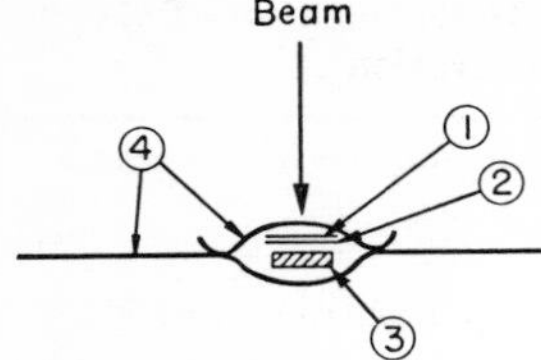

Fig. 19.7. Platinum seal. (1) Anodized Al beam monitor of 1.59 mm, 0.0625-in. diameter. (2) Plain Al absorber (same size as monitor). (3) Thick-target sample (different configuration and smaller than monitor). (4) Platinum seal.

TABLE 19.23. Analytical Procedures and Conditions for Determination of Oxygen in Actinide Metals by ^{3}He-Induced Nuclear Reactions

Type of sample	Fissionable actinides. Specifically thorium, plutonium, americium.
Nuclear reactions	$^{16}O(^3He, p)^{18}F$, and $^{16}O(^3He, n)^{18}Ne$
Hilac ^{3}He-beam energy	11.5 MeV collimated beam, degraded with Al foil. (Restricts reactions with elements of $Z > 36$, which excludes charged-particle fission in the actinide elements).
Target holder	Water- or helium-cooled to avoid melting or sublimation of the target material.
Irradiation time	110 min, to achieve half-saturation.
Counting	B^+ measurement with end-window gas-flow proportional counter, or γ detection of annihilation radiation with NaI detector connected to multichannel spectrometer.
Sample system	Thick-target samples of several milligrams weight, positioned with anodized Al beam monitor and plain Al absorber, all encapsulated with platinum, as shown in Fig. 19.7. (Platinum capsule removed in dry-box after irradiation; activated sample then sealed in plastic sheet prior to counting).

C. NITROGEN

1. Chemical Methods

The chemical dissolution–distillation techniques (often referred to as modified Kjeldahl procedures) are most frequently considered to yield the most reliable and accurate nitrogen results. A modified Kjeldahl procedure, which is satisfactory for the determination of nitrogen in the lanthanide elements and thorium, is described in Table 19.24. For very low nitrogen contents a colorimetric determination on the ammonia distillate [26] may be used in preference to the titrimetric measurement described.

TABLE 19.24. Reagents and Procedures for Determination of Nitrogen in Lanthanide and Actinide Metals by a Modified Kjeldahl Technique

Sample dissolution	Lanthanide metals: 1-g sample dissolved in hydrochloric or perchloric acid. Thorium: metal cleaned in warm solution of dilute (8*N*) nitric acid containing small amount of sodium fluosilicate, then rinsed under deionized water and dried. 5-g sample dissolved in 50 ml dilute (6*N*) hydrochloric acid containing 6 mg sodium fluosilicate; with heating if necessary. 20-ml aliquot taken from solution diluted to 100-ml volume.
Alkaline liberator	Sodium hydroxide solution: 1-lb pellets per liter deionized water.
Titrimetric indicator	Boric acid–methyl purple: 20-g boric acid crystals and 20-ml methyl purple indicator per liter deionized water. (Boric acid is sufficiently weak that the ammonium borate will titrate exactly as free ammonia with hydrochloric acid.)
Titrating acid	Hydrochloric acid, 0.01*N*. 0.83 ml concentrated HCl diluted to 1-liter with deionized water.
Acid standardization	Standardize against tris(hydroxymethyl)aminomethane [88], previously dried for 2-h at 100–103°C. 0.05 g of this standard in 50 ml deionized water is titrated to a pH of 4.70 with the (0.01*N*) HCl solution.
Procedure	The modified Kjeldahl apparatus is shown in Fig. 19.8. The apparatus is "conditioned" by collecting 100–150 ml redistilled water into the receiving flask from the steam generator. 25-ml of indicator solution is divided between vapor flask and receiving flask. The dissolved sample is placed in the distilling flask, to which 50-ml of the sodium hydroxide solution is added. About 50 ml distillate is collected in the receiving flask. The indicator solution in the vapor trap is drawn into the receiving flask when the boiling flask is cooled. The vapor and receiving flasks are rinsed with redistilled water, and the ammonia in the solution is titrated with the standardized (0.01*N*) hydrochloric acid to the grey endpoint. Blanks are determined in a similar fashion and should require 0.02 ml or less of the standard acid.

2. Vacuum Fusion

The criteria previously discussed for quantitative extraction of oxygen by the high-temperature fusion techniques may also be considered to apply to nitrogen, although the environmental conditions leading to complete extraction may not always be the same for both impurities in a given sample. The nitrogen results obtained by high-temperature fusion techniques, especially those obtained by vacuum-fusion procedures, have been subjected to critical scrutiny and are not as widely accepted as oxygen determinations.

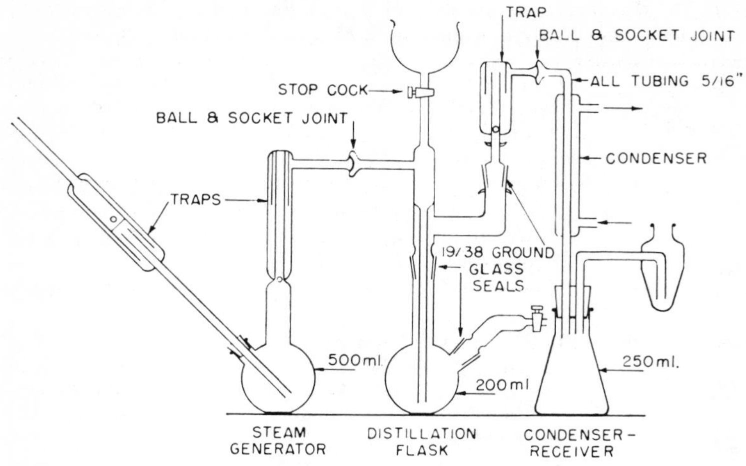

Fig. 19.8. Semi-micro Kjeldahl apparatus.

The vacuum-fusion operating conditions and procedures described in Table 19.17 appear to afford a reasonable degree of reliability for the determination of nitrogen in the lanthanide and actinide metals.

3. Carrier Gas Fusion

The criteria required for quantitative extraction and recovery of nitrogen for the lanthanide and actinide metals appear to be fulfilled to a maximal degree with the carrier-gas-fusion procedures and reaction conditions outlined in Table 19.19. Any carrier-gas-fusion facility that provides these conditions should be applicable to the determination of nitrogen in these metals.

4. DC Carbon Arc

The nitrogen contents of yttrium, terbium, thorium, and possibly of other lanthanide and actinide metals may be determined simultaneously with the oxygen content under the extraction conditions discussed in Table 19.21.

REFERENCES

1. C. L. Huffine and J. M. Williams, in F. H. Spedding and A. H. Daane, Eds., *The Rare Earths*, Wiley, New York, 1961, Chapter 11.
2. L. Grainger, *Uranium and Thorium*, Newnes, London, 1955.
3. R. M. Goldhoff, H. R. Ogden, and R. I. Jaffee, *U.S. At. Energy Comm. Rept.*, *BMI-776*, Battelle Memorial Institute, Columbus, Ohio, 1952.
4. H. W. Deem and R. A. Winn, *U.S. At. Energy Comm. Rept.*, *BMI-1052*, Battelle Memorial Institute, Columbus, Ohio, 1955.
5. C. D. Harrington and A. E. Ruehle, *Uranium Production Technology*, Van Nostrand, Princeton, N.J., 1959, Chapter 10.
6. H. A. Sloman, C. A. Harvey, and O. Kubaschewski, *J. Institute Metals*, **80,** 391 (1951–1952).
7. R. C. Tucker, E. D. Gibson, and O. N. Carlson, *Nucl. Metall. Ser.*, **10,** 315 (1964).
8. K. A. Gschneidner and J. T. Waber, *J. Less-Common Metals*, **6,** 354 (1964).
9. H. A. Wilhelm and P. Chiotti, *Trans. ASM*, **42,** 1295 (1950).
10. O. N. Carlson, D. W. Bare, E. D. Gibson, and F. A. Schmidt, Symposium on Newer Metals, *Spec. Tech. Publ. 272*, American Society for Testing and Materials, Philadelphia, 1959, p. 144.
11. C. R. Simmons, in F. H. Spedding and A. H. Daane, Eds., *The Rare Earths*, Wiley, New York, 1961, Chapter 18.
12. K. M. Bohlander, in F. H. Spedding and A. H. Daane, Eds., *The Rare Earths*, Wiley, New York, 1961, Chapter 12.
13. F. L. Cuthbert, *Thorium Production Technology*, Addison-Wesley, Reading, Mass., 1958.
14. D. T. Peterson, R. F. Russi, and R. L. Mickelson, Reprint No. 102, Nucl. Eng. and Sci. Congress, Cleveland, Ohio, 1955.
15. J. A. Milko, R. E. Adams, and W. O. Harms, in H. A. Wilhelm, Ed., *The Metal Thorium*, American Society for Metals, Cleveland, Ohio, 1958, Chapter 13.
16. M. D. Smith and R. W. K. Honeycombe, *J. Nucl. Mat.*, **1,** 345 (1959).
17. L. L. Marsh, G. T. Muehlenkemp, and G. K. Manning, *U.S. At. Energy Comm. Rept.*, *BMI-980*, Battelle Memorial Institute, Columbus, Ohio, 1955.
18. R. Eborall, *Spec. Rept. 68*, Iron and Steel Institute, London, 1960, 192.
19. J. T. Waber and E. S. Wright, *U.S. At. Energy Comm. Rept.*, *AECD-4280*, University of California, Los Alamos Sci. Lab., Los Alamos, N.M., 1955.
20. L. A. Fergason, D. E. Seizinger, and C. H. McBride, *U.S. At. Energy Comm. Rept.*, *MCW-1462*, Mallinckrodt Chem. Works, Weldon Spring, Mo., 1961.
21. Y. Morimoto and A. Hirayama, *Jap. Analyst*, **11,** 163 (1962).
22. U.K. Atomic Energy Authority, *PG-Report-412*, Springfields, Lancs, England, 1962.
23. A. Parker, *Spec. Rept. 68*, Iron and Steel Institute, London, 1960, p. 64.
24. W. E. Dallmann, "Comparative Studies on the Determination of Oxygen, Nitrogen, and Hydrogen in Some Rare Earth Metals," MS Thesis, Iowa State University, Ames, Iowa, 1964.

25. W. E. Dallmann, V. A. Fassel, C. C. Hill, and R. K. Winge, "Comparative Studies on the Determination of Oxygen, Nitrogen, and Hydrogen in Rare Earth Metals, and of Oxygen in Thorium," presented at Eighth Conf. on Anal. Chem. in Nuclear Tech., Gatlinburg, Tenn., Oct. 1964.
26. J. D. Hobson, *Spec. Rept. 68*, Iron and Steel Institute, London, 1960, p. 151.
27. Z. M. Turovtseva and L. L. Kunin, *Analysis of Gases in Metals*, Consultants Bureau Enterprises, New York, 1961.
28. R. T. vanSanten, private communication, Wah Chang Corp., Albany, Oreg., May 1967.
29. S. Kallmann, E. W. Hobart, H. K. Oberthin, and W. C. Brienza, *Anal. Chem.*, **40,** 332 (1968).
30. W. T. Elwell and D. F. Wood, *Analysis of the New Metals*, Pergamon, London, 1966, p. 76.
31. R. A. Yeaton, *Vacuum*, **2,** 115 (1952).
32. L. Singer, *Ind. Eng. Chem., Anal. Ed.*, **12,** 127 (1940).
33. W. G. Smiley, *Anal. Chem.*, **27,** 1098 (1955).
34. G. W. Goward, *Anal. Chem.*, **37,** 117R (1965).
35. W. E. Dallmann and V. A. Fassel, *Anal. Chem.*, **39,** 133R (1967).
36. J. N. Gregory, D. Mapper, and J. A. Woodward, *Analyst*, **78,** 414 (1953).
37. D. T. Peterson and D. J. Beerntsen, *Anal. Chem.*, **29,** 254 (1957).
38. M. W. Mallett, "Determination of Gases in Plutonium Nitride and Carbide and in Plutonium Metal," presented at Eighth Conf. on Anal. Chem. in Nuclear Tech., Gatlinburg, Tenn., Oct. 1964.
39. C. B. Griffith, W. M. Albrecht, and M. W. Mallett, *U.S. At. Energy Comm. Rept., BMI-1033*, Battelle Memorial Institute, Columbus, Ohio, 1955.
40. R. E. Taylor, *Anal. Chim. Acta*, **21,** 549 (1959).
41. Z. M. Turovtseva and N. F. Litvinova, *Proceedings of the Second U.N. Internat. Conf. on the Peaceful Uses of At. Energy*, **28,** 593 (1958).
42. E. Booth, F. J. Bryant, and A. Parker, *Analyst*, **82,** 50 (1957).
43. G. G. Glavin and Y. A. Karpov, *Ind. Lab*, **30,** 388 (1964).
44. U.K. Atomic Energy Authority, *PG-Report-411*, Springfields, Lancs, England, 1963.
45. V. A. Fassel, W. E. Dallmann, and C. C. Hill, *Anal. Chem.*, **38,** 421 (1966).
46. W. G. Smiley, *Spec. Tech. Publ. 222*, American Society for Testing and Materials, Philadelphia, 1958, p. 25.
47. B. D. Holt and H. T. Goodspeed, *Anal. Chem.*, **35,** 1510 (1963).
48. B. D. Holt and H. T. Goodspeed, "Manometric Measurement of Nitrogen, Oxygen, and Hydrogen Extracted from Plutonium Materials by Inert Gas Fusion," presented at Eighth Conf. on Anal. Chem. in Nuclear Tech., Gatlinburg, Tenn., Oct. 1964.
49. C. V. Banks, J. W. O'Laughlin, and G. J. Kamin, *Anal. Chem.*, **32,** 1613 (1960).
50. V. A. Fassel, W. A. Gordon, and R. W. Tabeling, *Spec. Tech. Publ. 222*, American Society for Testing and Materials, Philadelphia, 1958, p. 41.
51. R. K. Winge and V. A. Fassel, *Anal. Chem.*, **37,** 67 (1965).
52. V. A. Fassel and J. W. Goetzinger, *Spectrochim. Acta*, **21,** 289 (1965).

53. M. Chaput and J. N. Savarit, *Anal. Chim. Acta*, **31,** 563 (1964).
54. R. F. Coleman and J. L. Perkin, *Analyst*, **84,** 233 (1959).
55. R. G. Clark and W. A. Stensland, *U.S. At. Energy Comm. Rept.*, *IS-1064*, Iowa State University, Ames, Iowa, 1964.
56. R. F. Coleman, *Analyst*, **87,** 590 (1962).
57. S. S. Markowitz and J. D. Mahony, *Anal. Chem.*, **34,** 329 (1962).
58. A. C. Demildt, *Anal. Chem.*, **35,** 1228 (1963).
59. A. H. Daane, in F. H. Spedding and A. H. Daane, Eds., *The Rare Earths*, Wiley, New York, 1961, Chapter 8.
60. O. N. Carlson and F. A. Schmidt, in F. H. Spedding and A. H. Daane, Eds., *The Rare Earths*, Wiley, New York, 1961, Chapter 9.
61. H. A. Wilhelm, *Proceedings of the Internat. Conf. on the Peaceful Uses of At. Energy*, **8,** 162 (1955).
62. H. A. Wilhelm, in H. A. Wilhelm, Ed., *The Metal Thorium*, American Society for Metals, Cleveland, Ohio, 1958, Chapter 6.
63. D. T. Peterson, W. E. Krupp, and F. A. Schmidt, *J. Less-Common Metals*, **7,** 288 (1964).
64. N. D. Veigel, E. M. Sherwood, and I. E. Campbell, in H. A. Wilhelm, Ed., *The Metal Thorium*, American Society for Metals, Cleveland, Ohio, 1958, Chapter 7.
65. R. A. Noland, in H. A. Wilhelm, Ed., *The Metal Thorium*, American Society for Metals, Cleveland, Ohio, 1958, Chapter 9.
66. O. N. Carlson, F. A. Schmidt, and D. T. Peterson, *J. Less-Common Metals*, **10,** 1 (1966).
67. D. T. Peterson, F. A. Schmidt, and J. D. Verhoeven, *Trans. Metall. Soc. AIME*, **236,** 1311 (1966).
68. A. H. Daane, in F. H. Spedding and A. H. Daane, Eds., *The Rare Earths*, Wiley, New York, 1961, Chapter 13.
69. F. H. Spedding, J. J. Hanak, and A. H. Daane, *Trans. Metall. Soc. AIME*, **212,** 379 (1958).
70. H. T. Goodspeed, B. H. Holt, J. H. Marsh, and J. E. Stoessel, *Anal. Chem.*, **40,** 682 (1968).
71. E. Lassner and E. Wolfel, *Mikrochem. Acta*, **No. 3,** 394 (1960).
72. W. H. Smith, *Anal. Chem.*, **27,** 1636 (1955).
73. M. W. Mallett, *Talanta*, **9,** 133 (1962).
74. J. K. Stanley, J. von Hoene, and G. Weiner, *Anal. Chem.*, **23,** 377 (1951).
75. W. G. Guldner, *Talanta*, **8,** 191 (1960).
76. M. R. Everett and G. E. Thompson, *Analyst*, **87,** 515 (1962).
77. W. M. Albrecht and M. W. Mallett, *Anal. Chem.*, **26,** 401 (1954).
78. C. E. Habermann and A. H. Daane, *U.S. At. Energy Comm. Rept.*, *IS-789*, Iowa State University, Ames, Iowa, 1964.
79. F. H. Spedding, A. H. Daane, G. Wakefield, and D. H. Dennison, *Trans. Metall. Soc. AIME*, **218,** 608 (1960).
80. F. H. Spedding, R. J. Barton, and A. H. Daane, *J. Am. Chem. Soc.*, **79,** 5160 (1957).
81. R. J. Ackermann and E. G. Rauh, *J. Chem. Phys.*, **36,** No. 2, 448 (1962).

82. F. H. Spedding and A. H. Daane, *J. Metals*, **6,** 504 (1954).
83. F. H. Spedding and A. H. Daane, *Met. Rev.*, **5,** 297 (1960).
84. V. A. Fassel, W. E. Dallmann, R. Skogerboe, and V. M. Horrigan, *Anal. Chem.*, **34,** 1364 (1962).
85. W. G. Guldner and A. L. Beach, *Anal. Chem.*, **22,** 366 (1950).
86. W. E. Dallmann and V. A. Fassel, *Anal. Chem.*, **38,** 662 (1966).
87. F. M. Evens and V. A. Fassel, *Anal. Chem.*, **35,** 1444 (1963).
88. J. H. Fossum, P. C. Markunas, and J. A. Riddich, *Anal. Chem.*, **23,** 491 (1951).

CHAPTER

20

NUCLEAR AND ION BEAM TECHNIQUES FOR SURFACE AND NEAR-SURFACE ANALYSES

ELIGIUS A. WOLICKI

Nuclear Sciences Division
Naval Research Laboratory
Washington, D.C.

CONTENTS

I. INTRODUCTION

In the past few years the field of surface analysis has seen the development of a surprising number of promising new physical methods based on the use of high-energy ion beams. Improvements and advances continue to be reported at a rapid pace at the time of this writing, and the field can well be said to be scientifically most interesting and exciting. In addition to the extensive literature that is developing in this area, the reader may find Refs. 1–7 useful in that they emphasize areas of current research interest.

Ion beams will be taken here to be charged atomic species such as protons, deuterons, ^{3}He, α-particles and others, and the energies of interest, usually will be in the range from 1 million electron volts (MeV) upward. These particles are, for the most part, useful only for surface analyses, for the reason that their penetration ranges in solid matter are extremely short. At an energy of 2.0 MeV, for example, proton ranges in metals are typically about 50 μm and α-particle ranges are about 6 μm.

The very powerful and important photon- and electron-beam techniques for studying surfaces such as Auger electron spectroscopy, low-energy electron diffraction (LEED), and electron spectroscopy for chemical analysis (ESCA) will not be discussed, since they are outside the author's purview. Use of these techniques is widespread and growing, and commercial apparatus is available. Similar statements apply to the somewhat more recently developed technique of ion-beam mass spectrometry. In this technique a heavy ion beam with approximately 10 keV energy is used to sputter away the sample surface, the sputtered ions being mass-analyzed to provide not only a chemical analysis of the surface but also a depth profile for a selected element. Again, the use of this technique is growing rapidly, and the requisite equipment is available commercially.

While a few of the new techniques are close to routine application and others are clearly advancing in that direction, there are a number that are still being investigated and developed in the laboratory. In order that the text may remain germane to this changing area for a longer period than would otherwise be the case and be at the same time useful to the reader for anticipating future trends, the discussion will cover not only those techniques presently ready for application but others as well. The objective of the discussion will be to give the reader an overview of the present state of the art and, hopefully, sufficient information so that he can assess for his own problems whether a given technique may have advantage over more standard techniques.

Analysis for constituent chemical elements with ion beams from an accelerator consists of irradiating the sample with the ion beam and detecting the resulting characteristic radiations. The irradiations are usually performed in vacuum, and the ion beam is generally magnetically analyzed so that it consists of a single atomic species and has a well-defined energy. Emitted radiations may be detected with Geiger-Mueller counters, proportional counters, or scintillation or solid-state detectors, the pulses from these latter three detectors being analyzed typically with a multichannel pulse-height analyzer. The reactions that occur between the incident ions and the atoms of the sample (target) depend on the energy of the ions and the atomic species of both the beam and the sample. An incident ion beam, a sample, and some of the typical emitted radiations are shown schematically in Fig. 20.1.

The type of radiation and spectrum of pulse heights produced will be characteristic of the reactions induced by the incident beam in the constituent

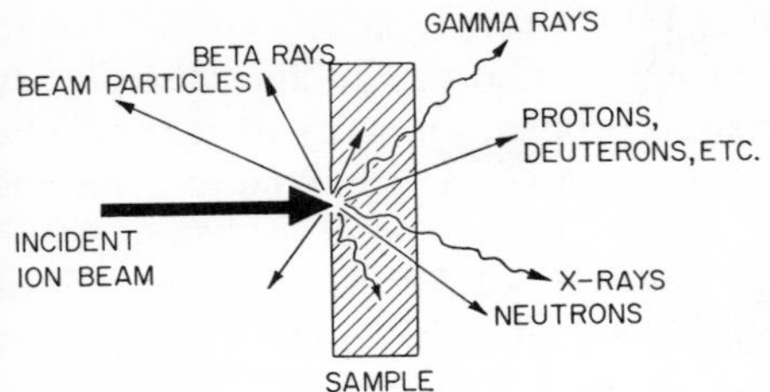

Fig. 20.1. A schematic diagram showing the interaction of an ion beam with a target sample and the various radiations that can be produced.

target elements. If the energy of the incident ion is great enough to overcome the electrostatic potential energy of repulsion, which builds up as the ion approaches a target atom nucleus, then a nuclear reaction can occur. In such a case, the emitted radiation will be nuclear in origin and will typically be a γ-ray, a charged nuclear particle that can be the same as or different from the incident ion, a neutron, or a β-particle. Because nuclear-reaction characteristics can differ markedly between two isotopes of the same chemical element, the reaction products are specific not only to the element but to a particular isotope of that element. This property, which is commonly utilized in stable and radioisotope tracer work, makes possible some important applications also in materials analysis. Nuclear reactions have the disadvantage, however, that with only a few exceptions that do not pertain to this discussion, they are not affected by chemical forces and so do not give information directly about the chemical compound form of the elements in a sample.

Radiations can be emitted from an irradiated sample also as a result of nonnuclear reactions. The two principal examples, for purposes of the present discussion, are heavy-ion-induced characteristic X-rays and Rutherford or elastic scattering. The energies used for X-ray analysis with such beam particles as protons and α-particles are most commonly in the MeV range. Rutherford scattering has been used for analysis with beam energies ranging from a few tenths of kiloelectron volts to megaelectron volts. Both of these techniques are of particular interest, because they can often be used to analyze for many elements in the sample with a single irradiation.

The fact that many of the techniques that will be listed here require highly sophisticated and expensive equipment in the form of small nuclear accelerators, radiation detectors, and pulse-height analyzing systems has not been overlooked. Neither is there any intent to underestimate the practical difficulties attendant to the wide use of these techniques, because of the high equipment costs involved. The view is taken rather that there may exist problems that cannot be satisfactorily solved in any other way and that are sufficiently important that technique costs will not be the dominant factor. In this connection it may be useful to call attention to the existence of a large number of nuclear-accelerator facilities [8], primarily at universities but in industry also; many of these are currently interested in interdisciplinary problems in general and materials analysis problems in particular. It may be possible, therefore, to test the application of a given technique to a specific analysis problem when a need arises.

The relative advantages of nuclear and ion-beam techniques, expressed for the cases where they can be applied, are that they can have excellent specificity, high sensitivity, and good absolute accuracy. The last named advantage, which is especially important in trace-element work, results from the fact that often little or no sample-processing is involved and the chances for introducing extraneous elements are thereby reduced. For metals in particular, but for many other materials as well, the irradiations are in addition nondestructive, so that elemental analysis on a particular sample does not preclude any other subsequent tests or measurements that may be desired.

The principal disadvantages are the high cost involved, as has previously been mentioned, and a low precision as compared to most standard analytical techniques. The disadvantage of high cost requires qualification. Sample analysis times typically are short, ranging from 1 to approximately 10 min. The techniques furthermore are such that they can often be made highly automatic. In situations where large numbers of similar samples are being routinely analyzed, the cost per sample may therefore compare favorably with other methods. If, in addition, multielement analysis is needed for every sample and can be provided by a nuclear or ion-beam technique, the cost per element per sample may turn out to be lower and on occasion even so much lower that one of these techniques may become the technique of choice because it is the least costly.

The low precision is caused by statistical errors associated with counting random events and by calibration problems. Calibrations, achieved most commonly through comparison standards, require special attention because the measurement result depends not only on the quantity of an element present but also on the depth distribution of that element in the sample-surface. The dependence on depth is caused by the facts that reaction cross sections depend on energy and that the incident ion loses energy as it penetrates into the sample. Precisions range commonly from 1% to approximately 10%. Precisions as good as 0.1% have been reported [9, 10] (for neutron-activation analysis), but these are achieved only with the utmost care. Chapter 15 of Ref. 7 deals specifically with accuracy, precision, and standards.

It is worth mentioning that ion beams at kiloelectron-volt energies can be used to produce accurate trace-element standards, including the noble gases, through ion implantation [11]. The number of implanted beam atoms is measured easily to a few percent or better through the integrated electrical charge that is deposited into the target. The sensitivity of the electrical measurement is such that amounts that are orders of magnitude below the sensitivity of the best chemical and physical analysis methods can be deposited quite accurately. The integrated charge can of course be controlled to produce whatever concentration is desired. The method is best suited to trace-element standards but can, with appropriately long irradiations, be used also to achieve concentrations of a fraction of a percent.

It is convenient to divide the ion-beam-analysis techniques into two categories: activation analysis and prompt-radiation analysis. Activation analysis with a charged-particle beam is analogous to neutron-activation analysis. The elements of interest in a sample are converted into radioactive species through nuclear reactions, and the characteristic radiations are measured after the irradiation. The half-life for decay frequently is used for identifying the isotope that is responsible for the radiation. *Prompt-radiation analysis* is the term used when the instantaneously emitted radiations are detected; the radiations in this case can be caused by either nuclear or nonnuclear reactions.

Charged-particle-activation analysis is covered in detail in Chapter 3; it is included here for completeness but will be discussed only briefly. The major emphasis will be on prompt-radiation analysis, a subject area that has not been treated earlier in this text and that is at the same time the one in which the most striking advances have recently been forthcoming.

II. CHARGED-PARTICLE-ACTIVATION ANALYSIS

In this method a stable element present in a sample is detected through the radioactivity produced as a result of nuclear reactions between that element and the irradiating beam. After irradiation, the sample containing the activated elements is placed in a low-background detector and the emitted radiations, usually γ-rays or β-rays, are measured. The sensitivity of the measurement is such that it is most commonly used for measuring elements present only in trace quantities. The trace elements to be measured may be present in the sample as extraneous impurities or they may be purposely introduced. The latter situation occurs, for example, when stable isotopes are used as tracers for studying some dynamic characteristic. Stable isotope tracers have the advantage that they can be used in many cases where the use of radioactive materials prior to analysis is precluded.

Although charged-particle-activation analysis is most commonly considered for trace-element analysis, its possibilities for minor and even major constituent analysis should not be overlooked. This comment applies particularly well to surface analyses, since the concentration of the detected element may be at the major or minor constituent level in the surface even though it is at the trace level in the bulk of the sample.

An excellent discussion and a list of references for charged-particle-activation analysis has been given by Tilbury [3]. Much interest has centered on ^{3}He particles, since their advantages for activating the light elements were pointed out by Markovitz and Mahony [12]. The proceedings from two series of conferences [1, 2, 6, 7] should be of particular interest because they contain descriptions of large numbers of specific examples. Other useful reports on charged-particle-activation analysis have appeared [12–24].

The charged bombarding particles of greatest interest for measuring gases in metals are protons, deuterons, and ^{3}He particles. Hydrogen and helium cannot practically be observed with activation analysis, although in principle, lithium-ion and other light-element beams can produce radioactivities upon interaction with the nuclei of these two elements. Very little charged-particle-activation analysis work has been reported for the noble gases Ne, Ar, Kr, and Xe [25, 26]. If needed, however, charged-particle-activation analyses could very probably be developed for these gases. The remaining gaseous elements, N, O, F, and Cl, have all been determined by charged-particle-activation analysis.

The problem of interferences from elements other than those whose determination is desired must be carefully considered for charged-particle activation analysis. That surface analyses are especially susceptible to contamination from the ambient atmosphere is well known.

Although the most efficient approach to a specific problem will usually be to consult the literature, the following general comments may be helpful. The magnitude of the cross section for a particular desired nuclear reaction will often be different from that of possible interfering reactions and in general will have a strongly different dependence on the energy of the incident ion. Both of these factors may be useful for minimizing or evaluating the contributions from interfering elements. Thus, for example, because of Coulomb-barrier effects, as the bombarding energy is lowered, the interference from higher-Z (atomic number) elements typically is reduced relative to lower-Z elements. An even simpler situation results if one of the interfering reactions in question has a negative Q-value, or energy release. Such a reaction will not proceed at all until the incident beam energy exceeds an easily calculable threshold energy; it is possible, then, by irradiation at an energy below this value, to eliminate the interference from this reaction completely. A changing from one bombarding particle species to another will usually also result in different relative interferences such that an evaluation of the various interfering contributions may be made possible. The above comments are directed particularly at those situations where no chemical separations are performed on the sample, since separations are to be avoided whenever possible. In practice, interfering radioactivities are often also removed by chemical separations.

The number and variety of samples that have been studied and reported upon in the literature are such that it may be possible for the reader to find a discussion of problems that are similar, if not identical, to his own. A comprehensive coverage and extensive cross-indexing by subject [26] should prove particularly valuable for this purpose. A few examples of applications are given.

The application of ^{3}He-induced nuclear reactions to activation analysis for oxygen and other elements has been discussed in a short review article by Markowitz and Mahony [12]. Analyses for oxygen were performed in beryllium, thorium and Mylar samples through detection of the radioactive ^{18}F isotope produced by the $^{16}O(^{3}He, p)^{18}F$ reaction; ^{18}F is a positron emitter with a 110-min half-life. For a 1-μA beam of 7.5-MeV ^{3}He ions incident on a 0.0005-in. Al foil, an ultimate sensitivity of 1 ppb is estimated; for higher-intensity beams and water-cooled target samples (to prevent melting), sensitivities of fractions of a part per billion are indicated. Actual analyses were performed with a 31-MeV ^{3}He beam from the Lawrence Radiation Laboratory Hilac accelerator. Procedures for determining the overall detection efficiency are described. Precision and accuracy are estimated to be 5% and 10%, respectively. Suggestions are made for evaluating possible interferences from sodium and fluorine as a result of the $^{23}Na(^{3}He, 2\alpha)^{18}F$ and $^{19}F(^{3}He, \alpha)^{18}F$ reactions. For determination of oxygen in thorium, thin foils of thorium (0.001 in.) were mechanically cleaned and washed in a nitrogen atmosphere, sealed in a 0.001-in. polystyrene envelope so that oxygen was excluded from the sample's surroundings, and irradiated and counted while in this envelope. The article also lists all of the common nuclear reactions of ^{3}He ions with nuclides from ^{6}Li through ^{48}Ca along with product half-lives and reaction Q-values.

The variation of the reaction cross section with energy, the same factor that can be useful in eliminating or evaluating interferences, complicates the problem of measuring the quantity of an element present in a sample. The reason for this is that the incident beam ion loses energy as it penetrates into the sample, and the magnitude of the cross section is therefore different at different positions along the penetration path. Some nuclear reactions exhibit such sharp "resonances"—that is, increased cross sections at specific incident energies—that the reaction cross section can change by an order of magnitude if the beam ion energy changes by only a few hundred parts per million. Such reactions are important when prompt nuclear-radiation measurements are made but are not commonly used for charged-particle-activation analysis; these reactions will therefore not be treated here but will be discussed instead in the prompt-radiation-analysis section.

For nonresonant nuclear reactions the cross section changes relatively more slowly with energy. In this case, therefore, when the target thickness is small compared to the penetration range of the incident ion, the reaction yield Y may be written as

$$Y = nN\,\Delta x\sigma(E), \tag{1}$$

where Y is the number of reactions produced, n is the number of incident beam particles, N is the number of reactant target atoms per cubic centimeter, Δx is the target thickness in centimeters, E is the incident particle energy, and $\sigma(E)$ is the reaction cross section in square centimeters at energy E. If the target is thick the formula becomes

$$Y = nN \int_0^E \frac{\sigma(E')\,dE'}{(dE'/dx)(E')} \tag{2}$$

where E' is the variable energy of the incident particle as it penetrates into the target and $(dE'/dx)(E')$ is the stopping power dE'/dx as a function of E'. For simplicity, N is assumed to be uniform in the sample; if it is not, its depth distribution must also be taken into account under the integral sign in Eq. 2. Normally, the dependences of the cross section $\sigma(E')$ and the stopping power $(dE'/dx)(E')$ on E' are obtained from experimental data or calculations [27] and are used to evaluate the integral over $\sigma(E')$.

Ricci and Hahn [15, 16] have greatly simplified this calculation by introducing a thick-target average cross section so defined that once this cross section is evaluated, it is possible to use range energy data without integration to calculate the reaction yield. The two articles cited also contain discussions of measurement calibrations and other experimental procedures as well as extensive lists of sensitivities for activation analysis of elements from Be through Ca with 18-MeV ^{3}He particles. The sensitivities for nitrogen and oxygen are among the highest. On a relative scale the sensitivities listed for nitrogen, oxygen, fluorine, and chlorine are, respectively, 725, 704, 114, and 11. The corresponding reactions and product half-lives are ^{14}N(^{3}He, d)^{15}O, 2.07 min; ^{16}O(^{3}He, p)^{18}F, 110 min; ^{19}F(^{3}He, α)^{18}F, 110 min; and ^{35}Cl(^{3}He, α)^{34m}Cl, 32.4 min. The notation ^{34m}Cl signifies that the product radioactive nucleus has been left in a metastable state. The radiation from this state, which decays with a 32.4-min half-life, is more convenient to measure than that from the ground state of ^{34}Cl, which has a half-life of 1.5 sec.

Deuteron-activation analysis has been used for determination of C, N, and O contamination of high-purity metallic surfaces by Benjamin et al. [17]. The energy of the deuterons used was 2.6 MeV, and the metals analyzed were beryllium, molybdenum, and tungsten. A particularly interesting technique was that in which a vacuum chamber for the irradiations was separated from the Van de Graaff accelerator vacuum system by a 0.0001-in.-thick molybdenum foil in order to avoid carbon contamination. The reactions used and the corresponding product half-lives were ^{12}C(d, n)^{13}N, 9.96 min;

$^{14}N(d, n)^{15}O$, 2.07 min; and $^{16}O(d, n)^{17}F$, 1.1 min. The authors found that the relatively long ^{13}N half-life makes the carbon analysis insensitive to the presence of oxygen and nitrogen. The sensitivities for nitrogen and oxygen, however, depend on the concentrations of carbon and oxygen, and carbon and nitrogen, respectively. Achievable detection limits were estimated to be 1 ppm for carbon, 2 ppm for oxygen, and 5–10 ppm for nitrogen.

Debrun et al. [18] have investigated the determination of carbon, oxygen, and sulfur in high-purity iron, nickel, and chromium samples by irradiation with ^3He and ^4He particles. Interferences from nitrogen, sodium, magnesium, and fluorine are discussed. While interference from the first three elements can be satisfactorily limited by lowering bombarding energy, that from fluorine cannot be so eliminated. The ratios of fluorine to oxygen production-cross sections for radioactive ^{18}F were therefore studied as a function of energy for both ^3He and ^4He particles. In the case of ^3He irradiations, the ^{18}F contribution per atom from fluorine is shown to be less than 8% of that from oxygen. Reactions interfering with carbon and sulfur determinations were also investigated and data for a number of interference ratios are presented. Activation analyses by irradiation with deuterons and protons are also discussed.

Tritons can be used to analyze for oxygen in the presence of fluorine, because they do not induce any radioactivity that can interfere with the ^{18}F produced by the $^{16}O(t, n)^{18}F$ reaction. A compact discussion of the use of tritons for oxygen analysis in metal surfaces has been given by Barrandon and Albert [19]. Excitation functions in the energy range from 0.5 to 3.0 MeV are presented. The investigations included a study of the variation of surface oxygen on pure zirconium and aluminum as a function of chemical and mechanical polishing. The method can be used, with no chemical separations being necessary, for aluminum, iron, chromium, nickel, and zirconium. Sensitivities lie in the range from 100 to 10^{-3} $\mu g/cm^2$.

Activation analyses for oxygen and carbon on the surfaces of highly purified gold and platinum have been reported by Butler and Wolicki [22]. Of particular interest are the tests made to determine whether contaminants were introduced onto the target surfaces from the accelerator vacuum system. These tests showed that carbon contamination could be kept below 10^{-2} atomic layers for the duration of an irradiation by surrounding the target as completely as possible with trapping surfaces at liquid-nitrogen temperature. One atomic layer is defined simply as $N^{2/3}$, where N is the number of host material atoms per cubic centimeter. Ultimate sensitivities for oxygen and carbon are estimated to be 10^{-3} and 10^{-4} atomic layers respectively. The reactions used were $^{16}O(^3He, p)^{18}F$ and $^{12}C(^3He, \alpha)^{11}C$, respectively.

The $^{37}Cl(d, p)^{38}Cl$ reaction has been used by Knudson and Dunning [24] to detect trace amounts of chlorine on aluminum. The sample was bombarded with 5.0-MeV deuterons and the resulting γ-ray activity was measured with a 5-in.-diameter, 5-in.-thick well-type NaI(Tl) scintillation detector. Carbon, nitrogen, and oxygen were not found to present an interference to the observation of ^{38}Cl. Measurements on 125-μm-thick commercial aluminum foil indicated considerable interference with the chlorine γ-ray spectrum as observed with a NaI(Tl) detector. Later measurements with a high-resolution Ge(Li) solid-state detector showed that the interference was caused by the presence of manganese and gallium in the commercial foils. Radioactive ^{28}Al, produced by the ^{27}Al(d, p)^{28}Al reaction could interfere with the ^{38}Cl. Fortunately, however, because the 2.27-min half-life of ^{28}Al is much shorter than the 37.2-min half-life of ^{38}Cl, counting can be delayed until the ^{28}Al activity has decayed to a tolerable level. A sensitivity of 0.2 μg/cm^2 was estimated. A substantially higher sensitivity should be achievable for chlorine on high-purity high-Z metals.

An interesting application of charged-particle-activation analysis has been described by Holm et al. [28]. In that work, the microscopic distribution of oxygen and carbon in metals was observed by bombardment of metallographically polished samples with 6.5-MeV ^{3}He ion and autoradiography of the activated areas. Potitional resolution of 13 μm was obtained in samples having bulk carbon concentrations of about 250 ppm.

Dynamic processes can be studied through the use of stable isotope tracers and subsequent charged-particle-activation analysis. Ritter et al. [29] and, more recently, Carosella and Comas [30] have used enriched ^{18}O to study the oxidation of silicon single crystals. The amounts of ^{18}O present on the sample-surface after various exposures of the sample were measured through the ^{18}O(p, n)^{18}F reaction. The sensitivity of the method was such that oxide layers as thin as 0.01 atomic layer were measurable.

III. PROMPT-RADIATION ANALYSIS

In this method the presence of an element in a sample is detected through the radiation that is emitted instantaneously when the incident-beam ion interacts with an atom of that element. Detection limits can be quite good; typically they are not as good as can be achieved under ideal conditions with charged-particle-activation analysis. Chemical separations between the irradiation and radiation detection are precluded, although, of course, prompt-radiation analysis can be performed on chemically separated sample-constituents. Specificity and accuracy of prompt-radiation analysis can be high; precision suffers from the same statistical counting limitations and calibration difficulties that were mentioned earlier. Even though chemical separations are not possible, the problem of interferences is sometimes simpler to solve for the case of prompt-radiation analysis than for charged-particle-activation analysis, because of the greater diversity of reactions and reaction types that can be considered for a specific problem.

The reaction between the two interacting atoms can be either nonnuclear or nuclear. The two principal examples of nonnuclear reactions induced by positive ion beams are elastic scattering and atomic excitation or ionization. Strictly speaking, if the incident ion energy is greater than, or comparable to, the Coulomb-barrier energy for the interaction, then nuclear reactions can contribute to elastic scattering. For simplicity, however, elastic scattering will be considered here as nonnuclear in origin; effects due to nuclear interactions can always be taken into account through the use of the experimentally measured elastic-scattering cross sections for the atoms and energies in question.

In elastic scattering, the incident-beam ion is scattered from an atom in the sample by a collision in which energy and momentum are both conserved. For a fixed incident energy and scattering angle the energy of the scattered beam ion depends on the mass of the struck atom and therefore can be used to identify that mass. The characteristic radiations from excited or ionized atoms in solids that are most useful for analysis are the X-rays. In distinction to nuclear reactions, which typically allow analysis for a few elements at most in a single irradiation, elastic scattering and ion-induced X-rays can often be used to measure a large number of elements in one irradiation.

Nuclear reactions may be categorized according to three basic mechanisms, each of which may exhibit either resonant or nonresonant dependence on the incident particle energy. The first of these is the simple capture of the incident particle by the struck nucleus to form a compound nucleus, which then decays by the emission of a neutron, a charged particle, or one or more γ-rays. The second is the inelastic scattering of the incident particle, part of the incident energy being absorbed by the struck nucleus, which is then left in an excited state. The third mechanism is that of *rearrangement collisions*. This term is applied to all reactions in which the particles that separate instantaneously after a collision are different from those that entered into the collision. For many reactions, in all three of the above categories, the radiations emitted have discrete and well-defined energies that are characteristic of the reaction that has occurred between the incident and the struck nuclei. These radiations can be used therefore to identify the nuclear species and, therefrom, the chemical elements in a sample being irradiated.

An important advantage of prompt-radiation analysis is that information can be obtained about the depth distribution of elements near the sample-surface. The dependence of prompt-radiation characteristics on depth results from the energy losses suffered by incident ions as they penetrate into the sample and, in the case of the emitted radiation being ionized particles, also from the energy losses suffered by these particles as they emerge from within the sample. Etching and other surface-removal techniques are sometimes used for depth measurements, but these are laborious and often do not approach the resolutions possible with prompt-radiation analysis. Nuclear resonant reactions are especially powerful in this regard.

Another advantage of prompt radiation analysis is that spatial information in the plane of the surface can be obtained about chemical composition, the spatial resolution of such measurements being determined by the diameter of the incident ion beam. The pioneering work of the Harwell Laboratories in England on the use of small-diameter beams for microprobe analysis deserves particular mention. Cookson and Pilling [31] have recently developed a 3-MeV, 6-nA proton beam of less than 4 μm in diameter. Pierce [32] has discussed the analysis of samples utilizing a scanning of this beam across the sample either mechanically or by electrostatic beam deflection. The need for spatial information about elemental composition in the plane of the surface will undoubtedly spur additional work and development in this area.

For the various categories of prompt-radiation-analysis reactions, a general description of the reaction characteristics and experimental techniques will be followed by examples of actual applications of a given technique. Whenever possible, examples will be discussed that pertain to the problems of gases in metals. The literature on prompt radiation analysis is, however, rather more limited than is the case for charged-particle-activation analysis. In order to show adequately how the various techniques are applied, examples dealing with elements other than gases and metals have therefore had to be used. The cross-indexed activation-analysis bibliography [26] should again be useful for consulting the literature. The proceedings for the four charged-particle-activation-analysis conferences [1, 2, 6, 7] also contain a number of articles that discuss prompt-radiation-analysis applications.

A very interesting physical phenomenon called *channeling* is the basis of a powerful technique for analyzing single crystals. This technique, which depends on accurate alignment of an incident well-collimated beam with one of the low-index or close-packed crystal axes, can utilize any one of the types of ion-beam interactions being considered here. Channeling can, under proper conditions, give information not only about elemental composition but also about the substitutional or interstitial character of a chemical impurity. It also provides a sensitive measurement of crystal surface disorder. This technique will be discussed in the elastic-scattering section, since most channeling measurements are performed by means of this type of reaction. It is interesting to note that when the beam diameter is small compared to the grain size, it should, in principle at least, be possible to perform channeling measurements on the individual crystals of a polycrystalline material such as a metal.

A. ELASTIC SCATTERING REACTIONS

The application of elastic scattering to elemental surface analysis was discussed in 1957 in an excellent article by S. Rubin et al. [33]. In a famous experiment, Turkevich et al. [34] used α-particle backscattering to perform the first direct chemical analysis of the moon. (Later, chemical analyses on lunar matter brought back to earth by the Apollo XI mission showed excellent agreement with the elastic-scattering measurements.) A useful review of the principles involved in elastic-scattering experiments has been given also by Anders [35]. Other contributions to this subject area have appeared [36–52].

In the past few years the elastic-scattering technique has received a great deal of interest. It has been improved in mass and depth resolution and has been applied at energies ranging from 0.5 keV to 10 MeV and above and used with beam-ion masses ranging from the proton to ^{40}Ar and above. The references cited contain more detailed discussions of the principles that apply to the elastic-scattering mechanism than can be included here. The following brief introduction to the subject may, however, be useful.

An elastic-scattering event is, first of all, a collision between two particles in which energy and momentum are both conserved. For such a collision the following relation may easily be derived for the energy E of the scattered (beam) ion:

$$E = E_0 \frac{m_1^2}{(m_1 + m_2)^2} \left[\cos\theta \pm \left(\frac{m_2^2}{m_1^2} - \sin^2\theta \right)^{1/2} \right]^2, \tag{3}$$

where E_0 is the incident beam ion energy, m_1 is the mass of the incident ion, m_2 is the mass of the struck atom, and θ is the angle of scattering as measured to the incident direction. This equation shows that for a fixed angle of detection, the energy of the scattered ion depends on the mass of the scattering atom. Conversely, if the scattered energy spectrum is measured, Eq. 3 may be used to calculate the various masses present in the sample.

For targets that are thin compared to the penetration range of the incident beam, the number of particles, S, scattered by atoms of mass m_2 into solid angle Ω is given by

$$S = nN(m_2)\,\Delta x \Omega \sigma(E_0, \theta, m_1, Z_1, m_2, Z_2), \tag{4}$$

where n is the number of incident beam atoms, $N(m_2)$ is the number of scattering atoms of mass m_2 per cubic centimeter, Δx is the target thickness in centimeters, Z_1 is the atomic number of incident atoms, Z_2 is the atomic number of scattering atoms, and σ is the elastic-scattering cross section in square centimeters per steradian. At incident energies below a few mega-electron volts and above about 50 keV, the elastic-scattering cross section is usually purely Rutherford and is given by

$$\sigma_R = \frac{Z_1^2 Z_2^2 e^4}{16 E_0^2} \times (k+1)^2 \csc^4\left(\frac{\varphi}{2}\right) \times W(k, \theta), \tag{5}$$

where e is the electronic charge; $k = m_1/m_2$;

$$\varphi = [\theta + \sin^{-1}(k \sin\theta)] = \text{scattering angle in the center of mass system} \tag{6}$$

and

$$W(k, \theta) = \frac{[k\cos\theta + (1 - k^2\sin^2\theta)^{1/2}]^2}{[1 - k^2\sin^2\theta]^{1/2}}. \tag{7}$$

Values of σ_R calculated from these equations are given in millibarns (10^{-27} cm^2) per steradian in Table 20.1 for the example $E_0 = 3.0$ MeV, $\theta = 165°$, $m_1 = 4$, and $Z_1 = 2$. When the scattering is not Rutherford, the appropriate experimental or calculated cross section must be used in Eq. 4. The number of beam particles, n, is usually measured as the integrated charge that is incident on the target; the detector solid angle is a constant for a particular experimental arrangement and can also easily be measured. A measurement of S can yield, therefore, $N(m_2)\,\Delta x$, the number of scattering atoms of mass m_2 per square centimeter, or, if Δx is known, $N(m_2)$, the number of such atoms per cubic centimeter. Frequently, as is done in other methods, the simplest and most practical way to obtain $N(m_2)\,\Delta x$ is to perform a comparison measurement with a standard sample for which this quantity is known.

In principle, for very thin targets, well-defined beam energy, and high-resolution detection systems, it should be possible to identify all masses in the target differing by at least 1 amu, each scattering mass being observed as a sharp peak in the energy spectrum of the scattered particles. The mass resolution of the measurement depends on the narrowness of the observed peaks, which in turn depends on the incident beam energy spread, the detector solid angle subtended, the intrinsic detector-system resolution, and the target thickness. Typically, beam energy spread is small (a few parts per thousand or less) because most accelerator beams are magnetically analyzed before they strike the target. Resolution is most often determined, therefore, by target thickness and detector-system resolution. The subtended solid angle affects resolution simply because the backscattered energy changes with the angle of scattering. The peak broadening produced by this effect is commonly termed *reaction kinematic broadening*, or sometimes just *kinematic broadening*. This term is applied also to the spectra of particles emitted from nuclear reactions. Detector resolutions of about 15 keV are conveniently obtainable with commercially available solid-state detectors. If resolutions significantly better than this value are desired, resort must be made to

TABLE 20.1 Values of E, E/E_0 and the Rutherford Scattering Cross Section σ_R Calculated for $E_0 = 3.0$ MeV, $m_1 = 4$, $Z_1 = 2$, and $\theta = 165°$

Element	m_2	E(MeV)	E/E_0	σ_R(mb/sr)
Pb	208	2.781	0.927	4005
Au	197	2.770	0.923	3716
W	184	2.754	0.918	3261
Ce	144	2.690	0.897	2002
Xe	129	2.656	0.885	1735
Ag	109	2.597	0.866	1313
Mo	96	2.546	0.849	1048
Kr	84	2.487	0.829	769.2
Zn	66	2.363	0.788	532.7
Ni	60	2.307	0.769	463.3
Fe	56	2.264	0.755	399.0
Cr	52	2.216	0.739	339.5
Ti	48	2.160	0.720	284.7
Ca	40	2.022	0.674	233.9
Cl	37	1.958	0.653	168.4
Cl	35	1.910	0.637	167.9
S	32	1.830	0.610	148.0
P	31	1.801	0.600	129.8
Si	28	1.704	0.568	112.3
Al	27	1.668	0.556	96.5
Mg	25	1.590	0.530	81.6
Na	23	1.503	0.501	68.0
Ne	22	1.456	0.485	55.9
Ne	20	1.352	0.451	55.1
F	19	1.294	0.431	44.2
O	16	1.098	0.366	33.7
N	14	0.944	0.315	24.8
C	12	0.767	0.256	17.1
B	11	0.670	0.223	11.32
Be	9	0.457	0.152	6.22
Li	7	0.232	0.077	2.49

magnetic or electrostatic analysis of the charged particles being detected. The backscattering technique is, however, subject to a basic ambiguity in the case of targets sufficiently thick that the beam particles can lose appreciable energy before the scattering event takes place; the corresponding scattered particles then also lose energy as they emerge from the target. In such case, it is not possible to distinguish between a scattering from a light atom close to the surface and a scattering from a heavy element deeper in the target.

In practice, for thin targets and solid-state detectors it is possible to resolve peaks in the energy spectrum that are about 60 keV apart. The mass resolution corresponding to this value can be estimated from Table 20.1, which shows the energy (calculated from Eq. 3) that a 3.0-MeV α-particle possesses after scattering through an angle of 165° from various masses. Mass resolution improves for a given incident particle and scattering angle as the scattering mass decreases. The column labeled E/E_0 can be used for calculating the energy E for other values of E_0, since the fraction of energy lost in a scattering does not depend on the incident energy. It is easy to see from this column that the scattered-energy difference between the peaks corresponding to two given scattering masses increases as the incident energy increases; mass resolution therefore will be better at higher energies, all other factors being equal.

Because mass and depth resolution in general are better at large or backward scattering angles than at forward angles, most experiments are performed in a backscattering geometry. The arrow labeled *beam particles* in Fig. 20.1 corresponds, for example, to a backscattering event. The example discussed in the following section is of this type. Included in the examples are experiments on high- and low-energy backscattering, channeling, and forward-scattering coincidence measurements. The discussions emphasize the measurement technique itself; the results of the various experiments are considered only to the extent that they display the potentialities of the scattering measurements.

1. High-Energy Backscattering

A number of interesting features of the backscattering technique can be illustrated by the results obtained recently by Hirvonen et al. [52] in an experiment that utilized a relatively simple geometry. In this experiment, thin films of chromium and gold were deposited on 1-mm-thick sapphire substrates, and the 165° scattering of 3-MeV α-particles was used to study the migration of the chromium as a function of annealing temperature. Figure 20.2 shows schematically the geometry of the experiment. Alpha-particles of 3.0 MeV energy were obtained from a Van de Graaff accelerator and were detected after scattering by means of a surface-barrier detector whose resolution was less than 15 keV full width at half maximum. The energy spread in the beam was approximately 3 keV.

Figure 20.3 shows the data obtained with chromium and gold film thicknesses of 500 and 2000 Å respectively. The lower part of the figure shows the energy spectrum of the scattered α-particles as obtained before annealing. Reference to the rays a, b, c, and d in Fig. 20.2 will be helpful toward explaining the features of this spectrum. Path ab first of all represents scattering from the topmost surface of the gold. For this path the scattered α-particle loses energy only in the elastic collision with a gold atom and

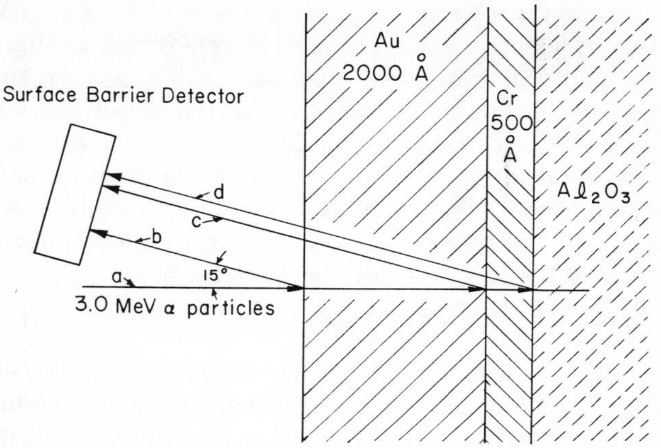

Fig. 20.2. A schematic diagram showing the backscattering of α-particles from thin films of Cr and Au on a sapphire substrate.

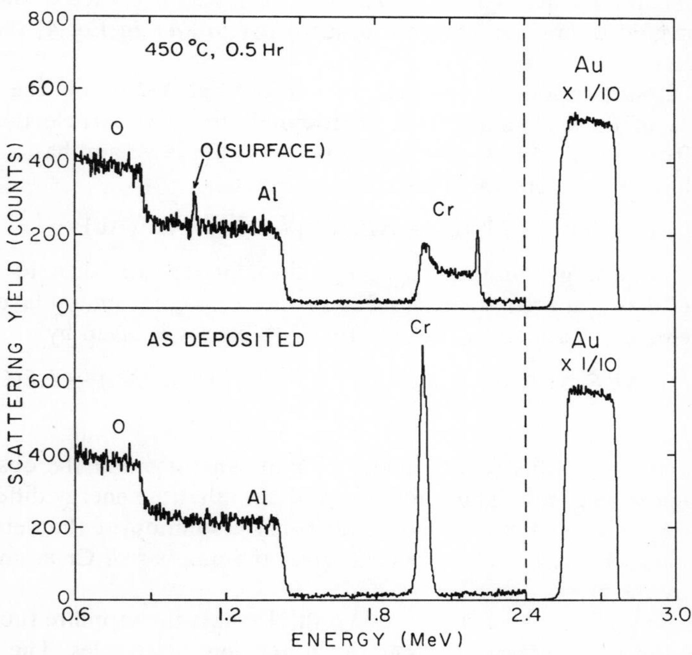

Fig. 20.3. Alpha-particle scattering yield at 165° from a Au–Cr–Al_2O_3 target before and after anneal. (Reproduced by permission of *Applied Physics Letters* [52].)

arrives at the detector with an energy (see Table 20.1) of $E_{ab}(\text{Au}) = 0.923\ E_0 = 2.770$ MeV. The high-energy side of the gold peak in Fig. 20.3 corresponds to path ab. For scattering events beneath the surface, the α-particle loses energy both on the way in and, after scattering, on the way out. The low-energy side of the gold peak in Fig. 20.3 corresponds to path ac—that is, to a scattering from a gold atom at the gold–chromium interface. If the assumption is made that the energy loss of the α-particle per centimeter of path in gold does not change rapidly with energy, the minimum energy on the low-energy side of the gold peak will be given by

$$E_{ac}(\text{Au}) = 0.923[E_0 - \Delta E(\text{Au})] - \sec 15^\circ\, \Delta E(\text{Au}), \tag{8}$$

where $\Delta E(\text{Au})$ is the energy lost by the α-particle in traversing the full thickness of the gold film. (In practice, for best accuracy, account should be taken of the fact that the energy loss per centimeter of path does change with energy.) For scattering points intermediate between the front and back surfaces of the gold, the α-particle will arrive at the detector with energies between 2.770 MeV and $E_{ac}(\text{Au})$, thus giving rise to a broad peak whose energy width is related to the thickness of the gold film from which the scattering is taking place. The total area under the peak is related to the total number of gold atoms per square centimeter [$N(m_2)\,\Delta x$ in Eq. 4] that the beam traverses.

Similar considerations show that the maximum energy on the high-energy side of the chromium peak corresponds to an α-particle that has scattered from the frontmost surface of the chromium layer and has traveled along path ac to the detector. This energy is given by

$$E_{ac}(\text{Cr}) = 0.739[E_0 - \Delta E(\text{Au})] - \sec 15^\circ\, \Delta E(\text{Au}). \tag{9}$$

The minimum energy on the low-energy side corresponds then to an α-particle that has scattered from the back surface of the chromium layer and has traveled along path ad to the detector. This energy is given by

$$E_{ac}(\text{Cr}) = 0.739[E_0 - \Delta E(\text{Au}) - \Delta E(\text{Cr})] - \sec 15^\circ[\Delta E(\text{Au}) + \Delta E(\text{Cr})], \tag{10}$$

where $\Delta E(\text{Cr})$ is the energy loss in the Cr film. The fact that the observed chromium peak does not exhibit a flat top implies that the energy difference $E_{ac}(\text{Cr}) - E_{ad}(\text{Cr})$ is comparable to the energy resolution of the detection system. Again, the area under the peak gives the number of Cr atoms/cm^2 traversed by the beam.

Finally, the plateaus labeled Al and O result because the sapphire substrate is thicker than the penetration range of the incident α-particles. The sharp rises or steps at approximately 1.45 and 0.89 MeV correspond to scattering along path ad from Al and O, respectively.

The upper portion of Fig. 20.3 shows the results obtained after the sample had been subjected to a half-hour anneal in argon at 450°C. (This sample had also previously received a number of heat-treatments at 150°C and 250°C; at 250°C only a relatively small amount of diffusion was observed.) The Cr peak has first of all broadened toward higher energies. The narrow peak on the high-energy side of the Cr now corresponds to scattering from Cr along path ab in Fig. 20.2—that is, to Cr situated at the front surface of the gold. This peak shows clearly that for these conditions of anneal, the Cr has migrated through the Au film and has built up in concentration at the surface. The flat "valley" portion of the Cr broad peak corresponds to a distribution of Cr throughout the rest of the Au layer, which is approximately uniform. It is important to note that the elastic scattering measurement performs an average over the beam spot area and cannot distinguish within this area between a homogeneous distribution of Cr in the gold and a nonuniform distribution consisting, for example, of thin Cr filaments having the same averaged density. The peak labeled O (surface) corresponds in energy to oxygen that is now present at the front surface of the sample, present there presumably because of the oxidation of the Cr that has reached this surface. The change in shape of the rear edge of the Au peak indicates interdiffusion of the Au and the Cr at the previously well-defined Au–Cr interface. It is of interest also to note that peak shapes can be different for the case when one interface element is predominantly diffusing into another, from those obtained when the two elements are diffusing equally into each other.

The qualitative features of the curves in the upper portion of Fig. 20.3 can be converted into quantitative depth distributions of elemental concentrations if the energy loss per centimeter of the incident and outgoing particles is known in the target. The reason for this is that it is then possible to convert observed energies to depths and observed energy intervals to layer thicknesses Δx. Counts per unit energy interval are used instead of peak areas to obtain the number of atoms of mass m_2 per cubic centimeter—that is, $N(m_2)$—at a given depth. It should be noted, however, that the conversion of the energy scale along the abscissa to a depth scale in the target depends on the mass from which the scattering is taking place, so it is necessary to correct for this dependence both in calculating the depth that corresponds to a given energy in the spectrum and in calculating the layer thickness Δx that corresponds to a given energy interval in the observed spectrum.

When the ratio of two elemental concentrations is desired, the situation simplifies somewhat in that the ratio depends less strongly on an accurate knowledge of the stopping power of the sample than do the individual concentrations. Thus, for example, the depth distribution of the stoichiometry of the aluminum oxide substrate could be estimated (if it were not known) from the heights of the Al and O "steps" in Fig. 20.3 to a better accuracy than that to which the stopping power for this material is known. For targets between 300 and 2000 Å thick, Meyer et al. [40] and Mitchell et al. [41] discuss a method that can measure the ratio of the concentrations of two elements to good accuracy even if the stopping power for the material is not known. The method utilizes the energy widths of the observed peaks to obtain a relative depth scale in the sample.

Picraux and Vook [44] have discussed multilayer thin-film analysis for Er, Sc, or V layers, 500–2000 Å thick, on Kovar or sapphire substrates. One particularly useful observation made by these authors is that interfacial diffusion at an upper surface—that is, one closer to the front surface of the target—will not decrease the energy sharpness of the energy edges due to lower interfaces, whereas upper film thickness nonuniformities will contribute directly to the lack of sharpness of the energy edges of lower interfaces. The article also treats the case when a narrow overlap between two broad peaks in the energy spectrum gives rise to a narrow peak in the region of overlap. The sharpness or lack thereof of this overlap peak also can be used to obtain information on interfacial diffusion.

In the case of thick targets, Eq. 4 must be restated in an integral form similar to that of Eq. 2. In practice, however, the need for integration can frequently be avoided through the use of simplifying approximations regarding target composition and the dependence of the energy loss on energy and on target composition. The ambiguity, mentioned earlier, between light elements close to the surface and heavy elements deeper in cannot be simply overcome. Similarly, a measurement on a given element cannot be performed if the corresponding backscattering results cannot be resolved from the rest of the spectrum. Outside of these two constraints, extensions of the measurements from the simple sample discussed here to more complex samples should be possible; a number of references [36–52] should prove useful for this purpose.

Proton and helium backscattering measurements have been carefully investigated at 100 keV by Buck and Wheatley [50]. This energy is interesting because the ion accelerator required is much less expensive than those which produce MeV energies. The authors cited have shown that the scattering at this energy is essentially Rutherford and that the sputtering away of metal target surfaces is negligible. Because the penetration range at 100 keV is small, this technique samples a thinner surface layer than higher energy experiments. This can be an advantage in that it may reduce the interfering background from the bulk of the sample. The penetration range is not so short, however, as to be greatly affected by monolayer films, so, in common with higher energy measurements, the technique does not require ultrahigh-vacuum techniques.

The method was used to detect surface impurities on silicon and graphite.

A mass resolution of better than one unit in the low-Z region around oxygen and carbon was obtained through the use of a small cylindrical electrostatic analyzer in front of a solid-state detector. The electrostatic analyzer introduced the complication that it does not transmit neutral particles. Since charge neutralization is not negligible for 100 keV protons and He^+ particles moving in matter, a correction for this effect must be made.

2. Channeling

Thus far, backscattering has been described for amorphous or polycrystalline materials. When the material being bombarded is a single crystal and when the beam is collimated so that its angular divergence is on the order of 0.1° or so, an interesting phenomenon can be observed that has implications for backscattering measurements. This phenomenon, which is called *channeling*, is evidenced by a large decrease in the yield from any reaction which requires a close ($\lesssim$ 0.2Å) interaction between the beam atom and the target atom when one of the close packed or low-index symmetry axes of the crystal is aligned closely with the incident beam direction. Typically the yield can decrease by factors of 20–50. High-energy backscattering is a close-interaction reaction; so also are ion induced X-ray production and inner-shell ionization, and nuclear reactions.

For simplicity, channeling will be discussed in succeeding sections in terms of backscattering only; it should be understood, however, that similar remarks will obtain for the other close-interaction reactions. In the channeling mode, just as in ordinary backscattering and subject to the same constraints, measurements can be made on a desired element through observations on the corresponding peak or structure in the backscattered energy spectrum. Thus, measurements are frequently possible on either the host crystal elements or on impurity elements or both.

To see how a reduced probability for close-interaction reactions can result when the beam is incident on a single crystal along a low-index direction, consider the schematic diagram shown in Fig. 20.4. A single crystal target is shown in cross section with the atoms located on a square array having spacing d. For convenience the crystal is shown in fixed position and beam rays A, B, and C are shown at various angles of incidence with respect to the crystal. In practice, of course, the beam direction is fixed in space and the crystal is rotated to the desired orientation.

Ray A, first of all, enters along a random direction into the crystal; the probability for a close interaction with an atom is therefore the same as it would be if the target were amorphous. The backscattering spectrum corresponding to ray A incident on a thick crystal is shown schematically in Fig. 20.5. The increase in yield at lower energies is caused in part by the inverse dependence of the Rutherford cross section on energy.

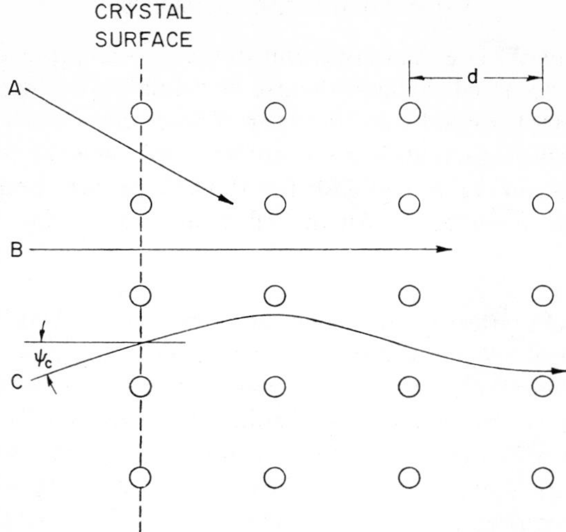

Fig. 20.4. A schematic diagram showing ion beams incident on a single crystal for three different inclinations relative to the crystal axes.

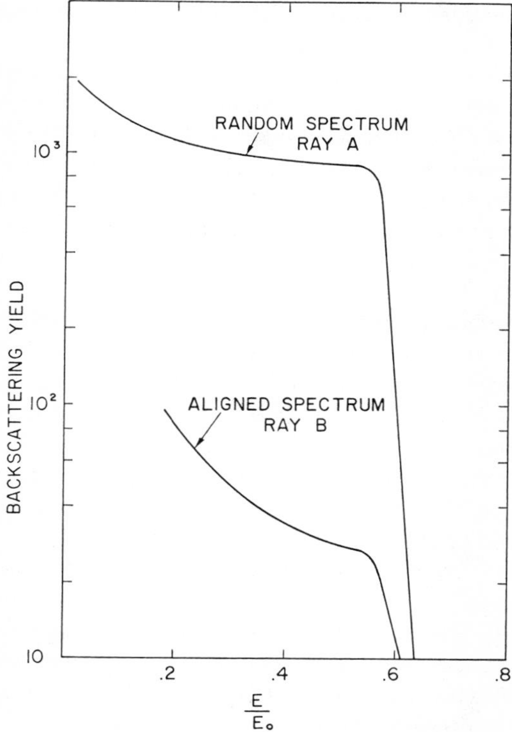

Fig. 20.5. Schematic backscattering spectrum corresponding to rays A and B incident on a crystal along random and aligned directions, respectively.

Ray B in Fig. 20.4 is aligned exactly with an atom row or plane. It is shown entering midway between two rows of atoms, but the entry points will in actuality be distributed uniformly over the crystal surface. Ray B will have a significant probability of undergoing a backscattering only if it enters the crystal at a point that is within about 0.2 Å of a surface atom center. Projected area considerations show that this should happen only about 2% of the time. In the ideal aligned case, therefore, the backscattering yield might be expected to decrease by a factor of about 50. In actuality, lattice vibrations, crystal imperfections, and the like combine to decrease this factor somewhat. Ray B gives rise to the reduced intensity spectrum labeled in Fig. 20.5 as the *aligned spectrum*. The increase in intensity of this curve at lower energies reflects the increasing dechanneling of the beam as it penetrates into the crystal.

A beam need not enter perpendicularly to the surface, as does ray B, in order to become channeled. As long as the momentum transverse to the crystal atom row is smaller than that required for approaching an atom site to within about 0.2 Å, the beam ion will suffer only small angle deflections and will not be scattered out of the channel. There exists a calculable critical angle below which the beam will be channeled and show low yield for backscattering and beyond which it will undergo backscattering reactions and approach the random direction yield. Ray C is shown schematically at the critical angle. In practice, the transition through the critical angle is not sharp, the critical angle being measured as the half width of the yield curve at the point where the yield is half that of the difference between the random yield and the minimum yield. Figure 20.6 shows a backscattering-yield curve

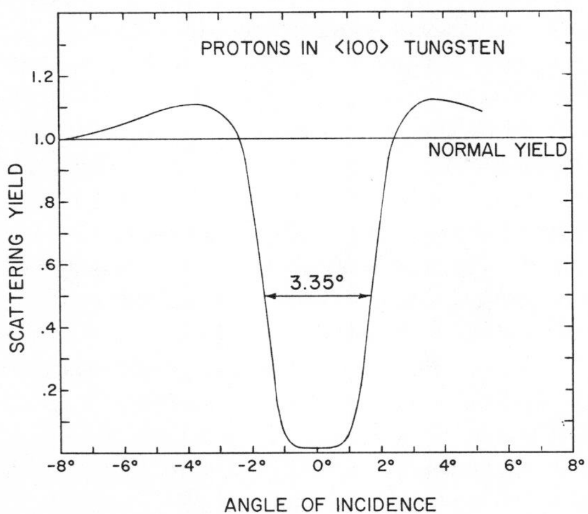

Fig. 20.6. Backscattering yield of protons as a function of angle of incidence relative to the ⟨100⟩ axis of single-crystal tungsten.

as a function of angle of incidence, which was obtained by Andersen [53] for 0.5 MeV protons on single-crystal tungsten; 0° incidence is parallel to the ⟨100⟩ crystal axis. The critical angle in this case was measured to be 1.68° and the yield at minimum was down by approximately a factor of 40 from the random yield.

If one of the atoms in Fig. 20.4 is displaced by more than about 0.2 Å from its lattice site, it is exposed to close interactions with an aligned beam. The decrease in backscattering yield occurs therefore only for atoms that are located on crystal lattice sites; backscattering from atoms that are interstitial or that are present in thin amorphous surface films are not affected. This means that if the host crystal lattice is presenting a background problem in the detection of a given interstitial or surface film element, alignment of the crystal so that the channeling mode is achieved can greatly reduce this problem. In consequence, a much higher sensitivity for detection of that element can be obtained than can be achieved if the entire sample is amorphous or polycrystalline [40].

The minimum backscattering yield from the crystal elements is strongly dependent on the perfection of the crystal—that is, on the extent to which all reactant atoms are in their proper lattice locations. Through this dependence, channeling has been used to study crystal-structure imperfections [54–60].

Impurity atoms in a crystal can be either substitutional or interstitial. If they are substitutional, the backscattering yield from the impurity atoms will show a decrease in yield similar to that from the host crystal element. If the impurity atoms are interstitial the backscattering yield will usually show no dependence on the angle of incidence. Sometimes an interstitial atom lies accurately in one or two crystal atom planes. In such case, backscattering measurements for several different channeling directions can be used to triangulate on the impurity atom's location in the lattice [61a–c]. More usually, impurity atoms are only partly substitutional; the factor by which the random yield decreases can be used to determine what fraction is so situated [61a–c, 62]. The atomic concentration of impurity atoms must be about 10^{-4} or greater before channeling measurements can be applied to such atoms [63].

Channeling has been of great interest in solid state physics and particularly in the field of semiconductors. The references listed here constitute merely an introduction to the literature. Excellent survey articles on channeling have been prepared by Gibson [64] and by Davies [65]. Much discussion pertinent to this subject may be found also in several conference proceedings [66–68] and in a book by Mayer et al. [69].

Oxide and nitride films on silicon and gallium arsenide single crystals have been studied by channeling and backscattering measurements [40–42, 70, 71]. In the case of silicon [40], measurements were made on the stoichiometry of the films as a function of distance from the interface between the bulk silicon and the surface films. The decrease in the aligned backscattering spectrum was used to reduce the silicon background under the oxygen and nitrogen peaks and thereby to improve the accuracy of the measurements. The ions used were 1–2-MeV ^{4}He. The energy resolution of the surface barrier detectors that were used provided a depth resolution of between 100 and 150 Å. Film thicknesses studied ranged from 400 to about 2000 Å. Stoichiometry profiles obtained with backscattering alone gave good agreement with measurements that combined surface-etching and backscattering. For gallium arsenide, backscattering and channeling were used to measure diffusion outward through silicon oxide and silicon nitride layers for various annealing conditions. Clear evidence was obtained for diffusion of gallium and for its accumulation at the surface of Si_3N_4 or SiO_2 layers on GaAs substrates after anneals at 700–800°C.

In still another interesting set of experiments, Matzke et al. [62] used channeling combined with both backscattering and nuclear reactions to study the location of excess oxygen in the fluorite crystal structure of UO_2. During oxidation of UO_2 the structure remains unchanged until a composition of U_4O_9 is reached; further excess oxygen produces a phase change to the tetragonal structure U_3O_7. Experiments were performed therefore with crystals having the composition $U_4O_{8.995}$. Uranium was detected by helium and deuterium ion backscattering, and oxygen was detected through protons emitted from the $^{16}O(d, p)^{17}O$ nuclear reaction. Channeling was used to determine that the excess oxygen was not, as had been previously postulated, occupying the centers of the large interstitial holes of the fluorite structure but was instead being incorporated into locations removed from both the normal lattice sites and from the interstitial holes by more than 0.2 Å. The channeling data also indicated that in contrast to a high degree of disorder in the oxygen sublattice, the uranium lattice is only slightly disturbed by the presence of excess oxygen.

3. High-Energy Forward Scattering

The backscattering technique is best for thin layers of heavy elements on light-element substrates because in such cases the backscattered peak occurs at higher energies than result from backscattering from the substrate. As the mass of a given surface-element decreases relative to the substrate elements, the energy of the backscattering peak decreases relative to the substrate spectrum, and when the substrate elements are the heavier, the backscattered peak "sits" on top of a relatively large contribution from the substrate. For single crystal substrates, channeling can reduce the relative substrate contribution and improve the accuracy and sensitivity with which lighter-element measurements can be made. For thick amorphous materials, the substrate background cannot be reduced, and this therefore puts limits on the light-element mass and concentration that can be practically detected above the background.

If the target sample can be made sufficiently thin that the lowest backscattered energy from the substrate is higher than the light-element backscattering, then the light-element response will be clearly resolved. The separation between the gold and chromium peaks in Fig. 20.3 illustrates this point. The technique is difficult because very thin samples are required.

Meek and Gibson [51] have shown that the technique can be improved if forward scattering is observed and if the collected particles are scattered forward through the crystal instead of back from it. In these experiments the target was tilted to an angle of 45° with respect to the beam, and 1.8-MeV He^+ or 1.0-MeV protons were observed at 90° with respect to the incident beam direction. For carbon on both surfaces of a 9000-Å-thick silicon wafer they obtained a clear separation between the silicon peak and the carbon scattering peaks. The He^+ scattering also resolved the carbon peaks from the front and back surfaces of the silicon, showing that distribution information about carbon within the film could be obtained. The sensitivity was estimated to be a few monolayers of carbon.

The gases hydrogen and helium are subject to additional limitations. Equation 3 shows that no backscattering at all occurs when the beam particle, m_1, is equal to or exceeds in mass the element, m_2, being measured. Thus, helium-particle beams produce no backscattering from hydrogen or helium, and proton beams produce no backscattering from hydrogen. Forward scattering does occur, however, and has some characteristics that can be put to advantage.

In an interesting experiment performed with protons, Cohen et al. [49] made use of the fact that when equal-mass particles collide, the angle between the directions of the scattered particle and the recoil particle must be 90°. Thus, if an incident proton is deflected through 45° by a hydrogen nucleus (i.e., a proton) the target proton must be emitted at 45° on the other side of the beam. Moreover, in this case the two protons must have identical energies and, of course, are emitted in time coincidence. The experiments therefore imposed the 45° deflection, 90° included angle, time coincidence, and equal energy requirements and provided a measurement highly specific to hydrogen and quite sensitive, the sensitivity being estimated to be well below 1 ppm.

Figure 20.7 shows a schematic diagram of the arrangement that was used. It can be seen that protons originating from point A—that is, at the back surface of the target—arrive at the detectors with greater energy than protons originating from a scattering at point B. There are two reasons for this. One is that the path length through the target is longer from point B; the second is that because the proton energy after a 45° deflection is only half the original energy, the energy loss per unit path length in the target is increased. The difference in energies corresponding to collisions at points A and B was therefore used to identify the depth at which each event had occurred. For this purpose the pulse heights from the two coincident detectors were summed for each event. The technique is therefore capable of providing not only the amount of hydrogen present in a sample but also its distribution in depth.

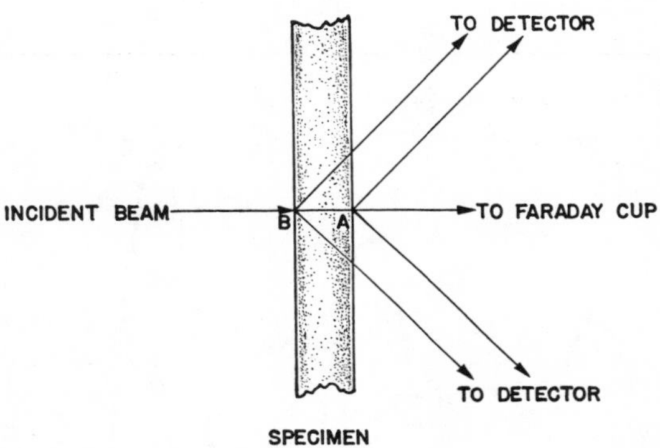

Fig. 20.7. Schematic diagram for incident beam, target, and detector geometry used in coincident measurements on scattering between two particles of like mass.

With a spatial resolution determined by the size of the beam spot, it can also provide information about hydrogen in the plane of the target. The technique furthermore is not limited to extremely thin samples. In the experiments cited here, 17-MeV protons from a three stage Van de Graaff accelerator were used to study foils of carbon, Teflon, aluminum, titanium, iron, and zirconium ranging in thickness from 7 mils for the lighter elements to 2 mils for the heavier.

Figures 20.8 and 20.9 [49] show some of the results that can be obtained with this technique. Figure 20.8 is the spectrum (summed coincident pulses) obtained from an Fe–Mylar sandwich target, which consisted of three 0.7-mil Fe foils interleaved with four 0.15-mil Mylar sheets. The peaks result from the hydrogen in the Mylar. The difference in peak areas is caused mostly by multiple scattering of the protons emerging from the target. The greater the path length a proton must traverse in the target, the more likely the probability that it will undergo scatterings that will cause it to miss the detector; peak area becomes progressively smaller, therefore, the closer a Mylar foil is to the target's front surface. These data show that multiple-scattering corrections must be made for thick targets before the true depth distribution can be obtained. Data such as shown in Fig. 20.8 can be used also to obtain empirical multiple-scattering corrections.

Figure 20.9 shows the hydrogen distribution measured in a 5-mil titanium foil. The dashed lines show the multiple-scattering corrections. The two spectra shown were obtained with the target being rotated through 180°. Hydrogen surface contamination on the foil shows up as the two peaks at the ends of the spectrum.

The same technique has been applied recently by Pieper and Theus [72] to measure the distribution of helium implanted into aluminum. In this case

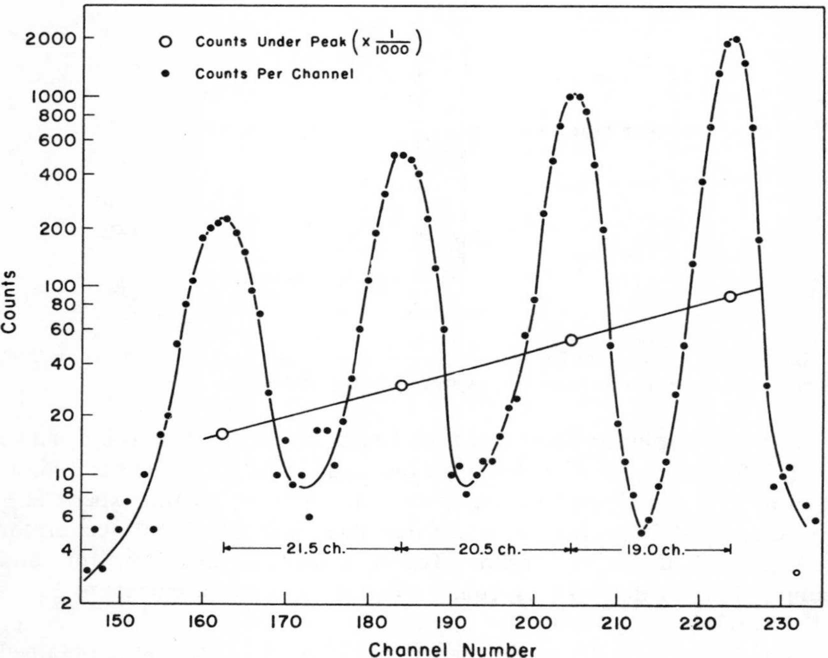

Fig. 20.8. Spectrum of summed coincident pulses obtained from hydrogen in a sandwich target consisting of three 0.7-mil Fe foils interleaved with four 0.15-mil Mylar sheets. (Reproduced by permission of *Journal of Applied Physics* [49].)

helium ions are used as the incident beam, and the reaction considerations are then identical with the case of protons on hydrogen. Figure 20.10 shows the results obtained with a 48.5-MeV helium ion beam from a cyclotron incident on a 3-mil-thick aluminum target implanted with two helium layers. The helium layers were implanted into opposite surfaces to a depth of 0.25 mils and were therefore separated in the foil by 2.5 mils. The two helium peaks were easily resolved. The helium concentration of the lower energy peak was 8000 ppm, and that of the higher energy peak was 3000 ppm. Again, the

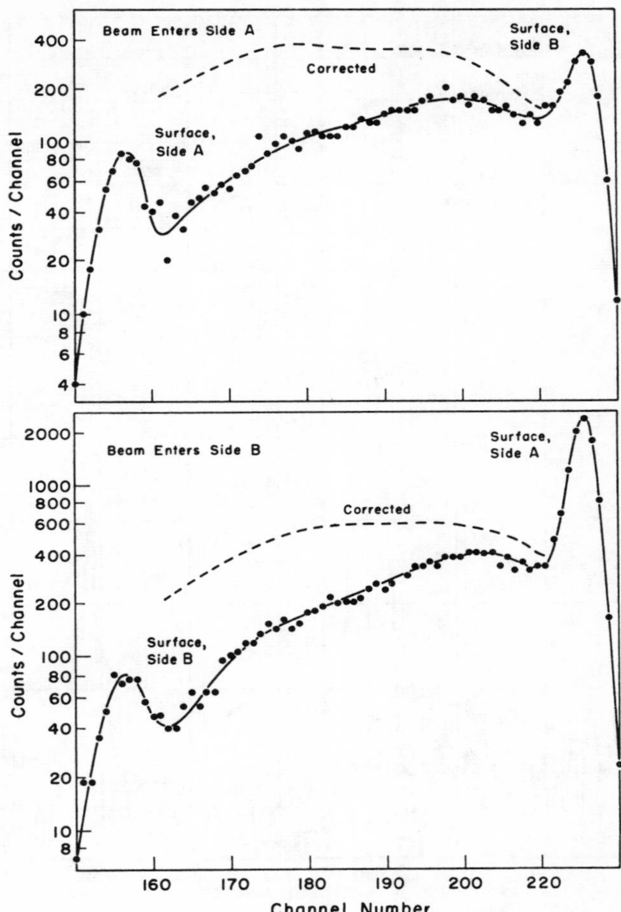

Fig. 20.9. Spectrum of summed coincident pulses obtained from hydrogen in a 5-mil titanium foil. (Reproduced by permission of *Journal of Applied Physics* [49].)

peak areas differ somewhat from the corresponding ratio because of multiple-scattering effects. Sensitivity is estimated to be 1 ppm or better. Interest in these experiments was associated with the helium embrittlement produced in nuclear reactor materials as a result of (n, α) reactions.

An interesting variation of the elastic-scattering technique has been reported by Cahill [73] in which recoil target nuclei are detected at forward angles instead of the beam particles. The technique is applicable to the analysis for light elements in thin samples and is quite sensitive. It is being applied to the study of light elements in atmospheric aerosols.

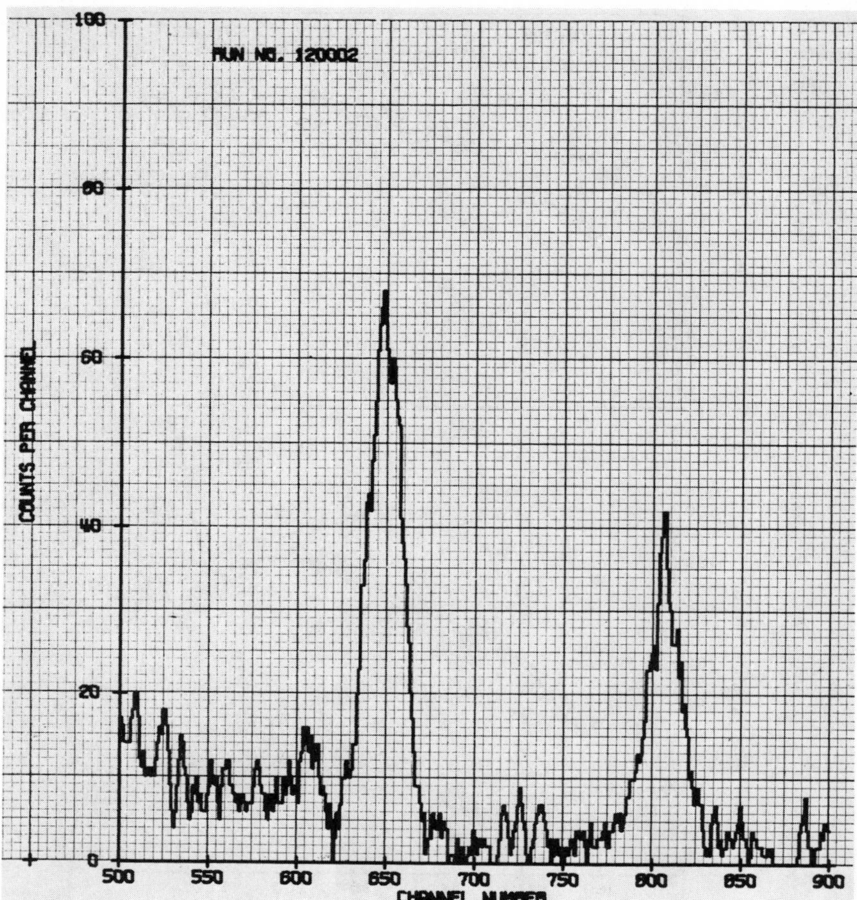

Fig. 20.10. Spectrum of pulses obtained from two layers of helium implanted into Al and separated by 2.5 mils of Al. (Reproduced by permission of the authors [72].)

4. Low-Energy Backscattering

When the incident ion beam energy is between 0.5 and about 3.0 keV, the characteristics of backscattering are different in several respects from the high-energy backscattering previously discussed. The importance of the applications made possible by these differences and the low cost of the equipment required as compared to that required for high-energy backscattering, and even to that required at 100 keV, are such that a separate section devoted to this low-energy range was deemed to be warranted. The region below 500 eV is excluded from the present discussion, because it is less well understood and because its utility for surface analyses is still open to question.

The application of low-energy ion scattering to surface analysis was pioneered in a very interesting set of experiments by Smith [38, 47]. Figure 20.11 shows a schematic diagram of the experimental arrangement used in this work. Subsequently a compact and relatively simple apparatus was

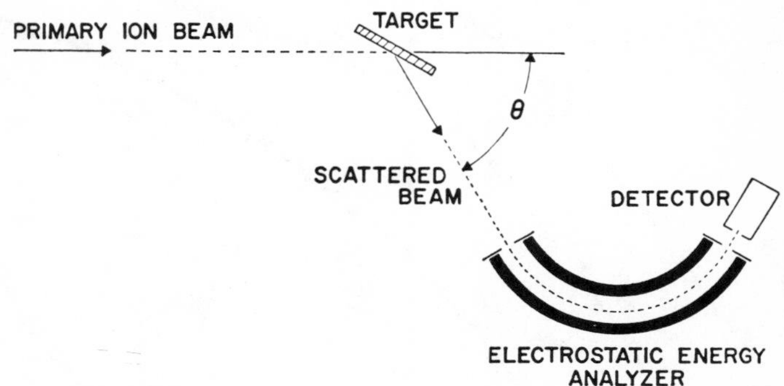

Fig. 20.11. Schematic diagram of the arrangement used for low-energy scattering studies of surfaces. (Reproduced by permission of *Surface Science* [47].)

developed by Smith and Goff [74] to facilitate application of the technique in other laboratories. More recently, Ball et al. [75] have investigated the factors which must be taken into account for quantitative measurements.

Two major differences exist between backscattering at low energy, $0.5 \lesssim E \lesssim 3.0$ keV, and at high energy, $E > 100$ keV. The first is that the cross section for the low-energy scattering is no longer Rutherford; it is not given therefore simply by Eq. 5. Instead, cross sections are derived from the screened Coulomb potential [75], a parameterized tabulation for which has been given by Everhart et al. [76].

The second and more important difference is that at low energies the backscattered particle has a significant probability of having its charge state neutralized. This probability increases as the path length of the particle in the target increases and as the energy of the particle decreases. Thus, for example, particles that are scattered from beneath the surface are more likely to be neutralized than those that scatter in the topmost atom layers of the target surface. This effect, when combined with an electrostatic analyzer that rejects neutral particles, produces the important consequence that low-energy-backscattering analysis can be made sensitive to just the first atom layer of the target surface [47, 75]. (The neutralization probability depends on the nature of the incident ion; reactive element ions show less of a tendency

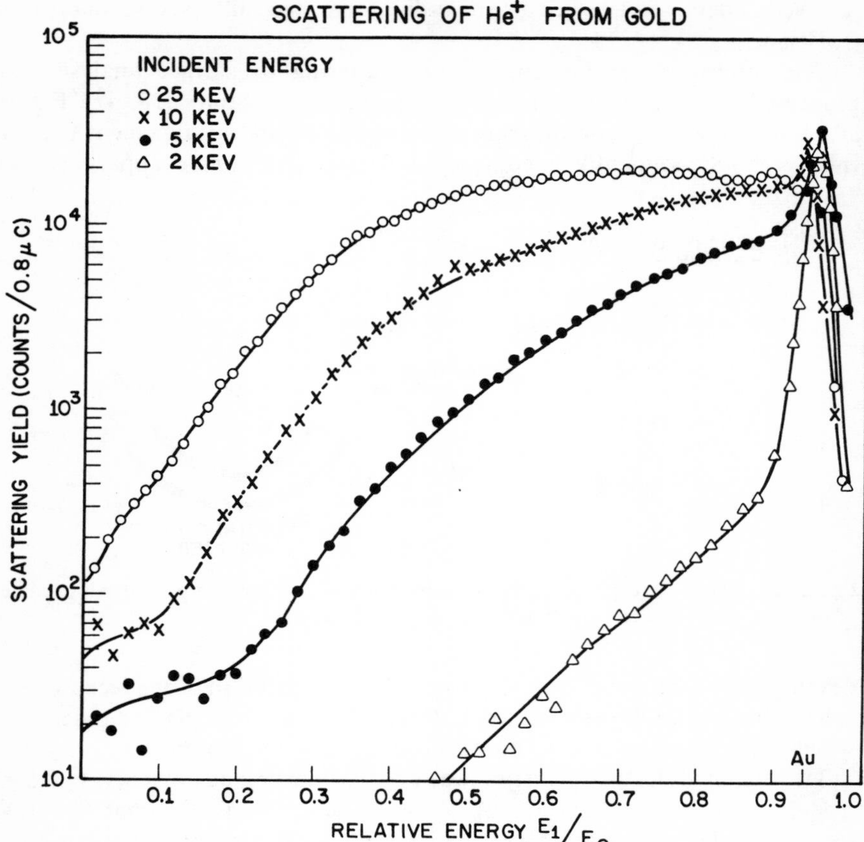

Fig. 20.12. Scattering yield of low-energy protons as a function of observed energy for various incident energies. (Reproduced by permission of *Surface Science* [75].)

for becoming neutralized after scattering from atoms beneath the surface and are therefore less useful for surface analyses [47].)

Figure 20.12 [75] illustrates the neutralization characteristic. The intensity on a semilog plot of α-particles backscattered from gold through 120° and at various incident energies is shown. For this angle of scattering, E/E_0 is expected to be 0.94. Note that at 25 keV, there is a plateau and an energy cutoff at $E_1/E_0 = 0.94$. This curve can be compared with the helium–gold backscatter peak in Fig. 20.3, where the lower-energy portion of the peak corresponds to alphas that have been scattered from progressively deeper portions of the gold layer. The target to obtain these data was thick enough so that the intensity of α-particles should be relatively flat down to the lowest energies. That the intensity falls off at lower energies is evidence that α-particles coming from beneath the surface are increasingly neutralized and do not get through the electrostatic analyzer to the detector. As the incident beam energy is lowered, this effect becomes more pronounced until at 2 keV a very sharp peak is observed, even though the gold target is thick to the beam. At this point the interfering contributions from the bulk of the target have been effectively eliminated and the technique can be used to measure surface impurities with high sensitivity.

As another illustration, the energy spectrum obtained by Smith [47] in the 90° scattering of 2-keV He^+ ions from polycrystalline copper is shown in Fig. 20.13 on a linear plot. Again, at higher energies the spectrum would not be a

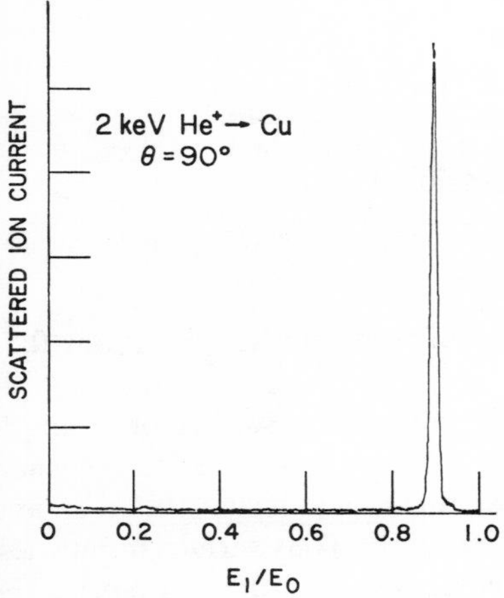

Fig. 20.13. Spectrum of 2-keV He^+ ions scattered through 90° from copper. (Reproduced by permission of *Surface Science* [47].)

single sharp peak but a flat plateau with a sharp cutoff at the peak position.

An interesting result of the technique's sensitivity to just the first layer of surface atoms is shown in Fig. 20.14 [47]. The different spectra were obtained when 2-keV He^+ ions were scattered from the two polar surfaces of single-crystal CdS. The difference is ascribed to the noncentrosymmetric crystal structure of the compound, which has one polar face comprised of Cd atoms and the other comprised of S.

The sensitivity of the low-energy-scattering technique to the structure of the top surface layer of atoms was exploited also by Heiland and Taglauer [77], who studied the sorption of oxygen on the ⟨100⟩ face of single-crystal nickel. By varying the angle of incidence and the crystal orientation, these authors were able to show that every second place in the ⟨110⟩ row of nickel is occupied by oxygen. From a shadow effect at small angles of incidence, they

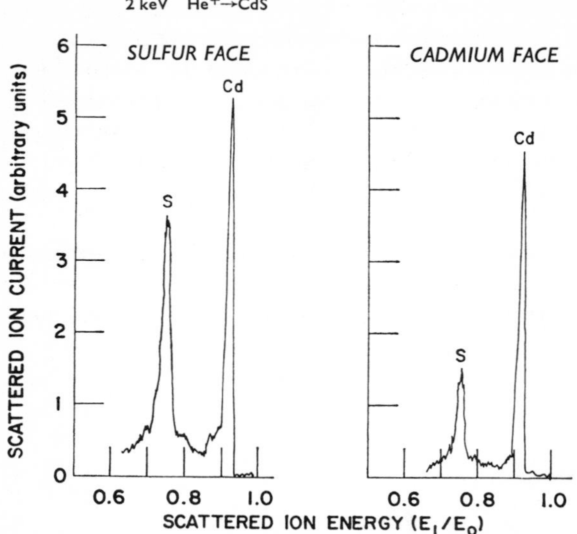

Fig. 20.14. Spectra obtained when 2-keV He^+ ions are scattered from the two polar surfaces of single-crystal CdS. (Reproduced by permission of *Surface Science* [47].)

also concluded that the oxygen protrudes from the surface by about 0.5–1.0 Å.

Because a single atom layer of an impurity film on the surface of a sample can completely mask the underlying surface, low-energy-scattering measurements must be performed in ultrahigh vacuum—that is, at 10^{-9} torr and better, and great care must be taken to remove extraneous surface impurities. For particles heavier than He^+ the sputtering away of the sample by the incident beam must also be taken into account. Calibration problems involve, in addition to these two factors, evaluation of the charge neutralization probability in collisions with the first layer of atoms, and the dependence of this probability on the elemental composition of the surface.

Low-energy-scattering measurements can detect both light and heavy elements on amorphous and crystalline surfaces with sensitivities ranging in some cases down to 10^{-4} atom layer. The technique can be used to study the first surface layer only and is relatively inexpensive. It should therefore come into increasing use for surface studies along with such techniques as LEED, ESCA, and Auger spectroscopy.

B. HEAVY-ION-INDUCED X-RAYS

Considerable interest has developed in the last two years in the use of energetic protons, alpha particles and other heavy ions for elemental analysis of samples through the characteristic X-rays induced by bombardments with these ions. The principal advantage of this technique comes about because heavy ions produce bremsstrahlung background that is orders of magnitude lower than that produced by electrons. Birks et al. [78] compared line–background ratios for 15 and 34 keV electrons and 0.7–2.0 MeV protons and found the ratios improved by at least one to two orders of magnitude and possibly more. Poole and Shaw [79], recognizing the advantages of proton X-ray excitation, developed a 100-keV, 10-μm beam diameter proton microprobe for surface analysis. They achieved a sensitivity of 10 ppm and estimated that 1 ppm could also be achieved.

Recent interest has resulted in part from the advent of high-resolution silicon X-ray detectors, which make possible efficient multielement analysis. Johansson et al. [80] in an important paper have given a new impetus to the field by pointing out that the sensitivity of the technique could extend to the 10^{-12}-g level and beyond. For samples prepared by the evaporation of droplets of a solution, this number translates into a parts-per-billion sensitivity.

Until recently the thin beryllium window covering commercially available lithium-drifted silicon X-ray detectors limited the use of these detectors to elements with $Z \geq \sim 12$. Presently, however, detectors are available with windows sufficiently thin that elements down to oxygen and possibly below can be measured. Specially constructed proportional counters [81] have been used to detect even the ultrasoft $K\alpha$ radiation from boron. Recent experimentation has occurred at incident beam energies from 20 keV–50 MeV and above with incident ion masses ranging from the proton to krypton ions and above [73, 78–89].

The bremsstrahlung radiation produced by heavy ions is caused mainly by the secondary electrons scattered by the passage of the ion through the target material. Highest sensitivity for surface analysis is obtained, therefore, if the sample is made as thin as possible. Even for thick samples, however, parts-per-million sensitivities are easily achievable with protons and α-particles of a few megaelectron volts in energy. The number of characteristic X-rays produced can again be given by Eq. 1, the reaction yield formula, provided that the cross section $\sigma(E)$ contains both the cross section for ionizing the electron shell in question and the fluorescence yield for the particular X-ray peak being used in the measurement.

Calibration of a measurement is simplest for thin targets, since the incident ion energy does not change appreciably in the target and since absorption of the emitted X-rays in the target can be neglected. For thick targets, calibration standards and procedures similar to those used for electron-microprobe work are required.

The effects of background, counter resolution, type of incident ion, and incident energy on detection sensitivity have been analyzed by a number of authors [82–86]. The energy at which a maximum is achieved in the X-ray excitation cross section increases with increasing target atomic number and with increasing beam ion mass. Figure 20.15 shows K- and L-X-ray production cross sections calculated for protons by Gordon and Kraner [82]. The corresponding maxima for α-particles occur at higher energies than is the

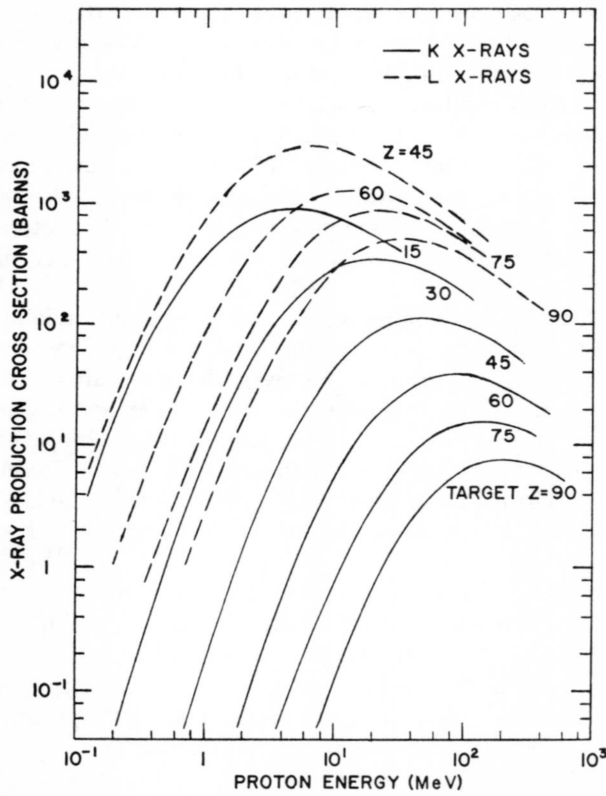

Fig. 20.15. Calculated cross sections for X-ray production by protons.

case for protons. At Van de Graaff accelerator facilities, therefore, protons are typically used for X-ray analysis; at cyclotrons α-particles are used.

An interesting application to X-ray analysis of the heavy ions K, V, and Kr with energies in the range 20–220 keV has been made by Cairns et al. [88]. These authors show that it is possible in certain cases to excite characteristic X-rays in a selective manner by adjusting the bombarding energy. Thus, for example, when a potassium beam is incident on copper, the threshold for copper L-X-ray production occurs at 40 keV while the threshold for potassium L-X-rays occurs at approximately 20 keV. It is therefore possible below 40 keV to excite the latter X-rays preferentially. This technique, which is particularly sensitive to elements at or near the surface, has been applied [89] in conjunction with anodic stripping to the determination of depth distributions of antimony implanted into silicon.

An example of the kind of spectrum that can easily be achieved with heavy-ion-induced X-rays and lithium-drifted silicon detectors is shown in Fig. 20.16. These data were obtained by Knudson and Wilkniss [90] from the

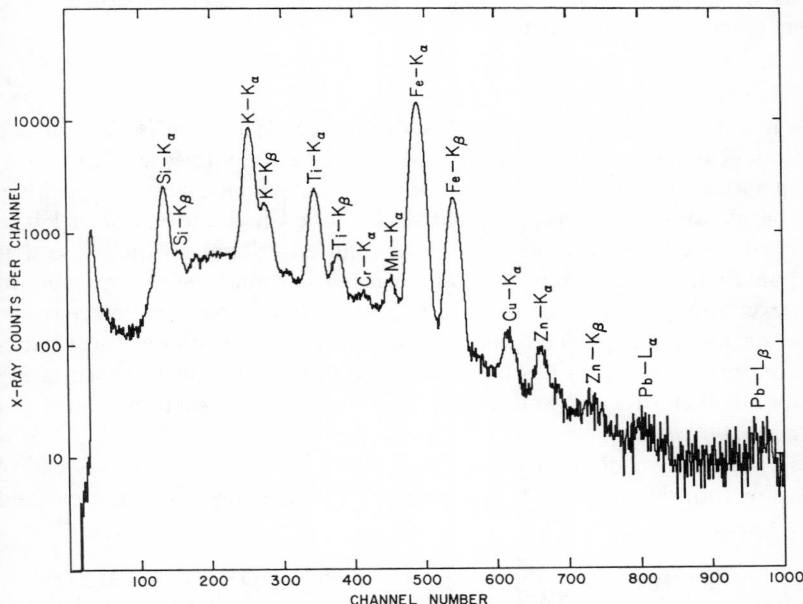

Fig. 20.16. Yield of X-rays obtained from the irradiation of sediment from a municipal water reservoir with 5.0-MeV protons. (Reproduced by permission of the authors [90].)

irradiation of a thin layer of sediment that had been collected from a municipal water reservoir. The sample, consisting of approximately 30 $\mu g/cm^2$ of material deposited in a thin self-supporting Zapon lacquer film, was bombarded for about 7.5 min with a 44-nA beam of 5.0-MeV protons. The lower energy cutoff in the spectrum is caused by two beryllium windows, with a total thickness of 9 mils, that was present between the target and the detector.

At the present stage of development heavy-ion-induced X-ray analysis is best suited to multielement measurements in thin samples. The recent availability of ultrathin window lithium-drifted silicon detectors should, however, promote research investigations on the gaseous elements nitrogen, oxygen, and fluorine. Another interesting possibility for the future is the recent discovery that chemical bonding can affect heavy-ion-induced K-X-ray spectra [91].

C. NUCLEAR REACTIONS

When the incident beam particle has enough energy to penetrate through the electrostatic Coulomb repulsion that develops as it approaches the nucleus of a target atom, a nuclear reaction can occur. Symbolically a nuclear reaction may be written

$$a + T \rightarrow b + R + Q, \tag{11}$$

where a is the incident nucleus, T is the target nucleus, b is the emitted radiation, R is the residual nucleus, and Q is the energy released (absorbed) in the reaction. Nuclear reactions can be exoergic, in which case Q is positive, or endoergic, in which case Q is negative. If the residual nucleus, R, is left in an excited state the Q for the reaction is reduced relative to the value that would obtain if the residual nucleus were left in its ground state by the amount of the excitation energy. The emitted radiation, b, can be a nuclear particle or one or more γ-rays. The case $a = b$ and $Q = 0$ is simply the elastic-scattering reaction, which has been discussed previously. Inelastic scattering corresponds then to $a = b$ and $Q < 0$. When $a \neq b$, the reaction is termed a *rearrangement collision*.

The conservation of total energy and momentum lead to the following expression (nonrelativistic) for the energy E_b of a nuclear particle emitted from a reaction:

$$E_b = E_a \frac{M_a M_b}{(M_b + M_R)^2} \left\{ 2\cos^2\theta + \frac{M_R(M_b + M_R)}{M_a M_b} \left[\frac{Q}{E_a} + \left(1 - \frac{M_a}{M_R}\right) \right] \right. \\ \left. \pm 2\cos\theta \left[\cos^2\theta + \frac{M_R(M_b + M_R)}{M_a M_b} \left\{ \frac{Q}{E_a} + \left(1 - \frac{M_a}{M_R}\right) \right] \right]^{1/2} \right\}. \tag{12}$$

E and M are energy and mass, respectively, and the subscript identifies the corresponding particle. In this formula θ is the angle between the emitted particle and the incident beam direction in the laboratory frame of reference. For elastic scattering, Eq. 12 reduces to Eq. 3.

Equation 12 shows that for a given reaction, incident energy E_a and observation angle θ, the energy spectrum E_b will depend on the Q-values possible to the reaction. In a manner quite analogous to optical spectroscopy, if the residual nucleus can be left in any of several sharply defined energy states and if the target is thin so the incident energy has a single well-defined value, the energy spectrum of the emitted particles will exhibit a series of narrow peaks, each of which corresponds to one of the excited states of the residual nucleus. Such a spectrum will then be highly specific to the reaction that has occurred and can be used to identify the presence of a given elemental isotope. The positions of the peaks in the energy spectrum are easily calculable from Eq. 12. The intensity ratios of the various peaks in the spectrum are also characteristic of the reaction and provide additional useful information for identifying the reaction responsible.

In the case of thin targets the amount of an isotope present can be calculated from the observed reaction yield, $Y_b(\theta)$, from the equation

$$Y_b(\theta) = nN\,\Delta x\sigma(E, \theta)\Omega, \tag{13}$$

where n is the total number of incident beam particles, N is the number of reactant atoms per cubic centimeters, Δx is the target thickness in centimeters, $\sigma(E, \theta)$ is the differential cross section in square centimeters per steradian at energy E and at angle θ, and Ω is the solid angle subtended by the detector. This equation is of the same form as Eq. 4 and, when integrated over all angles, as Eq. 1.

For thick targets, the energy-dependence of the cross section and any dependence of the concentration on depth in the sample must be taken into account. The thick target formula is then simply

$$Y_b(\theta) = n\Omega \int_0^{\mathscr{R}} N(x)\, \sigma(x, \theta)\, dx \tag{14}$$

where $\mathscr{R}$ is the range (maximum penetration distance) of the incident beam particle in the target, and the relationship between beam energy and depth x in the target is given by

$$E(x) = E_0 - \int_0^x \frac{dE}{dx'}\,(x')\, dx'. \tag{15}$$

In this equation E_0 is the beam energy just before it enters the target, and $(dE/dx)(x')$ is the energy loss of the beam per centimeter of target matter at depth x'; a number of stopping power tables exist that can be used to obtain this quantity (see, for example, Ref. 27.) Because penetration depth and beam energy are uniquely related, integration over target thickness can be performed either over distance or energy; Eqs. 2 and 14 are therefore basically equivalent.

The problems of target thickness, which were discussed in connection with charged-particle-activation analysis and Eq. 2, apply also here. Most often quantitative measurements are obtained by means of comparison with standard samples. If calculations are used, they are frequently based on approximations that avoid the full-scale integration called for in Eq. 14.

In principle, it should be possible through a series of yield measurements at various incident energies to unfold the depth dependence of $N(x)$. In practice, this is difficult to do when the reaction cross section varies only slowly with energy. The special and important case of reactions that exhibit a strong dependence on energy, the so-called *resonance reactions*, will be discussed for such application in a following section. For nonresonant reactions, information on depth distribution may be obtained in some cases, as is done in the backscattering technique, through careful measurement of the energy losses suffered by the observed particles b.

An experiment performed by Wolicki and Knudson [92] illustrates in a simple way how a nuclear reaction can be applied to surface analysis. In this work it was desired to detect sulfur nondestructively in thin films on a copper–nickel alloy in the presence of carbon and oxygen. The $^{32}S(d, p)^{33}S$ reaction was selected for the measurement because the Q-value (6.41 MeV) for this reaction is considerably higher than that for the (d, p) reaction on either carbon or oxygen. Interference from (d, p) reactions on copper and nickel was expected, but there existed the possibility that it would be small.

Initially a 4.0-MeV deuteron beam was used to bombard the samples. At this energy, however the intensity of protons from reactions produced in the copper and nickel isotopes was so high that the protons due to the thin sulfur film could not be observed. The deuteron beam energy was then reduced to 2.0 MeV. Because of the different Coulomb barriers involved, this reduction in beam energy produced a very large preferential reduction of the (d, p) yields from copper ($Z = 29$) and nickel ($Z = 28$) relative to those from sulfur ($Z = 16$). The discrete spectrum due to the thin sulfur film was thereby rendered easily identifiable.

The spectrum of protons obtained with a test sulfide film is shown in Fig. 20.17. The arrows labeled with symbols p_0 through p_8 show the energies calculated from Eq. 12 for protons emitted from the $^{32}S(d, p)^{33}S$ reaction. The symbol p_0 labels protons that correspond to ^{33}S being left in its ground state; p_1 corresponds to the first excited state of ^{33}S and so on. The correspondence of so many peaks with the expected energies provides a positive identification of sulfur on the irradiated sample. Not only the energy positions of the peaks but also the relative heights and, for some reactions, the widths of the peaks are all characteristic of the reaction that is responsible for the emitted particles. The sensitivity of the measurement was estimated to be easily 10^{-7} g/cm^2 with the possibility existing for considerable improvement beyond this value.

An excellent article on the analysis for oxygen by means of nuclear reactions has been written by Amsel and Samuel [93]. These authors discuss the reactions $^{18}O(p, \alpha)^{15}N$; $^{18}O(d, \alpha)^{16}N$; $^{16}O(d, \alpha)^{14}N$; $^{18}O(d, p)^{19}O$; and $^{16}O(d, p)^{17}O$. In each case the reaction is observed through the emitted charged particles. Relatively low bombarding energies (up to 600 keV) can be used, and the measurements determine either the absolute amount of each isotope of oxygen or the ratio ^{18}O–^{16}O. The reactions are highly specific, and amounts of the order of 10^{-12} g of ^{18}O can be detected. The article also

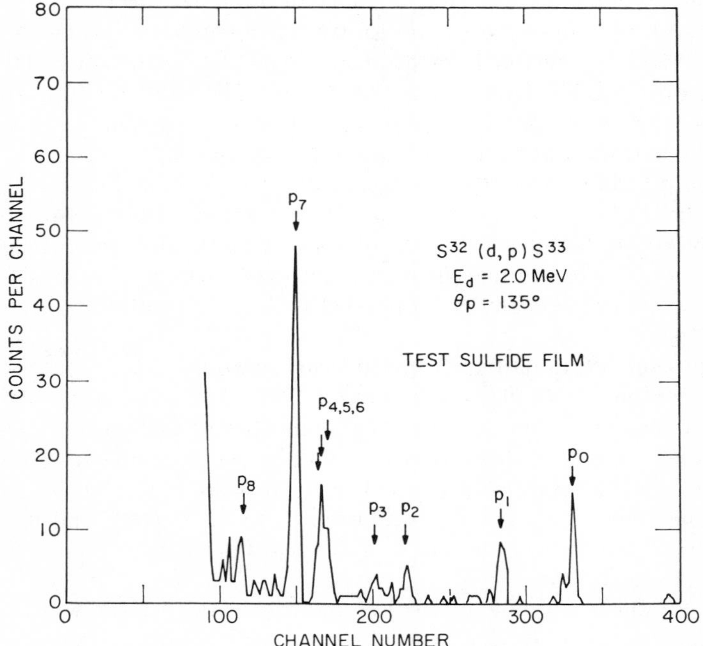

Fig. 20.17. Spectrum of protons emitted from the $^{32}S(d, p)^{33}S$ reaction in a 2.0-MeV deuteron irradiation of a thin sulfide film on a nickel alloy. (Reproduced by permission of *International Journal of Applied Radiation and Isotopes* [92].)

discusses the advantages of these measurements over all others, including nonnuclear analyses, and gives methods for correcting for targets of finite thickness.

The use of a 2-MeV Van de Graaff accelerator for nuclear reaction analysis for carbon, nitrogen, oxygen, fluorine, and other elements has been discussed by Amsel et al. [94]. The article includes a list of the most useful reactions along with discussions of the principles involved, particle detection instrumentation, concentration profile measurements with both resonant and nonresonant nuclear reactions, and several interesting applications.

Barrandon et al. [95] have studied (d, p) reactions for determining carbon and oxygen in the surfaces of metal samples. Results from the prompt-radiation analysis were compared with charged-particle-activation analysis through the $^{12}C(d, n)^{13}N$ and $^{16}O(t, n)^{18}F$ reactions and were found to be in agreement. Through these analyses it was discovered and quantitatively demonstrated that whatever the metal, the sample-surface always contains a certain quantity of carbon and oxygen after cleaning. It was also shown that this surface contamination causes an error in bulk carbon and oxygen determinations by combustion and fusion reduction and in oxygen determinations by 14-MeV neutron bombardment. With charged-particle-activation analysis, on the other hand, since the ^{13}N and ^{18}F radioactivities used are relatively long, it is possible to eliminate this error by cleaning the samples after radioactivation but before radiation measurements.

Oxygen diffusion in quartz has been studied by Palmer [96] through the use of enriched ^{18}O and the $^{18}O(p, \alpha)^{15}N$ reaction. Information on the concentration profile of the ^{18}O was obtained through measurements on the energies of the observed α-particles. The experimental α-particle energy resolution was 100 keV and corresponded to a depth resolution of approximately 2500 Å.

Prompt-neutron radiation from nuclear reactions has been used for analysis by Moller et al. [97] and by Peisach [98]. The former author discusses the use of (d, n) reactions to determine impurities of carbon, nitrogen, and oxygen on and below steel surfaces. Samples were irradiated by 3-MeV deuterons, and the spectra of emitted neutrons were measured by the time of flight method. Sensitivity was estimated to be 0.1 $\mu g/cm^2$, and the depth resolution, obtained from the shape of the peaks in the neutron spectra, was estimated to be 4500 Å.

Although the preceding discussion has dealt exclusively with emitted nuclear particles and Eq. 12 is derived for this case, prompt γ-radiation from nuclear reactions can also be used for analysis. The energy spectrum of the γ-rays is then characteristic of the nuclear reaction that has been produced. A reaction mode is also possible, called *direct capture*, for which the emitted γ-ray energy depends on the incident beam energy. By combining the excellent resolution of Ge(Li) detectors with this mode in the $^{16}O(p, \gamma)^{17}F$ reaction, Joy and Barnes [99] were able recently to achieve depth distribution measurements on oxygen in metals.

D. NUCLEAR RESONANCE REACTIONS

Many nuclear reactions have the property that the reaction yield exhibits resonances at one or more sharply defined incident ion energies. Experimentally such a yield curve is obtained one point at a time by a variation of the incident beam energy in small increments and a measurement of the quantity of emitted radiation per unit incident beam fluence (number of particles incident on a unit cross-sectional area) at each energy. The resonances and their energy widths correspond to the energy levels and level widths, respectively, in the compound nucleus that is formed. With reference to Eq. 11 the compound nucleus formed is $(a + T)$; the emitted radiation, b, can be one or more γ-rays or a nuclear particle. In somewhat simplified form, the resonant-reaction cross section can be written

$$\sigma(E) = \sigma_0 \frac{(\Gamma^2/4)}{(E - E_R)^2 + (\Gamma^2/4)}, \tag{16}$$

where E_R is the resonance energy, Γ is the full width of the resonance at half-maximum, and σ_0 is the peak cross section at resonance.

From the standpoint of elemental analysis, the resonance properties of a nuclear reaction are of interest for two reasons. The first is that the resonance energy, E_R, is characteristic of a particular nuclear reaction and can be used for identifying that reaction and for making a selective measurement on one element (actually just one isotope of that element) in the presence of many others. The second is that the energy dependence of the cross section, when combined with the energy loss of the incident beam as it penetrates into the target, makes possible depth distribution measurements on the element in question. Of particular importance is the fact that nuclear-reaction resonances can be sufficiently sharp so that depth resolutions better than 100 Å can be achieved in some cases. For such high resolution the resonance width and the energy loss of the beam in the target must be accurately taken into account.

The width, Γ, of the reaction yield curve that is observed experimentally is composed of three parts; the intrinsic width of the level in which the compound nucleus is formed, Doppler broadening due to the thermal motion of the reactant target atoms, and the energy inhomogeneity of the incident beam particles. The intrinsic width of a resonance depends only on the characteristics of the corresponding compound-nucleus energy level, so it needs to be evaluated only once for a particular resonance. The Doppler broadening can be calculated to a useful accuracy from the Debye temperature of a material. The energy inhomogeneity of the beam depends not only on the way in which the beam is prepared but also on the variations in energy loss that occur because of the statistical nature of the energy loss process as a charged particle traverses matter. Because of this *energy loss straggling*, as it is called, the energy inhomogeneity is a function of the depth of the beam as it penetrates into the target. When all the above factors are known and provided the beam energy inhomogeneity is not overly large, an experimentally observed yield curve $Y(E)$ can be unfolded to give the depth distribution $N(x)$ of the reactant atoms. The calculation, which involves a triple integration, requires the use of a large digital computer program [100].

A recent set of measurements on the depth distribution of a thin layer of aluminum within a silicon dioxide surface [101] provides a simple example for illustrating how the nuclear-reaction-resonance technique is used in practice and what kinds of results can be achieved with it. This case is a good example for discussion because the $^{27}Al(p, \gamma)^{28}Si$ reaction used for these measurements has an unusually sharp resonance and a relatively large peak cross section. The depth resolution and quantitative sensitivity of the technique are therefore both good.

Consider a target sample as shown in the top portion of Fig. 20.18. The sample consists of SiO_2 in which there are contained two thin layers of Al, one at the very surface and another, with a lower density of Al atoms in it, at some distance in from the surface.

For the ideal case of an infinitely narrow resonance width, a perfectly monoenergetic proton beam, and no straggling in energy loss as the beam

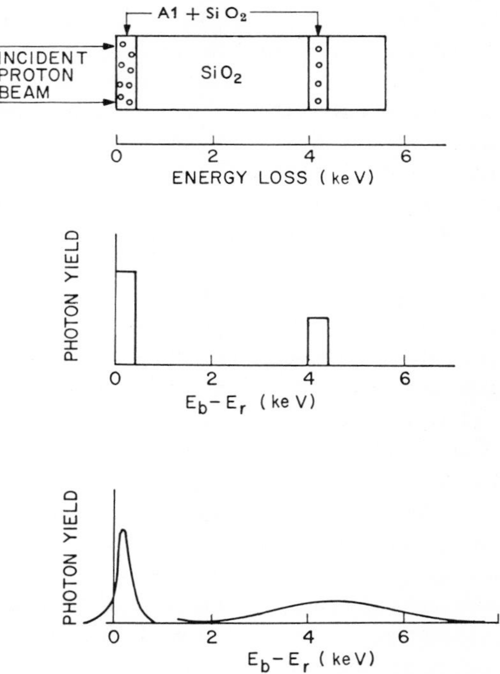

Fig. 20.18. Schematic diagrams showing a SiO_2 sample containing two thin layers of Al, and the corresponding idealized and real yield curves for the $^{27}Al(p, \gamma)^{28}Si$ reaction. (Reproduced by permission of the author [101].)

penetrates into the target, the γ-ray-yield curves that would be obtained are shown in the middle portion of the figure. Along the abscissa is plotted the difference between the incident beam energy and the reaction-resonance energy. For the surface layer the yield starts at the resonance energy and the width of the yield curve, in energy units, is proportional to the thickness, in atoms per square centimeter, of the Al–SiO_2 layer. The width of the curve is caused by the energy loss the protons undergo as they traverse the Al–SiO_2 layer. Thus, while the SiO_2 does not contribute to the γ-ray yield, it does cause the protons to lose energy and so must be taken into account along with the Al where the width of the yield curve is concerned. The area under the curve is proportional to the number of Al atoms per square centimeter in the layer.

For the internal layer, as it is shown, the resonance is displaced to a higher proton energy by 4 keV because the protons have lost this amount of energy in traversing the material between the surface and the internal layer. Since the density of Al atoms is somewhat lower in this layer than in the surface layer, the yield is correspondingly lower. The width of the yield curve is again proportional to the thickness of the Al–SiO_2 layer, and the area is proportional to the amount of Al.

The real situation is shown in the bottom part of the figure. The yield curves shown now reflect the real width of the resonance, Doppler broadening due to the thermal motion of the Al atoms, inhomogeneity in the beam energy, and beam-energy-loss straggling. It is this last factor that so greatly broadens the yield curve obtained for the internal layer. Given the distribution of the Al in the SiO_2 as shown in the top of the figure, the intrinsic width of the resonance, the beam energy inhomogeneity, and the computer program [100], it is possible to predict the yield curves shown at the bottom of the figure. Conversely, if a yield curve is observed as shown, the program can, by iteration, be used to obtain the Al distribution.

Results obtained by Dunning [101], for a SiO_2 sample into which 60-keV Al atoms were implanted to a fluence of 1×10^{15} atoms/cm^2, are shown in Fig. 20.19. The open triangles are the theoretical fit to the data, and the crosses are the Al depth distribution required to fit the data. The measurements showed that while the bulk of the Al had the depth distribution that would be expected for a 60-keV implantation, there was a small concentration of Al very close to the surface. The reasons why some of the Al came to rest at that depth are not known. The number of iterations required to obtain the depth distribution was four, a relatively small number. The depth resolution obtained in the measurements was estimated to be between 50 and 100 Å. For this work the resonance in the $^{27}Al(p, \gamma)^{28}Si$ reaction that occurs at 992 keV was used. This resonance has an intrinsic width of approximately 100 eV [102]. With a magnetic or electrostatic beam energy analysis

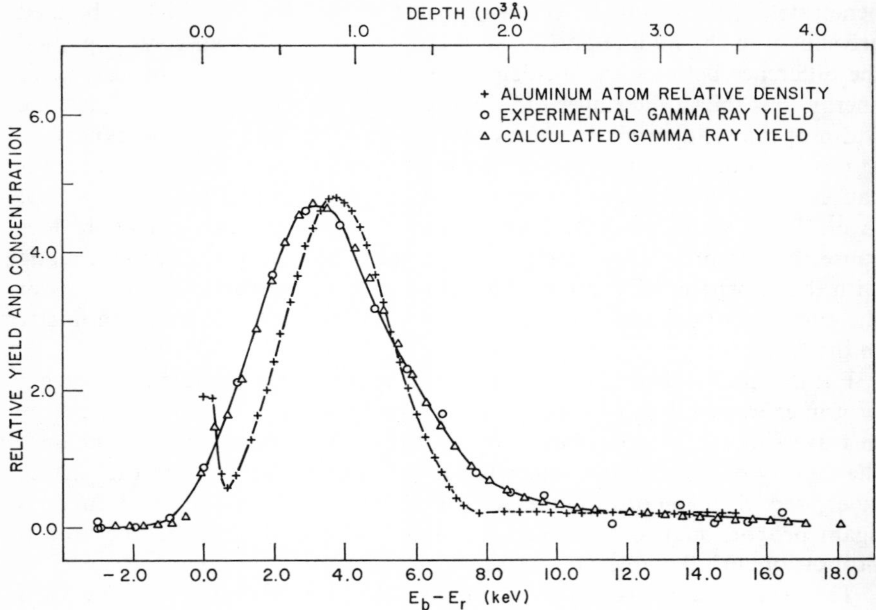

Fig. 20.19. Experimental and calculated γ-ray-yield curves and a depth distribution profile obtained for a thin layer of Al implanted into SiO_2. (Reproduced by permission of the author [101].)

system and a suitably thin Al target, an experimental yield curve width of 1 keV or less is readily achievable for this reaction. The depth resolution of 50–100 Å comes about because the center of the resonance can be determined with a relative accuracy that is better than the full width of the resonance.

Results from a somewhat different application of the same resonance reaction are shown in Fig. 20.20. In these experiments, which were performed by Bernett et al. [103], the object was to measure the amount of aluminum oxide powder of 0.3-μ diameter grit size, which remained in metals of various hardnesses after these metals had been polished with this powder and then rinsed clean.

The top figure shows the results obtained for zirconium. The resonance was observed at the correct energy for the grit particles to be located right on the surface of the sample, and the width of the yield curve corresponds closely to the 0.3-μ diameter of these particles. The area under the curve is proportional to the average number of Al atoms per square centimeter.

The middle figure shows the results of a similar measurement for gold. This curve has a greater area and is broader than the first, these features

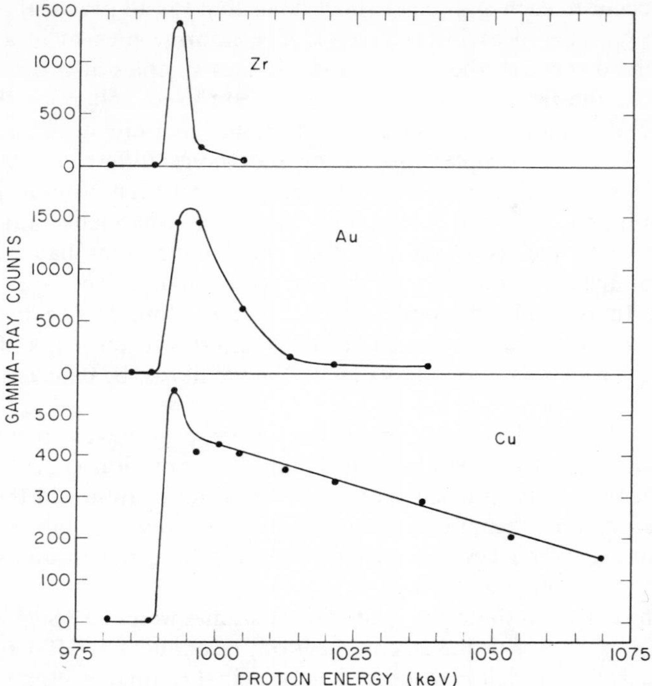

Fig. 20.20. Yield curves for the $^{27}Al(p, \gamma)^{28}Si$ reaction for Al_2O_3 polishing compound retained in Zr, Au, and Cu surfaces. (Reproduced by permission of the authors [103].)

showing, respectively, that more polishing compound has been retained in the sample and that some of the particles are embedded in the gold. The bottom curve shows finally that in copper there is still more polishing compound retained in the sample and that the grit particles are embedded still more deeply. In this latter case the interesting finding was that some of the particles are buried at least 1 μm deep and are covered over by the copper even though the polishing was performed by hand on a relatively slow-moving polishing surface. The thickness of copper covering the Al_2O_3 particles is evidenced by the fact that the reaction yield is still appreciable at 1070 keV, an energy that is over 70 keV above the resonance energy.

Measurements on fluorine contamination on and below the surface of Zircaloy were performed by Moller and Starfelt [104] by means of a resonance at 1375 keV in the $^{19}F(p, \alpha\gamma)^{16}O$ reaction. The notation in this case means that the compound nucleus ^{20}Ne, formed by the proton's capture in ^{19}F, decays by α-emission to an excited energy state of ^{16}O, which then emits γ-rays as it decays to its ground state. The α- and γ-emissions, although sequential, occur in such extremely short times that for all practical purposes they can be considered as instantaneous. The fluorine present in a sample can be observed through the γ-rays, and for thin samples also through the α-particles. In the present case, the γ-rays were used and both the total amount and the depth distribution of the fluorine were measured for a number of Zircaloy samples that had undergone different treatments. Altogether 29 samples were studied. The samples had been ground, pickled, or electropolished and had undergone autoclave treatments varying in temperature and time. It was found that the samples that had not been oxidized contained fluorine only on the surface or in a surface layer thinner than 0.1 μm. In the oxidized samples, the fluorine was found to be distributed down to depths of several microns. The yield from this reaction is sufficiently high that a quantity of fluorine less than 0.01 $\mu g/cm^2$ can be detected by this method.

Chemin et al. [105] have used resonances in (p, p$'\gamma$) nuclear reactions to measure trace amounts of Si and S in GaSb. The notation (p, p$'\gamma$) means that after the compound nucleus is formed by proton capture in the target nucleus, the proton is reemitted with a reduced energy, leaving the target nucleus in an excited energy state, which then decays to the ground state by the emission of one or more γ-rays.

The specific reactions that were used in these studies were the $^{28}Si(p, p'\gamma)^{28}Si$ and the $^{32}S(p, p'\gamma)^{32}S$. For the silicon reaction, resonances at 3100 and 3335 keV were used; for the sulfur reaction, the principal resonance used was that at 3379 keV. These reactions provided the excellent detection sensitivities of 0.12×10^{-9} g for sulfur and 0.16×10^{-9} g for silicon. Comparison was made with detection sensitivities for spectrophotometry, fluorescence, atomic absorption, flame spectrophotometry, neutron-activation analysis, and spark-source mass spectroscopy. Only spark-source mass spectroscopy with a sensitivity of 0.03×10^{-9} g for both silicon and sulfur exceeded the sensitivity of the (p, p$'\gamma$) resonance-reaction technique. The remaining techniques in general were 2–4 orders of magnitude less sensitive than the resonance technique.

As a final example, Whitton et al. [106] used the 639-keV resonance in the $^{18}O(p, \alpha)^{15}N$ reaction to measure the depth distribution of oxygen implanted into gallium phosphide at implantation energies of 20, 40, and 60 keV. These experiments differ from the previous examples cited in that the emitted radiation observed consisted not of γ-rays but of α-particles. The quantity of ^{18}O implanted in various samples varied from approximately 2×10^{15} atoms/cm^2 to 7×10^{15} atoms/cm^2; these values correspond, respectively, to 6×10^{-8} and 21×10^{-8} g/cm^2 of ^{18}O. Depth distribution curves for the implanted ^{18}O were measured before and after annealing for 15 min in vacuo at 600°C. The shift and broadening of the depth distribution after annealing were found to be consistent with thermal diffusion. The measurements showed that the $^{18}O(p, \alpha)^{15}N$ resonance reaction can be used to distinguish differences in the range and range straggling of oxygen implanted into GaP to depths that are of interest in the formation of electroluminescent devices. The reaction was judged to be useful also for oxygen-profiling in a large number of host materials.

REFERENCES.

1. J. P. Guinn, Ed., Proceedings of the 1965 International Conference on Modern Trends in Activation Analysis, College Station, Texas, April 19–22, 1965, Texas A & M University, College Station, Texas, 1965.
2. *Proceedings of the First Conference on Practical Aspects of Activation Analysis with Charged Particles* (Euratom, EUR 2957 d-f-e), Grenoble, France, June 23, 1965.
3. R. S. Tilbury, *Activation Analysis with Charged Particles*, Nat. Acad. Sci.–Nat. Res. Council Nuclear Science Series, NAS-NS3110 Clearinghouse for Federal Sci. and Tech. Info., Springfield, Va., 1966.
4. A. A. Smales, "Radioactivity Techniques in Trace Characterization," in W. W. Meinke and B. F. Scribner, Eds., *Trace Characterization, Chemical and Physical, Monograph 100* National Bureau of Standards, Washington, D.C., 1967, p. 307.
5. V. P. Guinn, "Nuclear Methods," in W. W. Meinke and B. F. Scribner, Eds., *Trace Characterization, Chemical and Physical, Monograph 100* National Bureau of Standards, Washington, D.C., 1967, p. 337.
6. H. G. Ebert, Ed., *Proceedings of the 2nd Conference on Practical Aspects of Activation Analysis with Charged Particles*, (Euratom, EUR 3896 d-f-e), September 21–22, 1967, Liege, Belgium, 1968.
7. J. R. DeVoe, Editor, *Modern Trends in Activation Analysis*, Special Publication 312, Vols. I and II, National Bureau of Standards, Washington, D.C., 1969.
8. E. H. Krause, "Particle Accelerators," Section 8 in *American Institute of Physics Handbook*, McGraw-Hill, New York, 1957, Tables 8i-1 and 8i-2.
9. R. H. Marsh and W. Allie, Jr., "High Precision Activation Analysis of Sodium Using an Internal Standard Technique," In Ref. 7, Vol. II, p. 1285.
10. F. F. Dyer, L. C. Bate, and J. E. Strain, "Three-Dimensionally Rotating Sample Holder for 14-Million Electron Volt Neutron Irradiation," *Anal. Chem.*, **39,** 1907 (1967).
11. J. Comas and E. A. Wolicki, "Argon Content in (111) Silicon for Sputtering Energies Below 200 eV," *J. Electrochem. Soc.*, **117,** 1197 (1970).
12. S. S. Markowitz and J. D. Mahony, "Activation Analysis for Oxygen and Other Elements by Helium-3-Nuclear Reactions," *Anal. Chem.*, **34,** 329 (1962).
13. J. D. Mahony, "Reactions of He^3 with Light Elements: Applications to Activation Analysis," *Rept. UCRL 11780*, Lawrence Radiation Lab, 1965.
14. E. L. Steel, "Helium-3 Charged Particles in Activation Analysis," *General Atomic Rept. GA-6568*, 1965.

15. E. Ricci and R. L. Hahn, "Theory and Experiment in Rapid, Sensitive Helium-3 Activation Analysis," *Anal. Chem.*, **37,** 742 (1965).
16. E. Ricci and R. L. Hahn, "Sensitivities for Activation Analysis of 15 Light Elements with 18-MeV Helium-3 Particles," *Anal. Chem.*, **39,** 794 (1967).
17. R. W. Benjamin, L. D. England, K. R. Blake, I. L. Morgan, and C. D. Houston, "Deuteron Activation Analysis for C, N, and O in High-Purity Metallic Surfaces," *Trans. Amer. Nucl. Soc.*, **9,** 104 (1966).
18. J.-L. Debrun, J.-N. Barrandon, and Ph. Albert, "Contribution to Activation Analysis by Charged Particles; Determination of Carbon and Oxygen in Pure Metals, Possibilities of Sulphur Determination," in Ref. 7, Vol. II, p. 774.
19. J.-N. Barrandon and Ph. Albert, "Determination of Oxygen Present at the Surface of Metals by Irradiation with 2 MeV Tritons," in Ref. 7, Vol. II, p. 794.
20. G. Revel and Ph. Albert, "A Study of the Possibilities of Measuring Oxygen in Zirconium, Molybdenum, Hafnium and Tungsten by Irradiation with ^{3}He and ^{4}He Particles," in Ref. 6.
21. P. Albert, "Nuclear Methods for Determination of Oxygen and Carbon on Metal Surfaces," *Vide*, **24,** 149 (1969).
22. J. W. Butler and E. A. Wolicki, "Surface Analysis of Gold and Platinum Disks by Activation Methods and by Prompt Radiation From Nuclear Reactions," in Ref. 7, Vol. II, p. 791.
23. C. Engelmann, J. Gosset, M. Loeuillet, A. Marschal, P. Ossart and M. Boissier, "Examples of Determination of Light Elements in Various High Purity Materials by Gamma Photon and Charged Particle Activation," in Ref. 7, Vol. II, p. 819.
24. A. R. Knudson and K. L. Dunning, "Detection of Chlorine on Aluminum by Means of Nuclear Reactions," *Anal. Chem.*, **44,** 1053 (1972).
25. R. C. Koch, *Activation Analysis Handbook*, Academic, New York, 1960.
26. G. J. Lutz, R. J. Boreni, R. S. Maddock, and W. W. Meinke, Eds., *Activation Analysis, A Bibliography, Tech. Note 467*, Pts. 1 and 2 National Bureau of Standards, Washington, D.C., 1971.
27. C. Williamson, J. Borojot, and J. Picard, *Tables of Charged Particle Energy Losses, CEA-R3042*, 1966.
28. D. M. Holm, J. A. Basmajian, and W. M. Sanders, "Observations of the Microscopic Distribution of Oxygen and Carbon in Metals by He^3 Activation," *Rept. LA-3515*, Los Alamos Scientific Laboratory, 1966.
29. J. C. Ritter, M. N. Robinson, B. J. Faraday and J. I. Hoover, "Room Temperature Oxidation of Silicon During and After Etching," *J. Phys. Chem. Solids*, **26,** 721 (1965).
30. C. A. Carosella and J. Comas, "Oxygen Sticking Coefficients on Clean (111) Silicon Surfaces," *Surface Sci.*, **15,** 303 (1969).
31. J. A. Cookson and F. D. Pilling, "A 3 MeV Proton Beam of Less Than Four Microns Diameter," *AERE-R-6300*, Harwell, 1970.
32. T. B. Pierce, "The Examination of Surfaces By Scanning With Charged Particle Beams," *Proc. Soc. Anal. Chem.*, **7,** 59 (1970).

33. S. Rubin, T. O. Passel, and L. E. Bailey, "Chemical Analysis of Surfaces by Nuclear Methods," *Anal. Chem.*, **29,** 736 (1957).
34. A. L. Turkevich, E. J. Franzgrote and J. H. Patterson, "Chemical Analysis of the Moon at the Surveyor V Landing Site," *Science*, **158,** 635 (1967).
35. O. U. Anders, "Use of Charged Particles From a 2 Megavolt Van de Graaff Accelerator for Elemental Surface Analysis," *Anal. Chem.*, **38,** 1442 (1966).
36. M. Peisach and D. O. Poole, "The Use of Semiconductor Detectors for Surface Analysis by Elastic Scattering of Accelerated Charged Particles," in Ref. 1, p. 206.
37. E. Bøgh and E. Uggerhøj, "Experimental Investigation of Orientation Dependence of Rutherford Scattering Yield in Single Crystals," *Nucl. Inst. Methods*, **38,** 216 (1965).
38. D. P. Smith, "Scattering of Low-Energy Noble Gas Ions from Metal Surfaces," *J. Appl. Phys.*, **38,** 340 (1967).
39. D. A. Thompson, H. D. Barber, and W. D. Mackintosh, "The Determination of Surface Contamination on Silicon by Large Angle Ion Scattering," *Appl. Phys. Lett.*, **14,** 102 (1969).
40. O. Meyer, J. Gyulai and J. W. Mayer, "Analysis of Amorphous Layers on Silicon by Backscattering and Channeling Effect Measurements," *Surface Sci.*, **22,** 263 (1970).
41. I. V. Mitchell, M. Kamoshida and J. W. Mayer, "Channeling-Effect Analysis of Thin Films on Silicon: Aluminum Oxide," *J. Appl. Phys.*, **42,** 4378 (1971).
42. J. Gyulai, J. W. Mayer, I. V. Mitchell and V. Rodriguez, "Outdiffusion Through Silicon Oxide and Silicon Nitride Layers on Gallium Arsenide," *Appl. Phys. Lett.*, **17,** 332 (1970).
43. T. B. Pierce, P. F. Peck and D. R. A. Cuff, "Examination of Contamination on the Surface of an Alumina Insulator by Elastic α-Particle Scattering," *J. Inorg. Nucl. Chem.*, **33,** 1963 (1971).
44. S. T. Picraux and F. L. Vook, "Multilayer Thin-Film Analysis by Ion Backscattering," *Appl. Phys. Lett.*, **18,** 191 (1971).
45. M. Croset, S. Rigo and G. Amsel, "Investigation of the Composition of Sputtered Silicon Nitride Films by Nuclear Microanalysis," *Appl. Phys. Lett.*, **19,** 33 (1971).
46. A. Van Wijingaarden, B. Miremadi and W. E. Baylis, "Energy Spectra of keV Backscattered Protons as a Probe for Surface-Region Studies," *Can. J. Phys.*, **49,** 2440 (1971).
47. D. P. Smith, "Analysis of Surface Composition with Low-Energy Backscattered Ions," *Surface Sci.*, **25,** 171 (1971).
48. A. Hiraki, M.-A. Nicolet, and J. W. Mayer, "Low-Temperature Migration of Silicon in Thin Layers of Gold and Platinum," *Appl. Phys. Lett.*, **18,** 178 (1971).
49. B. L. Cohen, C. L. Fink, and J. H. Degnan, "Non-destructive Analysis for Trace Amounts of Hydrogen," *J. Appl. Phys.*, **43,** 19 (1972).
50. T. M. Buck and G. H. Wheatley, "Studies of Solid Surfaces with 100 keV He^+ and H^+ Ion Beams," *Surface Sci.*, **33,** 35 (1972).
51. R. L. Meek and W. M. Gibson, "Determination of Lighter Impurities on

Silicon by 90° Forward Ion Scattering Through Thin Targets," *Nucl. Inst. Methods*, **98,** 375 (1972).

52. J. K. Hirvonen, W. H. Weisenberger, J. E. Westmoreland, and R. A. Meussner, "Backscattering Investigation of Low Temperature Migration of Chromium Through Gold Films," *Appl. Phys. Lett.*, **21,** 37 (1972).
53. J. U. Andersen, "Axial and Planar Dips in Reaction Yield for Energetic Ions in Crystal Lattice," *Mat. Fys. Medd. Dan Vid Selsk*, **36** (7) (1967).
54. E. Bøgh, "Interaction of Radiation with Solids," A. Bishay, Ed., *Proceedings of Cairo Solid State Conference, American University in Cairo, 1966*, Plenum, New York, 1967.
55. J. A. Davies, J. Denhartog, L. Eriksson, and J. W. Mayer, "Atom Location and Lattice Disorder by Means of 1.0-MeV Helium Ion Scattering," *Can. J. Phys.*, **45,** 4053 (1967).
56. F. H. Eisen, B. Welch, J. E. Westmoreland, and J. W. Mayer, "Lattice Disorder Produced in Silicon by Boron Ion Implantation," *Proceedings of International Conference on Atomic Collision Phenomena in Solids*, University of Sussex, Brighton, England, 1969, North-Holland Publ. Co., 1970, p. 111.
57. R. R. Hart, H. L. Dunlap and O. J. Marsh, "Disorder Produced in SiC by Ion Bombardment," *Radiation Effects*, **9,** 261 (1971).
58. J. W. Mayer, L. Eriksson, S. T. Picraux, and J. A. Davies, "Ion Implantation of Silicon and Germanium at Room Temperature. Analysis by Means of 1.0-MeV Helium Ion Scattering," *Can. J. Phys.*, **46,** 663 (1968).
59. W. H. Weisenberger, S. T. Picraux, and F. L. Vook, "Low Temperature Channeling Measurements of Ion Implantation Lattice Disorder in GaAs," *Radiation Effects*, **9,** 121 (1971).
60. J. E. Westmoreland, J. W. Mayer, F. H. Eisen, and B. Welch, "Analysis of Disorder Distributions in Boron Implanted Silicon," *Proceedings of the International Conference on Ion Implantation in Semiconductors*, Thousand Oaks, Calif., 1970, *Radiation Effects*, **6,** 161 (1970).

61a. L. Eriksson, J. A. Davies, N. G. E. Johansson, and J. W. Mayer, "Implantation and Annealing Behavior of Group III and V Dopants in Silicon as Studied by the Channeling Technique," *J. Appl. Phys.*, **40,** 842 (1969).

61b. F. H. Eisen, "New Aspects of Atom Location: Flux Peaking," in Ref. 68 p. 287.

61c. R. B. Alexander, G. Dearnaley, D. V. Morgan, J. M. Poate, and D. Van Vliet, "Lattice Site Location Measurements and Their Interpretation," in *Proceedings of European Conference on Ion Implantation*, Sept. 7–9, 1970, Reading, England, published by Peter Peregrinus Ltd, Stevenage, Herts., England, 1970, p. 181.

62. Hj. Matzke, J. A. Davies, and N. G. E. Johansson, "A Channeling Study of the Structure of U_4O_9," *Can. J. Phys.*, **49,** 2215 (1971).
63. J. A. Davies, "Channeling—A New Look in Crystallography and Solid State Investigations," *Phys. Can.*, **23** (3), 13 (1967).
64. W. M. Gibson, "Particle Channeling Experiments Using Small Accelerators," in J. L. Duggan, Ed., *Proceedings of the Conference on the Use of Small Accelerators For Teaching and Research*, Oak Ridge Associated Universities, Oak Ridge Tennessee, April 8–10, 1968.

65. J. A. Davies, "Rutherford Scattering and Channeling: A Useful Combination for Studying Crystal Surfaces," *J. Vac. Sci. Technol.*, **8,** 487 (1971).
66. *European Conference on Ion Implantation*, Reading, England, September 7–9, 1970, Peregrinus, Stevenage, Herts, England, 1970.
67. F. H. Eisen and L. T. Chadderton, Eds., *Ion Implantation*, Gordon Breach Science, New York, 1971.
68. I. Ruge and J. Graul, Eds., *Proceedings of the II International Conference on Ion Implantation in Semiconductors*, Garmisch-Partenkirchen, Bavaria, Germany, Springer Verlag, New York, 1971.
69. J. W. Mayer, L. Eriksson, and J. A. Davies, *Ion Implantation in Semiconductors*, Academic, New York, 1970.
70. J. Gyulai, O. Meyer, J. W. Mayer, and V. Rodriguez, "Analysis of Silicon Nitride Layers on Silicon by Backscattering and Channeling Measurements," *Appl. Phys. Lett.*, **16,** 232 (1970).
71. I. V. Mitchell, M. Kamoshida, and J. W. Mayer, "Lineshape Extraction from MeV He^+ Backscattering Energy Spectra: Aluminum Oxide on Silicon," *Phys. Lett.*, **35A** (1), 21 (1971).
72. A. G. Pieper and R. B. Theus, "Cyclotron Helium Implantation," in *NRL Memorandum Report 2394, Cooperative Radiation Effects Simulation Program*, Naval Research Laboratory, Washington, D.C., 1972, p. 69.
73. T. A. Cahill, "Use of Alpha-Particle Excited X-Ray Emission for Analysis of Smog Aerosols," *Bull. Amer. Phys. Soc. Ser. II*, **17,** 505 (1972).
74. R. F. Goff and D. P. Smith, "Surface Composition Analysis by Binary Scattering of Noble Gas Ions," *J. Vac. Sci. Technol.*, **7,** 72 (1970).
75. D. J. Ball, T. M. Buck, D. MacNair, and G. H. Wheatley, "Investigation of Low Energy Ion Scattering as a Surface Analytical Technique," *Surface Sci.*, **30,** 69 (1972).
76. E. Everhart, G. Stone, and R. J. Carbone, "Classical Calculation of Differential Cross Section for Scattering from a Coulomb Potential with Exponential Screening," *Phys. Rev.*, **99,** 1287 (1955).
77. W. Heiland and E. Taglauer, "Investigation of Surface Topography of Oxygen on Nickel Single Crystals by He Ion Backscattering," International Conference on Solid Surfaces, American Vac. Society, Boston, Mass., October 11-15, 1971.
78. L. S. Birks, R. E. Seebold, A. P. Bolt, and J. S. Grosso, "Excitation of Characteristic X-Rays by Protons, Electrons, and Primary X-Rays," *J. Appl. Phys.*, **35,** 2578 (1964).
79. S. M. Poole, J. S. Shaw, "Microanalysis with a Proton-Probe," in G. Möllenstedt and K. H. Gaukler, Eds., Vth International Congress on X-Ray Optics and Microanalysis, Tübingen, 1968, Springer-Verlag New York.
80. T. B. Johansson, R. Akselsson, and S. A. E. Johansson, "X-Ray Analysis: Elemental Trace Analysis at the 10^{-12} g Level," *Nucl. Inst. Methods*, **84,** 141 (1970).
81. J. A. Cairns, C. L. Desborough, and D. F. Holloway, "A New End Window Variable Geometry X-Ray Proportional Counter," *Nucl. Inst. Methods*, **88,** 239 (1970).

82. B. M. Gordon and H. W. Kraner, "Development of a System for Trace Element Analysis in the Environment by Charged Particle X-Ray Fluorescence," *BNL-16182*, 1971.
83. R. G. Flocchini, P. J. Feeney, R. J. Sommerville, and T. A. Cahill, "Sensitivity Versus Target Backings For Elemental Analysis by Alpha Excited X-Ray Fluorescence," *Nucl. Inst. Methods*, **100,** 397, (1972).
84. R. L. Watson, J. R. Sjurseth, and R. W. Howard, "An Investigation of the Analytical Capabilities of X-Ray Emission Induced by High Energy Alpha Particles," *Nucl. Inst. Methods*, **93,** 69 (1971).
85. J. W. Verba, J. W. Sunier, B. T. Wright, I. Slaus, A. B. Holman, and J. G. Kulleck, "Limitations and Improvements of Trace Element Analysis With Proton Induced X-Rays," paper presented at the International Conference on Chemical Analysis by Charged Particle Bombardment, L.A.R.N. Nemur Belgium, September 5–8, 1971.
86. P. K. Mueller, A. Alcocer, T. A. Cahill, R. Sommerville, and R. Flocchini "Elemental Analysis by Alpha Excited X-Ray Fluorescence," paper presented at the American Chemical Society Meeting, Los Angeles, Calif., March 1972.
87. J. M. Khan, D. S. Potter, and R. S. Worley, "Proposed Method for Microgram Surface Density Measurements by Observation of Proton-Produced X-Rays," *J. Appl. Phys.*, **37,** 564 (1966).
88. J. A. Cairns and R. S. Nelson, "Selective X-Ray Generation By Heavy Ions, Part 1—The Use of Energetic Heavy Ions to Generate Characteristic X-Rays from Elements in a Selective Manner," *Radiation Effects*, **7,** 163 (1971).
89. J. A. Cairns, D. T. Holloway, and R. S. Nelson, "Selective X-Ray Generation by Heavy Ions Part 2—Measurement of the Concentration Distribution of Ion Implanted Antimony in Silicon by the Use of Selective Heavy Ion X-Ray Excitation," *Radiation Effects*, **7,** 167 (1971).
90. A. R. Knudson and P. E. Wilkniss, private communication, February 1972.
91. P. G. Burkhalter, A. R. Knudson, D. J. Nagel, and K. L. Dunning, "Chemical Effects in Ion-Excited Aluminum K X-Ray Spectra," NRL *Memo Report 2438*, 1972, p. 13.
92. E. A. Wolicki and A. R. Knudson, "A Nuclear-Detection Method For Sulfur in Thin Films," *Int. J. Appl. Radiat. Isotope.*, **18,** 429 (1967).
93. G. Amsel and D. Samuel, "Microanalysis of the Stable Isotopes of Oxygen by Means of Nuclear Reactions," *Anal. Chem.*, **39,** 1689 (1967).
94. G. Amsel, J. P. Nadai, E. D'Artemare, D. David, E. Girard, and J. Moulin "Microanalysis by the Direct Observation of Nuclear Reactions Using a 2 MeV Van de Graaff," *Nucl. Inst. Methods*, **92,** 481 (1971).
95. J. N. Barrandon, L. Quaglia, J. L. Debrun, M. Cuypers, and G. Robaye, "Utilization of Low Energy Deuterons and Tritons for Measuring Carbon and Oxygen Present at the Surface of Metal Samples," *J. Radioanal. Chem.*, **4,** 115 (1970).
96. D. W. Palmer, "Oxygen Diffusion in Quartz Studied by Proton Bombardment," *Nucl. Instr. Methods*, **38,** 187 (1965).
97. E. Moller, L. Nilsson, and M. Starfelt, "Microanalysis of Light Elements by Means of (d, n) Reactions," *Nucl. Inst. Methods*, **50,** 270 (1967).

98. M. Peisach, "Analytical Use of Prompt Neutrons Produced by Pulses of Charged Particles," in Ref. 6, p. 65.
99. T. Joy and D. G. Barnes, "Some Practical Applications of the Observation of the ^{16}O-Direct Proton Capture Reaction With a Ge(Li) Detector," *Nucl. Inst. Methods*, **95,** 199 (1971).
100. K. L. Dunning, "Fortran Programs for (p, γ) Yield Calculations Based on Vavilov's Theory of Energy Loss Distributions," *NRL Report 7230*, 1971.
101. K. L. Dunning, "Aluminum in SiO_2; Depth Profiles by Means of a Nuclear Resonance Reaction," *NRL Memorandum Report 2438*, 1972, p. 35.
102. R. O. Bondelid and C. A. Kennedy, "Precise Determination of Nuclear Reaction Energies and Measurements of Resonance Widths," *Phys. Rev.*, **115,** 1601 (1959).
103. M. K. Bernett, J. W. Butler, E. A. Wolicki, and W. A. Zisman, "Relation Between Residual Polishing Agent and Material Hardness Determined by a Narrow Nuclear Reaction Resonance Technique," *J. Appl. Phys.*, **42,** 5826 (1971).
104. E. Moller and N. Starfelt, "Microanalysis of Fluorine in Zircaloy by the Use of the $^{19}F(p, \alpha\gamma)^{16}O$ Reaction," *Nucl. Inst. Methods*, **50,** 225 (1967).
105. J. F. Chemin, J. Roturier, B. Saboya, and G. Y. Petit, "Microanalysis of Si and S in GaSb by Direct Observation of Resonant Nuclear Reactions," *Nucl. Inst. Methods*, **97,** 211 (1971).
106. J. L. Whitton, I. V. Mitchell, and K. B. Winterbon, "Depth Distribution of Implanted Oxygen in Gallium Phosphide Measured by Nuclear Reaction Techniques," *Can. J. Phys.*, **49,** 1225 (1971).

INDEX

Where single pagination is given, the subject may extend continuously beyond the page designated. Inclusive pagination is used to indicate that the specific subject matter is broken up by figures or tables.